z	$P(Z < z)$	z	$P(Z < z)$	z	$P(Z < z)$	z	$P(Z < z)$	z	$P(Z < z)$	z	$P(Z < z)$
0.00	.5000	0.72	.7642	1.44	.9251	2.16	.9846	2.88	.9980	3.60	.9998
0.01	.5040	0.73	.7673	1.45	.9265	2.17	.9850	2.89	.9981	3.61	.9999
0.02	.5080	0.74	.7704	1.46	.9279	2.18	.9854	2.90	.9981	3.62	.9999
0.03	.5120	0.75	.7734	1.47	.9292	2.19	.9857	2.91	.9982	3.63	.9999
0.04	.5160	0.76	.7764	1.48	.9306	2.20	.9861	2.92	.9983	3.64	.9999
0.05	.5199	0.77	.7794	1.49	.9319	2.21	.9864	2.93	.9983	3.65	.9999
0.06	.5239	0.78	.7823	1.50	.9332	2.22	.9868	2.94	.9984	3.66	.9999
0.07	.5279	0.79	.7852	1.51	.9345	2.23	.9871	2.95	.9984	3.67	.9999
0.08	.5319	0.80	.7881	1.52	.9357	2.24	.9875	2.96	.9985	3.68	.9999
0.09	.5359	0.81	.7910	1.53	.9370	2.25	.9878	2.97	.9985	3.69	.9999
0.10	.5398	0.82	.7939	1.54	.9382	2.26	.9881	2.98	.9986	3.70	.9999
0.11	.5438	0.83	.7967	1.55	.9394	2.27	.9884	2.99	.9986	3.71	.9999
0.12	.5478	0.84	.7995	1.56	.9406	2.28	.9887	3.00	.9987	3.72	.9999
0.13	.5517	0.85	.8023	1.57	.9418	2.29	.9890	3.01	.9987	3.73	.9999
0.14	.5557	0.86	.8051	1.58	.9429	2.30	.9893	3.02	.9987	3.74	.9999
0.15	.5596	0.87	.8079	1.59	.9441	2.31	.9896	3.03	.9988	3.75	.9999
0.16	.5636	0.88	.8106	1.60	.9452	2.32	.9898	3.04	.9988	3.76	.9999
0.17	.5675	0.89	.8133	1.61	.9463	2.33	.9901	3.05	.9989	3.77	.9999
0.18	.5714	0.90	.8159	1.62	.9474	2.34	.9904	3.06	.9989	3.78	.9999
0.19	.5753	0.91	.8186	1.63	.9485	2.35	.9906	3.07	.9989	3.79	.9999
0.20	.5793	0.92	.8212	1.64	.9495	2.36	.9909	3.08	.9990	3.80	.9999
0.21	.5832	0.93	.8238	1.65	.9505	2.37	.9911	3.09	.9990	3.81	.9999
0.22	.5871	0.94	.8264	1.66	.9515	2.38	.9913	3.10	.9990	3.82	.9999
0.23	.5910	0.95	.8289	1.67	.9525	2.39	.9916	3.11	.9991	3.83	.9999
0.24	.5948	0.96	.8315	1.68	.9535	2.40	.9918	3.12	.9991	3.84	.9999
0.25	.5987	0.97	.8340	1.69	.9545	2.41	.9920	3.13	.9991	3.85	.9999
0.26	.6026	0.98	.8365	1.70	.9554	2.42	.9922	3.14	.9992	3.86	.9999
0.27	.6064	0.99	.8389	1.71	.9564	2.43	.9925	3.15	.9992	3.87	1.0000
0.28	.6103	1.00	.8413	1.72	.9573	2.44	.9927	3.16	.9992	3.88	1.0000
0.29	.6141	1.01	.8437	1.73	.9582	2.45	.9929	3.17	.9992	3.89	1.0000
0.30	.6179	1.02	.8461	1.74	.9591	2.46	.9931	3.18	.9993	3.90	1.0000
0.31	.6217	1.03	.8485	1.75	.9599	2.47	.9932	3.19	.9993	3.91	1.0000
0.32	.6255	1.04	.8508	1.76	.9608	2.48	.9934	3.20	.9993	3.92	1.0000
0.33	.6293	1.05	.8531	1.77	.9616	2.49	.9936	3.21	.9993	3.93	1.0000
0.34	.6331	1.06	.8554	1.78	.9625	2.50	.9938	3.22	.9394	3.94	1.0000
0.35	.6368	1.07	.8577	1.79	.9633	2.51	.9940	3.23	.9994	3.95	1.0000
0.36	.6406	1.08	.8599	1.80	.9641	2.52	.9941	3.24	.9994	3.96	1.0000
0.37	.6443	1.09	.8621	1.81	.9649	2.53	.9943	3.25	.9994	3.97	1.0000
0.38	.6480	1.10	.8643	1.82	.9656	2.54	.9945	3.26	.9994	3.98	1.0000
0.39	.6517	1.11	.8665	1.83	.9664	2.55	.9946	3.27	.9995	3.99	1.0000
0.40	.6554	1.12	.8686	1.84	.9671	2.56	.9948	3.28	.9995	4.00	1.0000
0.41	.6591	1.13	.8708	1.85	.9678	2.57	.9949	3.29	.9995		
0.42	.6628	1.14	.8729	1.86	.9686	2.58	.9951	3.30	.9995		
0.43	.6664	1.15	.8749	1.87	.9693	2.59	.9952	3.31	.9995		
0.44	.6700	1.16	.8770	1.88	.9699	2.60	.9953	3.32	.9996		
0.45	.6736	1.17	.8790	1.89	.9706	2.61	.9955	3.33	.9996		
0.46	.6772	1.18	.8810	1.90	.9713	2.62	.9956	3.34	.9996		
0.47	.6808	1.19	.8830	1.91	.9719	2.63	.9957	3.35	.9996		
0.48	.6844	1.20	.8849	1.92	.9726	2.64	.9959	3.36	.9996		
0.49	.6879	1.21	.8869	1.93	.9732	2.65	.9960	3.37	.9996		
0.50	.6915	1.22	.8888	1.94	.9738	2.66	.9961	3.38	.9996		
0.51	.6950	1.23	.8907	1.95	.9744	2.67	.9962	3.39	.9997		
0.52	.6985	1.24	.8925	1.96	.9750	2.68	.9963	3.40	.9997		
0.53	.7019	1.25	.8944	1.97	.9756	2.69	.9964	3.41	.9997		
0.54	.7054	1.26	.8962	1.98	.9761	2.70	.9965	3.42	.9997		
0.55	.7088	1.27	.8980	1.99	.9767	2.71	.9966	3.43	.9997		
0.56	.7123	1.28	.8997	2.00	.9773	2.72	.9967	3.44	.9997		
0.57	.7157	1.29	.9015	2.01	.9778	2.73	.9968	3.45	.9997		
0.58	.7190	1.30	.9032	2.02	.9783	2.74	.9969	3.46	.9997		
0.59	.7224	1.31	.9049	2.03	.9788	2.75	.9970	3.47	.9997		
0.60	.7257	1.32	.9066	2.04	.9793	2.76	.9971	3.48	.9998		
0.61	.7291	1.33	.9082	2.05	.9798	2.77	.9972	3.49	.9998		
0.62	.7324	1.34	.9099	2.06	.9803	2.78	.9973	3.50	.9998		
0.63	.7357	1.35	.9115	2.07	.9808	2.79	.9974	3.51	.9998		
0.64	.7389	1.36	.9131	2.08	.9812	2.80	.9974	3.52	.9998		
0.65	.7422	1.37	.9147	2.09	.9817	2.81	.9975	3.53	.9998		
0.66	.7454	1.38	.9162	2.10	.9821	2.82	.9976	3.54	.9998		
0.67	.7486	1.39	.9177	2.11	.9826	2.83	.9977	3.55	.9998		
0.68	.7517	1.40	.9192	2.12	.9830	2.84	.9977	3.56	.9998		
0.69	.7549	1.41	.9207	2.13	.9834	2.85	.9978	3.57	.9998		
0.70	.7580	1.42	.9222	2.14	.9838	2.86	.9979	3.58	.9998		
0.71	.7611	1.43	.9236	2.15	.9842	2.87	.9980	3.59	.9998		

STATISTICS
AN INTRODUCTION

STATISTICS
AN INTRODUCTION

Robert N. Goldman / Joel S. Weinberg

Simmons College

Simmons College

Prentice-Hall, Inc., Englewood Cliffs, New Jersey 07632

Library of Congress Cataloging in Publication Data

Goldman, Robert, (date).
Statistics, an introduction.

Includes index and bibliography.
1. Statistics. I. Weinberg, Joel. II. Title.
QA276.12.G65 1985 519.5 84-16098
ISBN 0-13-845918-5

Editorial/production supervision: Paula J. Martinac
Cover and interior designs: Suzanne Behnke
Manufacturing buyer: John Hall

Printed in the United States of America
10 9 8 7 6 5 4

ISBN 0-13-845918-5 01

Prentice-Hall International, Inc., *London*
Prentice-Hall of Australia Pty. Limited, *Sydney*
Editora Prentice-Hall do Brasil, Ltda., *Rio de Janeiro*
Prentice-Hall Canada Inc., *Toronto*
Prentice-Hall Hispanomericana, S.A., *Mexico*
Prentice-Hall of India Private Limited, *New Delhi*
Prentice-Hall of Japan, Inc., *Tokyo*
Prentice-Hall of Southeast Asia Pte. Ltd., *Singapore*
Whitehall Books Limited, *Wellington, New Zealand*

CONTENTS

16 NONPARAMETRIC METHODS 616

PREFACE

Statistics: An Introduction is designed for use in a first course in statistics. This course is generally taken by a wide range of students usually as a requirement for their majors. The only mathematical prerequisite is high school algebra which we have found is sufficient for an accurate and thorough treatment of modern statistics. The combination of our backgrounds—one of us is a statistician, the other an educator specializing in reading—has, we believe, helped us to provide clear explanations throughout the text. In particular, we have tried to avoid making the student "fill in the gaps" or "read between the lines" in our explanations. In extensive classroom trials the text was praised for its readability. We believe it will prove ideal for use as a regular class text.

We intend this text to fill the middle ground between the more traditional beginning statistics texts and the new and innovative texts such as those by Freedman[1] and others and by Koopmans[2]. We cover most of the topics discussed in the newer books but do so largely withing the structure of the traditional text. We feel that our approach will benefit both the student and the instructor.

Chapters 1 and 2 provide an introduction to univariate and bivariate data analysis. After a transitionary section, Chapters 3 through 7 cover probability, discrete and continuous distributions, and the sampling distribution of $\overline{X}$. Chapter 8 deals with elements of design in statistical studies. (More on this in a moment.) Standard large and small sample inferences are discussed in Chapters 9 through 12. The analysis of variance—and its relationship to experimental design is the subject of Chapter 13. The inferential aspects of regression are discussed in Chapter 14. The final two chapters cover the analysis of count data and non-parametric methods. We have found that it is possible to complete Chapters 1 through 10, the first half of Chapter 12 and most of one other chapter—generally linear regression—in one semester.

[1]Freedman, E., Pisani, R., and Purves, R. *Statistics*. New York: W. W. Norton, 1978.
[2]Koopmans, L. H. *An Introduction to Contemporary Statistics*. Boston: Duxbury, 1981.

The text has a number of features worthy of special note.

- In addition to a problem set at the end of each section, there is an extensive set of review problems at the end of each chapter. Because these problems are not keyed to specific sections, the student is required to choose the appropriate technique or approach. In many cases he or she must pull together ideas from several sections and, at times, several chapters. These review problems can serve as an excellent preparation for examinations and indeed may often serve as questions on examinations. Numerous "real life" applications are frequently presented through these review problems as well.
- A numerical example is used to introduce techniques or concepts and to motivate the students. After a technique has been explained, at least two further applications are provided. There is, in fact, an average of 14 examples per chapter.
- The solution to the first problem at the end of each section is worked out at the end of the problem set. This provides still another example and eases the student into the type of problems to be encountered in the set.
- Where appropriate a section illustrates how many of the techniques presented in the chapter can be performed by the statistical package Minitab. The simplicity and scope of Minitab make it ideal for an introductory course. Through these sections the student can also gain valuable experience in reading computer output without necessarily having access to the package.
- An innovative aspect of the book, and one we hope will represent a trend for the future, is the inclusion of a chapter (Chap. 8) devoted to the design of statistical studies. This chapter—which is *not* a list of the classical experimental designs—introduces many of the practical aspects of surveys and experiments. It focuses on the role that bias plays in drawing conclusions from studies.
- One of the most obscure points for the student to understand in an introductory statistics course in the transition from descriptive statistics to probability. In many texts this is made quite abruptly and leaves the student at a loss as to the connection between the two. To avoid this problem we devote a section (3.1) to a discussion of the reasons for this transition.
- At the end of each chapter there is a summary and review section which synthesizes and highlights the major facets of the chapter.

We owe thanks to a great many people. Many Simmons College students have provided insights from the perspective of the user. Karen Flora, T.E. Raghunathan, and Chang Park checked the answers to the problems in the text. Meg Burchnell, Norma Mosby, and Maryann Fidler typed earlier drafts fo the book. Dorothy Thayer did a superb job of typing the final draft. George Kimball typed the answers. Robert Goldman would especially like to express his thanks to his colleagues David Browder, John Garberson, and Margaret Menzin for their friendship and encouragement over the years.

The following professors reviewed the manuscript and offered comments and recommendations during the development of the project: Harvey Arnold, Oakland University; Louis Bush, San Diego City College; Damon Disch, Rose-Hulman Institute of Technology; Larry Haugh, University of Vermont; Michael Karelius, American River College; William Koellner, Montclair State College; and Larry Ringer, Texas A & M University. The authors would especially like to acknowledge the invaluable advice given to us by these professors: Robert Brown, U.C.L.A.; John Kellermeier, Northern Illinois University; Austin Lee,

Boston University; Robert Stephenson, Iowa State University; and Virginia Taylor, University of Lowell. Throughout the development of the book we have been fortunate to have the advice and patience of our editor Robert Sickles. It has been a singular pleasure working with him. Our production editor Paula Martinac and designer Sue Behnke did a marvelous job under a tight schedule.

Without the support, encouragement and patience of our wives Mary Glenn Vincens and Edith Weinberg this book would never have been completed. In addition Mary Glenn provided invaluable editorial assistance through much of the development of the text. We dedicate this text to our wives with our love and appreciation.

Robert N. Goldman
Joel S. Weinberg

STATISTICS
AN INTRODUCTION

STATISTICS: AN INTRODUCTION

WHAT IS COMMON TO ALL THESE QUESTIONS?

Can wearing seat belts significantly reduce the risk of serious injury in automobile accidents?

Does the Head Start program for preschoolers produce lasting benefits?

Can an opinion poll based on only 1500 voters accurately reflect the view of the entire American electorate?

Is cloud seeding more effective than nature in producing rain?

Which of the disputed *Federalist* papers were written by James Madison and which by Alexander Hamilton?

Does the use of interferon in the treatment of cancer live up to the claims of its proponents?

The questions on the preceding page may appear to have little or nothing in common. In fact they all have one thing in common—they have all been answered to one degree or another by the use of statistics.

In addition to its being used to solve problems such as these, statistics is used in areas from archaeology to zoology and in activities from the study of subatomic particles to the analysis of the farthest reaches of the universe.

The word "statistics" has an interesting origin. It is derived from the Latin for "state affairs." Historically it has been governments which have compiled and reported data (the statisticians' word for numerical information). The first reported use of statistics is believed to have been about 5000 years ago when Herodotus, a Greek historian, referred to figures on the wealth and population of Egypt, probably obtained in connection with the planning and the construction of the pyramids.

To many people the word "statistics" conjures up images of charts, of tables, of batting averages, test scores, rising crime rates, or details of the consumer price index. These images are correct as far as they go, and they illustrate that, to be useful, numerical information (i.e., data) must usually be organized and summarized, often graphically, but also by using averages, ranges, percentages, and other measures. For example, it is not very helpful to know that the 10 persons who received a new treatment for cancer of the larynx survived for the following numbers of months:

10.7 49.4 27.4 38.9 42.7 69.7 74.2 15.6 50.4 83.4

It is more useful to know that the average survival period is a little over 46 months (46.24 to be precise) or that a proportion, 3/10 or .3, survived longer than five years (60 months).

In Chapters 1 and 2 we outline the traditional procedures used by statisticians for organizing and summarizing data. We also present new and innovative methods for doing this. All of these procedures are included under the broad title **descriptive statistics.** This branch of statistics now forms but a small part of the entire discipline.

In the seventeenth century John Graunt, an English haberdasher, used the information in a sample of birth and death records from the London parish registers to estimate the total number of people living in London. This may have been the first instance of what is known as **inferential statistics.** This branch of statistics (now by far the more important) involves the use of information on a small and generally carefully selected part of a group in order to make a judgment about some aspect of the entire group. The small group is usually called the *sample* and the entire group the *population*. As used in statistics, however, the word "population" is not limited to people. It may refer to *all* the members of a group, for example all the schools in Nebraska, all cars sold last year, all TV sets now in use, or all the redheads in Roberts High School.

With this background in mind, we offer the following definition of statistics:

Statistics involves the collection, organization, presentation, and interpretation of numerical information.

Interpreting numerical information involves the use of inferential statistics, referred to above. There are two issues of special concern that arise when making such inferences. The first is the practical one of proper selection of the sample. For instance, inadequate sampling and follow-up methods probably resulted in the severe embarrassment and failure of the national magazine, the *Literary Digest,* when on the basis of an immense sample they incorrectly predicted the overwhelming victory of Alfred Landon over Franklin Roosevelt in the 1936 presidential election. The issue of sampling, and related questions that arise in the collection of data such as these, are discussed in Chapter 8.

The second issue in statistical inference can be illustrated by the well-known Nielsen ratings of TV programs. These ratings report the percentage of the viewing public watching each program during a specified time period. For example, the Nielsen organization may report that the latest miniseries was watched by 63% of viewers last evening. This figure will typically be based on reports from perhaps 1800 homes. With this information Nielsen will conclude that about 63% of *all* viewers were doing the same. But what does "about" mean? Statistical methods enable Nielsen to find out just *how likely* it is that 63% is within, say, 3% of the actual percentage of all viewers. The phrase "how likely" suggests the means for addressing this second issue. It indicates that a degree of uncertainty is involved in making inferences about a population (all viewers in this case) based on only a sample. The study of uncertainty is called *probability* and it is this subject which provides the foundation of inferential statistics. We begin the study of this topic in Chapter 3.

Once the basic principles of probability and sampling are understood, the theory and practice of inferential statistics can be properly explored. Such exploration will make it clear that, as used today, statistics is not an abstract subject. On the contrary, it is an exciting, ever-changing, and immensely practical field. In this text we have tried to convey the scope, elegance, and utility of the subject.

REFERENCES

BRYSON, M. C. "The Literary Digest Poll: Making of a Statistical Myth," *The American Statistician,* Vol. 30, No. 4 (Nov. 1976), 184–185.

FURTHER READING

The following books focus on either the application or the interpretion of statistical methods. Any one of the three would complement the material in this text.

FOLKS, J. L. *Ideas of Statistics*. New York: Wiley, 1981.

MOORE, D. S. *Statistics: Concepts and Controversies*. San Francisco: Freeman, 1979.

TANUR, J. M., et al., eds. *Statistics: A Guide to the Unknown*. San Francisco: Holden–Day, 1978.

1 DATA ANALYSIS

ARE THERE BETTER WARRANTIES?

Every consumer knows what it is like to have an appliance or automobile that starts to fall apart as soon as the warranty expires. One proposal has been made to protect consumers against such frustrations. This would be especially helpful to owners of automobiles faced with the expense and headache of costly repairs. The proposal suggests replacing the arbitrary warranty period with an *average life expectancy* of a product or major part of the product. During this period repairs would be made without charge. You are asked to analyze a more complicated version of this proposal in Review Problem 34 on page 65.

SECTION 1.1

DATA AND VARIABLES

DATA The information in Table 1 was collected as a small part of a study of pregnant women.

TABLE 1

Data on 25 new mothers

WOMAN IDENTIFICATION NUMBER	ETHNIC BACKGROUND [WHITE (W), BLACK (B), ORIENTAL (O)]	AGE AT DELIVERY	INFANT BIRTHWEIGHT (GRAMS)	NUMBER OF VISITS TO CLINIC
1	W	29	3310	12
2	B	33	2650	11
3	B	17	2450	9
4	B	19	2900	15
5	B	27	3100	17
6	W	26	3500	12
7	B	16	2400	4
8	O	31	3540	11
9	B	20	3770	12
10	B	19	3000	8
11	B	22	2970	10
12	B	25	2770	4
13	W	23	3250	14
14	B	19	3060	12
15	W	20	3260	12
16	W	32	2590	10
17	B	26	2960	12
18	B	19	2790	9
19	O	23	3570	12
20	W	36	3770	17
21	B	33	3870	14
22	B	19	2460	9
23	W	32	2380	10
24	W	35	2930	16
25	W	20	3800	9

Information such as that in Table 1 is called **data.** We can obtain data about almost anything: hospitals, companies, stores, automobiles, schools, molecules, and so on, as well as people. The data in Table 1 provide information on four characteristics of 25 women. Each characteristic is called a **variable** because its values may vary from person to person. Each number or letter in Table 1 (except the identification numbers) is called an **observation.**

When the data for a variable consist of numbers, as in the case of numbers of visits, birthweight, and age, they are called **quantitative variables.** But let us look at the three quantitative variables in Table 1 more closely. In each case the values are whole numbers: for instance, 27 years, 7 visits, or 2970 grams. But the number of visits can only be expressed as a whole number: 0, 1, 2, 3 visits—no woman could make 7.7 visits. Variables such as this are referred to as **discrete variables.** On the other hand, the women's ages could have been 27.5 or even 27.48 years (or theoretically any number of decimal places). Similarly, with sufficiently sensitive scales infant birthweights could have been obtained more precisely, say 2970.4 or 2970.46 grams. Variables such as these are referred to as **continuous variables.**

In general, we may distinguish between discrete and continuous variables as follows: A **discrete variable** is one that has gaps between possible values (e.g., 7 and 8 visits). A **continuous variable** is one that has no gaps between possible values. Thus it is always possible to find another amount between any two weights, e.g., 2970.46 between 2970.4 and 2970.5 grams and 2970.4384563 between 2970.438456 and 2970.438457 grams.

Having drawn this distinction we must add that there are some variables—generally those that can take a very large number of possible values—that are technically discrete but which we shall often treat as continuous. Examples include dollar amounts, test scores and grade point averages. We shall have more to say about such variables in Section 4.1 and in Chapter 6.

Some variables, such as ethnicity, involve only descriptive words or names. We refer to these as **qualitative variables.** The "values" they can take are called **levels.** The levels of the variable "ethnicity" used in Table 1 are white, black, and oriental. Similarly, the levels of the variable "patient's condition" may be listed as critical, poor, fair, and so on. But there is a difference between the levels of "ethnicity" and "patient's condition." In the latter case the levels are ranked or ordered and the variable "patient's condition" is therefore said to be **ordinal.** Other examples of ordinal variables are "degree of support for a proposal" (the levels may be: strongly support, support, neutral, and so on) and the classification of a piece of equipment as "perfect," "second," or "defective." A qualitative variable in which the levels are *not* ordered, as in the case of ethnicity, sex, nationality, or religion, is called a **nominal variable.**

Sometimes a quantitative variable is treated as being ordinal. For example, in a survey voters may be classified only as young, middle-aged, or old. Similarly, researchers often classify the variable "amount of education" as graduate degree, some graduate work, college degree, some college work, and so on.

PROBLEMS 1.1

1. The following data were collected on seven households:

PREVIOUS MONTH'S FOOD EXPENDITURE	OWN (H) OR RENT (R) THEIR HOME	NUMBER OF CHILDREN
$149.27	H	3
117.62	H	2
92.30	R	1
90.41	R	1
124.87	H	2
139.90	R	4
104.27	H	1

Classify each of the three variables as (a) discrete, (b) continuous, (c) ordinal, or (d) nominal.

2. The following data were obtained on eight automobile accident reports:

SIZE OF CAR	AGE OF CAR	NUMBER OF OCCUPANTS	DRIVER WEARING SEAT BELTS	NUMBER OF PREVIOUS ACCIDENTS	INSURANCE SETTLEMENT
Small	3	1	Yes	0	$ 431.04
Full size	4	2	No	1	238.41
Full size	8	4	No	3	38.92
Small	2	2	No	1	732.34
Midsize	2	2	Yes	0	1211.50
Full size	1	1	No	1	528.33
Midsize	1	3	No	0	92.51
Small	5	2	No	2	194.00

Classify each of the six variables as (a) discrete, (b) continuous, (c) ordinal, or (d) nominal. If ordinal or nominal, specify the levels.

3. Identify each of the following variables as (i) discrete, (ii) continuous, (iii) ordinal, or (iv) nominal:
   (a) The population of a city block
   (b) The country of origin
   (c) A person's height
   (d) Annual income
   (e) A student's area of concentration
   (f) The size of a school's graduating class
   (g) The time a person spends commuting in a day
   (h) The temperature of a sick child
   (i) The number of students absent from school on a specific day

4. Which of the following qualitative variables are nominal and which are ordinal?
   (a) Occupation
   (b) Socioeconomic status
   (c) The quality of seating chosen by an airline passenger (first class, business class, or economy class)
   (d) Eye color

Solution to Problem 1

Food expenditure is a continuous variable; home ownership, a nominal variable; and the number of children, a discrete variable.

SECTION 1.2
ORGANIZING AND DISPLAYING DATA

The values for each variable in Table 1 are listed according to the women's identification numbers. This may be a useful way to compile the data initially, but there are more useful ways of presenting data. For instance, the 25 ages can be summarized in five successive groups or intervals, as follows:

AGE	NUMBER OF WOMEN
14–18	2
19–23	11
24–28	4
29–33	6
34–38	2

Organizing and displaying data in this and other ways can make them more helpful and informative. The next example illustrates this in greater detail.

EXAMPLE 1 The following data were obtained as part of a study of the effectiveness of a new neurosurgical procedure. The numbers are the times, to the nearest one-tenth of an hour, used in performing 60 brain operations using the new procedure.

9.5	11.0	8.4	12.2	9.6	9.7	7.7	8.6	10.1	9.4
9.9	8.3	12.7	10.2	10.9	9.2	9.6	7.8	10.5	11.8
8.8	8.6	8.1	11.3	10.7	9.6	9.4	11.0	13.2	8.6
9.0	9.6	11.1	12.6	9.1	9.7	8.9	7.3	10.1	10.1
11.7	15.7	10.3	9.5	10.7	10.8	8.8	8.6	11.5	13.6
9.2	9.5	8.5	12.4	12.4	10.3	10.9	9.6	9.4	8.0

Table 2 represents one useful way of grouping these 60 observations. We begin by listing, in the first column, nine equal-size intervals. The first interval contains any times from 7.0 up to and including 7.9, the second 8.0 to 8.9, and so on, ending with 15.0 to 15.9. We shall have more to say later about these intervals. In the second column we have tallied the number of observations falling in each interval with vertical lines for the first four of each five tallies and a cross line for the fifth: I, II, III, IIII, ~~IIII~~. Any other method of tallying that avoids errors would also be appropriate.

TABLE 2

Frequency distribution for the surgery data

INTERVAL	TALLY	FREQUENCY	RELATIVE FREQUENCY	CUMULATIVE FREQUENCY
7.0–7.9	III	3	.050	3
8.0–8.9	~~IIII~~ ~~IIII~~ II	12	.200	15
9.0–9.9	~~IIII~~ ~~IIII~~ ~~IIII~~ III	18	.300	33
10.0–10.9	~~IIII~~ ~~IIII~~ II	12	.200	45
11.0–11.9	~~IIII~~ II	7	.117	52
12.0–12.9	~~IIII~~	5	.083	57
13.0–13.9	II	2	.033	59
14.0–14.9		0	.000	59
15.0–15.9	I	1	.017	60
		60	1.000	

The number of observations that fall in each interval is called its **frequency.** The frequency for the interval 9.0–9.9, for instance, is 18. The sum of all the frequencies is 60, the total number of observations. The list of the nine frequencies in the third column is called the **frequency distribution.** In this case the frequency distribution indicates that the times are concentrated between 8 and 11 hours with a peak between 9 and 10 hours.

Two other computations are useful. If we divide the frequency for a specific interval by 60 (the total number of observations) we obtain the **proportion** of all the observations that fall in that interval. For instance, the proportion of observations that are in the interval 9.0–9.9 is $18/60 = .3$. Each such proportion is called the **relative frequency** for the interval. The sum of the relative frequencies is 1 (can you see why?). Finally, **cumulative frequencies** are obtained by adding to the frequency of each interval the frequency of all the preceding intervals. Thus the cumulative frequency for the interval 9.0–9.9 is $3 + 12 + 18 = 33$, the number of observations that are less than 9.9. The list of the nine relative frequencies is called the **relative frequency distribution,** and the list of the nine cumulative frequencies is called the **cumulative frequency distribution.**

■

When constructing intervals, it is useful to keep three points in mind.

1. The larger the number of observations, the more intervals may be used; for most purposes between 5 and 15 are sufficient.

2. To make comparisons between intervals easier, they should be of equal length. Once the number of intervals has been decided on, their approximate width can be obtained by using the simple formula

$$\text{width of interval} = \frac{\text{largest observation} - \text{smallest observation}}{\text{number of intervals}}$$

Applying this formula to the surgery data, using nine intervals, we obtain

$$\text{width} = \frac{15.7 - 7.3}{9} = .93 \text{ hours}$$

For convenience we rounded this figure up to 1 hour. Notice that we also began the first interval at 7.0 rather than 7.3 so that each interval would begin with a whole number.

Occasionally, it makes sense to use intervals of different sizes. When incomes are grouped, for instance, we may want to use wider intervals for the larger incomes than for the smaller incomes.

3. Each observation should fit into one and only one interval. To ensure this, check that the intervals do not overlap. For example, the following are overlapping intervals: 7.0–8.0, 8.0–9.0, 9.0–10.0. Notice that the values 8.0 and 9.0 each occur in two different intervals. This problem can be avoided by using intervals such as 7.0–7.9, 8.0–8.9, and 9.0–9.9 if the observations are given to one decimal place or 7.00–7.99, 8.00–8.99, 9.00–9.99 if two decimal

places, and so on. You can also define the ends of intervals as 7.0–under 8.0, 8.0–under 9.0, and so on, regardless of the number of decimal places used.

RELATIVE FREQUENCY HISTOGRAMS

The relative frequency distribution in Table 2 can also be represented pictorially by a graph, as shown in Figure 1. The horizontal axis is divided into the nine intervals. The height of each bar is the relative frequency for the interval. The resulting display is called a **relative frequency histogram.** A similar histogram can be constructed using the frequencies themselves. This would be called a **frequency histogram.** We prefer to use relative frequencies, for reasons discussed in Chapter 4.

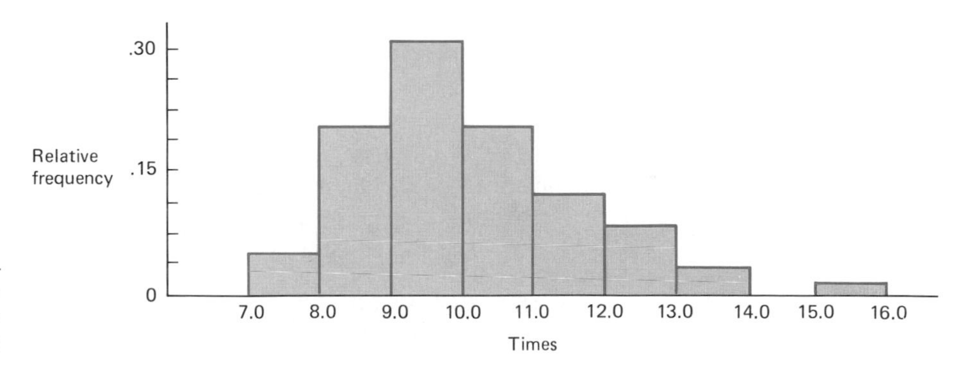

FIGURE 1

Relative frequency histogram for the surgery data

The histogram in Figure 1 highlights several features of the data:

1. Most of the times lie between 8.0 and 11.0 minutes.
2. The right end of the histogram (called the *right tail*) is longer than the left, illustrating the fact that a few operations took much longer than 9 or 10 hours.

If the right tail of a histogram is longer than the left, the histogram is said to be *skewed* to the right. If the left tail is longer, the histogram is said to be skewed to the left and if the tails are about the same, the histogram is *symmetric*. Examples of right-skewed, left-skewed, and symmetric histograms are shown on page 28.

You may wonder about the effect of grouping data into too few or too many intervals. The relative frequency histograms for the 60 times grouped in 4 and 18 intervals are shown in Figures 2(a) and (b), respectively.

Although the general shape of the histogram in Figure 2(a) is similar to that in Figure 1, much of the detail seen with nine intervals is lost. Conversely, the histogram in Figure 2(b) shows too much detail. It is too ragged, with the 18

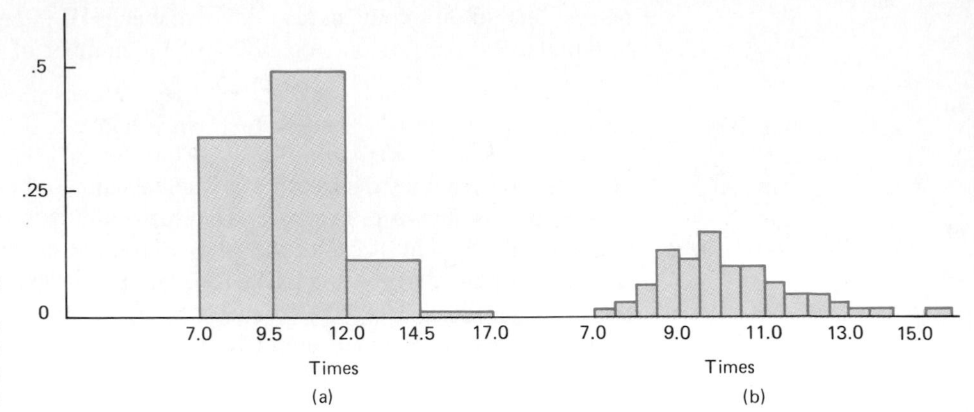

FIGURE 2

Relative frequency histogram for the surgery data using (a) 4 intervals and (b) 18 intervals

intervals reflecting some of the chaos of the ungrouped data. The histogram with nine intervals offers a good compromise.

STEM-AND-LEAF DISPLAYS

An alternative to the histogram is the **stem-and-leaf display.** This procedure was popularized by the statistician John Tukey.

There are three steps involved in constructing a stem-and-leaf display. We follow them using the 60 surgery times in Example 1.

1. Decide on the stem values. We have used the whole-number parts of the 60 times. The result is nine stems, as shown in Figure 3(a). The asterisk here stands for the decimal part of each time, which is then written separately.

FIGURE 3

The three steps involved in constructing a stem-and-leaf display

(a)	(b)		(c)	
7.*	7.*	783	7.*	378
8.*	8.*	463861698650	8.*	013456666889
9.*	9.*	5674926640617525б4	9.*	0122444555666667779
10.*	10.*	129571137839	10.*	111233577899
11.*	11.*	0830175	11.*	0013578
12.*	12.*	27644	12.*	24467
13.*	13.*	26	13.*	26
14.*	14.*		14.*	
15.*	15.*	7	15.*	7

2. Write the decimal part of *each* of the 60 times to the right of the corresponding stem. For example, for the number 9.5, place a 5 to the right of the stem 9.*. For the number 11.0, place a 0 to the right of the stem 11.*. Continue in the same way with the other 58 numbers. The display then looks as shown here in Figure 3(b). The numbers to the right of the stem are the leaves. Actually, the first two steps are combined, Figure 3(a) being shown here only for purposes of clarity.

3. Finally, order the leaves for each stem from smallest to largest, as shown in the stem-and-leaf display in Figure 3(c).

It is important to align the leaves vertically. For instance, the third decimal number of the stem 9.* (a 2) is immediately above the third decimal number of the stem 10.* (a 1). With this alignment you can compare the number of leaves with the different stems by just looking at the length of the leaves. For instance, there are four times as many leaves with the stem 8.* as with the stem 7.*. The actual numbers are 12 and 3. The stem-and-leaf display in Figure 3 thus reflects the distribution of the data exactly as the histogram did in Figure 1. The advantage of the stem and leaf display, however, is that all the original observations are retained and are *in order*. Indeed, the observations themselves indicate the shape of their distribution.

In general, the nature of the stems and leaves will vary with the types of numbers involved and the number of intervals required. For example, if the 60 numbers were between 7.0 and 9.9, we might use the following stems:

7.*

7.+

8.*

8.+

9.*

9.+

In this case the asterisk indicates leaves with the values 0, 1, 2, 3, or 4. The symbol + indicates leaves with the values 5, 6, 7, 8, or 9.

Other formats for stem-and-leaf displays will be illustrated in the examples and problems that follow.

DOT DIAGRAMS

When the number of observations is 15 or less, a **dot diagram** is an effective way of displaying the data. The numbers below are the cost estimates of seven contractors for the repair of damaged gutters.

\$305 \$251 \$295 \$260 \$287 \$365 \$270

A dot diagram representing the data consists of the seven numbers ordered and plotted along a straight line as shown in Figure 4. The distances between the points on the line should at least approximately reflect the differences between

FIGURE 4

Dot diagram for the contractor data

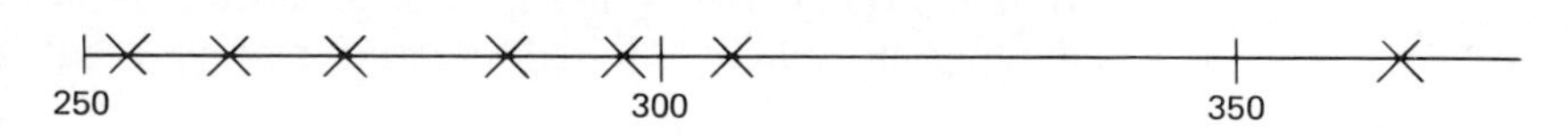

the corresponding values. These seven estimates fall into three groups, one centered at $260, another centered at $295, and a single outlying estimate at $365.

DISPLAYING DISCRETE DATA

The procedures described above are more relevant to the organization and display of continuous data. We now consider techniques for use with discrete data.

EXAMPLE 2 A group of 66 students studying for a professional examination were asked how many times they had taken the exam before. The results were as follows:

3 2 0 0 2 0 1 1 0 0 0 1 0 1 2 0 1 2 3 0 0 2

1 1 0 0 4 1 1 1 2 0 3 0 3 0 1 0 1 0 2 1 4 1

2 2 0 0 3 1 1 0 2 2 1 5 3 0 0 5 1 1 0 1 2 0

In this case a few observations are repeated many times. It is natural to summarize such data by recording the number of times each value occurred rather than grouping the values. The frequency distribution for the number of previous attempts is shown in Table 3.

TABLE 3

Summarized examination data

NUMBER OF PREVIOUS ATTEMPTS	TALLY	FREQUENCY	RELATIVE FREQUENCY	CUMULATIVE FREQUENCY
0	~~IIII~~ ~~IIII~~ ~~IIII~~ ~~IIII~~ IIII	24	.37	24
1	~~IIII~~ ~~IIII~~ ~~IIII~~ ~~IIII~~	20	.30	44
2	~~IIII~~ ~~IIII~~ II	12	.18	56
3	~~IIII~~ I	6	.09	62
4	II	2	.03	64
5	II	2	.03	66
		66	1.00	

From the frequency column we can see that 24 of the students had not taken the exam before, whereas 12 had made two previous attempts. From the relative frequencies we note that a proportion of .18 (12/66 = .18) of the students had taken the exam twice before. From the cumulative frequencies we can see that 56 students had taken the exam two *or fewer* times.

Unlike the distribution of times in Table 2, the entries in Table 3 refer to specific values, so that none of the original data are lost with this summary. Figure 5 illustrates a method for representing the relative frequency distribution for these discrete data. The distribution is represented by lines with heights which are the relative frequencies. This emphasizes that the relative frequencies apply *only* to the values 0, 1, 2, 3, 4, and 5. Where there are a large number of different discrete values they may be grouped into intervals and the histograms and other procedures referred to earlier may be used.

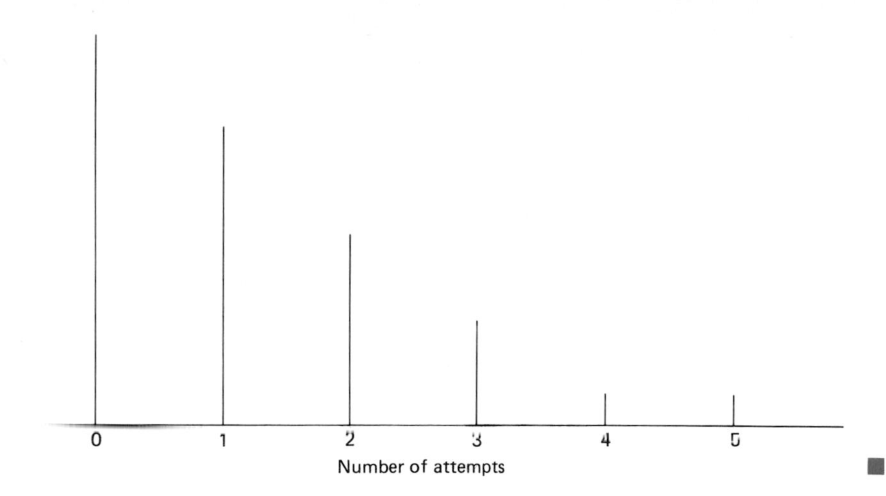

FIGURE 5

Mothod for representing the relative frequencies in the examination data

ARRANGING AND DISPLAYING QUALITATIVE DATA

We now turn to the organization of qualitative data.

EXAMPLE 3

As part of a report to the trustees of a college, the registrar classifies the 530 members of the senior class by area of concentration. The categories (levels) are (1) natural sciences, (2) social sciences, (3) humanities, and (4) professional. In Table 4 we have recorded the frequency and relative frequency for each area of concentration. The table shows that 176 of the 530 majors are social scientists and make up one-third (.33) of the total. Notice that Table 4 has no column of cumulative frequencies. The reason for this is that the variable—area of concentration—is nominal, so it makes no difference in what order we list the areas of concentration. We could as easily have placed humanities at the top rather than in third place.

TABLE 4

Distribution of seniors by area of concentration

AREA OF CONCENTRATION	FREQUENCY	RELATIVE FREQUENCY
Natural sciences	80	.15
Social sciences	176	.33
Humanities	211	.40
Professional	63	.12
All students	530	1.00

Data like those in Table 4 can be displayed in a number of ways. One popular device is a **bar diagram,** shown in Figure 6. The bar diagram is similar to the histogram in Figure 1. As before, the height of each block is a relative frequency—in this case, that with which each concentration occurs. Here, how-

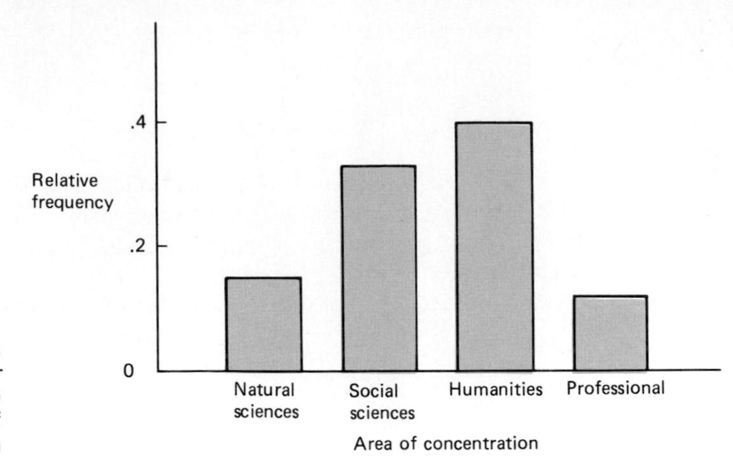

FIGURE 6

Bar diagram for the distribution of areas of concentration

ever, since the variable is qualitative and not quantitative the bars may be separated. There is also no significance to the width of the bars or to the distance between them.

■

PROBLEMS 1.2

5. The following data are the amounts (in dollars) spent by the first 36 persons going through the checkout line of a supermarket:

9.76	9.80	21.00	34.85	19.33	33.42	13.43	12.62	3.94
22.90	56.56	10.33	9.13	42.37	11.16	5.54	11.00	5.23
12.02	27.09	8.83	25.56	2.17	3.68	2.30	38.64	9.60
51.57	3.93	26.29	18.83	11.27	27.72	26.82	36.52	37.26

(a) Arrange these data into six intervals of the form \$0.00–\$9.99, \$10.00–\$19.99, and so on.

(b) Construct the columns of frequencies, relative frequencies, and cumulative frequencies.

(c) How many of these amounts are less than \$40?

(d) Construct the relative frequency histogram for these data.

(e) Construct a stem-and-leaf display using the last three digits as the leaves.

6. The temperatures (in degrees Fahrenheit) at midday on July 4 in a small New England town for the last 70 years were as follows:

71	77	80	74	85	87	85	73	83	78	82	80	85	92
81	80	84	69	80	81	84	84	67	60	84	75	73	91
88	80	76	91	80	81	83	83	81	79	83	72	67	93

80 78 84 82 79 81 84 94 72 74 79 84 81 76

84 76 75 77 70 84 78 79 82 84 76 73 87 71

(a) Construct a frequency distribution for these data using seven intervals of the form 60°–64°, 65°–69° and so on.

(b) Compute the columns of relative and cumulative frequencies.

(c) In how many years was the temperature below 75°F?

(d) Construct a stem-and-leaf display for these data using the same seven intervals.

(e) Use the stem-and-leaf display to answer the following questions:

(i) What is the lowest temperature? The highest?

(ii) For how many years was the temperature 80°F or above?

(iii) For what proportion of years was the temperature below 70°F?

7. Following are the monthly salaries (in dollars) of the 80 middle managers in a company:

1690	1358	1586	1117	1136	1012	1055	1392	1147	1761
1173	1960	1781	1829	905	879	1263	1182	1373	1263
1586	1477	1209	1472	1667	1353	1624	1243	1333	1500
1157	1904	1576	936	1210	1660	1231	1365	1344	1177
1490	1278	1211	1289	1783	2098	1735	1597	1329	1343
1013	1615	1295	1027	1267	995	1453	1122	1487	1352
1800	1334	1571	1917	1713	1490	1480	1728	1093	1561
1343	1661	1218	1383	1475	1192	1176	1282	1223	1697

(a) Arrange these salaries into nine intervals of the form $850–$999, $1000–$1149, and so on.

(b) Construct the columns of relative and cumulative frequencies.

(c) Sketch the relative frequency histogram for these data.

(d) How many of the managers earn less than (i) $1750; (ii) $2050?

(e) What proportion of managers earn (i) between $1150 and $1299; (ii) between $1600 and $1899?

(f) Construct the stem-and-leaf display for these data. The stems should be 8**, 9**, 10**, . . . , 20**, and the leaves the last two digits.

(g) Is the stem and leaf display symmetric? If not, is it skewed to the left or to the right?

(h) What advantage does the stem and leaf display have over the corresponding relative frequency histogram?

8. The following observations are the speed (in mph) of each of 40 cars recorded by police radar on the Pennsylvania Turnpike:

54	67	58	66	54	56	54	66	59	57
53	54	67	65	50	67	54	53	51	50
57	66	66	66	58	66	64	67	64	56
55	55	70	56	59	72	66	59	68	62

(a) Arrange these data into a frequency distribution using intervals of the form 50–54, 55–59, ...

(b) Construct the relative frequency and the cumulative frequency distributions.

(c) How many motorists were traveling at (i) 59 mph or less; (ii) 54 mph or less?

(d) What proportion of the motorists were traveling at between (i) 60 and 64 mph (both limits included); (ii) 55 and 64 mph (both limits included)?

(e) Construct a relative frequency histogram for these data.

9. A commuter recorded the travel time (in minutes) to travel to work on each of eight successive Mondays:

35 39 31 29 37 36 22 36

Arrange these data in a dot diagram.

10. Following are the grade-point averages (GPAs) for 10 college seniors:

3.14 1.98 3.35 2.72 2.47 3.42 2.14 3.90 2.96 3.57

Arrange these data in a dot diagram.

11. Following is the distribution of the number of goals per game scored by the Boston Bruins in an 82-game season:

NUMBER OF GOALS	0	1	2	3	4	5	6	7	8	9	10
NUMBER OF GAMES IN WHICH THIS NUMBER OF GOALS WERE SCORED	2	4	8	16	21	13	10	5	2	0	1

(a) Construct the relative frequency and the cumulative frequency distributions for these data.

(b) Sketch the relative frequency histogram.

(c) In what proportion of games did the Bruins score four goals?

(d) In how many games did the Bruins score (i) four or fewer goals; (ii) seven or more goals?

12. The number of courses being taken by each of the 43 students in a course on government are

3 4 4 4 3 3 5 4 4 2 4 4 4 4 3 3 4 3 5 4 4 4

4 4 3 4 4 4 4 2 4 3 4 4 5 3 4 3 4 4 4 3 4

(a) Arrange the data into a frequency distribution.

(b) Construct the columns of relative and cumulative frequencies.

(c) Sketch the relative frequency histogram.

(d) Does the shape surprise you?

(e) What proportion of students take three courses?

(f) What proportion take *other* than the usual load of four courses?

13. In a batch of 35 income tax returns the Internal Revenue Service (IRS) notes the following numbers of dependents claimed:

2 3 4 4 6 2 2 4 2 6 8 3 3 4 3 2 2 8

2 3 3 1 5 5 7 8 6 4 5 6 4 3 3 4 5

(a) Arrange the data in a frequency distribution.

(b) Construct the columns of relative and cumulative frequencies.

(c) How many families claimed (i) four dependents; (ii) more than four dependents?

(d) What proportion of families claimed five dependents?

14. At an interfaith gathering on a college campus the religion of each participant was recorded (RC, Roman Catholic; P, Protestant; J, Jewish; B, Buddhist; M, Moslem):

P RC J RC RC P RC P P P M RC P

P RC J J B RC RC P P RC P J J

P P RC P J M P RC J P RC M

(a) Compute the relative frequency for each religion.

(b) Display these relative frequencies in a bar diagram.

(c) Is the variable "religion" ordinal or nominal?

15. The number of visitors to a museum on each of five weekdays was recorded as follows:

DAY	NUMBER OF VISITORS
Monday	192
Tuesday	124
Wednesday	152
Thursday	96
Friday	64

(a) Compute the total number of visitors and the relative frequency for each day.

(b) Display the relative frequencies in a bar diagram.

(c) Is the variable "day of the week" ordinal or nominal?

16. In one year a consumer organization rates 390 movies as E (excellent), G (good), F (fair), or P (poor). The results were as follows:

RATING	NUMBER OF MOVIES
E	15
G	95
F	100
P	180

(a) Compute the relative frequency for each rating.

(b) Display the relative frequencies in a bar diagram.

(c) Is the variable "rating" a nominal or ordinal variable?

Solution to Problem 5

(a) and (b) The frequency, relative frequency, and cumulative frequency distributions are as follows:

INTERVAL	TALLY	FREQUENCY	RELATIVE FREQUENCY	CUMULATIVE FREQUENCY
$ 0.00–under $10.00	~~IIII~~ ~~IIII~~ II	12	.333	12
$10.00–under $20.00	~~IIII~~ IIII	9	.250	21
$20.00–under $30.00	~~IIII~~ II	7	.194	28
$30.00–under $40.00	~~IIII~~	5	.139	33
$40.00–under $50.00	I	1	.028	34
$50.00–under $60.00	II	2	.056	36
		36	1.000	

(c) From the column of cumulative frequencies, 33 of the 36 amounts are less than $40.00.

(d)

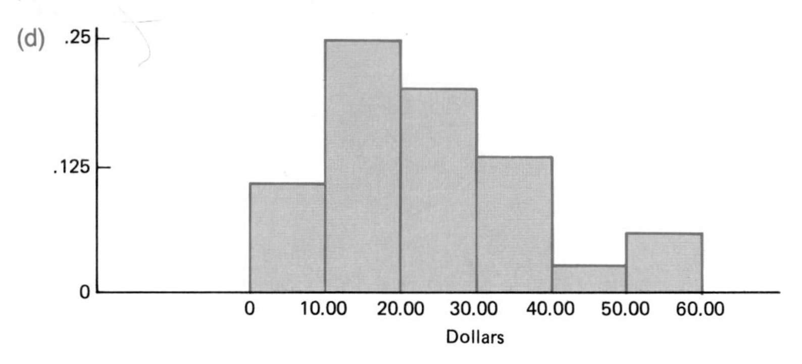

(e) The three-step procedure for obtaining the stem-and-leaf display is as follows:

(i)

Stem	Leaves
*.**	
1*.**	
2*.**	
3*.**	
4*.**	
5*.**	

(ii)

Stem	Leaves
*.**	9.76 9.80 3.94 9.13 5.54 5.23 8.83 2.17 3.68 2.30 9.60 3.93
1*.**	9.33 3.43 2.62 0.33 1.16 1.00 2.02 8.83 1.27
2*.**	1.00 2.90 7.09 5.56 6.29 7.72 6.82
3*.**	4.85 3.42 8.64 6.52 7.26
4*.**	2.37
5*.**	6.56 1.57

(iii)

Stem	Leaves
*.**	2.17 2.30 3.68 3.93 3.94 5.23 5.54 8.83 9.13 9.60 9.76 9.80
1*.**	0.33 1.00 1.16 1.27 2.02 2.62 3.43 8.83 9.33
2*.**	1.00 2.90 5.56 6.29 6.82 7.09 7.72
3*.**	3.42 4.85 6.52 7.26 8.64
4*.**	2.37
5*.**	1.57 6.56

SECTION 1.3

MEASURES OF LOCATION

THE MEAN Perhaps the most widely known method for summarizing quantitative data numerically is the arithmetic mean, illustrated in the next example.

EXAMPLE 4 The reading scores for nine tenth graders are

58 76 75 66 90 90 71 76 80

The simplest way to summarize these scores is to compute their average. To do this, add up all nine scores and divide the total by 9:

$$\text{average score} = \frac{58 + 76 + 75 + 66 + 90 + 90 + 71 + 76 + 80}{9}$$

$$= \frac{682}{9} = 75.778 = 75.8$$

■

The value 75.8 is called the arithmetic mean or more simply, the **mean.** The mean is referred to as a "measure of central tendency" or "location" because it indicates the location of the center (or middle) of a set of data.

> DEFINITION 1
>
> The **mean** of a set of observations is their sum divided by the number of observations

Two other common measures of central tendency are the median and the mode, both defined a bit later.

To provide a general formula for the mean, we need the following notations:

1. The letter n stands for the number of observations.
2. The first observation is represented by x_1, the second by x_2, and so on, to the last observation, x_n. The order in which the designations $x_1, x_2, \ldots$ are assigned does not matter.
3. The mean of the x's is represented by an x with a bar on top, $\bar{x}$, and called "x-bar." Thus

$$\bar{x} = \frac{x_1 + x_2 + \cdots + x_n}{n}$$

where the three dots stand for all the numbers between x_2 and x_n. It is more common, however, to write $\overline{x}$ in summation notation as in Equation 1.1 below. (You can review summation notation in Appendix B.)

$$\overline{x} = \frac{\sum_{i=1}^{n} x_i}{n} \qquad \text{(Eq. 1.1)}$$

EXAMPLE 5 The weights of five 2-year-old girls (in pounds) are 25.4, 21.9, 31.3, 24.4, and 28.6. Find their mean weight.

Solution In this case $n = 5$, $x_1 = 25.4$, $x_2 = 21.9$, $x_3 = 31.3$, $x_4 = 24.4$, and $x_5 = 28.6$. From Equation 1.1,

$$\overline{x} = \frac{25.4 + 21.9 + 31.3 + 24.4 + 28.6}{5} = \frac{131.6}{5} = 26.32 \text{ pounds}$$

■

ROUNDING OFF In Example 5 we computed the mean to two decimal places whereas in Example 4 we used only one decimal place. This is a good place to establish rules for deciding how many decimal places to use during computations and in final answers.

1. Retain as many decimal places as practical *during* any computation. For example, six or more decimal places are quite practical with a calculator.
2. The final answer should be rounded off to one more decimal place than the data on which the answer was based. In Example 4 the reading scores were whole numbers and $\overline{x} = 75.8$ was recorded to one decimal place. In Example 5 the five weights were recorded to one decimal place, and their mean, $\overline{x} = 26.32$ pounds, to two places.
3. If the last digit to the right of a decimal point is a 5, round so as to leave an *even* digit. For example, if we are to leave two decimal places, 5.635 should be rounded up to 5.64 and 5.645 rounded down to 5.64.

THE MEDIAN The mean is not always the best measure of location for a set of data. For example, suppose that the monthly after-tax incomes reported by each of seven households are

$1540 $1360 $15,200 $1730 $1480 $1690 $1580

The mean monthly income for the seven households is

$$\overline{x} = \frac{1540 + 1360 + 15{,}200 + 1730 + 1480 + 1690 + 1580}{7}$$

$$= \frac{24{,}580}{7} = \$3511.43$$

This mean is more than twice as large as six of the seven incomes and is completely uncharacteristic of the entire group. The problem, of course, is that the \$15,200 income distorts the mean. In cases such as this, the **median** is a useful alternative to the mean.

The first step in computing the median is to order the seven numbers from smallest to largest. The result of this step is

1360 1480 1540 (1580) 1690 1730 15,200

The median of this ordered list is the middle amount or \$1580. We denote the median of a set of numbers by m. Here $m = \$1580$, which is certainly more characteristic of the group than was the mean.

Notice that the same number of observations (three in this case) are below as are above m.

EXAMPLE 6 Compute the median weight for the five girls in Example 5.

Solution The five ordered weights are

21.9 24.4 (25.4) 28.6 31.3

and the median weight is the third smallest (or largest), 25.4 pounds. ■

Suppose that in Example 6 the list had included a sixth weight, 23.1 pounds. The ordered list would then have been

21.9 23.1 (24.4 25.4) 28.6 31.3

Since there is now no single middle value, the median is the mean of the two middle numbers:

$$m = \frac{24.4 + 25.4}{2} = 24.90 \text{ pounds}$$

In general, to obtain the median:

1. Order the observations from smallest to largest.
2. If the number of observations, n, is *odd:*
   (a) Obtain the value for $\frac{n+1}{2}$.
   (b) Count off the observations from either end until you reach the $\left(\frac{n+1}{2}\right)$th. This observation is the median.
3. If the number of observations, n, is *even:*
   (a) Obtain the values for $\frac{n}{2}$ and for $\frac{n}{2} + 1$.

(b) Count off the observations from either end until you reach the $\left(\frac{n}{2}\right)$th and the $\left(\frac{n}{2} + 1\right)$th. The median is the mean of these two observations.

> DEFINITION 2
>
> The **median** of a set of observations is the middle observation if n is odd and the mean of the two middle ones if n is even.

In Example 6, $n = 5$, $(n + 1)/2 = 3$ and m, the third smallest or third largest (they will always be the same observation) = 25.4 pounds. If $n = 6$, $n/2 = 3$, $(n/2) + 1 = 4$, and m is the mean of the third and fourth smallest observations. Remember that the observations must be ordered before the median can be located.

Both the mean and the median are sometimes referred to as averages. In this text we use the term "average" only as an alternative term for the mean.

THE MODE

One other measure that is sometimes helpful in describing the center of a set of observations is the **mode.**

> DEFINITION 3
>
> The **mode,** designated m_o, is that value which occurs most frequently.

EXAMPLE 7 The number of children in eight families are 4, 2, 3, 0, 2, 1, 3, and 2. Find m_o.

Solution The modal number of children is $m_o = 2$, because 2 occurs more frequently (three times) than any other value.

If you examine the nine reading scores in Example 4, you will notice that there are two modes, 76 and 90, both of which occur twice. In this case the data are described as **bimodal.** A set of data with a single mode is called **unimodal.** ■

MEASURES OF LOCATION WITH DISCRETE DATA

EXAMPLE 8 The following observations are the number of visits made to 15 prisoners in a cell block of a state prison during one weekend:

3 0 2 1 1 2 1 1 2 2 0 0 1 2 1

Compute the mean, median, and modal number of visits per prisoner.

Solution It will be helpful in this case to order the data and then construct a frequency distribution.

0 0 0 1 1 1 1 1 1 2 2 2 2 2 3

NUMBER OF VISITS	NUMBER OF PRISONERS	CUMULATIVE FREQUENCIES
0	3	3
1	6	9
2	5	14
3	1	15
	15	

We compute the mean from Equation 1.1:

$$\bar{x} = \frac{0 + 0 + 0 + 1 + 1 + 1 + 1 + 1 + 1 + 2 + 2 + 2 + 2 + 2 + 3}{15}$$

This can be rewritten as

$$\frac{(3 \times 0) + (6 \times 1) + (5 \times 2) + (1 \times 3)}{15} = 1.2667 \quad \text{or} \quad 1.3 \text{ visits}$$

The median number of visits can be found from the ordered data or equivalently from the cumulative frequencies above. From Definition 1 the median is the

$$\left(\frac{n + 1}{2}\right)\text{th} = \left(\frac{15 + 1}{2}\right)\text{th} = \text{eighth smallest observation}$$

Notice from the cumulative frequencies that the smallest three numbers are 0 and the fourth smallest to the ninth smallest (which includes the eighth) are 1. Thus the eighth smallest $= m = 1$.

The modal number of visits can also be obtained from the frequency distribution by searching for the value with the largest frequency. In this case $m_o = 1$.

More generally, when discrete data consisting of only whole numbers are in frequency form as in Example 8, the mean, median, and mode can be found using the frequencies and the cumulative frequencies. Call f_0 the frequency of 0's, f_1 the frequency of 1's, f_2 the frequency of 2's, and so on. The mean $\bar{x}$ can then be computed as

$$\bar{x} = \frac{(0 \times f_0) + (1 \times f_1) + (2 \times f_2) + \cdots + k \times f_k}{n}$$

or in summation notation,

$$\bar{x} = \frac{\sum_{i=0}^{k} if_i}{n} \qquad \text{(Eq. 1.2)}$$

where k is the largest value in the set of observations.

EXAMPLE 9 The examination data from Example 2 are summarized as follows:

NUMBER OF PREVIOUS ATTEMPTS, i	FREQUENCY, f_i	CUMULATIVE FREQUENCIES
0	24	24
1	20	44
2	12	56
3	6	62
4	2	64
5	2	66
	66	

Find the mean, median, and modal number of previous attempts.

Solution From Equation 1.2 the mean is

$$\bar{x} = \frac{(0 \times 24) + (1 \times 20) + (2 \times 12) + (3 \times 6) + (4 \times 2) + (5 \times 2)}{66}$$

$$= \frac{80}{66} = 1.2121 \quad \text{or} \quad 1.2 \text{ previous attempts}$$

Since $n = 66$ is even, the median number of previous attempts is the mean of the $(n/2)$th $= 33$rd and $[(n/2) + 1]$th $= 34$th observations in the ordered list. From the column of cumulative frequencies the 24 smallest observations are all zero and the 25th to the 44th smallest (which include the 33rd and 34th) are all 1. Therefore, $m = (1 + 1)/2 = 1$.

The modal number of previous attempts is $m_o = 0$ because this value occurs more frequently ($f_0 = 24$) than any other.

■

QUANTIFYING QUALITATIVE VARIABLES Occasionally there are advantages to quantifying artificially what are essentially qualitative data and then computing a mean. Suppose, for instance, that a group of 30 executives was classified by sex:

1	2	3	4	5	6	7	8	9	10	11	12	13	14	15
M	M	M	M	F	M	F	M	M	M	M	M	M	F	F

16	17	18	19	20	21	22	23	24	25	26	27	28	29	30
M	M	M	M	M	M	F	M	M	F	M	F	M	M	M

Since 7 of the 30 executives are women, the proportion of women is $7/30 = .23$.

One way to quantify these data is to assign the number 1 to each female and 0 to each male as follows:

M	M	M	M	F	M	F	M	M	M	M	M	M	F	F
0	0	0	0	1	0	1	0	0	0	0	0	0	1	1

M	M	M	M	M	M	F	M	M	F	M	F	M	M	M
0	0	0	0	0	0	1	0	0	1	0	1	0	0	0

The mean of these ones and zeros is

$$\bar{x} = \frac{23(0) + 7(1)}{30} = \frac{7}{30} = .23$$

which is again the proportion of women in the sample. Where the data consist of zeros and ones, we use the symbol $\bar{p}$ in place of $\bar{x}$. This is a reminder that in this case the mean is also a proportion.

Many of you will be familiar with another example of this kind. Suppose that a batter in baseball has had 14 hits in 45 "at bats." His "average" is $14/45 = .311$, which is the proportion of "at bats" in which he gets a hit. The .311 is also the mean ($\bar{p}$) of 14 ones (one for each hit) and 31 zeros (one for each at bat when he did not get a hit).

WHICH MEASURE OF LOCATION TO USE

The mode is most useful with discrete data. It is usually not very useful with continuous data since repeated observations are not common. If every observation occurs once, there is no mode. Even if one or two of the observations are repeated, the mode can be misleading. Consider the following 13 IQ scores, for instance:

92 97 102 81 109 88 111 101 106 88 107 116 103

Here $\bar{x} = 100.1$, $m = 102$, and $m_o = 88$. But $m_o = 88$ is not indicative of the center of the data in this case.

When the relative frequency distribution is close to being symmetric as in Figure 7(a), the mean and the median will be close to each other. When the distribution is skewed to the right as in Figure 7(b), the mean will be larger than the median. In this case the mean is "pulled up" by extremely large numbers. On the other hand, when the distribution is skewed to the left as in Figure 7(c), the mean will be "pulled down" below the median. An extreme example of case (b) involves the income data on page 22, where one of a group of seven incomes was roughly 10 times the other six. In this case the median, \$1580, was more characteristic of the data than the mean, \$3511.43. However, we could argue that the median ignores the presence of the one very large observation. The

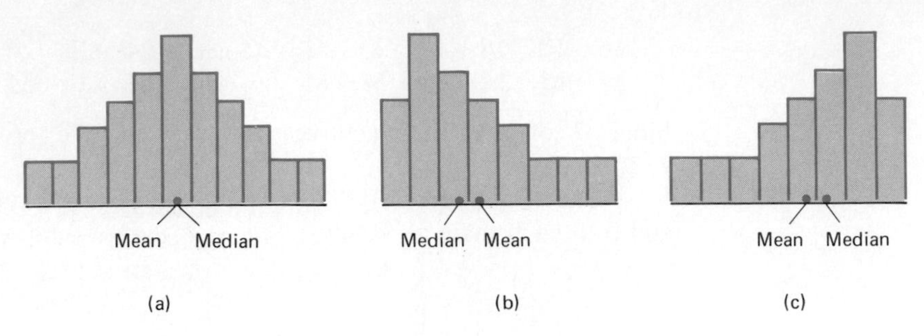

FIGURE 7

Effect of the "shape" of the data on the location of the mean and the median

answer, of course, is to report *both* the mean and the median. The fact that these two measures of location, neither of which is wrong, are so different is in itself an indication that the data are extremely skewed.

Despite its sensitivity to extreme values, the mean is the most widely used measure of location. This is in part because it is well understood. There are also some theoretical reasons for its preeminence in statistics, as you will see in Chapter 7.

PROBLEMS 1.3

17. Following are the scores on the first hour exam by the 12 students in a physics course:

    76 84 86 95 93 76 75 85 48 73 81 77

    Compute the (a) mean, (b) median, and (c) modal scores.

18. The ages at appointment of the nine members of the U.S. Supreme Court are

    61 50 43 44 59 61 64 47 55

    (a) Compute the (i) mean, (ii) median, and (iii) modal age at appointment.
    (b) Plot these ages on a dot diagram and mark the positions of $\bar{x}$, m, and m_o.

19. Ten persons enroll in a weight-reducing program. After two weeks the weight loss is recorded for each person:

    20.7 10.8 4.9 −1.4 7.3 8.9 12.4 −4.0 9.4 6.9

    (Two persons actually gained weight. These are negative losses.)
    (a) Compute the (i) mean, and (ii) median weight loss.
    (b) Why is the mode not a useful measure of location in this case?

20. Refer to the supermarket data in Problem 5 on page 16. Compute the (a) mean and (b) median amounts. (*Hint:* The stem-and-leaf display in the solution of Problem 5 will be helpful in computing the median.)

21. Refer to the speed data in Problem 8 on page 17. Compute the (a) mean, (b) median, and (c) modal speed. (*Hint:* A stem-and-leaf display will be helpful in computing the median and the mode.)

22. The number of misprints on each of the 20 pages of a booklet are

    0 2 2 1 2 2 2 1 1 2 2 1 1 0 1 0 1 3 2 2

(a) Arrange these data in a frequency distribution.

(b) Compute the (i) mean, (ii) median, and (iii) modal number of misprints per page.

23. For the goals per game data in Problem 11 on page 18, compute the (a) mean, (b) median, and (c) modal number of goals per game.
24. For the number of courses data in Problem 12 on page 18, compute the (a) mean, (b) median, and (c) modal number of courses per student.
25. Of the 25 hospitals in a city, 16 are public and 9 are private. Each of the public hospitals is given a score of 1 and each of the private hospitals a score of 0.
    (a) Find the mean ($\overline{p}$) of these scores.
    (b) How would you interpret this mean?
26. A sample of 1200 voters are interviewed. A score of 1 is assigned to each of the 730 in the sample who approve of the U.S. president's handling of the economy. A score of 0 is assigned to others in the sample.
    (a) What is the mean ($\overline{p}$) of these scores?
    (b) How would you interpret this mean?
27. The ages at death for 25 people were obtained from a local paper. The mean age was 65.4 years, and the median age was 74.8 years. What do these values suggest about the shape of the distribution of these 25 ages?

Solution to Problem 17

(a) The mean score is

$$\overline{x} = \frac{76 + 84 + 86 + 95 + 93 + 76 + 75 + 85 + 48 + 73 + 81 + 77}{12}$$

$$= 79.0833 \quad \text{or} \quad 79.1$$

(b) The ordered scores are

48 73 75 76 76 77 81 84 85 86 93 95

From Definition 1 the median is the mean of the $n/2$ = sixth and $(n/2) + 1$ = seventh smallest values. Thus

$$\frac{77 + 81}{2} = 79.0$$

(c) The modal score is 76, the value occurring most frequently and in this case the only value that occurs more than once.

SECTION 1.4

MEASURES OF SPREAD

Measures of location such as the mean or median are often not sufficient to describe a set of data adequately. Let us see why not.

EXAMPLE 10 The prices of a particular tape deck at the five retail outlets in Alton are \$110, \$121, \$117, \$123, and \$114. The mean price, $\overline{x}_A$ = \$117, and the median price, m_A = \$117, are the same. The town of Belton has both discount stores and high-priced department stores. The prices of the tape deck at the five retail outlets in Belton are \$59, \$45, \$117, \$169, and \$195. In this town the mean and median prices ($\bar{x}_B$ and m_B) are also \$117. But note that although the tape deck prices in the two towns have the same means and medians, there is a great difference in their **variability** or **spread.** This is emphasized by the two dot diagrams in Figure 8. Some means of indicating how spread out these two sets of observations are is clearly needed.

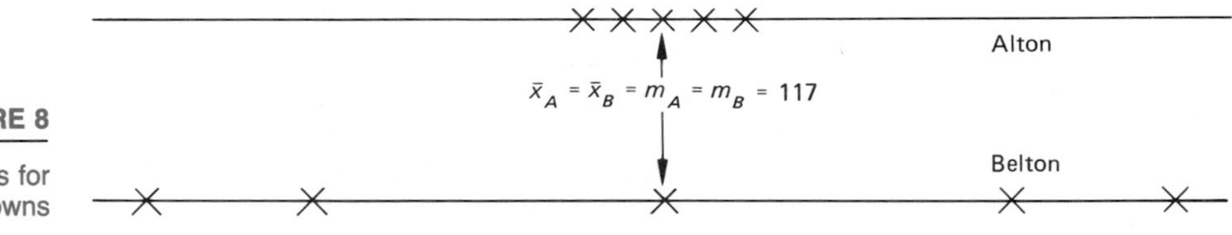

FIGURE 8

Dot diagrams of prices for a tape deck in two towns

■

THE RANGE

The simplest indicator of the degree of spread of a set of data is their **range** (denoted by R). This is the difference between the highest and the lowest value in a set of data. For the two towns the values of R are:

$$\text{Alton:} \quad R = \$123 - \$110 = \$13$$

$$\text{Belton:} \quad R = \$195 - \$\ 45 = \$150$$

The range of prices for Belton is over 10 times that for Alton.

The range is based on only the largest and the smallest observations. But what of the observations in between? They could all be clustered around the mean with just a few large or small ones as in Figure 9(a), or they could be spread out evenly as in Figure 9(b). Two distributions may have the same mean

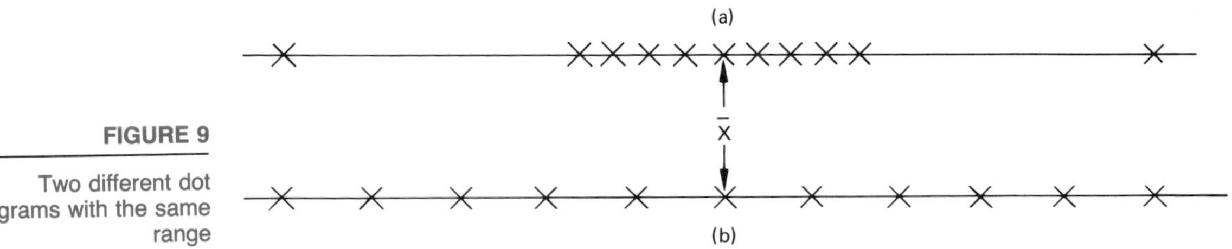

FIGURE 9

Two different dot diagrams with the same range

and range, yet be quite different in the way the scores are spread around the mean. Is there a measure more sensitive to such differences than the range?

One approach that suggests itself is to (1) obtain the mean of the set of observations, then (2) find the difference between each observation and this mean, and finally (3) compute the mean of these differences. For example, the difference between the first two prices and the mean for Alton are $110 - 117 = -\$7$ and $121 - 117 = \$4$. Differences such as these are called **deviations from the mean,** and we are to find the mean of these deviations. The computations for doing so in the case of our two towns are shown in Table 5. In both cases the negative and positive deviations cancel each other out, resulting in the sum of the deviations being zero. In fact, no matter what the data set, the sum of the deviations from the mean will always be zero. This procedure, therefore, offers no help in comparing the spread of prices in the two towns.

TABLE 5

Deviations and their means for two towns

ALTON		BELTON	
	DEVIATION		DEVIATION
110 − 117 =	−7	59 − 117 =	−58
121 − 117 =	4	45 − 117 =	−72
117 − 117 =	0	117 − 117 =	0
123 − 117 =	6	169 − 117 =	52
114 − 117 =	−3	195 − 117 =	78
Sum =	0	Sum =	0

There are other ways of using deviations to measure spread or variability. One such measure is based on the absolute values of the deviations. In case you do not remember, the absolute value of a number x, written $|x|$, is the magnitude of the number regardless of sign. Thus $|-5| = |5| = 5$. Unfortunately, manipulating absolute values is particularly complicated. There is, however, a more useful measure of spread, which we turn to next.

THE VARIANCE

The **variance** is approximately the mean of the *squared* deviations from the mean. Denoted s^2, the variance has proved useful in both descriptive and inferential statistics. The steps in computing s^2 for the prices in Alton and Belton are shown in Table 6 on the next page.

To obtain s^2 we follow four steps: (1) we find the five deviations from the mean, (2) we square each of these deviations, (3) we then add up the squared deviations, and (4) we divide this sum by one less than the number of observations $(n - 1)$. The variance of prices in Belton is over 150 times as large as that in Alton $(4334/27.5 = 157.6)$. Incidentally, the reason the variance is expressed as "s squared" (s^2) instead of just s is that s itself has a separate significance, which we will explain momentarily.

Translating this process into more general terms, we again denote the n

observations as $x_1, x_2, \ldots, x_n$ and their mean by $\bar{x}$. The deviations are then $(x_1 - \bar{x}), (x_2 - \bar{x}), \ldots, (x_n - \bar{x})$ and the variance is

$$s^2 = \frac{(x_1 - \bar{x})^2 + (x_2 - \bar{x})^2 + \cdots + (x_n - \bar{x})^2}{n - 1}$$

or in summation notation,

$$s^2 = \frac{1}{n-1} \sum_{i=1}^{n} (x_i - \bar{x})^2 \qquad \text{(Eq. 1.3)}$$

TABLE 6

Computing the variance of prices in Alton and Belton

PRICES	DEVIATIONS	SQUARED DEVIATIONS
	ALTON	
110	110 − 117 = −7	$(-7)^2 =$ 49
121	121 − 117 = 4	$4^2 =$ 16
117	117 − 117 = 0	$0^2 =$ 0
123	123 − 117 = 6	$6^2 =$ 36
114	114 − 117 = −3	$(-3)^2 =$ 9
		110

$$s^2 = \frac{110}{5-1} = \frac{110}{4} = 27.5$$

PRICES	DEVIATIONS	SQUARED DEVIATIONS
	BELTON	
50	59 − 117 = −58	$(-58)^2 =$ 3,364
45	45 − 117 = −72	$(-72)^2 =$ 5,184
117	117 − 117 = 0	$0^2 =$ 0
169	169 − 117 = 52	$52^2 =$ 2,704
195	195 − 117 = 78	$78^2 =$ 6,084
		17,336

$$s^2 = \frac{17{,}336}{5-1} = \frac{17{,}336}{4} = 4334$$

THE STANDARD DEVIATION

The variance, s^2, is useful when we need to compare the degree of spread in two sets of data, but you may be wondering how to interpret a single variance—for instance 27.5 for the prices in Alton. A variance is hard to interpret largely because it is a squared measure. In the example above the observations are in dollars, so the squared deviations (and hence the variance) are in "squared dollars." More useful and somewhat easier to interpret than s^2 is its square root, s, which we call the **standard deviation.** From Table 6 the standard deviation of prices in Alton is $s = \sqrt{s^2} = \sqrt{27.5} = \$5.20$ and in Belton $s = \sqrt{s^2} = \sqrt{4334} = 65.8331$ or \$65.8.

How shall we interpret the standard deviation and its relation to the mean? To answer this we note that, if we disregard the sign, a deviation can be represented as a distance from the mean. For example, the deviations of the tape deck prices for the town of Alton (−7, 4, 0, 6, −3) are shown as such distances in Figure 10. The standard deviation of these prices, 5.2, is also shown as such a distance. As you can see from the diagram, the standard deviation is approx-

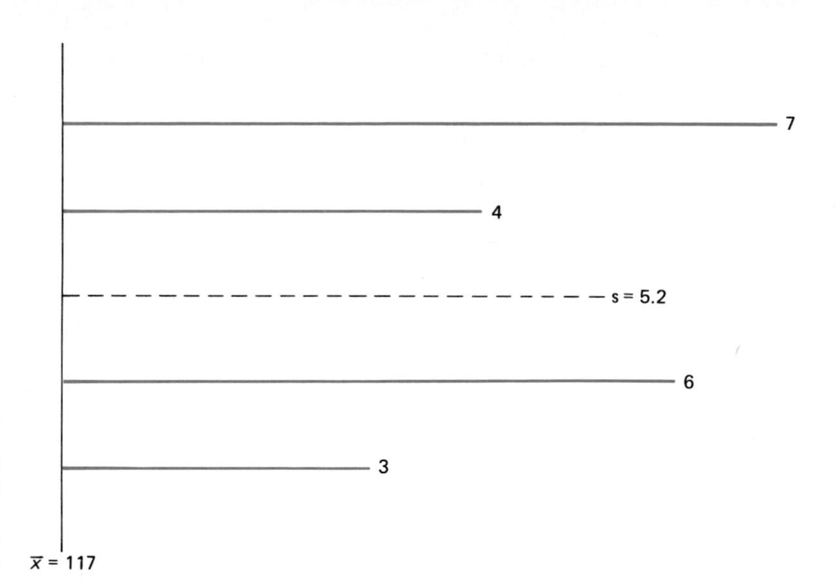

FIGURE 10

Viewing the standard deviation as approximately the average distance a set of observations are from their mean

imately the average of these distances. In general, you may find it helpful to view the standard deviation as approximately the average distance that a set of observations are from their mean.

EXAMPLE 11 Find the standard deviation of the nine reading scores in Example 4. The scores are

58 76 75 66 90 90 71 76 80

Solution From Example 4, $\bar{x} = 75.7778$. The details of the computations used in obtaining s are as follows:

x_i	$x_i - 75.7778$	$(x_i - 75.7778)^2$
58	−17.7778	316.05017
76	.2222	.04937
75	−.7778	.60497
66	−9.7778	95.60537
90	14.2222	202.27097
90	14.2222	202.27097
71	−4.7778	22.82737
76	.2222	.04937
80	4.2222	17.82697
		857.55553

$$s^2 = \frac{857.55553}{8}$$

$$= 107.19444$$

Thus $s = \sqrt{107.19444} = 10.35347$ or 10.4. ■

As Example 11 shows, computing s^2 or s involves many steps which are somewhat messy, even with a calculator. Computing s^2 can be simplified considerably by using a computational form for s^2. More specifically, it can be shown that s^2 can be written as

$$s^2 = \frac{1}{n(n-1)}[n(x_1^2 + x_2^2 + \cdots + x_n^2) - (x_1 + x_2 + \cdots + x_n)^2]$$

or in summation notation,

$$s^2 = \frac{1}{n(n-1)}\left[n \sum_{i=1}^{n} x_i^2 - \left(\sum_{i=1}^{n} x_i\right)^2\right] \qquad \text{(Eq. 1.4)}$$

For clarity in the following discussions, we shall use simply Σ in place of $\sum_{i=1}^{n}$. Using Equation 1.4, we proceed as outlined in Table 7.

TABLE 7

Simplified procedure for computing the variance s^2

1. Square each of the observations: x_i^2.
2. Sum the squared values in step 1: Σx_i^2.
3. Multiply the sum of the squares from step 2 by n: $n \Sigma x_i^2$.
4. Sum the observations: Σx_i.
5. Square the sum in step 4: $(\Sigma x_i)^2$.
6. Subtract the result in step 5 from that in step 3: $n \Sigma x_i^2 - (\Sigma x_i)^2$.
7. Divide the result in step 6 by $n(n - 1)$:

$$s^2 = \frac{1}{n(n-1)}\left[n \sum x_i^2 - \left(\sum x_i\right)^2\right]$$

Notice that step 2 produces Σx_x^2, which is a *sum of squares,* whereas step 5 produces $(\Sigma x_i)^2$, which is the *square of a sum.*

Let us apply Table 7 to the reading scores. The components of Equation 1.4 for these scores are conveniently supplied in the following table:

x_i	x_i^2
58	3,364
76	5,776
75	5,625
66	4,356
90	8,100
90	8,100
71	5,041
76	5,776
80	6,400
$\Sigma x_i = 682$	$\Sigma x_i^2 = 52{,}538$

$$s^2 = \frac{9(52{,}538) - 682^2}{9(9-1)}$$

$$= \frac{472{,}842 - 465{,}124}{72}$$

$$= \frac{7718}{72} = 107.19444$$

$$s = \sqrt{107.19444} = 10.4$$

As you can see, Equation 1.4 involves fewer and simpler steps for computing s^2 than does Equation 1.3.

EXAMPLE 12 Compute the range and the standard deviation of the surgery data from Example 1 given that $\Sigma x_i = 605.3$ and $\Sigma x_i^2 = 6259.69$.

Solution From the stem-and-leaf display for these data in Figure 3, the range of times is $R = 15.7 - 7.3 = 8.4$ hours

Applying Equation 1.4 for s^2 with $n = 60$ yields

$$s^2 = \frac{1}{60(59)}[60(6259.69) - (605.3)^2] = \frac{9193.31}{3540}$$

$$= 2.59698$$

$$s = \sqrt{2.59698} = 1.6115 \quad \text{or} \quad 1.61 \text{ hours}$$

■

COMPUTING THE VARIANCE FOR DISCRETE DATA

With discrete data, when a few values are repeated many times a slightly different version of Equation 1.4 is helpful, as seen in the following example.

EXAMPLE 13 The following table shows the distribution of the number of visits per prisoner for the 15 prisoners in Example 8.

NUMBER OF VISITS, i	NUMBER OF PRISONERS, f_i
0	3
1	6
2	5
3	1
	15

Compute the standard deviation of the number of visits.

Solution From Equation 1.4 we need the total of the squared numbers of visits, Σx_i^2, and the total number of visits, Σx_i. Using the frequencies above, $\Sigma x_i^2 = x_1^2 + x_2^2 + \cdots + x_{15}^2 = (0^2 \times 3) + (1^2 \times 6) + (2^2 \times 5) + (3^2 \times 1) = 35$ and $\Sigma x_i = (0 \times 3) + (1 \times 6) + (2 \times 5) + (3 \times 1) = 19$. Thus

$$s^2 = \frac{1}{15(14)}[15(35) - 19^2] = .78095$$

and $s = .8837$ or .9 visit.

■

To generalize: Call f_0 the frequency of the zeros, f_1 the frequency of the 1's, f_2 the frequency of the 2's, and so on. Then

$$s^2 = \frac{1}{n(n-1)}\left[n[(0^2 \times f_0) + (1^2 \times f_1) + (2^2 \times f_2) + \cdots + (k^2 \times f_k)] - [(0 \times f_0) + (1 \times f_1) + (2 \times f_2) + \cdots + (k \times f_k)]^2\right]$$

or, using summation notation,

$$s^2 = \frac{1}{n(n-1)}\left[n \sum_{i=0}^{k} i^2 f_i - \left(\sum_{i=0}^{k} i f_i\right)^2\right] \qquad \text{(Eq. 1.5)}$$

where k is the largest value in the set of observations.

EXAMPLE 14 The examination data from Example 2 are summarized as follows:

NUMBER OF PREVIOUS ATTEMPTS, i	NUMBER OF STUDENTS, f_i
0	24
1	20
2	12
3	6
4	2
5	2
	66

Compute the standard deviation of the number of previous attempts at the examination.

Solution From Equation 1.5, with $n = 66$ and $k = 5$

$$\sum_{i=1}^{5} i^2 f_i = (0^2 \times 24) + (1^2 \times 20) + (2^2 \times 12) + (3^2 \times 6) + (4^2 \times 2) + (5^2 \times 2)$$

$$= 204$$

$$\sum_{i=1}^{5} i f_i = (0 \times 24) + (1 \times 20) + (2 \times 12) + (3 \times 6) + (4 \times 2) + (5 \times 2)$$

$$= 80$$

$$s^2 = \frac{1}{66(65)}[66(204) - 80^2] = \frac{7064}{66(65)} = 1.64662$$

and $s = \sqrt{1.64662} = 1.2832$ or 1.3 previous attempts. ■

THE EMPIRICAL RULE

We can represent s, the standard deviation of a set of data, as a distance along the base of the relative frequency histogram. Let us see how this can be done with the surgery data. In Example 12 we noted that $\Sigma x_i = 605.3$, so $\bar{x} = 605.3/60 = 10.0883$ or 10.09 hours. In Example 12 we also computed $s = 1.61$ hours. The relative frequency histogram for these data from Figure 1 is reproduced in Figure 11.

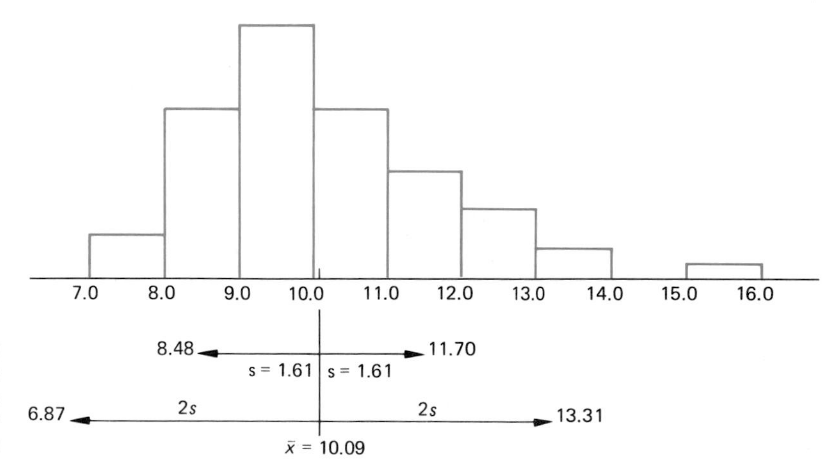

FIGURE 11

Displaying the standard deviation as a distance along the base of the relative frequency histogram

A distance of 1 standard deviation above the mean brings us to $10.09 + 1.61 = 11.70$ hours. A similar distance below the mean brings us to 8.48 hours. In the same way, a distance of 2 standard deviations above and below the mean brings us to the values 6.87 hours and 13.31 hours.

When a large number of observations have been obtained, there is a useful rule—called the **empirical rule**—for predicting what proportion of the observations will occur within 1, 2, or 3 standard deviations of the mean. The rule is summarized as follows:

> DEFINITION 5
>
> 1. Approximately 68% of the observations will occur within 1 standard deviation of the mean (i.e., between $\bar{x} - s$ and $\bar{x} + s$).
> 2. Approximately 95% of the observations will occur within 2 standard deviations of the mean (i.e., between $\bar{x} - 2s$ and $\bar{x} + 2s$).
> 3. Approximately 99.75% of the observations will occur within 3 standard deviations of the mean (i.e., between $\bar{x} - 3s$ and $\bar{x} + 3s$).

The rule is illustrated in Figure 12.

The empirical rule works best when the histogram has the symmetric bell

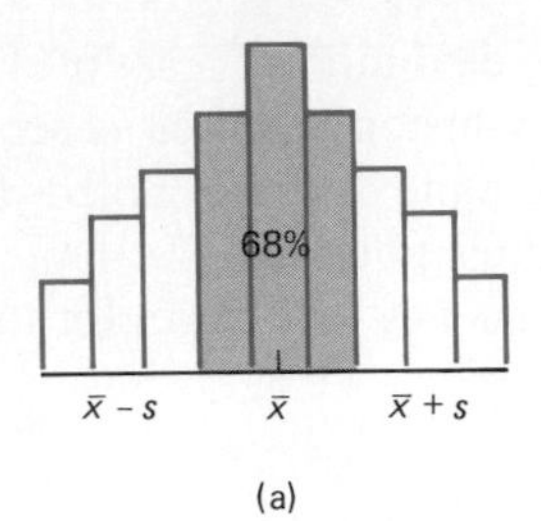

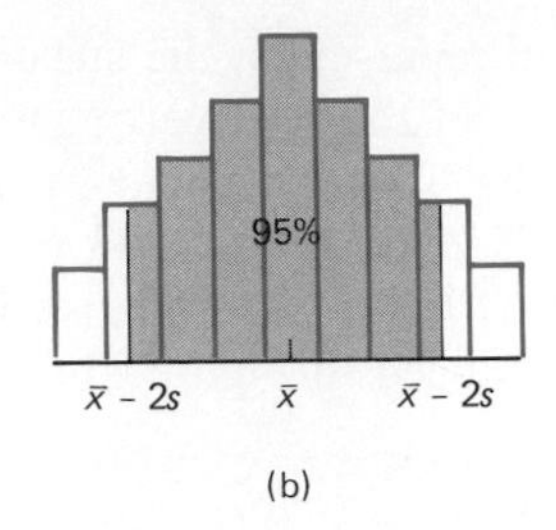

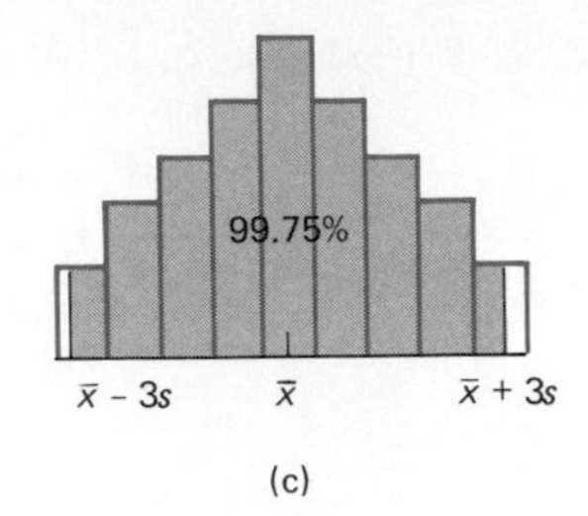

FIGURE 12

Empirical rule

shape shown in Figure 12, but it will also work well if the histogram is not too far from this shape. For example, the histogram for the surgery data (in Figure 11) is skewed to the right, but the empirical rule does well. Referring to the stem-and-leaf display for these data in Figure 3, you can check that 42, or 70% of the times occur between 8.48 and 11.70 hours, and 58, or 97%, of the times occur between 6.87 and 13.31 hours. These percentages are close to the predicted 68% and 95%.

AN APPROXIMATE RELATIONSHIP BETWEEN THE RANGE AND THE STANDARD DEVIATION

The empirical rule can also be used to provide a simple way of relating the range, R, to the standard deviation, s, of a set of data. The empirical rule predicts that most (95%) of the observations will lie within 2 standard deviations above and below the mean. This suggests that the range of observations, R, should be roughly 4 standard deviations. This approximate relationship can be summarized as

$$R \doteq 4s \qquad \text{(Eq. 1.6)}$$

The notation "$\doteq$" means "approximately equal to." This relationship between R and s enables us to obtain a "rough" or "ballpark" figure for s using the easier-to-compute range. If the range of a set of data is 12.6 for instance, the standard deviation s, will be *about* $12.6/4 = 3.15$. In fact, s may be 4.05 or 2.69. But if s is computed as 40.5 or .269, an arithmetic mistake has almost certainly occurred.

This approximation, like the empirical rule, works best when the data have a bell shape as in Figure 12, but it, also, works fairly well if the data are not too far from symmetric. For the surgery data, for instance, $R = 8.4$ hours, so a ballpark estimate for s is $s \doteq 8.4/4 = 2.1$ hours. In fact, in Example 12 we computed $s = 1.61$ hours.

We recommend that before you compute s, you first use Equation 1.6 to obtain a rough estimate. Remember, though, that the rough estimate is not a substitute for the computed value of s. Rather, it is a check for errors in computing s.

PROBLEMS 1.4

28. The six students in a seminar record the time spent (in hours) preparing their final papers as follows:

$$14 \quad 17 \quad 25 \quad 18 \quad 12 \quad 20$$

Compute the (a) range and (b) standard deviation of these times.

29. The eight weather stations in a state report the following temperatures (in degrees Fahrenheit) at midnight on February 3:

$$-12 \quad -16 \quad -14 \quad -22 \quad -18 \quad -20 \quad -9 \quad -12$$

Compute the (a) range and (b) standard deviation of these temperatures.

30. Compute the range and standard deviation of age at appointment for the members of the U.S. Supreme Court. The ages are:

$$61 \quad 50 \quad 43 \quad 44 \quad 59 \quad 61 \quad 64 \quad 47 \quad 55$$

31. Compute the (a) range, (b) variance, and (c) standard deviation for the weight-loss data in Problem 19 on page 28.

$$20.7 \quad 10.8 \quad 4.9 \quad -1.4 \quad 7.3 \quad 8.9 \quad 12.4 \quad -4.0 \quad 9.4 \quad 6.9$$

32. (a) Using the supermarket data in Problem 5 on page 16, compute (i) the range of dollars spent per person; (ii) a rough estimate for s based on the range; (iii) the actual standard deviation of outlays. (Use the fact that $\Sigma\, x_i = 702.47$ and $\Sigma\, x_i^2 = 20{,}980.601$.)

    (b) What proportion of the outlays lie within (i) 1 standard deviation of $\bar{x}$; (ii) 2 standard deviations of $\bar{x}$?

33. (a) Using the speed data in Problem 8 on page 17, compute (i) the range of speeds; (ii) a rough estimate for s based on the range; (iii) the standard deviation of the speeds. (Use the fact that $\Sigma\, x_i = 2407$ and $\Sigma\, x_i^2 = 146{,}343$.)

    (b) What proportion of the speeds lie within (i) 1 standard deviation of $\bar{x}$; (ii) 2 standard deviations of $\bar{x}$?

34. From Problem 22 on page 28, the number of misprints for 20 pages are

$$0 \quad 2 \quad 2 \quad 1 \quad 2 \quad 2 \quad 2 \quad 1 \quad 1 \quad 2 \quad 2 \quad 1 \quad 1 \quad 0 \quad 1 \quad 0 \quad 1 \quad 3 \quad 2 \quad 2$$

Compute the standard deviation of the number of misprints.

35. Using the goals data in Problem 11 on page 18, compute (a) the range of the number of goals; (b) a rough estimate for s; (c) the standard deviation of the

36. Among a group of 400 teachers the mean income is \$20,500 and the standard deviation of income is \$3750. The distribution of income is approximately bell-shaped. Use the empirical rule to compute how many of the 400 incomes you might expect to be within (a) 1, (b) 2, and (c) 3 standard deviations of \$20,500.

37. In a study of 30 first graders a test score for each child was recorded. The range of scores was 17 points.

    (a) What would you use as a rough estimate for s, the standard deviation of test scores?

    (b) What kind of mistake would you expect if a value $s = .684$ were computed?

38. If the standard deviation of the number of children in 50 families is $s = 2.05$, what figure would you use as an estimate for the range?

Solution to Problem 28

(a) $R = 25 - 12 = 13$ hours.

(b) For these data,

$$\sum x_i = 14 + 17 + 25 + 18 + 12 + 20 = 106$$

$$\sum x_i^2 = 14^2 + 17^2 + 25^2 + 18^2 + 12^2 + 20^2 = 1978$$

and

$$s^2 = \frac{n \Sigma x_i^2 - (\Sigma x_i)^2}{n(n-1)} = \frac{6(1978) - 106^2}{6(5)} = \frac{632}{30} = 21.06667$$

$$s = \sqrt{21.06667} = 4.5898 \quad \text{or} \quad 4.6 \text{ hours}$$

SECTION 1.5

QUARTILES AND BOX-AND-WHISKER DISPLAYS

QUARTILES In section 1.3 we noted that the median divides an ordered set of data into two equal parts. It is often useful to go one step further and divide the data into four approximately equal parts. The point below which approximately 25% of the data lie is then called the **first quartile** and designated Q_1. The median, m, the point below which 50% of the data lie, is sometimes referred to as the **second quartile,** Q_2, and the point below which approximately 75% of the data lie is called the **third quartile,** Q_3.

The quartiles Q_1 and Q_3 are computed in the following examples.

EXAMPLE 15 Spot checks of the amount of fluoride in the water at nine different locations of a town reservoir result in the following observations (in parts of fluoride per million gallons):

1.31 .94 1.40 1.26 1.05 1.17 .98 1.25 1.09

First we order the data:

.94 .98 1.05 1.09 (1.17) 1.25 1.26 1.31 1.40

The median of the ordered data is 1.17.

We next divide the entire set of numbers into two equal parts. Since there are nine numbers, we include the median in each part to make them equal. The two parts are then

.94 .98 (1.05) 1.09 1.17 and 1.17 1.25 (1.26) 1.31 1.40

Q_1 is the median of the set of five smaller numbers = 1.05 and Q_3 is the median of the five larger numbers = 1.26. In summary,

$$Q_1 = 1.05 \qquad Q_2 = m = 1.17 \qquad Q_3 = 1.26$$

■

EXAMPLE 16 We now reconsider the fluoride data with the observation 1.40 removed so that there are only eight (ordered) observations:

$$.94 \quad .98 \quad 1.05 \quad 1.09 \quad 1.17 \quad 1.25 \quad 1.26 \quad 1.31$$

Q_2, the median of these eight observations is $(1.09 + 1.17)/2 = 1.13$. With eight observations, we can divide the entire set of numbers into two equal parts with four each, as follows:

$$.94 \quad .98 \quad 1.05 \quad 1.09 \qquad \text{and} \qquad 1.17 \quad 1.25 \quad 1.26 \quad 1.31$$

Q_1, the median of the four smaller numbers $= (.98 + 1.05)/2 = 1.015$, and Q_3, the median of the four larger numbers $= (1.25 + 1.26)/2 = 1.255$.

The three quartiles are shown on the dot diagram in Figure 13. In this case the quartiles divide the data into four groups of two observations each. Notice that the term "quartile" refers to the *three values* or *points* that divide the numbers into four groups.

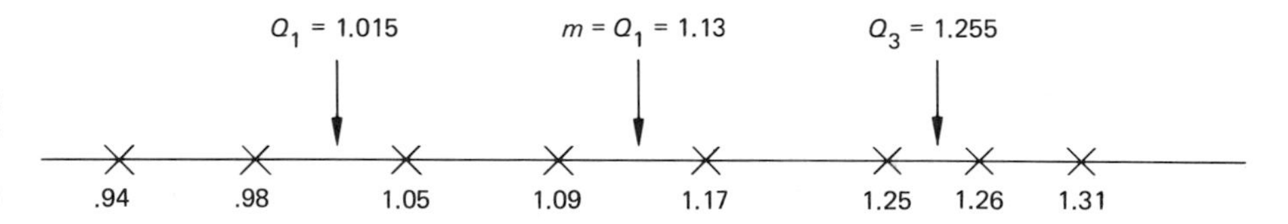

FIGURE 13

The three quartiles in Example 16

The general procedure for finding the three quartiles is outlined in Table 7.

TABLE 7

Procedure for finding the three quartiles

1. Order the n observations from smallest to largest.
2. The second quartile, Q_2, is the median, m, of the n observations.
3. If n is even, Q_1 is the median of the smallest $n/2$ observations and Q_3 is the median of the largest $n/2$ observations.
4. If n is odd, Q_1 is the median of the smallest $(n + 1)/2$ observations, and Q_3 is the median of the largest $(n + 1)/2$ observations. (Both groups will contain Q_2.)

EXAMPLE 17 Find the three quartiles of the 60 surgical times in Example 1. For convenience the stem-and-leaf display representing the data is reproduced in Figure 14.

```
 7.* | 3 7 8
 8.* | 0 1 3 4 5 6 6 6 6 8 8 9
 9.* | 0 1 2 2 4 4 4 5 5 5 6 6 6 6 6 7 7 9      Q1 = 8.95
10.* | 1 1 1 2 3 3 5 7 7 8 9 9                 Q2 = m = 9.65
11.* | 0 0 1 3 5 7 8                           Q3 = 10.95
12.* | 2 4 4 6 7
13.* |  2 6
14.* |
15.* | 7
```

FIGURE 14

Stem-and-leaf display for the 60 surgical times

Solution Following the steps in Table 7 with $n = 60$, the median is the mean of the 30th and 31st observations, 9.6 and 9.7. Thus $m = Q_2 = 9.65$.

Since $n = 60$ is even, Q_1 is the median of the smallest 30 observations. This in turn is the mean of the 15th and 16th numbers, 8.9 and 9.0 so that $Q_1 = 8.95$. Similarly, Q_3 is the median of the largest 30 observations which is the mean of the 45th and 46th numbers, 10.9 and 11.0. Thus $Q_3 = 10.95$. To summarize, $Q_1 = 8.95$, $Q_2 = m = 9.65$, and $Q_3 = 10.95$. These three quartiles divide the 60 numbers into four groups, each with 15 observations.

■

THE INTERQUARTILE RANGE

A useful measure of spread, called the **interquartile range** (IR), is the difference, $Q_3 - Q_1$. It is less affected by unusually small or large observations than are the range and the standard deviation. The IR is frequently used with ungrouped data as a measure of spread in conjunction with the use of the median as a measure of location.

EXAMPLE 18 Compute the range and the interquartile range for the fluoride data in Example 15.

Solution From the nine observations, $R = 1.40 - .94 = .46$. In Example 15 we obtained $Q_1 = 1.05$ and $Q_3 = 1.26$. Thus IR $= Q_3 - Q_1 = 1.26 - 1.05 = .21$.

Had the largest observation been 1.80 instead of 1.40, the range would be .86, an increase of almost 100% due to a change in a single observation. By contrast, the interquartile range would not be affected by this change.

■

THE FIVE-NUMBER SUMMARY AND BOX-AND-WHISKER DISPLAYS

Still another convenient way to summarize a set of data is the **five-number summary.** The five numbers are the three quartiles, Q_1, Q_2, and Q_3, plus the smallest and largest observations. For instance, the five-number summary for the surgery data from Example 1 is

15.70
10.95
9.65
8.95
7.30

A useful way to represent the five-number summary graphically is a **box-and-whisker display.** This is shown for the surgery data in Figure 15.

A box-and-whisker display has the following characteristics:

1. It represents the values of the five-number summary on a vertical scale, in this case "time" in hours.
2. The box is a rectangle with the height = the interquartile range. Here IR $= 10.95 - 8.95 = 2.00$ hours.

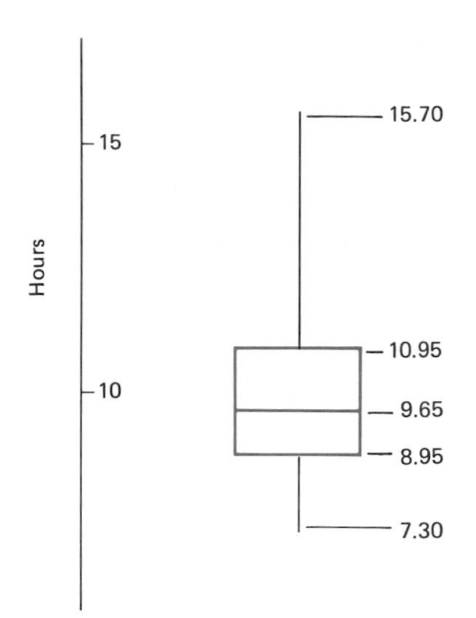

FIGURE 15

Box-and-whisker display for the surgical data

3. The width of the box is arbitrary and, therefore, not significant.
4. The horizontal line through the box represents the median value, here 9.65 hours.
5. The whiskers are the two lines stretching respectively from Q_3 up to the largest observation and from Q_1 down to the smallest observation. The lengths of the whiskers are then:

$$\text{upper} = \text{largest} - Q_3 \qquad \text{lower} = Q_1 - \text{smallest}$$

The box-and-whisker display also reflects the following:

6. One-fourth (here 15) of all the observations fall between successive numbers in the five-number summary: that is, smallest $- Q_1$, $Q_2 - Q_1$, $Q_2 - Q_3$, and $Q_3 -$ largest.
7. The *distances* between successive numbers in a five-number summary are *not* necessarily the same, as we emphasize in this hypothetical dot diagram:

Bearing the last two points in mind, the shape of the box-and-whisker display in Figure 15 suggests that in the case of the surgery data the distribution of the observations within the interquartile range is reasonably symmetrical around the median. On the other hand, the *length* of the upper whisker $(15.7 - 10.95 = 4.75)$ is almost three times the length $(8.95 - 7.3 = 1.65)$ of the lower whisker. This indicates that the right tail of the relative frequency distribution will be longer than the left tail. The stem-and-leaf display in Figure 14 verifies this.

In the following example, box-and-whisker displays are used to compare two sets of data.

EXAMPLE 19 The salaries of the 30 male and 23 female members of the faculty of a small college are as follows (in thousands of dollars), ordered for convenience:

Men:	12.2	12.5	12.7	13.2	13.6	14.1	14.6	14.8	15.2	15.8
	16.2	16.2	17.1	18.1	19.3	19.5	19.5	19.9	20.6	21.5
	22.3	23.4	23.8	24.4	25.1	25.4	26.1	27.3	28.0	30.4
Women:	12.3	12.6	13.0	13.1	13.4	13.4	13.6	13.9	14.2	14.7
	15.3	16.0	16.8	17.4	18.5	19.0	20.4	20.4	21.3	21.4
	23.1	24.1	26.9							

(a) Compute the five-number summary for male and for female faculty members. (b) Draw the corresponding box-and-whisker displays and point out any striking differences between the two displays.

Solution (a) There are 30 men. Their median salary is therefore the mean of the two ordered observations: $n/2 = 15$th and $(n/2) + 1 = 16$th. The values of these observations are 19.3 and 19.5, so that $m = Q_2 = (19.3 + 19.5)/2 = 19.4$.

Q_1 is the median of the lower $n/2$ observations $= 30/2 = 15$, so Q_1 is the eighth smallest observation $= 14.8$. Similarly, Q_3 is the median of the upper 15 observations = the eighth largest = 23.8.

There are 23 women; therefore, Q_2, the median salary for women, is the 12th smallest = 16.0.

Q_1 is the median of the women's 12 smallest salaries. Since 12 is even, this is the mean of the sixth and seventh smallest salaries $= (13.4 + 13.6)/2 = 13.5$. Similarly, Q_3 is the median of the women's 12 largest salaries $= (20.4 + 20.4)/2 = 20.4$.

The five-number summary for each group is, therefore,

	MEN	WOMEN
LARGEST	30.4	26.9
Q_3	23.8	20.4
m	19.4	16.0
Q_1	14.8	13.5
SMALLEST	12.2	12.3

(b) The corresponding box-and-whisker displays are shown in Figure 16. We see from Figure 16 that, overall, women faculty members tend to earn less than their male counterparts. Four of the five summary numbers for women are smaller than the corresponding numbers for men.

Notice also how much closer to symmetric the distribution for men is than that for women. For the women the range of incomes for the top 50% is $26.9 - 16.0 = 10.9$. But for the lower 50% of women, the range of incomes

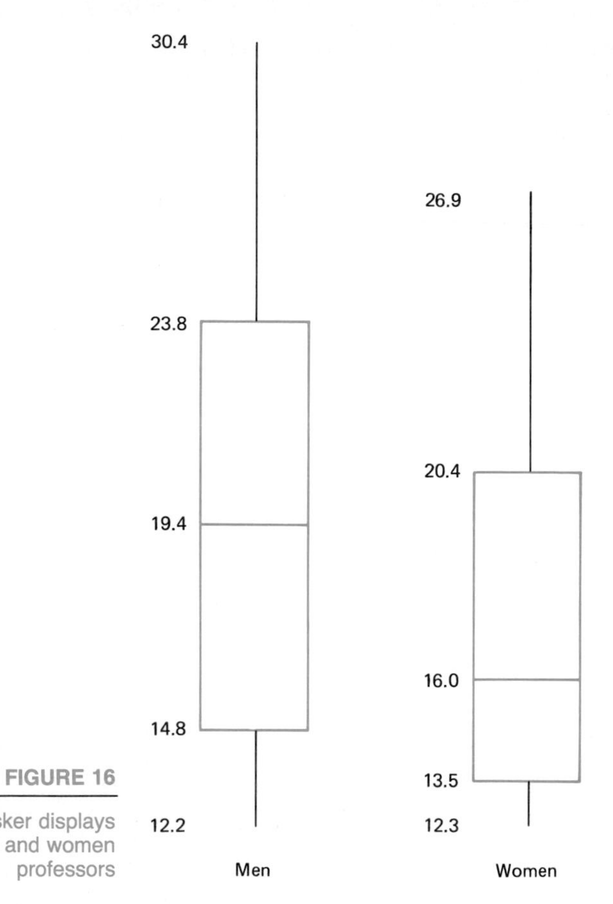

FIGURE 16

Box-and-whisker displays for men and women professors

is only $16.0 - 12.3 = 3.7$. By contrast, the corresponding ranges for the men, $30.4 - 19.4 = 11.0$ and $19.4 - 12.2 = 7.2$, are much closer to one another. ■

PROBLEMS 1.5

39. An orchestra performs each of the 10 symphonies of a composer. The length of the symphonies (in minutes) are

    100 95 107 86 99 68 92 112 80 77

    Compute (a) the three quartiles; (b) the interquartile range; (c) the five-number summary, for the data.

40. The price of the same item in 16 different stores in a city is recorded (in dollars) as

    8.60 2.10 9.40 10.82 11.51 5.85 9.32 10.31
    10.10 6.25 5.91 1.99 8.42 7.85 8.73 5.86

(a) Compute (i) the three quartiles; (ii) the interquartile range; (iii) the five-number summary.

(b) Construct the corresponding box-and-whisker display. Comment on its shape.

41. Construct the (a) five-number summary and (b) box-and-whisker display for the temperature data in Problem 6 on page 16.

42. The Scholastic Aptitude Test (SAT) verbal scores are recorded for each senior in a school district. The five-number summary is

620
474
420
366
222

(a) Compute (i) the range; (ii) the interquartile range.

(b) Construct and comment on the shape of the box-and-whisker display.

(c) Interpret the five numbers in the five-number summary for a person who does not know any statistics.

43. A biologist collects 100 specimens of each of two species of salamander and records the length of each creature (in inches). The five-number summary for each species is as follows:

SPECIES A	SPECIES B
3.40	4.29
2.92	3.35
2.65	2.63
2.33	1.89
1.93	1.32

(a) Compute the (i) range and (ii) interquartile range for each species.

(b) Construct the box-and-whisker display for each species and comment on differences between the two displays.

44. The amount of annual rainfall (in inches) recorded in 25 sites in Alabama and Louisiana was recorded as follows:

Alabama:	6.3	5.6	3.1	4.3	6.9	6.7	5.9	3.7	6.6	5.7
	4.4	4.8	7.0	4.3	3.6	6.9	8.1	5.3	5.0	2.9
	4.4	6.0	7.5	6.6	6.5					
Louisiana:	5.6	4.5	6.4	6.6	7.9	7.0	3.1	3.5	5.4	5.8
	6.2	4.6	6.4	6.8	6.6	7.2	6.2	5.5	4.8	4.2
	4.3	3.1	3.8	5.1	7.6					

(a) Construct a five-number summary and the corresponding box-and-whisker display for each state.

(b) Comment on any similarities or differences between the two sets of rainfall.

45. In a comparative study at a major maternity hospital, the age at birth for the first 40 first-pregnancy mothers of 1955 was compared with the age at birth of the first 40 first-pregnancy mothers of 1975. The 80 ages were as follows:

1955:	21	29	25	32	37	30	24	36	23	19	37	22	26	
	31	26	27	19	21	33	17	24	21	22	36	22	25	
	31	20	18	20	16	21	25	26	34	27	18	39	24	21

1975:	24	22	22	32	17	28	38	20	30	39	19	33	44	
	24	18	27	24	33	29	23	26	18	28	32	43	28	
	26	22	41	28	25	35	31	23	19	46	20	23	34	29

(a) Construct a five-number summary and the corresponding box-and-whisker display for each set of data.

(b) Comment on the differences between the two sets of data.

Solution to Problem 39

The 10 ordered times are

68 77 80 86 92 95 99 100 107 112

(a) The median $m = Q_2 = 93.5$ is the mean of the two middle times 92 and 95. Following the rules in Table 7 with $n = 10$, $Q_1 = 80$ is the median of the five shortest times and $Q_3 = 100$, the median of the five longest times.

(b) The interquartile range is $\text{IR} = Q_3 - Q_1 = 100 - 80 = 20$ minutes.

(c) The five-number summary is

112
100
93.5
80
68

SECTION 1.6

USING COMPUTER OUTPUT: MINITAB AND DATA ANALYSIS

INTRODUCTION TO MINITAB

With the widespread availability of computers most of the drudgery of excessive statistical calculations can be avoided by using a statistical package. Such packages are simply collections of standardized computer programs designed to perform a range of statistical analyses on data. In this section we indicate how one such package, Minitab, performs many of the computations covered in this chapter. The commands in Minitab are so close to conversational English that the package can be used by students with no previous experience with computers or programming.

Minitab may be used in **batch** mode or **interactively.** In **batch mode** the data and Minitab commands are punched on cards which are later fed into the computer. The printed output is usually available to the user some hours later. The better way of using Minitab is **interactively.** The user enters the data and the commands from a terminal, often remote from the actual computer, and the results are printed out almost instantaneously. One advantage of this approach is that errors can be corrected immediately. All the output on the following pages were obtained interactively.

THE NUTRITION DATA

For small amounts of data most routine analyses can be performed with a calculator. However, if data are collected on more than two or three variables and for more than 20 or so persons, a statistical package is almost a necessity. To illustrate Minitab, we shall make extensive use throughout the text of the data set in Table 8. These data were collected as part of a recent study of low-income pregnant women. Table 8 contains the data for a subsample of 68 of these women.

TABLE 8

Nutrition Data

(C1) ETH	(C2) AGE	(C3) SMOK	(C4) PRWT	(C5) DVWT	(C6) BSTFD	(C7) BTWT	(C8) BTLH	(C9) TIM
1	29	1	115	140	0	3310	45	99
2	33	1	112	126	0	2650	48	64
2	19	1	125	145	0	2900	49	60
1	26	2	108	146	1	3500	51.5	102
1	35	1	112	133	1	2600	51	77
2	20	1	115	137	0	3770	52	110
2	22	1	99	135	1	2970	49	125
1	23	2	140	178	0	3250	51	50
1	20	3	115	150	0	3260	50	67
2	26	1	135	172	0	2960	48	32
3	23	1	103	138	1	3575	52.5	90
2	33	1	191	215	0	3870	55	100
1	32	2	125	156	0	2380	47	87
1	20	1	112	140	0	3800	50.5	83
2	21	1	140	165	0	3460	53	86
3	41	1	170	202	1	2000	45.5	87
1	20	1	115	138	1	3206	52	76
1	24	1	130	160	1	3780	51	91
3	24	1	107	146	0	3350	52	70
1	21	1	180	206	0	2980	47	102
1	27	1	125	163	1	3300	50	98
1	25	2	140	160	1	3120	52	96
3	26	1	120	141	1	3210	50	105
2	34	1	189	200	1	3600	48	81
3	22	1	120	155	0	2800	46	63
2	28	1	140	182	1	3760	50	110
1	25	1	133	164	1	3850	53	100
3	28	1	92	117	1	2950	50	96
1	25	1	165	224	1	4458	60	128
2	27	1	125	148	1	3860	52	102
3	22	1	118	161	1	3493	49	82
1	23	3	135	173	1	2650	49.5	48
3	27	3	127	148	1	3200	48	65
1	20	3	87	163	1	3300	50	98
2	28	2	165	187	0	3640	51	57
1	21	1	112	138	0	4080	54	79
3	31	1	138	148	0	2950	50	66
1	17	2	120	139	1	3560	53	102
2	24	3	125	150	0	2760	45	82
2	24	2	125	150	0	2760	45	126
2	25	1	175	190	1	3664	47	77
2	20	2	128	159	0	3070	48	84
3	35	1	140	174	0	3940	48	105
1	25	3	144	180	0	3050	51	80
1	29	1	125	160	1	3350	48	85
2	27	1	150	183	1	3360	52	90
1	24	1	105	129	1	3170	52	99

TABLE 8 *(continued)*

(C1) ETH	(C2) AGE	(C3) SMOK	(C4) PRWT	(C5) DVWT	(C6) BSTFD	(C7) BTWT	(C8) BTLH	(C9) TIM
1	22	1	128	172	1	2850	53	90
1	25	1	90	125	1	3465	53.5	85
2	29	1	148	191	1	3730	53	72
1	22	1	155	162	0	3750	51	65
3	30	1	128	160	1	3460	48	125
2	16	1	113	138	0	2360	47	72
1	21	3	115	142	1	3030	50	84
2	26	1	98	129	1	2820	49	71
3	23	1	115	129	0	2950	45	93
2	17	3	99	124	0	3300	51	88
2	28	3	152	168	0	3000	41	71
1	25	1	148	173	1	2880	52	92
1	20	1	105	152	1	3460	50	69
1	36	1	145	180	1	2826	51	105
1	23	1	264	250	1	3380	51	115
1	20	3	190	255	1	3124	50	85
1	17	1	85	125	0	3240	48	67
2	19	2	137	172	0	2400	46	60
2	19	1	102	147	0	3000	47	105
2	25	1	125	152	0	3050	52	100
2	19	1	132	156	0	3360	51	84

The variables listed in Table 8 are identified in Table 9.

TABLE 9 The nine variables in Table 8

VARIABLE	DESCRIPTION
ETH	Ethnic background (1 = white, 2 = black, 3 = Hispanic)
AGE	Age of mother
SMOK	Smoking habits of mother (1 = nonsmoker, 2 = light smoker, 3 = heavy smoker)
PRWT	Weight of mother (in pounds) at conception
DVWT	Weight of mother (in pounds) at delivery
BSTFD	Whether or not mother is breastfeeding (1 = breastfeeding, 0 = not breastfeeding)
BTWT	Infant birthweight (in grams)
BTLH	Length of the infant (in centimeters)
TIM	The time (in minutes) that mother spent with nutritionist during pregnancy

DATA ANALYSIS WITH MINITAB

We begin by illustrating one method for entering data and obtaining frequency histograms.

EXAMPLE 20

Minitab allows the user 50 columns in which to store data. In Table 10 we show the command "READ DATA INTO C1-C9" on the first line followed by the nutrition data in the appropriate columns, using a separate line for each person. The command above will place the ETH values in column 1 (C1), the AGEs in column 2 (C2), and so on. The command "HISTOGRAM OF C7" produces a frequency histogram of the infant birthweights (BTWT) in column 7 (C7). Instead of writing each interval in the form 1900–under 2100, 2100–under 2300, and so on, Minitab prints the *midpoint* of each interval: 2000, 2200, and so on. The frequency histogram is formed by rows of asterisks, the number in each row corresponding to the frequency of observations in the interval. The histogram

TABLE 10

Entering the nutrition data and obtaining a frequency histogram of BTWT (the output is shaded)

```
MTB > READ DATA INTO C1-C9
DATA> 1,29,1,115,140,0,3310,45,99
DATA> 2,33,1,112,126,0,2650,48,64
DATA> 2,19,1,125,145,0,2900,49,60
DATA> 1,26,2,108,146,1,3500,51.5,102
DATA> 1,35,1,112,133,1,2600,51,77
DATA> 2,20,1,115,137,0,3770,52,110
DATA> 2,22,1,99,135,1,2970,49,125
DATA> 1,23,2,140,178,0,3250,51,50
DATA> 1,20,3,115,150,0,3260,50,67
DATA> 1,20,3,190,255,1,3124,50,85
DATA> 1,17,1,85,125,0,3240,48,67
DATA> 2,19,2,137,172,0,2400,46,60
DATA> 2,19,1,102,147,0,3000,47,105
DATA> 2,25,1,125,152,0,3050,52,100
DATA> 2,19,1,132,156,0,3360,51,84
```

```
MTB > HISTOGRAM OF C7

  C7

  MIDDLE OF    NUMBER OF
  INTERVAL     OBSERVATIONS
     2000          1     *
     2200          0
     2400          3     ***
     2600          3     ***
     2800          7     *******
     3000         13     *************
     3200          9     *********
     3400         14     **************
     3600          6     ******
     3800          9     *********
     4000          2     **
     4200          0
     4400          1     *
```

indicates that the distribution of infant birthweight is slightly skewed to the left, with a few more low-weight than high-weight infants.

■

EXAMPLE 21 One aspect of considerable interest in the nutrition study was the increase in the mothers' weights during pregnancy. This increase is DVWT–PRWT. In Table 11 we indicate how this increase is obtained for each woman by using the command "SUBTRACT C4 FROM C5, PUT INTO C10." With this command Minitab computes the 68 increases in weight and inserts them in column 10 (C10). In Table 11 we also indicate the commands necessary to obtain the frequency histogram for the weight increases (C10), their mean ("AVERAGE OF C10"), median ("MEDIAN OF C10"), and standard deviation ("STANDARD DEVIATION OF C10").

TABLE 11

Summarizing the data on the increase in the mothers' weights

```
MTB > SUBTRACT C4 FROM C5, PUT INTO C10
MTB > HISTOGRAM OF C10

 C10

   MIDDLE OF    NUMBER OF
   INTERVAL     OBSERVATIONS
       -10          1     *
         0          0
        10          5     *****
        20         15     ***************
        30         22     **********************
        40         20     ********************
        50          2     **
        60          1     *
        70          1     *
        80          1     *

MTB > AVERAGE OF C10
   MEAN     =       29.779
MTB > MEDIAN OF C10
   MEDIAN =        29.000
MTB > STANDARD DEVIATION OF C10
   ST.DEV. =        13.074
MTB > MINIMUM OF C10
   MINIMUM =       -14.000
```

The negative midpoint of the interval on the histogram of C10 prompted one of the authors to use the command "MINIMUM OF C10" to obtain the (surprising) result that one of the women actually *lost* 14 points during her pregnancy. This proved not to be a typing error; nor had the values for PRWT and DVWT been inadvertently reversed. The woman in question proved to have been seriously overweight. On the advise of her nutritionist she conscientiously dieted and, in fact, lost weight during her pregnancy.

■

Some other features of Minitab are indicated in the next example.

EXAMPLE 22 An automatic bank teller is located in a large hospital. The following data are the numbers of bank customers who use the teller on 42 successive days:

92	85	91	106	84	99	72	84	101	80	83	63	84	99
80	92	78	92	62	114	63	95	72	93	95	63	90	77
83	81	80	98	80	112	86	87	83	119	83	99	75	101

In Table 12 we indicate how these data are entered and a frequency histogram obtained with Minitab. The command "SET DATA IN C1" can be used when all the values are for the same variable and therefore can go into one column. In this case any number of values can be entered on a line as opposed to the format for the READ DATA command, where the number of values per row must correspond to the number of variables.

Instead of obtaining the summary measures for these data one at a time as we did in Table 11, Minitab provides nine summary measures with the single command "DESCRIPTION OF C1." Two of the nine measures will be new to the reader. TMEAN stands for the "trimmed" mean and is obtained by removing the largest and smallest values and finding the mean of the remaining $n - 2$ values. The trimmed mean is a useful measure of location when the data include an unusually large value, an unusually small value, or both. SEMEAN is short

TABLE 12

Printout for the automatic teller data

```
MTB > SET DATA IN C1
DATA> 92,85,91,106,84,99,72,84,101,80,83,63,84,99
DATA> 80,92,78,92,62,114,63,95,72,93,95,63,90,77
DATA> 83,81,80,98,80,112,86,87,83,119,83,99,75,101
DATA> HISTOGRAM OF C1

 C1

  MIDDLE OF    NUMBER OF
  INTERVAL     OBSERVATIONS
       60          1    *
       65          3    ***
       70          2    **
       75          2    **
       80          6    ******
       85         10    **********
       90          5    *****
       95          3    ***
      100          6    ******
      105          1    *
      110          1    *
      115          1    *
      120          1    *
```

TABLE 12 *(continued)*

```
MTB > DESCRIPTION OF C1

                C1
N               42
MEAN          87.0
MEDIAN        84.5
TMEAN         86.8
STDEV         13.4
SEMEAN         2.1
MAX          119.0
MIN           62.0
Q3            95.8
Q1            80.0
```

for the standard error of the mean, which is simply $s/\sqrt{n}$, where s is the standard deviation of the number of customers per day in the sample. The significance of this measure is explained in Chapter 7.

■

PROBLEMS 1.6

46. A researcher is analyzing some data using Minitab. She has taken the following summary output home:

```
MTB > DESCRIPTION OF C1

                C1
N               20
MEAN           9.4
MEDIAN         9.5
TMEAN          9.4
STDEV         [illegible]
SEMEAN        [illegible]
MAX           32.0
MIN          -12.0
Q3            22.2
Q1            -3.5
```

Unfortunately, the STDEV and SEMEAN terms were unclear. Explain how the remaining output can be used to obtain a ballpark estimate for s.

47. The following output is for C6 in the nutrition data (Table 8):

```
MTB > AVERAGE OF C6
   MEAN    =        0.52941
```

How would you interpret this value?

48. Use the summary data in Table 12 to verify that the *trimmed mean* (TMEAN) is 86.8. (*HINT:* Use $\bar{x}$ and n to find the sum of all n values.)
49. Each of the 38 members of a sixth-grade class filled out a questionnaire designed to indicate how ambitious they are. Ambition scores were calculated for each student. The boys' scores were entered into C1, the girls' into C2.

```
DATA> DESCRIPTION OF C1,C2

                C1        C2
N               20        18
MEAN         30.15     29.44
MEDIAN       29.50     27.50
TMEAN        29.89     29.38
STDEV         8.85      9.02
SEMEAN        1.98      2.13
MAX          49.00     46.00
MIN          16.00     14.00
Q3           36.50     36.50
Q1           23.00     22.00
```

(a) Use the summary data to construct a box-and-whisker display for each of the groups.
(b) Comment on differences between the two groups.

50. Indicate whether each of the following sets of results indicate evidence of (a) symmetry; (b) skewness (specify in which direction).

(i)
```
DATA> AVERAGE OF C1
   MEAN    =        15.333
MTB > MEDIAN OF C1
   MEDIAN =        22.000
```

(ii)
```
DATA> AVERAGE OF C1
   MEAN    =        27.333
MTB > MEDIAN OF C1
   MEDIAN =        27.000
```

(iii)
```
DATA> AVERAGE OF C1
   MEAN    =        33.333
MTB > MEDIAN OF C1
   MEDIAN =        27.500
```

51. Use the output in Table 12 to compute the interquartile range.

SECTION 1.7
SUMMARY AND REVIEW

DATA AND VARIABLES

Information obtained on a group of persons or other units is called **data.** Each characteristic on which information is obtained is called a **variable.** A **quantitative variable** can take various values and can be either discrete or continuous. If there are gaps between the possible values the variable can take, it is referred to as **discrete.** When there are no gaps between the possible values the variable can take, it is referred to as **continuous. Qualitative variables** involve only descriptive words and phrases called **levels.** When the levels can be ordered, the variable is referred to as **ordinal.** When the levels cannot be ordered, the variable is **nominal.**

ORGANIZING AND DISPLAYING DATA

A large number of observations can be divided into a sequence of nonoverlapping **intervals.** The intervals should preferably be of equal length and should number between 5 and 15. The number of observations falling in each interval is called the **frequency;** the list of frequencies is called the **frequency distribution.** The proportion of observations that fall in each interval is called the **relative frequency** for the interval. Adding to the frequency of each interval the frequency of all the preceding intervals results in a **cumulative frequency distribution.** Discrete data are often organized by recording the frequency and relative frequency with which each *value* occurs.

A relative frequency distribution can be represented pictorially by a **relative frequency histogram.** This consists of adjacent blocks, one block for each interval (or value). The height of each block is the relative frequency for the interval (or value) below the block. If one tail of the histogram is longer than the other, the data are said to be skewed to the longer side.

A **stem-and-leaf display** is one in which the observations themselves are used to indicate the "shape" of the distribution. The advantage of this type of display over the histogram is that all the observations are retained and ordered.

Small ($n \leq 15$) amounts of data can be illustrated graphically by representing the values as points on a line. The result is known as a **dot diagram.**

Qualitative data may be represented graphically by a **bar diagram.** Here each level of the variable is represented by a bar or block whose height is the relative frequency for that level. In contrast to the histogram, there is no significance to the *width* of each bar or to the distances between bars.

MEASURES OF LOCATION

Measures of location indicate the middle of a set of observations. There are three commonly used measures.

The **mean** of a set of observations is their sum divided by the number of observations (Definition 1). The mean ($\overline{x}$) is found by using the equation

$$\overline{x} = \frac{\sum_{i=1}^{n} x_i}{n} \qquad \text{(Eq. 1.1)}$$

Where discrete data consisting of only whole numbers are arranged by frequencies, we can use the form

$$\bar{x} = \frac{\sum_{i=0}^{k} i f_i}{n} \qquad \text{(Eq. 1.2)}$$

where f_i is the number of times the value i occurs and k is the largest value in the set of observations.

The **median** (m) is the middle observation if n is odd and the mean of the two middle observations if n is even (Definition 2).

The **mode** (m_o) is the value that occurs most frequently (Definition 3).

MEASURES OF SPREAD

Measures of spread indicate the extent to which the observations are spread out or vary—generally around the mean.

The **range** (R) is the difference between the largest and the smallest value.

The **variance** (s^2) is the sum of the squared deviations around the mean divided by $n - 1$. That is,

$$s^2 = \frac{1}{n-1} \sum_{i=1}^{n} (x_i - \bar{x})^2 \qquad \text{(Eq. 1.3)}$$

A more convenient form of s^2 is given by

$$s^2 = \frac{1}{n(n-1)} \left[n \sum_{i=1}^{n} x_i^2 - \left(\sum_{i=1}^{n} x_i \right)^2 \right] \qquad \text{(Eq. 1.4)}$$

When discrete data are arranged by frequencies, Equation 1.4 can be written

$$s^2 = \frac{1}{n(n-1)} \left[n \sum_{i=0}^{k} i^2 f_i - \left(\sum_{i=0}^{k} i f_i \right)^2 \right] \qquad \text{(Eq. 1.5)}$$

where, as before, f_i is the number of times the value i occurs and k is the largest value in the set.

The **standard deviation,** s, is the square root of the variance. It may be helpful to think of s as *approximately* the mean distance that the observations are from $\bar{x}$.

A rough relationship between the range (R) and the standard deviation (s) is

$$R \doteq 4s \qquad \text{(Eq. 1.6)}$$

This approximate relationship can be used as a check of the computed value for s.

QUARTILES

The three **quartiles,** Q_1, $Q_2 = m$, and Q_3, divide a set of ordered observations into four groups each having approximately the same number of values. If n is even, Q_1 is the median of the smallest $n/2$ observations and Q_3 the median of the largest $n/2$ observations. If n is odd, Q_1 is the median of the smallest $(n + 1)/2$ observations and Q_3 the median of the largest $(n + 1)/2$ observations. When n is odd, both groups include the median, m.

The **interquartile range** is the difference between Q_3 and Q_1 ($\text{IR} = Q_3 - Q_1$) and is a useful measure of spread.

The **five-number summary** consists of the smallest observation, Q_1, $Q_2 = m$, Q_3, and the largest observation. A **box-and-whisker display** is a pictorial representation of the five-number summary.

REVIEW PROBLEMS

GENERAL

1. An airline records the time taken (in hours) to fly from New York to London on 25 successive flights. The times are

7.1	7.0	7.1	7.1	7.1
7.2	7.1	7.1	6.8	7.8
7.1	7.1	7.1	7.7	7.9
6.9	7.0	7.2	7.1	6.9
7.2	6.9	7.0	7.8	7.1

Compute the (a) mean and (b) standard deviation of the time taken for the flights.

2. Following are the college GPAs for 60 students:

2.21	2.82	2.50	3.13	3.12	2.48	3.07	2.91	2.60	2.87
2.81	3.03	2.70	2.86	2.98	2.55	2.48	2.42	3.14	2.81
2.53	2.63	2.37	3.20	3.38	2.13	2.70	2.38	3.40	3.29
2.11	3.07	2.36	3.09	2.79	2.68	2.71	3.39	3.00	2.59
3.06	2.83	2.84	2.74	2.90	2.68	2.78	3.22	2.67	2.30
2.80	2.46	2.83	2.73	3.00	2.96	2.71	2.10	2.24	3.88

(a) Arrange the data into eight intervals of the form 2.01–2.25, 2.26–2.50 and so on. Construct the relative frequency distribution and histogram.

(b) Find the mean and the standard deviation of the GPAs if $\Sigma\, x_i = 167.02$ and $\Sigma\, x_i^2 = 472.3250$.

3. Following are the ages at inauguration of the first 40 presidents of the United States:

57	61	57	57	58	57	61	54	68	51	49	64	50	48
65	52	56	46	54	49	50	47	55	55	54	42	51	56
55	51	54	51	60	62	43	55	56	61	52	69		

(a) Construct a stem-and-leaf display for the ages.

(b) Is the distribution of ages symmetric? If not, in which direction is it skewed?

(c) Compute the mean and standard deviation of age if $\Sigma x_i = 2193$ and $\Sigma x_i^2 = 121{,}701$.

(d) Compute Q_1, Q_2, Q_3, and the interquartile range.

4. The number of reported road accidents per week in a town for one year are summarized as follows:

NUMBER OF ACCIDENTS	2	3	4	5	6	7	8	10	12	15
NUMBER OF WEEKS	1	4	7	17	14	4	2	1	1	1

(a) Construct the relative frequency distribution and the histogram.

(b) What were the mean, median, and modal number of weekly accidents?

(c) Compute the range and standard deviation of the weekly number of accidents.

(d) Compute the three quartiles and the interquartile range.

(e) In how many weeks were there four or fewer accidents?

(f) In what proportion of weeks were there six or more accidents?

5. In an analysis of the poem "The Expiration" by John Donne, a count is made of the number of letters in each word. These numbers are:

2 2 6 3 4 4 9 5 5 5 3 6 3 6 4 4 5 4
5 4 3 3 3 3 5 4 3 3 3 6 7 3 8 3 2 4
4 5 2 4 3 4 2 3 3 2 6 1 5 2 6 3 3 3
2 4 4 4 3 5 4 4 4 3 4 5 2 6 3 3 3 3
2 2 2 4 3 2 4 4 2 3 3 1 4 6 2 1 8 3
6 2 2 3 4 2 4 2 2 5 6 4 5 3 6

(a) Arrange the data in a frequency distribution.

(b) Compute the relative and cumulative frequencies.

(c) How many words contain (i) fewer than five letters; (ii) three or fewer letters?

(d) Compute the mean, median, and modal number of letters per word.

(e) Compute the standard deviation of the number of letters per word.

6. A group of 800 people were asked to name their favorite sports. The response were as follows:

Baseball	261
Basketball	158
Football	222
Ice hockey	42
Soccer	85
Tennis	24
Other	8

(a) Compute the relative frequency of people favoring each sport.

(b) Display these relative frequencies in a bar diagram.

(c) Is the variable "favorite sport" nominal or ordinal?

7. At a certain university the number of men and of women faculty members at each rank is recorded. The results are as follows:

	MEN	WOMEN
INSTRUCTOR	30	16
ASSISTANT PROFESSOR	75	49
ASSOCIATE PROFESSOR	60	24
PROFESSOR	43	8
TOTAL	208	97

(a) Compute the relative frequency of each rank for each sex.

(b) How do these two sets of relative frequencies differ?

(c) Is the variable "rank" nominal or ordinal?

8. The weight (in pounds) of 30 male college students is recorded. Each student is also given a score of 1 if he exercises regularly and a 0 otherwise.

WEIGHT	139	163	148	146	143	155	147	154	158	169
EXERCISES	0	1	0	0	1	1	1	0	0	0

WEIGHT	138	145	144	137	153	137	157	181	169	138
EXERCISES	1	1	0	0	0	1	1	1	1	0

WEIGHT	153	175	154	157	184	167	160	137	152	133
EXERCISES	0	0	0	0	0	1	0	1	1	1

(a) Find the mean and median weight for the 30 students. ($\Sigma x_i = 4593$ pounds.)

(b) Compute the standard deviation of weight for the 30 students. ($\Sigma x_i^2 = 708{,}507$.)

(c) Compute the mean, $\overline{p}$, of the 0 and 1 scores. How would you interpret this mean?

(d) Compute the mean and median weight separately for (i) those who exercise regularly; (ii) those who do not exercise regularly. What do these numbers suggest about the effect of exercise on weight?

9. Of the 55 runners who started a marathon race, the 33 who finished received (among other things) a score of 1, while the remainder received a score of 0. Compute and interpret the mean of these ones and zeros.

10. There are 14 counties in Arizona. In one year the mean number of accidents per county was 95 and the median number of accidents was 24. What do these numbers tell you about the distribution of the number of accidents per county?

HEALTH SCIENCES

11. The following data are the systolic blood pressures for each of 80 patients who have suffered heart attacks:

84	183	115	116	114	112	135	117	144	116
123	106	95	141	92	146	140	139	138	184
146	140	139	138	184	146	124	134	126	141
114	130	135	111	118	89	82	168	132	106
139	129	127	77	115	131	143	105	102	111
104	101	134	116	131	125	104	132	103	139
100	144	127	150	85	120	150	112	129	112
119	106	132	105	107	127	165	121	116	122

(a) Arrange the data in eight equal-length intervals of the form 70–84, 85–99, and so on.

(b) Compute the relative and cumulative frequencies.

(c) Construct the relative frequency histogram for the data.

(d) Is the histogram symmetric or skewed?

(e) Hazard a guess as to the mean blood pressure.

12. For the blood pressure data in Problem 11:

(a) Compute the mean blood pressure. ($\Sigma\, x_i = 9960$.)

(b) Compute the range of blood pressure and obtain a rough estimate for the standard deviation of blood pressure based on that range.

(c) Compute the standard deviation of blood pressure. ($\Sigma\, x_i^2 = 1{,}278{,}020$.) Does your result agree with the rough estimate in part (b)?

13. Following are the hemoglobin readings for each of 48 healthy adult women:

15.08	13.03	14.10	15.53	14.06	14.10	15.08	12.79	15.18	13.09
13.28	13.87	15.16	13.03	14.32	15.49	14.85	13.32	15.78	14.67
12.63	13.39	15.11	16.16	13.42	13.51	17.03	13.82	15.15	15.88
15.32	14.64	13.80	16.38	15.05	12.57	12.53	12.08	14.16	15.07
13.96	13.51	15.30	13.54	16.37	15.49	15.67	13.42		

(a) Construct a stem-and-leaf display for these readings using the two digits to the left of the decimal place as stems and the two digits to the right as leaves.

(b) Comment on the shape of the data.

14. For the hemoglobin readings in Problem 13:

(a) Compute the three quartiles Q_1, Q_2, and Q_3.

(b) Construct the box-and-whisker display.

(c) Does this display convey the same impression of the data as that of the stem-and-leaf display in Problem 13?

(d) Compute the mean and standard deviation of the hemoglobin readings. ($\Sigma\, x_i = 690.77$, $\Sigma\, x_i^2 = 10{,}006.7499$.)

15. A medical laboratory obtains the amount of nicotine (in milligrams) in each of 12 cigarettes of a particular brand. The amounts are

.789	.802	.781	.857	.779	.779
.788	.783	.854	.780	.856	.797

(a) Compute the mean and median amounts of nicotine per cigarette.

(b) Compute the range of nicotine amounts and obtain a rough estimate of the standard deviation.

(c) Compute the standard deviation of nicotine amount. ($\Sigma x_i^2 = 7.763511$.) Does your result agree with the rough estimate in part (b)?

16. The number of hospitalizations within the last 10 years was recorded for each of 40 patients enrolled in a mental rehabilitation program:

3	1	0	2	2	4	0	1	1	3
4	0	2	2	1	1	3	1	3	0
0	0	4	1	1	2	2	3	2	3
1	0	3	6	4	1	2	0	2	3

(a) Arrange the data in a frequency distribution. Compute the relative and cumulative frequencies.

(b) What proportion of patients were hospitalized (i) twice; (ii) at least once?

(c) Display the data in a relative frequency histogram.

(d) Compute the mean, median, and modal number of hospitalizations.

(e) Compute the range and standard deviation of the number of hospitalizations.

17. Following are the deaths (in ten thousands) associated with leading cancer sites as reported by the American Cancer Society for 1979:

SITE	ESTIMATED DEATHS	SITE	ESTIMATED DEATHS
Breast	35	Kidney/bladder	18
Colon/rectum	52	Larynx	3.5
Lung	98	Prostate	21
Oral	9	Stomach	14
Skin	6	Leukemia	15
Uterus	11	Lymphomas	20

(a) Compute the relative frequency of deaths associated with each site.

(b) Display these relative frequencies in a bar diagram.

18. At a maternity hospital 25 newborns were assigned a score of 1 if they were delivered by caesarian section and 0 if delivered by a normal vaginal delivery. The data were

0 0 1 0 0 0 1 0 0 0 0 0 0

0 0 1 0 1 0 0 0 0 0 0 0

Compute and interpret the mean, $\overline{p}$, of these zeros and ones.

GOVERNMENT AND SOCIOLOGY

19. A state legislature orders an investigation into the distribution of Aid to Families with Dependent Children (AFDC). As part of the investigation the monthly amount received by each of 60 families was recorded. The results were (in dollars)

289.08	252.38	274.48	226.17	266.27
248.26	253.39	222.93	278.82	229.20

230.90	285.34	239.00	209.70	266.32
273.08	231.39	253.17	241.30	333.39
226.48	231.42	255.92	333.72	238.26
246.93	255.36	288.36	265.17	224.27
259.05	237.04	263.13	285.88	233.08
258.66	214.02	241.05	248.52	284.31
279.46	232.46	235.88	201.28	233.19
314.81	224.54	264.52	259.22	264.58
305.37	203.33	289.58	280.42	301.89
290.11	260.38	299.57	197.79	184.92

(a) Arrange these data into a frequency distribution containing eight intervals of the form 180.00–199.99, 200.00–219.99, and so on. Compute the relative and cumulative frequencies.

(b) What proportion of the 60 families receive between \$280 and \$300 a month?

(c) How many families receive less than \$260 a month?

(d) Sketch the relative frequency histogram for the data. Comment on its shape.

20. Using the AFDC data in Problem 19:

(a) Compute the mean monthly payments. ($\Sigma\, x_i = 15{,}318.5$.)

(b) Compute the range of payments and obtain a rough estimate for the standard deviation of payment.

(c) Compute the standard deviation of payment. ($\Sigma\, x_i^2 = 3{,}971{,}827.76$.) Does your answer agree with the rough estimate in part (b)?

(d) Compute the three quartiles Q_1, Q_2, and Q_3.

(e) Construct the box-and-whisker display for these payments.

21. A political pressure group measures the performance of members of Congress by means of a score which runs from 0 to 100. A score of 0 implies that the member has consistently voted against the aims of the pressure group. A score of 100 implies a voting record in complete accord with the group's aims. The scores by party are as follows:

Democrats:	26	33	47	44	67	10	33	34	49	41	57	59	30
	52	58	67	67	46	40	42	42	30	7	36	51	47
	64	77	14	41	75	45	46	21	59	59	58	27	50
	56	39	50	57	55	42	60	57	57				
Republicans:	77	85	88	63	62	66	69	75	86	94	71	77	35
	76	39	79	71	77	74	95	86	84	81	47	82	94
	87	87	74	57	58	67	62	64	64	86	97	63	69
	36	76	72	76	96	30	95	75	68	69	64	75	75

(a) For the entire 100 scores construct a frequency distribution using 10 intervals of the form 0–9, 10–19, ... Compute the relative and cumulative frequency distributions.

(b) What proportion of all members have scores (i) between 20 and 29; (ii) 80 or greater?

(c) How many members have scores less than or equal to 49?

(d) Construct the relative frequency histogram for the 100 scores and comment on its shape.

22. Referring to the data in Problem 21:

(a) Construct a stem-and-leaf display for (i) the 48 Democratic scores; (ii) the 52 Republican scores. In each case use the first digit as the stem.

(b) Find the three quartiles Q_1, Q_2, and Q_3 for the two sets of scores.

(c) Construct a box-and-whisker display for the Democratic scores and for the Republican scores. Comment on any differences between them.

23. Refer again to the scores in Problem 21.

(a) Compute the mean score for each party. (For the Democrats, $\Sigma x_i = 2224$; for the Republicans, $\Sigma x_i = 3775$.)

(b) Compute the range of scores for each party and obtain a rough estimate for the standard deviation of scores in each party.

(c) Find the standard deviation of scores for each party. (For the Democrats, $\Sigma x_i^2 = 114{,}758$; for the Republicans, $\Sigma x_i^2 = 286{,}487$.)

24. As part of a pilot study of charitable deductions the IRS lists the amount of such deductions for 16 federal tax returns. The amounts are (in dollars)

80 25 50 10 150 25 78 80

225 28 50 40 95 400 85 40

(a) Plot the data on a dot diagram.

(b) Find the mean and the median deduction.

(c) Compute the range of deductions and obtain a rough estimate of the standard deviation.

(d) Compute the standard deviation of the deductions. Does your answer agree with the rough estimate in part (c)?

(e) Find the first and third quartiles for these data.

(f) Sketch the box-and-whisker display and comment on its shape.

25. The number of years actually served in prison by eight convicted felons sentenced to seven years are

6.2 4.9 3.7 5.2 5.8 4.8 4.5 4.0

(a) Plot the data on a dot diagram.

(b) Compute the mean and median years served.

(c) Compute the standard deviation of the number of years served.

26. The number of automobile deaths recorded in one year in each of the 14 Arizona counties are

30 15 183 9 7 39 28

30 76 13 21 18 723 3

(a) Compute the mean and the median number of deaths per county. Why are these two measures so different?

(b) Compute the standard deviation of the number of deaths per county.

(c) Find the first and third quartiles for these data.

27. Four hundred and fifty elderly couples were asked whether they received Supplemental Security Income (SSI) payments. Two hundred and eleven couples receiving SSI payments received a score of 1. The other couples received a score of 0. Compute and interpret the mean of these zeros and ones.

28. In a survey of 628 registered voters in a city, each person reported the number of elections in which he or she had voted during the past three years. The distribution of responses was as follows:

NUMBER OF ELECTIONS	0	1	2	3	4	5	6
NUMBER OF VOTERS	72	109	245	117	56	28	1

(a) Construct the relative and cumulative frequency distributions.

(b) What proportion of voters voted in four elections? In one or fewer elections?

(c) Sketch the relative frequency distribution for the data. Comment on its shape.

(d) Compute the mean, median, and modal number of elections in which the voters had cast votes.

(e) Compute the range and the standard deviation of the number of elections.

29. The 628 voters referred to in Problem 28 were also asked how long they had lived in their present homes. The mean and standard deviation of the 628 lengths of times were 10.4 and 3.8 years, respectively. How many of the voters would you expect to have lived in their homes (a) between 6.6 and 14.2 years; (b) between 2.8 and 18 years?

ECONOMICS AND MANAGEMENT

30. A retailer has 50 outstanding accounts due. The values of these accounts are as follows:

77.66	68.14	13.77	41.66	30.09	49.18	37.69	49.41	55.35	34.66
52.17	25.53	36.17	38.61	54.32	71.72	25.66	33.54	67.86	29.28
59.90	74.21	43.83	37.12	40.89	28.54	54.11	25.70	45.14	60.51
56.57	46.15	85.98	25.34	15.08	59.02	52.42	35.08	38.86	60.91
43.39	29.51	15.50	69.19	18.89	27.55	32.39	58.84	87.65	28.79

(a) Arrange these data in a stem-and-leaf display. (For display purposes, round each value to the nearest dollar and use the first digit as the stem and the second as the leaf.)

(b) Using these rounded numbers, compute the three quartiles.

(c) Construct the box-and-whisker display.

31. Compute the mean and the standard deviation of the (unrounded) accounts data in Problem 30. ($\Sigma x_i = 2249.53$, $\Sigma x_i^2 = 117{,}924.4$.)

32. A total of 1044 small businesses reported their number of employees to the Department of Labor. The returns are summarized as follows:

EMPLOYEES	FREQUENCY (NUMBER OF BUSINESSES)
1–20	17
21–40	43
41–60	85

EMPLOYEES	FREQUENCY (NUMBER OF BUSINESSES)
61–80	144
81–100	228
101–120	224
121–140	164
141–160	103
161–180	30
181–200	6

(a) Compute the relative frequency distribution and the cumulative frequency distribution.

(b) How many of the businesses have (i) 40 or fewer employees; (ii) 120 or fewer employees; (iii) between 81 and 140 employees?

(c) What proportion of businesses have between (i) 61 and 80 employees; (ii) between 121 and 160 employees?

(d) Sketch the relative frequency histogram for the data. Comment on its shape.

33. For the data in Problem 32, suppose that the mean and the standard deviation of the number of employees per company are 118 and 34, respectively. How many of the 1044 companies would you expect to have (a) between 84 and 152 employees; (b) between 50 and 186 employees?

34. A manufacturer of automobile parts has developed a spark plug which it hopes will last considerably longer than standard plugs. It is considering guaranteeing the plugs under a scheme based on the average life span of a plug. Plugs lasting less than the average lifespan are replaced at half-price except that plugs lasting a period of time shorter than one standard deviation below the average are replaced without charge. The mean and standard deviation are based on the lifetimes of a sample of 50 such plugs. The summary data for the 50 lifetimes (in months) are:

$$\sum x_i = 1021 \qquad \sum x_i^2 = 21{,}386$$

For what length of life will the manufacturer replace the plugs without charge?

35. A plumber notes the length of 18 consecutive home visits with the following results (in hours):

1.3	1.8	.9	2.3	1.7	1.5	1.4	1.1	.8
1.5	1.6	1.4	2.0	1.4	1.6	.9	1.4	1.0

(a) Sketch a dot diagram for the data.

(b) Compute the mean and the median length of visit.

(c) Compute the range and the standard deviation of the length of visit.

(d) Compute the first and third quartiles of the data.

(e) Construct the box-and-whisker display. Comment on its shape.

36. A bank makes 10 mortgage loans in one two-week period. The following figures are the proportion of gross income that each mortgage payment represents:

.28 .34 .22 .32 .35 .29 .29 .30 .33 .35

(a) Sketch a dot diagram for the data.

(b) Compute the mean and the median proportion of gross income represented.

(c) Compute the range and the standard deviation of this proportion.

(d) Compute the first and third quartiles and the interquartile range of the data.

37. Each of the 70 supermarkets in a large town is assigned a score of 1 if they are using computerized checkouts and 0 otherwise. Twenty-six of the supermarkets receive a 1, the remainder a 0. Compute and interpret $\overline{p}$, the mean of these zeros and ones.

38. An airline records the number of "no-shows" for a particular flight for 29 consecutive days:

8	7	10	11	8	12	8	9	9	7
13	5	15	8	8	9	10	11	11	10
4	9	11	9	7	8	9	10	12	

(a) Arrange these data in a frequency distribution (do not use intervals).
(b) Compute the relative and cumulative frequencies.
(c) On what proportion of days do seven no-shows occur?
(d) On how many days do 10 or fewer no-shows occur?
(e) Compute the mean, median, and modal number of no-shows.
(f) Compute the range and the standard deviation of the number of no-shows.

39. A brand of sauna bath is marketed in five southwestern states. Analysis shows the following breakdown of sales in the five states:

Arizona	2494
California	4892
Nevada	984
New Mexico	1702
Utah	1682

(a) Compute the relative frequency of sales for each state.
(b) Illustrate these figures with a bar diagram.

EDUCATION AND PSYCHOLOGY

40. The following data are the SAT math scores obtained by the 65 members of a high school senior class.

489	462	482	541	432	372	345	460	479	452
474	541	479	561	561	468	496	593	456	457
570	528	597	424	549	559	576	657	364	605
364	427	592	517	434	402	473	588	500	318
424	459	461	542	663	287	411	310	462	443
460	362	431	440	504	523	463	513	530	669
448	463	349	405	582					

(a) Arrange the data into a frequency distribution using intervals of the form 250–299, 300–349, ... Construct the relative frequency and the cumulative frequency distributions.
(b) What proportion of the students fall (i) in the interval 350–399; (ii) in the interval 500–549?
(c) How many of the students score (i) below 400; (ii) below 600; (iii) 600 or above?
(d) Construct the relative frequency histogram for the data.

41. For the SAT scores in Problem 40:

(a) Compute the mean SAT score. ($\Sigma x_i = 31{,}248$.)

(b) Compute the range of SAT scores and obtain a rough estimate for the standard deviation.

(c) Compute the standard deviation of SAT scores. ($\Sigma x_i^2 = 15{,}484{,}060$.)

42. Following are the IQ scores for 50 third graders:

122	107	110	115	120	109	119	105	109	103
112	114	112	109	120	122	118	102	110	110
104	127	108	104	119	117	113	118	113	104
108	101	112	126	117	119	104	100	101	114
112	117	105	117	117	103	109	113	108	104

(a) Arrange these scores in a stem-and-leaf display using stems of the form 10*, 10+, 11*, and so on, where * indicates that the third digit is a 0, 1, 2, 3, or 4 and + that the third digit is a 5, 6, 7, 8, or 9.

(b) Compute the five-number summary.

(c) Interpret Q_1 and Q_2 and $Q_3 - Q_1$ in this case.

(d) Compute the mean IQ. ($\Sigma x_i = 5582$.)

(e) Compute the range of IQ scores and obtain a rough estimate of the standard deviation.

(f) Compute the standard deviation of IQ scores. ($\Sigma x_i^2 = 625{,}468$.)

43. The age at which each of 10 children began walking is (in months)

10.9	13.2	12.4	12.9	12.1
14.2	11.1	12.5	13.0	18.8

(a) Plot these ages on a dot diagram.

(b) Compute the mean age at which walking began.

(c) Compute the median age at which walking began. Why are the mean and the median so different?

(d) Find the first and third quartiles, Q_1 and Q_3.

(e) Construct the box-and-whisker display for the data and comment on its shape.

(f) Compute the standard deviation, s, of age at which walking began.

44. In an experiment 54 young children were shown how to perform a simple task. They were then encouraged to practice performing it themselves. The number of practice trials it took each child before the task was performed correctly are as follows:

5	6	6	5	5	4	5	7	5	5	6	4	4	7	5	6	5	4
5	4	3	6	5	8	5	6	4	5	7	6	4	5	3	7	5	4
4	6	6	5	11	5	4	8	3	6	5	3	7	5	6	4	6	4

(a) Arrange the data in a frequency distribution. Compute the relative and the cumulative frequency distributions.

(b) What proportion of the children took (i) five trials; (ii) seven trials; (iii) eight or more trials?

(c) How many of the children took five or fewer trials?

(d) Draw the relative frequency histogram for the data and comment on its shape.

(e) Compute the mean and the median number of trials per child.
(f) Find the standard deviation, s, of the number of trials per child.

45. The raw scores for 11 fifth graders on a certain reading test are

42 35 46 40 29 33 41 51 44 35 45

(a) Plot the data on a dot diagram.
(b) Compute the mean score.
(c) Compute the standard deviation of scores.

46. The 27 colleges in a state are given a score of 1 if they are private and a score of 0 if they are public. Seven of the colleges were scored 0. Compute and interpret the mean of those zeros and ones.

47. In a large calculus class, the distribution of grades was as follows:

GRADE	NUMBER GETTING GRADE
A	10
B	52
C	43
D	7
F	3

(a) Compute the relative frequency for each grade.
(b) Plot these relative frequencies on a bar diagram.
(c) Is the qualitative variable "grade" ordinal or nominal?

48. The 200 top scorers in a nationwide high school statistics contest are classified by region of the country as follows:

New England	37
Mideast	55
Southeast	12
Great Lakes	20
Plains	18
Southwest	12
Rocky Mountains	5
Far West	41

(a) Compute the relative frequency for each region.
(b) Plot these relative frequencies on a bar diagram.
(c) Is the qualitative variable "region" ordinal or nominal?

SCIENCE AND TECHNOLOGY

49. The resistance (in ohms) of each of 50 resistance boxes is as follows:

7382	7538	8764	7511	7599
7300	6106	6729	7341	8564
7161	7086	6130	6447	7710

6530	7125	6767	6179	6125
7430	7054	7648	7646	6975
6089	7974	6531	7093	5365
7486	5954	7893	6790	6299
6925	7150	7427	6860	7577
8657	6867	5594	7284	6345
5522	7314	7388	7522	8239

(a) Arrange the data into eight intervals of the form 5000–under 5500, 5500–under 6000, ... Construct the frequency and the relative frequency distributions.

(b) What proportion of boxes have resistances between (i) 6000 and 6500 ohms; (ii) between 7000 and 8000 ohms?

(c) Construct the relative frequency histogram for the data. Comment on its shape.

50. For the resistance data in Problem 49:

(a) Order the data from smallest to largest and compute Q_1, Q_2, and Q_3 and the five-number summary.

(b) Construct the box-and-whisker display for the data. Comment on its shape.

51. The following data are the amounts of carbon in 40 samples of lunar soil (in parts of carbon per million):

152.7	151.4	169.4	171.7	138.5
159.1	149.1	169.0	160.2	37.0
147.7	165.1	166.7	56.8	52.2
162.4	162.8	159.4	170.0	157.5
163.5	150.0	147.7	151.6	140.6
150.1	153.3	146.6	164.5	136.8
148.2	167.9	138.6	146.1	151.3
159.4	152.6	140.1	147.7	150.9

(a) Round these numbers to the nearest whole number and construct the stem-and-leaf display. Use the first two digits as stems (some stems will have no leaves).

(b) Compute the mean and median amount of carbon. ($\Sigma\, x_i = 5866.2$.) Why are the two quite different?

(c) Compute the standard deviation of the amount of carbon. ($\Sigma\, x_i^2 = 895{,}147.04$.)

52. The following data represent the tensile strength (in pounds per square inch) for 10 equal-size blocks for each of two varieties of cement, A and B:

A	325	335	378	315	346	369	350	272	453	407
B	354	377	370	376	356	348	355	358	331	348

(a) Compute the mean tensile strength for each variety of cement.

(b) Compute the standard deviation, s, of tensile strength for each variety of cement.

(c) Using your results in parts (a) and (b), comment on any differences between the two varieties of cement.
(d) Compute the five-number summary for both groups.
(e) Construct the box-and-whisker display for each group on the same scale.
(f) Do these displays agree with your conclusions in part (c)?

53. The sea-level barometric pressure (in millibars of mercury) is recorded in eight cities along the eastern seaboard as follows:

1015.2 1025.5 1030.2 1021.1 1012.8 1027.0 1026.1 1017.9

(a) Plot these data on a dot diagram.
(b) Compute the mean pressure over the eight cities.
(c) Compute the median pressure.
(d) Find the standard deviation of pressure.

54. At 48 randomly chosen points during a week, a computer is classified as functioning (designated by a score of 1) or "down" (designated by a score of 0). A total of five zeros were recorded. Compute and interpret the mean of these 48 zeros and ones.

55. The number of alpha particles emitted by a piece of radioactive material per second for 200 consecutive seconds is recorded and summarized as follows:

NUMBER OF PARTICLES	0	1	2	3	4	5	6	7	8	9	10	11	12	13	14
NUMBER OF TIMES THAT EACH NUMBER OF ARTICLES WERE EMITTED IN A SECOND	1	0	11	18	25	40	30	26	20	14	10	4	0	0	1

(a) Construct the relative frequency distribution for these data.
(b) Construct the relative frequency histogram.
(c) Compute the mean number of alpha particles per second.
(d) Find the range of the number of particles per second and obtain a rough estimate for the standard deviation of the number of particles per second.
(e) Compute the standard deviation of the number of particles emitted per second. Does your answer agree with that in part (d)?

56. The number of doctorates awarded in the United States in the major science areas in a year are as follows:

Biology	2098
Engineering	1436
Chemistry	1525
Geology	367
Mathematical sciences	305
Physics	873

(a) Compute the relative frequency for each area.
(b) Construct a bar diagram for the data.

SUGGESTED READING

KOOPMANS, L. H. *An Introduction to Contemporary Statistics*. Boston: Duxbury, 1981.

Two excellent books on recent developments in data analysis are:

MCNEIL, D. R. *Interactive Data Analysis*. New York: Wiley, 1977.

VELLEMAN, P. F. and HOAGLIN, D. C. *Applications, Basics, and Computing of Exploratory Data Analysis*. Boston: Duxbury, 1981.

The best guide to Minitab is:

RYAN, T. A., et al. *Minitab Student Handbook*. Boston: Duxbury, 1976.

A more thorough introduction to the use of statistical computer packages is:

BERENSON, M. L., et al. *Intermediate Statistical Methods and Applications: A Computer Package Approach*. Englewood Cliffs, N.J.: Prentice-Hall, 1983.

2 BIVARIATE DATA

HOW MUCH DOES LATITUDE AFFECT TEMPERATURES?

It is common knowledge, especially to those who have moved from North Dakota to Florida, that a location's average temperature is determined to a large extent by its latitude. In other words, in general, the higher the latitude (i.e., the farther north the location), the lower the average temperature. The following data substantiate this "fact." These figures consist of the average temperature (in degrees Fahrenheit) and the latitude (in degrees north of the equator) for 13 cities in the United States.

CITY	LATITUDE, x	TEMPERATURE y
New York	40	54.1
Los Angeles	34	61.8
Chicago	42	49.9
Philadelphia	40	54.6
Houston	29	68.2
Miami	25	75.5
Boston	42	50.3
Atlanta	33	61.5
Detroit	42	49.2
San Francisco	37	56.2
Minneapolis	45	44.9
New Orleans	30	69.0
Denver	39	50.2

This rule is not perfect. Clearly, temperatures are affected by altitude, wind currents, and whether a city is located on the western side, on the eastern side, or in the interior of the continent. For example, New Orleans is farther north than Houston but has a higher average temperature. In Problem 15 on page 89 you are asked to use these figures to compute a measure of just how closely latitude and temperature *are* related.

SECTION 2.1

DISPLAYING BIVARIATE DATA

BIVARIATE DATA

Sometimes we obtain information on more than one variable for the same person or unit. Table 1 of Chapter 1 (page 6), for instance, contained observations on each of four variables for 25 women. Of particular importance are those cases in which we have information on two variables for each person or unit. Such data are called **bivariate.** Since each variable in a bivariate set may be either quantitative or qualitative, there are three combinations of bivariate data which can arise. They are:

1. Both variables are quantitative.
2. One variable is quantitative and the other qualitative.
3. Both variables are qualitative.

In this section we describe some methods for displaying bivariate data.

TWO QUANTITATIVE VARIABLES

EXAMPLE 1

Toward the end of a management course each of the eight students rates the instructor on a scale of 0 to 20; the higher the rating, the greater the regard for the instructor. Table 1 contains these ratings (y) together with each student's final course score (x).

TABLE 1

Ratings of their instructor and final course scores for eight students

	STUDENT							
VARIABLE	1	2	3	4	5	6	7	8
Final course score, x	84	77	68	94	76	80	72	88
Rating of instructor, y	12	8	7	13	15	18	10	16

Where both variables are quantitative the values for one variable are identified as $x_1, x_2, \ldots, x_n$ and the values for the other variable as $y_1, y_2, \ldots, y_n$. Thus (x_1, y_1) represents the two values for the first student, (x_2, y_2) the two values for the second student, and so on. For the data in Table 1 these pairs are $(84, 12), (77, 8), \ldots, (88, 16)$.

In Figure 1 we illustrate how these data can be represented graphically. In accordance with custom, the horizontal axis in Figure 1 is labeled x and the vertical axis is labeled y. The hatches to the left of the x axis and at the bottom of the y axis are reminders that the sections of the axes shown do not extend to the natural meeting point of the two axes at $x = 0$ and $y = 0$.

We plot the "position" of each student beginning with a dot for the first student at $x = 84$, $y = 12$; another dot for the second student at $x = 77$, $y = 8$; and so on for each student. The result in Figure 1 is called a **scatterplot.**

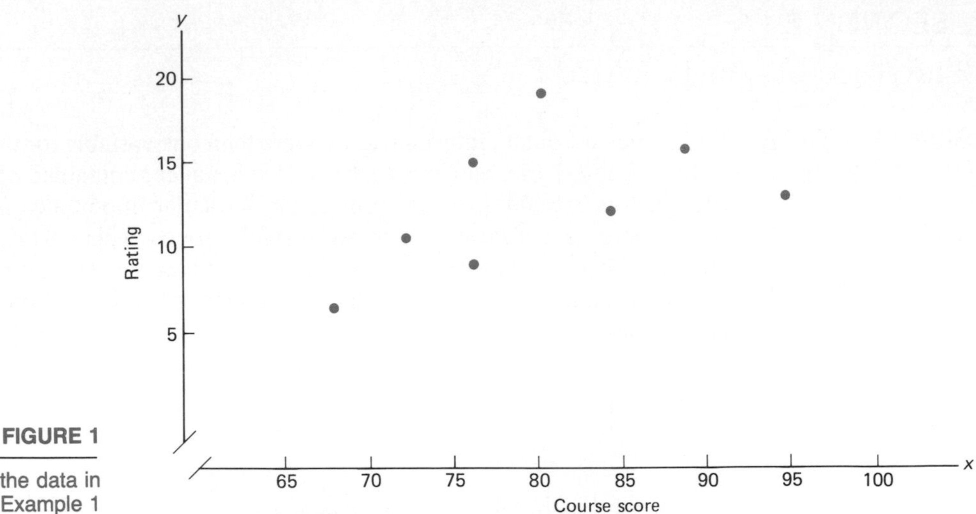

FIGURE 1

Scatterplot for the data in Example 1

Looking at the placement of dots, note that they arrange themselves in a pattern generally going from lower left to upper right. This suggest that students who give the lower ratings tend to have the lower scores, and those who give the higher ratings tend to have higher scores. ■

ONE QUANTITATIVE AND ONE QUALITATIVE VARIABLE

EXAMPLE 2 Table 2 contains bivariate data in which one variable is quantitative and the other qualitative. The data, taken from the *Boston Sunday Globe* newspaper, consists of the advertised prices (to the nearest $1000) for houses in three towns: Milton, Brookline, and Newton. For convenience the prices are already ordered. The price for each house is the quantitative variable and the town in which it is located is the qualitative variable, with three levels.

TABLE 2

Advertised price of houses in three communities (in thousands of dollars)

MILTON (n = 26)			BROOKLINE (n = 29)			NEWTON (n = 33)		
56	88	155	48	94	142	55	110	150
60	89	175	50	95	148	70	114	170
60	94	195	54	100	150	75	115	179
61	94	200	55	105	160	80	120	179
65	95		60	108	186	82	125	180
69	95		60	110	189	85	129	190
70	96		62	115	225	92	130	199
79	110		62	125		103	133	200
82	115		78	125		106	140	210
85	118		80	129		108	145	225
87	150		80	130		110	148	269

The distribution of prices for each town can be represented by a stem–and–leaf display. For purposes of clarity and comparison we have used the same stem for each community and have written each one as a range of values. Each leaf is a specific value.

FIGURE 2

Stem-and-leaf display of house prices in three communities (in thousands of dollars)

	Milton	Brookline	Newton
40–59	56	48 50 54 55	55
60–79	60 60 61 65 69 70 79	60 60 62 62 78	70 75
80–99	82 85 87 88 89 94 94 95 95 96	80 80 94 95	80 83 85 92
100–119	10 15 18	00 05 08 10 15	03 06 08 10 10 14 15
120–139		25 25 29 30	20 25 29 30 33
140–159	50 55	42 48 50	40 45 48 50
160–179	75	60	70 79 79
180–199	95	86 89	80 90 99
200–219	00		00 10
220–239		25	25
240–259			
260–279			69

All three distributions are skewed to the right because of a few unusually large prices. But the distributions also differ in some respects. In Milton the prices generally seem to fall into two clusters, one centered around $90,000 and another in the high hundred thousands. The distribution of prices in Brookline is "flatter" than Milton's but not quite as symmetric as Newton's.

The median price and the range of prices for each town are as follows:

	MILTON	BROOKLINE	NEWTON
MEDIAN	91.5	105	129
RANGE	144	177	214

Newton has the highest median price and Milton the lowest.

Much of the same kind of information just described can be obtained in a starker form by drawing the box-and-whisker display for each town. The results appear in Figure 3 on the next page together with the corresponding five-number summaries.

TWO QUALITATIVE VARIABLES

We now consider the case of bivariate data in which both variables are qualitative.

EXAMPLE 3 Each of the 1200 students at a small liberal arts college was classified by sex and by area of concentration. The results are summarized in Table 3. The levels of the variable "area of concentration" are natural science (NS), social science (SS), and humanities (H). The array in Table 3 is called a **contingency table** or more precisely a 2 × 3 (read "2 by 3") contingency table.

Each of the six boxes delineated by single lines in Table 3 is called a **cell.** The number in each cell is the number of students with the combination of sex and area of concentration for the row and column of that cell. For example, there are 291 male social science concentrators and 83 female natural science concen-

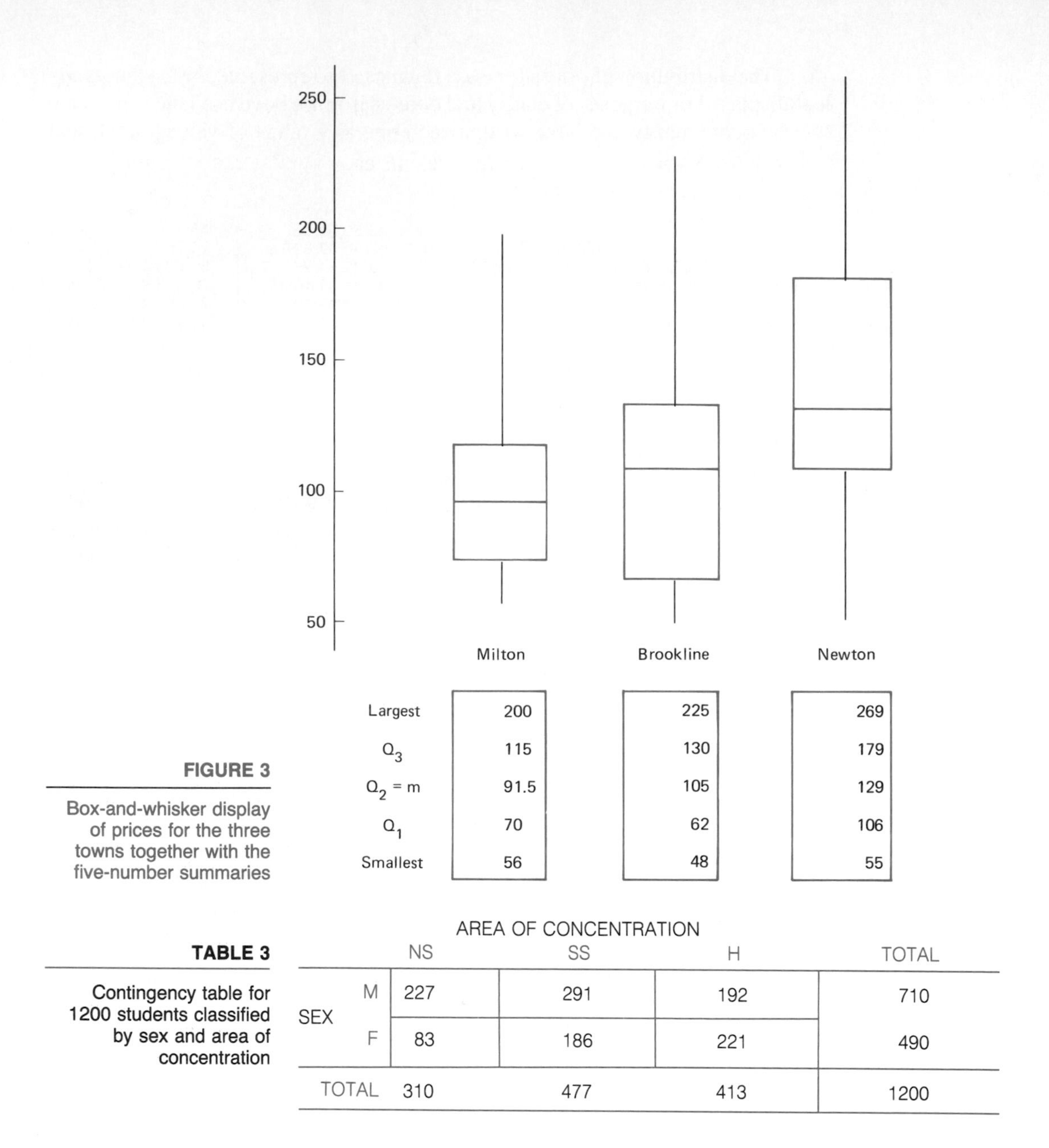

	Milton	Brookline	Newton
Largest	200	225	269
Q_3	115	130	179
$Q_2 = m$	91.5	105	129
Q_1	70	62	106
Smallest	56	48	55

FIGURE 3

Box-and-whisker display of prices for the three towns together with the five-number summaries

TABLE 3

Contingency table for 1200 students classified by sex and area of concentration

		AREA OF CONCENTRATION			
		NS	SS	H	TOTAL
SEX	M	227	291	192	710
	F	83	186	221	490
TOTAL		310	477	413	1200

trators. The number of students in each cell is not a measurement (i.e., a price, a weight, or an age, etc.) but a count of people. The numbers around the border of the table are the total numbers of students in each row or column. Thus there are in all 310 natural science concentrators, 490 female students, and so on. The number in the bottom right-hand corner is the total number of students in the study (1200). For clarity we have omitted many of the lines in subsequent tables.

In Table 4 we indicate the *proportion* of each sex in each area of concentration. The proportion of males who are social science concentrators, for instance, is 291/710 = .41. The number in parentheses beside each proportion is the corresponding *percentage* of that sex in each concentration (obtained by multiplying the proportion by 100).

TABLE 4

Proportion (and percentage) of each sex that are in each area of concentration

		AREA OF CONCENTRATION			
		NS	SS	H	TOTAL
SEX	M	227/710 = .32 (32)	291/710 = .41 (41)	192/710 = .27 (27)	1.00 (100)
	F	83/490 = .17 (17)	186/490 = .38 (38)	221/490 = .45 (45)	1.00 (100)

In Figure 4 the percentage of each sex concentrating in each area is represented by the height of an appropriately labeled block. Notice, in particular, the quite different distribution of concentrations for the two sexes. Almost a third (32%) of males are natural science concentrators, whereas only 17% of women concentrate in this area. Figure 4 is called a **block diagram.**

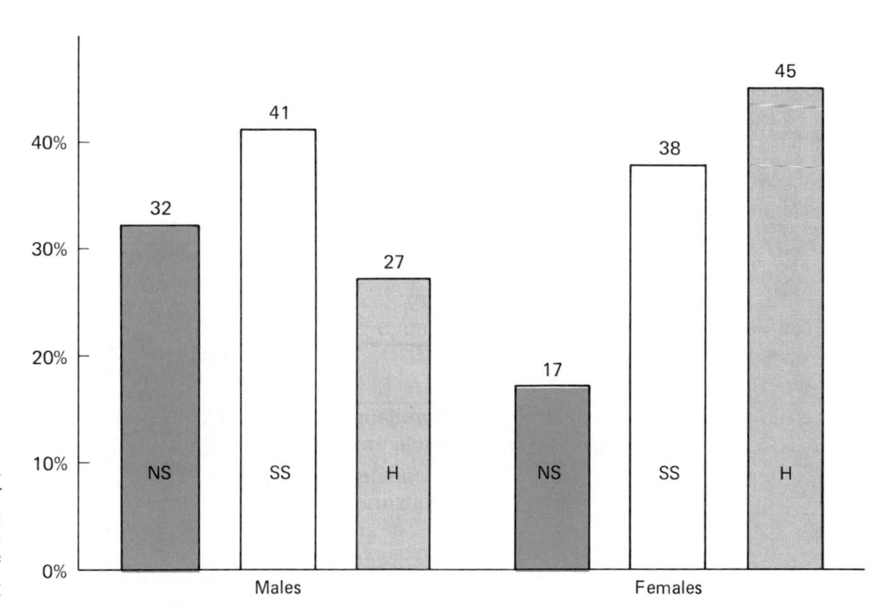

FIGURE 4

Block diagram showing the distribution (in percent) of type of concentration by sex

Conversely, the percentage of humanities concentrators among the women (45%) is almost twice that for men (27%). Approximately the same percentage of both sexes are social science concentrators. Notice that we have included the 0% point in Figure 4. Omitting this baseline can produce a misleading impression. Examine Figure 5, for example. Here we have plotted the same percentages but used 15% as the baseline instead of 0. Figure 4 accurately shows that slightly more than twice the percentage of females are social scientists

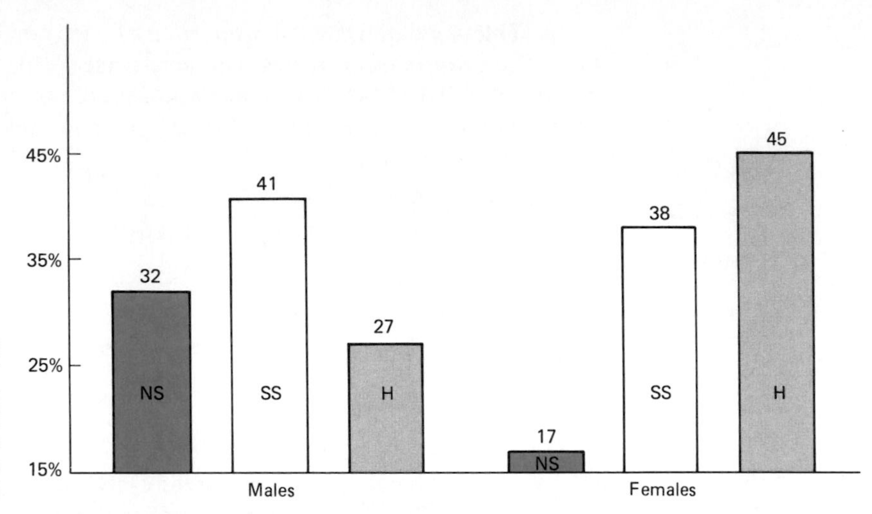

FIGURE 5

Block diagram showing the distribution of type of concentration by sex (using 15% as the baseline)

(38%) than natural scientists (17%). But the blocks in Figure 5 give the impression that the percentage of women who are social scientists is almost 10 times that of women who are natural scientists.

It is also possible to provide a breakdown of the areas of concentration by sex: that is, the percentages of natural scientists who are men and those who are women. But these percentages are easy to misinterpret as the next example will illustrate.

■

EXAMPLE 4 There are 500 natural science concentrators at a college, 400 men and 100 women. This means that 80% of the natural science concentrators are men and appears to suggest that women are not very interested in this area. However, the college has only recently begun to admit women and of the 2000 undergraduates, only 200 are women. So, in fact, 50% of the women are natural science concentrators, whereas only 400 of the 1800 men, or 22%, are natural science concentrators.

This example, together with our comments on Figure 5, illustrates the care that must be taken in interpreting statistical findings.

■

PROBLEMS 2.1

1. The time taken (in hours) by six groups of workers to assemble a car (y) is recorded, together with the number of weeks the group have worked together (x):

x	4	23	12	18	7	14
y	2.7	1.6	2.0	1.8	2.5	2.3

(a) Plot these data on a scatterplot.

(b) Comment on the appearance of your scatterplot.

2. The height (in inches), x, and the weight (in pounds), y, for each of eight male college students are as follows:

x	66	70	68	69	71	69	67	72
y	152	145	162	172	160	158	140	184

(a) Plot these data on a scatterplot.

(b) Does the appearance of your plot agree with your intuition?

3. The following data are the number of years of full-time education (x) and the annual salary in thousands of dollars (y) for 15 persons.

x	20	17	18	18	13	18	9	18	16	12	12	19	16	14	13
y	35.2	34.6	23.7	33.2	24.4	33.4	11.2	32.3	25.1	22.1	18.9	37.8	25.9	28.4	29.6

Plot these data on a scatterplot and comment on its appearance.

4. The unemployment rate (in percent), y, and the rate of inflation (in percent), x, for eight Western European countries are as follows:

x	17.2	8.1	4.7	12.0	6.5	9.2	8.4	10.7
y	5.4	10.0	12.9	6.3	10.2	7.0	8.4	6.4

(a) Plot these data on a scatterplot.

(b) What does your plot suggest about the relationship between inflation and unemployment?

5. The following data are the number of persons *not* reporting to work at a plant for each of 15 consecutive Mondays, Tuesdays, Wednesdays, Thursdays, and Fridays:

M	18	17	16	18	20	22	19	17	16	18	19	20	17	16	19
T	12	14	15	16	12	16	13	13	10	12	15	12	11	11	13
W	9	9	11	13	12	9	8	8	9	10	10	11	9	7	8
TH	14	12	15	12	16	13	10	15	12	12	11	12	14	11	10
F	18	17	18	16	16	15	14	19	18	18	15	14	16	17	13

(a) Construct the five-number summary for each day.

(b) Draw the corresponding box-and-whisker displays.

(c) What differences in attendance do these displays suggest for the different days?

6. Thirty cars for each of three models, A, B, and C, are run over a 200-mile course and the mileage per gallon computed. The results are as follows:

A			B			C		
31.1	32.5	31.4	35.7	34.5	35.3	29.4	30.6	32.6
32.6	34.5	31.8	34.1	34.2	36.3	31.0	31.4	29.9
32.3	32.9	33.0	35.0	37.1	35.0	31.8	28.9	29.4
32.6	32.4	31.7	35.0	34.9	32.8	33.1	31.1	31.7
30.5	28.5	30.0	33.9	35.1	34.6	32.9	31.7	31.9
33.1	31.1	31.4	33.1	37.2	35.5	32.1	29.2	29.1
31.8	31.8	32.8	35.2	36.1	32.2	26.7	29.8	30.4
32.3	31.3	32.3	30.2	36.4	34.6	31.6	29.5	28.4
32.9	32.9	31.7	31.2	33.3	28.3	27.2	29.7	27.6
33.3	29.6	31.8	34.1	34.2	30.5	29.8	30.8	27.8

(a) What are the two variables of interest in this example?

(b) Compare the distribution of mileage for the three models by constructing three stem-and-leaf displays.

(c) Comment on the differences among the three models.

(d) If mileage were the only factor of concern to you, which model car would you purchase?

7. Six hundred men and 600 women were asked whether they approved, disapproved, or were neutral toward, the U.S. president's economic program. The results are summarized as follows:

	ATTITUDE			
	APPROVE	NEUTRAL	DISAPPROVE	TOTAL
MEN	310	54	236	600
WOMEN	241	81	278	600

(a) Interpret the numbers 241 and 54.

(b) Construct a table similar to Table 4 showing the percentage of each sex falling in each attitude category.

(c) Construct a block diagram similar to that in Figure 4 showing the breakdown of attitude by sex.

(d) How, if at all, do the attitudes vary by sex?

8. The number of full-time students enrolled in public and private schools in four regions of the country is as follows:

		TYPE OF SCHOOL	
		PUBLIC	PRIVATE
REGION	NEW ENGLAND	81,529	73,336
	MID-ATLANTIC	282,276	130,412
	GREAT LAKES	381,973	79,139
	WEST	372,749	30,897

(a) Construct a table similar to Table 4 showing the percentage of students in each region who are in public and private schools.

(b) Construct a block diagram which shows the percentage breakdown for each region.

(c) Do the data indicate any trends which run from East Coast to West Coast?

9. A group of 800 people who watched TV one evening recorded the network they were watching at 9:15. The results, classified by household income, were as follows:

		NETWORK			
		ABC	CBS	NBC	PBS
INCOME	HIGH	70	45	30	65
	MEDIUM	110	125	65	50
	LOW	67	73	90	20

(a) Construct a table showing the percentage of each income group that watched each network.

(b) Represent these percentages in a block diagram.

(c) Comment on the differences among income groups.

Solution to Problem 1

(a) The scatter plot is as follows:

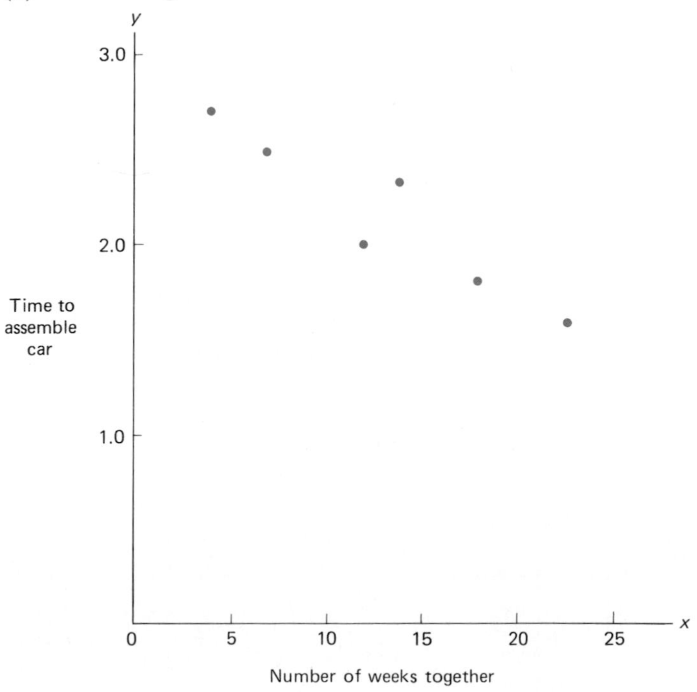

(b) The scatterplot suggests that generally the longer the group members have worked together, the less time it takes the group to assemble a car.

SECTION 2.2
THE CORRELATION COEFFICIENT

In addition to displaying quantitative bivariate data, it is useful to summarize the relationship between the two variables numerically. We shall use the ratings data from Example 1 to illustrate how such a numerical summary can be obtained. The scatterplot of the data is reproduced in Figure 6. It indicates that students with lower scores tend to give their instructors lower ratings, whereas students with higher scores tend to give higher ratings. The result is that the points lie relatively close to a straight line, as shown in Figure 6.

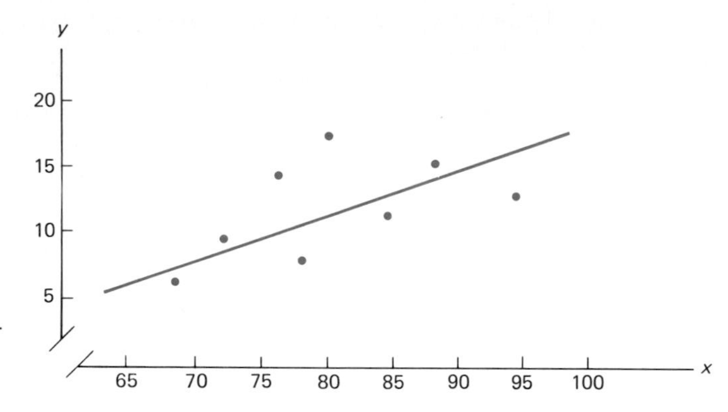

FIGURE 6

Scatterplot for the ratings data in Example 1

When the points representing bivariate data lie exactly in a straight line, the two variables are said to have a **perfect linear relationship.** If the points lie close to a straight line, the two variables are said to have a **strong** linear relationship. If the points hardly approach a straight line, the variables are said to have a **weak** or almost no linear relationship. It is possible to represent the strength of a linear relationship numerically. A measure for doing so was developed by Karl Pearson, a pioneering statistician. This measure, designated by the letter r, is known as the Pearson product moment correlation or, more commonly, as the **correlation coefficient.**

THE CORRELATION COEFFICIENT

The formula for r is

$$r = \frac{\Sigma (x_i - \bar{x})(y_i - \bar{y})}{\sqrt{\Sigma (x_i - \bar{x})^2 \, \Sigma (y_i - \bar{y})^2}} \qquad \text{(Eq. 2.1)}$$

This formula is not very convenient for computing r but will be helpful when we examine the sign of r a bit later. A more convenient form of Equation 2.1 is

$$r = \frac{n \sum x_i y_i - (\sum x_i)(\sum y_i)}{\sqrt{[n \sum x_i^2 - (\sum x_i)^2][n \sum y_i^2 - (\sum y_i)^2]}} \qquad \text{(Eq. 2.2)}$$

We used sums like $\sum x_i$ and $\sum y_i^2$ in Chapter 1. The only component of Equation 2.2 that is new is the sum of the products, $\sum x_i y_i$. We illustrate how this sum and the value for r are computed, using the ratings data in Example 1. The necessary computations are shown in Table 5.

TABLE 5

Computations necessary to compute r for the ratings data

x	x^2	y	y^2	xy	
84	7056	12	144	1008	
77	5929	8	64	616	$(=77 \times 8)$
68	4624	7	49	476	
94	8836	13	169	1222	
76	5776	15	225	1140	
80	6400	18	324	1440	
72	5184	10	100	720	
88	7744	16	256	1408	
$\sum x_i = 639$	$\sum x_i^2 = 51549$	$\sum y_i = 99$	$\sum y_i^2 = 1331$	$\sum x_i y_i = 8030$	

The numbers in the last column were obtained by multiplying together the x and y values for each student and adding the products over all eight students to obtain $\sum x_i y_i = 8030$.

Entering these various sums into Equation 2.2 and recalling that $n = 8$, we obtain

$$r = \frac{8(8030) - (639)(99)}{\sqrt{[8(51{,}549) - 639^2][8(1331) - 99^2]}}$$

$$= \frac{979}{\sqrt{(4071)(847)}} = .5272$$

Before interpreting the value $r = .5272$, some general observations about the correlation coefficient will be helpful.

INTERPRETING r

The value for r will always lie between -1 and $+1$ (inclusive); or symbolically, $-1 \leq r \leq 1$. Examples of scatterplots showing the relationships between x and y when the value for r is 1, 0, and -1 are shown in Figure 7.

As these scatterplots show, $r = 1$ when (1) the (x, y) values (i.e., points) lie exactly along a straight line; and (2) the larger the x value, the larger the y value. In this case x and y are said to have an exact *positive* linear relationship.

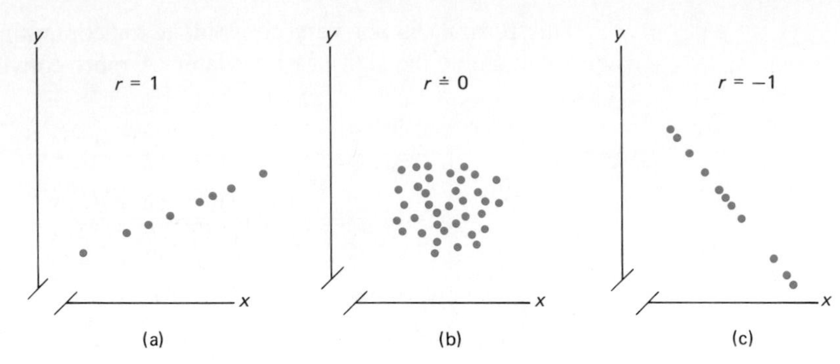

FIGURE 7

Scatterplots showing the situations when (a) $r = 1$, (b) $r = 0$, and (c) $r = -1$.

The value $r = -1$ also occurs where (1) the (x, y) values lie exactly along a straight line, but now (2) the larger the x value, the smaller the y value. In this case x and y have an exact *negative* linear relationship. As the middle scatterplot indicates, r is close to zero when there is no apparent linear relationship at all between x and y [see Figure 7(b)].

In practice the value for r indicates the degree to which the points lie *close* to a straight line. Returning to the ratings example, the value $r = .5272$ suggests only a slight linear relationship between ratings and test scores.

In Figure 8 we show the scatterplot for the ratings data again, but this time we have drawn two, new (dashed) axes with their origin at the point representing the two *means* $\bar{x} = 79.875$ and $\bar{y} = 12.375$. These axes define four quadrants labeled A, B, C, and D. In quadrant A values of both x and y are above their respective means. In quadrant C values of both x and y are below their respective means. By contrast, in quadrant B values of x above $\bar{x}$ occur with values for y *below* $\bar{y}$. In quadrant D values of x below $\bar{x}$ occur with values of y *above* $\bar{y}$. When points lie primarily in quadrants A and C, as in Figure 8, r will be positive. When points lie primarily in quadrants B and D, r will be negative.

It will be helpful to relate these ideas to the first equation for r (Equation 2.1). The denominator of r is always positive since both sums of squares under the square root are positive. Thus the sign of r will depend only on the numerator of r:

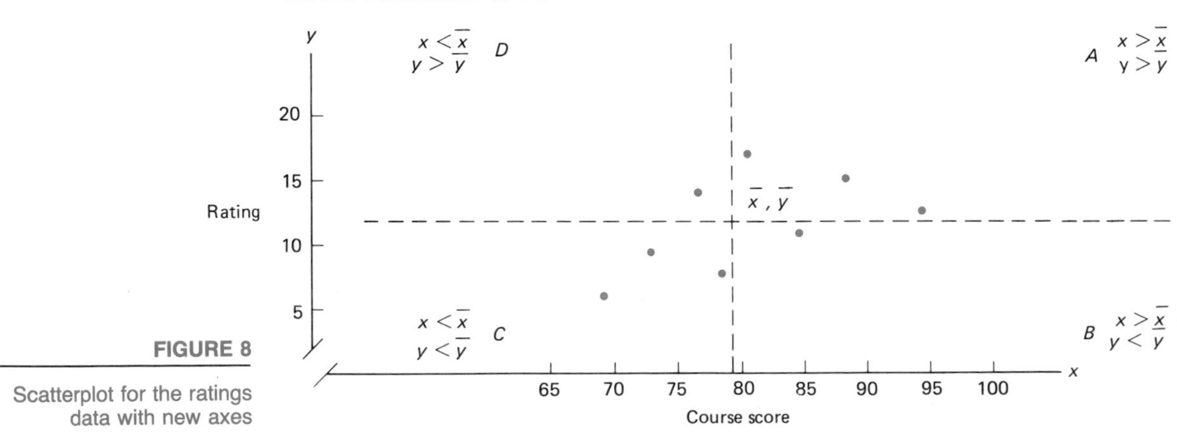

FIGURE 8

Scatterplot for the ratings data with new axes

$$\Sigma (x_i - \bar{x})(y_i - \bar{y}) = (x_1 - \bar{x})(y_1 - \bar{y}) + (x_2 - \bar{x})(y_2 - \bar{y}) + \cdots + \cdots + (x_n - \bar{x})(y_n - \bar{y})$$

When many of the n points are such that $(x_i - \bar{x})$ and $(y_i - \bar{y})$ have the same sign (i.e., lie in quadrants A and C), the values for both $\Sigma (x_i - \bar{x})(y_i - \bar{y})$ and r are likely to be positive. By contrast, when many of the n points are such that $(x_i - \bar{x})$ and $(y_i - \bar{y})$ have opposite signs (i.e., lie in quadrants B and D), $\Sigma (x_i - \bar{x})(y_i - \bar{y})$ and hence r is likely to be negative.

In Table 6 we suggest terms that may be used to describe the strength of the linear relationship between x and y for different values of r. Thus (a) the value $r = -.15$ indicates a weak negative linear relationship between x and y; (b) the value $r = -.62$, a moderate negative linear relationship; and (c) the value $r = .93$, a strong positive linear relationship. Scatterplots appropriate to each of these values for r are shown in Figure 9.

TABLE 6

Describing the linear relationship between x and y

IF, REGARDLESS OF SIGN:	THE LINEAR RELATIONSHIP BETWEEN x AND y IS DESCRIBED AS:
$0 \le r < .3$	Weak
$.3 \le r < .55$	Slight
$.55 \le r < .8$	Moderate
$.8 \le r < 1$	Strong

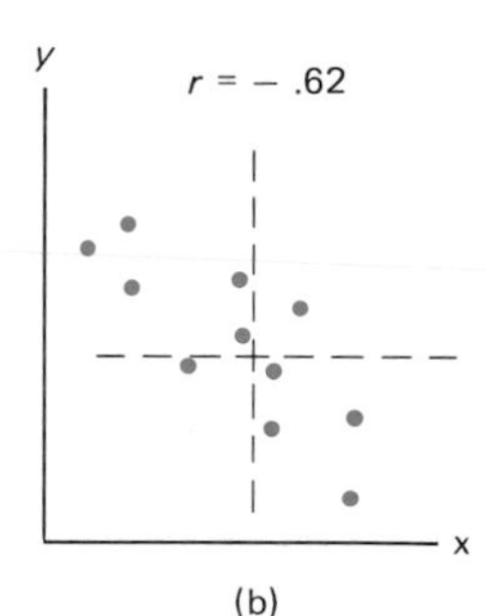

FIGURE 9

Three scatterplots representing data for which (a) $r = -.15$, (b) $r = -.62$, and (c) $r = .93$

One more example will help clarify the computation and interpretation of r.

EXAMPLE 5 A new product has been test-marketed in four similar supermarkets, being sold at a different price in each market. The prices charged and the numbers of units sold are as follows:

MARKET	PRICE, x	UNITS SOLD, y
A	\$41	435
B	45	375
C	50	410
D	54	354

(a) Plot these data on a scatterplot, and (b) compute the correlation coefficient, r, between price and sales.

Solution (a) The scatterplot is shown in Figure 10. It suggests a negative linear relationship between price and sales. (b) The computations necessary to obtain r are shown in Table 7.

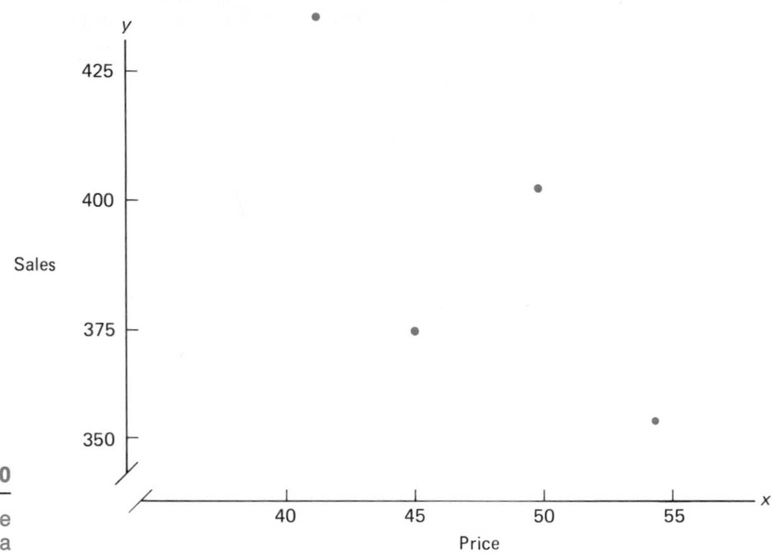

FIGURE 10

Scatterplot for the prices/sales data

TABLE 7

Components of r for the prices/sales data

x	x^2	y	y^2	xy
41	1681	435	189,225	17,835
45	2025	375	140,625	16,875
50	2500	410	168,100	20,500
54	2916	354	125,316	19,116
$\Sigma x_i = 190$	$\Sigma x_i^2 = 9122$	$\Sigma y_i = 1574$	$\Sigma y_i^2 = 623{,}266$	$\Sigma x_i y_i = 74{,}326$

Using Equation 2.2 with $n = 4$, we have

$$r = \frac{4(74{,}326) - (190)(1574)}{\sqrt{[4(9122) - 190^2][4(623{,}266) - 1574^2]}}$$

$$= \frac{-1756}{\sqrt{(388)(15{,}588)}} = -.714$$

This suggests a moderate, negative relationship between price and sales. ■

A study of Equation 2.2 will show that it is symmetric with respect to x and y. Thus if the variable labeled x (price in Example 5) were to be labeled y and the one that had been y (sales) became x, the value for r would remain the same. Because of the change in labeling, the arrangement of points on the scatterplot will change. However, the extent to which they lie close to a line—which is what r measures—will not change.

MISINTERPRETING r

It is important to realize that the value for r must be interpreted with caution, as we indicate in the following example.

EXAMPLE 6

The following data are the total monthly electric bills (in dollars) and the average monthly temperatures (in degrees Fahrenheit) for an all-electric home in Central City.

MONTH	JAN.	FEB.	MAR.	APR.	MAY	JUNE	JULY	AUG.	SEPT.	OCT.	NOV.	DEC.
AVERAGE TEMPERATURE, x	38	34	48	55	64	68	75	86	75	61	47	39
ELECTRIC BILL, y	153	130	71	58	50	35	78	102	33	43	48	100

Given the following sums, compute and interpret the correlation between average temperature and the bill for electricity.

$$\sum x_i = 690 \qquad \sum x_i^2 = 42{,}746 \qquad \sum y_i = 901$$

$$\sum y_i^2 = 84{,}169 \qquad \sum x_i y_i = 48{,}288$$

Solution

Using Equation 2.2 with $n = 12$, we have

$$r = \frac{12(48{,}288) - (690)(901)}{\sqrt{[12(42{,}746) - 690^2][12(84{,}169) - 901^2]}} = -.494$$

This value for r suggests only a slight negative linear relationship between temperature and amount of electricity. The scatterplot for these data, shown in Figure 11, verifies that there is just a slight linear relationship. But r measures

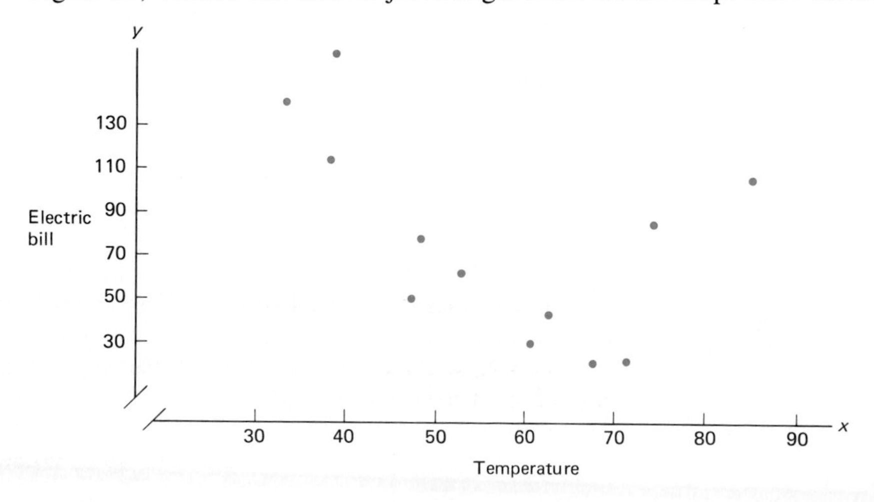

FIGURE 11

Scatterplot for the temperature/electricity use data

only the extent to which x and y have a *linear* relationship. The arrangement of points in Figure 11 suggests that there is, in fact, quite a strong relationship between x and y but it is more complex than a linear one. Indeed, consideration of the variables in this problem suggests that the cost of electricity will increase in extremely cold weather (for heating) *and* in extremely hot weather (for cooling), a situation reflected in the scatterplot in Figure 11. It is not difficult to imagine that if similar data were obtained for three successive years, a scatterplot showing cost and time might look something like Figure 12. In this case x and y exhibit a strong *cyclical* relationship.

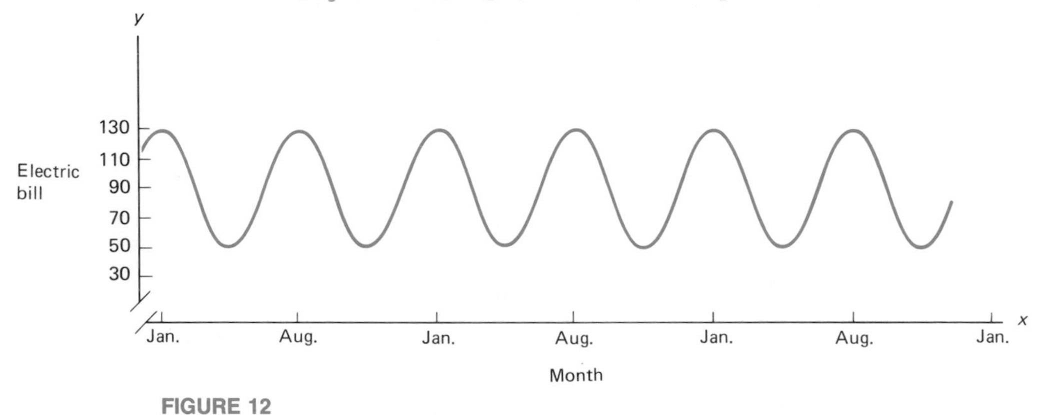

FIGURE 12

Long-term cyclical relationship between date and electric bill

Another source of confusion in interpreting r can be illustrated by the following example. It is said that in New York City there is a high correlation between increases in the number of cracks in the sidewalk and the hospitalization of elderly people. It might be assumed that the cracked sidewalks caused elderly people to trip, resulting in serious injury. Actually, it is more likely that it is the summer heat which causes both the cracks in the sidewalks and also heart attacks and other heat-related illnesses among the elderly. Thus a high correlation may, but does not necessarily, imply a *causal* relationship between two variables.

Only an understanding of the mechanisms that cause two variables to change their values can determine whether or not a causal relationship exists between them. In the case of the price/sales data, for example, an economist might well conclude—based on an understanding of economic forces—that, in view of the value $r = -.714$, higher prices *do* in fact cause a reduction in sales, and vice versa.

One further aspect of the correlation coefficient is helpful to keep in mind. As you saw, means or standard deviations are in minutes, grams, dollars, numbers of children, or similar units. But r is merely a number between -1 and 1; that is, it has *no units* just as in one sense we may say that a fraction (e.g., 1/2) in and of itself has no units.

PROBLEMS 2.2

Save your calculations. You will need them in Section 2.3.

10. Compute the correlation coefficient, r, for the following automobile assembly data from Problem 1 on page 78.

NUMBER OF WEEKS THE GROUP MEMBERS HAVE WORKED TOGETHER, x	4	23	12	18	7	14
TIME TO ASSEMBLE CAR, y	2.7	1.6	2.0	1.8	2.5	2.3

11. Compute the correlation coefficient, r, between height and weight using the data in Problem 2 on page 79.

HEIGHT, x	66	70	68	69	71	69	67	72
WEIGHT, y	152	145	162	172	160	158	140	184

$$\sum x_i^2 = 38{,}116 \qquad \sum y_i^2 = 203{,}977 \qquad \sum x_i y_i = 87{,}956$$

12. The following data, given first in Problem 3 on page 79, are the number of years of full-time education (x) and the annual salary (y) for 15 persons. Compute r for these data.

x	20	17	18	18	13	18	9	18	16	12	12	19	16	14	13
y	35.2	34.6	23.7	33.2	24.4	33.4	11.2	32.3	25.1	22.1	18.9	37.8	25.9	28.4	29.6

$$\sum x_i^2 = 3761 \qquad \sum y_i^2 = 12{,}237.78 \qquad \sum x_i y_i = 6725.6$$

13. Does your answer in Problem 12 imply that more education is responsible for a higher salary?
14. Compute the correlation coefficient between unemployment rate (y) and rate of inflation (x) using the following data from Problem 4 on page 79.

x	17.2	8.1	4.7	12.0	6.5	9.2	8.4	10.7
y	5.4	10.0	12.9	6.3	10.2	7.0	8.4	6.4

$$\sum x_i^2 = 839.48 \qquad \sum y_i^2 = 599.82 \qquad \sum x_i y_i = 579.85$$

15. For each of 13 cities in the United States, the latitude (in this case the angular distance, in degrees, north of the equator) and the average annual temperature (in degrees Fahrenheit) were recorded, with the following results:

CITY	LATITUDE, x	TEMPERATURE, y
New York	40	54.1
Los Angeles	34	61.8
Chicago	42	49.9
Philadelphia	40	54.6
Houston	29	68.2
Miami	25	75.5
Boston	42	50.3
Atlanta	33	61.5
Detroit	42	49.2
San Francisco	37	56.2
Minneapolis	45	44.9
New Orleans	30	69.0
Denver	39	50.2

$$\sum x_i^2 = 18{,}018 \qquad \sum y_i^2 = 43{,}757.18 \qquad \sum x_i y_i = 26{,}746.5$$

(a) Plot these data on a scatterplot.
(b) Compute the correlation coefficient, r, between latitude and temperature.
(c) Does a causal relationship seem appropriate here? Explain.

16. Discuss the value of the correlation coefficient that you would expect to find for the following pairs of variables:
(a) Speed and gasoline mileage at speeds over 40 m.p.h.
(b) IQ and height
(c) Poverty and crime for 25 communities
(d) Altitude and ZIP code for 25 communities
(e) Age and income for working persons

17. Which of the following values for r indicates the strongest linear relationship? the weakest?

$$.59 \quad -.18 \quad -.77$$

Solution to Problem 10

The calculations we need are as follows:

x	x^2	y	y^2	xy
4	16	2.7	7.29	10.8
23	529	1.6	2.56	36.8
12	144	2.0	4.00	24.0
18	324	1.8	3.24	32.4
7	49	2.5	6.25	17.5
14	196	2.3	5.29	32.2
$\Sigma x_i = 78$	$\Sigma x_i^2 = 1258$	$\Sigma y_i = 12.9$	$\Sigma y_i^2 = 28.63$	$\Sigma x_i y_i = 153.7$

Inserting these various sums into Equation 2.2 for r, we get

$$r = \frac{6(153.7) - (78)(12.9)}{\sqrt{[6(1258) - 78^2][6(28.63) - 12.9^2]}}$$

$$= \frac{-84}{\sqrt{(1464)(5.37)}} = -.947$$

SECTION 2.3
THE LEAST SQUARES LINE

As we have indicated, the correlation coefficient, r, measures the extent to which a set of bivariate quantitative data have a linear relationship (i.e., the data points seem to lie along a straight line). In many instances it is helpful to find the equation of the line that best represents the path the data points appear to follow. This equation helps the investigator better understand how the two variables are related and can be used to make predictions about one variable (y) given some value for the other (x). For instance, using available sales/price data, a sales manager might predict the effect of changes in the price of a company's product on sales.

We begin with some comments about the equation of a straight line. The general equation of a straight line is

$$y = a + bx \qquad \text{(Eq. 2.3)}$$

The constant a is called the ***y* intercept** and is the value for y when $x = 0$. The constant b is called the **slope** of the line and is the change in y corresponding to a one-unit increase in x. The definitions of a and b are illustrated in Figure 13.

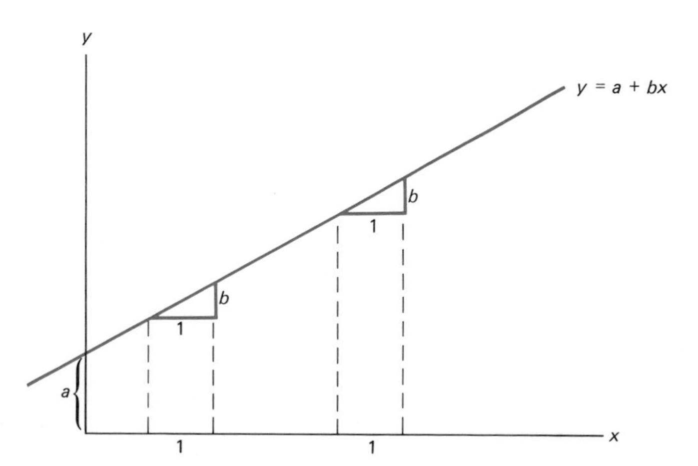

FIGURE 13

Showing a as the y intercept and b as the slope of the line $y = a + bx$

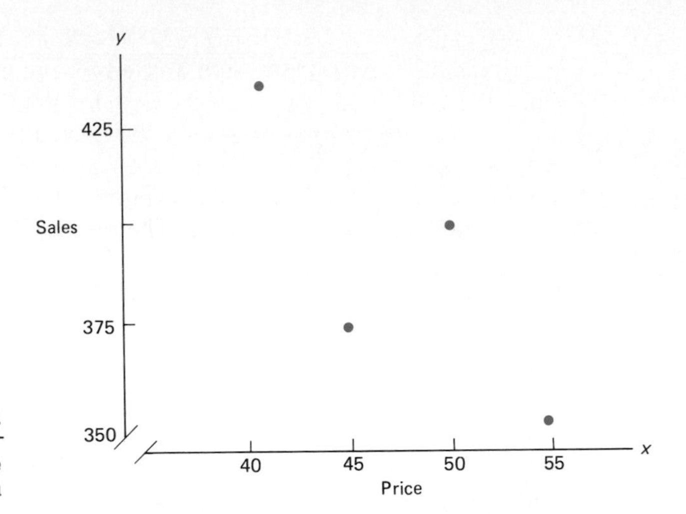

FIGURE 14

Scatterplot for the price/sales data

Returning to our discussion of the line that best fits the data, consider again the price/sales data from Example 5 as represented in the scatterplot reproduced in Figure 14. The value $r = -.714$ suggests a moderate negative relationship between prices and sales. How are we to find the line that passes as close as possible to all four points? A useful approach is to specify some criterion that the line should meet. Such a criterion, developed by the French mathematician André Legendre in 1806, is the principle of least squares. This principle involves taking the vertical distance from each data point to a hypothetical line as the measure of its closeness to the line. These distances, shown in Figure 15, are labeled D_1, D_2, D_3, and D_4. The value of the D's will vary for different lines. According to the principle of least squares, the line that is the best fit to the data

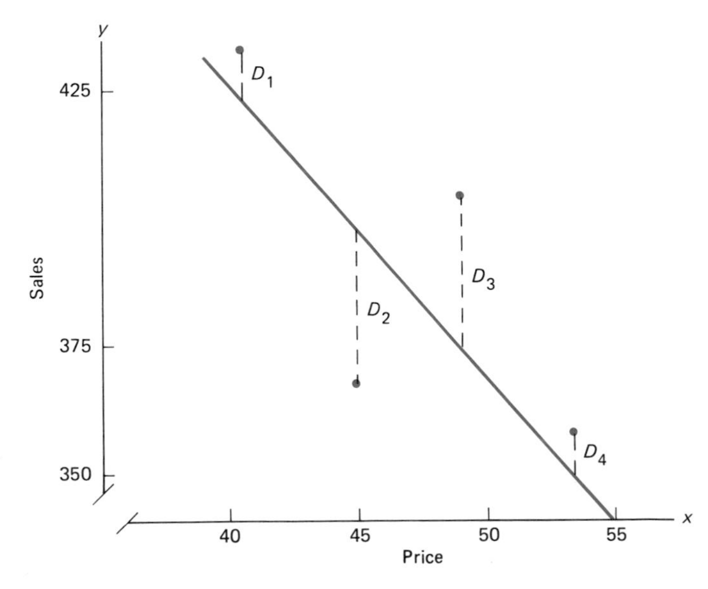

FIGURE 15

Scatterplot for the price/sales data showing the four vertical distances from a hypothetical line

is that for which the sum $D_1^2 + D_2^2 + D_3^2 + D_4^2$ is as small as possible. This line is referred to as the **least squares line** or the **regression line.**

More generally, with n data points we label the corresponding distances $D_1, D_2, \ldots, D_n$. The values for a and b associated with the least squares or best-fitting line are now those which make the sum $D_1^2 + D_2^2 + \cdots + D_n^2$ as small as possible. The procedure used to determine the values of a and b is quite lengthy and therefore has been omitted. The resulting values for a and b can, however, be computed as follows:

The equation for b, the slope of the best-fitting line, is

$$b = \frac{\Sigma\,(x_i - \bar{x})(y_i - \bar{y})}{\Sigma\,(x_i - \bar{x})^2} \qquad \text{(Eq. 2.4)}$$

A computationally more helpful form of Equation 2.4 is

$$b = \frac{n\,\Sigma\,x_i y_i - (\Sigma\,x_i)(\Sigma\,y_i)}{n\,\Sigma\,x_i^2 - (\Sigma\,x_i)^2} \qquad \text{(Eq. 2.5)}$$

The equation for a, the intercept of the best-fitting line, is

$$a = \frac{1}{n}\left(\sum y_i - b\sum x_i\right) \qquad \text{(Eq. 2.6)}$$

The equation $y = a + bx$ now becomes the least squares or regression line. Notice that the numerator of b in Equation 2.5 is the same as the numerator of r in Equation 2.2. The denominator of b is one of the two components under the square root in the denominator of Equation 2.2.

Now let us apply Equations 2.5 and 2.6 to find the least squares line for the price/sales data.

From Table 7 the components for Equations 2.5 and 2.6 are

x	x^2	y	xy
41	1681	435	17,835
45	2025	375	16,875
50	2500	410	20,500
54	2916	354	19,116
$\sum x_i = 190$	$\sum x_i^2 = 9122$	$\sum y_i = 1574$	$\sum x_i y_i = 74{,}326$

Entering these various sums in Equation 2.5, we have

$$b = \frac{4(74{,}326) - (190)(1574)}{4(9122) - 190^2}$$

$$= \frac{-1756}{388} = -4.52577$$

The value for a is then

$$a = \frac{1}{4}[1574 - (-4.52577)(190)]$$

$$= \frac{1}{4}(1574 + 859.8963) = 608.4741$$

The resulting least squares line is thus:

$$y = 608.4741 - 4.52577x$$

or, more usefully,

$$y = 608.47 - 4.53x$$

This least squares line is plotted in Figure 16.

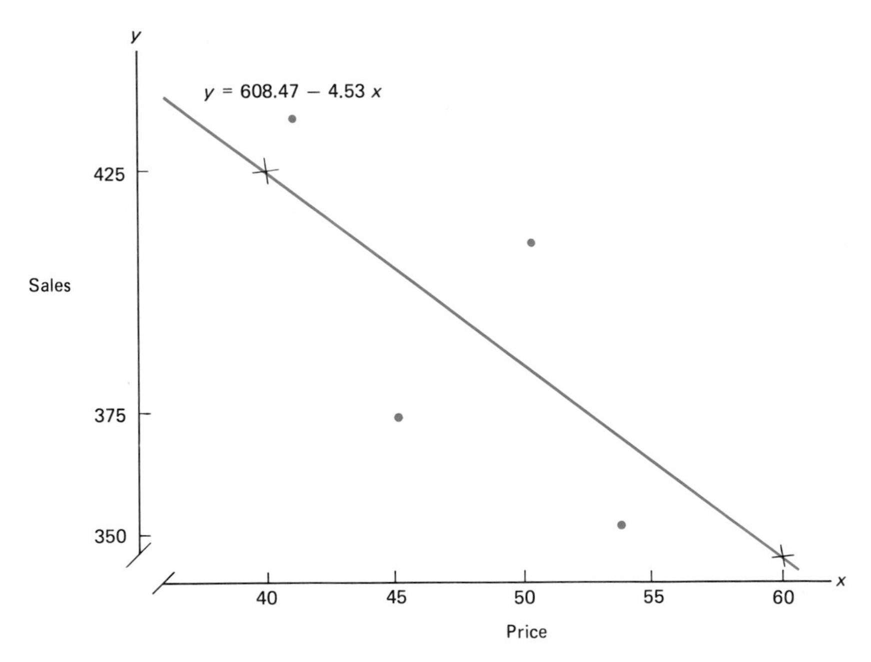

FIGURE 16

Least squares line predicting sales (y) from price (x)

The method for sketching in the line is straightforward:

1. Select two values for x which are far apart. In this case we choose $x = 40$ and $x = 60$.

2. Compute the corresponding values for y. In this case when $x = 40$, $y = 608.47 - 4.53(40) = 427.27$, and when $x = 60$, $y = 608.47 - 4.53(60) = 336.67$.
3. Mark the points (40, 427) and (60, 337) and join the points to obtain the line.

The line $y = 608.47 - 4.53x$ is the best fit to the data in that there is no other line for which the sum of the squared vertical distances from the four points is smaller.

The intercept of the least squares line, $a = 608.47$, is not very meaningful. It is the predicted number of units sold (y) when the price (x) is zero. The slope of the line, $b = -4.53$, is more useful. An increase of one dollar in the price (x) is associated with a decrease of 4.53 in the number of units sold.

PREDICTION WITH THE LEAST SQUARES LINE At the beginning of this section we implied that the least squares line can be used to make predictions about one variable (y) given some value for another variable (x). We illustrate this by an example.

EXAMPLE 7 Use the least squares line obtained above, $y = 608.47 - 4.53x$, to predict the number of units that will be sold if the price per unit is (a) $x = \$47$; (b) $x = \$60$; (c) $x = \$140$.

Solution (a) If $x = \$47$, the predicted value for y is $y = 608.47 - 4.53(47) = 395.56$. The number of items sold has to be a whole number, so it makes sense to round off the predicted value to the nearest integer—in this case, 396 (items).

(b) If $x = \$60$, the predicted value for y is $y = 608.47 - 4.53(60) = 336.67(337)$ items.

(c) If $x = \$140$, the predicted value for y is $y = 608.47 - 4.53(140) = -25.73$.

The last result clearly makes no sense. It is impossible to have a negative number of sales. This result points up the danger of using a least squares line to make a prediction far beyond the range of the data on which it is based. There is no guarantee at all that the line that is the best fit to a set of data within which x varies from approximately 40 to approximately 55 will be appropriate outside that range.

■

We next obtain the least squares line for the ratings data, and then use the line to predict ratings from scores.

EXAMPLE 8 (a) Use the computations in Table 5 to find the least squares line relating a student's rating of the instructor to the student's course score; (b) interpret b in this case; and (c) use the line to predict the rating by a student with a score of 85.

Solution (a) From Table 5 we obtain the values $n = 8$, $\Sigma\, x_i = 639$, $\Sigma\, x_i^2 = 51{,}549$, $\Sigma\, y_i = 99$, and $\Sigma\, x_i y_i = 8030$. Entering these values in Equation 2.5 gives us

$$b = \frac{8(8030) - (639)(99)}{8(51{,}549) - 639^2} = \frac{979}{4071} = .2405$$

Using Equation 2.6 for a, results in

$$a = \frac{1}{8}[99 - (.2405)639] = -6.835$$

The least squares line is thus

$$y = -6.835 + .2405x$$

or, more conveniently,

$$y = .24x - 6.84$$

(b) The value $b = .24$ indicates that with each increase of 1 point on a student's course score there is an increase of .24 in the student's rating of the instructor. More usefully, with an increase of 4 points on the student's score there is an increase of .96 (almost 1) point in the rating.
(c) When $x = 85$, the predicted value for y is

$$y = .24(85) - 6.84 = 13.56$$

The least squares line $y = .24x - 6.84$ is sketched on the corresponding scatterplot in Figure 17.

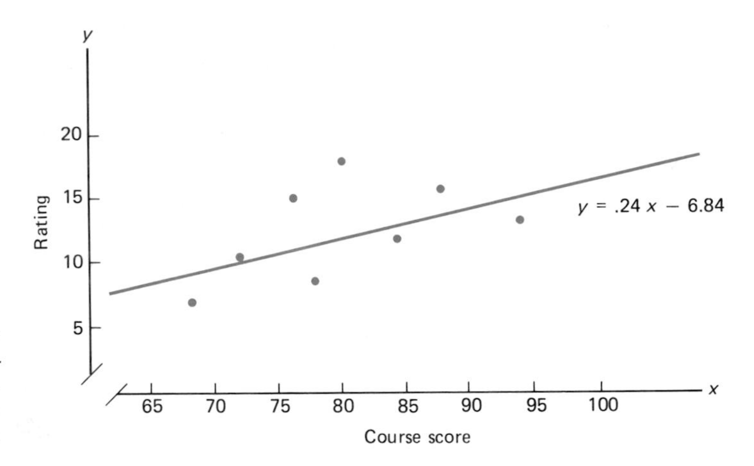

FIGURE 17

Least squares line $y = .24x - 6.84$ for the ratings data

■

The least squares line is frequently referred to as the regression line. This terminology originated with the early application of least squares in genetic studies. Throughout the remainder of this text we shall use the terms "regression line" and "least squares line" interchangeably.

CORRELATION, LINEAR RELATIONSHIPS, AND CAUSATION

In Chapter 1 we discussed measures of spread or variability in a set of quantitative data (s, R etc.). To the extent that two such variables x and y have a linear relationship, it can be shown (we do so in Section 14.1) that some part of the variability in y can be associated with, or determined from, that in x. More specifically:

DEFINITION 1

The **proportion** of the variability in y that can be accounted for by its linear relationship to x is r^2.

Multiplying r^2 by 100, we can say that:

DEFINITION 2

The **percentage** of the variability in y that can be accounted for by its linear relationship to x is $100r^2$.

Thus, if $r = -.8$, for instance, $100r^2 = 100\ (.64) = 64$ and we can conclude that sixty-four percent of the variability in y can be accounted for by its linear relationship to x.

Notice that, as usual, these definitions say nothing about the *cause* of the linear relationship. The words "accounted for" here refer only to the degree to which changes in y can be predicted from values for x in light of the linear relationship between them. As we stated earlier (see page 88), the existence of even a strong linear relationship between two variables, and therefore the ability to predict one from the other, does not imply a *causal* relationship between them.

In this context it is important to realize that in addition to any linear relationship to x, y is generally related to factors other than x, whether these relationships are causal or not. For instance, in addition to its relation to price (x), sales of a product in any month (y), may be related to such factors as the method or degree of promotion, competitors' prices, and pent-up demand. The farther the data points representing x and y are from a straight line, the more the values of y are related to factors *other* than x.

The next two examples illustrate the preceding discussion.

EXAMPLE 9 The value of r for the ratings data in Section 2.2 was $r = .5272$ and thus $r^2 = .5272^2 = .278$. The value $100(.278) = 27.8$ indicates that only about 28% of the variability in the ratings can be accounted for by its linear relationship to course scores. As important, over 72% of the variability in ratings is related to variables other than the student's course score.

■

The value for r^2 is sometimes called the **coefficient of determination.** We stress that Definitions 1 and 2 refer to r^2, *not* r.

EXAMPLE 10 In Example 5 we found that the value for r for the prices/sales data was $-.714$. In this case $r^2 = (-.714)^2 = .51$. This number indicates that just over one-half of the variability in sales between supermarkets can be accounted for by the linear relationship between sales and price.

■

PROBLEMS 2.3

18. (a) Using the data in Problem 10 on page 89, find the least squares line relating the time to assemble a car (y) to the length of time the team has been together (x).
    (b) Interpret b in this case.
    (c) Compute and interpret the value of r^2 in the context of this problem.
    (d) Use the least squares line to predict the value of y when (i) $x = 10$; (ii) $x = 20$.
19. (a) Use the data in Problem 11 on page 89 to find the least squares line relating weight (y) to height (x).
    (b) Interpret the value of b.
    (c) Compute and interpret the value of r^2.
    (d) Predict the weight of a man who is (i) 65 inches tall; (ii) 70 inches tall.
    (e) Plot these data on a scatterplot and draw the least squares line.
20. (a) Use the data in Problem 12 on page 89 to find the least squares line relating annual salary (y) to years of full-time education (x).
    (b) Compute and interpret r^2 in this case.
    (c) Interpret b.
    (d) Predict the annual income of a person who has 15 years of education.
21. (a) Using the data in Problem 14 on page 89, obtain the least squares line relating unemployment rate (y) to the inflation rate (x).
    (b) Compute and interpret r^2 in this case.
    (c) Interpret b,
    (d) Predict the unemployment rate for a country whose inflation rate is (i) 10%; (ii) 5%.
    (e) Plot these data on a scatterplot and draw the least squares line.
22. (a) Use the data in Problem 15 on page 89 to obtain the least squares line relating mean temperature (y) to latitude (x).
    (b) Interpret b.
    (c) Compute and interpret r^2 for these data.
    (d) Predict the average annual temperature in Kansas City, Missouri, latitude 39°.

Solution to Problem 18

(a) In the solution to Problem 10 on page 90, we obtained the following summary data:

$$n = 6 \qquad \sum x_i = 78 \qquad \sum x_i^2 = 1258 \qquad \sum y_i = 12.9$$

$$\sum y_i^2 = 28.63 \qquad \sum x_i y_i = 153.7$$

Using Equation 2.5, the slope of the regression line is

$$b = \frac{6(153.7) - (78)(12.9)}{6(1258) - (78)^2} = \frac{-84}{1464} = -.0574$$

and the y intercept is

$$a = \frac{1}{6}[12.9 - (-.0574)(78)]$$

$$= 2.896$$

The least squares line is thus

$$y = 2.896 - .0574x$$

(b) A one-week increase in the time the team has spent together is associated with a .0574-hour decrease in the time to assemble a car.

(c) From Problem 10 on page 90, $r^2 = (-.947)^2 = .897$. Approximately 90% of the variability in the time it takes a group to assemble a car is accounted for by the linear relationship between the time to assemble a car and the time the group members have been together.

(d) (i) If $x = 10$, $y = 2.896 - .0574(10) = 2.32$ hours.
   (ii) If $x = 20$, $y = 2.896 - .0574(20) = 1.75$ hours.

SECTION 2.4
USING COMPUTER OUTPUT: MINITAB AND BIVARIATE DATA

In this section we consider some of the capabilities of Minitab in describing and summarizing bivariate data where both variables are quantitative.

EXAMPLE 11 Returning to the nutrition data in Table 8 of Chapter 1 (page 48) it is interesting to examine the relationship between infant birthweight (BTWT) in C7 and birth length (BTLH) in C8. The command "PLOT OF C7 VS C8" produces a scatterplot of these data with BTWT on the vertical (y) axis and BTLH on the horizontal (x) axis. This command and the scatterplot are shown in Figure 18.

FIGURE 18

Scatterplot for infant birthweight (C7) and birth length (C8)

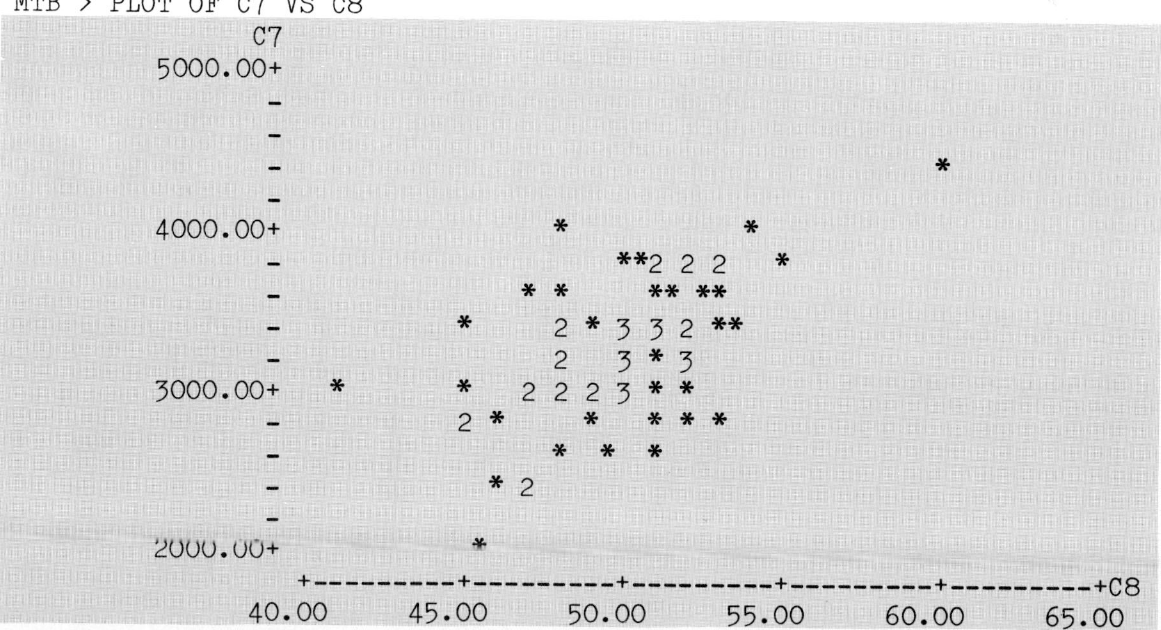

The points on the scatterplot are indicated by asterisks and numbers. The numbers are used to indicate points where there is more than one (x, y) pair. For example, there are three infants who weighed approximately 3000 grams and who were 50 centimeters in length. The numbers make it a bit difficult to interpret the scatterplot, but the impression is created of a moderate positive relationship between the two variables.

To investigate further we obtained the correlation coefficient r between C7 and C8. The command is "CORRELATION C7, C8" and the resulting output, shown in Table 8, indicate that the correlation between BTWT and BTLH is $r = .58$, indicating a moderate positive linear relationship between the two variables. To obtain the least squares or regression line relating BTWT (C7) to BTLH (C8), we use the command "REGRESS C7 ON 1 PREDICTOR IN C8" (see Table 8). The "1" in this command needs to be inserted because it is possible to obtain a regression line relating y (C7) to two or more predictor variables $(x_1, x_2, \ldots)$. In this case the equation of the regression line is C7 = −1233 + 89.6 C8 or in the more usual form, $y = -1233 + 89.6x$. As we have seen, the regression line can be used for prediction. For instance, the predicted weight for an infant who is 48 centimeters long is $-1233 + 89.6(48) = 3067.8$ grams.

TABLE 8

Correlation coefficient and the regression line relating BTWT to BTLH

```
MTB > CORRELATION C7,C8

   CORRELATION OF        C7 AND C8        = 0.580

MTB > REGRESS C7 ON 1 PREDICTOR IN C8

 THE REGRESSION EQUATION IS
 C7 = - 1233 + 89.6 C8
```

■

Actually, as we indicate in Chapter 14, the "REGRESS" command above produces a good deal more information about the least squares line than simply its equation.

EXAMPLE 12

How does the average speed of long-distance runners vary with age? The following data are the ages and the average speeds in mph of a random sample of 18 runners who completed a recent marathon.

TABLE 9

Scatterplot, correlation coefficient, and regression line for the age/speed data

AGE	AVERAGE SPEED	AGE	AVERAGE SPEED	AGE	AVERAGE SPEED
31	7.3	44	6.7	23	8.0
29	7.7	37	7.0	33	6.6
24	8.1	35	7.5	33	7.2
29	7.6	32	7.2	25	7.4
23	8.2	37	7.0	34	7.0
38	6.9	41	6.8	31	7.9

```
MTB > READ DATA INTO C1,C2
DATA> 31,7.3
DATA> 29,7.7
DATA> 24,8.1
DATA> 29,7.6
DATA> 23,8.2
DATA> 38,6.9
DATA> 44,6.7
DATA> 37,7
DATA> 35,7.5
DATA> 32,7.2
DATA> 37,7
DATA> 41,6.8
DATA> 23,8
DATA> 33,6.6
DATA> 33,7.2
DATA> 25,7.4
DATA> 34,7
DATA> 31,7.9
DATA> PLOT OF C2 VS C1
    18 ROWS READ
              C2
          8.40+
               -          *
               -            *
               -          *
               -                             *
          7.70+                        *
               -                       *           *
               -              *
               -                             *
               -                              * *
          7.00+                                  *    2
               -                                        *    *
               -                                                 *
               -                                *
               -
          6.30+
                 +---------+---------+---------+---------+---------+C1
               18.00     24.00     30.00     36.00     42.00     48.00

MTB > CORRELATION C1,C2

   CORRELATION OF        C1 AND C2      =-0.835

MTB > REGRESS C2 ON 1 PREDICTOR IN C1

THE REGRESSION EQUATION IS
C2 = 9.53 - 0.0680 C1
```

In Table 9 we indicate the process of entering the data through the "READ" command and obtaining the scatterplot, the correlation coefficient, and the regression line. The scatterplot suggests an inverse (negative) close-to-linear relationship between age (C1) and average speed (C2). This impression is confirmed by the value of the correlation coefficient: $r = -.835$.

■

PROBLEMS 2.4

23. An additive is expected to increase mileage when it is combined with unleaded gasoline. Seven mixtures are made up, each containing a different percentage of the additive. The percentages are given in C1 in the accompanying diagram and the mileages obtained on a standard test run using the mixture, with each percentage, in C2. The scatterplot, correlation coefficient, and regression line relating C2 to C1 are also given.

```
MTB > READ DATA INTO C1,C2
DATA> 1,46.4
DATA> 1.5,46.7
DATA> 2,47.4
DATA> 2.5,48
DATA> 3,48.3
DATA> 3.5,48.2
DATA> 4,48.5
DATA> PLOT C2 VS C1
```

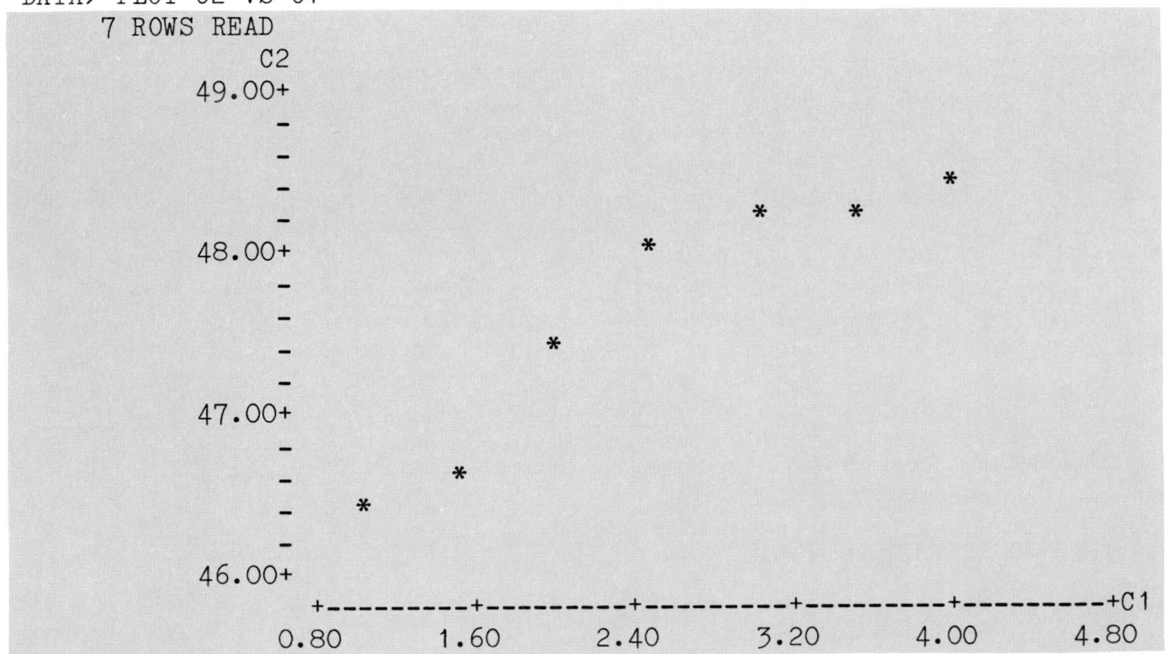

```
MTB > CORRELATION C1,C2

   CORRELATION OF          C1 AND C2         = 0.952

MTB > REGRESS C2 ON 1 PREDICTOR IN C1

THE REGRESSION EQUATION IS
C2 = 45.8 + 0.729 C1
```

(a) Interpret the slope of the regression line in this case.

(b) Predict the mileage if a mixture containing 2.75% additive is used.

(c) What proportion of the variability in mileage can be accounted for by its linear relationship to the percentage of the additive?

24. The heights (in inches) of 30 fathers and their 30 adult sons were entered into C1 and C2, respectively. The scatterplot, correlation coefficient, and regression line predicting C2 from C1 were obtained.

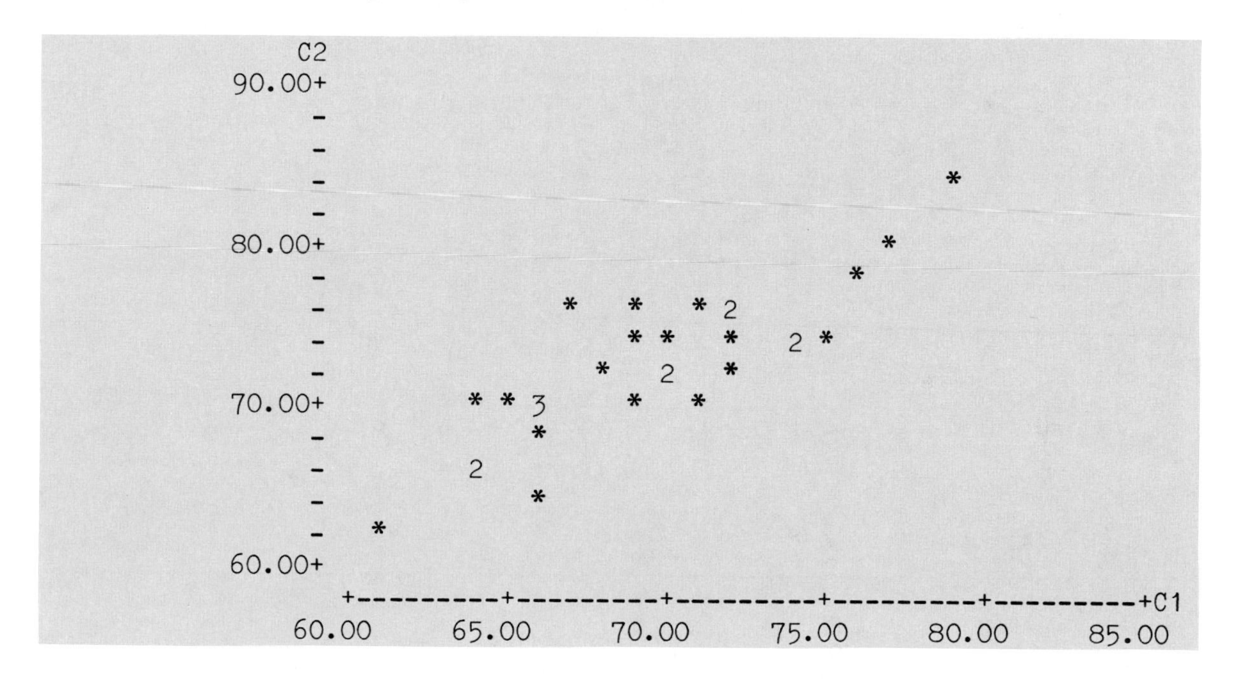


```
MTB > CORRELATION C1,C2

   CORRELATION OF          C1 AND C2         = 0.835

MTB > REGRESS C2 ON 1 PREDICTOR IN C1

THE REGRESSION EQUATION IS
C2 = 10.0 + 0.887 C1
```

(a) Interpret the slope of the regression line in this case.

(b) Predict the height of the son of a man (i) 6 feet tall; (ii) 66 inches tall.

(c) What proportion of the variability in the heights of the sons can be accounted for by their linear relationship to the heights of the fathers?

25. Refer to the nutrition data in Table 8 of Chapter 1 (page 48). It was of interest to investigate the relationship between the mother's weight gain (in C10) and her prepregnancy weight (in C4). The scatterplot for these two variables is shown in the accompanying plot. Why would it be inappropriate to compute and use the regression line relating C10 to C4?

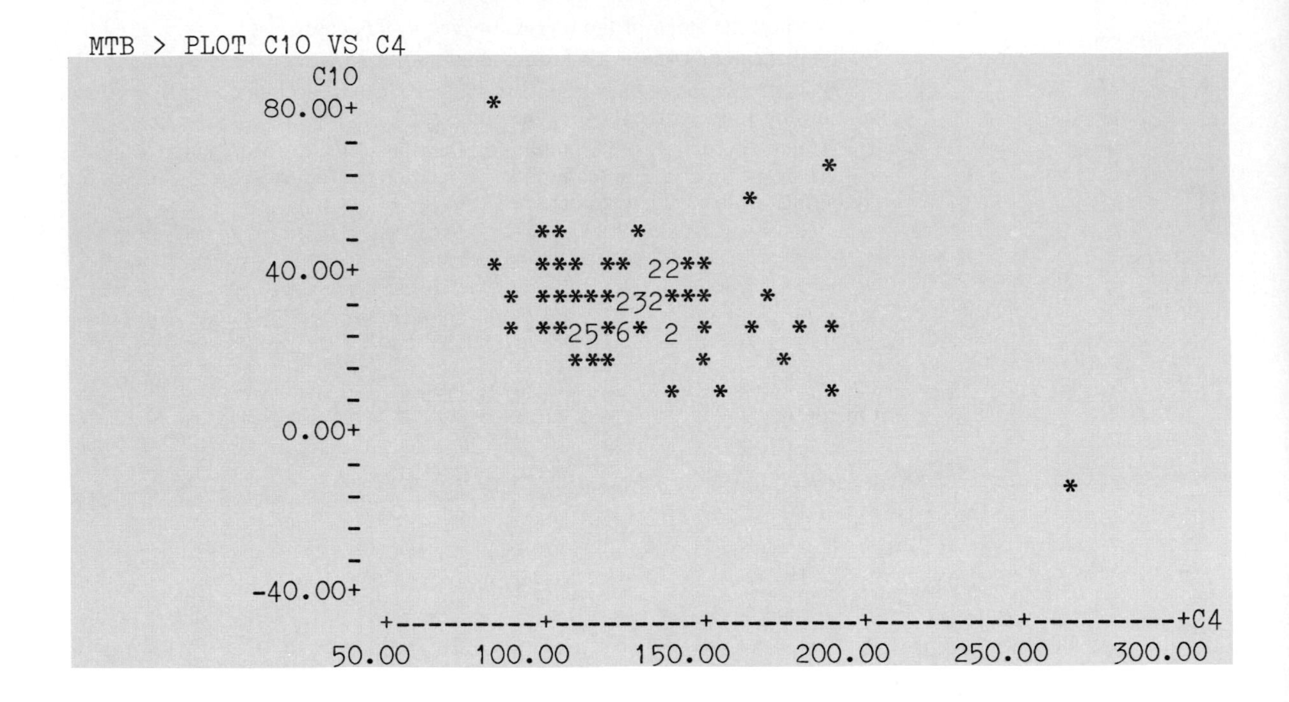

SECTION 2.5
SUMMARY AND REVIEW

BIVARIATE DATA

Data on two variables are often obtained for each person or unit in a group. Such data are called **bivariate data.**

A number of graphic options are available if one variable is quantitative and the other qualitative. The observations should first be grouped on the basis of the different levels of the qualitative variables. Then a histogram, a stem-and-leaf display, or a box-and-whisker display showing the distribution of the quantitative variable can be drawn for each level.

When both variables are qualitative, the data are usually represented as counts in a **contingency table.** The rows in the table represent one variable and the columns the other. Such data can also be represented graphically by a sequence of blocks, one for each level of one of the variables.

When both variables are quantitative, it is customary to display the data on a **scatterplot,** a two-dimensional display in which the axes are x and y. Each (x, y) pair is represented by a point on the plot.

THE CORRELATION COEFFICIENT

The **correlation coefficient,** r, measures the extent to which the points on a scatterplot tend to lie on or near a straight line. The computationally useful formula for r is

$$r = \frac{n \sum x_i y_i - (\sum x_i)(\sum y_i)}{\sqrt{[n \sum x_i^2 - (\sum x_i)^2][n \sum y_i^2 - (\sum y_i)^2]}} \qquad \text{(Eq. 2.2)}$$

The value for r will always lie between -1 and 1 (inclusive). The closer r is to -1 or 1, the stronger the linear relationship between x and y (i.e., the closer the points lie to a straight line). Depending on the value of r we describe the extent of the linear relationship between x and y as weak, slight, moderate, or strong.

It is important to note that, although a small value for r indicates a weak linear relationship, this does not necessarily mean that the two variables are unrelated. They may have a cyclical, quadratic, or some other relationship which is more sophisticated than a linear one.

On the other hand, a large value for r does not necessarily mean there is a causal relationship between x and y. There may, for instance, be a third "intervening" variable (whether known or not) that affects both x and y.

THE LEAST SQUARES LINE

If we are interested in the linear relationship between two variables, we compute the **least squares** or **regression line.** This is that line for which the sum of the squared vertical distances from the data points to the line is as small as possible. The resulting line is of the form $y = a + bx$, where

$$b = \frac{\sum (x_i - \bar{x})(y_i - \bar{y})}{\sum (x_i - \bar{x})^2} \qquad \text{(Eq. 2.4)}$$

A computationally simpler version for b is

$$b = \frac{n \sum x_i y_i - (\sum x_i)(\sum y_i)}{n \sum x_i^2 - (\sum x_i)^2} \qquad \text{(Eq. 2.5)}$$

and

$$a = \frac{1}{n}\left(\sum y_i - b \sum x_i\right) \qquad \text{(Eq. 2.6)}$$

As defined earlier, a is the y *intercept* (i.e., the value for y when $x = 0$); b is the *slope* of the line and measures the change in y corresponding to an increase of one unit in x. By entering a new value for x into the equation of the least squares line, we can obtain a *predicted* value for y.

In the context of the least squares line, r^2 has an interesting interpretation. The value r^2 (not r) is the proportion of the variability in y that can be accounted for by its linear relationsip to x. This can be stated more usefully as: $100r^2$ is the *percentage* of the variability in the y's that can be accounted for by its linear relationship to x (Definition 2). Whatever the value of r^2, however, no conclusions as to a causal relationship between x and y should be drawn without further knowledge of the reasons for changes in their values.

Great care needs to be taken when using the least squares line for prediction beyond the range of x values on which it is based.

REVIEW PROBLEMS

GENERAL

1. (a) Plot the following data on a scatterplot:

x	−5	−4	−2	0	2	4
y	34	32	25	19	10	4

(b) Compute the correlation coefficient.

2. The men's world record times (in seconds) for races of various lengths (in meters) are as follows:

LENGTH, x	100	200	400	800	1000	1500
TIME, y	9.95	19.72	43.86	102.40	133.40	211.40

(a) Plot these data on a scatterplot.
(b) Compute the regression line relating time to length. Draw this line on the scatterplot.
(c) Interpret b in this case.
(d) Predict the world record time for a race of 600 meters.

$$\sum x_i^2 = 4{,}100{,}000 \qquad \sum y_i^2 = 75{,}382.861 \qquad \sum x_i y_i = 554{,}903$$

3. A random sample of nine used-car sales involving the same model was taken. For each sale the age of the car and the price paid (in hundreds of dollars) were recorded, with the following results:

AGE, x	4	5	1	5	3	1	7	4	2
PRICE, y	27	24	50	20	39	55	10	30	43

(a) Plot these data on a scatterplot.
(b) Compute the regression line relating price to age.

(c) Interpret b in this case.

(d) Predict the difference in prices for two cars that differ in age by three years.

4. The following data are the ages, at the time of their marriage, of seven married couples:

AGE OF HUSBAND, x	28	20	36	25	28	35	29
AGE OF WIFE, y	25	20	32	19	29	34	29

(a) Compute the regression line relating the age of the wife to the age of the husband.

(b) Compute the regression line relating the age of the husband to the age of the wife.

(c) Are these two lines the same?

$$\sum x_i^2 = 5955 \qquad \sum y_i^2 = 5248 \qquad \sum x_i y_i = 5570$$

5. Two variables x and y are related by the equation $y = 2x + 1$. The values of y corresponding to four values for x are as follows:

x	2	4	6	8
y	5	9	13	17

Find (a) the regression line relating y to x; (b) the correlation coefficient, r, between x and y. Do your answers surprise you? Explain.

6. Two variables, x and y, are related by the equation $y = 8 + 2x - x^2$. The values of y corresponding to six different values for x are as follows:

x	−4	−2	0	2	4	6
y	−16	0	8	8	0	−16

(a) Plot these data on a scatterplot.

(b) Compute the correlation coefficient r.

(c) Why is the value for r so small when the relationship between x and y is so well defined?

7. Show that the correlation between a variable x and *itself* is $r = 1$.

8. Show that the regression line always goes through the point $(\bar{x}, \bar{y})$ on the scatterplot. (*Hint:* Compute the value of y when $x = \bar{x}$.)

9. The midterm and the final scores for each of the 20 students in a physics class are as follows:

MIDTERM SCORE, x	95	82	98	100	61	73	86	88	84	65
FINAL SCORE, y	90	71	86	100	57	75	79	84	71	61

MIDTERM SCORE, x	84	86	100	76	81	86	95	81	71	78
FINAL SCORE, y	83	79	92	53	84	90	85	81	72	74

(a) Plot these data on a scatterplot.
(b) Compute the regression line relating the final exam score to the midterm score.
(c) Interpret b in this case.

$$\sum x_i = 1670 \qquad \sum x_i^2 = 141{,}720 \qquad \sum y_i = 1567$$

$$\sum y_i^2 = 125,475 \qquad \sum x_i y_i = 132{,}925$$

10. Does offense or defense win games? The following data are the total number of points (y) for each of the 21 teams in the National Hockey League (NHL) in 1979–1980, together with the goals scored by them (x_1) and against them (x_2). (Points are based on games won or tied.) Compute the correlation between x_1 and y and between x_2 and y. Are goals for or goals against a team the best predictor of performance in terms of the number of points obtained in a season?

TEAM	x_1	x_2	y	TEAM	x_1	x_2	y
Philadelphia	327	254	116	Montreal	328	240	107
New York Islanders	281	247	91	Los Angeles	290	313	74
New York Rangers	308	284	86	Pittsburgh	251	303	73
Atlanta	282	269	83	Hartford	303	312	73
Washington	261	293	67	Detroit	268	301	63
Chicago	241	250	87	Buffalo	318	201	110
St. Louis	266	278	80	Boston	310	234	105
Vancouver	256	281	70	Minnesota	311	253	88
Edmonton	301	322	69	Toronto	304	327	75
Colorado	234	308	51	Quebec	248	313	61
Winnipeg	214	314	51				

$$\sum x_{1i} = 5902 \qquad \sum x_{1i}^2 = 1{,}680{,}108 \qquad \sum x_{1i} y_i = 481{,}376$$

$$\sum x_{2i} = 5897 \qquad \sum x_{2i}^2 = 1{,}679{,}667 \qquad \sum x_{2i} y_i = 461{,}185$$

$$\sum y_i = 1680 \qquad \sum y_i^2 = 141{,}110$$

11. Thirty angel cake batters are divided into three groups, differing only in the sugar content. In group A, the standard amount, (1/3 cup) of sugar, is used. In group B, no sugar is used, while in group C, 1/3 cup of confectioners' sugar is used. After preparation each angel cake is tested for texture (on a scale of 0 to 60—the higher the measurement, the less tender the cake). The scores are as follows:

A	30	32	28	40	35	36	33	35	29	33
B	40	36	38	42	37	42	45	33	41	40
C	33	40	36	34	36	33	30	30	35	34

(a) Compute the five-number summary for each of the three groups.
(b) Construct the corresponding box-and-whisker displays.

(c) Comment on differences among the distributions of texture levels for the three batters.

12. To examine the effect of the seasons on gas mileage, 18 cars, all of the same make and year, were included in an experiment. Each car was driven over the same distance for the same time in each of the four seasons. The mileage per gallon was recorded on each occasion, with the following results:

FALL		WINTER		SPRING		SUMMER	
25.8	26.9	22.9	25.0	28.9	30.0	25.5	26.0
26.5	27.2	25.0	22.2	28.1	27.6	25.9	28.3
26.4	26.9	22.8	22.9	28.7	29.8	25.3	23.7
27.5	26.7	24.1	23.8	28.3	27.7	26.1	27.1
26.7	27.4	22.5	23.5	29.4	27.6	27.4	29.1
27.7	27.4	20.9	22.3	20.1	20.1	23.8	25.5
26.4	27.1	25.7	24.2	28.6	28.8	26.7	25.2
26.0	26.1	23.5	25.3	28.5	29.3	25.4	27.0
27.2	26.3	26.0	26.1	28.0	29.1	24.4	26.3

(a) Compute the five-number summary for each season.

(b) Construct the corresponding box-and-whisker displays.

(c) Comment on differences among the distributions of mileages for the four seasons.

13. Each county in a sample of 400 was classified by (i) whether it was predominantly urban, suburban, or rural, and by (ii) whether the majority of those voting in the last presidential election voted Democratic or Republican. The data are summarized as follows:

	URBAN	SUBURBAN	RURAL
REPUBLICAN	47	91	34
DEMOCRATIC	133	79	16

(a) Draw a block diagram showing the proportion in each party by region and briefly summarize the information on the chart.

(b) Draw a block diagram showing the proportion in each region by party and briefly summarize the information on the chart.

HEALTH SCIENCES

14. A zoologist records the average number of different behaviors exhibited by a baby gibbon at different ages. The data are as follows:

NUMBER OF MONTHS, x	1	2	3	4	5	6
NUMBER OF BEHAVIORS, y	3	8	24	32	42	69

(a) Compute the regression line relating the number of behaviors to the age in months.

(b) Interpret b in this case.

(c) Why would it be unwise to use your line to predict the number of different behaviors exhibited by a 10-month-old gibbon?

15. As male patients come to a clinic for a routine physical examination, they are asked the average number of cigarettes they smoke per day. For 20 cigarette-smoking patients these data are matched with their systolic blood pressure (SBP) to obtain the following bivariate data:

NUMBER OF CIGARETTES, x	20	35	20	10	16	5	50	20	35	20
SBP, y	167	182	170	150	172	150	189	160	175	149

NUMBER OF CIGARETTES, x	60	30	40	10	40	40	20	15	15	20
SBP, y	204	180	185	150	183	194	139	142	161	159

(a) Plot these data on a scatterplot.

(b) Compute the correlation between these two variables.

(c) Find the regression line relating y to x.

(d) Predict the SBP for a male smoking an average of (i) 25 cigarettes per day; (ii) 10 cigarettes per day.

$$\sum x_i = 521 \qquad \sum x_i^2 = 17{,}581 \qquad \sum y_i = 3361$$

$$\sum y_i^2 = 571{,}197 \qquad \sum x_i y_i = 91{,}992$$

16. Ten diabetic patients enroll in a supervised weight-loss program. Their initial weights and six-month weight losses are as follows:

INITIAL WEIGHT, x	142	223	175	235	240	154	193	154	200	185
WEIGHT LOSS, y	5	37	31	44	40	14	21	13	11	24

(a) Compute the regression line relating weight loss to initial weight.

(b) What is an objective measure of the extent to which the least squares line "fits" these data?

(c) Predict the weight loss for a person who enters the program weighing 180 pounds.

(d) Predict the *difference* in weight loss for two persons whose initial weights differed by 25 pounds.

$$\sum x_i = 1901 \qquad \sum x_i^2 = 372{,}249 \qquad \sum y_i = 240$$

$$\sum y_i^2 = 7394 \qquad \sum x_i y_i = 49{,}177$$

17. Forty patients suffering from terminal cancer of the prostate gland are randomly divided into two treatment regimens, A and B. The number of months until death is recorded for each of the 40 patients. The results are as follows:

A				*B*			
7.0	3.2	5.5	1.2	2.3	5.3	5.2	7.0
1.8	7.4	5.5	4.8	4.6	1.6	4.8	8.6
5.3	5.9	6.8	5.6	2.6	5.6	4.8	6.7
3.5	8.0	6.4	5.8	5.0	5.4	6.0	5.0
2.2	2.3	6.1	8.0	5.2	3.4	5.7	6.3

(a) What are the two variables of interest here?
(b) Compute the five-number summary for each group.
(c) Sketch the corresponding box-and-whisker displays for each group.
(d) Comment on the relationship between treatment regimen and survival time.

18. A large group of rats is divided into 10 subgroups. Members of each subgroup are given a specific dose of a new drug. The proportion dying is recorded for each dose. These proportions, together with the corresponding doses (in milligrams), are as follows:

DOSAGE, x	.5	1.0	1.5	2.0	2.5	3.0	3.5	4.0	4.5	5.0
PROPORTION DYING, y	.22	.29	.38	.41	.54	.63	.61	.75	.90	.93

(a) Plot these data on a scatterplot. (The result is called a *response curve*.)
(b) Compute the regression line and draw it on the scatterplot.
(c) Interpret b.
(d) Comment on how well the line "fits" the data.

$$\sum x_i = 27.5 \qquad \sum x_i^2 = 96.25 \qquad \sum y_i = 5.66$$

$$\sum y_i^2 = 3.743 \qquad \sum x_i y_i = 18.865$$

19. A doctor treats 45 patients suffering pain with one of two drugs (C and D) or a *placebo* pill (a placebo pill is usually a sugar pill with no pain-relieving ability whatsoever). The time until the patient is relieved of the pain is recorded. The times (in minutes) are as follows:

DRUG *C*					DRUG *D*					PLACEBO				
10	11	13	13	16	14	7	5	13	9	9	6	13	11	9
10	10	11	8	14	16	17	8	12	4	13	8	14	4	10
6	2	14	9	10	11	5	11	12	10	5	15	11	19	4

(a) What are the two variables of interest here?
(b) Construct a stem-and-leaf display of the recovery times for the two drugs and for the placebo.
(c) Comment on the effect of the placebo drug.

20. One of the earliest medical experiments was performed by a British surgeon, Joseph Lister. He performed amputations on 75 persons. Each patient was

classified according to whether or not an antiseptic (carbolic acid) had been used on the bandages and whether or not the patient survived the operation. The results were as follows:

	ANTISEPTIC USED	
	YES	NO
PATIENT SURVIVED	34	19
PATIENT DIED	6	16

(a) Draw a block diagram showing the percentage surviving and not surviving for for those having the antiseptic and those not having the antiseptic.

(b) Comment on the relationship between these two qualitative variables.

GOVERNMENT AND SOCIOLOGY

21. The following data are the number of murders (y) and the number of police officers per 100,000 population (x) for eight U.S. cities:

CITY	MURDER RATE (PER 100,000)	POLICE OFFICERS (PER 100,000)
San Francisco	11.9	250
Miami	15.6	206
Denver	7.0	288
Detroit	14.1	434
New Orleans	19.2	275
Memphis	13.2	186
Fresno	21.9	182
Los Angeles	16.0	266

(a) Plot these data on a scatterplot.

(b) Compute the correlation coefficient between x and y. Comment on the magnitude and sign of r.

$$\sum x_i = 2087 \qquad \sum x_i^2 = 590{,}337 \qquad \sum y_i = 118.9$$

$$\sum y_i^2 = 1911.27 \qquad \sum x_i y_i = 30{,}301$$

22. The following figures represent the incidence of suicide by sex (per 100,000 persons of that sex) for each of 15 countries:

COUNTRY	MALES, x	FEMALES, y
Austria	36.7	14.7
Switzerland	34.5	15.4
Denmark	31.8	19.8
West Germany	30.1	15.1
Sweden	28.3	12.9

COUNTRY	MALES, x	FEMALES, y
France	23.3	9.9
Poland	22.8	4.2
Japan	22.6	13.6
Canada	21.2	7.3
United States	19.0	6.3
Norway	17.2	7.1
Australia	16.6	6.7
England and Wales	10.7	6.5
Israel	6.6	4.7
Spain	6.1	2.1

(a) Plot these data on a scatterplot.
(b) Compute the correlation coefficient between the two sets of scores.

$$\sum x_i = 327.5 \qquad \sum x_i^2 = 8405.87 \qquad \sum y_i = 146.3$$

$$\sum y_i^2 = 1797.35 \qquad \sum x_i y_i = 3774.98$$

23. Each of 13 fathers and his oldest adult child are given a questionnaire designed to indicate political preference on a scale from 0 to 100. (The lower the score, the more liberal; the higher the score, the more conservative.) The results are summarized as follows:

FATHERS' VIEWS, x	86	74	55	35	64	49	57
CHILDREN'S VIEWS, y	25	70	50	40	25	42	10

FATHERS' VIEWS, x	34	71	84	66	70	25
CHILDREN'S VIEWS, y	29	66	50	54	55	17

(a) Plot these data on a scatterplot. Without doing any computations, summarize any relationship between a father and child's political views.
(b) Compute the correlation coefficient between x and y.
(c) Compute the regression line relating the child's score to that of the father.

$$\sum x_i = 770 \qquad \sum x_i^2 = 50{,}002 \qquad \sum y_i = 533$$

$$\sum y_i^2 = 26{,}041 \qquad \sum x_i y_i = 33{,}419$$

24. The following data are the annual per capita income, in dollars, (x) and the mortality rate per 1000 (y) for 20 of the poorest nations.

COUNTRY	INCOME	MORTALITY	COUNTRY	INCOME	MORTALITY
Bhutan	70	20	Chad	120	24
Laos	70	23	Afghanistan	130	21
Upper Volta	90	26	Guinea	130	23
Mali	90	26	Benia	140	23
Burandi	100	20	Pakistan	140	15
Ethiopia	100	18	India	150	13
Republic of Maldives	100	23	Malawi	150	24
Somalia	100	22	Timor	150	23
Bangladesh	110	20	Haiti	180	16
Burma	110	16	Madagascar	200	21

Compute the correlation coefficient between income and mortality. Does your result substantiate the claim that a higher per capita income results in a lower mortality rate?

$$\sum x_i^2 = 317{,}500 \qquad \sum y_i^2 = 8945 \qquad \sum x_i y_i = 49{,}950$$

25. A researcher chooses and interviews two groups, each composed of 15 male workers in an automobile plant. All the workers in group A were middle-aged and had migrated to the United States up to 10 years earlier. Group B were also middle-aged but were at least third-generation Americans. Each of the 30 workers received a score ranging from 1 to 20 on a scale of patriotism—the higher the score, the more patriotic the person appeared to be. The results were as follows:

A	10	14	15	17	17	18	18	19	16	18	20	12	14	9	17
B	15	12	19	8	9	16	17	6	11	14	12	11	10	12	6

(a) Construct the five-number summary for each group.
(b) Sketch the box-and-whisker display for each group.
(c) Comment on differences between the two sets of scores.

26. Three hundred families were included in a study of family size. Each set of parents was classified according to two variables: the size of their family, small or large (four or more children), and whether one, both, or neither parent was part of a large family. The results were as follows:

		SIZE OF PARENTS' FAMILIES		
		BOTH LARGE (A)	ONE LARGE (B)	NEITHER LARGE (C)
OWN	LARGE	38	89	27
FAMILY	SMALL	18	83	45

(a) Draw a block diagram showing the percentage of large and small families for each group, A, B, and C, separately.
(b) Comment on the relationship between these two variables.

27. One hundred and fifty top executives were classified by sex and birth order: (i)

firstborn or only child and (ii) other. The data are shown below. Do these data appear to substantiate the claim that a much greater proportion of successful women than men are firstborn or only children?

	BIRTH ORDER	
	FIRSTBORN OR ONLY CHILD	OTHER
MEN	47	61
WOMEN	27	15

ECONOMICS AND MANAGEMENT

28. The median number of school years completed and the unemployment rate for 12 census tracts are as follows:

MEDIAN SCHOOL YEARS, x	15.3	16.1	12.9	13.4	8.7	8.2
UNEMPLOYMENT RATE, y	3.8	1.2	4.7	3.1	7.5	7.6

MEDIAN SCHOOL YEARS, x	13.4	8.7	8.1	12.4	11.2	11.3
UNEMPLOYMENT RATE, y	.3	10.8	4.8	3.9	5.3	5.6

(a) Plot these data on a scatterplot.

(b) Compute the correlation coefficient, r.

(c) Compute the regression line relating unemployment rate to median number of school years completed. Draw your line on the scatterplot.

(d) Interpret b in this case.

(e) Predict the unemployment rate in an area where the median number of school years is (i) 10; (ii) 12.

$$\sum x_i = 139.7 \qquad \sum x_i^2 = 1709.95 \qquad \sum y_i = 58.6$$

$$\sum y_i^2 = 376.02 \qquad \sum x_i y_i = 615.06$$

29. During one week a company records the sales (in thousands of dollars) and the number of sales representatives for each of 20 sales areas. The results are as follows:

SALES, y	437	477	462	237	497	466	348	330	426	394
REPRESENTATIVES, x	15	14	19	7	12	10	10	16	12	20

SALES, y	491	327	345	458	258	294	357	434	387	370
REPRESENTATIVES, x	13	11	9	17	8	9	5	14	8	11

(a) Compute the regression line predicting sales from the number of sales representatives.

(b) Interpret b in this case.

(c) Predict the *difference* in sales for two sales areas in which the number of representatives differ by 4.

$$\sum x_i = 240 \qquad \sum x_i^2 = 3186 \qquad \sum y_i = 7795$$

$$\sum y_i^2 = 3{,}151{,}265 \qquad \sum x_i y_i = 96{,}654$$

30. A bank records the number of mortgage applications and its own prevailing interest rate (in the middle of the month) for each of 16 consecutive months. The results are as follows:

INTEREST RATE, x	12.5	12.75	13	13.5	14	14.5	14	15
MORTGAGE APPLICATIONS, y	27	29	25	25	19	20	17	13

INTEREST RATE, x	15	15.5	16	16.5	16	15.5	14.5	14.5
MORTGAGE APPLICATIONS, y	15	10	10	6	5	5	11	14

(a) Plot these data on a scatterplot.

(b) Compute the correlation coefficient between the bank's interest rate and number of mortgage applications.

$$\sum x_i = 232.75 \qquad \sum x_i^2 = 3407.5625 \qquad \sum y_i = 251$$

$$\sum y_i^2 = 4867 \qquad \sum x_i y_i = 3517.75$$

31. The families of 10 randomly chosen college professors in each of three countries are included in a study of saving habits. The percentage of each family's annual income that is saved is recorded. The results are as follows:

U.S.	6.2	9.4	5.2	4.6	8.1	6.7	7.5	3.2	4.4	1.9
BRITAIN	8.3	1.3	13.3	13.4	9.4	6.4	10.8	10.5	2.2	4.7
W. GERMANY	11.7	10.0	4.9	11.9	8.3	9.3	11.5	16.3	9.7	3.9

(a) What are the two variables of interest here?

(b) Compute the five-number summary for each country.

(c) Construct the box-and-whisker display for each country.

(d) Comment on the differences between countries.

32. For advertising purposes an automobile company obtains the ages of 20 recent purchasers of each of its three models, the Skipper, the Mate, and the Crew. The data are as follows:

SKIPPER				MATE				CREW			
44	45	54	51	34	28	49	32	41	32	35	27
38	59	37	60	25	50	44	32	27	30	38	27
33	54	61	63	41	23	34	51	26	25	27	19
44	44	41	52	18	45	32	29	33	17	26	31
47	44	49	44	39	30	32	33	19	25	37	36

Construct the stem-and-leaf display for each model of car and comment on differences in the respective age distributions.

33. Each of 2000 adults were asked whether they viewed unemployment, high interest rates, or inflation as the greatest economic problem facing the nation. The responses, broken down by income, are summarized as follows:

		GREATEST PROBLEM		
		UNEMPLOYMENT	HIGH INTEREST RATES	INFLATION
INCOME	LESS THAN $25,000	312	288	650
	$25,000 OR MORE	83	360	307

(a) For both income groups, compute the percentage indicating each problem as the most serious.

(b) For each income group, draw a block diagram showing the percentage indicating each problem as the most serious.

(c) Comment on differences between the two income groups.

EDUCATION AND PSYCHOLOGY

34. The following data are the reading comprehension score and a creativity index for each of 10 students:

READING SCORE, x	31	25	14	47	42	25	30	43	28	35
CREATIVITY INDEX, y	13	9	18	15	13	10	16	20	12	11

(a) Plot these data on a scatterplot.

(b) Compute the correlation coefficient, r.

(c) Compute the regression line relating creativity to reading score.

(d) Why would you caution a psychologist who wanted to use the relationship obtained in part (c) to predict creativity indices from reading scores?

$$\sum x_i^2 = 11{,}138 \qquad \sum y_i^2 = 1989 \qquad \sum x_i y_i = 4442$$

35. A psychologist develops two alternative versions of the same test (T_1 and T_2). The two versions are incorporated into a single test by the frequently used strategy of using T_1 as the odd questions and T_2 as the even questions. The composite test is then given to 20 subjects. The scores on T_1 and T_2 are as follows:

T_1	40	59	63	84	71	22	59	90	25	67
T_2	43	62	58	77	74	24	54	87	20	62

T_1	39	48	75	85	60	44	55	52	51	42
T_1	49	42	75	82	58	48	56	58	43	45

Compute the correlation between T_1 and T_2. (This is a measure of the *reliability* of the test.)

36. The expenditure per student on education, in dollars, (x) and the number of juvenile crimes per 10,000 persons under 18 (y) for 12 U.S. counties are as follows:

EDUCATIONAL EXPENDITURE, x	1073	2004	1432	1831	1222	1549
JUVENILE CRIME RATE, y	79.2	21.0	29.3	24.1	63.6	30.7

EDUCATIONAL EXPENDITURE, x	2216	1336	1914	992	1515	1241
JUVENILE CRIME RATE, y	22.3	49.9	30.0	59.3	36.2	42.2

(a) Compute the correlation coefficient and comment on its sign and magnitude.
(b) Compute and plot the regression line relating y to x.
(c) Interpret b.
(d) Predict the juvenile crime rate for a county spending \$1700 per student.
(e) Predict the *difference* in juvenile crime rates for two counties that differ in expenditure per student by (i) \$400; (ii) \$900.

$$\sum x_i = 18{,}325 \qquad \sum x_i^2 = 29{,}641{,}533 \qquad \sum y_i = 487.8$$

$$\sum y_i^2 = 23{,}635.46 \qquad \sum x_i y_i = 677{,}965.8$$

37. A self-assessment test was given to 100 college students in each of three countries: the United States, France, and Greece. Scores on the test varied from 0 to 50—the higher the score, the more self-confident the person. The five-number summary for each country is as follows:

U.S.	7	24	30	38	47
FRANCE	8	18	24	34	42
GREECE	12	27	34	40	46

(a) Sketch the three box-and-whisker displays for these data.

(b) Comment on the relationship between self-assessment and country evidenced by the data.

38. The 36 students in a statistics class are randomly divided into two sections, each taught by the same person. One section is taught in standard fashion, while the other is taught with the aid of computer graphic displays. The final scores for each of the students are as follows:

STANDARD METHOD						COMPUTER GRAPHICS CLASS					
57	88	78	57	58	76	79	63	73	66	55	87
68	67	63	60	59	59	68	69	57	79	77	75
57	70	70	67	70	70	85	74	78	68	56	61

(a) What are the two variables of interest here?

(b) Construct a stem-and-leaf display for each section.

(c) Compute the mean, median, and standard deviation for each section.

(d) Comment on differences between the two sections.

39. Three mental hospitals use different therapeutic techniques. The 280 patients in the three hospitals were asked whether they felt they had improved since entering the hospital. The results are summarized as follows:

	HOSPITAL		
	A	*B*	*C*
IMPROVED	35	76	55
NOT IMPROVED	25	49	40

(a) Draw a block diagram showing the percentages improved and not improved, by hospital.

(b) Comment on the differences between hospitals.

40. A mail questionnaire is to be sent to 500 persons. A dollar bill is included in 100 of the letters. A five-dollar bill is included in 50 of the letters. The breakdown of the number of respondents to the questionnaire is given in the following table:

	$5 PEOPLE	$1 PEOPLE	NO-MONEY PEOPLE
RESPONDED	43	82	219
DID NOT RESPOND	7	18	131

(a) Compute the percentage responding and not responding for each money group.

(b) Construct a block diagram showing these percentages.

(c) Would you recommend including $1 in future questionnaires?

(d) Would you recommend including $5?

SCIENCE AND TECHNOLOGY

41. The solubility (x) and tensile strength (y) in nine samples of a polymer are as follows:

SOLUBILITY, x	6.32	6.32	6.36	6.29	6.30	6.31	6.35	6.30	6.35
TENSILE STRENGTH, y	6260	6251	6262	6234	6247	6244	6252	6246	6250

(a) Plot these data on a scatterplot.
(b) Compute the correlation coefficient r.
(c) Compute the regression line relating tensile strength to solubility.
(d) Interpret b in this case.

$$\sum x_i = 56.9 \qquad \sum x_i^2 = 359.7396 \qquad \sum y_i = 56{,}246$$

$$\sum y_i^2 = 351{,}513{,}066 \qquad \sum x_i y_i = 355{,}600.94$$

42. The number of bacteria (y) per unit area on a piece of barley after various hours (x) is given below, together with the natural logarithm of y (call it Y).

HOURS, x	0	1	2	3	4	5	6	7
NUMBER OF BACTERIA, y	27	84	104	149	228	429	817	2142
LOG $y = Y$	3.30	4.43	4.64	5.00	5.43	6.06	6.71	7.67

(a) Plot the (x, y) data on a scatterplot. On a different scatterplot draw the (x, Y) data.
(b) Which data seem to lie closest to a straight line?
(c) Compute (i) the correlation between x and y; (ii) the correlation between x and Y. Do your answers support your impressions in part (b)?

$$\sum y_i = 4010 \qquad \sum y_i^2 = 5{,}661{,}900$$

$$\sum Y_i = 43.24 \qquad \sum Y_i^2 = 247.106$$

$$\sum x_i y_i = 23{,}902 \qquad \sum x_i Y_i = 174.68$$

43. In a cloud-seeding experiment 28 clouds were seeded with silver nitrate pellets in the hope that this would increase the chances of their producing rain. A control group of 28 similar clouds were not seeded. The results were as follows:

	PRODUCED MEASURABLE RAIN	
	YES	NO
CLOUDS SEEDED	12	16
CLOUDS NOT SEEDED	9	19

(a) Draw a block diagram showing the breakdown of the percentages of clouds producing and not producing measurable rain for both the seeded and the nonseeded group.

(b) Does cloud seeding seem to be beneficial in producing rain?

SUGGESTED READING

There is a wide ranging but largely intuitive discussion of correlation and regression in:

FREEDMAN, E., et al. *Statistics*. New York: W.W. Norton, 1978.

A more thorough treatment of this topic appears in:

WEISBERG, S. *Applied Linear Regression*. New York: Wiley, 1980.

3 PROBABILITY

CONVICTED BY PROBABILITY

Arguments based on probability are being introduced increasingly in criminal and civil trials. An early misuse of probability, however, occurred in the trial of a Los Angeles couple in 1964.

A witness saw a young woman rob an elderly woman and leave the scene in a car driven by a man. Subsequently, police arrested a married couple who matched the description provided by the elderly woman and a witness.

The prosecutor, using computations made by a mathematics instructor, suggested that there was only a 1-in-12 million chance that there was another couple in Los Angeles who possessed these characteristics. Faced with this "evidence," the jury convicted the couple.

On appeal the convictions were overturned, in large part because the mathematician who had testified had misused elementary probability. The mathematician's error is explained in Example 24 on page 148.

SECTION 3.1

STATISTICS AND UNCERTAINTY

It is often necessary to use information about a part of a group to answer questions about the entire group. The next two examples illustrate this type of problem.

EXAMPLE 1 In Example 4 of Chapter 1 (page 21) the mean reading score was computed for nine seventh graders who had completed an innovative remedial reading course. Reading specialists might be interested in the extent to which the results for these nine students could be applicable to *all* seventh graders who might benefit from the course. Often referred to as the issue of *generalizability,* this leads to such questions as:

1. How representative were the nine students of *all* seventh graders who might benefit from the course?
2. How different is the mean reading score of the nine students likely to be from that of all such seventh graders if they were to take the course?

■

EXAMPLE 2 Over the past five years a difficult surgical procedure has been successful 60% of the time. A surgeon has developed a new technique which it is believed will have a success rate significantly higher than 60%. Since the new technique is more expensive, it will be adopted only if it can be shown to have a success rate significantly higher than 60%. The new technique is used in the next 20 operations and 15 are successful, for a success rate of 75%. At first glance this suggests that the new technique is indeed better than the old one. But we might ask: Isn't it possible that in a short run of 20 operations, there would sometimes be 15 successes even if the long-run success rate were still only 60%?

■

In Example 1 it was clear that we had data on *part* of a larger group and wanted to use this to make a judgment about the larger group. Although it may not be quite as apparent, the same is also true in Example 2. We had information on the success rate for only 20 operations and were to use this to judge whether the success rate would still be substantially above 60% if the new technique were used in the larger group of all succeeding operations of this type.

POPULATION AND SAMPLE

We are now ready to define two important terms.

> DEFINITION 1
>
> Statisticians use the word **population** to refer to a *whole group* of interest.

DEFINITION 2

The word **sample** refers to a part (usually a small part) of the population.

In Example 1 the population was all seventh graders eligible for the remedial reading course. The sample was the nine students to whom the course was given. In Example 2 the population consisted of all future operations in which the new technique might potentially be used. The sample consisted of the 20 operations in which it was used.

The statistician uses the word "population" to refer to the members of any group: for example, all cars, not just people. In each case, however, it includes all members of the group. On many occasions the term "population" is also used to refer to a (usually) numerical *characteristic* of members of a group rather than to the members themselves. For instance, a list of the grade-point averages for each student at Downstate U. would be a population of grade-point averages. Similarly, we may refer to the population of reading scores rather than to the population of seventh graders who achieved these scores.

INFERENCE AND UNCERTAINTY

The themes of Examples 1 and 2 can both be summarized in the question: How can we use the sample data to draw conclusions about the population of interest? Statisticians refer to the use of sample data in this way as a form of *statistical inference*.

Before we can begin to answer the question above, we need some idea of the extent to which the sample is representative of the population. This, in turn, depends on the way the sample was selected. It is generally assumed that the selection procedure was "random." We shall have more to say about this in Chapter 8.

Once the sample is selected, we follow two steps in making inferences about the population.

1. First, the sample data must be arranged and summarized in a form that readily permits us to generalize. Methods for doing this were discussed in Chapters 1 and 2.
2. The second step is new. In Examples 1 and 2 we used such phrases as "likely to be" and "isn't it possible that" These phrases suggest degrees of uncertainty as to the extent to which information about the sample can be generalized to the population. By providing the tools for measuring uncertainty the theory of probability enables us to address this problem. We therefore consider the subject of probability first and then apply it in generalizing from a sample to a population.

SECTION 3.2
PROBABILITY AND EQUALLY LIKELY OUTCOMES

Although the ancient Egyptians were familiar with some of the ideas of probability, the study of this subject really began with the popular games of chance in the seventeenth century. The theory of probability has far outgrown these origins, but the basic rules can still be illustrated by reference to such games.

We begin the study of probability by introducing some common words which are used with very specialized meanings in this context.

> DEFINITION 3
>
> A **statistical experiment** denotes any phenomenon in which the outcome is uncertain.

Using this definition, examples of experiments include (1) rolling a die, (2) drawing a card from a well-shuffled deck, (3) selecting a sample of 100 voters for an opinion poll, (4) a baseball game, (5) a baseball season, (6) drawing the winners in a state lottery, and (7) selecting 12 jurors from a pool of 300 persons.

> DEFINITION 4
>
> Associated with an experiment is a list of all its possible outcomes. This list is called the **sample space** (denoted SS).

EXAMPLE 3 An experiment consists of selecting a single card from a well-shuffled deck of 52 cards. The sample space consists of 52 possible outcomes corresponding to the values of the individual cards.

$$SS = \begin{array}{llllllllllllll} \text{C—Ace} & 2 & 3 & 4 & 5 & 6 & 7 & 8 & 9 & 10 & \text{J} & \text{Q} & \text{K} \\ \text{D—Ace} & 2 & 3 & 4 & 5 & 6 & 7 & 8 & 9 & 10 & \text{J} & \text{Q} & \text{K} \\ \text{H—Ace} & 2 & 3 & 4 & 5 & 6 & 7 & 8 & 9 & 10 & \text{J} & \text{Q} & \text{K} \\ \text{S—Ace} & 2 & 3 & 4 & 5 & 6 & 7 & 8 & 9 & 10 & \text{J} & \text{Q} & \text{K} \end{array}$$

■

EXAMPLE 4 The initials of the 12 persons serving on a jury are placed in a hat. A name is then selected. The person whose initials are selected is to be the foreman. The sample space is

$$SS = [\text{RNG}, \text{JSW}, \text{MGV}, \text{DSB}, \text{MSM}, \text{DWN}, \text{JKG}, \text{WSC}, \text{WH}, \text{CSM}, \text{PE}, \text{KBE}]$$

■

EXAMPLE 5 The sample space for the rolling of a six-sided die is

$$SS = [1, 2, 3, 4, 5, 6]$$

■

EXAMPLE 6 Which one of the 26 major league baseball teams will win the World Series?

Viewing the baseball season as an experiment, the sample space is simply a list of the 26 teams.

		NATIONAL LEAGUE	AMERICAN LEAGUE
SS =	EAST	Chicago, Montreal, New York, Philadelphia, Pittsburgh, St. Louis	Baltimore, Boston, Cleveland, Detroit, Milwaukee, New York, Toronto
	WEST	Atlanta, Cincinnati, Houston, Los Angeles, San Diego, San Francisco	California, Chicago, Kansas City, Minnesota, Oakland, Seattle, Texas

Note: The sample space could have been represented as a single list of the 26 individual teams. However, grouping the outcomes as we have here will prove useful, as you will see a bit later.

■

EXAMPLE 7 A mathematics department can afford to send only two of the five members, call them A, B, C, D, and E, to a professional meeting. The chairperson will select the two at random. The sample space for this experiment consists of all possible *pairs* of instructors.

$$\text{SS} = [AB, AC, AD, AE, BC, BD, BE, CD, CE, DE]$$

■

EXAMPLE 8 A weather forecaster is sure that the weather tomorrow will consist of one of the outcomes in the following sample space:

SS = [no precipitation, rain, snow, rain turning to snow]

■

Although we have so far referred to individual outcomes, we shall be interested primarily in *collections* of outcomes.

> DEFINITION 5
>
> The occurrence of any one of a specific collection of outcomes is referred to as an **event.**

In Example 3, selecting a king would be an event, since it includes four outcomes: the kings of clubs, diamonds, hearts, and spades. Similarly, selecting a diamond would be an event made up of the 13 diamonds.

In Example 6 the World Series being won by a National League East team is an event consisting of six outcomes—can you name them? Two events that might be of interest in Example 7 are (1) B goes to the meeting, which consists of the four outcomes AB, BC, BD, and BE; and (2) C goes to the meeting with either D or E—which consists of the pairs CD and CE.

Any specific outcome of an experiment is also an event, consisting of a collection of one outcome. Thus (1) selecting the king of diamonds, (2) the World Series is won by the Los Angeles Dodgers, and (3) the pair CD go to the meeting are each events associated with Examples 3, 6, and 7, respectively. The sample space is equivalent to that event which consists of all the possible outcomes of an experiment.

We shall often use a letter, or a combination of letters, to refer to an event. For instance, we might refer to the event "a king" by the letter A, or the event "an American League West team wins the World Series" by the letter C.

A sample space and related events can be represented pictorially by means of a **Venn diagram.** The Venn diagram for the sample space in Example 7 is shown in Figure 1. The sample space is represented by a rectangle and each outcome by a dot. The events G: instructor B attends the meeting, and H: instructor A goes with either D or E, are indicated.

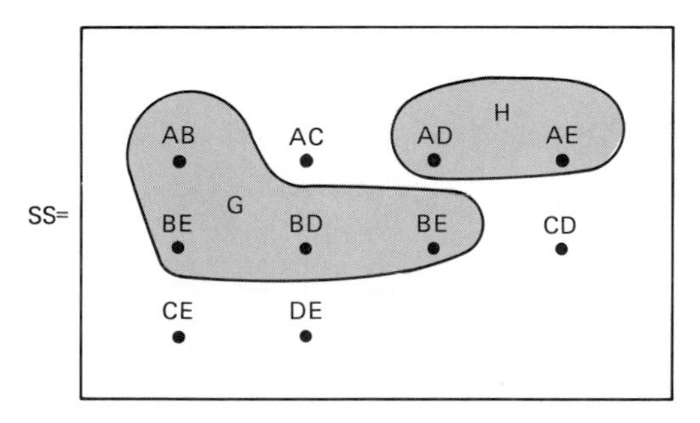

FIGURE 1

Venn diagram representing the sample space and the events H and G in Example 7

Prior to performing an experiment, we are interested in the likelihood or probability that some event A will occur. We denote this probability $P(A)$, but how should we assign *values to* $P(A)$? We begin by assigning the value 1 to the probability of the event "one of the outcomes in the sample space occurs"; that is $P(\text{SS}) = 1$. This may appear somewhat arbitrary but we will explain in the next section why it is appropriate. We shall be interested primarily in experiments in which all the outcomes in the sample space can be considered as equally likely. This assumption is reasonable in the case of Examples 3, 4, 5, and 7.

By contrast, in Example 6, although there are 26 teams, there is good reason to believe that they are *not* all equally likely to win the World Series. Similarly, in Example 8 there is no reason to suppose that all four outcomes (weather forecasts) are equally likely to occur.

Let us consider Example 3 again. If the probability of the entire sample space is 1 and the selection of each of the 52 cards is equally likely, we should assign a probability of $1/52$ to each outcome. For example, if A is the event, "the 7 of hearts," $P(A) = 1/52$. Suppose that we call A the event "the chosen card is a heart." What is $P(A)$? Since 13 of the 52 equally likely outcomes are hearts, it seems reasonable to assign the value $13/52$ to $P(A)$; that is, $P(A) = 13/52$.

Generalizing from this example, assume that the sample space contains M equally likely outcomes. Assume also that an event A occurs if any one of M_o outcomes occurs. Then

The probability of each outcome is $1/M$. (Eq. 3.1)

The probability of the event A is

$$P(A) = M_o\left(\frac{1}{M}\right) = \frac{M_o}{M} \qquad \text{(Eq. 3.2)}$$

In the last example, $M = 52$ and for the event A, $M_o = 13$ and $P(A) = M_o/M = 13/52 = .25$.

EXAMPLE 9 Consider again the sample space for Example 7.

$$\begin{array}{c} SS = [AB, AC, AD, AE, BC, BD, BE, CD, CE, DE] \\ \quad G \qquad\quad H \quad H \quad G \quad G \quad G \\ \qquad\qquad\qquad\qquad I \end{array}$$

Compute the probability of the following events:
(a) G: instructor B goes to the meeting
(b) H: instructor A goes with either D or E
(c) I: instructors B and C go

Solution (a) Using Equation 3.2 with $M = 10$ and $M_o = 4$, $P(G) = 4/10 = .4$.
(b) In this case $M_o = 2$, so $P(H) = 2/10 = .2$.
(c) The event I is one of the outcomes in the sample space, so $P(I) = 1/10 = .1$.

■

EXAMPLE 10 A total of 254 tickets numbered 1 through 254 are sold in a raffle. After the 254 stubs are mixed in a large barrel, a single prize-winning ticket is selected. Find the probability of the event A: one of the numbers 20–30 (inclusive) is selected.

Solution Here the sample space is simply a list of the 254 numbers from 1 to 254, that is,

$$SS = [1, 2, 3, \ldots, 253, 254]$$

In this case $M = 254$ and $M_o = 11$ since there are 10 numbers in the twenties, plus 30. Using Equation 3.2, $P(A) = M_o/M = 11/254 = .043$.

■

EXAMPLE 11 A street has 80 voters classified as shown in the following table:

		POLITICAL AFFILIATION			
		DEMOCRAT	REPUBLICAN	INDEPENDENT	TOTAL
SEX	M	20	8	16	44
	F	21	7	8	36
TOTAL		41	15	24	80

We see that, for instance, 41 of the voters are Democrats. Of these, 20 are males and 21 females. The names of the 80 voters are written on slips of paper, the slips are mixed in a hat, and one is selected. Find the probability of the following events:

(a) R: the voter selected is a Republican
(b) F: the voter selected is a female
(c) A: the voter selected is a male independent

Solution Each of the $M = 80$ voters is equally likely to be selected. There are 15 Republicans, 36 females, and 16 male independents. Thus (a) $P(R) = 15/80 = .1875$; (b) $P(F) = 36/80 = .45$; (c) $P(A) = 16/80 = .20$.

In computing the probabilities, the table above is more useful than a sample space listing all 80 voters with their sex and political affiliation individually.

■

When computing probabilities, rounding to three decimal places is generally sufficient. For example, .46748 should be rounded to .467. However, where this would result in a probability of zero or one, it is best to round to one *significant* figure. For instance, .0004238 should not be rounded to zero but to .0004. Similarly, .99976 should not be rounded to 1 but to .9998.

PROBLEMS 3.2

1. A perfectly balanced coin is to be tossed twice. Write down the four equally likely outcomes in the sample space. Compute the probability of the following events:
   (a) A: both tosses result in a tail
   (b) B: the first toss results in a head, the second in a tail
   (c) C: one toss is a head, the other a tail
2. An experiment consists of rolling a perfectly balanced die. Find the probability of the following events:
   (a) A: a 5
   (b) B: an even number
   (c) C: a number greater than 3
   (d) D: a number divisible by 3
3. The distribution of final grades for the 125 students in a statistics course is as follows:

	FINAL GRADE				
	A	B	C	D	F
FRESHMAN	1	4	8	1	1
SOPHOMORE	4	10	12	3	1
JUNIOR	15	28	9	5	3
SENIOR	6	8	4	2	0

An experiment consists of selecting a student at random. Find the probability of the following events:

(a) J: a junior

(b) L: a student with a B

(c) S: a sophomore with a C

(d) U: an upperclassman (a junior or senior)

(e) T: a lowerclassman with an A

4. An experiment consists of selecting a card at random from a well-shuffled deck of 52 cards. What is the probability of selecting:

(a) The 7 of diamonds

(b) An ace

(c) A spade

(d) A red card

(e) A number less than 5

(f) A face card (a jack, queen, or king)

5. The president of a company wishes to select a subcommittee of two at random from among the six vice-presidents, A, B, C, D, E, and F.

(a) Write down the sample space corresponding to this experiment.

(b) What is the probability that (i) C and F are selected; (ii) A is chosen with either D or F; (iii) at least one of A and B are chosen?

6. Suppose that in a certain city the same number of voters were born on each of the 365 days of the year. Nobody was born on February 29. What is the probability that the birthday of a voter selected at random:

(a) Is November 20?

(b) Is in November?

(c) Falls between January 15 and February 15—not including either day?

(d) Is in the first three months of the year?

(e) Is not on April 15?

(f) Is not in July?

7. Suppose that in Example 7 three of the five instructors are to be selected.

(a) Write down the sample space for this experiment.

(b) What is the probability that (i) A, B, and C are selected; (ii) D and E are included in the group?

8. An experiment is equally likely to result in one of the 40 numbers 1, 2, 3, . . . , 40. What is the probability that the number chosen:

(a) Is 20?

(b) Is less than 8?

(c) Lies between 15 and 20 (inclusive)?

(d) Is at least as large as 30?

9. Consider the following argument. A baby is (approximately) equally likely to be a boy or a girl. Among families with two children the three possibilities are two boys, two girls, or one of each. Therefore, roughly one-third of such families should contain two boys, one-third should contain two girls, and one-third should contain one of each. Is this argument correct? If so, explain why. If not, explain why not.

Solution to Problem 1

With the obvious notation the sample space is

$$SS = [\mathrm{HH}, \mathrm{HT}, \mathrm{TH}, \mathrm{TT}]$$

(a) $P(\mathrm{TT}) = M_o/M = \frac{1}{4}$
(b) $P(\mathrm{HT}) = M_o/M = \frac{1}{4}$
(c) The event "one of each" consists of the two outcomes HT and TH. Therefore,

$$P(\text{one of each}) = \frac{M_o}{M} = \frac{2}{4} = \frac{1}{2}$$

SECTION 3.3
INTERPRETING PROBABILITIES

An experiment consists of tossing a coin that is chipped, so that we are not willing to assume that it is equally likely to come up heads and tails. What does $P(\mathrm{H})$ (the probability of a toss coming up heads) mean in this case? How can we obtain $P(\mathrm{H})$? To answer these questions, imagine tossing this coin repeatedly and recording the ratio of the number of heads to the number of tosses. The results might look as shown in Figure 2.

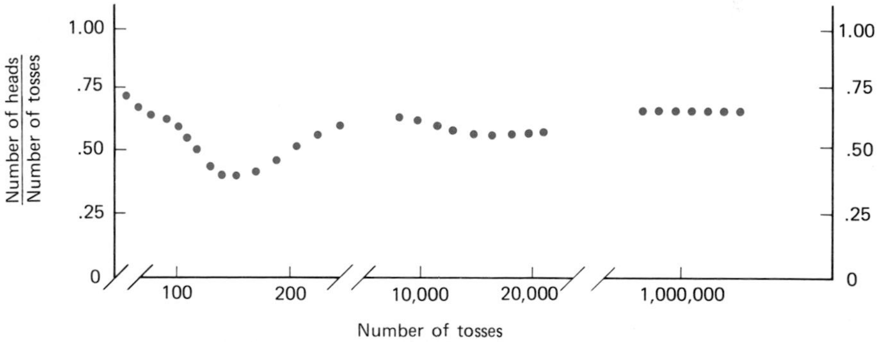

FIGURE 2

Graph of the ratio $\frac{\text{number of heads}}{\text{number of tosses}}$ for different numbers of tosses

We can view $P(\mathrm{H})$ as the value of this ratio for an infinite number of tosses. At first, when the number of tosses is small, the ratio fluctuates considerably, but as the number of tosses increase, the ratio begins to stabilize at approximately .55. We therefore say that $P(\mathrm{H})$ is approximately .55.

Generalizing from this example, we define what is known as the **relative frequency interpretation** of probability.

DEFINITION 6

If an experiment can be repeated, we interpret the probability of an event A as the proportion of times the event A would occur if the experiment were repeated an infinite number of times.

Since we cannot repeat an experiment an infinite number of times, Definition 6 cannot actually be used to compute $P(A)$. However, you saw in Section 3.2 that where we can regard each outcome in the sample space as equally likely then, $P(A) = M_o/M$. Where this is not true, as with the chipped coin, Definition 6 suggests a method for obtaining at least an estimate for $P(A)$: namely, performing the experiment a large number of times and computing the proportion of these times that the event A occurs.

Two extreme cases of Definition 6 are of special interest. If an event A must occur, it will occur each time the experiment is performed. In such a case, $P(A)$, the proportion in Definition 6, will be 1. Since one of the outcomes in the sample space must always occur, this is a justification for assigning a probability of 1 to the entire sample space in the preceding section. If an event A *cannot* occur, $P(A)$, the proportion in Definition 6, will be zero. These results suggest that all probabilities must lie between 0 and 1.

Definition 6 does not apply at all in the case of an experiment that cannot be repeated. Take, for instance, a presidential election in which the candidates Arnold, Barry, and Charles are running. Designate the events: Arnold wins, Barry wins, and Charles wins by A, B, and C, respectively. This is an example of an experiment that *cannot* be repeated—usually referred to as a **unique** experiment. Presidential elections are sufficiently different so that past elections are of little value in determining $P(A)$, $P(B)$, and $P(C)$. Nor can we argue that $P(A) = P(B) = P(C) = \frac{1}{3}$ simply because there are three candidates. One expert may assess these probabilities as $P(A) = .5$, $P(B) = .4$, and $P(C) = .1$; another may assess them as $P(A) = .45$, $P(B) = .4$, and $P(C) = .15$. In the case of a unique experiment, the probability of an event is a subjective assessment of the chance of that event occurring. There need be no single correct $P(A)$.

DEFINITION 7

In general, for unique experiments, $P(A)$ is a subjective assessment of the chance of the event A occurring.

EXAMPLE 12 Sports experts are frequently asked to assess or predict the performance of baseball teams before the season begins. Such an expert may quote odds of 19 to 2 against the Chicago White Sox winning the World Series. This is simply another way of saying that in this expert's opinion the probability of this event is

$$\frac{2}{19+2} = \frac{2}{21} = .095$$

A second expert may of course (and probably will) quote different odds.

No matter how $P(A)$ is viewed, it is important to note that it will always lie between 0 and 1 (inclusive). That is

DEFINITION 8

For any event A,

$$0 \leq P(A) \leq 1$$

PROBLEMS 3.3

10. An experiment consists of rolling a thumbtack that can land either point up (U) or point down (D). Of interest is $P(U)$. The experiment is repeated 400 times and the tack lands point up 78 of those times. How would you use this information to estimate $P(U)$?
11. A new test for pregnancy is developed. Call $P(A)$ the probability that the test will be positive (i.e., indicate a pregnancy) when a woman is, in fact, pregnant. The test is given to 250 women known to be pregnant. The test proves positive on 229 of these occasions. How would you use this information to estimate $P(A)$?
12. An industrial engineer is interested in estimating $P(D)$, the probability that a new machine will produce cylinder blocks with diameters beyond the accepted tolerances. An examination of 2000 blocks shows that 154 are unacceptable. Use this information to estimate $P(D)$.

Solution to Problem 10

An estimate for $P(U)$ is $78/400 = .195$.

SECTION 3.4

THE ADDITION OF PROBABILITIES

Sometimes we are interested in the probability of a *combination* of events. In this section we introduce procedures for computing such probabilities.

THE ADDITION RULE

We begin by reconsidering the details of Example 11. The distribution of the 80 voters in that example is reproduced in Table 1. An experiment consists of selecting one of these 80 voters at random. We are interested in the two events,

TABLE 1

Distribution of voters by sex and party affiliation

		POLITICAL AFFILIATION			
		DEMOCRAT	REPUBLICAN	INDEPENDENT	TOTAL
SEX	M	(20)	(8)	(16)	44
	F	(21)	7	8	36
	TOTAL	41	15	24	80

A: a Democrat is selected and *B*: a male is selected. We can combine these events to produce new events in a number of ways. The event designated (*A and B*) corresponds to selecting a voter who is *both* a male and a Democrat (i.e., a male Democrat). The event designated (*A or B*) corresponds to the selection of a voter who is either a Democrat or a male or a male Democrat. The numbers of voters corresponding to the event (*A* or *B*) are circled in Table 1.

The Venn Diagrams in Figure 3 illustrate the following events: (a)(i) event A and (ii) event B, (b) the event (A or B) and (c) the event (A and B).

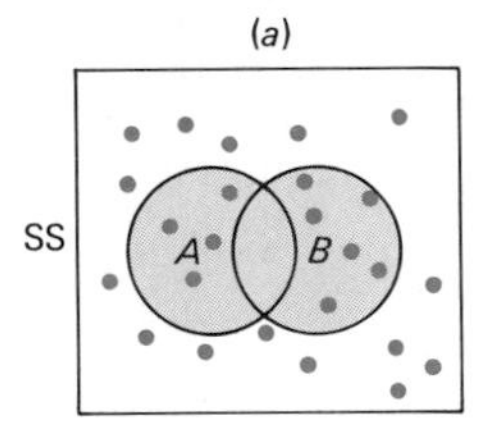

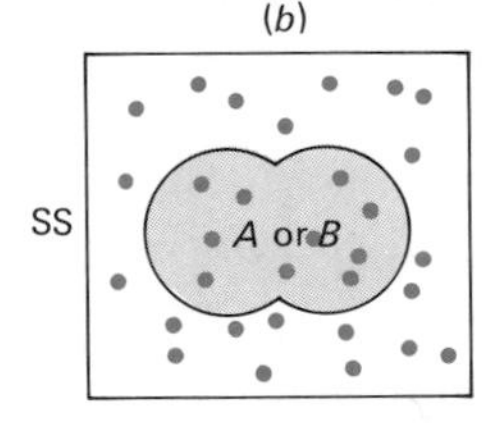

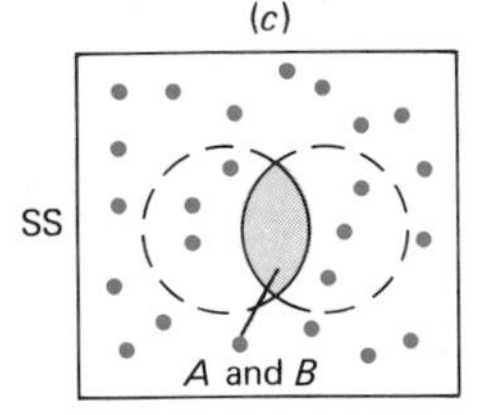

FIGURE 3

Venn diagrams highlighting the events (a) *A*, *B*, (b) (*A* or *B*), and (c) (*A* and *B*)

Using Equation 3.2, the probability of the event (*A* and *B*) is the ratio $M_o/M = 20/80 = .25$. Now how should we compute $P(A \text{ or } B)$? Adding the circled numbers in Table 1, we compute $M_o = 21 + 20 + 8 + 16 = 65$; therefore, $P(A \text{ or } B) = 65/80 = .8125$. Can we express this probability in terms of $P(A)$ and $P(B)$? If we add $P(A) = 41/80$ and $P(B) = 44/80$, we obtain

$$P(A) + P(B) = \frac{41 + 44}{80} = \frac{85}{80} = 1.0625$$

which is impossible. The problem is that in adding $P(A)$ and $P(B)$ we counted the probability of selecting a male Democrat in $P(A)$ and again in $P(B)$, thus effectively double counting these 20 voters. We indicate this by the double circle around the 20 in Table 1. We can, however, obtain $P(A \text{ or } B)$ by adding $P(A)$ and $P(B)$ and substracting $P(A \text{ and } B)$ from the total. This yields

$$P(A \text{ or } B) = \frac{41}{80} + \frac{44}{80} - \frac{20}{80} = \frac{65}{80} = .8125$$

as before.[1]

[1]If we wanted $P(A \text{ or } B$ but *not* both), we would add $P(A)$ and $P(B)$ and subtract *twice* $P(A$ and $B)$. For example,

$$P(\text{Democrat or a male but not both}) = P(A) + P(B) - 2P(A \text{ and } B)$$
$$= \frac{41}{80} + \frac{44}{80} - 2\left(\frac{20}{80}\right) = \frac{45}{80} = .5625$$

In general, for any two events A, B:

> DEFINITION 9
> (A and B) is the event "both A and B occur."

Expressing $P(A \text{ and } B)$ in terms of $P(A)$ and $P(B)$ is quite complicated and will be discussed later. For the present we will determine it, as we did above, using the ratio M_o/M in Equation 3.2.

> DEFINITION 10
> (A or B) is the event "any one of the following occur: A, or B, or (A and B)."

$$P(A \text{ or } B) = P(A) + P(B) - P(A \text{ and } B) \qquad \text{(Eq. 3.3)}$$

Equation 3.3 is called the **addition rule.** Those familiar with set theory will recognize (A and B) as equivalent to ($A \cap B$) and (A or B) as ($A \cup B$). Now let us apply the addition rule again to the same data.

EXAMPLE 13 Refer again to Table 1. This time let us call the events A: an independent is selected and B: a female is selected. What is $P(A \text{ or } B)$?

Applying Equation 3.3, we have

$$P(A \text{ or } B) = P(A) + P(B) - P(A \text{ and } B)$$
$$= \frac{24}{80} + \frac{36}{80} - \frac{8}{80} = \frac{52}{80} = .65$$

■

MUTUALLY EXCLUSIVE EVENTS

Continuing with the experiment of selecting one voter, suppose that we refer to the events A: a Democrat is selected and B: a Republican is selected. What is $P(A \text{ or } B)$? In this case a voter cannot be both a Democrat *and* a Republican, so

$$P(A \text{ and } B) = 0 \text{ and } P(A \text{ or } B) = P(A) + P(B) - 0$$
$$= \frac{41}{80} + \frac{15}{80} - 0 = \frac{56}{80} = .70$$

When, as above, the events A and B cannot both occur, so that $P(A \text{ and } B) = 0$, we say that they are mutually exclusive.

> DEFINITION 11
>
> The event A and the event B are **mutually exclusive** if $P(A \text{ and } B) = 0$.

> Where A and B are mutually exclusive the addition rule becomes
>
> $$P(A \text{ or } B) = P(A) + P(B) \qquad \text{(Eq. 3.4)}$$

This situation is represented in the Venn diagram in Figure 4.

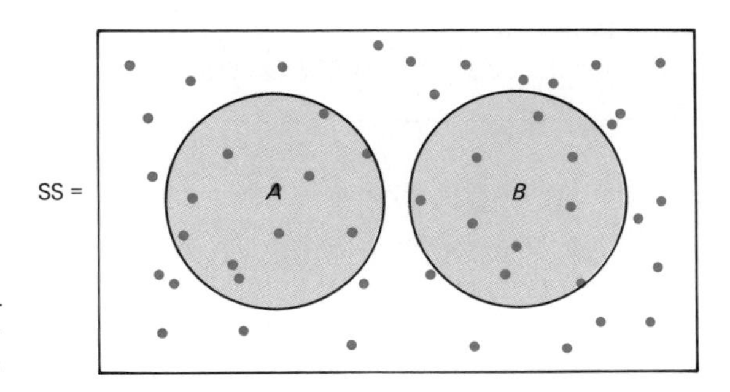

FIGURE 4

Representing two mutually exclusive events

We apply Equations 3.3 and 3.4 in the following two examples.

EXAMPLE 14 A card is drawn at random from a well-shuffled deck of 52 cards. The events of interest are H: a heart, C: a club, T: a 3, and F: a face card. These four events are outlined on the sample space as follows:

SS: (T outlines the 3 column; F outlines the J, Q, K columns; C outlines the C row; H outlines the H row)

C	Ace	2	3	4	5	6	7	8	9	10	J	Q	K
D	Ace	2	3	4	5	6	7	8	9	10	J	Q	K
H	Ace	2	3	4	5	6	7	8	9	10	J	Q	K
S	Ace	2	3	4	5	6	7	8	9	10	J	Q	K

Find (a) $P(C \text{ or } T)$; (b) $P(C \text{ or } H)$; (c) $P(T \text{ or } F)$; (d) $P(F \text{ or } H)$.

Solution (a) From the addition rule, we have

$$P(C \text{ or } T) = P(C) + P(T) - P(C \text{ and } T) = \frac{13}{52} + \frac{4}{52} - \frac{1}{52} = \frac{16}{52} = .308$$

(b) The events C and H are mutually exclusive. From Equation 3.4,

$$P(C \text{ or } H) = P(C) + P(H) = \frac{13}{52} + \frac{13}{52} = \frac{26}{52} = .50$$

(c) The events T and F are also mutually exclusive, so

$$P(T \text{ or } F) = P(T) + P(F) = \frac{4}{52} + \frac{12}{52} = \frac{16}{52} = .308$$

(d) $P(F \text{ or } H) = P(F) + P(H) - P(F \text{ and } H) = \frac{12}{52} + \frac{13}{52} - \frac{3}{52}$

$$= \frac{22}{52} = .423$$

[The three outcomes corresponding to the event (F and H) are the jack, queen, and king of hearts.]

■

EXAMPLE 15 The medical literature suggests that the probability of a child contracting chickenpox is .18; of contracting German measles, .12; and of contracting both, .07. (a) Are the two events C: contracting chickenpox and G: contracting German measles, mutually exclusive? (b) Compute the probability that a child will contract at least one of the two diseases.

Solution (a) C and G are *not* mutually exclusive since both C and G can occur. In fact $P(C \text{ and } G) = .07$, *not* 0.

(b) The event "at least one of C or G" is equivalent to the event (C or G). Thus, using the addition rule, Equation 3.3, we have

$$P(C \text{ or } G) = P(C) + P(G) - P(C \text{ and } G) = .18 + .12 - .07 = .23$$

■

The addition rule for two mutually exclusive events in Equation 3.4 can readily be generalized to three or more mutually exclusive events. For example, if A, B, and C are mutually exclusive, then $P(A \text{ or } B \text{ or } C) = P(A) + P(B) + P(C)$.

EXAMPLE 16 On one evening 30% of the patients in a hospital are listed as being in good condition (G), 30% are listed as being in fair condition (F), 25% in poor condition (P), and 15% in critical condition (C). If a patient is selected at random, what is (a) $P(G \text{ or } F)$; (b) $P(F \text{ or } P \text{ or } C)$?

Solution Since a patient can be in only one of the four categories, the events G, F, P, and C are mutually exclusive. Thus (a) $P(G \text{ or } F) = P(G) + P(F) = .3 + .3 = .6$ and similarly, (b) $P(F \text{ or } P \text{ or } C) = P(F) + P(P) + P(C) = .3 + .25 + .15 = .7$.

■

COMPLEMENTARY EVENTS

Suppose that A is the event "drawing a king from a well-shuffled deck of cards." Then the event "any card *except* a king" is known as the complement of A and is denoted as A'. The event A' is read as "A complement."

> DEFINITION 12
> A' is the event "not A."

In effect, A and A' are a special case of mutually exclusive events, as shown in Figure 5.

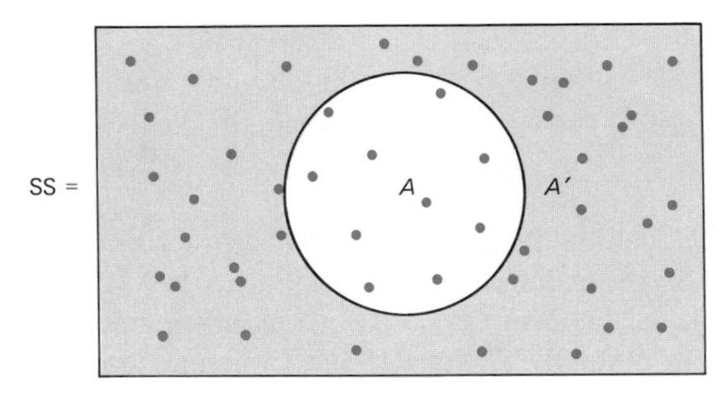

FIGURE 5

Relationship between A and A'

The events A and A' are distinguished from other such events by an important relationship, which is derived as follows:

1. $P(\text{SS}) = 1$.
2. Since A and A' are mutually exclusive, SS can be thought of as the event (A or A').

As a consequence,

3. $P(\text{SS}) = P(A \text{ or } A') = P(A) + P(A') = 1$

Thus

$$P(A) = 1 - P(A') \qquad \text{(Eq. 3.5)}$$

This result can be very useful if $P(A)$ is difficult to compute but $P(A')$ is relatively simple. The following example suggests how Equation 3.5 is used in practice.

EXAMPLE 17 Referring to Example 16, call A the event (F or P or C); then $P(A) = 1 - P(A') = 1 - P(G) = 1 - .3 = .7$, as before.

■

PROBLEMS 3.4

13. Referring to the experiment in Example 7 of selecting two instructors from five, we are interested in new events G: E goes with either A or B, H: B goes to the meeting, and I: C and E go. Compute (a) $P(G \text{ and } H)$; (b) $P(H \text{ and } I)$; (c) $P(G \text{ or } H)$; (d) $P(H \text{ or } I)$; (e) $P(G \text{ or } I)$; (f) $P(G')$; (g) $P(I')$.
14. Three students, Lyman, Michaels, and Nichols, are taking a psychology course. We call L, M, and N the events "Lyman passes," "Michaels passes," and "Nichols passes," respectively. The instructor assesses the following probabilities: $P(L) = .6$, $P(M) = .7$, $P(N) = .5$, $P(L \text{ and } M) = .45$, $P(L \text{ and } N) = .35$, and $P(M \text{ and } N) = .40$.

    (a) Are the events L, M, and N mutually exclusive?

    (b) Compute the probability that (i) at least one of Lyman and Michaels passes; (ii) at least one of Lyman and Nichols passes; (iii) at least one of Michaels and Nichols passes; (iv) Nichols fails.
15. An automobile manufacturer produces cars at the six plants listed below. Beside each plant name is the proportion of the company's total output that is produced at the plant.

Akron (.1)	Binghamton (.15)	Columbus (.2)
Dayton (.2)	Evanston (.05)	Flint (.3)

    An experiment consists of selecting an automobile from a retail outlet of the company. A is the event "the car selected was produced in Akron." Events B, C, D, E, and F are defined similarly.

    (a) Are events A, B, C, D, E, and F mutually exclusive?

    (b) Compute (i) $P(A \text{ or } B)$; (ii) $P(C \text{ or } F)$; (iii) $P(A \text{ or } B \text{ or } D)$; (iv) $P(E \text{ or } F \text{ or } C)$; (v) $P(B \text{ or } C \text{ or } D \text{ or } E)$.

    (c) What is the probability that the car selected was produced in a plant (i) other than Dayton; (ii) other than Evanston?
16. A and B are two mutually exclusive events such that $P(A) = .7$ and $P(B) = .2$. Find (a) $P(A \text{ and } B)$; (b) $P(A \text{ or } B)$.
17. A and B are two events such that $P(A) = .3$, $P(B) = .4$, and $P(A \text{ or } B) = .6$. Compute $P(A \text{ and } B)$.
18. The 65 middle managers in a company are classified by age and income as follows:

		AGE		
		THIRTIES (T)	FORTIES (F)	FIFTIES OR SIXTIES (S)
	LOW (L)	13	4	1
INCOME	MODERATE (M)	8	10	3
	HIGH (H)	2	16	8

    A manager is selected at random from this population. Compute (a) $P(L)$; (b) $P(S)$; (c) $P(F \text{ and } M)$; (d) $P(T \text{ or } S)$; (e) $P(F \text{ or } M)$; (f) $P(L')$; (g) $P(T')$; (h) $P[(T \text{ and } L)']$.
19. A card is to be selected at random from a well-shuffled deck of 52 cards. The events D, R, F, T, and K are D: a diamond, R: a red card, F: a face card, T: a 3, and K: the king of diamonds. Compute (a) $P(D \text{ and } T)$; (b) $P(D \text{ or } T)$; (c) $P(R \text{ and } F)$; (d) $P(R \text{ or } F)$; (e) $P(F \text{ or } T)$; (f) $P(T \text{ or } R)$; (g) $P(K')$; (h) $P(T')$; (i) $P(D')$; (j) $P(D' \text{ and } T)$.

20. In a recent survey of senior citizens 80% favored giving greater powers of arrest to policemen, 65% favored longer sentences for convicted persons, and 50% favored both propositions.
    (a) What percentage favored at least one of the two proposals? (*Hint:* Interpret these percentages as probabilities and apply Equation 3.3.)
    (b) What percentage favored neither proposal?
21. At a college in California with 2294 undergraduates, the number of students from each home state was as follows:

HOME STATE	NUMBER OF STUDENTS
Arizona (A)	215
California (C)	1452
Nevada (N)	68
Oregon (O)	318
Washington (W)	198
Other states (OS)	43
	2294

If an undergraduate is selected at random, compute (a) $P(A)$, (b) $P(N \text{ and } W)$; (c) $P(N \text{ or } O)$; (d) $P(C \text{ or } O \text{ or } W)$; (e) $P(C')$; (f) $P(A')$; (g) $P[(O \text{ or } W)']$.

22. A, B, C, D, and E are five mutually exclusive events each having probability .15 of occurring. Find (a) $P(A \text{ or } B)$; (b) $P(B \text{ or } C \text{ or } D)$; (c) $P(A \text{ or } B \text{ or } C \text{ or } D \text{ or } E)$; (d) $P[(A \text{ or } B \text{ or } C \text{ or } D \text{ or } E)']$.

Solution to Problem 13

The sample space with the events G, H, and I marked looks like this:

$$\begin{array}{rcccccccccc} \text{SS} = [& AB, & AC, & AD, & AE, & BC, & BD, & BE, & CD, & CE, & DE] \\ & H & & & G & H & H & G & & & I \\ & & & & & & & H & & & \end{array}$$

(a) $P(H \text{ and } G) = P(\text{BE}) = 1/10 = .1$.
(b) The events H and I are mutually exclusive so $P(H \text{ and } I) = 0$.
(c) $P(G \text{ or } H) = P(G) + P(H) - P(G \text{ and } H) = 2/10 + 4/10 - 1/10 = .5$.
(d) $P(H \text{ or } I) = P(H) + P(I) = 4/10 + 1/10 = .5$.
(e) Events G and I are mutually exclusive, so $P(G \text{ or } I) = P(G) + P(I) = 2/10 + 1/10 = .3$.
(f) $P(G') = 1 - P(G) = 1 - 2/10 = .8$.
(g) $P(I') = 1 - P(I) = 1 - 1/10 = .9$.

SECTION 3.5
THE MULTIPLICATION OF PROBABILITIES

CONDITIONAL PROBABILITY

The distribution of voters in a street, as seen in Example 11, is as follows:

		POLITICAL AFFILIATION			
		DEMOCRAT (D)	REPUBLICAN (R)	INDEPENDENT (I)	TOTAL
SEX	M	20	8	16	44
	F	21	7	8	36
	TOTAL	41	15	24	80

The probability that a randomly selected voter is an independent is $24/80 = .3$. But suppose that a selected voter is known to be a female. What is the probability that she is an independent? We denote this probability $P(I \mid F)$, which should be read as "the probability of a randomly selected voter being an independent *given* that this voter is female." The vertical line between I and F means "given" or "conditional on." Of the 36 females, eight are independent, so $P(I \mid F) = 8/36 = .222$. We call this value the **conditional probability** of I given F.

Following the same logic, we may compute, for example, $P(D \mid M) = 20/44 = .455$, $P(M \mid D) = 20/41 = .488$ and $P(D \mid R) = 0$. The last result is true since, once the voter is known to be a Republican, he or she cannot also be a Democrat.

Our original conditional probability, $P(I \mid F) = 8/36$, can be expressed in terms of two more familiar probabilities, as follows. Divide both top and bottom of the fraction $8/36$ by 80 (the total number of voters); then

$$P(I \mid F) = \frac{8/80}{36/80}$$

But $8/80$ is $P(I \text{ and } F)$ and $36/80$ is $P(F)$. Thus we can write $P(I \mid F) = P(I \text{ and } F)/P(F)$.

More generally, for two events A and B,

$$P(B \mid A) = \frac{P(A \text{ and } B)}{P(A)} \qquad \text{(Eq. 3.6)}$$

provided that $P(A) \neq 0$.

For example,

$$P(D \mid M) = \frac{P(D \text{ and } M)}{P(M)}$$

$$= \frac{20/80}{44/80} = .455 \quad \text{as above.}$$

EXAMPLE 18 Forty-five percent of adults in a town read the *Globe*, 30% read the *Herald*, and 20% read both newspapers. (a) What proportion of those who read the *Globe* also read the *Herald*? (b) What proportion of those who read the *Herald* also read the *Globe*?

Solution Treating these percentages as probabilities and with the obvious notation, we have:

(a) $P(H|G) = P(H \text{ and } G)/P(G) = .2/.45 = .44$

(b) $P(G|H) = P(G \text{ and } H)/P(H) = .2/.3 = .67$

■

INDEPENDENCE

The following table shows 200 students classified by sex and whether they have completed a statistics course.

	M	*F*	TOTAL
STATISTICS COURSE (S)	50	30	80
NO STATISTICS COURSE (S')	75	45	120
TOTAL	125	75	200

From the table we see that if a student is selected at random, then $P(S)$, the probability that he or she has completed a statistics course, is $80/200 = .4$. Notice that $P(S|M) = 50/125$ is also .4. Knowing that the student is a male does not change the probability of selecting a student who has completed a statistics course. In the situation where $P(S|M) = P(S)$ we say that the events S and M are independent. More generally:

> DEFINITION 13
>
> Two events A and B are **independent** if the occurrence of one of them does not change the probability of the other occurring, that is, if $P(B|A) = P(B)$ and $P(A|B) = P(A)$. Otherwise, A and B are **dependent.**

The following is an intuitive explanation for the independence of M and S. Since the proportion of males who have taken a statistics course $(50/125 = .4)$ is the same as that for females $(30/75 = .4)$ and for all students, $80/200 = .4$, knowing that the student is a male does not change the value of $P(S)$. We note that since $P(S|F) = 30/75 = .4, = P(S)$ the events F and S are also independent.

By contrast, suppose that the following table applied:

	M	F	TOTAL
STATISTICS COURSE (S)	50	15	65
NO STATISTICS COURSE (S')	75	60	135
TOTAL	125	75	200

In this case S and M are dependent since $P(S) = 65/200 = .325$ but $P(S|M) = 50/125 = .4$. The overall proportion of students who have completed a statistics course is $65/200 = .325$, but the proportion is greater for males, $50/125 = .4$, and less for females, $15/75 = .2$.

THE MULTIPLICATION RULE

Up to this point we have obtained probabilities of the form $P(A \text{ and } B)$ by computing M_o/M, the proportion of outcomes in which both events occur. In the table above, for instance, $P(S \text{ and } M) = 50/200 = .25$. On occasions, however, it will be helpful to compute $P(A \text{ and } B)$ using a variation of Equation 3.6:

$$\frac{P(A \text{ and } B)}{P(A)} = P(B|A)$$

If we multiply both sides of this Equation by $P(A)$, we obtain

$$P(A \text{ and } B) = P(A) \cdot P(B|A) \qquad \text{(Eq. 3.7)}$$

Equation 3.7 is known as the **multiplication rule** of probability.

This rule is useful when an experiment consists of two separate operations and we want the probability that each operation results in a specified event. For instance, if an experiment consisted of selecting two cards without replacement from a deck, we might be interested in the probability that the first card is an ace and the second a king, or the probability that both cards are diamonds. (See Problem 38 at the end of this section.)

The following two examples illustrate how useful this rule can be.

EXAMPLE 19

The New York Yankees have won the American League (AL), Eastern Division. A sports writer estimates that the probability they will win the AL championship (the event A) is .6. Further, if they win the championship, the probability that they will win the World Series (the event $B|A$) is .7. What is the probability that the Yankees will (a) win the AL championship and the World Series; (b) win the AL championship but lose the World Series?

Solution

It will be helpful in applying the multiplication rule to represent the possibilities in this experiment by the tree diagram in Figure 6. The two complementary

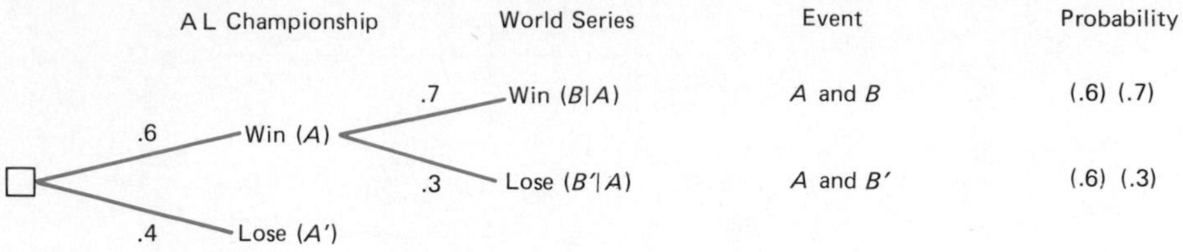

	AL Championship		World Series	Event	Probability
	.6 Win (A)	.7	Win ($B\|A$)	A and B	(.6) (.7)
		.3	Lose ($B'\|A$)	A and B'	(.6) (.3)
	.4 Lose (A')				

FIGURE 6

Tree diagram for Example 19

events A and A' (win and lose the AL championship) are represented by two branches. Alongside each branch we have written the probability of that event. The conditional events $B|A$: win the World Series and $B'|A$: lose the World Series are also shown with their conditional probabilities as branches beginning at the end of the event A branch. Of course, if the Yankees lose the AL championship, their season ends and there are no more branches.

Now to answer the questions.

(a) Using Equation 3.7 yields $P(A \text{ and } B) = P(A) \cdot P(B|A) = (.6)(.7) = .42$. This is the product of the probabilities associated with each of the "win–win" sequences of branches.

(b) The event (A and B') represents the "win–lose" sequence of branches. Notice that the events $B|A$ and $B'|A$ are complementary. That is, having won the AL championship (A) the Yankees can either win ($B|A$), or lose ($B'|A$), the World Series. We know $P(B|A) = .7$. Therefore $P(B'|A) = 1 - P(B|A) = 1 - .7 = .3$. Using Equation 3.7 with B' in place of B gives us $P(A \text{ and } B') = P(A) \cdot P(B'|A) = (.6)(.3) = .18$.

■

EXAMPLE 20 A student liaison committee consists of three seniors and two juniors. Two officers are to be chosen at random. The first to be selected will be the chairperson and the second will be the treasurer. Assuming that the same student cannot occupy both posts, what is the probability that (a) the chairperson will be a senior and the treasurer a junior; (b) both will be juniors? Answer the same questions if the same student *can* occupy both posts.

Solution The appropriate tree diagram is shown in Figure 7. At the first selection the probability of a junior is 2/5 and of a senior is 3/5. The conditional probabilities at the second selection are also given in the tree diagram. For example, if the first student is a junior, there will be one junior out of four students left and the probability that the next choice will also be a junior will, therefore, be $P(J_2|J_1) = 1/4$. Similarly, if the first student is a senior there will be two seniors out of four students remaining and $P(S_2|S_1) = 2/4$.

Now turning to the question:

(a) From Equation 3.7, $P(S_1 \text{ and } J_2) = P(S_1) \cdot P(J_2|S_1) = (3/5)(2/4) = .3$.

(b) $P(J_1 \text{ and } J_2) = P(J_1) \cdot P(J_2|J_1) = (2/5)(1/4) = .1$.

In the case where the same student may hold both posts, we replace the name of the student selected as chairperson in the pool. In this case, the choice

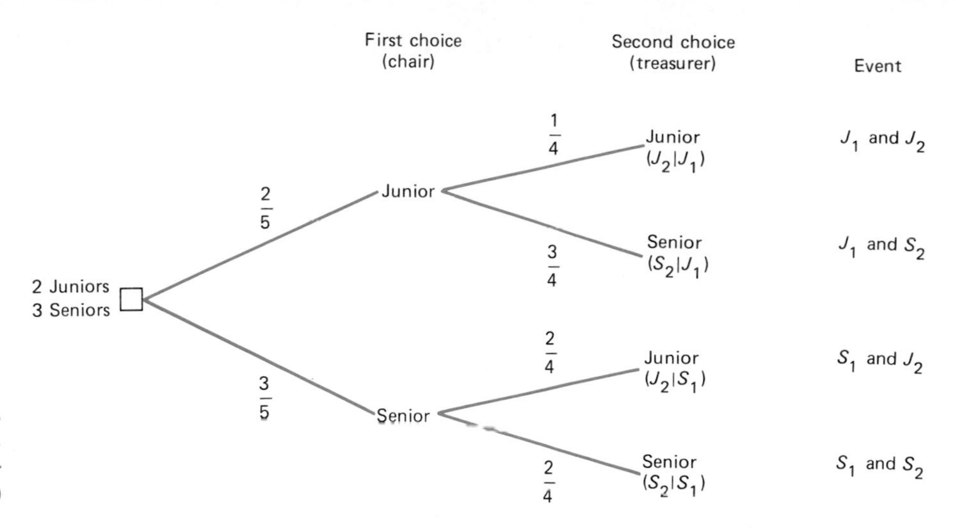

FIGURE 7

Tree diagram for Example 20

of treasurer is independent of the choice of chairperson. Thus the solutions in this case are:

(a) $P(J_2 | S_1) = P(J_2) = 2/5$, so $P(S_1 \text{ and } J_2) = P(S_1) \cdot P(J_2) = (3/5)(2/5) = .24$. Similarly,

(b) $P(J_2 | J_1) = P(J_2) = 2/5$ and $P(J_1 \text{ and } J_2) = P(J_1) \cdot P(J_2) = (2/5)(2/5) = .16$.

■

THE MULTIPLICATION RULE FOR INDEPENDENT EVENTS

The last part of Example 20 was an application of the multiplication rule (Equation 3.7) *for two independent events*. Since the events A and B are independent, $P(B | A) = P(B)$ and the multiplication rule becomes as follows:

> Where A and B are **independent** events
>
> $$P(A \text{ and } B) = P(A) \cdot P(B) \qquad \text{(Eq. 3.8)}$$

We may also note here that if A and B are independent events, the following pairs are also independent: A and B', A' and B, and A' and B'.

Applications of Equation 3.8 occur in many contexts other than sampling with replacement. The following are examples of two such contexts.

EXAMPLE 21 A student applies to two banks for a college loan. A is the event "the first bank grants the loan," B the event "the second bank grants the loan." The student judges that $P(A) = .7$ and $P(B) = .4$ and that A and B are independent. Compute the probability that (a) both banks grant the loan; (b) the first bank does not grant the loan but the second does.

Solution The tree diagram for this problem is shown in Figure 8.

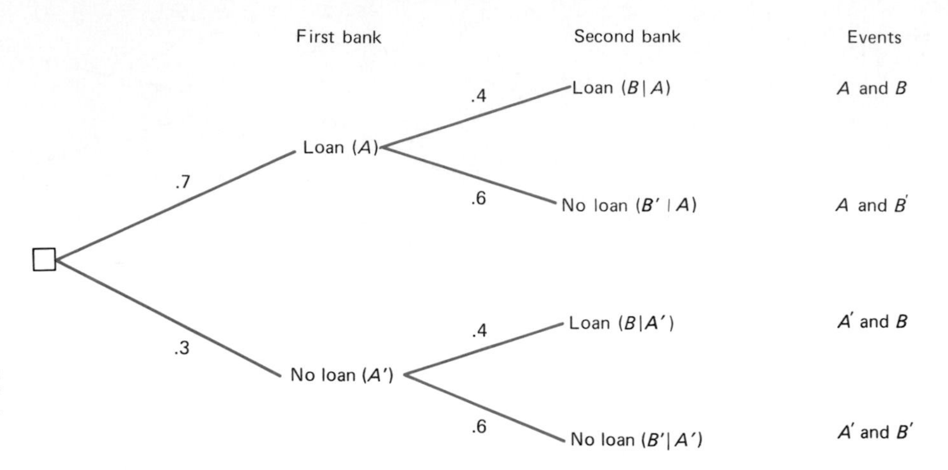

FIGURE 8

Tree diagram for Example 21

(a) Applying Equation 3.8, we multiply the probabilities for the two branches and obtain $P(A \text{ and } B) = P(A) \cdot P(B) = (.7)(.4) = .28$.

(b) Similarly, with A' replacing A in Equation 3.7, $P(A' \text{ and } B) = P(A') \cdot P(B) = (.3)(.4) = .12$. ■

EXAMPLE 22 Two of the six sides of a fair die are green, three are blue, and one is white. If the die is rolled twice, what is the probability that (a) both throws result in a green side; (b) both throws result in the same color?

Solution The tree diagram is shown in Figure 9. By the nature of the experiment the color obtained on the second roll is independent of that obtained on the first. Therefore:

(a) $P(G_1 \text{ and } G_2) = P(G_1) \cdot P(G_2) = (2/6)(2/6) = .111$.

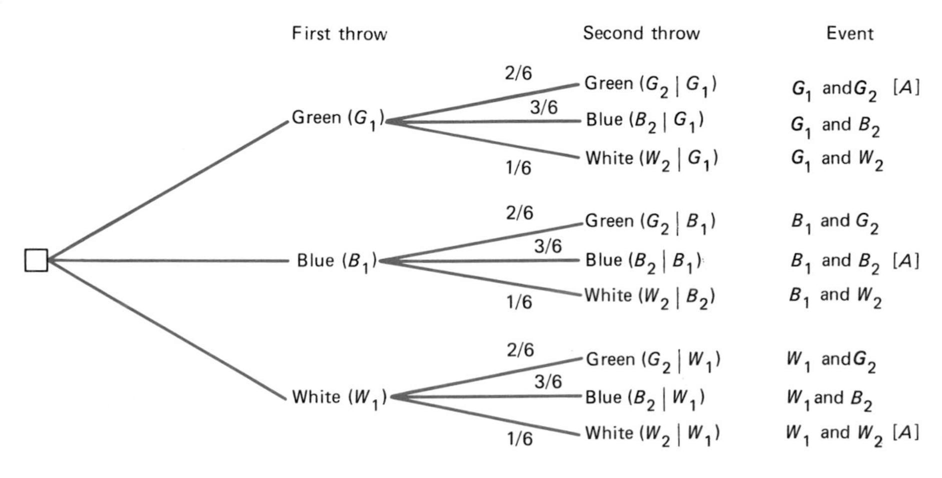

FIGURE 9

Tree diagram for Example 22

(b) The event A "both throws result in the same color" occurs if *either* both colors are green *or* both are blue *or* both are white. That is, $A = (G_1 \text{ and } G_2)$ or $(B_1 \text{ and } B_2)$ or $(W_1 \text{ and } W_2)$. This is shown on the right of Figure 9. Since these three events are mutually exclusive (if one color appears twice, the others cannot), we can write

$$\begin{aligned} P(A) &= P(G_1 \text{ and } G_2) + P(B_1 \text{ and } B_2) + P(W_1 \text{ and } W_2) \\ &= P(G_1) \cdot P(G_2) + P(B_1) \cdot P(B_2) + P(W_1) \cdot P(W_2) \\ &= \frac{2}{6} \cdot \frac{2}{6} + \frac{3}{6} \cdot \frac{3}{6} + \frac{1}{6} \cdot \frac{1}{6} = \frac{14}{36} = .3889 \end{aligned}$$

■

It may be useful to emphasize here the difference between mutually exclusive and independent events. If A and B are mutually exclusive, both *cannot occur* and thus

DEFINITION 11
The event A and the event B are **mutually exclusive** if

$$P(A \text{ and } B) = 0$$

If A and B are independent, both *may occur* but if one does, it does not affect the probability of the other occurring and thus

Where A and B are **independent** events

$$P(A \text{ and } B) = P(A) \cdot P(B) \qquad \text{(Eq. 3.8)}$$

Comparing Equation 3.8 with Definition 11, note that if A and B are independent and both $P(A) > 0$ and $P(B) > 0$, then $P(A \text{ and } B) = P(A) \cdot P(B) > 0$ and therefore A and B *cannot* be mutually exclusive, i.e., both may occur.

The multiplication rule for two independent events (Equation 3.8) can now be generalized to three or more independent events. For example, if A, B, and C are independent events, then $P(A \text{ and } B \text{ and } C) = P(A) \cdot P(B) \cdot P(C)$.

EXAMPLE 23 The probability is .2 that the traffic lights at a particular intersection will malfunction on a given day. Whether the lights malfunction on one day is independent of how they function on any other day. What is the probability that in four successive days: (a) a malfunction does not occur on any of the four days; (b) they malfunction on at least one day?

Solution We will use the notation F for functioning and, for convenience, M (instead of F') for malfunctioning. From above, $P(M) = .2$ and $P(F) = .8$.

(a) We require the probability of the event (F_1 and F_2 and F_3 and F_4). For simplicity we write this as $P(FFFF)$. Since the four events are independent, we can write $P(FFFF) = P(F) \cdot P(F) \cdot P(F) \cdot P(F) = (.8)^4 = .4096$.

(b) Designate the event "a malfunction on at least one day" as A. The complement of the event A is A': no malfunctions at all. We have computed $P(A')$ above as $P(A') = .4096$. Therefore, from Equation 3.5, $P(A) = 1 - P(A') = 1 - .4096 = .5904$.

■

EXAMPLE 24 An interesting but inappropriate application of the multiplication of probabilities occurred in the trial of a couple in California [see Fairley and Mosteller (1977)]. In 1964 the couple was accused of stealing a woman's purse. A witness could not identify the defendants but did testify that the couple he saw leaving the scene had the following characteristics—shared by the defendants.

CHARACTERISTIC	PROBABILITY
A: partly yellow automobile	$P(A) = 1/10$
B: man with mustache	$P(B) = 1/4$
C: white girl with ponytail	$P(C) = 1/10$
D: white girl with blonde hair	$P(D) = 1/3$
E: black man with beard	$P(E) = 1/10$
F: interracial couple in car	$P(F) = 1/1000$

Beside each characteristic is its probability as estimated by a mathematician called by the prosecution. It was not made clear at the trial how accurate these probabilities were, but this is not our main point. The mathematician proceeded to "demonstrate" the "unlikelihood" of there being another couple having the same combination of characteristics by multiplying the probabilities together to obtain the probability of this occurring—approximately 1/12,000,000. In trying to obtain $P(A \text{ and } B \text{ and } C \text{ and } D \text{ and } E \text{ and } F)$ by multiplying together the individual probabilities, the mathematician treated the six events as if they were independent. But were they? For instance, are you not more likely to find a bearded man among mustached men than among all men? To see what we mean, consider the following population, consisting of 1000 men classified according to their possession of a beard and/or a mustache.

		MAN HAS BEARD		
		YES (B)	NO (B')	TOTAL
MAN HAS MUSTACHE	YES (E)	80	20	100
	NO (E')	170	730	900
	TOTAL	250	750	1000

If events B and E were independent, $P(B \mid E)$ would equal $P(B)$. But in this case $P(B)$, the proportion of men with beards, is $250/1000 = .25$ and $P(B \mid E)$, the proportion of men with mustaches who have beards, is $80/100 = .8$. These two characteristics are *dependent*. Suppose, however, that we followed the court mathematician and (incorrectly) computed $P(B$ and $E)$ by multiplying $P(B) = .25$ and $P(E) = .1$. Then we would obtain the value $P(B$ and $E) = .025$. The *actual* value for $P(B$ and $E)$, however, as obtained directly from the table above, is $80/1000 = .08$. By multiplying $P(B)$ and $P(E)$ our mathematician would have *underestimated* $P(B$ and $E)$ by a factor of more than 3, $(.08/.025 = 3.2)$. By multiplying the six individual probabilities the mathematician probably underestimated $P(A$ and B and C and D and E and $F)$ by a factor much greater than 3. The mathematician's arguments were accepted and the couple were convicted, but this decision was reversed on appeal in large part based on the argument outlined above.

■

PROBLEMS 3.5

23. The contents of a box of 90 light bulbs may be classified as follows:

	GOOD (G)	DEFECTIVE (D)	TOTAL
40-WATT (A)	26	4	30
60-WATT (B)	33	7	40
80-WATT (C)	16	4	20
TOTAL	75	15	90

(a) If a light bulb is selected at random, what is the probability that it is (i) defective if it is a 60-watt bulb; (ii) good if it is an 80-watt bulb; (iii) a 40-watt bulb if it is known to be defective?
(b) If two light bulbs are selected at random and without replacement, what is the probability that both will be good?

24. A fair die is to be rolled. Define the events A, B, C, and D as follows; A: a 4, B: a 6, C: a 4 or smaller number, and D: an even number. Compute (a) $P(A \mid C)$; (b) $P(A \mid D)$; (c) $P(B \mid C)$; (d) $P(B \mid D)$.

25. Forty-three percent of foreign visitors to New York City visit the Statue of Liberty, 64% visit the United Nations, and 31% visit both landmarks.
(a) What proportion of those who visit the United Nations visit the Statue of Liberty?
(b) What proportion of those who visit the Statue of Liberty visit the United Nations?

26. The 180 members of a high school graduating class are classified by sex and whether or not they went to college.

		WENT TO COLLEGE YES (C)	NO (C')
SEX	M	60	20
	F	75	25

(a) If a student is selected at random, compute (i) $P(M|C)$; (ii) $P(C|F)$.
(b) Are the events F and C independent? Explain.

27. A bookstore is featuring the works of three mystery writers, J. Arrow, S. Blake, and T. Chan, on one shelf. Each book has been rated by a literary magazine as excellent (E), good (G), or unexceptional (U) as follows:

		AUTHOR ARROW (A)	BLAKE (B)	CHAN (C)
RATING	EXCELLENT (E)	4	2	6
	GOOD (G)	4	7	1
	UNEXCEPTIONAL (U)	2	3	1

A person selects a mystery at random from the shelf.
(a) If the book is by Blake, what is the probability that it will be excellent?
(b) If the book is by Chan, what is the probability that it will be (i) good; (ii) unexceptional?
(c) Are the events B and E independent? Explain.
(d) What about the events A and G?

28. A and B are events such that $P(A) = .4$ and $P(B|A) = .6$. Find (a) $P(A')$; (b) $P(B'|A)$; (c) $P(A \text{ and } B)$; (d) $P(A \text{ and } B')$.

29. A tennis player will participate first at Wimbledon and then at the U.S. Open. Whether she wins or loses at Wimbledon is independent of her performance at the U.S. Open. The probability of her winning Wimbledon is .3, and of her winning the U.S. Open, .45. Construct a tree diagram for this experiment. What is the probability that she (a) wins both tournaments; (b) wins Wimbledon but loses the U.S. Open; (c) wins only one of the tournaments?

30. Department of Transportation data indicate that seat belts were worn by the drivers in 65% of reported accidents. When seat belts were worn, 8% of the accidents were fatal. When seat belts were not worn, 20% of the accidents were fatal. What is the probability that in any accident (a) the driver is wearing a seat belt and is not killed; (b) the seat belt is not worn and the accident is fatal? (Construct a tree diagram.)

31. In each of two soccer games the home team has a probability of .6 of winning. Assume that the outcome of one game has no effect on the outcome of the other. What is the probability that (a) the home team wins both games; (b) the home team loses both games?

32. A purse contains five nickels and three dimes. Two coins are to be selected at random and without replacement.
(a) Construct a tree diagram for this experiment.
(b) What is the probability that (i) the first coin will be a dime and the second a nickel; (ii) both will be nickels; (iii) both will be dimes; (iv) the first will be a nickel and the second a dime?
(c) What is the sum of these four probabilities? Explain.

33. Suppose that a woman has a probability of .65 of giving birth to a boy and that the sex of one child is independent of the sex of another. If the woman has two children, what is the probability that (a) both will be girls; (b) the first will be a boy and the second a girl; (c) she will have a child of each sex?
34. Suppose that the woman referred to in Problem 33 has three children. What is the probability that (a) all will be boys; (b) she will have at least one girl?
35. The probability that a plane flown by Skyhigh Airlines is on time is .75. What is the probability that of four successive flights (a) all are late; (b) the first two are on time but the last two are late; (c) at least one is late? You may assume independence. A tree diagram will help.
36. A total of 240 tickets are sold in a lottery. You hold 24 of them. Five prizes are to be awarded. At the drawing the five winning tickets are to be selected *with* replacement. What is the probability that you win (a) no prizes; (b) at least one prize; (c) all five prizes?
37. Example 24 contained a table in which the events E and B were *dependent*. Fill in the four missing entries in the following table so as to make the events E and B *independent*. [*Hint:* What should be the relationship between $P(B|E)$ and $P(B)$?]

		MAN HAS BEARD		
		YES (B)	NO (B')	TOTAL
MAN HAS MUSTACHE	YES (E)			100
	NO (E')			900
	TOTAL	250	750	1000

38. Two cards are drawn without replacement from a deck of 52 cards. What is the probability that:
(a) the first is an ace and the second a king
(b) both are diamonds
(c) neither is a diamond
(d) exactly one of the two is a club?

Solution to Problem 23

(a) Refer to the events indicated in the table.
(i) $P(D|B) = 7/40 = .175$; (ii) $P(G|C) = 16/20 = .8$; (iii) $P(A|D) = 4/15 = .267$.
(b) $P(G_1 \text{ and } G_2) = P(G_1) \cdot P(G_2|G_1)$

$$= \frac{75}{90} \cdot \frac{74}{89} = .693$$

SECTION 3.6
SUMMARY AND REVIEW

The term **population** is used to refer to an entire group of interest. Frequently, the term is used to refer to numerical characteristics of members of the group. That part of the population for which data have been obtained is called the

sample. Probability provides the tools necessary to draw conclusions, or **inferences,** about the population by using the data in a sample.

Other terms also have a special meaning in probability. The word **experiment** denotes any phenomenon in which the outcome is uncertain. The list of all possible **outcomes** of an experiment is the **sample space** and is denoted SS. The occurrence of any one of a specific collection of outcomes is called an **event.** The notation $P(A)$ denotes the probability that event A will occur.

In experiments in which we can treat all the M outcomes in the sample space as equally likely and if M_0 is the number of outcomes in which A occurs, then

$$\text{The probability of each outcome is } \frac{1}{M} \qquad \text{(Eq. 3.1)}$$

$$\text{The probability of the event } A \text{ is } P(A) = \frac{M_0}{M} \qquad \text{(Eq. 3.2)}$$

If an experiment can be repeated, one interpretation of $P(A)$ is the proportion of times that A occurs when the experiment is repeated an infinite number of times.

If, by contrast, an experiment is unique (i.e., nonrepeatable), then $P(A)$ may be regarded as a subjective assessment of the chances of A occurring.

Regardless of the nature of the experiment,

$$\text{for any event } A,\ 0 \leq P(A) \leq 1 \qquad \text{(Def. 8)}$$

The event (A *and* B) occurs when the event A and the event B both occur. The event (A *or* B) occurs when either the event A, or the event B, or (A and B) occur.

The **addition rule** of probabilities is

$$P(A \text{ or } B) = P(A) + P(B) - P(A \text{ and } B) \qquad \text{(Eq. 3.3)}$$

The events A and B are said to be mutually exclusive if an experiment cannot result in the event (A and B), that is, if $P(A \text{ and } B) = 0$. The **addition rule for mutually exclusive events** is

$$P(A \text{ or } B) = P(A) + P(B) \qquad \text{(Eq. 3.4)}$$

The complement of the event A is the event "*not* A" and is designated A'. The probabilities of the events A and A' sum to 1, and therefore

$$P(A) = 1 - P(A') \qquad \text{(Eq. 3.5)}$$

The **conditional probability,** $P(B \mid A)$, is the probability that the event B will occur given that A has already occurred. We can compute $P(B \mid A)$ using the equation

$$P(B \mid A) = \frac{P(A \text{ and } B)}{P(A)} \qquad \text{(Eq. 3.6)}$$

Equation 3.6 can be rearranged to provide a means of computing $P(A$ and $B)$. The result

$$P(A \text{ and } B) = P(A) \cdot P(B \mid A) \qquad \text{(Eq. 3.7)}$$

is called the **multiplication rule.** It is useful to keep in mind that when we are interested in $P(A$ or $B)$, we use the addition rule, and in $P(A$ and $B)$, the multiplication rule.

Two events are **independent** if the occurrence of one of them does not change the probability that the other will occur. This can be expressed as

$$P(B \mid A) = P(B)$$

When A and B are independent, the multiplication rule becomes

$$P(A \text{ and } B) = P(A) \cdot P(B) \qquad \text{(Eq. 3.8)}$$

Events that are not independent are said to be **dependent.**

The terms "mutually exclusive" and "independent" are easy to confuse. If A and B are mutually exclusive events, both *cannot* occur. If A and B are independent, both may occur, but if one does occur it does not affect the probability of the other occurring.

REVIEW PROBLEMS

GENERAL

1. Eighty tourists at the United Nations are classified by nationality and sex.

	AMERICAN (*A*)	JAPANESE (*J*)	BRITISH (*B*)
M	20	10	6
F	30	10	4

A tourist is to be selected at random to win a prize. Compute (a) $P(B$ and $M)$; (b) $P(A$ and $J)$; (c) $P(J$ or $F)$; (d) $P(A)$; (e) $P[(F$ and $A)']$; (f) $P[(A$ and $M)$ or $(B$ and $F)]$.

2. For some purposes the families registered with a certain health center are arranged by the number of children under 18 in the family. The proportion of each size of family is as follows:

NUMBER OF CHILDREN	0	1	2	3	4	5	6	7+
PROPORTION OF FAMILIES WITH THIS NUMBER OF CHILDREN	.25	.1	.27	.21	.1	.04	.02	.01

An experiment consists of selecting one family at random from all registrants and counting the number of children in the family.

(a) What is the probability that the chosen family will have (i) exactly three children; (ii) fewer than three children; (iii) more than three children; (iv) either fewer than two children or six or more children?

(b) What is the most likely number of children in the family chosen?

3. The books on a shelf are classified by fiction (F) or nonfiction (N) and by author (A, B, or C).

	A	B	C
F	1	10	15
N	29	10	10

If a book is selected at random from the shelf, compute (a) $P(F)$; (b) $P(F \text{ and } B)$; (c) $P(C \text{ or } N)$; (d) $P(A \text{ or } C)$; (e) $P(A|F)$; (f) $P(F|A)$; (g) $P[(C \text{ or } B|N)]$.

4. A and B are events such that $P(A) = .6$, $P(B) = .3$, and $P(A \text{ and } B) = .3$.

(a) Are A and B mutually exclusive?

(b) Find (i) $P(A')$; (ii) $P(B')$; (iii) $P(A \text{ or } B)$.

5. Two events A and B are such that $P(A) = .4$, $P(B) = .3$, and $P(A \text{ and } B) = .2$. Find (a) $P(A \text{ or } B)$; (b) $P(A')$; (c) $P[(A \text{ and } B)']$.

6. A coin is known to be weighted in favor of heads. We wish to get some idea of the probability of the tossed coin showing heads. The coin is tossed 2000 times. The following figures record the number of heads after various numbers of tosses (these were the only records kept):

NUMBER OF TOSSES	100	500	1000	2000
NUMBER OF HEADS	62	361	712	1404

(a) Based on the 2000 tosses, what would be your best guess as to the probability of the coin coming up heads?

(b) How would this guess change if the coin had been tossed only 500 times?

(c) Suppose that you found out that immediately after the 100th toss, a chip had come off the coin. This would presumably affect its probability of its coming up heads. Use the data above to obtain the best estimate of the new probability of the coin coming up heads after it had been chipped.

7. A die is weighted so that the probabilities of each of the digits 1, 2, 3, 4, 5, and 6 are as follows:

NUMBER	1	2	3	4	5	6
PROBABILITY	$\frac{1}{21}$	$\frac{2}{21}$	$\frac{3}{21}$	$\frac{4}{21}$	$\frac{5}{21}$	$\frac{6}{21}$

This die is to be rolled twice. What is the probability that the sum of the two values is (a) equal to 4, (b) equal to 6. (Notice that in this case the 36 pairs of values in the sample space are not equally likely.)

8. A specially constructed roulette wheel contains 36 numbers "colored" red or black as follows:

RED						BLACK					
1	2	3	4	5	10	6	7	8	9	14	15
11	12	13	24	25	26	16	17	18	19	20	21
27	32	33	34	35	36	22	23	28	29	30	31

If the wheel is spun, what is the probability of:
(a) A black number?
(b) A number greater than 20?
(c) A black number less than 16?
(d) A black even number or a red odd number?
(e) A red number greater than 20?
(f) An even red number?

9. If the roulette wheel in Review Problem 8 is spun twice, what is the probability that:
(a) Both spins result in a red number?
(b) Both spins result in an even number?
(c) One of the spins result in a number greater than 20?
(d) One of the spins result in black number less than 16?
(e) At least one of the two spins result in a black number?

10. A student applies to two graduate schools, A and B. She estimates that the probability of getting into A is .6, into B .5, and of getting into both .2. (a) Are the events "She gets into A" and "She gets into B" independent? Find the probability that (b) She gets into at least one of the two schools? (c) She does not get into either school? (d) She gets into B if she has been accepted by A?

11. (a) Two events A and B are such that $P(A) = .4$, $P(B) = .6$, and $P(A \text{ and } B) = .3$. Find $P(A|B)$ and $P(B|A)$. Are A and B independent? Explain your answer.

(b) Two other events, C and D, are such that $P(C) = .4$, $P(D) = .25$, and $P(C \text{ and } D) = .1$. Find $P(C|D)$ and $P(D|C)$. Are C and D independent? Explain your answer.

12. In each of the following cases explain whether or not A and B are independent events.
(a) $P(A) = .75$, $P(A|B) = .6$, $P(B|A) = .75$
(b) $P(A) = .8$, $P(A \text{ and } B) = .2$, $P(B) = .25$
(c) $P(A|B) = .4$, $P(B|A) = .7$, $P(A) = .4$
(d) $P(B) = .2$, $P(A \text{ and } B) = .2$, $P(A) = .9$
(e) $P(B|A) = .6$, $P(A|B) = .3$, $P(A) = .3$
(f) $P(A) = .5$, $P(B) = .5$, $P(A \text{ and } B) = .25$

13. We are interested in families with two children. With the obvious notation, assume that each of the four events BB, BG, GB, and GG are equally likely. Compute the probability of two girls if a family with two children is known to have at least one girl.

14. The probability that it will rain tomorrow is .6. If it rains tomorrow, the probability that it will rain on the following day is .7. If it does not rain tomorrow, the probability that it will rain on the following day is .2. What is the probability that:
    (a) It rains tomorrow and the next day?
    (b) It rains tomorrow but not the next day?
    (c) It does not rain on either day?
15. An athlete is to run in the 100-meter race on one day and in the 200-meter race on the next day. She estimates that the probability that she will win the 100 meters is .2, that she will win the 200 meters is .4, and that she will win both races is .1. What is the probability that she will win:
    (a) Either the 100 meters or the 200 meters or both?
    (b) Neither of the two races?
16. If seven-digit telephone numbers are assigned so that the first digit is equally likely to be any of the digits 1, 2, . . . , 9 and the remaining six are all equally likely to be any of the digits 0, 1, 2, . . . , 9, what is the probability of the number 123-4567 being assigned? The number 482-0541? (Assume digits are assigned independently.)
17. A fair coin is to be tossed six times. It can be shown (we will indicate how in a later chapter) that the probability of exactly four heads is .234 and that the probability of less than four heads is .656.
    (a) Are these two events mutually exclusive?
    (b) Compute the probability of (i) four or fewer heads; (ii) five or more heads; (iii) any number of heads except four.
18. The probability that a particular basketball player misses a foul shot is .15. If he takes six shots in a game, what is the probability that he misses at least one of the six?
19. Two teams, A and B, play in the baseball World Series. The first team to win four games wins the series and is world champion. Suppose that the probability that A wins each game is .6 and that the outcome of each game is independent of any other game. What is the probability that:
    (a) A wins the first three games, loses the fourth, and wins the fifth?
    (b) A wins the first two games, loses the third game, and wins the fourth and fifth games?
    (c) A wins the World Series in five games.

HEALTH SCIENCES

20. A man and a woman decide to get married. Assume that the probability that each will have a specific blood group is as follows:

BLOOD GROUP	O	A	B	AB
PROBABILITY	.50	.35	.10	.05

Assume also that the blood group of one partner is independent of the other. What is the probability that:
    (a) The husband is (i) group A; (ii) not group AB?
    (b) The husband is group O and the wife group B?
    (c) Both are group A?
    (d) The wife is group AB and the husband O?

(e) One of the two is group A?
(f) Neither is group AB?
(g) At least one is group A or group B?

21. The following is an extract from a life table for the U.S. population for 1959–1961.

AGE INTERVAL	PROBABILITY OF SURVIVING THROUGH THE PERIOD
0–1	.9741
1–5	.9958
5–10	.9976
10–15	.9978
15–20	.9954
20–25	.9938
25–30	.9936
30–35	.9920
35–40	.9885
40–45	.9819
45–50	.9713
50–55	.9544
55–60	.9333
60–65	.8992
65–70	.8554
70–75	.7915
75–80	.6970
80–85	.5522

The value .9978, for example, is the probability that a person aged 10 will live through age 15.
(a) What is the probability that a person aged 50 will live to age 55?
(b) What is the probability that a person aged 40 will *not* survive to age 45?
(c) What is the probability that a person aged 50 will live to age 60?
(d) What is the probability that a person aged 50 will live to age 55 but will not survive to age 60?
(e) What is the probability that two persons aged 25 will both live to age 30?

22. One person in 300 has a particular form of cancer. There is a diagnostic test for this cancer but it is not perfect. Specifically, if a person has this type of cancer, the test will be positive (i.e., indicate the presence of the cancer) with probability .9. Also, if a person does not have the cancer, the test will be positive with probability .2. If a randomly chosen person is given the test, what is the probability that:
(a) The cancer is present but the test is negative?
(b) The cancer is not present and the test is negative?
(c) The cancer is present and the test is positive?

23. A biologist is interested in cross-fertilizing two varieties of rose. She is interested in the color of the petal and the size of the leaf in the resulting progeny. Genetic theory suggests that the probability of a white petal is .4, the probability of a large leaf is .6, and the probability of both these characteristics together is .2. What is the probability that the progeny have:
(a) At least one of these two characteristics?
(b) Neither of these two characteristics?

24. In the past a particular operation has involved a rather long recovery period, 30 days on average for men and 40 days on average for women. A new technique for performing this operation, which it is hoped will reduce the average recovery time, is carried out on the next 80 men and 120 women requiring this operation. The results of these 200 operations are summarized in the following table. For example, 30 of the men took between 20 and 29 days to recover.

RECOVERY TIME (DAYS)	MEN (M)	WOMEN (W)
0–9 (A)	4	0
10–19 (B)	17	6
20–29 (C)	30	24
30–39 (D)	13	65
40–49 (E)	7	17
50–59 (F)	3	7
60–69 (G)	3	0
70–79 (H)	2	1
80–89 (I)	1	0

One of the 200 patients is to be selected at random for a follow-up study.
(a) What is the probability that the selected person (i) is a woman; (ii) took more than 49 days to recover; (iii) took 19 or fewer days to recover; (iv) is a man who took more than 49 days to recover?
(b) Are the nine events A, B, C, D, E, F, G, H, and I equally likely? Explain.
(c) Compute (i) $P(A \text{ or } B \text{ or } E)$; (ii) $P(D \text{ or } G \text{ or } M)$.

25. In Review Problem 24, suppose that a man and a woman are to be selected separately. What is the probability that:
(a) The woman recovered in less than 40 days?
(b) The man recovered in less than 30 days?
(c) both recovered in less than 30 days?

26. The 85 subjects volunteering for a medical study are classified by sex and blood pressure (high, normal, and low):

	H	N	L
M	8	22	10
F	11	22	12

If a subject is selected at random, compute (a) $P(N)$; (b) $P(F \text{ and } H)$; (c) $P(F \text{ or } H)$; (d) $P(F \mid L)$; (e) $P(H' \text{ and } M)$; (f) $P(H' \mid F)$.

27. In a classic experiment the German botanist K. E. Correns crossed a white-flowered plant (W) with a red-flowered plant (R) of the same genus. All the first generation offspring of these unions were pink, but of the 565 second-generation offspring of these R-W unions, only 292 had pink flowers. Based on this information, what would Correns's estimate be of the probability of getting a second-generation pink from an R-W union? Hazard a guess as to the true probability of this event.

28. A surgical procedure can be classified as (i) unsuccessful (U), (ii) partially successful (P), or (iii) completely successful (C). The probability of these various

outcomes are $P(U) = .2$, $P(P) = .3$, and $P(C) = .5$. If four patients receive this procedure in turn and their outcomes are independent, compute the probability that:

(a) The first two procedures are unsuccessful and the last two completely successful.
(b) All are partially successful.
(c) At least one is partially successful.

GOVERNMENT AND SOCIOLOGY

29. The jury of 12 persons in a murder trial is constituted as follows:

	FAVORS CAPITAL PUNISHMENT	OPPOSES CAPITAL PUNISHMENT
MEN	3	4
WOMEN	4	1

A jury foreman is to be selected at random from the jury members. What is the probability that the person selected is:
(a) A woman?
(b) A man favoring capital punishment?
(c) A person favoring capital punishment?

30. The 100 members of the state senate are classified as follows:

		BACKGROUND			
		BUSINESS (B)	LAWYER (L)	OTHER (O)	TOTAL
PARTY	REPUBLICAN (R)	22	10	2	34
	DEMOCRAT (D)	17	28	7	52
	INDEPENDENT (I)	5	1	8	14
	TOTAL	44	39	17	100

(a) If a senator is selected at random, compute (i) $P(B)$; (ii) $P(O \text{ and } I)$; (iii) $P(B \text{ and } L)$; (iv) $P(L \text{ or } O)$; (v) $P(R \text{ or } L)$; (vi) $P[(B \text{ and } R)']$.
(b) If the 34 Republicans meet to randomly select a chairperson, what is the probability that the chairperson's background will be (i) business; (ii) either business or as a lawyer?

31. A recent poll indicated that in a particular state 45% of the voters favored capital punishment, 30% were opposed to capital punishment, and 25% had no opinion. Assume that all voters are eligible for, and are equally likely to be picked for, jury duty. If a jury of 12 people are selected, what is the probability that:
(a) All will favor capital punishment?
(b) At least one will be opposed to capital punishment?
(c) All will have some opinion on the subject?
(d) At least one will have no opinion on the subject?

32. In the primary elections for a seat in the U.S. Senate, the Democratic incumbent is unopposed. There are three candidates for the Republican nomination: Bolan, Dolan, and Nolan. The probabilities that each will win the nomination are judged

to be .5, .3, and .2, respectively. If she wins the primary, the probability that Bolan will win election to the Senate is .3. The corresponding probabilities for Dolan and Nolan are .6 and .3, respectively. What is the probability that:
(a) Bolan wins the primary and the election?
(b) Dolan wins the primary and the election?
(c) Dolan wins the primary but loses the election?
(d) Nolan wins the primary but loses the election?

33. For a certain society the likelihood of intergenerational movement between social classes is summarized by the following probabilities:

		SOCIAL CLASS OF PARENTS			
		A	*B*	*C*	*D*
SOCIAL CLASS	*A*	.7	.3	.1	.1
OF CHILD	*B*	.2	.6	.3	.1
(AS AN	*C*	.08	.07	.5	.2
ADULT)	*D*	.02	.03	.1	.6

For example, the probability that the child of a couple in class C will end up in class B is .3. The probability that the child of a couple in class B will end up (stay) in B is .6. Use the 16 probabilities above to compute the probability that:
(a) A child of a couple in C will end up in D and a grandchild of the couple will end up in C.
(b) A child and grandchild of a couple in B will both stay in B.
(c) A child of a couple in D will end up in C, a grandchild of the couple, in B, and a greatgrandchild of the couple, in A.
(Assume that marriages are within the same social class)

ECONOMICS AND MANAGEMENT

34. Every car that comes off the production line is inspected. Ten percent of a certain make of car are defective. The probability that a defect in one of these cars will be discovered during inspection is .4.
(a) What is the probability that (i) a given car has a defect and it is discovered; (ii) a given car has a defect but it is not discovered?
(b) If two cars are selected at random, what is the probability that both have defects which have not been discovered? (Assume that defects occur independently and their discovery also occurs independently.)

35. A supermarket has four manned checkout counters. A person is in a hurry and will leave without making a purchase if all the counters are busy. At that time of day the probability of each counter being free is .25. What is the probability that the person will make a purchase? (Assume that whether or not one counter is busy is independent of any other counter.)

36. Airline records show that the probability of a passenger losing an item of baggage is .001. When an item is lost the probability of its being recovered in less than 24 hours is .6. Compute the probability that:
(a) A passenger loses an item and it is recovered within 24 hours.
(b) A passenger loses an item but it is not recovered within 24 hours.
(c) A passenger does not lose an item.

37. Television sets coming off a production line are numbered sequentially. During one day's production, sets are numbered 4875 to 5346. Unfortunately, those (and only those) numbered 4925 to 4968 have defects. If one of the sets produced that day is selected at random, what is the probability that it will be a defective one?

38. Sixty workers in a factory are classified by sex and seniority as shown below (G_1, G_2, G_3, with G_3 the most senior). A single worker is selected at random.

	M	F
G_1	10	12
G_2	12	8
G_3	13	5

Compute (a) $P(G_1)$; (b) $P(G_2 \text{ and } F)$; (c) $P(G_2 \text{ or } F)$; (d) $P(G_2 | F)$; (e) $P(F | G_2)$; (f) $P(G_2')$.

39. A company places an advertisement in two magazines, A and B. They estimate that of all potential customers a proportion .05 will see the ad in A, a proportion .1 will see the ad in B, and a proportion .02 will see both ads.
(a) What proportion of potential customers will see (i) at least one of the two ads; (ii) neither of the ads?
(b) (i) Of those potential customers who see the ad in A, what proportion will see the ad in B? (ii) Of those potential customers who see the ad in B, what proportion will see the ad in A?

40. The company referred to in Review Problem 39 estimates that a proportion .2 of potential customers who see at least one ad will purchase the product. What proportion of potential customers will:
(a) See at least one of the two ads and purchase the product?
(b) See at least one of the two ads but not purchase the product?

41. Three women and nine men apply for a position with a company. All are equally qualified and the position is filled by selecting from among the 12 persons at random. What is the probability that the position is filled by (a) a woman; (b) a man?

42. A company that explores for oil in the ocean is considering two new sites, A and B. The exploration will take the entire year. The company estimates the probability of finding oil is .6 at site A and .4 at site B. The sites are far enough apart so that, whether or not oil is found at one site, is independent of its being found at the other. What is the probability that:
(a) Oil is found at both sites?
(b) Oil is found at site A but not at B?
(c) Oil is found at only one of the sites?

EDUCATION AND PSYCHOLOGY

43. A monkey is being taught a simple task. If it completes the task correctly six times in a row it is said to have learned the task. Suppose that, in fact, the monkey merely guesses how to perform the task on each occasion and that the probability of getting the task correct on each occasion is .4. What is the probability that the monkey will be (falsely) declared to have learned the task?

44. A multiple-choice exam contains k questions. Each question has t choices. If a person simply guesses the answer to each question, what is the probability that he gets all the questions wrong? That he gets at least one correct?

45. It is estimated that approximately 8% of persons seeking treatment at a mental health clinic have a particular disorder. There is a test to detect whether or not a person has this disorder but it is not foolproof. It is estimated that of all persons having the disorder the test will be positive, indicating the presence of the disorder, in only 80% of the cases. Moreover, the test also proves positive on 10% of

persons not having the disorder. If a particular person seeks treatment, what is the probability that:
(a) The person has the disorder and the test is positive?
(b) The person does not have the disorder and the test is positive?
(c) The person does not have the disorder and the test is negative?

46. The institutions of higher education in a region are classified by type of control and highest degree awarded.

	B.S./B.A. (B)	M.S./M.A. (M)	Ph.D. (P)
PRIVATE (T)	8	8	6
PUBLIC (L)	2	7	9

If an institution is selected at random, compute (a) $P(B)$; (b) $P(M \text{ and } T)$; (c) $P(P \text{ or } T)$; (d) $P(B \text{ or } P)$; (e) $P(B' \text{ and } L)$; (f) $P(M|L)$; (g) $P(B|M)$.

47. It is estimated that 20% of students in a school district have a learning disability, A, 15% have a learning disability, B, and 5% have both.
(a) What percentage of students have (i) at least one of the two disabilities; (ii) neither of the disabilities?
(b) Of those students who have disability A, what percentage have disability B?

48. Referring to Review Problem 47, suppose that a sample of four students is selected from all those in the district. What is the probability that:
(a) All four have learning disability A?
(b) All four have at least one of the learning disabilities?
(c) At least one of the four has learning disability B?

SCIENCE AND TECHNOLOGY

49. The probability that a calculator battery will last more than 25 hours is.75. The length of life of any one battery is independent of that of any other. If four such batteries are tested, what is the probability that:
(a) All last longer than 25 hours?
(b) At least one lasts less than 25 hours?

50. A bank has four independent security systems. Only one of them need function for the police to be alerted. The probability that each system will fail is .1, .2, .3, and .3, respectively. What is the probability that, where necessary, the police will be alerted?

51. Components of an electrical system are said to be in "parallel" if current will pass through the system if any one (or more) of the components is capable of carrying current. An example with two components in parallel is as follows:

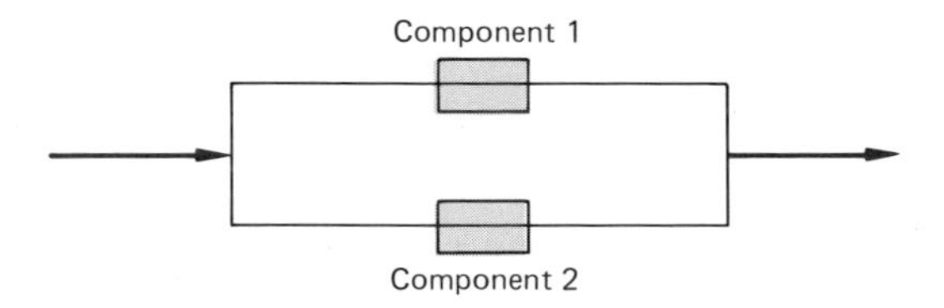

Components are said to be in "series" if current will pass through the system only if all the components are capable of carrying current. An example with two components in series is as follows:

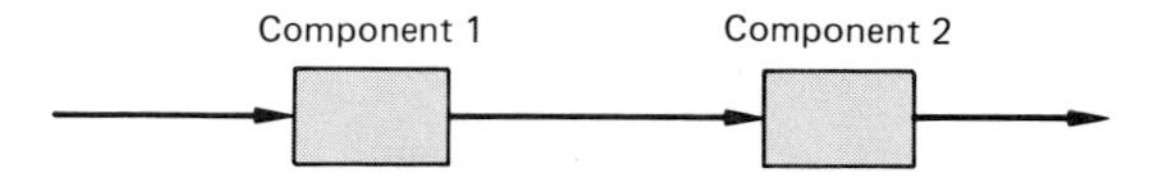

Suppose that each component is capable of carrying current with probability .9 and that whether or not each is so capable is independent of any other component. What is the probability that a two-component system will carry electricity if:
(a) The system is in parallel?
(b) The system is in series?

52. A component in a rocket has a probability .999 of working when switched on. A backup component, which is switched on automatically only when the first fails, has a probability of .99 of working when switched on. What is the probability that:
(a) The first component fails to work but the backup does work?
(b) Neither component works?
(c) One or the other of the components work? (A tree diagram will be helpful.)

REFERENCES

FAIRLEY, W. B. and MOSTELLER, F., eds. *Statistics and Public Policy*. Reading, Mass.: Addison-Wesley, 1977, pp. 350–397.

SUGGESTED READING

Some interesting applications of probability and statistics in law can be found in:

GRAY, M. W. "Statistics and the Law," *Mathematics Magazine,* Vol. 56 (March 1983), pp. 67–81.

Other excellent introductions to probability are:

AYER, A. J. "Chance"; WEAVER, W. "Probability"; and KAC, M. "Probability." Readings in *Mathematics in the Modern World*. San Francisco: Foreman, 1968.

MOORE, D. S. *Statistics: Concepts and Controversies*. San Francisco: Freeman, 1979, pp. 237–268.

STANCL, D. L. and STANCL, M. L. *Applications of College Mathematics*. Lexington, Mass.: D. C. Heath, pp. 323–414.

A more complete treatment of probability can be found in:

ROSS, S. *A First Course in Probability*. New York: Macmillan, 1976.

4 RANDOM VARIABLES

BEATING THE HOUSE!

However optimistic, Las Vegas gamblers are probably aware that the house will only offer a game if, in the long run, the house will win more than they lose. This is certainly true of the following game. A player draws a card from a deck and then draws another without replacing the first. If he picks cards with the same value (both are queens, for instance), he wins $10. If he picks cards of the same suit (both are diamonds, for instance), the player wins $2. For any other combination, the player pays the house $2. In this chapter we indicate how to compute the long-run gain to the house offering this game (see Problem 17 on page 177).

SECTION 4.1

RANDOM VARIABLES

Imagine the experiment of tossing a fair coin three times in which we are interested in the number of "runs" that occur with the three tosses. Here a run is a sequence of one or more of the same letter. For example, the outcome HHH contains only one run (three H's). The outcome THH contains two runs (T is a sequence of one letter).

Any series of three tosses can result in one of the following outcomes: HHH, HHT, HTH, THH, HTT, THT, TTH, TTT. The sample space for the experiment (together with the probability of each outcome) will then be as shown in Table 1.

TABLE 1

Sample space for tossing a fair coin three times and the probability of each outcome

SS	HHH	HHT	HTH	THH	HTT	THT	TTH	TTT
PROBABILITY	1/8	1/8	1/8	1/8	1/8	1/8	1/8	1/8

Let X stand for the number of runs in any outcome (i.e., any series of tosses). For example, X will be 2 if the series of tosses is THH. The value of X can vary from one series of tosses to another, so it is a variable. But the value of X for any particular series is also determined by chance, so X is called a random variable. The following is a list of the number of runs associated with each possible outcome:

HHH	1	HTH	3	HTT	2	TTH	2
HHT	2	THH	2	THT	3	TTT	1

Therefore, X can take one of the values 1, 2, or 3 and no others.

We now define a random variable.

> DEFINITION 1
>
> A **random variable** is a variable that can take any one of the values associated with the set of outcomes of an experiment.

We can now construct Table 2, showing the probability of X taking each of the possible values 1, 2, or 3 (runs). Beneath each number we have written the probability of that number of runs occurring. For example, four of the eight outcomes result in two runs, so the probability of the value 2 is $4/8(= .5)$.

TABLE 2

Table of probabilities of X taking the value 1, 2, or 3

NUMBER OF RUNS	1	2	3
PROBABILITY	2/8	4/8	2/8

TABLE 3

Probability distribution of the random variable X

x	1	2	3
$P(X = x)$	2/8	4/8	2/8

It is customary to write these results as shown in Table 3 and to refer to such a table as the probability distribution of the random variable X. The capital letter X in Table 3 stands for the random variable "the number of runs in three tosses of a fair coin." We use lowercase x to represent those particular values that uppercase X can take, in this case 1, 2, or 3. Thus the heading $P(X = x)$ should be read as "the probability of the random variable X taking one of these particular values." Since only the mutually exclusive values 1, 2, and 3 are possible, the corresponding probabilities must add to 1. We can now introduce the following definition:

> DEFINITION 2
>
> The **probability distribution** of a discrete random variable is an array containing (a) all the values the random variable may take together with (b) the corresponding probabilities of its taking each value.[1]

For the sake of clarity we shall use the simpler notation $P(x)$ for $P(X = x)$. For instance, in the last example $P(X = 3) = P(3) = 2/8$.

A probability distribution can also be represented by a probability line histogram. That for X, the number of runs, is shown in Figure 1. Notice how closely it resembles the line histogram for relative frequencies in Chapter 1. Here, however, the height of each line is the probability rather than the relative frequency of the value.

FIGURE 1

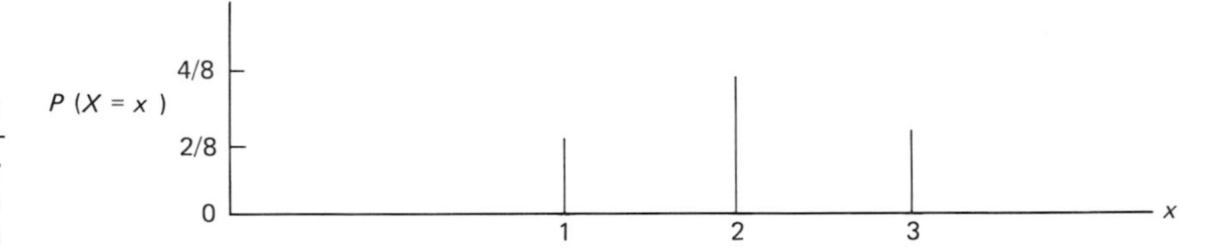

Probability histogram for the number of runs in three tosses of a fair coin

EXAMPLE 1 Census figures indicate that the distribution of the number of TV sets per household in a small town of 5000 households is that shown in Table 4. For example, 1500, or a proportion .3 of the households, have a single TV set; 250, or a proportion .05, have four sets; and so on. The values in the bottom row of the table are the proportion of all households with the specified number of TV sets. These are the relative frequencies of each number of TV sets. Assume now that an experiment consists of selecting one of the 5000 households at random

[1]We will explain the use of the term "discrete" later in the section.

and recording their number of TV sets. Call Y the random variable "the number of TV sets in the household selected." Y can then take one of the values 0 through 5. Notice that in this case each outcome of the experiment is a *number*, so that the value of the random variable coincides with the outcome. (This was not so in the last problem where for instance THH was an outcome and 2 the value of the random variable.)

TABLE 4

Distribution of the number of television sets per household

NUMBER OF TV SETS	0	1	2	3	4	5	TOTAL
Number of Households	200	1500	2000	1000	250	50	5000
Relative Frequency of Households	.04	.30	.40	.20	.05	.01	

The probability of Y taking any specific value is the relative frequency with which that value occurs. Thus the probability of selecting a household with three TV sets is .2. The probability distribution of Y is shown in Table 5. The corresponding probability histogram is shown in Figure 2.

■

TABLE 5

Probability distribution of Y, the number of TV sets in a randomly chosen household

y	0	1	2	3	4	5
$P(y)$	.04	.30	.40	.20	.05	.01

FIGURE 2

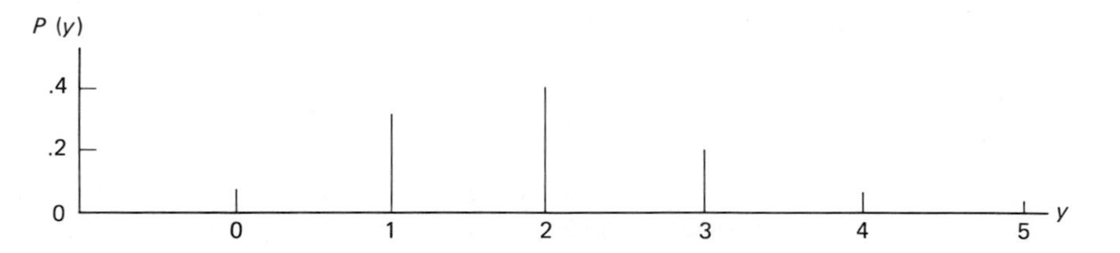

Probability histogram of Y, the number of TV sets in a randomly chosen household

EXAMPLE 2 Of the 200 prisoners in the county jail, 80 had only a single conviction, 60 had two, 40 had three, and 20 had four convictions. One of the prisoners is to be selected at random and his number of convictions (T) recorded. The probability distribution of T is shown in Table 6. For example, $P(2) = 60/200 = .3$. As in Example 1, the probability of T taking any value is the relative frequency of each value.

■

TABLE 6

Distribution of the number of convictions per prisoner in Example 2

t	1	2	3	4
$P(t)$	.4	.3	.2	.1

In Examples 1 and 2 the outcome of the experiment *is* the value of the random variable.

PROBABILITY DISTRIBUTIONS AS POPULATIONS

In Example 1 the experiment consisted of selecting a household at random and recording their number of TV sets (Y). The probability distribution of Y can be thought of as a convenient way of representing the population of "numbers of TV sets per household." A proportion, .04, of the population are zeros, .25 are ones, and so on. Similarly, the probability distribution of t in Example 2 is a convenient way of summarizing a population of 200 numbers of convictions.

It is not so clear that the distribution of the number of runs in three tosses of a fair coin also represents a population. In this case the population is not finite and real as in the other two examples but is infinite and hypothetical, consisting of all possible numbers of runs that result from repeating the experiment of tossing a fair coin three times over and over again. One-fourth of this population are ones, one-half are twos, and one-fourth are threes. Thus tossing the coin three times may be viewed as selecting a sample of 1 from this population.

Where a population is real and finite, the probabilities of any value being selected will be the same as the relative frequency or proportion of each value in the population, as seen in Examples 1 and 2. Where the population is not finite, as when tossing a coin, this will not be true.

DISCRETE AND CONTINUOUS RANDOM VARIABLES

As we did in Section 1.1 with variables in general, we may distinguish between continuous and discrete random variables. If there are gaps between the values that the random variable can take, the random variable is discrete. The most important group of discrete random variables are those that take only whole-number values. These are usually counts of things. All three examples in this section are of this kind. There are , however, discrete random variables which are not counts. The random variable—the *proportion* of heads in three tosses of a coin—which takes the values 0/3, 1/3, 2/3 or 3/3—is an example.

A continuous random variable may take *any* value in some specified interval, that is, there are no gaps in the interval at all. For instance, X, the age (in years) of a randomly chosen American, at some particular time is a continuous random variable. At least theoretically, X can take any value in the interval 0–120. Continuous random variables are discussed in Chapter 6.

As the general case, we shall refer to a discrete random variable X, without reference to any particular experiment, as one that takes one of the k values x_1, x_2 . . . , x_k with respective probabilities $P(x_1)$, $P(x_2)$, . . . , $P(x_k)$. The probability distribution of X is, therefore,

x	x_1	x_2	$\cdots$	x_k
$P(x)$	$P(x_i)$	$P(x_2)$	$\cdots$	$P(x_k)$

Since one of the k mutually exclusive values must occur, the corresponding probabilities $P(x_1)$, $P(x_2)$, . . . , $P(x_k)$ must sum to 1.

PROBLEMS 4.1

1. An experiment consists of rolling first a red and then a green die. Both are fair.
   (a) Write down the sample space of 36 equally likely pairs of numbers.
   (b) Beside each pair write down the sum of the two numbers.
   (c) Write down the probability distribution of S, the sum of the two numbers showing on two rolled dice.
2. The distribution of the number of children (18 years of age or younger) per family in a town of 400 families follows:

NUMBER OF CHILDREN	0	1	2	3	4	5
NUMBER OF FAMILIES WITH THIS NUMBER OF CHILDREN	60	80	100	80	60	20

   A family is chosen at random and the number of children (Y) in that family is recorded.
   (a) Write down the probability distribution for Y. (*Hint:* Compute the proportion of families that have each number of children.)
   (b) Sketch the probability histogram for Y.
3. The probability that a particular woman will give birth to a girl is .7. The outcome of one birth is independent of the outcome of any other. She gives birth to two children. We call the random variable G the number of girls. Write down the probability distribution of G.
4. The following numbers are the number of automobile accidents per week at a particular intersection for 30 consecutive weeks.

$$\begin{array}{ccccccccccccccc} 5 & 1 & 4 & 4 & 2 & 6 & 4 & 3 & 3 & 2 & 4 & 3 & 4 & 4 & 3 \\ 2 & 1 & 4 & 5 & 3 & 2 & 4 & 6 & 3 & 2 & 4 & 5 & 3 & 2 & 5 \end{array}$$

   The random variable X is the number of accidents in a randomly chosen week.
   (a) Find the probability distribution for X.
   (b) Sketch the probability histogram.
   (c) Compute $P(X \leq 3)$.
5. A person rolls first a red and then a green die. Both are fair. The person is paid according to the following scheme: $50 if both dice show a one, $5 if the sum of the two dice is 7, and $1 otherwise. The random variable of interest is X, the amount the person is paid. Write down the probability distribution of X. (*Hint:* You will find the sample space in Problem 1 helpful in computing the probabilities.)
6. The probability that a welfare worker makes an error in determining a client's eligibility is .1. A federal auditor selects three cases at random and records W, the number that contain an error.
   (a) Write down the probability distribution of W.
   (b) Sketch the probability histogram.
   (*Hint:* You will find it helpful to draw a tree diagram for this experiment.)
7. The number of courses being taken by each of the six students in a seminar are 4,

3, 4, 4, 3, 2. An experiment consists of selecting two of the six students at random and computing the value for $\overline{X}$, their mean number of courses.

(a) Write down the sample space of 15 equally likely pairs of numbers.

(b) Beside each pair write down the corresponding value for $\overline{X}$.

(c) Construct the probability distribution of the random variable $\overline{X}$.

8. Which of the following random variables are discrete and which are continuous?

(a) The rainfall in a particular location next year.

(b) The number of earthquakes during the next six months in Japan.

(c) The number of miners killed in mining accidents last year.

(d) The average number of courses taken by the two students referred to in Problem 7. (This is tricky.)

(e) The volume of helium left in a balloon after five hours.

(f) The time it takes to commute to work next Monday.

(g) The number of kings in a poker hand (which consists of five cards drawn without replacement from a deck of 52).

9. Explain which of the following arrays are probability distributions. Where you think an array is *not* a probability distribution, explain why.

(a)

x	1	2	3	4
$P(x)$	.3	.5	.1	.2

(b)

x	1.6	2.4	3.2
$P(x)$	.3	.4	.3

(c)

y	−2	−1	0	2
$P(y)$	.6	.1	.1	.2

(d)

z	6	12	18	24
$P(z)$	1.3	−.4	.7	.4

(e)

v	4	3	2	1
$P(v)$	.3	.4	.1	.1

(f)

y	105	110	115
$P(y)$	.3	.4	.3

Solution to Problem 1

(a) and (b) The sample space is shown below. Beside each pair of values we have recorded their sum(s). (c) The probability distribution of S is shown below the sample space. For example, $P(S = 4) = P(1, 3) + P(2, 2) + P(3, 1) = 3/36$.

	(R, G)	Sum		(R, G)	Sum		(R, G)	Sum
1.	(1, 1)	2	13.	(3, 1)	4	25.	(5, 1)	6
2.	(1, 2)	3	14.	(3, 2)	5	26.	(5, 2)	7
3.	(1, 3)	4	15.	(3, 3)	6	27.	(5, 3)	8
4.	(1, 4)	5	16.	(3, 4)	7	28.	(5, 4)	9
5.	(1, 5)	6	17.	(3, 5)	8	29.	(5, 5)	10
6.	(1, 6)	7	18.	(3, 6)	9	30.	(5, 6)	11
7.	(2, 1)	3	19.	(4, 1)	5	31.	(6, 1)	7
8.	(2, 2)	4	20.	(4, 2)	6	32.	(6, 2)	8
9.	(2, 3)	5	21.	(4, 3)	7	33.	(6, 3)	9
10.	(2, 4)	6	22.	(4, 4)	8	34.	(6, 4)	10
11.	(2, 5)	7	23.	(4, 5)	9	35.	(6, 5)	11
12.	(2, 6)	8	24.	(4, 6)	10	36.	(6, 6)	12

SS =

s	2	3	4	5	6	7	8	9	10	11	12
$P(s)$	1/36	2/36	3/36	4/36	5/36	6/36	5/36	4/36	3/36	2/36	1/36

SECTION 4.2
THE MEAN OF A RANDOM VARIABLE

In Chapter 1 we discussed the mean, $\bar{x}$, of *a set of data*. We use the notation μ_X or simply μ (the Greek lowercase letter mu) to denote the mean of *a random variable*. The principle behind each of the two means is identical.

EXAMPLE 3 Consider again the experiment of selecting a prisoner at random from the 200 referred to in Example 2. Of these, 80 had one conviction, 60 had two, 40 had three, and 20 had four convictions. The population of 200 numbers of convictions were summarized by the distribution of T.

t	1	2	3	4
$P(t)$	.4	.3	.2	.1

The mean value, μ_T of T is the mean number of convictions among all prisoners. To find μ_T we use the same procedure that we used to find $\bar{x}$ in Chapter 1.

$$\mu_T = \frac{\text{total number of convictions among all prisoners}}{\text{total number of prisoners}}$$

$$= \frac{(1 \times 80) + (2 \times 60) + (3 \times 40) + (4 \times 20)}{200}$$

which can be written as

$$\mu_T = 1 \times \frac{80}{200} + 2 \times \frac{60}{200} + 3 \times \frac{40}{200} + 4 \times \frac{20}{200}$$

$$= (1 \times .4) + (2 \times .3) + (3 \times .2) + (4 \times .1) = 2 \text{ convictions}$$

You can see from this last expression that μ_T can be computed directly as the sum of the products of the values that T can take multipled by their respective probabilities of occurring. The value $\mu_T = 2$ is marked on the probability histogram of T in Figure 3. We shall interpret this value in a moment. ■

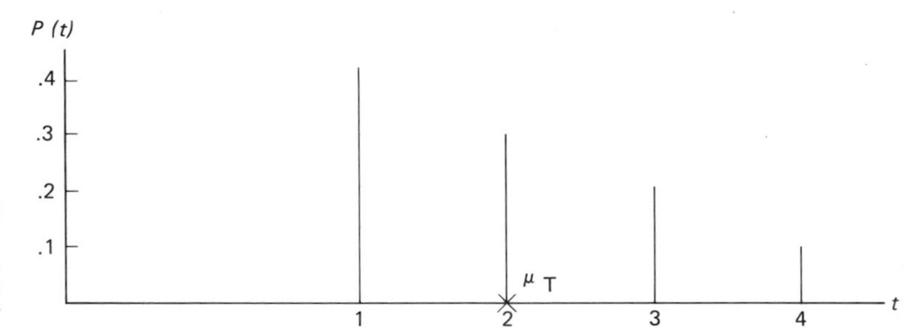

FIGURE 3

The mean $\mu_T = 2$ as a point on the probability histogram of T

THE GENERAL CASE FOR μ_x

We shall occasionally use the notation $E(X)$ in place of μ_X. This is short for *the expected value of* X and may be read as "E of X." Before interpreting $\mu_X = E(X)$, we give a general formula for computing its value.

If a random variable X has the probability distribution

x	x_1	x_2	$\cdots$	x_k
$P(x)$	$P(x_1)$	$P(x_2)$	$\cdots$	$P(x_k)$

then we compute $\mu_X = E(X) = x_1 P(x_1) + x_2 P(x_2) + \cdots + x_k P(x_k)$, or

$$\mu_X = E(X) = \sum_{i=1}^{k} x_i P(x_i) \qquad \text{(Eq. 4.1)}$$

EXAMPLE 4 From Section 4.1 the distribution of the number of runs, X, in three tosses of a fair coin is

x	1	2	3
$P(x)$	2/8	4/8	2/8

From Equation 4.1 the mean or expected value of X is

$$\mu_X = E(X) = 1\left(\frac{2}{8}\right) + 2\left(\frac{4}{8}\right) + 3\left(\frac{2}{8}\right) = \frac{16}{8} = 2$$

■

EXAMPLE 5 From Example 1 the population of the numbers of TV sets in 5000 households was summarized by the probability distribution of the random variable, Y, the number of TV sets in a randomly chosen household.

y	0	1	2	3	4	5
$P(y)$	.04	.30	.40	.20	.05	.01

The mean number of TV sets per household in this population is

$$\begin{aligned} \mu_Y = E(Y) &= 0(.04) + 1(.30) + 2(.40) + 3(.2) + 4(.05) + 5(.01) \\ &= 1.95 \text{ sets} \end{aligned}$$

The value $\mu_Y = 1.95$ is shown on the probability histogram of Y in Figure 4.

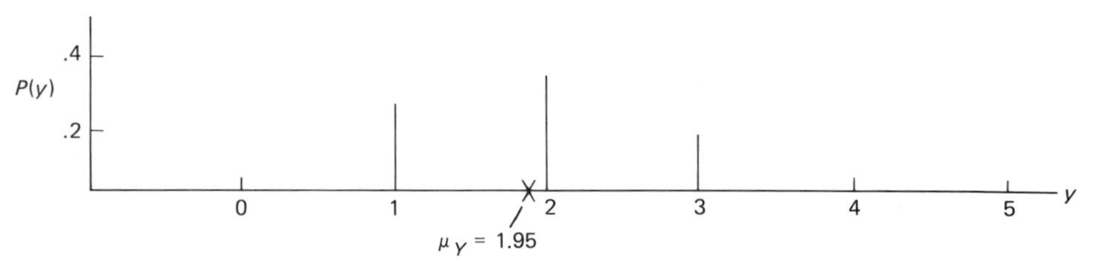

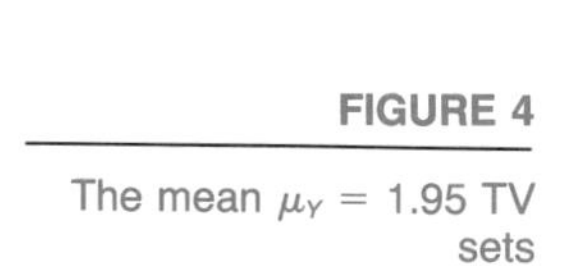

FIGURE 4

The mean $\mu_Y = 1.95$ TV sets

THE MEAN OF A RANDOM VARIABLE AS A POPULATION MEAN

In Section 4.1 we introduced the idea that the probability distribution for a random variable can be viewed as a convenient way of summarizing a numerical population, whether real or imaginary.

If the population is real, as with the number of convictions for 200 prisoners, or the numbers of TV sets in each of 5000 households, then the mean of the random variable is simply the mean value in the population of numbers and is generally referred to as μ. For instance, the mean number of convictions per prisoner is $\mu_T = 2$ and the mean number of TV sets per household is $\mu_Y = 1.95$.

If the population is imaginary, consisting of all the possible values that would occur if an experiment were performed an infinite number of times, then $E(X)$ is generally used to designate the (long-run) mean value of this infinite population. For instance, if the experiment of recording the number of runs in three tosses of a fair coin was repeated an infinite number of times, the long run mean number of runs would be $E(X) = 2.0$.

The mean of a real population can also be interpreted in a "long-run" sense. For instance, if the experiment of recording the number of TV sets in a randomly selected household was repeated (with the replacement of households) an infinite number of times then $E(X)$ will be the mean of this infinite number of numbers. Since, in the long run the proportions of households with 0, 1, 2, . . . , 5 TV sets will be the same as those in the 5000 households, $E(X) = \mu_X = 1.95$ sets.

In this and succeeding chapters we shall use the symbol μ, or less frequently $E(X)$, to denote the mean of a population, reserving the symbol $\bar{x}$ for use as the sample mean. To illustrate the difference, recall from Example 5 that the mean number of TV sets per household in a town is $\mu = 1.95$. Suppose that a sample of six households contained 2, 1, 1, 1, 3, and 2 sets, respectively; then the sample mean would be $\bar{x} = (2 + 1 + 1 + 1 + 3 + 2)/6 = 1.67$ sets.

The following example provides an interesting illustration of the notion of an expected value.

EXAMPLE 6 Extensive death records held by an insurance company indicate that a proportion .996 of 38-year-old men will live for at least one more year. Thirty-eight-year-old men wishing to have their lives insured by the company for \$20,000 will be charged a premium of \$100 for the first year. (a) What is the expected gain to the company on such a transaction? (b) What premium should the company charge such a man in order to have an expected gain of \$30?

Solution (a) The insurance company will regard an insured 38-year-old man's next year as an experiment. Associated with this experiment is a random variable, G, the gain to the company, which takes the value \$100 (with probability .996) if the man lives through the year and $-$\$19,900 (with probability $1 - .996 = .004$) if the man dies. The figure $-$\$19,900 = \$100 premium $-$ \$20,000 payment is a negative gain (i.e. a loss). The probability distribution of G is thus

g	100	−19,900
$P(g)$	.996	.004

and the company's expected gain is

$$E(G) = 100(.996) + (-19{,}900)(.004) = \$20$$

Of course the insurance company will insure many thousands of such men, each of whom may be viewed as a replication of this experiment. Most will live through the year; some will die. The value $E(G) = \$20$ may be interpreted as the mean gain for insuring all such men.

(b) What premium will provide an expected gain of \$30 to the company? Call the premium D. The probability distribution for G may now be written

g	D	$D - 20{,}000$
$P(g)$	.996	.004

Since $E(G) = 30$, we need to solve the equation $30 = .996(D) + .004(D - 20{,}000)$ for D. Gathering terms in D on one side of the equation, we obtain

$$.996D + .004D = 30 + 80 \quad \text{or} \quad D = \$110$$

■

ZERO/ONE POPULATIONS

At the end of Section 1.2 we illustrated how qualitative data can be artificially but usefully quantified. We can also do this with random variables. Consider, for instance, the following example.

EXAMPLE 7

An experiment consists of selecting at random one of the 60 employees in a small company. Only 24 of the 60, or a proportion $24/60 = .4$, are smokers. We can assign the value $X = 1$ to all 24 smokers and $X = 0$ to all 36 nonsmokers, thus creating a random variable X having the probability distribution

x	0	1
$P(x)$	.6	.4

The mean of X is the mean of these zeros and ones and is $\mu_X = 0(.6) + 1(.4) = .4$, the proportion of employees who are smokers.

■

Experiments such as the one in Example 7, where we are interested in only two complementary outcomes, are called **Bernoulli trials** after the Swiss mathematician James Bernoulli. For convenience we shall sometimes refer to these two outcomes as successes or failures. These loaded words will stand for such outcomes as smoker and nonsmoker, male and female, defective and nondefective item, and so on. If we designate the proportion of successes in the population as π (the Greek lowercase letter pi), the proportion of failures will be $1 - \pi$. If we also assign the value $X = 1$ to all successes, and $X = 0$ to all failures, in the population, we will create a 0/1 population represented by the probability distribution.

x	0	1
$P(x)$	$1 - \pi$	π

In this case X is the random variable "the value (0 or 1) of a randomly selected member of the population." The mean of this population is

$$\mu_X = 0(1 - \pi) + 1(\pi) = \pi$$

The value π is also the proportion of the population that are successes (as designated above) or, equivalently, the probability of selecting a success.

PROBLEMS 4.2

10. The distribution of the number of job offers received by the 80 graduates of a computer school is

x	0	1	2	3	4	5	6
$P(x)$	15/80	25/80	20/80	12/80	6/80	1/80	1/80

(a) Compute $\mu_X = E(X)$.
(b) Interpret your answer in two ways.

11. Find the mean of the random variables having the following probability distributions. In each case sketch the probability histogram marking the position of the mean.

(a)

y	−1	0	1
$P(y)$	.1	.3	.6

(b)

z	−3	−2	−1	1
$P(z)$	.25	.25	.25	.25

(c)

x	3	4
$P(x)$	.7	.3

(d)

x	4
$P(x)$	1

12. For a particular batter in baseball, the probability of getting 0, 1, 2, 3, or 4 bases each time at bat is as follows:

x	0	1	2	3	4
$P(x)$	.65	.25	.06	.01	.03

(a) What is the probability that the batter gets on base?
(b) What is the expected or mean number of bases per at bat? Interpret this value.

13. A random variable X is such that $P(4) = .2$, $P(5) = .6$, and $P(6) = .2$. Compute (a) $P(X \leq 5)$; (b) $E(X)$.

14. The distribution of the number of visits to a dental health center per person in a year is as follows:

x	1	2	3	4	5	6	8	10
$P(x)$	.38	.32	.18	.07	.03	.01	.005	.005

(a) How would you interpret the random variable X?
(b) What proportion of visitors made at least four visits?
(c) Compute $\mu_X = E(X)$
(d) Interpret your answer in part (c) in two ways.

15. A gambling house makes the following game available to its customers. A player

rolls two dice. If the sum of the dice is 11 or 12, the house pays the player \$10. Otherwise, the player pays the house \$1.

(a) What is the house's expected profit from this game? Interpret this number.

(b) What is the player's expected gain?

(c) What would the house have to pay out (instead of \$10) in order to make this a completely fair game, that is, a game in which its expected profit was zero?

16. For a \$5 monthly fee the gas company guarantees customers complete service. The company estimates the probability that a customer will require one call in a month as .05 and the probability of two calls as .01. (Ignore the possibility of more than two calls.) Each call costs the gas company \$40.

(a) What is the gas company's expected monthly gain from such a contract?

(b) Interpret your answer.

(c) Suppose that the cost of each call increased from \$40 to \$50. What monthly fee would the company have to charge in order to maintain the same expected gain as before?

17. The following is a simple card game that might be offered by a gambling house. Players draw two different cards from a deck without replacement. If both cards have the same value (i.e., both are 4's or both are queens, for example), the player wins \$10. If both cards are of the same suit (i.e., both are diamonds, for example), the player wins \$2. For any other combination the player pays the house \$2. What is the average gain to the house for this game?

Solution to Problem 10

(a) Using Equation 4.1, we have

$$E(X) = 0\left(\frac{15}{80}\right) + 1\left(\frac{25}{80}\right) + 2\left(\frac{20}{80}\right) + 3\left(\frac{12}{80}\right) + 4\left(\frac{6}{80}\right) + 5\left(\frac{1}{80}\right) + 6\left(\frac{1}{80}\right)$$

$$= \frac{136}{80} = 1.7 \text{ offers}$$

(b) (i) The mean number of job offers per graduate is 1.7.

(ii) If we were to continue to sample graduates, record their number(s) of offers and replace them, the long-run mean number of offers would be 1.7.

SECTION 4.3

THE VARIANCE OF A RANDOM VARIABLE

In Chapter 1 we noted that a measure of location alone, such as a mean, is insufficient to adequately summarize a set of data. We also need a measure of spread or variability. In the case of a random variable the mean, μ_X, provides a measure of location of its probability distribution, but now we need a measure of the extent to which the distribution is spread out around μ_X.

We have defined the variance of a set of observations $x_1, x_2, \ldots, x_n$ as

$$s^2 = \frac{1}{n-1}[(x_1 - \bar{x})^2 + (x_2 - \bar{x})^2 + \cdots + (x_n - \bar{x})^2]$$

We shall use a slight modification of this result to compute the variance of a random variable.

In Example 2 we referred to the 200 prisoners in the county jail. Eighty had a single conviction, 60 had two, 40 had three, and 20 had four convictions. The population of 200 numbers of convictions was summarized by the following distribution of T:

t	1	2	3	4
$P(t)$	.4	.3	.2	.1

The mean value of T (i.e., the mean of the population) is $\mu_T = 2$ convictions. To obtain the variance of T, we modify the formula for s^2 above by replacing $\bar{x}$ by μ and $n - 1$ by n in the denominator. Noting again that the value 1 occurs 80 times, 2 occurs 60 times, and so on, the variability of the number of convictions is

$$\frac{1}{200}[80(1-2)^2 + 60(2-2)^2 + 40(3-2)^2 + 20(4-2)^2]$$

$$= \frac{80}{200}(1-2)^2 + \frac{60}{200}(2-2)^2 + \frac{40}{200}(3-2)^2 + \frac{20}{200}(4-2)^2$$

$$= .4(1) + .3(0) + .2(1) + .1(4) = 1.0$$

When we refer to the variance of a random variable we use the notation $\text{Var}(T)$ or σ_T^2. The symbol σ is the Greek lowercase sigma (Σ is the uppercase sigma) and σ^2 should be read as "sigma squared." In this example, then, $\text{Var}(T) = \sigma_T^2 = 1.0$.

In the computations above we computed σ_T^2 as the sum of all the products of possible values for $(t - \mu)^2$ multiplied by the probability of getting each such value for t. In general, and with the usual notation,

$$\text{Var}(X) = \sigma_X^2 = (x_1 - \mu)^2 P(x_1) + (x_2 - \mu)^2 P(x_2) + \cdots + (x_k - \mu)^2 P(x_k)$$

or, in summation notation

$$\text{Var}(X) = \sigma_X^2 = \sum_{i=1}^{k} (x_i - \mu)^2 P(x_i) \qquad \text{(Eq. 4.2)}$$

The standard deviation of a random variable X is then

$$\text{SD}(X) = \sigma_X = \sqrt{\text{Var}(X)} = \sqrt{\sigma_X^2} \qquad \text{(Eq. 4.3)}$$

Equation 4.2 for Var(X) is not the simplest method for computing Var(X). A more convenient formula is

$$\sigma_X^2 = \text{Var}(X) = [x_1^2 P(x_1) + x_2^2 P(x_2) + \cdots + x_k^2 P(x_k)] - \mu^2$$

or

$$\sigma_X^2 = \text{Var}(X) = \sum x_i^2 P(x_i) - \mu^2 \qquad \text{(Eq. 4.4)}$$

EXAMPLE 8 Apply Equation 4.4 to find the variance and standard deviation of the number of convictions, T.

Solution From the distribution of T above, we have

$$\begin{aligned} \text{Var}(X) &= 1^2(.4) + 2^2(.3) + 3^2(.2) + 4^2(.1) - 2^2 \\ &= .4 + 1.2 + 1.8 + 1.6 - 4 \\ &= 5 - 4 \\ &= 1 \qquad \text{as before} \\ \text{SD}(X) &= \sqrt{1} = 1 \text{ conviction} \end{aligned}$$

■

EXAMPLE 9 The distribution of the number of misprints, M, per page in a textbook is

m	0	1	2	3
$P(m)$	.35	.45	.15	.05

Find the mean and the standard deviation of the number of misprints.

Solution Using Equation 4.1,

$$\begin{aligned} \mu_M &= 0(.35) + 1(.45) + 2(.15) + 3(.05) \\ &= .9 \text{ misprint} \end{aligned}$$

From Equation 4.4,

$$\begin{aligned} \sigma_M^2 &= 0^2(.35) + 1^2(.45) + 2^2(.15) + 3^2(.05) - .9^2 \\ &= 1.50 - .81 = .69 \end{aligned}$$

and thus

$$\sigma_M = \sqrt{.69} = .83 \text{ misprint.}$$

■

INTERPRETING σ

Just as we interpreted the mean of a random variable as a population mean, so we can interpret σ, the standard deviation of a random variable, as the standard deviation of a population, that is, as a measure of the extent to which the numbers in the population are spread out around their mean. The population may be real, as with the 200 numbers of convictions, or imaginary, as with all the possible results of repeating the experiment of tossing a fair coin three times an indefinite number of times. In Example 8, the number of convictions per prisoner differ from the mean number in the population (2.0) by an average of approximately 1.0. We will reserve the symbol s for the standard deviation of a sample.

THE VARIANCE OF A ZERO/ONE POPULATION

As we indicated earlier a population in which a proportion π are ones and a proportion $1 - \pi$ are zeros can be represented by the distribution

x	0	1
$P(x)$	$1 - \pi$	π

where X is the random variable, the value (0 or 1) of a randomly selected member. We indicated that the mean of this population is $\mu_X = 0(1 - \pi) + 1(\pi) = \pi$. Using Equation 4.4, the variance in such a population is

$$\sigma_X^2 = 0^2(1 - \pi) + 1^2(\pi) - \pi^2$$
$$= \pi - \pi^2 = \pi(1 - \pi)$$

and the standard deviation is

$$\sigma_X = \sqrt{\pi(1 - \pi)}$$

EXAMPLE 10

At the Panther Automobile plant the probability of a defective car is .15. (a) Write the probability distribution of the random variable associated with the experiment of selecting a car for inspection. Compute the (b) mean and (c) standard deviation of this random variable.

Solution

(a) If we assign a value $X = 1$ whenever a defective car is found, and $X = 0$ otherwise, the distribution of X is

x	0	1
$P(x)$	.85	.15

(b) The mean of X is $\mu_X = \pi = .15$.

(c) From above, the standard deviation of X is

$$\sigma_X = \sqrt{(.15)(1 - .15)} = \sqrt{.1275} = .357$$

■

For convenience, we now summarize the principal results for 0/1 populations. A 0/1 population can be represented by the probability distribution

x	0	1
$P(x)$	$1 - \pi$	π

where X is the random variable, the value (0 or 1) of a randomly selected member of the population, and π is the proportion of successes in the population. The mean, variance, and standard deviation of the zeros and ones in the population are

$$\mu_X = \pi \qquad \text{(Eq. 4.5)}$$

$$\sigma_X^2 = \pi(1 - \pi) \qquad \text{(Eq. 4.6)}$$

$$\sigma_X = \sqrt{\pi(1 - \pi)} \qquad \text{(Eq. 4.7)}$$

PROBLEMS 4.3

18. In Example 5 we gave the distribution of the number of TV sets as

y	0	1	2	3	4	5
$P(y)$	.04	.30	.40	.20	.05	.01

and computed the mean $\mu_X = 1.95$ sets. Compute the standard deviation of the number of sets. Interpret your answer.

19. A crossword puzzle enthusiast attempts the puzzle in the daily newspaper each day. The distribution of the number of words left unsolved each day is

x	0	2	3	4	5	6
$P(x)$	.05	.3	.3	.2	.1	.05

Compute (a) the mean number of unsolved words; (b) the standard deviation of the number of unsolved words; (c) the probability that the enthusiast will fail to complete a puzzle.

20. Three women and five men apply for two (identical) positions with a company. The selection subcommittee feels that all eight candidates are equally qualified and intends to select two at random to fill the positions.

    (a) Write down all the 28 equally likely pairs of applicants in the sample space.

    (b) Beside each pair write down the value for W, the number of women selected.

    (c) Construct the probability distribution of W.

    (d) Compute the (i) expected number and (ii) standard deviation of the number of women.

21. The distribution of the number of taxicabs available at a stand is

t	0	1	2	3	4
$P(t)$	1/20	1/20	4/20	5/20	9/20

    (a) Compute (i) the mean number of taxicabs available; (ii) the standard deviation of the number of taxicabs.

    (b) What percentage of the time is there (i) one cab; (ii) at most two cabs available?

22. An industrial process produces lengths of piping which are supposed to be 10 centimeters long, but the lengths actually vary according to the following distribution. (We are in effect making discrete a variable-length-which is essentially continuous.)

(l)	9.97	9.98	9.99	10.00	10.01	10.02	10.03
$P(l)$	.05	.1	.15	.4	.15	.1	.05

    (a) Compute the mean length of the piping.

    (b) Compute the standard deviation of length.

    (c) Pipe sections are rejected if they differ from 10 centimeters by *more* than .02 centimeter. What proportion of lengths will be rejected?

23. Sixty percent of a population of voters are Democrats.

    (a) Write down the probability distribution of the random variable associated with the experiment of selecting a voter at random from this population.

    (b) Compute the (i) mean and (ii) standard deviation of this population.

24. A medical procedure has a probability .8 of being successful.

    (a) Write down the probability distribution of the random variable associated with the experiment of performing this procedure.

    (b) Compute the (i) mean and (ii) standard deviation of this random variable.

Solution to Problem 18

Using Equation 4.4,

$$\sigma_Y^2 = 0^2(.04) + 1^2(.30) + 2^2(.40) + 3^2(.20) + 4^2(.05) + 5^2(.01) - (1.95)^2$$
$$= 4.75 - 3.8025 = .9475$$

and

$$\sigma_Y = \sqrt{.9475}$$
$$= .973 \text{ T.V. sets}$$

Approximately, the number of TV sets per household differs from the mean number (1.95) by an average of .973 (almost 1) set.

SECTION 4.4
SUMMARY AND REVIEW

A **random variable** is a variable that can take any one of the values associated with the set of outcomes of an experiment (Definition 1).

A **discrete random variable** has gaps between the values that the variable can take. A **continuous random variable** may take *any* value within a specified interval; that is, there are no gaps between the possible values. This chapter dealt only with discrete random variables.

The **probability distribution** of a discrete random variable is an array containing (a) all the values the random variable may take together with (b) the corresponding probabilities of its taking each value. (Definition 2). Such a distribution is a convenient way to represent a population of discrete values. Where the population is real and finite, the probabilities are the same as the relative frequencies of each value.

The probability histogram for a random variable X is similar to a relative frequency histogram for discrete data except that the height of each line is the probability rather than the relative frequency of the value of the random variable.

The *mean* of a random variable is denoted by either μ_X or $E(X)$ (the "expected value of X") and is computed as

$$E(X) = \mu_X = x_1 P(x_1) + x_2 P(x_2) + \cdots + x_k P(x_k)$$
$$= \sum_{i=1}^{k} x_i P(x_i) \qquad \text{(Eq. 4.1)}$$

If the probability distribution of X represents a real population of numbers, its mean value (μ) is simply the mean value in the population. Sometimes the probability distribution of X represents an imaginary population, consisting of all the possible values obtained by performing the experiment an infinite number of times. In this case $E(X)$ is the (long-run) mean of this population.

The variance of a random variable X is denoted Var (X) or σ_X^2 and may be computed as

$$\text{Var }(X) = \sigma_X^2 = \sum_{i=1}^{k} x_i^2 P(x_i) - \mu^2 \qquad \text{(Eq. 4.4)}$$

The standard deviation of X is

$$SD(X) = \sigma_X = \sqrt{\text{Var }(X)} \qquad \text{(Eq. 4.3)}$$

Experiments in which we are interested in only two complementary outcomes are called Bernoulli trials. We usually refer to the outcomes as successes and failures.

If the proportion of successes in a population is designated π, the proportion of failures is $1 - \pi$. Assigning the value $X = 1$ to all successes and $X = 0$ to all failures in the population, we create a 0/1 population represented by the probability distribution

x	0	1
$P(x)$	$1 - \pi$	π

The mean, variance, and standard deviation of such a 0/1 population are

$$E(X) = \mu_X = \pi \quad \text{(Eq. 4.5)}$$

$$\text{Var}(X) = \sigma_X^2 = \pi(1 - \pi) \quad \text{(Eq. 4.6)}$$

$$\text{SD}(X) = \sigma_X = \sqrt{\pi(1 - \pi)} \quad \text{(Eq. 4.7)}$$

REVIEW PROBLEMS

GENERAL

1. Only one of the following four arrays is a probability distribution. Which one is it? Explain why each of the other three arrays is not a probability distribution.

(a)

t	−1	0	2
$P(t)$	.4	.25	.25

(b)

x	3	4	9
$P(x)$	.3	−.2	.9

(c)

z	1.5	2.0	2.5
$P(z)$	9	.05	.05

(d)

y	−7	7
$P(y)$	.7	.7

2. A random variable V has the following probability distribution:

v	1	2	3	4	5
$P(v)$	c	$2c$	$3c$	$4c$	$5c$

Find (a) the constant c; (b) $E(V) = \mu_V$; (c) $P(V \geq \mu_V)$.

3. A random variable W has the following probability distribution:

w	1	2	3	4	5
$P(w)$	c	.4	.1	.2	.05

Find (a) the constant, c; (b) the most likely value for W; (c) the least likely value for W; (d) $E(W)$; (e) $P(W \geq 3)$; (f) $P(W < 4)$.

4. The random variable Z has the following probability distribution:

z	2	3	4	8
$P(z)$	.3	.2	.3	.2

(a) Sketch the probability histogram for Z.

(b) Find (i) $P(Z > 3)$; (ii) $E(Z)$; (iii) $P(Z < 3.5)$.

5. A soccer team gets two points for each game it wins, one point for each tie, and no points for each loss. The probabilities of winning, tying, or losing depend on whether it is playing at home or away and are as follows:

	WIN	TIE	LOSE
HOME	.5	.2	.3
AWAY	.4	.3	.3

(a) What is the expected number of points (i) per home game; (ii) per game away?

(b) If the team plays 25 home games and 15 games away in a season, how many total points would you expect it to have at the end of the season?

6. A game consists of a blindfolded person sticking a pin into a square board divided up and numbered as shown. The person's score, X, for the game is the number showing in the section of the board where the pin lands. Assume that the pin is equally likely to land at any point on the board.

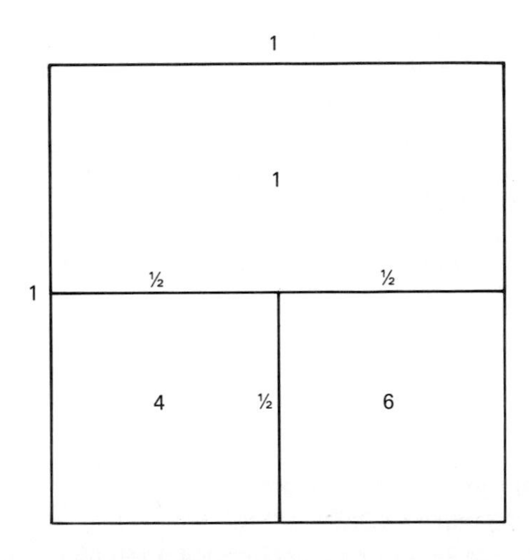

(a) Construct the probability distribution for X. (*Hint:* Consider the relative areas of the three divisions.)

(b) Compute $E(X)$.

7. A population consists of the five numbers 1, 2, 3, 4, and 5.

(a) Compute the mean of these numbers. Each of these five numbers is written on a slip of paper and put in a hat. An experiment consists of randomly selecting two of the numbers without replacement.

(b) Write down the sample space for this experiment. (It will consist of 10 pairs of numbers.)

(c) What is the probability of getting each pair?

(d) Beside each pair in the sample space write down the mean, M, of the two numbers.

(e) Use your answers in parts (c) and (d) to construct the probability distribution- for M.

(f) Compute $E(M)$. [It should correspond to your answer in part (a). Can you see why?]

8. X is the random variable "the number of misprints on a randomly selected page of a book." The distribution of X is

x	0	1	2	3	4	5
$P(x)$	.500	.350	.120	.020	.008	.002

(a) Compute and interpret $E(X) = \mu_X$.

(b) Compute SD(X).

(c) If the book contains 2000 pages, what is the total number of misprints in the book?

9. An experiment consists of rolling two fair dice.

(a) Write down the sample space of 36 pairs of numbers for this experiment.

(b) Beside each pair write down the value for X, the larger of the two numbers. [In the case of the pair (4, 4), 4 is the larger of the two.]

(c) Construct the probability distribution of X.

(d) Compute and interpret $E(X)$.

(e) Compute and interpret SD(X).

(f) A gambling house offers the following game. A player rolls two fair dice. If the larger of the two numbers is 3 or less, the player is paid $9. Otherwise, the player pays the house $4. What is the expected gain of the house with this game?

10. A study of car-riding patterns suggests that the distribution of the number of riders (R) per car entering a city on the expressway on a workday morning is

r	1	2	3	4	5	6
$P(r)$	.42	.25	.18	.10	.04	.01

(a) Compute and interpret $E(R)$.

(b) Compute SD(R).

(c) The fees (for automobiles) at a toll booth on the expressway run as follows:

Cars carrying one person pay a $1.00 toll.
Cars carrying two persons pay a $.50 toll.
Cars carrying three or more persons pay no toll.

If T is the random variable "the toll for a randomly selected car," (i) construct the probability distribution for T; (ii) compute and interpret $E(T)$.

HEALTH SCIENCES

11. The distribution of the number of days spent in the hospital by patients having a specific procedure is

d	3	4	5	6	7
$P(d)$	.15	.2	.35	.2	.1

(a) Compute and interpret $E(D) = \mu_D$.
(b) Compute and interpret SD(D).
(c) Find $P(D > \mu_D)$.
(d) Suppose that patients undergoing this procedure are charged $250 a day for the first three days in the hospital and then $200 a day thereafter. What is the mean cost of hospitalization for such patients?

12. The number of arrivals per hour at an emergency room has been observed to have the following probability distribution:

a	0	1	2	3	4	5	6	7	8
$P(a)$	.082	.205	.257	.214	.133	.067	.028	.010	.004

(a) Compute the mean number of arrivals per hour.
(b) Compute the standard deviation of the number of arrivals per hour.
(c) If, in any one hour more than five persons arrive, additional staff members have to be diverted to the emergency room. In what proportion of hours does this happen?

13. The probability that a woman will get a migraine headache in a day is .3. If she gets such a headache the probability that it will require three aspirins is .6, that it will require two, .3, and that it will require only one, .1. Find (a) the probability distribution of the number of aspirins taken per day; (b) the mean number of aspirins per day. (Assume that the woman never gets more than one headache a day and that she takes aspirins only for migraine headaches.)

SOCIOLOGY AND GOVERNMENT

14. The experience of federal auditors suggests that the number of errors (i.e., grant errors and eligibility errors) per welfare case has the following distribution:

e	0	1	2	3	4
$P(e)$	.5	.2	.15	.1	.05

(a) What is (i) the most likely number of errors per case; (ii) the least likely number of errors per case?
(b) What is the probability that a randomly chosen case will have (i) at least one error; (ii) at least two errors?

(c) What is (i) the expected number of errors; (ii) the standard deviation of the number of errors?

15. Based on extensive research a sociologist suggests that the number of close friends claimed by women will have the following distribution:

f	0	1	2	3	4	5
$P(f)$	.20	.25	.25	.15	.10	.05

and that the distribution for men will be

m	0	1	2	3	4	5
$P(m)$	.35	.30	.20	.10	.03	.02

(a) Compute $E(F)$ and $E(M)$.

(b) What is the probability that a randomly chosen woman will claim more than two close friends?

(c) If a man and woman meet, what is the probability that both have no close friends? (Assume that the numbers of "close friends" are independent.)

16. Fifty-five percent of students at State U have smoked marijuana.

(a) Write down the probability distribution of the (0/1) random variable associated with the experiment of selecting a student at random.

(b) Compute and interpret μ in this case.

(c) Compute σ in this case.

17. Each week a salesman has to visit a downtown store for two hours. For $.50 he can park at a meter for one hour and only one hour. Experience suggests that there is a probability of .4 that he will get a $10 ticket during the second hour.

(a) What is the salesman's expected cost of parking at the meter for two hours?

(b) A garage nearby charges $2.50 an hour. Would you advise the salesman to use the garage rather than the meter?

(c) Would your answer in part (b) change if the garage charged $2.00 an hour?

18. Assume a state allows the police to stop a random sample of two cars in a 15-minute period and to ask the driver to undergo a blood alcohol test. Suppose, for simplicity, that 30% of cars are driven by young people and 20% of such drivers have blood alcohol levels (BAL) above the legal limit.

(a) What is the probability that a randomly chosen driver will be young and have a BAL above the legal limit?

(b) Find the probability distribution of X, the number of young drivers per 15-minute period having a BAL above the legal limit.

(c) Find and interpret $E(X)$.

ECONOMICS AND MANAGEMENT

19. There are empirical grounds for assuming that the distribution of the number of complaints per day received by a certain manufacturer is

c	0	1	2	3	4
$P(c)$	.36	.42	.12	.07	.03

Compute and interpret both (a) $E(C)$ and (b) $SD(C)$.

20. A 45-year-old person wishes to obtain life insurance for five years. An insurance company agrees to insure the person's life for $50,000 for a five-year period in return for a one-time premium of $1500.

    (a) Use the probabilities in Review Problem 21 in Chapter 3 to determine the probability distribution of the random variable G, the company's gain on this transaction.

    (b) Compute $E(G)$.

    (c) What premium should the company charge in this case in order to have an expected gain of $80?

21. A company is planning to produce a new product that will cost the company $8 million in the first year. Their sales department pıojects that revenue from this product in the first year cannot be estimated exactly but will depend on various factors outside the control of the company. These factors include the reaction of the public and their competitors to the new product. Some reactıons are more lıkely than others, however, and the sales department is sure that revenues will take one of five values (in millions of dollars) with the probabilities

REVENUE, r	4	8	12	16	20
$P(r)$	.3	.2	.3	.1	.1

What is the company's expected first-year profits? (Use profit = revenue − costs.)

22. The distribution of Y, the number of persons in line at a supermarket checkout in midafternoon, is

y	0	1	2	3	4
$P(y)$	.35	.40	.10	.10	.05

    (a) Compute and interpret $E(Y)$.

    (b) Compute $SD(Y)$.

    (c) What is $P(Y \leq 2)$?

23. A manufacturer of automobiles produces four models A, B, C, and D. Their respective mileages per gallon of gasoline are 20, 25, 30, and 35. The profits made on cars of each model are $4000, $3000, $2000, and $1000, respectively. The manufacturer is considering the following four different production plans:

PLAN	A	B	C	D
I	.25	.25	.25	.25
II	.2	.2	.3	.3
III	.2	.2	.2	.4
IV	.2	.1	.35	.35

For instance, under plan III 20% of the cars produced would be model A, 20% model B, 20% model C, and 40%, model D. Federal legislation requires that the mean mileage per gallon over all models for this manufacturer must equal or exceed 29 mpg. Which of the four plans would satisfy this law and at the same time produce the greatest mean profit?

EDUCATION AND PSYCHOLOGY

24. People requiring fellowship standing in a professional society must pass two examinations. Approximately 80% of those taking the two exams pass exam A, and 60% exam B. Whether or not a person passes one exam is independent of whether or not they pass the other.

(a) Find the probability that a randomly selected student (i) passes both examinations; (ii) passes neither exam; (iii) passes only one of the exams.

(b) Write down the probability distribution for the number of examinations that a randomly selected student passes.

(c) What is the mean number of examinations passed per student?

25. The probability distribution of the total number of different "behaviors" exhibited by mice in the first minute after being exposed to a particular stimulus is

b	1	2	3	4	5	6
$P(b)$	.05	.2	.3	.25	.15	.05

Find (a) the mean and (b) the standard deviation of the number of behaviors.

SCIENCE AND TECHNOLOGY

26. The number of particles of radiation (X) hitting a shield in 1/100 of a second has approximately the following distribution:

x	0	1	2	3	4	5
$P(x)$	.20	.32	.26	.14	.06	.02

(a) Compute the mean number of particles hitting the shield per 1/100 second.

(b) Compute SD(X).

(c) Compute $P(X > \mu)$.

27. An old four-engine plane makes a weekly trip between two cities. The probability that each *specific* engine will fail is .05. The probability that each *specific pair* of engines will fail is .015. The probability that any *specific three* engines will fail is .0009. The probability that all four engines will fail is so small that it can be ignored. Call the random variable X "the number of engines that fail."

(a) Write down the probability distribution for X.

(b) Find $E(X)$.

5 THE BINOMIAL DISTRIBUTION

AIRLINE OVERBOOKING

The limited experience of a new commuter airline, Windblown, suggests that 5% of those who make reservations fail to show up for the flight. As a result, Windblown intends to follow the practice of older and larger airlines and overbook—that is, sell more tickets for their 38-seat planes than they have seats. But how many more should they sell? If they sell 40 tickets for instance, what is the probability that all those who show up for the flight can be accommodated? (See Problem 20 on page 203.)

SECTION 5.1

BINOMIAL EXPERIMENTS

In Chapter 4 we introduced Bernoulli trials. A Bernoulli trial is an experiment in which we are concerned with only two complementary outcomes, designated a success or a failure. Examples include tossing a coin, observing whether a machine is functioning, a randomly selected juror is black, or a student answers a question correctly.

We use the symbol π to indicate the probability of a success. If a success is rolling a 4 with a fair die, then $\pi = 1/6$. If 20% of voters are foreign-born and a success is selecting such a voter, then $\pi = .2$.

Experiments frequently involve not one but a series of Bernoulli trials. For instance, we might toss a coin 10 times, note whether each of six flashcubes work, or select a sample of 100 voters and check whether or not each of them approves of the president's foreign policy.

If such experiments share the characteristics listed in Table 1, they are called **binomial experiments.**

TABLE 1

Characteristics of binomial experiments

1. A fixed number, n, of Bernoulli trials.
2. The probability of success is the same (π) for each trial.
3. The outcome of each trial is independent of the outcome of any other trial.
4. We are interested in the random variable, Y, the number of trials that are successes.

Following are two examples of binomial experiments.

EXAMPLE 1 The probability is .95 that a telephone call coming into a particular exchange will be correctly connected. Checking whether or not each of 400 randomly chosen calls are completed may be regarded as a binomial experiment with $n = 400$ and $\pi = P(\text{completed call}) = .95$. Y is the number of completed calls.

■

EXAMPLE 2 If two fair dice are rolled, the probability that the numbers on the two dice will add to 6 is 5/36 (can you see why?). Rolling these dice 20 times, and recording whether or not the sum for each throw is 6 is a binomial experiment with $n = 20$ and $\pi = 5/36$. Y is the number of occasions on which the sum is 6.

■

SAMPLING PROCEDURES AS BINOMIAL EXPERIMENTS

In some cases selecting a sample can be considered a binomial experiment. Following is an example of such an experiment.

EXAMPLE 3 A street contains 20 voters of whom 12 (or 60%) are Democrats. An experiment consists of selecting 10 voters at random and with replacement; that is, when a voter's name is selected and recorded, it is replaced in the pool before the next voter is selected. With this scheme the probability of drawing a Democrat is the same (.6) for each selection and the selection of the entire sample is therefore a binomial experiment with $n = 10$ and $\pi = P(\text{Democrat}) = .6$.

Now assume that the selection is made without replacement (i.e., as the voters are selected, their names are *not* replaced in the pool). The probability of selecting a Democrat on the first pick will still be .6. But suppose that seven of the first nine voters selected are Democrats. Since only five of the remaining 11 voters are Democrats, the probability of choosing a Democrat on the tenth and final pick is $5/11 = .455$. Since the probability of a success varies from selection to selection, this is *not* a binomial experiment.

Going one step further, assume, more realistically, that a sample of 1000 voters is to be selected from a population of 2,000,000 voters of whom 1,200,000 (or 60%) are Democrats. If selection is made with replacement, the sampling procedure is a binomial experiment with $n = 1000$ and $\pi = .6$. But what if sampling were without replacement! The probability that the first voter selected is a Democrat is .6. But this time the number of voters is so large that removing this one voter, in fact even removing as many as 1000 voters, will not significantly change the probability of selecting a Democrat. For instance, suppose that *all* of the first 999 voters selected were Democrats. The probability of the 1000th voter selected being a Democrat is

$$\frac{1{,}200{,}000 - 999}{2{,}000{,}000 - 999} = .5998$$

which is very close to .6. This selection procedure is thus approximately a binomial experiment with $n = 1000$ and $\pi = .6$.

■

We can summarize the findings in Example 3 as follows:

Assume a population of size N of whom a number A are successes. A sample of size n is to be selected from the population.

1. If sampling is *with* replacement, selecting the sample is exactly a binomial experiment with $\pi = A/N$.
2. If sampling is *without* replacement but the sample size n is small compared to the population size N, as in the last part of Example 3, selecting the sample may be regarded as a binomial experiment with $\pi = A/N$. As a guide, this will be true if n is less than one-twentieth of N (i.e., if $n/N \leq .05$).
3. In all other cases, selecting the sample without replacement *cannot* be considered a binomial experiment.

In practice, most samples are taken without replacement. Accordingly, any selection of a sample described in this text is assumed to be without replacement unless otherwise specified. Consider, for instance, the following example.

EXAMPLE 4 Of the 14,500 students at a university, 5075, or 35%, participated in the January Program of Independent Activities (P.I.A.). Each of a random sample of 50 students will be asked whether or not they participated in the P.I.A. Since $n/N = 50/14{,}500 = .00345$ is much smaller than .05, selecting this sample may be regarded as a binomial experiment with $n = 50$ and $\pi = P$ (participant in P.I.A.) = .35.

■

PROBLEMS 5.1

In each of the following problems, indicate whether the experiment described may be classified (at least approximately) as a binomial experiment. Where you think an experiment is binomial, give the value for n and π. Where you think it is not binomial, explain your reasons. Where appropriate, assume that sampling is without replacement unless otherwise indicated.

1. Of the 20 colleges in a city, 70% are private. In a certain study the board of higher education selects a random sample of 10 colleges and counts the number of private colleges in the sample.
2. Forty-five percent of the millions of companies in the United States administer pension plans for their employees. The federal government selects a sample of 300 companies and records how many have such plans.
3. Referring to Problem 2, assume that the federal government randomly selects companies *until* the sample contains 100 companies with pension plans.
4. A basketball player has a probability of .8 of hitting each foul shot. His performance on any shot is not affected by his performance on previous shots. In one game he attempts 18 foul shots and the number he sinks is counted.
5. A sequence of 10 shots are made at a target. The shooter's initial probability of a bull's-eye is .4, but he expects his probability of obtaining a bull's-eye to go up by about 2% with each shot. He keeps a count of the results.
6. It is known that 75% of subscribers to a national news magazine are men. A sample of 20 subscribers is randomly selected and the number of men in the sample is recorded.
7. Of the thousands of junior colleges in the United States, 50% are in urban areas, 30% are in the suburbs, and 20% are in rural areas. A sample of 25 junior colleges is selected and the type of location of each is recorded.
8. Twelve jurors are to be selected from a pool of 15 men and 15 women. We are concerned with the number of women in the jury.
9. The number of visitors to a hospital's emergency room is recorded for one 12-hour period.
10. A coin having a probability of .7 of coming up heads is tossed 15 times and the number of tails counted.
11. A sample of 20 families is selected from a list of 150,000 families and the average income per family is computed for the sample.
12. A gambling house advertises the following game. A player rolls two dice. If the sum of the dice is 11 or 12, the house pays the player $10; otherwise, the player pays the house $1. A player plays this game until he has won four times.
13. A large state university has over 30,000 undergraduates, 10% of whom are on the dean's list. A sample of five undergraduates will be taken and the number that are on the dean's list recorded.

Solutions to Problems 1–4

1. This is not a binomial experiment. Since the sample is such a large proportion of the population and we assume that sampling is without replacement, the probability of selecting a private college will vary from one selection to another.
2. This is a binomial experiment. The number of companies with pension plans is so large compared to $n = 40$ that the probability of selecting such a company will remain at almost exactly $\pi = .45$ for each selection.
3. This is not a binomial experiment since it does not involve a fixed number of trials (selections). We do not know the number of trials necessary to obtain 100 companies with pension plans.
4. This is not a binomial experiment since the number of trials (foul shots in the game) is not fixed in advance.

SECTION 5.2
BINOMIAL PROBABILITIES

As we have indicated, our main interest in a binomial experiment is Y, the number of successes that occur. In Example 4, Y was the number of students who participated in the P.I.A. In general, we refer to Y as the **binomial random variable** and define it as follows:

> DEFINITION 1
>
> The **binomial random variable,** Y, is the number of successes in a binomial experiment.

The random variable, Y, may take one of the values $0, 1, 2, \ldots n - 1$ or n with corresponding probabilities $P(0), P(1), \ldots, P(n-1)$ and $P(n)$. In this section we develop a formula for computing these probabilities. Before doing so, however, we require some new notation.

FACTORIAL NOTATION

For any positive whole number, J, we define $J!$ as J multiplied by each whole number smaller than J:

> DEFINITION 2
>
> $J!$ (referred to as "J **factorial**")
>
> $$= J \times (J-1) \times (J-2) \times \cdots \times 2 \times 1$$

For instance, with $J = 3$, $3! = 3 \times 2 \times 1 = 6$. A few special features of $J!$ may prove helpful.

1. For consistency we define $0! = 1$.
2. We can write $J!$ in more than one way. For instance, $6!$ can be written as $6 \times 5 \times 4 \times 3 \times 2 \times 1$, or as $6 \times 5!$, or as $6 \times 5 \times 4!$, and so on.

EXAMPLE 5 Compute (a) $4!$; (b) $\frac{8!}{5!}$; (c) $\frac{6!}{3!}$; (d) $\frac{5!}{(3!)(2!)}$; (e) $\frac{8!}{(5!)(3!)}$; (f) $\frac{4!}{(0!)(4!)}$.

Solution (a) $4! = 4 \times 3 \times 2 \times 1 = 24$

(b) $$\frac{8!}{5!} = \frac{8 \times 7 \times 6 \times 5!}{5!} = 8 \times 7 \times 6 = 336$$

(c) $$\frac{6!}{3!} = \frac{6 \times 5 \times 4 \times 3!}{3!} = 6 \times 5 \times 4 = 120$$

(Note that $6!/3!$ is *not* $2!$.)

(d) $$\frac{5!}{(3!)(2!)} = \frac{5 \times 4 \times 3!}{(3!)(2!)} = \frac{5 \times 4}{2 \times 1} = 10$$

(e) $$\frac{8!}{(5!)(3!)} = \frac{8 \times 7 \times 6 \times 5!}{(5!)(3!)} = \frac{8 \times 7 \times 6}{3 \times 2 \times 1} = 56$$

(f) $$\frac{4!}{(0!)(4!)} = 1 \qquad \text{since } 0! = 1$$

■

BINOMIAL PROBABILITIES

Now we can go on to compute probabilities for Y. Suppose, for instance, that a variety of seed has a probability of germinating of $\pi = .7$. Assume that whether or not one seed germinates is independent of whether or not any other one will. If $n = 5$ seeds are randomly selected and planted, what is the probability that exactly three seeds will germinate, that is, $P(3)$? The tree diagram representing the possible outcomes of this experiment is shown in Figure 1. It is large because each of the five seeds can germinate or not. We have indicated a successful germination by the letter S and a failure by F. The probability of a failure is $P(F) = 1 - P(S) = 1 - .7 = .3$. Notice from Figure 1 that of the 32 outcomes (sequences of five letters) the event $Y = 3$ occurs 10 times. One of these is $SSSFF$. Since each of these five events are independent

$$\begin{aligned} P(SSSFF) &= P(S) \cdot P(S) \cdot P(S) \cdot P(F) \cdot P(F) \\ &= (.7)(.7)(.7)(.3)(.3) \\ &= (.7)^3(.3)^2 \end{aligned}$$

The probability of each of the other nine sequences is the same. For example,

$$\begin{aligned} P(FSFSS) &= P(F) \cdot P(S) \cdot P(F) \cdot P(S) \cdot P(S) \\ &= (.3)(.7)(.3)(.7)(.7) \\ &= (.7)^3(.3)^2 \end{aligned}$$

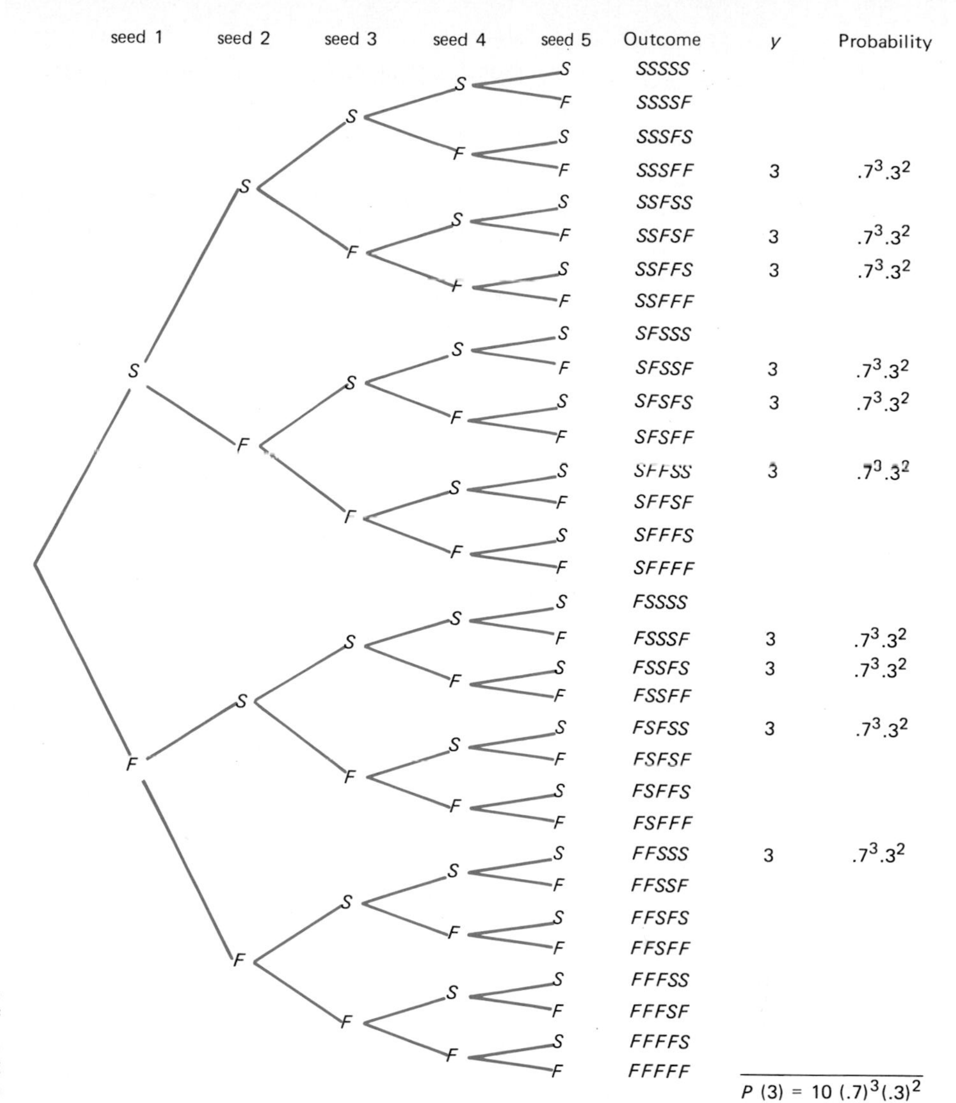

FIGURE 1

Tree diagram for the binomial experiment of planting five seeds and counting the number that germinate

Since the 10 sequences are mutually exclusive, $P(3)$ is the sum of the probabilities of the 10 sequences, or

$$\begin{aligned} P(3) &= P(SSSFF) + P(SSFSF) + \cdots + P(FFSSS) \\ &= (.7)^3(.3)^2 + (.7)^3(.3)^2 + \cdots + (.7)^3(.3)^2 \\ &= \boxed{10(.7)^3(.3)^2} = .3087 \end{aligned}$$

The expression for $P(3)$ which we have circled is, therefore, the product of two components:

The probability of each sequence involving three successes and $5 - 3 = 2$ failures $= (.7)^3(.3)^2$.

The *number* of sequences involving three successes and two failures = 10.

Keeping these points in mind, we now turn to the general case. Assume a binomial experiment consisting of n trials each having a probability of success, π. What is the probability $P(y)$ that exactly y of the trials are successes? Following our reasoning above, the probability of each sequence involving exactly y successes (and hence $n - y$ failures), for instance, is $\pi^y(1 - \pi)^{n-y}$.

$$\overleftarrow{\quad\quad\quad n \quad\quad\quad}\!\!\rightarrow$$

$$\overleftarrow{\quad y \quad}\!\!\rightarrow\overleftarrow{\; n - y \;}\!\!\rightarrow$$

$$SSS \cdot \cdot \cdot \cdot \cdot SFF \cdot \cdot \cdot \cdot F$$

Thus $P(y)$ will have the form

$$P(y) = \begin{pmatrix} \text{number of sequences} \\ \text{that involve } y \text{ successes} \\ \text{and } n - y \text{ failures} \end{pmatrix} \cdot \pi^y(1 - \pi)^{n-y}$$

But how can we find, in general, the number of sequences that include y successes (and $n - y$ failures)? We used a tree diagram with $n = 5$ above, but when n is large this is not very practical. Fortunately, there is a simple formula for determining this number.

DEFINITION 3

The number of sequences in which y successes can occur in n trials is $\dfrac{n!}{y!\,(n - y)!}$.

For instance, if $n = 5$ and $y = 3$,

$$\frac{n!}{y!\,(n - y)!} = \frac{5!}{3!\,(5 - 3)!} = \frac{5!}{3!\,2!} = 10$$

as we found earlier. The number of sequences,

$$\frac{n!}{y!\,(n - y)!}$$

is called the **binomial coefficient.** We can now write the general equation for $P(y)$.

If a binomial experiment consists of n trials each with probability of success π, then the probability of exactly y successes is

$$P(y) = \frac{n!}{y!\,(n-y)!} \cdot \pi^y(1-\pi)^{n-y} \qquad \text{for } y = 0, 1, 2, \ldots, n \qquad \text{(Eq. 5.1)}$$

Equation 5.1 is called the **binomial probability.** Some examples will clarify the use of Equation 5.1.

EXAMPLE 6 In our earlier example a variety of seed had a probability of germinating $\pi = .7$. If $n = 5$ such seeds are planted, what is the probability that (a) 0, (b) 1, (c) 2, (d) 3, (e) 4, and (f) 5 of the seeds germinate?

Solution Applying Equation 5.1 with $n = 5$, $\pi = .7$, and y equal to 0, 1, 2, 3, 4, and 5, in turn, we obtain:

$$\text{(a) } P(0) = \frac{5!}{0!\,5!}(.7)^0(.3)^5 = (1)(1)(.3)^5 = .00243$$

Remember that $0! = 1$ and any number raised to the power of zero is 1.

$$\text{(b) } P(1) = \frac{5!}{1!\,4!}(.7)^1(.3)^4 = 5(.7)(.0081) = .02835$$

$$\text{(c) } P(2) = \frac{5!}{2!\,3!}(.7)^2(.3)^3 = 10(.49)(.027) = .13230$$

$$\text{(d) } P(3) = \frac{5!}{3!\,2!}(.7)^3(.3)^2 = 10(.343)(.09) = .30870$$

$$\text{(e) } P(4) = \frac{5!}{4!\,1!}(.7)^4(.3)^1 = 5(.2401)(.3) = .36015$$

$$\text{(f) } P(5) = \frac{5!}{5!\,0!}(.7)^5(.3)^0 = (1)(.16807)(1) = .16807$$

We have, in effect, just computed the probability distribution of Y, the number of germinating seeds. This distribution is

y	0	1	2	3	4	5
$P(y)$	.00243	.02835	.1323	.3087	.36015	.16807

■

As always with a probability distribution, the probabilities add to 1.

The probability distribution in Example 6 is called the **binomial distribution** for $n = 5$ and $\pi = .7$. There is a *family* of binomial distributions with a different member for each combination of n and π.

EXAMPLE 7 Forty percent of the students at a college are commuters. If there are eight students in a seminar, what is the probability that (a) four, (b) six, (c) at most two, and (d) at least three of them are commuters?

Solution We regard the seminar as a binomial experiment with $n = 8$ and $P(\text{commuter}) = \pi = .4$.

(a) $P(4) = \dfrac{8!}{4!\,4!}(.4)^4(.6)^4 = 70(.0256)(.1296) = .2322$

(b) $P(6) = \dfrac{8!}{6!\,2!}(.4)^6(.6)^2 = 28(.004096)(.36) = .0413$

(c) We may write $P(\text{at most two commuters})$ as

$$P(Y \leq 2) = P(0) + P(1) + P(2)$$

Thus

$$P(0) = \frac{8!}{0!\,8!}(.4)^0(.6)^8 = (1)(1)(.0168) = .0168$$

$$P(1) = \frac{8!}{1!\,7!}(.4)^1(.6)^7 = 8(.4)(.028) = .0896$$

$$P(2) = \frac{8!}{2!\,6!}(.4)^2(.6)^6 = 28(.16)(.04666) = .2090$$

and

$$P(Y \leq 2) = .0168 + .0896 + .2090 = .3154$$

(d) We could compute $P(\text{at least three commuters}) = P(Y \geq 3)$ as $P(3) + P(4) + P(5) + P(6) + P(7) + P(8)$, but note that if the event $Y \geq 3$ occurs, the event $Y \geq 2$ will not occur, and vice versa. These events are thus complementary, so $P(Y \geq 3) + P(Y \leq 2) = 1$. Therefore, $P(Y \geq 3) = 1 - P(Y \leq 2) = 1 - .3154 = .6846$.

■

BINOMIAL TABLES

With the availability of electronic claculators, computing binomial probabilities has been made much easier. But computing $P(25)$ when $n = 40$ and $\pi = .6$, for instance, requires a lot of busy work even with a calculator. To make life easier, tables of binomial probabilities have been computed.

In Table A1 in Appendix A, binomial probabilities are tabulated for $n = 2$,

3, . . . , 19, 20, 25, 30, and 40 and for π = .05, .1, .2, .25, .3, .4, .5, .6, .7, .75, .8, .9, and .95. Suppose, for instance, we require $P(6)$ when $n = 8$ and $\pi = .4$. The relevant portion of Table A1 for $n = 8$ is shown in Figure 2. Looking at the left-hand column, you can see the value $n = 8$. Look down the next column for the value $y = 6$ and then across to the column headed $\pi = .4$. The entry for $P(6) = .041$. In Example 7(b) we used Equation 5.1 and obtained the same answer .0413, but to one more decimal place. The entire set of nine entries (probabilities) for $n = 8$ and $\pi = .4$ together form the binomial distribution for these values of n and π.

Table A2 contains *cumulative* binomial probabilities for the same combination of values of n and π as found in Table A1. These are the probabilities of getting y *or fewer* successes, that is, $P(Y \leq y)$. We show a portion of Table A2 in Figure 3 for $n = 8$ and $\pi = .4$ For instance, if $n = 8$ and $\pi = .4$, what is $P(Y \leq 2)$? Looking down the left-hand column of Table A2 to $n = 8$ and $y = 2$ and across to the column headed $\pi = .4$, we find that $P(Y \leq 2) = .315$. Again we obtained the same answer, .3154, to one more decimal place in Example 7(c).

A feature of Table A2 to bear in mind concerns the multiple .999's which occur in most columns. For any specific n and π, the first should be read as .999

FIGURE 2

The number circled, .041, is $P(6)$ when $n = 8$ and $\pi = .4$

n	y	π 0.05	0.1	0.2	0.25	0.3	0.4	0.5	0.6	0.7	0.75	0.8	0.9	0.95
8														
	0	.663	.430	.168	.100	.058	.017	.004	.001	.000	.000	.000	.000	.000
	1	.279	.383	.336	.267	.198	.090	.031	.008	.001	.000	.000	.000	.000
	2	.051	.149	.294	.311	.296	.209	.109	.041	.010	.004	.001	.000	.000
	3	.005	.033	.147	.208	.254	.279	.219	.124	.047	.023	.009	.000	.000
	4	.000	.005	.046	.087	.136	.232	.273	.232	.136	.087	.046	.005	.000
	5	.000	.000	.009	.023	.047	.124	.219	.279	.254	.208	.147	.033	.005
	6	.000	.000	.001	.004	.010	.041	.109	.209	.296	.311	.294	.149	.051
	7	.000	.000	.000	.000	.001	.008	.031	.099	.198	.267	.336	.383	.279
	8	.000	.000	.000	.000	.000	.001	.004	.017	.058	.100	.168	.430	.663

FIGURE 3

The number circled, .315, is $P(Y \leq 2)$ when $n = 8$ and $\pi = .4$

n	y	π 0.05	0.1	0.2	0.25	0.3	0.4	0.5	0.6	0.7	0.75	0.8	0.9	0.95
8														
	0	.663	.430	.168	.100	.058	.017	.004	.001	.000	.000	.000	.000	.000
	1	.943	.813	.503	.367	.255	.106	.035	.009	.001	.000	.000	.000	.000
	2	.994	.962	.797	.679	.552	.315	.145	.050	.011	.004	.001	.000	.000
	3	.999	.995	.944	.886	.806	.594	.363	.174	.058	.027	.010	.000	.000
	4	.999	.999	.990	.973	.942	.826	.637	.406	.194	.114	.056	.005	.000
	5	.999	.999	.999	.996	.989	.950	.855	.685	.448	.321	.203	.038	.006
	6	.999	.999	.999	.999	.999	.991	.965	.894	.745	.633	.497	.187	.057
	7	.999	.999	.999	.999	.999	.999	.996	.983	.942	.900	.832	.570	.337
	8	.999	.999	.999	.999	.999	.999	.999	.999	.999	.999	.999	.999	.999

and the others as 1. For example, if $n = 8$ and $\pi = .25$, $P(Y \leq 6) = .999$, but $P(Y \leq 7) = 1$ and $P(Y \leq 8) = 1$. To preserve the simple layout of the table, these last two values have not been rounded up but should be read as if they had been.

EXAMPLE 8 A darts player has a probability of .3 of getting a bull's-eye *each* time she throws her favorite dart. If she intends to throw the dart 25 times, what is the probability that she gets (a) 10 or fewer bull's-eyes; (b) eight or more bull's-eyes? Assume that there is no practice effect.

Solution In this example $n = 25$ and $\pi = .3$. Call the number of bull's-eyes Y.
(a) We require $P(Y \leq 10)$, which may be found directly from Table A2. For $n = 25$, $\pi = .3$ and $y = 10$, $P(Y \leq 10) = .902$.
(b) The required probability $P(Y \geq 8)$ cannot be found directly from Table A2 but we can find the probability of the event $Y \leq 7$, which is complementary to $Y \geq 8$. From Table A2, $P(Y \leq 7) = .512$ and $P(Y \geq 8) = 1 - P(Y \leq 7) = 1 - .512 = .488$.

■

In some cases the examples in the preceding section and indeed throughout this chapter may appear unrealistic in assuming that the value for π is known or that the probability of some number of successes is of interest. They will, however, prove very helpful in understanding the procedures involved in the more common situation (addressed in later chapters) when π is *not* known.

PROBLEMS 5.2

In each problem use Table A1 or A2 unless asked to use Equation 5.1 or unless the values for n and π are not in the table.

14. Sixty percent of graduating seniors in a school district go on to college within a year of graduation. If a random sample of 20 seniors is selected, what is the probability that the number going on to college is (a) 14; (b) more than 15; (c) less than 10?
15. Use Equation 5.1 to compute the following binomial probabilities. In each case verify your answer by referring to the appropriate entry in Table A1.
    (a) $P(4)$ if $n = 10$ and $\pi = .6$
    (b) $P(1)$ if $n = 3$ and $\pi = .2$
    (c) $P(4)$ if $n = 4$ and $\pi = .75$
    (d) $P(8)$ if $n = 10$ and $\pi = .5$
    (e) $P(2)$ if $n = 6$ and $\pi = .1$
    (f) $P(0)$ if $n = 4$ and $\pi = .2$
    (g) $P(12)$ if $n = 13$ and $\pi = .8$

16. A fair coin is to be tossed seven times.

(a) Use Equation 5.1 to compute the probability of (i) two heads; (ii) four heads; (iii) five heads.

(b) Verify your answers by referring to the appropriate entry in Table A1.

17. Use Table A2 to find the following binomial probabilities.

(a) $P(Y \leq 20)$ if $n = 30$ and $\pi = .75$

(b) $P(Y < 5)$ if $n = 9$ and $\pi = .6$

(c) $P(Y \geq 14)$ if $n = 40$ and $\pi = .4$

(d) $P(Y > 36)$ if $n = 40$ and $\pi = .95$

18. Suppose that internal studies indicate that approximately 20% of "next day" mail sent through the U.S. Postal Service is delivered late. If 25 such letters are sent to various places in the United States, what is the probability that:

(a) Exactly four letters are late?

(b) Three or fewer letters are late?

(c) One or more letters are late?

(d) Exactly 18 letters arrive on time?

(e) More than 15 letters arrive on time?

19. A distributor of breakfast cereal places coupons for a free carton in every fourth carton. A family purchases eight cartons.

(a) Use Table A1 to find the probability that the family will find (i) three coupons; (ii) no coupons.

(b) Find the probability distribution of the number of coupons in the eight cartons.

(c) What is the most likely number of coupons?

(d) Sketch the probability histogram.

20. A commuter airline flys 38-seat planes. The airline knows that on the average 5% of passengers with reservations do not show up. To compensate, the airline sells 40 tickets for each flight. What is the probability that for a particular flight:

(a) All 40 ticket holders show up?

(b) All those who show up for the flight can be accommodated? (Assume that whether one ticket holder shows up is independent of whether any other one does)

21. Assume the Internal Revenue Service audits a random sample of 10% of tax returns where the gross income exceeds $100,000. If a tax lawyer assists in the completion of 40 such returns, what is the probability that:

(a) None will be audited?

(b) At least one will be audited?

(c) At most two will be audited?

(d) At least three will be audited?

22. A student is given a multiple-choice examination containing 30 questions. The teacher suspects that this student has not prepared for the test and will guess the answer to each of the questions, in which case $P(\text{correct answer}) = .25$.

(a) If the student does guess at each question, what is the probability of (i) four or fewer correct answers; (ii) nine or more correct answers; (iii) 15 or more correct answers?

(b) Suppose that the student gets 15 correct answers. Recalling your answer in (a, iii), does this throw doubt on the presumption that the student guessed at each question? Explain.

23. Suppose it is equally likely that a randomly selected student was born on any one of the 365 days of the year. (Assume that nobody was born on February 29.) Those born between October 23 and November 21 (inclusive) have the sign of Scorpio,

the scorpion. A class has nine students. Use Equation 5.1 to write down an expression for the following probabilities (you do not need to evaluate them unless you wish to):

(a) The probability that *exactly* one of the nine students will be a Scorpio.

(b) The probability that *at least* one of the nine students will be a Scorpio.

Solution to Problem 14

We regard the selection of the 20 seniors as a binomial experiment with $n = 20$ and $\pi = P(\text{going to college}) = .6$. Designate the number in the sample going to college as Y.

(a) From Table A1, $P(14) = .124$.

(b) We require $P(Y > 15)$. The event $Y > 15$ is equivalent to $Y \geq 16$. Therefore, $P(Y \geq 16) = 1 - P(Y \leq 15) =$ (from Table A2) $1 - .949 = .051$.

(c) The event $Y < 10$ is equivalent to $Y \leq 9$. Using Table A2 directly, we have $P(Y < 10) = P(Y \leq 9) = .128$.

SECTION 5.3

THE MEAN AND STANDARD DEVIATION OF THE NUMBER OF SUCCESSES

In Chapter 4 we computed the mean or expected value of a discrete random variable X using the formula

$$E(X) = \mu_X = x_1 P(x_1) + x_2 P(x_2) + \cdots + x_k P(x_k)$$

We can use this formula to find the mean of a binomial random variable. In Example 6, for instance, where $n = 5$ and $\pi = .7$, the probability distribution of the number of germinating seeds was

y	0	1	2	3	4	5
$P(y)$	.00243	.02835	.13230	.30870	.36015	.16807

The "expected" number of seeds that will germinate will therefore be

$$E(Y) = 0(.00243) + 1(.02835) + \cdots + 5(.16807) = 3.5 \text{ seeds.}$$

Computing $E(Y)$ in this way when n is large, however, can be quite time consuming. Happily, there is a simple formula for both $E(Y)$ and $\text{Var}(Y)$ which can be used for any binomial random variable.

If Y is a binomial random variable, these formulas are

$$\mu_Y = E(Y) = n\pi \qquad \text{(Eq. 5.2)}$$

$$\sigma_Y^2 = \text{Var}(Y) = n\pi(1 - \pi) \quad \text{(Eq. 5.3)}$$

$$\sigma_Y = \text{SD}(Y) = \sqrt{n\pi(1 - \pi)} \quad \text{(Eq. 5.4)}$$

For instance, when $n = 5$ and $\pi = .7$, $E(Y) = n\pi = 5(.7) = 3.5$ seeds, as we found above; $\text{Var}(Y) = n\pi(1 - \pi) = 5(.7)(1 - .7) = 1.05$; and $\text{SD}(Y) = \sqrt{1.05} = 1.025$ seeds.

As the next two examples indicate, Equation 5.2 really accords with one's intuition.

EXAMPLE 9 A newspaper sponsors a random sample of 1000 voters. Each voter is to be asked whether he or she favors repeal of the state's death penalty. Suppose, unknown to the newspaper, that 35% of all voters favor repeal. Compute (a) the number in the sample that would be *expected* to favor repeal; (b) the standard deviation of the number favoring repeal.

Solution We regard the sampling as a binomial experiment with $n = 1000$ and $\pi = P(\text{favoring repeal}) = .35$. Here Y is the number in the sample favoring repeal. From Equations 5.2 and 5.4,

$$\mu_Y = E(Y) = n\pi = 1000(.35) = 350 \text{ voters}$$

and

$$\sigma_Y = \text{SD}(Y) = \sqrt{n\pi(1 - \pi)} = \sqrt{1000(.35)(.65)} = 15.083 \text{ voters}$$

We interpret $E(Y) = 350$ as the mean value for Y over all the different samples of 1000 voters that could be selected from this population of voters. It is unlikely that the value for Y actually obtained in any one sample will be exactly 350. Recalling the definition of the standard deviation in Chapter 1, we may interpret $\sigma_Y = 15.083$ as follows. Over all possible samples of 1000 votes the value for Y will differ from 350 by an average of approximately 15 voters. ■

EXAMPLE 10 An experiment consists of rolling a fair die 60 times. What are the mean and standard deviation of the number of fives that will be rolled?

Solution In this case $n = 60$ and $\pi = P(5) = 1/6$.

If we call Y the number of fives, then from Equations 5.2 and 5.4, we have

$$E(Y) = n\pi = 60\left(\frac{1}{6}\right) = 10 \text{ fives}$$

$$\text{SD}(Y) = \sqrt{n\pi(1 - \pi)} = \sqrt{60\left(\frac{1}{6}\right)\left(\frac{5}{6}\right)} = 2.89 \text{ fives}$$

Here $E(Y) = 10$ is the mean number of fives that would occur if this experiment were repeated an infinite number of times. The standard deviation, $\sigma_Y = 2.89$, is approximately the average "distance" that all the numbers of fives (Y) will be from 10, again if the experiment were repeated an infinite number of times.

For a better understanding of our interpretation of $E(Y)$ and $SD(Y)$, the authors actually performed this experiment 20 times, in each case recording the number of fives. The results were

13	11	8	14	13	9	8	10	7	12
11	15	10	7	5	12	10	9	14	13

$$\sum y_i = 211 \qquad \sum y_i^2 = 2367$$

Using the methods of Chapter 1, the mean of the 20 values, $\bar{y} = \Sigma y_i/20 = 211/20 = 10.55$ fives and the standard deviation is

$$s = \sqrt{\frac{20\,\Sigma y_i^2 - (\Sigma y_i)^2}{20(19)}} = \sqrt{\frac{20(2367) - 211^2}{380}} = 2.72 \text{ fives}$$

These values, although based on only 20 replications of the experiment rather than an infinite number, are close to the corresponding "population" values $\mu_Y = 10$ and $\sigma_Y = 2.89$.

■

We emphasize that Equations 5.2, 5.3, and 5.4 apply only to *binomial* random variables. More generally, it is necessary to use the methods of Chapter 4 when computing the mean and variance of a discrete random variable.

In Figure 4 we show the probability histograms for three binomial distributions. The top one, (a), is for the binomial distribution when $n = 10$ and

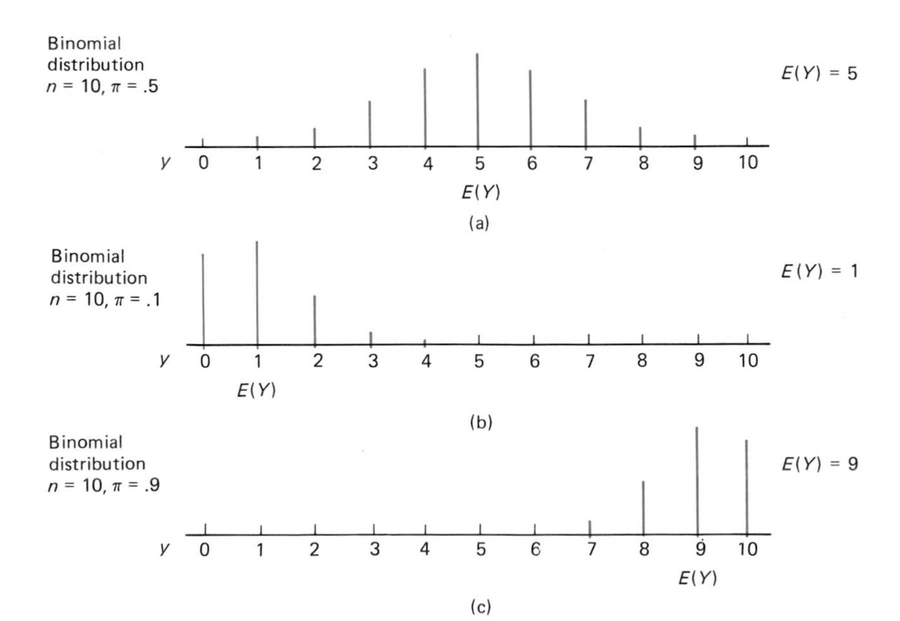

FIGURE 4

Probability histograms for three binomial distributions

$\pi = .5$. The height of each line is the probability of getting the baseline value. For example, the height of the line above $y = 3$ is $P(3) = .117$, from Table A1. When $\pi = .5$ the distribution is symmetric around the mean and, $E(Y) = 5$. In (b) and (c) $n = 10$, but in (b) $\pi = .1$ and in (c) $\pi = .9$. Notice that when π is small, as in (b), the distribution is skewed to the right. When π is large, as in (c), the distribution is skewed to the left.

PROBLEMS 5.3

24. A fair coin is to be tossed 25 times.
    (a) Compute the expected number and the standard deviation of the number of heads.
    (b) Interpret your answers.
25. An experiment consists of rolling a fair die twice. If this experiment is to be repeated 200 times, what is the expected number and the standard deviation of the number of occasions on which the sum of the two throws is 4?
26. A batter in baseball has a probability of .3 of getting a hit each time at bat.
    (a) If he comes to bat four times per game in each game of a 120-game season, how many hits can he expect to get?
    (b) What is the standard deviation of the number of hits?
27. It has been estimated that in the 1950s the probability that a murderer was caught, convicted and executed was 1/250. If there were 6500 murders in 1955, how many murderers could be expected to be executed?
28. Each of a random sample of 40 second-semester freshmen is asked whether he or she feels that their choice of college was correct. If, in fact, 65% of *all* such freshmen feel that they made the correct choice, how many in the sample can be expected to answer in this way?
29. Insurance company records indicate that the probability that an 80-year-old man will live through his next five years is .453. If the company insures 4382 such men, how many of them can be expected to live to 85?
30. A college admissions office knows that the probability that a student accepted by the school will attend is .75. How many students should the school accept so that they can expect to obtain a freshman class of 660?
31. A state's department of health and human services wants to select a sample of children under the age of 11 for a study of the factors associated with child abuse. Their judgment is that one child in 20 is abused. How large a sample should be selected in order that the number of abused children "expected" in the sample will be 250.

Solution to Problem 24

(a) With $n = 25$ and $\pi = P(\text{head}) = .5$, $E(Y) = n\pi = 25(.5) = 12.5$ and $\text{SD}(Y) = \sqrt{n\pi(1 - \pi)} = \sqrt{25(.5)(.5)} = 2.5$ heads.

(b) If this experiment were repeated an infinite number of times, the mean number of heads would be 12.5. The average distance that the number of heads thrown would be above or below 12.5 is approximately 2.5.

SECTION 5.4
SUMMARY AND REVIEW

A **Bernoulli trial** is an experiment in which we are concerned with only two complementary outcomes, designated a success and a failure. A series of Bernoulli trials with the following characteristics is called a **binomial experiment:**

1. A fixed number, n, of Bernoulli trials.
2. The probability of a success is the same (π) for each trial.
3. The outcome of each trial is independent of the outcome of any other.
4. We are interested in the random variable, Y, the number of trials that are successes.

In some cases *selecting a sample* can be considered a binomial experiment. Specifically:

Assume a population of size N of which a number A are successes. A sample of size n is to be selected from this population. Then:

1. If sampling is *with* replacement, selecting the sample is exactly a binomial experiment with $\pi = A/N$.
2. If sampling is *without* replacement but the sample size n is small compared to the population size N, then selecting the sample may be regarded as a binomial experiment with $\pi = A/N$. As a guide, this will be true if $n/N \leq .05$.
3. In all other cases, selecting the sample without replacement *cannot* be regarded as a binomial experiment.

The probability of exactly y successes in a binomial experiment (and hence $n - y$ failures) is given by

$$P(y) = \frac{n!}{y!\,(n-y)!}\pi^y(1-\pi)^{n-y} \qquad y = 0, 1, 2, \ldots, n \qquad \text{(Eq. 5.1)}$$

Equation 5.1 is called a **binomial probability.**

The **binomial probability distribution** is an array consisting of all the values that Y can take ($y = 0, 1, 2, \ldots, n-1, n$) together with the probability $P(y)$ of each value. The entries in Table A1 are the binomial probabilities $P(y)$ for all possible values of y for $n = 2, 3, \ldots, 19, 20, 25, 30$, and 40, and for $\pi = .05, .1, .2, .25, .3, .4, .5, .6, .7, .75, .8, .9$, and .95. For each combination of n and π, the values of $P(y)$ are arranged so as to form the probability distribution of Y. By contrast, the entries in Table A2 are **cumulative binomial probabilities** of the form $P(Y \leq y)$.

There are simple formulas for the mean, variance, and standard deviation of Y. They are:

$$\mu_Y = E(Y) = n\pi \qquad \text{(Eq. 5.2)}$$

$$\sigma_Y^2 = \text{Var}(Y) = n\pi(1-\pi) \qquad \text{(Eq. 5.3)}$$

$$\sigma_Y = \text{SD}(Y) = \sqrt{n\pi(1-\pi)} \qquad \text{(Eq. 5.4)}$$

It is important to recall that these formulas apply only to *binomial* random variables.

REVIEW PROBLEMS

GENERAL

1. Decide which of the following sampling schemes is *not* a binomial experiment and explain the reason for your conclusion:

   (a) A sample of 25 teachers is selected from the membership of the teachers' union. Each teacher selected reports his or her number of years of teaching experience.

   (b) A sample of 50 names is taken from a list of all registered voters in a city. Each of the 50 voters is asked whether or not he or she approves the mayor's veto of a specific piece of legislation.

   (c) The records of a large maternity hospital are to be sampled until an instance of triplets is discovered.

   (d) The records of a large maternity hospital are to be sampled until six instances of twins have been discovered.

2. Five cards are selected *with* replacement from a deck of 52 cards.

   (a) What is the probability that three are red and two are black?

   (b) What is the most likely number of red cards?

   (c) What is the expected number of red cards?

   (d) Explain the difference between the most likely and the expected value.

3. Suppose that a batter in baseball has a probability of .3 of getting a hit each time at bat. If he comes up to bat five times in a game, what is the probability that:

   (a) He will get two hits?

   (b) He will get no hits?

   (c) He will get at least one hit?

   (d) Twenty percent of his at-bats result in hits?

4. Y is a binomial random variable with $n = 7$ and $\pi = .5$.

   (a) Use the binomial formula to compute (i) $P(Y = 5)$; (ii) $P(Y = 7)$; (iii) $P(Y = 6)$. (b) Compute (i) $E(Y)$; (ii) $\text{SD}(Y)$.

5. X is a binomial random variable with $n = 20$ and $\pi = .8$. Use the binomial tables to compute (a) $P(X = 10)$; (b) $P[X = E(X)]$; (c) $P(X \geq 13)$.

6. A student estimates that the probability that her commute to school will take less than an hour is .86. She goes to school on each of five days. What is the probability that she will take longer than an hour to get to school (a) only on Tuesday; (b) on only one of the five school days.

7. A specially constructed roulette wheel contains the numbers 1 to 36. Half the numbers are white, half red. When the wheel is spun, both colors are equally likely as are all 36 numbers. The wheel is spun 180 times during a day. How many times would you expect to see (a) the number 4; (b) an odd number; (c) a white number?

8. Y is a binomial random variable with $n = 40$ and $\pi = .5$. Compute (a) $P(Y = 21)$; (b) $P(Y > 21)$; (c) $P(Y \leq \mu_Y)$.

9. An experiment consists of rolling a fair die four times. A success is getting a 1 or a 2. Call X the number of rolls on which a success occurs.

   (a) What are the possible values that X can take?

(b) Find the probability distribution for X.

(c) Use Equation 4.1 to verify that $E(X) = 4/3$ in this case.

10. A French gambler of the seventeenth century, the Chevalier de Méré, claimed that the chance of throwing at least one 1 when throwing four fair dice is equal to the probability of throwing at least one double 1 in 24 throws with two fair dice. Was he correct?

11. A binomial random variable has mean 30 and variance 21. Find n and π.

12. Thirty percent of the adults in a city saw a particular Olympic event on television. Suppose that each of a random sample of 40 adults are asked whether or not they watched the program.

    (a) What is the probability that the number of adults in the sample who saw the program (i) equals 14; (ii) is less than 12?

    (b) Find and interpret $E(X)$ in this case.

13. A test consists of 20 true–false questions. The passing score, K, must be set so that if someone guesses the answer to each question, there is at most a 5% chance of passing the test. What should K be?

HEALTH SCIENCES

14. When crossing two plants, each having a red/white gene pair, there is a probability of one-fourth that the progeny will have white flowers.

    (a) If five plants from such a cross are obtained, what is the probability that (i) two will be white; (ii) none will be white; (iii) at least one will be white?

    (b) If 20 plants from such a cross are obtained, what is the probability that (i) one-fourth of them are white; (ii) more than one-fourth are white; (iii) between (but, not including) one-fourth and one-half are white?

15. It is estimated that 20% of births at a particular hospital are by caesarian section. Assume that 30 babies are born in a given week.

    (a) What is the probability that (i) at least four are delivered by caesarian section; (ii) exactly 20% of the babies are delivered by caesarian section?

    (b) What is the most likely number of caesarian section deliveries during that week?

16. A test for a form of cancer is imperfect in that there is a probability of .05 that it will falsely indicate the presence of the cancer when, in fact, a person does not have the cancer. If 25 persons, none of whom have the cancer are tested, what is the probability that the test will wrongly indicate the presence of the cancer in (a) at least one case; (b) exactly two cases; (c) no cases?

17. Seventy-five percent of a particular population have a systolic blood pressure (SBP) greater than 112. A sample of 20 members of this population are selected. Call X the number in the sample having a SBP greater than 112. Compute (a) $P(X = 15)$; (b) $P(X > 15)$; (c) $P(X < 12)$; (d) $E(X)$; (e) $P[X \geq E(X)]$.

18. Using the traditional technique, the probability that a particular kind of medical operation will be successful is .6. A young surgeon claims to have invented a new technique for which the probability of a successful operation is .8. Her superiors decide to test this new technique on the next 20 patients needing the operation. The technique will be adopted for general use if 16 or more of the 20 operations are successful.

    (a) If, in fact, the new technique is no better than the old (i.e., the probability of a success is still .6), what is the probability that at least 16 of the 20 operations are successful (so that the new technique will, incorrectly, be adopted)?

    (b) If, in fact, the new technique has a probability of success of .8, what is the probability that fewer than 16 of the 20 operations will be successful (so that the new test will, incorrectly, *not* be adopted)?

SOCIOLOGY AND GOVERNMENT

19. A candidate for reelection commissions a small random survey of 40 voters.

(a) If exactly one-half of the population of voters intend to vote for her, what is the probability that (i) exactly one-half of the sample will favor her; (ii) 60% or more of the sample will favor her?

(b) If only 40% of the population of voters intend to vote for her, what is the probability that (i) exactly 40% of the sample will favor her; (ii) 50% or more of the sample will favor her?

20. It is known from federal audits that in a particular city approximately 30% of all welfare cases have an error in the grant payment. Suppose that a sample of 30 cases is selected at random.

(a) What is the probability that (i) at most 10 contain a grant error; (ii) more than 10 contain a grant error?

(b) Compute the (i) expected number, and (ii) standard deviation of the number of cases containing grant errors.

21. A doctor believes that 15% of the children who are brought to the health center where she works are abused to some extent. If five children are brought in for examination on one day, what is the probability that:

(a) None are abused?

(b) At least one is abused?

(c) Two are abused?

22. In a particular city 20% of the population of voters are black. If juries of 12 persons are selected at random from the voting population, what proportion of juries will have (a) no black persons; (b) at least one black person; (c) exactly two black persons?

23. Referring to Review Problem 22, what is the probability that three consecutive juries will have no black jurors? Explain why, if three consecutive juries were all without black jurors, you should doubt that the selection process was random.

ECONOMICS AND MANAGEMENT

24. Twenty men aged 45 apply for a 10-year term life insurance policy. Actuarial studies indicate that of all 45-year-old men, 5% will die before age 55.

(a) Find the expected number of deaths (in the next 10 years) among the 20 men. Interpret your result.

(b) What is the probability that (i) none of the 20 applicants will die before age 55; (ii) exactly one of the 20 will die before age 55; (iii) one or less, or three or more, of the men will die before age 55?

25. Smash Motor Company knows that approximately 20% of its cars are "lemons." If 10 of its cars are selected at random, what is the probability that:

(a) Two are lemons?

(b) At most two are lemons?

(c) All the cars are good?

(d) Nine are good?

26. The probability that an overseas telephone call will be completed is .9. Whether or not any call is completed is independent of the result of any other call. If a company attempts 20 overseas calls in a day, what is the probability that:

(a) All will be completed?

(b) All but two will be completed?

(c) Three-fourths of them will be completed?

(d) Three will not be completed?

27. A salesman judges that at 80% of the homes he calls somebody will be home. If there is somebody home, he will make a sale at approximately one home in four.

(a) What is the probability that when he visits a home he will make a sale?

(b) If he visits four houses on a street, what is the probability that (i) he will find somebody at home in each house; (ii) he will make only one sale; (iii) he will make at least one sale?

28. In a particular community 65% of households have an annual income greater than $14,000. Assume that a random sample of five households is selected.

(a) What is the probability that the number of households with an income greater than $14,000 is (i) equal to four; (ii) exceeds three?

(b) What is the expected number of such households in the sample?

EDUCATION AND PSYCHOLOGY

29. Approximately 34% of students score between 500 and 600 on the SAT math test. If a random sample of four students is selected, what is the probability that (a) only one; (b) an odd number of the four score between 500 and 600?

30. Madame X claims to have extrasensory perception. A test is arranged to investigate this claim. Madame X is seated behind a screen while in front the tester randomly selects one of five different shapes from a barrel. Madame X is asked to identify the selected shape. This procedure is repeated 15 times (replacing the selected shape each time).

(a) Suppose that Madame X just guesses on each of the 15 occasions. What is the probability that she will make (i) four correct guesses; (ii) seven correct guesses; (iii) six or more correct guesses?

(b) If she were guessing, what is the expected number of correct guesses?

(c) Suppose Madame X does have some extrasensory abilities and that the probability of her correctly naming the chosen shape is .5. Answer parts (a) and (b).

31. An examination consists of 25 multiple-choice questions. Each question has four possible answers. A total of 12 correct answers are required to pass the examination.

(a) If a student simply guesses at the answer to each question, what is the probability that the student passes?

(b) How many questions would you expect the student to get correct by guessing?

32. A test consists of n true–false questions. State regulations require that persons getting 80% or more of the questions correct must pass. Compute the probability of passing the test by simply guessing the answer to each question if (a) $n = 10$; (b) $n = 20$; (c) $n = 30$. On the basis of your answers which number of questions would you recommend.

33. It is known that approximately one child in 10 entering first grade has a particular learning disability. Suppose that a first-grade class consists of 20 children.

(a) What is the probability that (i) three of them have the specified disability; (ii) more than three have the disability?

(b) What is the most likely number of children with the disability?

SCIENCE AND TECHNOLOGY

34. In a small town there are 40 sets of traffic lights. Engineers with the state highway department estimate that there is a probability of .05 that any one of them will malfunction in any given week.

(a) In a specific week what is the probability that (i) at least three malfunctions will occur; (ii) no malfunctions will occur; (iii) at least one malfunction will occur?

(b) What is the most likely number of malfunctions?

(c) Compute the expected number of malfunctions and interpret your result.

35. A carton containing 1500 electronic components is delivered to a manufacturer. The quality control engineer will sample 10 of the components at random and

reject the carton if more than two of the components are defective. What is the probability that the carton will be rejected if 150 of the componenents in the carton are in fact defective?

36. Every minute a machine generates a signal with probability .7. Whether a signal appears in any one minute is independent of whether or not a signal occurs in any other minute. Assume that the machine is run for 25 minutes.

(a) What is the probability that (i) exactly 16 and (ii) more than 20 signals, are generated?

(b) What is the expected number of signals?

SUGGESTED READING

AYER, A. J. "Chance"; WEAVER, W. "Probability"; and KAC, M. "Probability." Readings in *Mathematics in the Modern World*. San Francisco: Foreman, 1968.

MOORE, D. S. *Statistics: Concepts and Controversies*. San Francisco: Freeman, 1979, pp. 237–268.

STANCL, D. L. and STANCL, M. L. *Applications of College Mathematics*. Lexington, Mass: D.C. Heath, 1983, pp. 323–414.

ROSS, S. *A First Course in Probability*. New York: Macmillan, 1976.

6 THE NORMAL DISTRIBUTION

FALSE ADVERTISING?

The owner of a new light van complained to the dealer that he was getting only 16.5 miles per gallon (mpg). Apparently, this make of van was prominently advertised as having "EPA estimated mpg 19." (The EPA is the Environmental Protection Agency.) The dealer pointed out the small print in the advertisement, which indicated that this figure was just a "guide" and that "actual mileage will vary." The owner is not satisfied and remains convinced that his experiences are not consistent with an EPA estimate of 19 mpg. Assuming that 19 *is* the *mean* mpg, is a figure as low as 16.5 really unusual? The normal probability model discussed in this chapter provides a basis for determining just how unusual such a figure is. You are asked to investigate this question further in Review Problem 24 on page 239.

SECTION 6.1

CONTINUOUS RANDOM VARIABLES AND THE NORMAL DISTRIBUTION

In Chapter 4 we introduced the idea of a continuous random variable. As you saw there, in contrast to a discrete random variable, there are *no* gaps between the values that a continuous random variable may take. As a consequence:

> DEFINITION 1
>
> **A continuous random variable:**
> (a) May take an *infinite* number of values.
> (b) Has no limit as to the accuracy with which it can be measured.

For instance, with increasingly more sensitive scales, we may measure a person's weight as 143 pounds, 143.3 pounds, 143.28 pounds, 143.281 pounds, 143.2808 pounds, and so on, without limit. In practice we are usually satisfied with rounding to the nearest pound.

Examples of continuous random variables include:

1. The diameter of a randomly selected mass-produced engine cylinder
2. The distance traveled by a car on a gallon of gasoline
3. The age of a randomly selected first grader
4. The waiting time to see a doctor at a clinic for a randomly chosen patient

PROBABILITY DISTRIBUTIONS FOR CONTINUOUS RANDOM VARIABLES

In the next example we illustrate the nature of the probability distribution for a continuous random variable.

EXAMPLE 1

Suppose that 4000 women gave birth at a large maternity hospital last year. We are interested in the weight of the women at the time of delivery. If the 4000 weights were known, we might summarize the population of weights by arranging them into a relative frequency distribution. Using the methods of Chapter 1, we can display this distribution by means of a relative frequency histogram, as in Figure 1 where we indicate that *all* the women have weights between 80 and 400 pounds. It is not easy to work with a distribution such as that in Figure 1, and statisticians usually go further and approximate the histogram with a smooth curve. In Figure 2 such a curve is shown superimposed on the original histogram of weights.

A feature of this curve and others used by statisticians is that areas under the curve correspond to probabilities. For example, if we call W the weight of a randomly chosen mother, then $P(W < 120)$ is the area above the baseline to

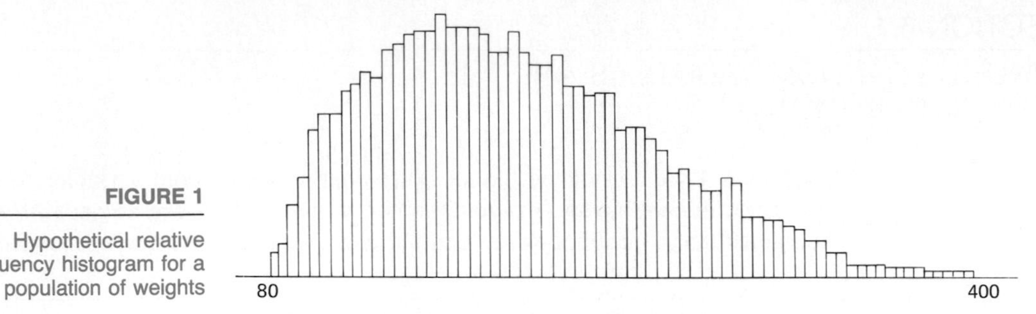

FIGURE 1

Hypothetical relative frequency histogram for a population of weights

the left of 120 lb. Similarly, $P(160 < W < 200)$ is the area above the interval $160 < W < 200$ (i.e., between 160 and 200 lb). These areas are shown shaded in Figure 3. The total area under the curve is $P(80 < W < 400)$, which is 1.

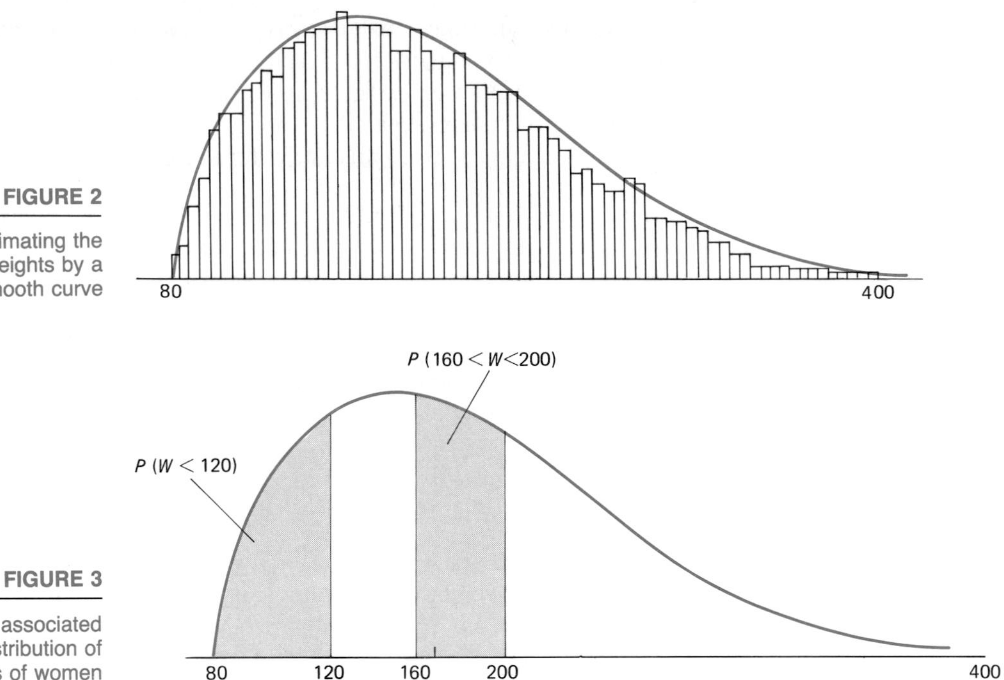

FIGURE 2

Approximating the histogram of weights by a smooth curve

FIGURE 3

Probabilities associated with the distribution of weights of women

With a continuous random variable we can only obtain probabilities associated with intervals since the probability that the variable takes any specific value is zero. For example, $P(W = 143) = 0$. Although a woman's weight may be close to 143 lb, as we pointed out above, with increasingly more sensitive scales her weight could be recorded as 143.3 pounds, 143.28 pounds, and so on, but it is never exactly 143 pounds.

■

To generalize these points:

> DEFINITION 2
>
> The **probability distribution** for a continuous random variable X can be represented by areas under a curve; the total area under the curve is 1; and the probability that X takes a specific value is zero.

Occasionally, we shall treat variables that are technically discrete as though they were continuous. This will occur where the variable may take so many values that it is not practical to treat it as discrete. Examples include variables such as income, populations, prices, and numbers of items sold.

Like their discrete counterparts, continuous random variables generally have a mean (μ) and a standard deviation (σ). For a population of weights, for instance, μ might be 158 pounds and σ, 49 pounds. The procedures for computing μ and σ for continuous random variables require calculus and will not be developed in this text.

In Chapter 4 we noted that the probability distribution of a discrete random variable can be thought of as a convenient means of summarizing how a population of numbers vary. The same is true for the probability distribution of a continuous random variable. The curve in Figure 3, for instance, summarizes how a population of weights vary.

Statisticians have developed families of curves, such as that shown in Figure 2, that are good approximations to the relative frequency histogram for a variety of continuous variables. Some, such as that in Figure 3, are skewed to the right. Others are skewed to the left. The most important continuous distribution in statistics is called the **normal distribution** and is symmetric, as shown in Figure 4.

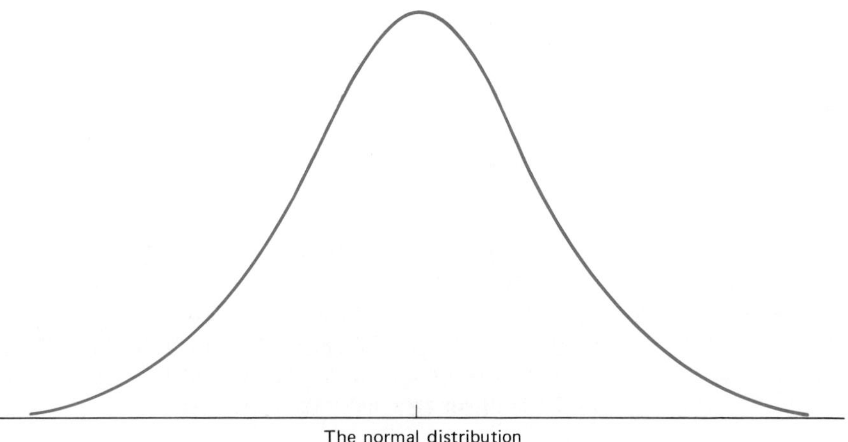

FIGURE 4

The normal distribution

THE NORMAL DISTRIBUTION

Although usually referred to as *the* normal distribution, the name actually applies to a family of distributions all of which are bell-shaped. Sometimes called the bell-shaped or Gaussian curve (after one of its discoverers Karl Friedrich Gauss), the normal distribution owes its importance in statistics in part to the fact that it closely approximates the distribution of many different measurements. These include heights, blood pressures, the lifetime of tires, brain sizes, and many others. The distribution of certain kinds of test and IQ scores, for example, are also normal.

Figure 5 shows three normal distributions. Two of them, (a) and (b), have the same mean, $\mu = 100$, but different standard deviations, and two of them, (b) and (c), have the same standard deviations but different means, $\mu_1 = 100$ and $\mu_2 = 110$. In general, the values for μ and σ determine the location and degree of spread of a specific normal distribution.

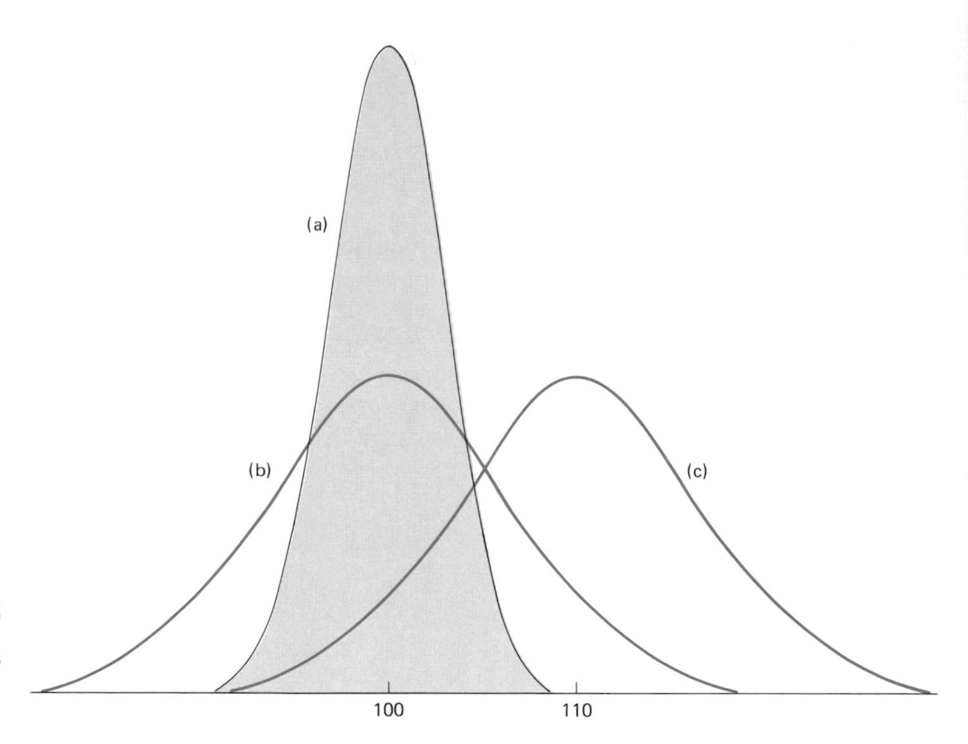

FIGURE 5

Three normal distributions

PROPERTIES OF THE NORMAL DISTRIBUTION

Every normal distribution has the following four properties:

1. It is symmetric around its mean μ.
2. The curve gets closer and closer to the horizontal axis but never touches or crosses that axis.
3. The total area under the curve is 1.

4. Approximately 68% of the area under the curve lies in the interval extending from a point one standard deviation below the mean to a point one standard deviation above the mean. In other words since the area under the curve is 1, .68 of the area lies between $\mu - \sigma$ and $\mu + \sigma$. The approximate areas that lie within one, two, and three standard deviations of the mean are listed below.

NUMBER OF STANDARD DEVIATIONS	PERCENT OF AREA
1	68
2	95
3	99.75 (i.e., practically all)

These four properties are illustrated in Figure 6 for the normal distribution with mean $\mu = 100$ and $\sigma = 10$.

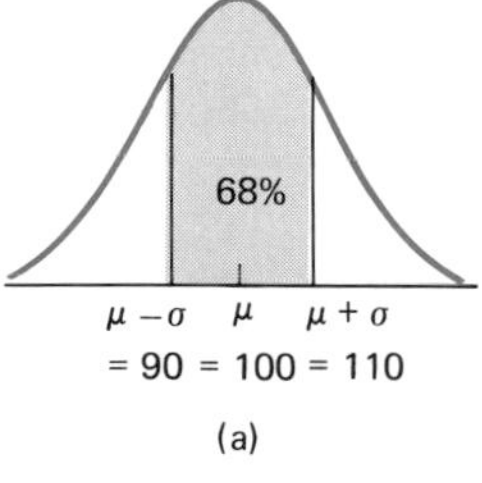

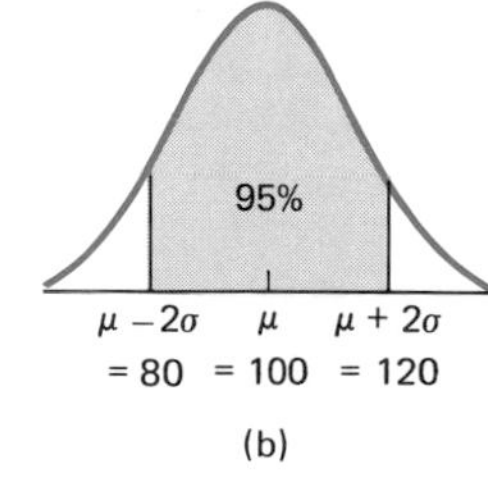

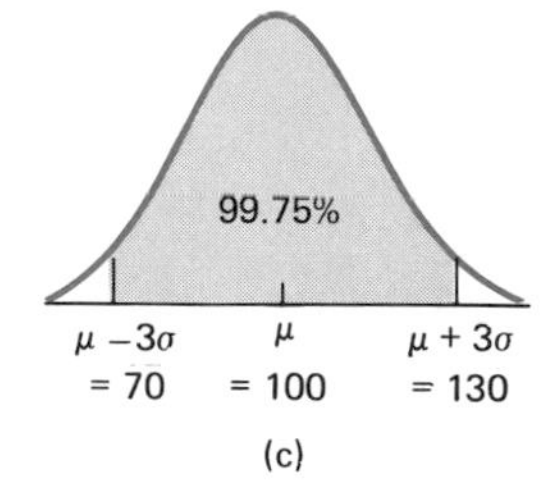

FIGURE 6

Illustrating the 68, 95, 99.75% property of the normal distribution

A continuous random variable X having a normal distribution is called a **normal random variable.** We signify this with the notation

$$X \text{ is } N[\mu, \sigma]$$

The N stands for normal, the first number inside the brackets stands for the mean of X, and the second number in the brackets stands for the standard deviation of X. This notation should be read as "X has the normal distribution with mean μ and standard deviation σ."

Probabilities associated with normal random variables, as with any continuous random variable, correspond to areas under the appropriate curve. We show this in the following example.

EXAMPLE 2 The distribution of IQ scores in a particular occupation is $N[100, 10]$. Call X the random variable "the IQ of a randomly selected person in this occupation." The three probabilities, $P(X < 95)$, $P(105 < X < 112)$, and $P(X > 120)$, are shown in Figure 7.

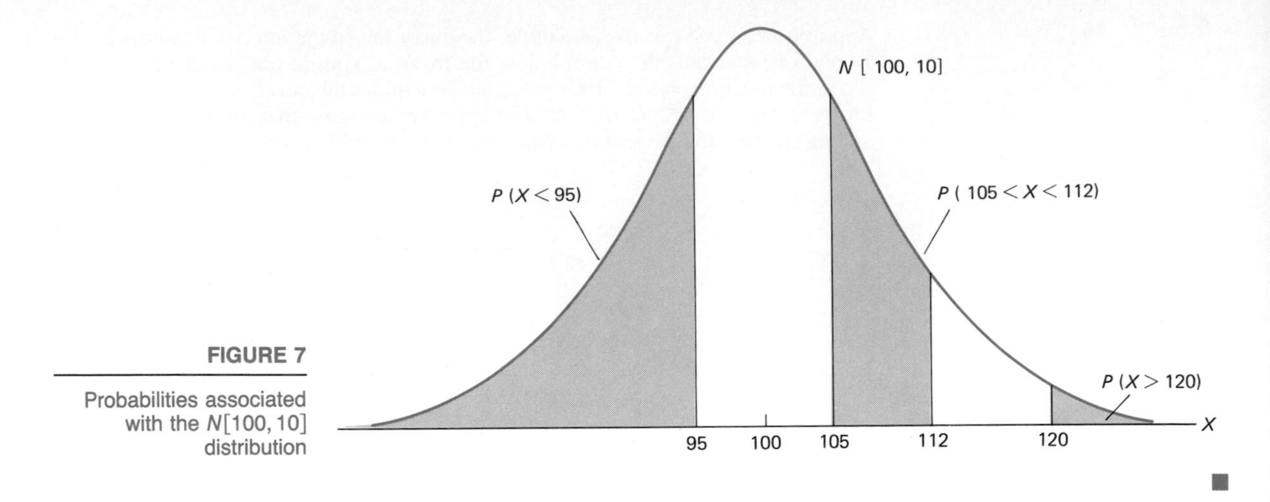

FIGURE 7

Probabilities associated with the $N[100, 10]$ distribution

■

PROBLEMS 6.1

1. The distribution of grade-point average (G) at a university is $N[2.6, .4]$. Use the properties of the normal distribution to compute the percentage of students having GPAs (a) between 1.8 and 3.4; (b) greater than 3.0.
2. Call the random variable "the height of a randomly selected adult male" H. Assume that H is $N[68, 3]$:

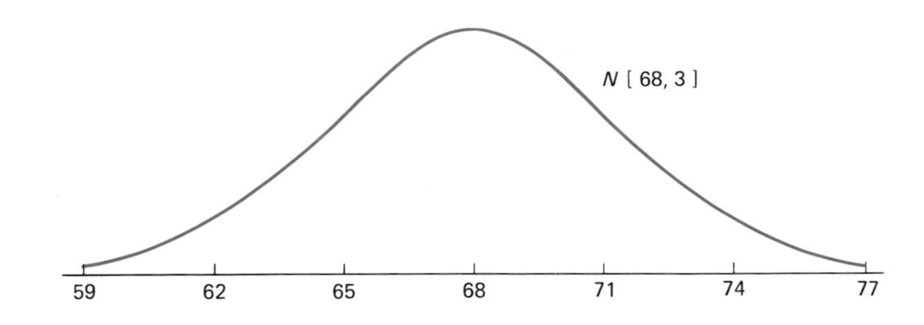

   (a) Copy this distribution and shade in the areas corresponding to the probabilities (i) $P(H < 63)$; (ii) $P(70 < H < 74)$.

   (b) Use the properties of the normal distribution to find the percentage of adult males having heights (i) between 65 and 71 inches; (ii) between 68 and 74 inches; (iii) between 59 and 71 inches.

3. A manufacturing process produces engine cylinders whose diameters (D) vary according to a normal distribution with a mean of 20 centimeters (cm) and a standard deviation of .02 cm:

   (a) Copy this distribution and indicate the values of the seven positions indicated along the D axis. (The distance between adjacent points is one standard deviation)

   (b) Shade in the areas corresponding to (i) $P(D > 20.01)$; (ii) $P(19.98 < D < 19.99)$.

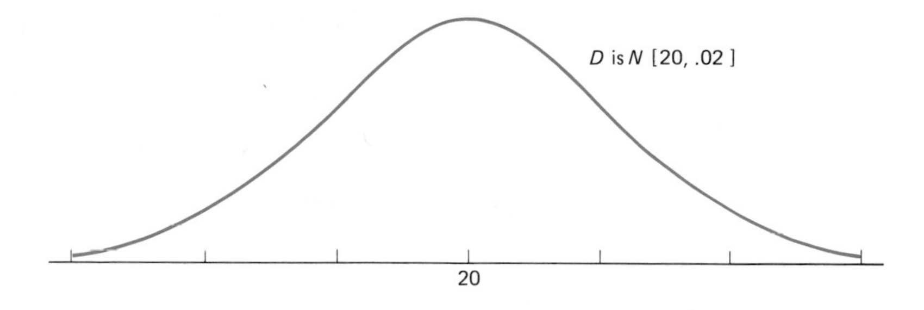

(c) Use the properties of the normal distribution to find the percentage of cylinders having diameters (i) between 19.98 and 20.02 cm; (ii) between 20.00 and 20.04 cm; (iii) greater than 20.04 cm; (iv) between 19.96 and 19.98 cm; (v) less than 20.02 cm.

Solution to Problem 1

(a) Since $\sigma = .4$, the values $G = 1.8$ and $G = 3.4$ are two standard deviations below and above $\mu = 2.6$, respectively. Accordingly, from property 4 approximately 95% of students have GPAs between 1.8 and 3.4.

(b) From property 1, 50% of the distribution lies above 2.6. We now require the percentage between 2.6 and 3.0. From property 4, approximately 68% of the distribution lies between 2.2 and 3.0. But again from property 1, since the distribution is symmetric around 2.6, half of the 68%, or 34%, lies between 2.6 and 3.0. Therefore, the percentage of students with GPAs greater than 3.0 is approximately 50% − 34% = 16%. This result is illustrated below:

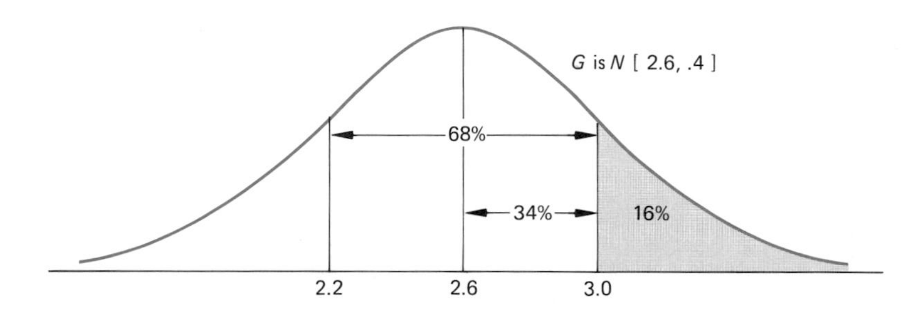

SECTION 6.2
THE STANDARDIZED NORMAL DISTRIBUTION

THE STANDARDIZED NORMAL RANDOM VARIABLE

As we said earlier, there is not just one, but a whole family of normal distributions. One member of this family is of special importance in statistics. In this section we consider properties of this distribution.

First, let us assume a random variable, Z, which has a normal distribution with mean $\mu = 0$ and standard deviation $\sigma = 1$. There is no natural phenomenon known to have this distribution, but the random variable Z can be derived from another (actually any other) normal random variable. For instance, suppose that we have a random variable X that is $N[16, 4]$. Then $(X - 16)/4$ is also a random variable since for every value of X there is a corresponding value of $(X - 16)/4$. It can be shown that the random variable $(X - 16)/4$ has a normal distribution with mean 0 and standard deviation 1 and is therefore $= Z$. In effect, $(X - 16)/4$ and Z are just two representations of the same random variable. More generally it is true, but difficult to prove, that

> DEFINITION 3
>
> If X is $N[\mu, \sigma]$, then $\dfrac{X - \mu}{\sigma} = Z$ and is $N[0, 1]$.

For reasons that are explained in the next section, Z is referred to as the **standardized normal random variable** and its $N[0, 1]$ distribution as the **standardized normal distribution.**

The distribution of Z is shown in Figure 8. Recall from Property 4 in Section 6.1 that most (actually 99.75%) of a normal distribution lies within three standard deviations of the mean. In this case since $\mu_Z = 0$ and $\sigma_Z = 1$, this will be between -3 and $+3$. In other words, $P(-3 < Z < 3) = .9975$.

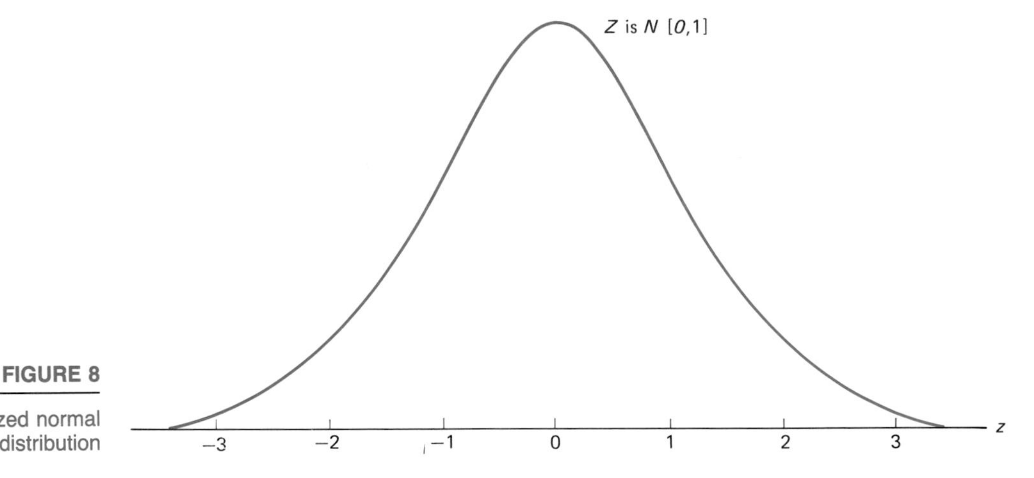

FIGURE 8

Standardized normal distribution

In Figure 9 we have shaded the areas corresponding to the probabilities $P(Z < 1.72)$ and $P(Z < -.54)$. These and other probabilities for Z can be found by using Table A3. This table provides the area *to the left of z*, that is, $P(Z < z)$, for values of z from -4.00 to $+4.00$ at intervals of .01. Two sections of Table A3 are shown in Table 1. There it can be seen by checking the value $z = 1.72$ that the corresponding area, $P(Z < 1.72)$, is .9573. Similarly, $P(Z < -.54)$ may be found by checking the entry corresponding to $z = -.54$. It is .2946.

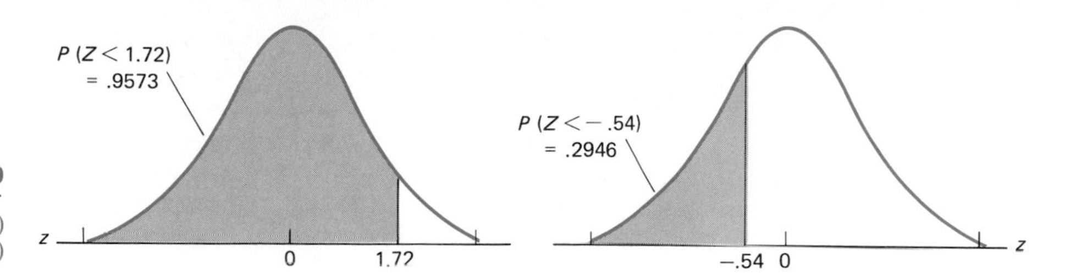

FIGURE 9

Probabilities $P(Z < 1.72)$ and $P(Z < -.54)$

TABLE 1

Portions of Table A3 showing that $P(Z < 1.72) = .9573$ and that $P(Z < -.54) = .2946$

z	$P(Z < z)$	z	$P(Z < z)$	z	$P(Z < z)$	z	$P(Z < z)$	z	$P(Z < z)$	z	$P(Z < z)$
−4.00	.00003	−3.33	.0004	−2.66	.0039	−1.99	.0233	−1.32	.0934	−.65	.2579
−3.99	.00003	−3.32	.0004	−2.65	.0040	−1.98	.0239	−1.31	.0951	−.64	.2611
−3.98	.00003	−3.31	.0004	−2.64	.0042	−1.97	.0244	−1.30	.0968	−.63	.2644
−3.97	.00004	−3.30	.0004	−2.63	.0043	−1.96	.0250	−1.29	.0985	−.62	.2676
−3.96	.00004	−3.29	.0005	−2.62	.0044	−1.95	.0256	−1.28	.1003	−.61	.2709
−3.95	.00004	−3.28	.0005	−2.61	.0045	−1.94	.0262	−1.27	.1020	−.60	.2743
−3.94	.00004	−3.27	.0005	−2.60	.0047	−1.93	.0268	−1.26	.1038	−.59	.2776
−3.93	.00004	−3.26	.0006	−2.59	.0048	−1.92	.0274	−1.25	.1057	−.58	.2810
−3.92	.00004	−3.25	.0006	−2.58	.0049	−1.91	.0281	−1.24	.1075	−.57	.2843
−3.91	.00005	−3.24	.0006	−2.57	.0051	−1.90	.0287	−1.23	.1094	−.56	.2877
−3.90	.00005	−3.23	.0006	−2.56	.0052	−1.89	.0294	−1.22	.1112	−.55	.2912
−3.89	.00005	−3.22	.0006	−2.55	.0054	−1.88	.0301	−1.21	.1131	−.54	.2946
−3.88	.00005	−3.21	.0006	−2.54	.0055	−1.87	.0307	−1.20	.1151	−.53	.2981
−3.87	.00005	−3.20	.0006	−2.53	.0057	−1.86	.0314	−1.19	.1170	−.52	.3015
0.24	.5948	0.96	.8315	1.68	.9535	2.40	.9918	3.12	.9991	3.84	.9999
0.25	.5987	0.97	.8340	1.69	.9545	2.41	.9920	3.13	.9991	3.85	.9999
0.26	.6026	0.98	.8365	1.70	.9554	2.42	.9922	3.14	.9992	3.86	.9999
0.27	.6064	0.99	.8389	1.71	.9564	2.43	.9925	3.15	.9992	3.87	1.0000
0.28	.6103	1.00	.8413	1.72	.9573	2.44	.9927	3.16	.9992	3.88	1.0000
0.29	.6141	1.01	.8437	1.73	.9582	2.45	.9929	3.17	.9992	3.89	1.0000
0.30	.6179	1.02	.8461	1.74	.9591	2.46	.9931	3.18	.9993	3.90	1.0000
0.31	.6217	1.03	.8485	1.75	.9599	2.47	.9932	3.19	.9993	3.91	1.0000
0.32	.6255	1.04	.8508	1.76	.9608	2.48	.9934	3.20	.9993	3.92	1.0000
0.33	.6293	1.05	.8531	1.77	.9616	2.49	.9936	3.21	.9993	3.93	1.0000
0.34	.6331	1.06	.8554	1.78	.9625	2.50	.9938	3.22	.9994	3.94	1.0000
0.35	.6368	1.07	.8577	1.79	.9633	2.51	.9940	3.23	.9994	3.95	1.0000
0.36	.6406	1.08	.8599	1.80	.9641	2.52	.9941	3.24	.9994	3.96	1.0000
0.37	.6443	1.09	.8621	1.81	.9649	2.53	.9943	3.25	.9994	3.97	1.0000

EXAMPLE 3 Using Table A3, compute the probabilities (a) $P(Z < 1.07)$; (b) $P(Z > .29)$; (c) $P(-1.91 < Z < .45)$.

Solution The three probabilities are shown in Figure 10.

(a) The $P(Z < 1.07)$ can be read directly from Table A3 as .8577.

(b) The $P(Z > .29)$ must be found indirectly. The events $Z > .29$ and $Z < .29$ are complementary. This can be seen in Figure 10(b). As a consequence, $P(Z > .29) = 1 - P(Z < .29)$. From Table A3, $P(Z < .29) = .6141$ and thus $P(Z > .29) = 1 - .6141 = .3859$.

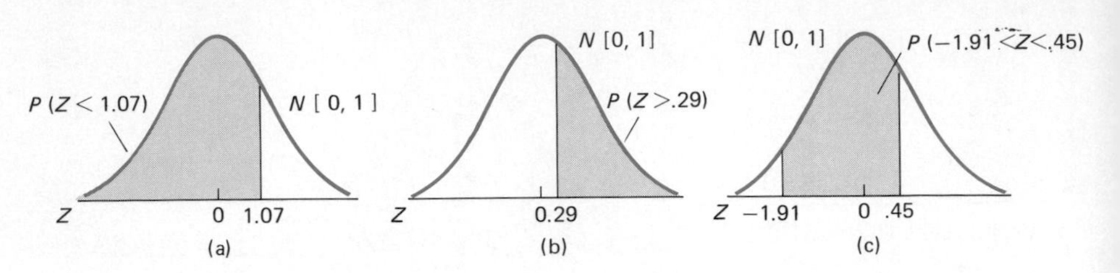

FIGURE 10

Probabilities $P(Z < 1.07)$, $P(Z > .29)$, and $P(-1.91 < Z < .45)$

(c) The $P(-1.91 < Z < .45)$ can be thought of as the difference between two areas: the total area to the left of .45, that is, $P(Z < .45)$, and the part of this which is the area to the left of -1.91, that is, $P(Z < -1.91)$. Thus $P(-1.91 < Z < .45) = P(Z < .45) - P(Z < -1.91) = .6736 - .0281 = .6455$.

■

EXAMPLE 4 Compute (a) $P(Z > 0)$; (b) $P(Z < -5.7)$; (c) $P(Z < 4.8)$.

Solution These three probabilities are shown in Figure 11.

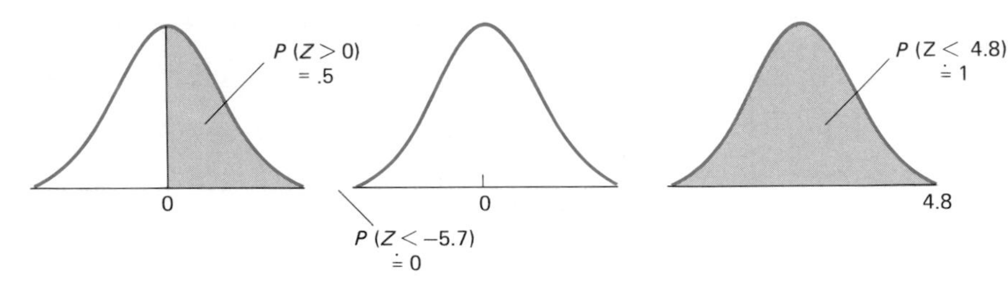

FIGURE 11

Probabilities $P(Z > 0)$, $P(Z < -5.7)$, and $P(Z < 4.8)$

(a) From Table A3 we find that $P(Z > 0) = 1 - P(Z < 0) = 1 - .5 = .5$. Actually, the symmetry of the standardized normal distribution around zero would be sufficient to indicate that $P(Z > 0) = P(Z < 0) = .5$.

(b) The value $z = -5.7$ is beyond the range of Table A3, but notice that the area to the left of -4.0, that is, $P(Z < -4.0) = .00003$. Since -5.7 is even farther out in the left tail, the area to the left of -5.7, that is, $P(Z < -5.7)$, is certainly less than .00003 and is, therefore, to four decimal places, zero.

(c) Again 4.8 is beyond the range of the table but in the opposite direction. Since the area to the left of 4.0, $P(Z < 4.0) = .99997$, the area to the left of 4.8 is certainly greater than .99997 and is, therefore, to four decimal places, equal to 1.

Notice that $P(Z \leq 1.07)$, for example, is simply $P(Z < 1.07)$ without the equal sign. The reason is that, as with any continuous random variable, the probability that Z takes a specific value is 0.

■

PERCENTILES OF Z It is sometimes useful to reverse the process described above. For example, if it is known that $P(Z < A) = .9$, what is A? As Figure 12 shows, A is that value

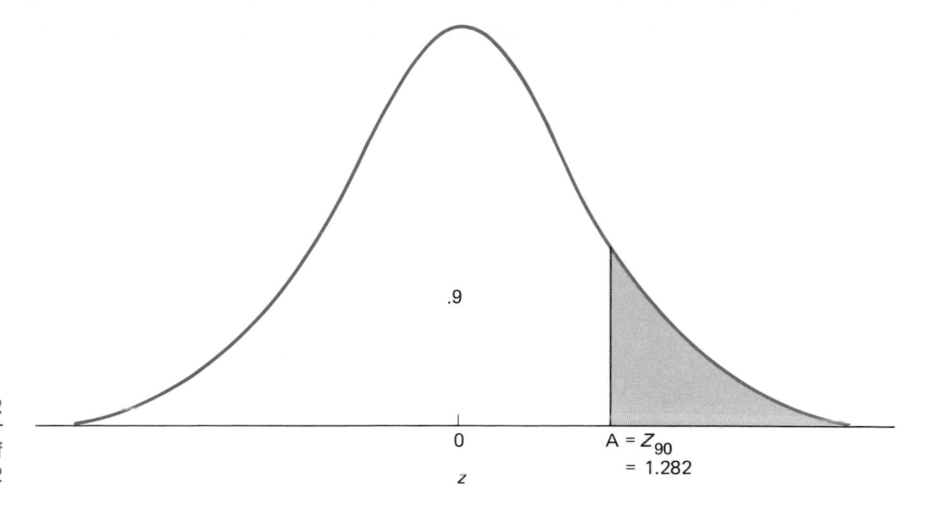

FIGURE 12

A is the 90th percentile of $Z = Z_{90} = 1.282$

of Z which has 90% of the distribution of Z to the left. Notice that A must be located to the right of 0 since the area to the left of 0 is .5 and we are concerned with an area greater than .5.

An approach to finding A is to use Table A3 in reverse; that is, locate the area and find the corresponding value for Z. The nearest area to .9 is .8997. The corresponding value is $z = 1.28$. Therefore, A is close to 1.28. We can do better than this, however, by using Table A4, which contains values for z corresponding to various areas (probabilities) of the form $P(Z < z)$. A portion of this table is shown in Table 2. Searching for the entry .9 in the column headed $P(Z < z)$, we see that the corresponding value for z is 1.282. We call 1.282 the 90th percentile of Z since 90% of the distribution of Z lies to the left of this value. This percentile is designated Z_{90}.

TABLE 2

Portion of Table A4 showing that $Z_{90} = 1.282$

K	$P(Z < Z_K) = \frac{K}{100}$	Z_K
50	.50	0
55	.55	.1256
90	.90	1.282
92.5	.925	1.440
95	.950	1.645
96	.960	1.751

Since the total area under a probability distribution is 1, K percent (%) of the area will be $[K/100](1) = K/100$. Thus, in the example above, the area to the left of 1.282, the 90th percentile is $90/100 = .9$.

The term **percentile** can be used in connection with any distribution. It refers to the *value* of a variable below which a specific percentage of its distribution lies.

We can define the Kth percentile of Z as follows:

> DEFINITION 4
>
> The Kth percentile of Z, designated Z_K, is that value of Z below which $K\%$ of the Z distribution lies.

EXAMPLE 5 Find Z_{40}, the 40th percentile of Z.

Solution Check either the first column of Table A4, headed "K" for 40 or the second column, headed "$P(Z < z)$," for .4. The corresponding value in the third column is $Z_{40} = -.2533$. This is shown in Figure 13.

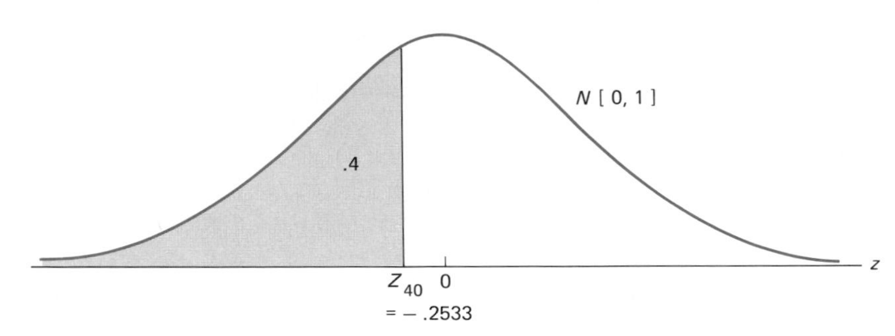

FIGURE 13

40th percentile of Z; $Z_{40} = -.2533$

■

Notice that Z_{50}, the value of Z below which 50% of the distribution lies, is 0. Thus percentiles *below* the 50th are negative. For instance, $Z_{25} = -.674$, whereas percentiles above the 50th are positive.

It is easy to confuse a percentile of Z with a probability or a percentage. These three concepts may be distinguished as follows:

EXPRESSION	TERM	MEANING
$Z_{75} = .674$	A percentile	A value of Z
75%	A percentage	Percentage of the area under the Z or $N[0, 1]$ distribution below the value $Z_{75} = .674$
$P(Z < .674) = .75$	A probability	The probability that Z is less than .674

Tables A3 and A4 are used extensively in later chapters, and we urge you to become proficient in their use. In working through the following problem set, graph each problem as we have done throughout the section.

PROBLEMS 6.2

In all problems, Z is the standardized normal random variable; that is, Z is $N[0, 1]$.

4. (a) Compute (i) $P(Z < -.46)$; (ii) $P(.05 < Z < .15)$.
   (b) Find Z_{25} and Z_{75}.
5. Compute (a) $P(Z < .13)$; (b) $P(Z < -.54)$; (c) $P(Z < 4.7)$; (d) $P(Z < 1.74)$.
6. Compute (a) $P(Z > .36)$; (b) $P(Z \geq -.36)$; (c) $P(Z > -1.24)$; (d) $P(Z > -7.1)$.
7. Compute (a) $P(-.91 < Z < .61)$; (b) $P(-.91 < Z < -.61)$; (c) $P(.75 < Z < 1.75)$; (d) $P(-1.2 < Z < 1.2)$; (e) $P(0 < Z < .72)$; (f) $P(-5.7 < Z < 1.64)$; (g) $P(.36 < Z < 4.8)$.
8. Compute (a) $P(Z < -.8 \text{ or } Z > .8)$; (b) $P(Z < -2.4 \text{ or } Z > .49)$.
9. Find the (a) 15th (b) 99th (c) 1st (d) 55th (e) .5th and (f) 99.75th percentiles of Z.
10. Find (a) A such that $P(Z < A) = .55$; (b) B such that $P(Z < B) = .04$; (c) C such that $P(Z > C) = .55$.

Solution to Problem 4

(a) (i) Directly from Table A3 we obtain $P(Z < -.46) = .3228$.
(ii) The $P(.05 < Z < .15)$ is the difference between two probabilities (areas), that is,

$$P(.05 < Z < .15) = P(Z < .15) - P(Z < .05)$$
$$= .5596 - .5199 = .0397$$

These two probabilities are shown as follows:

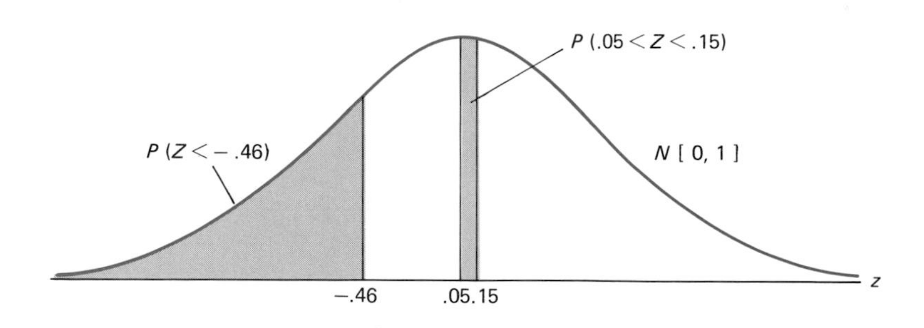

(b) From Table A4, $Z_{25} = -.674$ and $Z_{75} = .674$. The fact that $Z_{25} = -Z_{75}$ follows from the symmetry of the distribution around 0. Although 75% of the distribution lies *below* Z_{75}, 25% lies above this point, and this is the same percentage of the distribution that lies below Z_{25}. This is shown as follows:

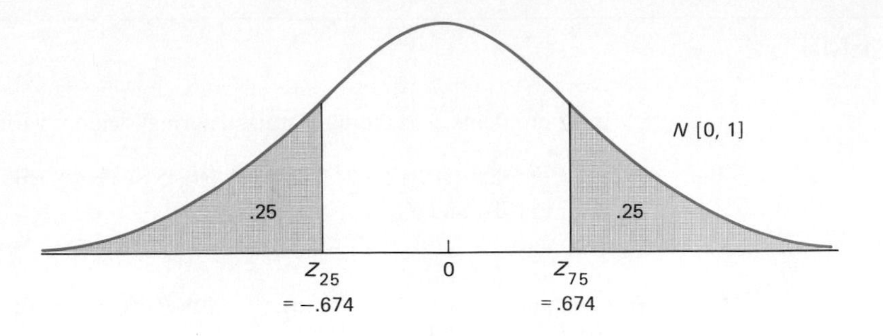

Section 6.3

FINDING PROBABILITIES AND PERCENTILES FOR NORMAL DISTRIBUTIONS

In Definition 3 on page 222 we noted that if X has the $N[\mu, \sigma]$ distribution, then $(X - \mu)/\sigma$ has the $N[0, 1]$ distribution and is therefore equivalent to Z. This relationship is illustrated in Figure 14 with the $N[16, 4]$ distribution of X. Although this is only an approximate visual representation, it shows that the Z distribution is much narrower than that for X since the standard deviation of Z is 1, while that of X is 4.

There is an equivalent correspondence between the *values* for X and those for Z. For instance, the value $x = 14$ will be $(14 - 16)/4 = -2/4 = -.5$ (or 1/2 standard deviation) below 16 and $x = 22$ will be $(22 - 16)/4 = 1.5$ standard deviations above 16. The values of Z corresponding to these x values will also be $(14 - 16)/4 = -.5$ and $(22 - 16)/4 = 1.5$ (see Figure 14). Thus in each case the Z value turns out to be the number of standard deviations the corresponding value for X is above or below μ. We therefore call $z = -.5$ and $z = 1.5$ the standardized values for $x = 14$ and $x = 22$.

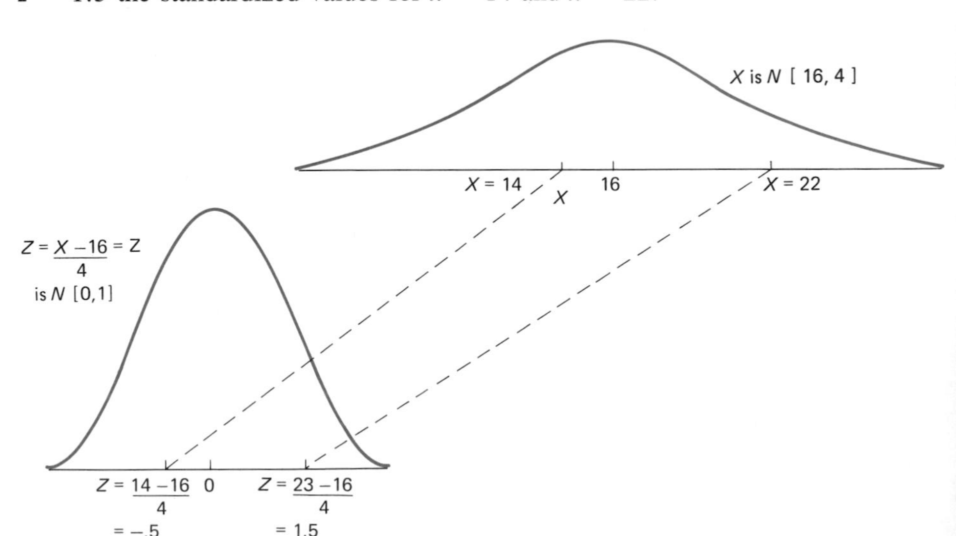

FIGURE 14

Showing the relationship between $x = 14$ and $z = -.5$ and between $x = 22$ and $z = 1.5$

More generally:

> DEFINITION 5
>
> For any specific value (x) for a normal random variable X, there is a corresponding value, $z = \dfrac{x - \mu}{\sigma}$, which indicates the number of standard deviations x is above or below μ. This is referred to as the **standardized value** for x.

COMPUTING NORMAL PROBABILITIES

The relationship in Definition 5 enables us to use the probabilities about Z in Table A3 to compute probabilities associated with *any* normal random variable. An example will show how this is done.

EXAMPLE 6

During a particular week the distribution of the number of hours of nonacademic work performed by students at a university is $N[16, 4]$. If X is the random variable, the number of hours worked by a randomly chosen student in this week, find (a) $P(X < 14)$; (b) $P(X > 22)$.

Solution

(a) The area corresponding to $P(X < 14)$ is shown in Figure 15. We compute this probability by translating it into a probability about Z. Since the value $x = 14$ corresponds to the value $z = (14 - 16)/4 = -.5$, the events $X < 14$ and $Z < -.5$ are equivalent. As a result, $P(X < 14) = P(Z < -.5)$. Finally, from Table A3, $P(Z < -.5) = .3085 = P(X < 14)$. The equivalence of $P(Z < -.5)$ and $P(X < 14)$ is illustrated roughly in Figure 15.

(b) The $P(X > 22)$ is the area to the right of 22 under the $N[16, 4]$ distribution in Figure 15. Since the value $x = 22$ corresponds to the value $z = (22 - 16)/4 = 1.5$, the events $X > 22$ and $Z > 1.5$ are equivalent. As a result, $P(X > 22) = P(Z > 1.5)$. From Table A3, $P(Z > 1.5) = 1 -$

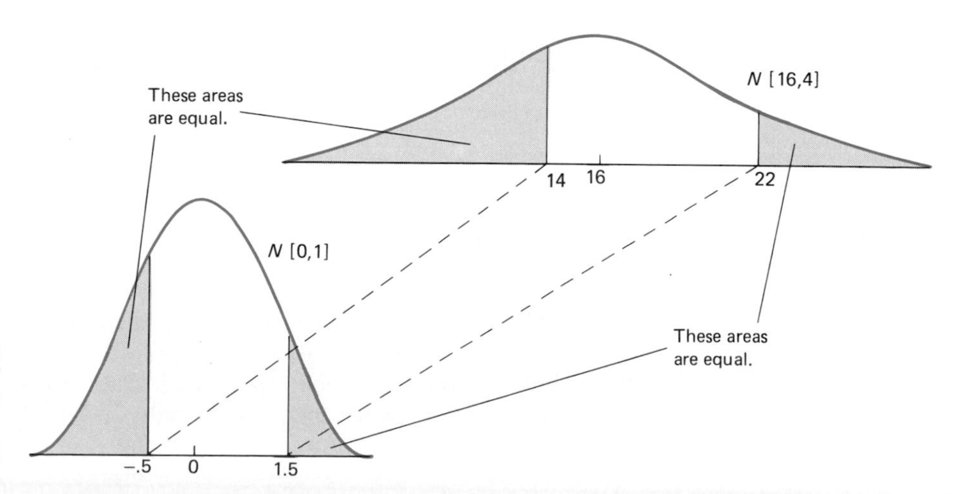

FIGURE 15

Translating probabilities about a $N[16, 4]$ random variable into probabilities about Z

$P(Z < 1.5) = 1 - .9332 = .0668 = P(X > 22)$. The equivalence of $P(X > 22)$ and $P(Z > 1.5)$ is also shown in Figure 15.

Each of these probabilities can be interpreted in two ways. For instance, $P(X < 14) = .3085$ is the probability that a randomly selected student will work less than 14 hours during the week. This same value, .3085, can also be thought of as the *proportion* of all students who work fewer than 14 hours in that week.

■

The general procedure for obtaining probabilities about X, an $N[\mu, \sigma]$ random variable, by standardizing it is outlined in Table 3.

TABLE 3

Procedure for computing probabilities about X, a $N[\mu, \sigma]$ random variable[a]

(a) $P(X < A) = P\left(Z < \frac{A - \mu}{\sigma}\right)$

(b) $P(X > A) = P\left(Z > \frac{A - \mu}{\sigma}\right) = 1 - P\left(Z < \frac{A - \mu}{\sigma}\right)$

(c) $P(A < X < B) = P\left(\frac{A - \mu}{\sigma} < Z < \frac{B - \mu}{\sigma}\right)$

$= P\left(Z < \frac{B - \mu}{\sigma}\right) - P\left(Z < \frac{A - \mu}{\sigma}\right)$

[a] A and B are constants.

Let us apply this standardizing procedure to another example.

EXAMPLE 7 Among pregnancies that go to full term the length of the human gestation period is approximately normal with a mean of 266 days (just about nine months) and a standard deviation of 12 days. What percentage of mothers carry their children between 260 and 270 days?

Solution If X is the random variable "the gestation period for a randomly chosen birth," then X is $N[266, 12]$. We require the $P(260 < X < 270)$. Applying procedure (c) in Table 3, with $A = 260$ and $B = 270$ days, we obtain

$$P(260 < X < 270) = P\left(\frac{260 - 266}{12} < Z < \frac{270 - 266}{12}\right)$$

$$= P(-.5 < Z < .33) = P(Z < .33) - P(Z < -.5)$$

$$= .6293 - .3085 = .3208$$

Approximately 32% of women carry their children between 260 and 270 days. These calculations are illustrated in Figure 16. In actuality, the difference in the curves would be much greater than that shown here since the standard deviation of X is 12 while that of Z, is only 1.

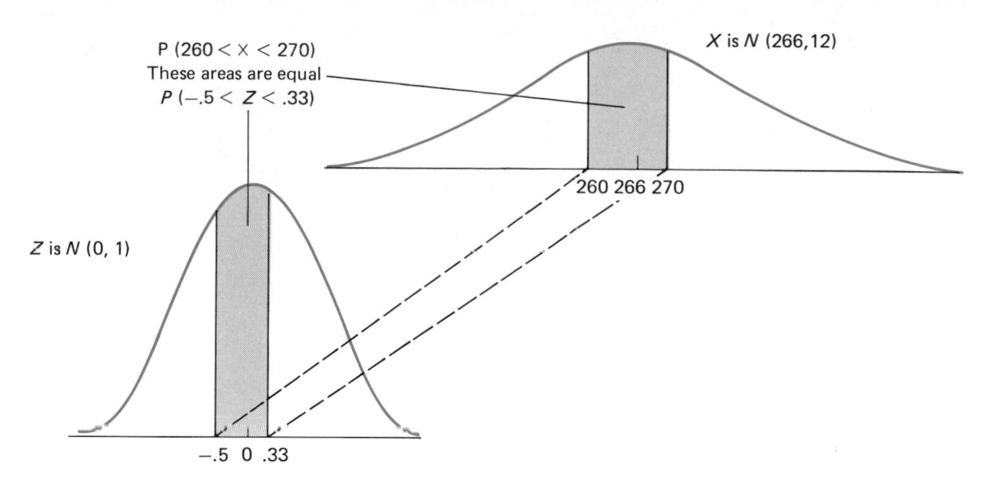

FIGURE 16

Equivalence of $P(260 < X < 270)$ and $P(-.5 < Z < .33)$

The $N[266, 12]$ distribution is an idealization of how the human gestation period varies, and careless use can lead to absurd results. For example, following procedure (b) in Table 3,

$$P(Z > 314) = P\left(Z > \frac{314 - 266}{12}\right) = P(Z > 4) = 1 - P(Z < 4)$$

$$= 1 - .99997 = .00003$$

Although this probability is not zero, the event $X > 314$ is impossible since pregnancies are terminated long before 314 days (10.5 months).

■

PERCENTILES OF A NORMAL DISTRIBUTION

The relationship in Definition 5 can also be used to find percentiles for any normal distribution. The next example illustrates the procedure.

EXAMPLE 8 Assume that the distribution of SAT math scores are normal with a mean of 450 and a standard deviation of 80. What is the 90th percentile of SAT math scores, that is, the score below which 90% of scores lie?

Solution Call the score below which 90% of the scores lie X_{90}, that is, the 90th percentile. The value for Z corresponding to X_{90} is $Z_{90} = (X_{90} - 450)/80$, that is, the 90th percentile of the standardized normal distribution. The correspondence between X_{90} and Z_{90} is indicated in Figure 17. From Table A4, $Z_{90} = 1.282$. Therefore, $(X_{90} - 450)/80 = Z_{90} = 1.282$.

We can now find X_{90} by solving the equation $(X_{90} - 450)/80 = 1.282$ for X_{90}. Multiplying both sides of the equation by 80 and transposing terms, we obtain

$$X_{90} = 450 + 80(1.282) = 552.56$$

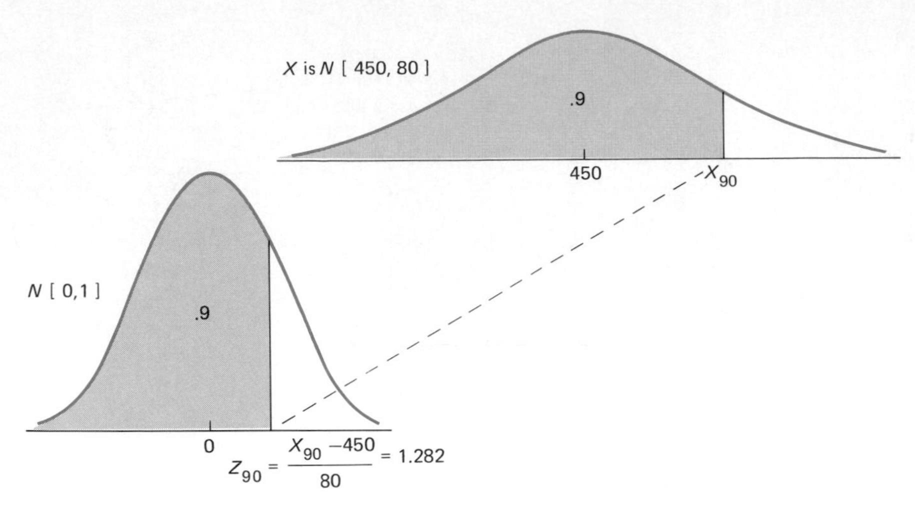

FIGURE 17

Correspondence between X_{90} and Z_{90}

The 90th percentile of SAT math scores is approximately 553. A student obtaining a math SAT score of 553 has attained a score higher than 90% of those taking the test.

■

In general, if X is $N[\mu, \sigma]$, we denote the Kth percentile of X as X_K. Recall that this is the value of X such that $K\%$ of the population are below X_K. The standardized value for X_K is $Z_K = (X_K - \mu)/\sigma$. Proceeding as we did in Example 8, we obtain the result

$$X_K = \mu + \sigma(Z_K) \qquad \text{(Eq. 6.1)}$$

Equation 6.1 indicates that the Kth percentile of X is Z_K standard deviations above or below μ. Z_K, in turn, is the value of the Kth percentile of Z and can be read from Table A4. X_K is above μ if Z_K is positive and below μ if Z_K is negative.

EXAMPLE 9 The distribution of the height of adult males is approximately $N[68, 3]$. Find the (a) 15th, (b) 65th, and (c) 98th percentile of adult male heights.

Solution Using Table A4 and Equation 6.1 with $K = 15$, 65, and 98, respectively, we have

(a) $X_{15} = \mu + \sigma(Z_{15}) = 68 + 3(-1.036) = 64.9$ inches
(b) $X_{65} = \mu + \sigma(Z_{65}) = 68 + 3(.3854) = 69.2$ inches
(c) $X_{98} = \mu + \sigma(Z_{98}) = 68 + 3(2.054) = 74.2$ inches

These percentiles are shown in Figure 18.

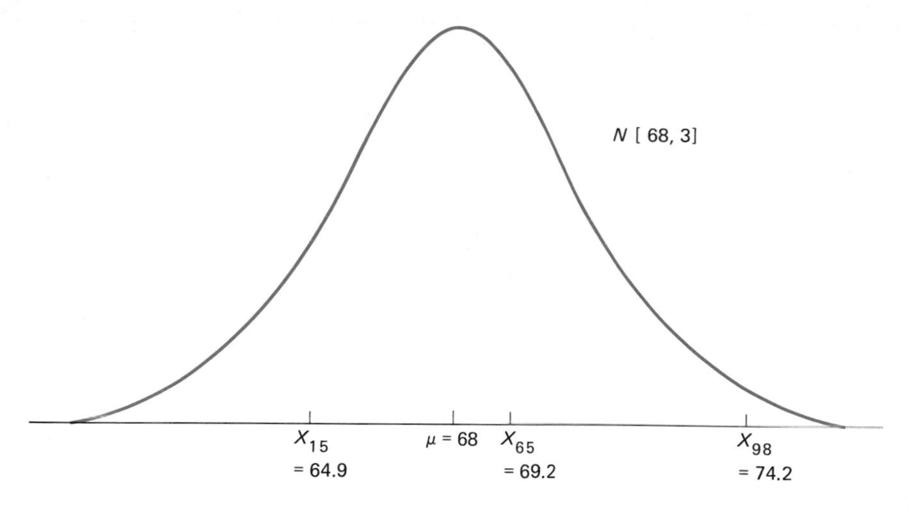

FIGURE 18

15th, 65th, and 98th percentiles of the height of adult males

EXAMPLE 10 At a large university, students attaining the top 5% of grade-point averages (GPAs) make the dean's list. Assume that the distribution of GPAs at the school is $N[2.5, .5]$. Above what GPA do students make the dean's list?

Solution The problem is represented in Figure 19. We require the specific GPA such that 5% of the distribution lies to the right. But that means that 95% of the distribution lies to the left. We therefore require X_{95}. Using Table A4 and Equation 6.1 with $K = 95$, we have

$$X_{95} = \mu + \sigma(Z_{95}) = 2.5 + (.5)(1.645) = 3.3225$$

Students with a GPA above 3.32 will make the dean's list.

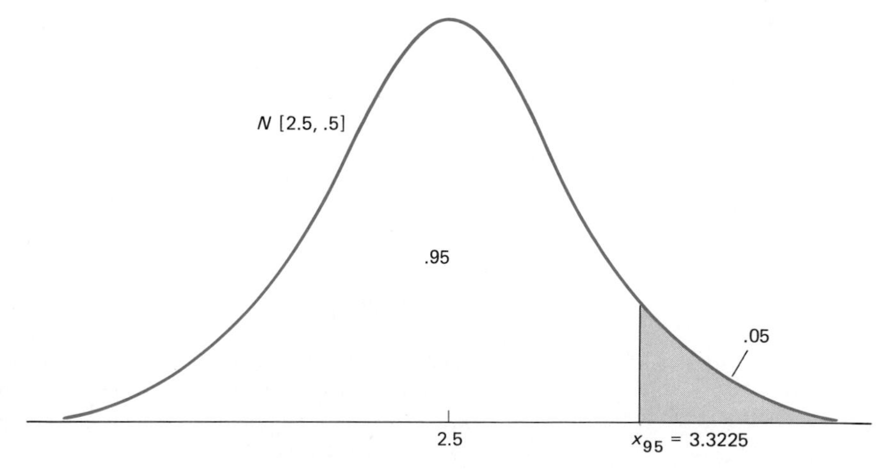

FIGURE 19

95th percentile of the $N[2.5, .5]$ distribution

PROBLEMS 6.3

11. The daily maximum temperature (T) on a Caribbean Island has approximately a normal distribution with a mean of 78°F and a standard deviation of 5°F. On what proportion of days is the maximum temperature (a) below 70°F; (b) greater than 85°F?
12. The distribution of systolic blood pressure (SBP) among undergraduates is approximately normal with a mean of 120 mm Hg and a standard deviation of 14 mm Hg. What proportion of undergraduates have a SBP (a) greater than 125; (b) between 110 and 120; (c) less than 140; (d) between 108 and 112?
13. A commuter estimates that the length of the train ride to downtown varies according to a normal distribution with a mean of 50 minutes and a standard deviation of 4.5 minutes. What is the probability that on a randomly chosen day the length of the journey will (a) be less than 45 minutes; (b) lie between 46 and 54 minutes; (c) exceed an hour?
14. The distribution of final scores in freshman calculus is approximately $N[74, 8]$.

    (a) What proportion of students score (i) over 90; (ii) less than 60?

    (b) The instructor gives A's to students receiving the top 15% of scores. What score must a student exceed to earn an A?
15. The distribution of income tax rebates claimed by doctors is approximately $N[6400, 1600]$. (a) All of the returns claiming the largest 2.5% of rebates will be audited. Above what rebate amount will a doctor's return be audited?

    (b) None of the returns claiming the smallest 7.5% of returns will be audited. Below what rebate amount need a doctor not be concerned about an audit?
16. The distribution of the age of college graduates at graduation is $N[21.6, .3]$.

    (a) Find the (i) 70th, (ii) 50th, and (iii) 25th percentiles of age at graduation.

    (b) What proportion of graduates is (i) older than 22 years; (ii) younger than 21 years?
17. Using the information in Problem 16, suppose that two graduates are randomly selected. What is the probability that:

    (a) Both are older than 22?

    (b) Only one of them exceeds the 70th percentile of age?
18. The breaking strength of a particular variety of worsted yarn has approximately a normal distribution with a mean of 16.5 ounces and a standard deviation of 1.2 ounces.

    (a) If X is the breaking strength of a randomly selected length of the material, compute (i) $P(X > 16.8)$; (ii) $P(X \leq 16.42)$.

    (b) Find the (i) 20th, (ii) 35th, and (iii) 95th percentiles of the distribution.
19. The distribution of SAT math scores is approximately $N[450, 80]$.

    (a) If a student is told that she has scored at the 90th percentile, what score did she get?

    (b) Answer the same question if the student scored at the (i) 30th, and (ii) the 99th percentiles.
20. The distribution of the amount of coffee dispensed by a machine is normal with a mean of 6 ounces and a standard deviation of .2 ounce.

    (a) If a 6-ounce cup actually holds 6.36 ounces without spilling, on what proportion of times will the coffee spill?

    (b) On what proportion of times will the amount of coffee dispensed (i) exceed 5.9 ounces; (ii) be between 5.9 and 6.1 ounces?

(c) If a person gets two cups of coffee, what is the probability that (i) both cups will contain less than 6 ounces; (ii) the coffee will spill over at least one of the cups?

21. In a small town the distribution of family income has a normal distribution with a mean of $18,900. If the 60th percentile of income is $19,810, what is the 40th percentile?

Solution to Problem 11

(a) We require $P(T < 70)$. Following procedure (a) in Table 3 with $A = 70$ (but using T instead of X) yields

$$P(T < 70) = P\left(Z < \frac{70 - 78}{5}\right) = P(Z < -1.6) = .0548$$

(b) Following procedure (b) in Table 3 with $A = 85$,

$$P(T > 85) = P\left(Z > \frac{85 - 78}{5}\right) = P(Z > 1.4)$$

$$= 1 - P(Z < 1.4) = 1 - .9192 = .0808$$

SECTION 6.4
SUMMARY AND REVIEW

The probability distribution for a continuous random variable, X, can be represented by areas under a curve; the total area under the curve is 1, and the probability that X takes a specific value is zero (Definition 2).

Many random variables, such as income or populations, are technically discrete because there are gaps between the possible values that they may take, but because of the large number of such values they are treated as continuous random variables.

The most important distribution in all of probability and statistics is the *normal* distribution. Usually referred to as *the* normal distribution, this is really a family of distributions all of which are bell-shaped and symmetric around their mean. We indicate that a random variable X has a normal distribution with mean μ and standard deviation σ with the notation X is $N[\mu, \sigma]$.

The total area under a normal curve is 1 and the following percentages of this area lie between 1, 2, or 3 standard deviations above and below the mean:

NUMBER OF STANDARD DEVIATIONS	PERCENTAGE OF AREA
1	68
2	95
3	99.75

A special member of the family of normal distributions is the standardized normal distribution. The standardized normal random variable Z is $N[0, 1]$; that is, Z is normal with mean 0 and standard deviation (and variance) equal to 1.

Table A3 provides the area to the left of values of Z; that is, $P(Z < z)$ from -4 to $+4$ at intervals of .01. This table, together with properties of the normal distribution, may be used to find any probability associated with Z.

The Kth percentile of Z, designated Z_K, is that value of Z below which K% of the Z distribution lies (Definition 4).

If any random variable, X, has the $N[\mu, \sigma]$ distribution, the random variable $(X - \mu)/\sigma$ has the standardized normal distribution. We summarize this result as

$$\text{If } X \text{ is } N[\mu, \sigma], \text{ then } \frac{X - \mu}{\sigma} = Z \text{ is } N[0, 1] \qquad \text{(Definition 3)}$$

Each value, x, has a corresponding value z that represents the number of standard deviations that x is above (or below) μ, the mean of the X distribution.

Probabilities associated with X can be found by using Definition 3 to translate these probabilities into the corresponding probabilities for Z and then referring to Table A3. This procedure, known as standardizing, is outlined in Table 3.

The Kth percentile, X_K, of an $N[\mu, \sigma]$ distribution can be found by using the equation

$$X_K = \mu + \sigma(Z_K) \qquad \text{(Eq. 6.1)}$$

where the value for Z_K is found from Table A4.

REVIEW PROBLEMS

GENERAL

1. As usual, Z is the standardized normal random variable. Find (a) $P(Z < 1.25)$; (b) $P(Z < -1.42)$; (c) $P(Z < 5.6)$; (d) $P(-2.3 < Z < .23)$; (e) $P(-1.08 < Z < 0)$; (f) $P(-5.2 < Z < .06)$; (g) $P(.41 < Z < 4.1)$.
2. The random variable Z has the standardized normal distribution.
   (a) Find (i) Z_{15}; (ii) Z_{65}.
   (b) Find the values for A, B, and C that satisfy the following equations: (i) $P(Z < A) = .4$; (ii) $P(Z > B) = .8$; (iii) $P(Z < C) = .025$.
3. The number of hours per week spent watching television by a young boy is approximately normal with a mean of 15 hours and a standard deviation of 2 hours.
   (a) In what proportion of weeks does the boy watch television (i) more than 16 hours; (ii) between 15 and 19 hours; (iii) more than 12.5 hours?
   (b) What are the (i) 90th, (ii) 50th, and (iii) the 85th percentile of the time per week spent watching television?
4. The distribution of the end-of-season batting averages for major league baseball players is approximately normal with a mean of .270 and a standard deviation of .018.
   (a) Find the (i) 90th, (ii) 60th, and (iii) 20th percentile of batting averages.
   (b) Interpret the value you obtained in part (a, i).

5. The number of apples per tree in a large commercial orchard follows approximately a normal distribution with a mean of 2500 apples and a standard deviation of 400 apples.

   (a) What proportion of trees have (i) more than 3000 apples; (ii) more than 4000 apples; (iii) between 2000 and 2500 apples?

   (b) Find the (i) 35th, (ii) 50th, (iii) 65th, and (iv) 80th percentiles of the distribution of the number of apples per tree.

6. The number of minutes past the hour that a student arrives in class has a normal distribution with mean 0 (on average she arrives on time) and standard deviation 3 minutes. In what proportion of classes does she arrive (a) more than 5 minutes late; (b) within 5 minutes of the start of class? (The class starts right on the hour.)

7. In a large statistics class the distribution of students' final test scores is approximately normal. The instructor assigns grades as follows: an A to those who score more than 1.5 standard deviations above the mean, a B to those who score between .5 and 1.5 standard deviations above the mean, a C to those who score within .5 standard deviation of the mean, a D to those who score between .5 and 1.5 standard deviations below the mean, and an F to those who score more than 1.5 standard deviations below the mean.

   (a) What proportion of students achieve each grade?

   (b) If the class consists of 200 students, how many students would you expect to get each grade?

8. The amount dispensed by a soft-drink dispensing machine has a normal distribution with mean μ and standard deviation .25 ounce. The mean amount dispensed, μ, can be controlled. At what value should μ be set so that a 9-ounce cup will overflow with probability .02?

9. The distribution of the annual salary of professors at a large university is approximately normal. The 25th percentile is $21,500. The 90th percentile is $35,400. Use these figures to compute the mean and the standard deviation of salary at the university.

10. A particular measurement X has a normal distribution with a mean of 7.5. If the 85th percentile of X is 9.25, what is the 15th percentile? Do *not* compute the SD(X); use properties of the normal distribution.

11. The distribution of the height of adult females is approximately normal with a mean of 64 inches and a standard deviation of 3 inches. What proportion of adult females are (a) taller than 60 inches; (b) between 60 and 64 inches tall; (c) shorter than 6 feet?

12. Referring to Review Problem 11, if four females are randomly selected, what is the probability that:

    (a) All four are taller than 60 inches?

    (b) Three of the four are taller than 60 inches?

13. A random variable X is normal with a mean of 50. If $P(X > 40) = t$, use the various properties of the normal distribution to express the following probabilities in terms of t: (a) $P(X < 40)$; (b) $P(X < 60)$; (c) $P(40 < X < 60)$; (d) $P(50 < X < 60)$; (e) $P(X < 40 \text{ or } X > 50)$.

HEALTH SCIENCES

14. The distribution of heart rate for patients suffering from heart disease is approximately normal with a mean of 97 (beats per minute) and a standard deviation of 18.

    (a) What proportion of such persons have heart rates (i) below 120; (ii) between 100 and 110?

    (b) Compute and interpret the 75th percentile of heart rate.

15. Researchers have found that when anesthetic X is used on patients undergoing a particular surgical operation, the recovery time is approximately normally distributed with a mean of 5.5 hours and a standard deviation of 45 minutes.

(a) What proportion of such patients recover in (i) less than 5 hours; (ii) between 5 and 6 hours?

(b) What is the 95th percentile of recovery time?

16. The number of tapeworms found in a hamster 10 days after 10 such tapeworms were introduced into its food seems to have approximately a normal distribution with a mean of 75 and a standard deviation of 11. What proportion of hamsters so infected would contain (a) more than 100 tapeworms; (b) less than 70 or more than 90 tapeworms?

17. Referring to Review Problem 16, suppose that five hamsters were infected as described. What is the probability that after 10 days:

(a) Only one of them had more than 100 tapeworms?

(b) At least one of them had more than 100 tapeworms?

18. A doctor has developed a test which, it is hoped, will discriminate well between persons who do and do not have a particular disease. Research suggests that persons scoring above 18.8 on the test be classified as having the disease. Scores for persons who are later found to have the disease follow a normal distribution with a mean of 20 and a standard deviation of 1.5. The scores for those persons who are later found *not* to have the disease follow a normal distribution with a mean of 17 and a standard deviation of 1.

(a) What proportion of people who, in fact, have the disease will be classified as not having it?

(b) What proportion of persons who do not, in fact, have the disease will be classified as having it?

SOCIOLOGY AND GOVERNMENT

19. Federal officials estimate that the distribution of Supplementary Security Income (SSI) payments to those eligible is approximately normal with a mean of $120 per month and a standard deviation of $35 per month.

(a) What proportion of SSI recipients received (i) less than $100 per month; (ii) either less than $100 per month or more than $200 per month?

(b) What is the 85th percentile of the distribution of SSI payments?

(c) Interpret the number found in part (b).

20. The length of waiting time for trial for a criminal offense in a state is approximately normal with a mean of 95 days and a standard deviation of 25 days.

(a) What proportion of persons wait (i) longer than 100 days; (ii) fewer than 75 days?

(b) Compute the (i) 25th, (ii) 50th, and (iii) 75th percentile of waiting times.

(c) Interpret the value in part (b, iii).

21. Referring to Review Problem 20, suppose that 10 persons awaiting trial are selected at random. What is the probability that:

(a) All wait longer than 75 days?

(b) Exactly three wait less than the 25th percentile of waiting time?

(c) Exactly eight wait less than the 75th percentile of waiting time?

ECONOMICS AND MANAGEMENT

22. Records kept by a producer of automobile tires suggest that the distribution of the mileage to be expected from these tires is approximately normal with a mean of 35,000 miles and a standard deviation of 4000 miles.

(a) What proportion of this company's tires last (i) less than 30,000 miles; (ii) less than 40,000 miles?

(b) The company's advertising manager wishes to claim that "90% of our tires last longer than A miles." What should A be?

23. In Review Problem 22, if two tires are selected at random, what is the probability that:

(a) Both will last longer than 40,000 miles?

(b) Only one of them will last longer than 40,000 miles?

24. The owner of a light van complained to the dealer that he was getting only 16.5 miles per gallon (mpg). The dealer pointed out that the "19 mpg" referred to in an advertisement for this make of van was "just a guide and that actual mileage will vary." Suppose that 19 *is* the mean mpg for this make of van and that the distribution of mileage per van is normal with a standard deviation of $\sigma = .75$ mpg.

(a) How unlikely is a van that averages only 16.5 mpg or less?

(b) What does your answer suggest about the manufacturer's claim?

25. The distribution of the amount of overdue accounts for a large public utility is approximately normal with a mean of $29 and a standard deviation of $12.

(a) What proportion of accounts were overdue by more than $30?

(b) If three accounts are selected at random, what is the probability that all are overdue by more than $30?

26. The distribution of the time it takes for an automobile inspection to be completed is approximately normal with a mean of 20 minutes and a standard deviation of 5 minutes.

(a) What proportion of inspections take (i) less than 15 minutes; (ii) between 15 and 30 minutes; (iii) longer than 18 minutes?

(b) The inspector charges $4 if the inspection lasts less than 15 minutes, $6 if it lasts between 15 and 30 minutes, and $10 if it lasts longer than 30 minutes. What is the mean payment made to the inspector?

EDUCATION AND PSYCHOLOGY

27. The distribution of IQ scores in a population is normal with a mean of 100 and a standard deviation of 15.

(a) If a person is chosen at random, what is the probability that that person will have an IQ (i) greater than 110; (ii) between 95 and 105; (iii) between 110 and 120?

(b) Why are your answers in part (a, ii and iii) different though the intervals 95–105 and 110–120 are of equal lengths?

(c) A high-IQ club accepts as members only persons with IQs above the 99.5th percentile. Above what score are you eligible for membership in this club?

28. A psychologist is testing the reaction time of people to simulated traffic crises. Her studies suggest that reaction time is approximately normally distributed with a mean of .3 second and a standard deviation of .04 second. What proportion of people react in (a) less than .2 second; (b) more than .35 second?

29. The distribution of grade-point averages (GPAs) at a large college is approximately normal with a mean of 2.4 and a standard deviation of .5. Students having a GPA below 1.5 are automatically placed on probation. Students having a GPA of 3.3 or better make the dean's list.

(a) What proportion of students are put on probation?

(b) What proportion fall between the top and bottom groups, that is, those on the dean's list and those on probation?

30. Referring to Review Problem 29, suppose that two students are selected at random. What is the probability that:

(a) Both fall between the top and bottom group?

(b) One of them is on the dean's list?

SCIENCE AND TECHNOLOGY

31. Each day random samples of a city's water supply are analyzed and the concentration of fluoride (in parts per million) measured. It is found that the distribution of this concentration is approximately normal with a mean of 1.3 parts per million and a standard deviation of .15 part per million.

(a) On what proportion of days is the concentration (i) over 1 part per million; (ii) over 1.5 parts per million?

(b) What are the (i) 80th, and (ii) 40th percentiles of the distribution?

(c) Interpret the two numbers obtained in part (b).

32. Laboratory experiments suggest that the amount of weight loss (in grams) of a particular compound held in a controlled environment for 2 days has approximately a normal distribution with a mean of 120 grams and a standard deviation of 25 grams. On what proportion of occasions will the weight loss (a) exceed 100 grams; (b) be less than 180 grams; (c) be between 100 and 150 grams?

33. A particular section of piping is mass produced. The distribution of the width of the piping produced is approximately normal with a mean of 10 centimeters and a standard deviation of .025 centimeter. Piping sections are discarded if the width exceeds 10.045 cm or is less than 9.955 cm.

(a) What proportion of piping sections have to be discarded?

(b) How would this proportion decline if the standard deviation of the width were reduced to .015 centimeter?

34. The distribution of voltage (V) in a circuit is approximately normal with a mean of 110 volts and a standard deviation of 3 volts.

(a) Compute (i) $P(V > 110)$; (ii) $P(V < 100)$.

(b) What is the 80th percentile of voltage?

SUGGESTED READING

AYER, A. J. "Chance"; WEAVER, W. "Probability"; and KAC, M. "Probability." Readings in *Mathematics in the Modern World,* San Francisco; Foreman, 1968.

MOORE, D. S. *Statistics: Concepts and Controversies,* San Francisco; Freeman, 1979, pp. 237–268.

STANCL, D. L. and STANCL, M. L. *Applications of College Mathematics,* Lexington, Mass.: D. C. Heath, 1983, pp. 323–414.

ROSS, S. *A First Course in Probability,* New York: Macmillan, 1976.

7 THE SAMPLING DISTRIBUTION OF $\overline{X}$

THE ARMY VERSUS BRASH

The U.S. Army orders 10,000 advanced four-wheel drive vehicles from Brash, Inc. On average the vehicles must travel 2500 miles before a breakdown.

Before accepting the entire shipment the Army will test drive a sample of 35 vehicles. However, the Army and Brash differ on the terms on which the vehicles will be acceptable. The Army wants to accept the entire order only if the average mileage before breakdown in the sample exceeds 2500 miles. The company argues that even if the mean over all 10,000 cars is 2500 miles in a sample of only 35, the mean may, by chance, be somewhat less than 2500 miles. After some discussion the Army agrees to accept the entire order if the mean of the sample of 35 vehicles exceeds 2250 miles.

This agreement raises some interesting and important possibilities. For example: (1) if over the 10,000 vehicles, μ, the mean mileage until a breakdown is, in fact, 2500 as specified, what is the probability that by chance the mean in a sample of only 35 will be *less* than 2250 miles (so that the Army will not accept the entire order), or (2) if in fact μ is only 2100 miles, what is the probability that, by chance, the sample mean *will exceed* 2250 miles (so that the Army *will* accept the entire order)? You are asked to compute these probabilities in Review Problem 13 on page 264.

SECTION 7.1

THE BEHAVIOR OF $\overline{X}$ IN REPEATED SAMPLES

In this chapter we begin to tie together the descriptive statistics of Chapter 1 and the probability theory of the preceding chapters.

THE NOTATION OF SAMPLING

In Section 4.1 we assumed a population of numbers of TV sets in 5000 households. This population can be represented by the probability distribution of the random variable Y, the number of TV sets in a randomly selected household:

y	0	1	2	3	4	5
$P(y)$	.04	.30	.40	.20	.05	.01

The mean, variance, and standard deviation of this population were $\mu = 1.95$ TV sets, $\sigma^2 = .9475$, and $\sigma = .9734$ TV sets, respectively.

Suppose that instead of selecting a single household we select a random sample of 12 households, or more precisely, a sample of 12 "numbers of TV sets." We denote the members of the sample as follows.

Let X_1 represent the number of TV sets in the first household, X_2 the number in the second, and so on to X_{12}. Prior to selecting the sample, $X_1, X_2, \ldots, X_{12}$ are random variables.

We now introduce yet another random variable, $\overline{X}$, the mean number of TV sets over the 12 households. From Chapter 1, $\overline{X} = (X_1 + X_2 + \cdots + X_{12})/12$.

Once the sample has been selected we have *observed* values for $X_1, X_2, \ldots, X_{12}$. These might be 1, 2, 2, 1, 2, 3, 1, 3, 2, 3, 3, 2. The corresponding observed value of $\overline{X}$ will then be $\overline{x} = (1 + 2 + \cdots + 3 + 2)/12 = 2.08$ TV sets per household. This is a little higher than the population mean $\mu = 1.95$.

We now generalize this process as follows: We begin with a population of numbers, either discrete or continuous, with mean μ and standard deviation σ. Now consider the selection of a random sample of size n from such a population. As we have said earlier, we shall assume that sampling is without replacement unless the contrary is stated. Denote the sample numbers as the random variables $X_1, X_2, \ldots, X_n$ and the sample mean as the random variable $\overline{X} = (X_1 + X_2 + \cdots + X_n)/n$. Once the sample has been selected, we have the corresponding *observed* values $x_1, x_2, \ldots, x_n$ and $\overline{x} = (x_1 + x_2 + \cdots + x_n)/n$. The following two examples will illustrate these ideas.

EXAMPLE 1

The distribution of the ages of all army recruits over the past five years is $N[19.7, 2.1]$. A random sample of 25 recruits is to be selected. The sample mean is represented by the random variable $\overline{X} = (X_1 + X_2 + \cdots + X_{25})/25$. Assume that the 25 observed ages are

23.4	22.7	22.0	21.6	21.3	21.2	20.8	20.6	20.3
20.1	19.9	19.7	19.5	19.3	19.1	18.8	18.6	
18.4	18.1	17.8	17.4	17.4	17.2	17.0	16.8	

The observed value for $\overline{X}$ will then be $\overline{x} = (23.4 + 22.7 + \cdots + 16.8)/25 = 19.56$ years, a little smaller than the population mean $\mu = 19.7$ years.

■

EXAMPLE 2 For advertising purposes a national news magazine is interested in the proportion of subscribers who own their homes. The editor of the magazine intends to use a random sample of 50 subscribers to obtain the relevant information. Here we are not concerned with a number for each subscriber but whether or not they own their homes.

This sampling plan can be formulated as before, but this time assign the number 1 to every subscriber in the population owning his or her home and a 0 to all others. If, in fact, 42% of all subscribers own their homes (i.e., are 1's), the population can be represented by the probability distribution

x	0	1
$P(x)$	.58	.42

where X is the random variable, the value (0 or 1) associated with a randomly chosen subscriber. From Equation 4.5, the mean of this population is $\mu = \pi = .42$, the proportion of subscribers in the population who own their homes.

Assume that the sample results in 50 observed values consisting of 18 ones and 32 zeros, somewhat as follows:

$$0 \quad 1 \quad 0 \quad 0 \quad 1 \quad 0 \quad 0 \quad \cdots \quad 0 \quad 1 \quad 1 \quad 1 \quad 0 \quad 0 \quad 1$$

The sample mean is $\overline{x} = (0 + 1 + \cdots + 0 + 1)/50$. Since 18 of those sampled reported owning their homes (i.e., are ones), the 32 other subscribers are zeros. The sample mean is, therefore, $[(18 \times 1) + (32 \times 0)]/50 = 18/50 = .36$. Recall from Section 1.3 that when the sample consists of zeros and ones, the mean is written $\overline{p}$ instead of $\overline{x}$. In fact, $\overline{p}$ is simply the proportion of ones in the sample, which in this case is .36. As before, we should not be surprised that the sample proportion ($\overline{p} = .36$) differs from the population proportion ($\pi = .42$).

You may have realized that in this example, since the population of subscribers is large, the selection of the 50 subscribers is approximately a binomial experiment with $n = 50$ and $\pi = .42$.

■

REPEATED SAMPLES For convenience we record below the distribution of the number of TV sets referred to earlier.

y	0	1	2	3	4	5
$P(y)$	.04	.3	.4	.2	.05	.01

$$\mu = 1.95 \qquad \sigma^2 = .9475 \qquad \sigma = .9734$$

Consider now the following three circumstances:

1. Twelve households are selected at random and the number of TV sets in each household is recorded. For convenience we view this as 12 *samples* each containing $n = 1$ household rather than as a single sample of 12 households. The results are given in the first column of Table 1.
2. Twelve samples, each sample of size $n = 10$ households, are randomly selected. We note first the number of TV sets in each household and then compute the sample mean for each of the 12 samples. These 12 means are given in the second column of Table 1.
3. We again select 12 samples, but this time each sample contains $n = 100$ households. The mean number of TV sets in each of the 12 samples is listed in the third column of Table 1.

Beneath each column we have recorded the mean, variance, and standard deviation of the 12 means in that column. For example, as noted above, the single observations in the first column are considered the means in samples of size $n = 1$. The mean of these means is then 2.083 sets. The mean of the 12 means in the second column is 1.947 and that for the third column is 1.954.

TABLE 1

Means of 12 samples of size $n = 1$, $n = 10$, and $n = 100$ numbers of TV sets

	MEANS IN 12 SAMPLES OF SIZE $n = 1$[a]	MEANS IN 12 SAMPLES OF SIZE $n = 10$	MEANS IN 12 SAMPLES OF SIZE $n = 100$
	1	1.92	2.02
	2	2.20	1.84
	2	1.75	1.98
	1	2.05	1.80
	2	1.41	1.90
	3	2.33	2.08
	1	1.84	1.94
	3	1.64	1.87
	2	1.98	2.04
	3	2.40	2.06
	3	2.14	1.99
	2	1.70	1.93
Mean of the 12 means	2.083	1.947	1.954
Variance of the 12 means	.629	.087	.008
Standard deviation of the 12 means	.793	.295	.090

[a]That is, a single observation.

In Figure 1 we have plotted the three sets of 12 sample means, identifying each mean by an asterisk. The horizontal line through the plot marks the position of the *population mean* $\mu = 1.95$. There are two important features that an examination of these plots show:

1. In each case the sample means exhibit the same tendency to distribute themselves around the population mean $\mu = 1.95$. Some are above and some below 1.95, with the mean of each set of means somewhere near 1.95 (i.e., 2.083, 1.947, and 1.954).
2. The plots also suggest that the larger the sample size, the more closely the 12 sample means cluster around $\mu = 1.95$. This is supported by an examination of the standard deviations of the three sets of means. These are .793 when $n = 1$, .295 when $n = 10$, and .090 when $n = 100$ (see Problem 8 at the end of this section).

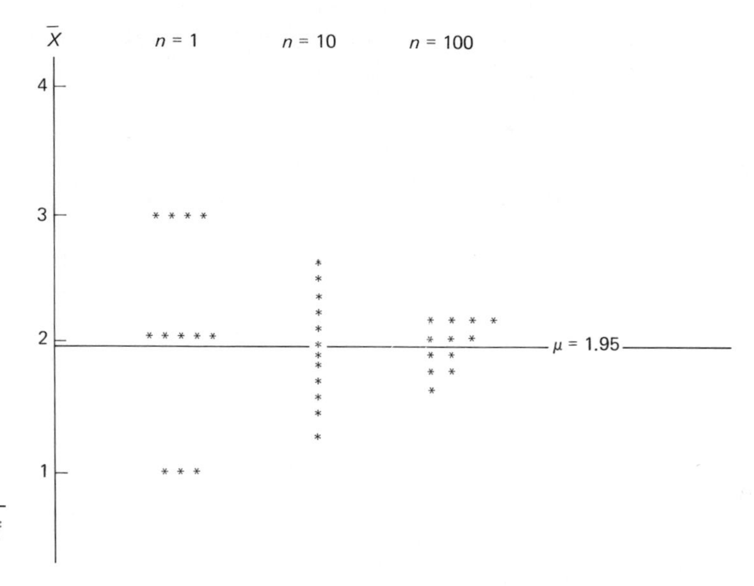

FIGURE 1

Plots of the three sets of 12 means in Table 1

THE MEAN AND VARIANCE OF THE SAMPLE MEAN

The intuitive ideas about the behavior of $\bar{X}$ in repeated samples which have just been described can be formalized. Suppose that repeated samples of size n are selected from a population with mean μ and standard deviation σ. For each successive sample the *sample mean* will take different values $\bar{x}_1, \bar{x}_2, \ldots$. Instead of taking just 12 samples, as we did earlier, however, we imagine taking *all possible* samples of size n. Then over all such samples the random variable $\bar{X}$, the sample mean will itself have a mean value which we designate $E(\bar{X})$, a variance, Var $(\bar{X})$, and a standard deviation SD$(\bar{X})$. It can be shown that

$$\mu_{\bar{X}} = E(\bar{X}) = \mu \qquad \text{(Eq. 7.1)}$$

$$\sigma_{\bar{X}}^2 = \text{Var}(\bar{X}) = \frac{\sigma^2}{n} \qquad \text{(Eq. 7.2)}$$

$$\sigma_{\bar{X}} = \text{SD}(\bar{X}) = \frac{\sigma}{\sqrt{n}} \qquad \text{(Eq. 7.3)}$$

These three results accord closely with the insights gained in Figure 1. The 12 means for each set of samples tended to distribute themselves around $\mu = 1.95$, with the mean of each set not too far from 1.95. Equation 7.1 indicates that over all possible samples of size n the mean value of $\bar{X}$ will be precisely μ.

We also noted from Figure 1 that the larger the sample size, the more closely the sample means tended to cluster around $\mu = 1.95$. Equations 7.2 and 7.3 quantify this impression. The variance of $\bar{X}$, Var $(\bar{X})$, is equal to the variance in the population, σ^2, divided by n. As you saw in Figure 1, as n increases Var $(\bar{X}) = \sigma^2/n$ and $\text{SD}(\bar{X}) = \sigma/\sqrt{n}$ both get smaller.

Applying Equations 7.2 and 7.3 to the TV example with $\sigma^2 = .9475$ and $\sigma = .973$:

1. With $n = 1$:

$$\text{Var}(\bar{X}) = \frac{\sigma^2}{1} = \sigma^2 = .9475 \quad \text{and} \quad \text{SD}(\bar{X}) = \frac{\sigma}{\sqrt{1}} = \sigma = .9734$$

2. With $n = 10$:

$$\text{Var}(\bar{X}) = \frac{\sigma^2}{10} = \frac{.9475}{10} = .09475 \quad \text{and} \quad \text{SD}(\bar{X}) = \frac{\sigma}{\sqrt{10}} = \frac{.9734}{\sqrt{10}} = .3078$$

3. With $n = 100$:

$$\text{Var}(\bar{X}) = \frac{\sigma^2}{100} = \frac{.9475}{100} = .009475 \quad \text{and} \quad \text{SD}(\bar{X}) = \frac{\sigma}{\sqrt{100}} = \frac{.9475}{10} = .09734$$

These variances and standard deviations agree fairly well with the variances and standard deviations over just 12 samples at the foot of Table 1.

EXAMPLE 3 The mean tuition per credit hour at private four-year colleges is $\mu = \$240$. The standard deviation is $\sigma = \$40$. If $\bar{X}$ is the mean cost in a sample of 50 such colleges, find $E(\bar{X})$ and $\text{SD}(\bar{X})$. Interpret these two values.

Solution From Equations 7.1 and 7.3, $E(\overline{X}) = \mu = \$240$ and $SD(\overline{X}) = \sigma/\sqrt{n} = 40/\sqrt{50} = \$5.66$. Over all possible samples of 50 colleges the mean value of $\overline{X}$ is precisely $\mu = \$240$. The $SD(\overline{X}) = \$5.66$ can be interpreted as approximately the mean distance that all the $\overline{x}$'s will be from $\mu = \$240$.[1]

■

THE MEAN AND VARIANCE OF THE SAMPLE PROPORTION

We can adapt Equations 7.1, 7.2, and 7.3 to obtain the mean, variance, and standard deviation of the random variable $\overline{P}$, the proportion of success in the sample.

Assume a population made up of successes and failures with a proportion of successes $= \pi$. As usual, each success is assigned the value 1 and each failure a 0. From Section 4.3 the mean of this population of zeros and ones is $\mu = \pi$ and the variance is $\sigma^2 = \pi(1 - \pi)$. As always, we write $\overline{P}$ in place of $\overline{X}$ for the random variable, the mean of the zeros and ones in a sample. Replacing $\overline{X}$ by $\overline{P}$, μ by π, and σ^2 by $\pi(1 - \pi)$ in Equations 7.1, 7.2, and 7.3, we obtain the following results:

$$\mu_{\overline{P}} = E(\overline{P}) = \pi \qquad \text{(Eq. 7.4)}$$

$$\sigma^2_{\overline{P}} = \text{Var}(\overline{P}) = \frac{\pi(1 - \pi)}{n} \qquad \text{(Eq. 7.5)}$$

$$\sigma_{\overline{P}} = SD(\overline{P}) = \sqrt{\frac{\pi(1 - \pi)}{n}} \qquad \text{(Eq. 7.6)}$$

EXAMPLE 4 In a particular city 20% of teenagers not in school are unemployed. A sample of 1000 teenagers not in school is to be selected. Call $\overline{P}$ the random variable "the proportion of unemployed in the sample of 1000." Compute the mean and standard deviation of $\overline{P}$. Interpret your answers.

Solution Using Equations 7.4 and 7.6 with $\pi = .2$ and $n = 1000$,

$$E(\overline{P}) = \pi = .2 \qquad \text{and} \qquad SD(\overline{P}) = \sqrt{\frac{\pi(1 - \pi)}{1000}} = \sqrt{\frac{(.2)(.8)}{1000}} = .013$$

We interpret these results as follows. Over all possible samples of 1000 teenagers the mean value of $\overline{P}$ is $\pi = .2$, the proportion unemployed in the city.

[1]Some textbooks refer to $SD(\overline{X}) = \sigma/\sqrt{n}$ as the **standard error** of $\overline{X}$.

Over all samples, the $\bar{p}$'s will differ from .2 by an average of approximately .013.

■

The results $SD(\bar{X}) = \sigma/\sqrt{n}$ in Equation 7.3 and $SD(\bar{P}) = \sqrt{\pi(1 - \pi)/n}$ in Equation 7.6 are based on the assumption that the sample size n is negligible compared to the size of the population. It can be shown that if this assumption is not appropriate, Equations 7.3 and 7.6 overestimate the standard deviation. As you will see later, it is far preferable in statistical work to overestimate rather than underestimate standard deviations. Accordingly, we shall continue to use Equations 7.3 and 7.6 in all cases. For more details on this issue see our discussion at the end of Section 9.5.

PROBLEMS 7.1

1. The mean height of adult deer of a certain species is $\mu = 40.5$ inches. The standard deviation is $\sigma = 3.2$ inches.
   (a) If $\bar{X}$ is the random variable, the mean of a random sample of 25 such deer, compute $E(\bar{X})$, $\text{Var}(\bar{X})$, and $SD(\bar{X})$.
   (b) Interpret $E(\bar{X})$ and $SD(\bar{X})$.
2. Among adults in a particular social group the mean and standard deviation of the number of hours spent watching television per week are, respectively, $\mu = 15.0$ and $\sigma = 4.5$. Call $\bar{X}$ the random variable, "the mean number of hours spent watching TV by a random sample of n such adults."
   (a) Compute $E(\bar{X})$ and $SD(\bar{X})$ if (i) $n = 25$; (ii) $n = 400$.
   (b) Interpret your answers in part (a, i).
3. In a large city 60% of adults watch public television at least once a week. Call $\bar{P}$ the random variable "the proportion of adults who watch public television at least once a week in a sample of 600 adults."
   (a) Compute (i) $E(\bar{P})$; (ii) $SD(\bar{P})$.
   (b) Interpret these two values.
4. Five percent of telephone calls attempted through the college exchange are not completed. If (a sample of) 50 telephone calls are attempted, find (a) the mean and (b) the standard deviation of the proportion $\bar{P}$ that are completed.
5. Suppose that both the mean and the standard deviation of the number of visits per member made to a health plan clinic in a year are $\mu = \sigma = 2.4$. Suppose also that 20% of members make no visits at all. The plan administrator intends to select a random sample of 60 members. Call the random variables $\bar{X}$ and $\bar{P}$, respectively, the mean number of visits and the proportion making no visits in the sample. Compute (a) $E(\bar{X})$; (b) $SD(\bar{X})$; (c) $E(\bar{P})$; (d) $SD(\bar{P})$.
6. A population consisting of just four numbers 3, 4, 6, and 7, can be represented by the probability distribution where X is the random variable "a number chosen at random from this population."

x	3	4	6	7
$P(x)$	$1/4$	$1/4$	$1/4$	$1/4$

(a) Show that the population mean is $\mu = 5$.

(b) What is $E(\overline{X})$?

(c) List all the six possible samples of size $n = 2$ that can be selected without replacement from this population. For each sample record the corresponding sample mean $\overline{X}$.

(d) Arrange these six sample means into a probability distribution for $\overline{X}$.

(e) From this distribution compute the mean of $\overline{X}$, that is, $E(\overline{X})$. Your answer should correspond to that in part (b).

(f) Compute $E(\overline{X})$ by applying the formula $E(\overline{X}) = \sum_{i=1}^{6} x_i/6$ to the six sample means.

7. An opinion pollster intends to select a sample of n voters in Massachusetts. The random variable $\overline{P}$ is the proportion of voters in the sample who favor a 200-mile fishing limit for foreign ships. The pollster assumes that the corresponding population proportion is approximately $\pi = .6$. How large a sample should she select to ensure that $SD(\overline{P})$ will be at most .01?

8. Refer to the second column in Table 1. Using the equation

$$s = \sqrt{\frac{n \sum x_i^2 - (\sum x_i)^2}{n(n-1)}}$$

show that the standard deviation among the 12 sample means is .295.

Solution to Problem 1

(a) Using Equations 7.1, 7.2, and 7.3, $E(\overline{X}) = \mu = 40.5$, $\text{var}(\overline{X}) = \sigma^2/n = 3.2^2/25 = .4096$, and $SD(\overline{X}) = \sigma/\sqrt{n} = 3.2/\sqrt{25} = 3.2/5 = .64$.

(b) Over all possible samples of 25 deer, the mean value of $\overline{X}$ will be $\mu = 40.5$ inches. The average distance that the $\overline{x}$'s will be from $\mu = 40.5$ is approximately .64 inch.

SECTION 7.2

THE CENTRAL LIMIT THEOREM

THE DISTRIBUTION OF $\overline{X}$ WHEN THE POPULATION IS NORMAL

In Section 7.1 we paid considerable attention to the mean and standard deviation of $\overline{X}$. We are now concerned with the *shape* of the distribution of $\overline{X}$. We begin with the statement that *if the population is normal*, the *distribution of $\overline{X}$* is also normal with mean μ and standard deviation $\sigma/\sqrt{n}$, or more formally:

DEFINITION 1

If $\overline{X}$ is the mean of a sample of size n from an $N[\mu, \sigma]$ population, then the distribution of $\overline{X}$ is $N[\mu, \sigma/\sqrt{n}]$. Standardizing the random variable $\overline{X}$, we obtain

$$\frac{\overline{X} - \mu}{\sigma/\sqrt{n}} = Z$$

which is $N[0, 1]$.

Definition 1 can be used to compute probabilities about $\bar{X}$ in samples from a normal population. An example will illustrate the procedure.

EXAMPLE 5 The distribution of grade-point averages (GPAs) at a large university has the $N[2.5, .5]$ distribution. A random sample of nine students is to be selected. What is the probability that $\bar{X}$, the sample mean GPA, is (a) less than 2.3; (b) greater than 3.0?

Solution Following Definition 1, since the population is $N[2.5, .5]$, the distribution of $\bar{X}$ is $N[2.5, .5/3]$ or $N[2.5, .1667]$. The population and the (dashed) distribution of $\bar{X}$ are both shown in Figure 2. Since $\bar{X}$ has a normal distribution, we can use the standardizing procedure of Chapter 6. Thus

$$\text{(a) } P(\bar{X} < 2.3) = P\left(Z < \frac{2.3 - 2.5}{.1667}\right) = P(Z < -1.2) = .1314$$

$$\text{(b) } P(\bar{X} > 3.0) = P\left(Z > \frac{3.0 - 2.5}{.1667}\right) = P(Z > 3.0) = .0013$$

There is only a small chance of getting a sample mean above 3.0. The areas corresponding to the required probabilities are also shown in Figure 2.

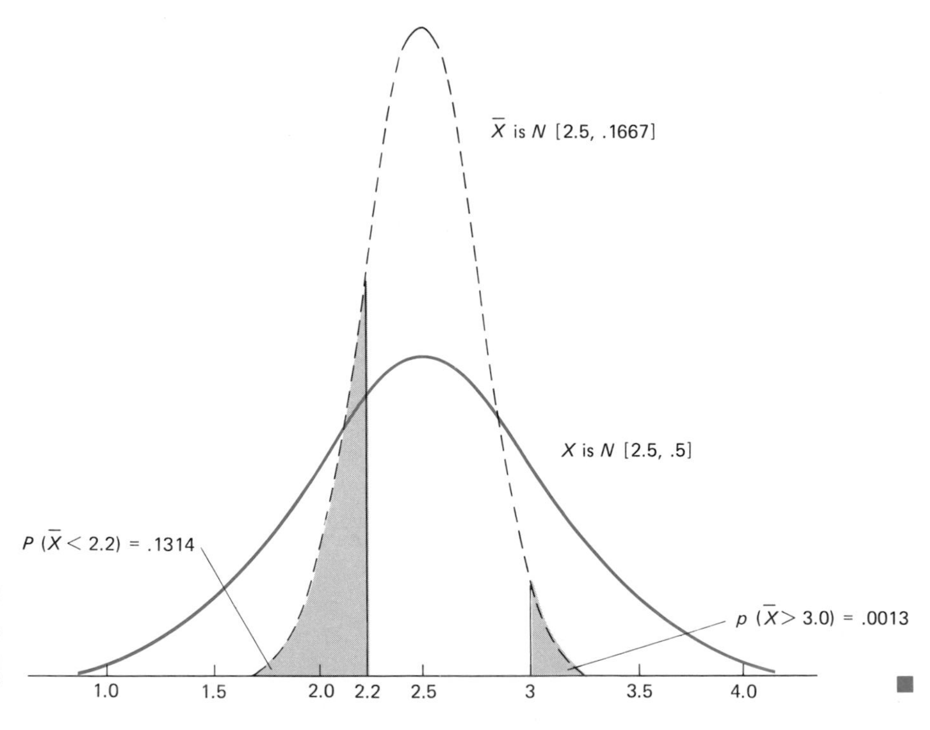

FIGURE 2

$N[2.5, .5]$ distribution of GPA and $N[2.5, .1667]$ distribution of $\bar{X}$

■

THE CENTRAL LIMIT THEOREM

You have seen that if the population is normal, the distribution of the sample mean is also normal. But what if the population is not normal? In reality we frequently know very little about the shape of the population being sampled.

However, it can be shown that regardless of the shape of the population, if the value of n is sufficiently large, the distribution of $\overline{X}$ will be approximately normal with a mean μ and standard deviation $\sigma/\sqrt{n}$. This result is known as the **central limit theorem (CLT)** and can be stated as follows.

> DEFINITION 2
>
> Regardless of the shape of the population, if the size of a random sample, n, is sufficiently large, $\overline{X}$ is approximately $N[\mu, \sigma/\sqrt{n}]$. Standardizing the random variable $\overline{X}$, we obtain $(\overline{X} - \mu)/(\sigma/\sqrt{n})$, which is approximately $N[0, 1]$.

Although the proof of the CLT is beyond the scope of this text, we can illustrate its utility in a number of ways. In Figure 3 we show two populations, one discrete and skewed to the left, and the other continuous and skewed to the right. Both populations have mean μ and standard deviation σ. The dashed curve is the distribution of $\overline{X}$ for large samples from the population. Like the population(s), the distribution of $\overline{X}$ has mean μ, but the standard deviation of $\overline{X}$, $\sigma/\sqrt{n}$, is much smaller than the population standard deviation, σ. More important is the fact that the distribution of $\overline{X}$ is practically normal even though the two populations are quite skewed. The distribution of $\overline{X}$ is often referred to more explicitly as the **sampling distribution of $\overline{X}$.**

As we suggest in Figure 3, the population may be discrete or continuous, symmetric, or skewed. No matter! Provided that the sample size is sufficiently large, the sampling distribution of $\overline{X}$ will be approximately normal.

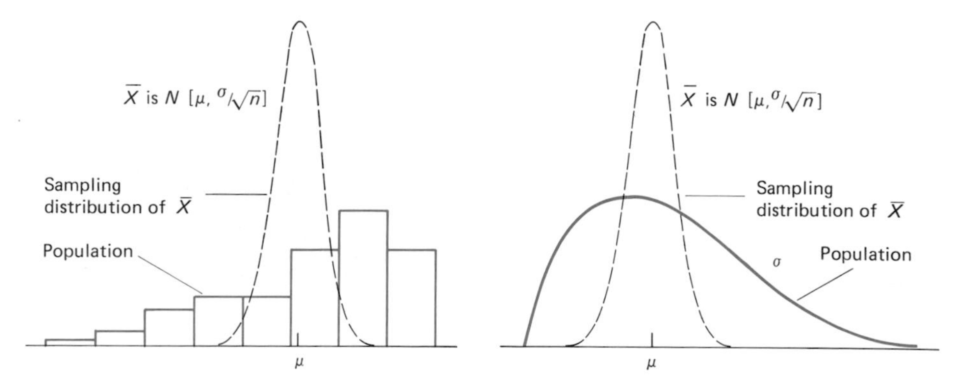

FIGURE 3

The (dashed) distribution of $\overline{X}$ is approximately normal

SAMPLE SIZE FOR APPLYING THE CLT

How large a value for n is sufficiently large? First, the closer the distribution of the population itself is to being normal, the smaller n need be for the CLT to hold. Conversely, the more skewed the population, the larger n will need to be. The problem is that in practice we rarely know or are willing to assume much about the shape of the population. We overcome this difficulty by adopting the following rule:

> DEFINITION 3
>
> Apply the central limit theorem for $\bar{X}$ when the sample size n is *30 or more*.

The following examples illustrate how the CLT is applied.

EXAMPLE 6 A new type of electronic flash for cameras will last an average of 5000 hours with a standard deviation of 500 hours. A company quality control engineer intends to select a random sample of 100 of these flashes and use them until they fail. What is the probability that the mean lifetime of the 100 flashes will be (a) less than 4928 hours; (b) between 4950 and 5050 hours (i.e., within 50 hours of μ)?

Solution Since $n = 100$ exceeds 30, we may use the CLT with $\mu = 5000$ and $\sigma = 500$ hours. From the CLT, $\bar{X}$ is approximately $N[5000, 500/\sqrt{100}]$ or $N[5000, 50]$. The required probabilities (a) $P(\bar{X} < 4928)$ and (b) $P(4950 < \bar{X} < 5050)$ are shown in Figure 4.

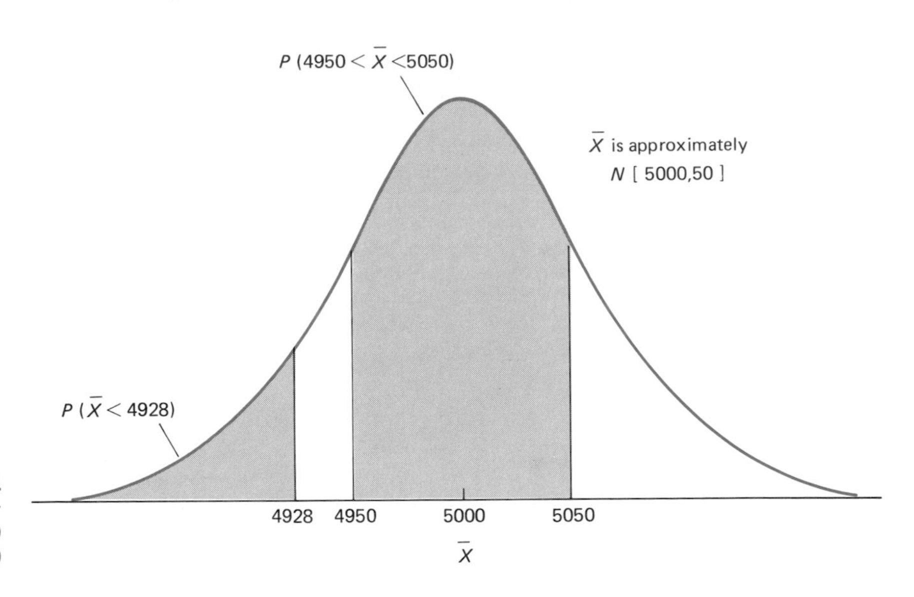

FIGURE 4

Probabilities $P(\bar{X} < 4928)$ and $P(4950 < \bar{X} < 5050)$

(a) We standardize $P(\bar{X} < 4928)$ as before, that is,

$$P(\bar{X} < 4928) = P\left(\frac{\bar{X} - 5000}{50} < \frac{4928 - 5000}{50}\right) \doteq P(Z < -1.44) = .0749$$

There is only a small probability of getting a sample mean below 4928 hours.

(b) We compute $P(4950 < \bar{X} < 5050)$ by standardizing as in part (a):

$$P(4950 < \bar{X} < 5050)$$

$$\doteq P\left(\frac{4950 - 5000}{50} < Z < \frac{5050 - 5000}{50}\right) = P(-1 < Z < 1)$$

$$= P(Z < 1) - P(Z < -1) = .6826$$

There is approximately a two-thirds chance that the sample mean will be within 50 hours of the population mean.

■

In Table 2 we summarize the general procedure for computing probabilities relating to $\bar{X}$ using the CLT. The procedures are similar to those of Table 3 of Chapter 6 (page 230) except that in Table 2, $SD(\bar{X}) = \sigma/\sqrt{n}$ replaces σ. Recall that σ measures the spread among all numbers in the population. By contrast, $SD(\bar{X}) = \sigma/\sqrt{n}$ measures the spread among the *means of all samples* of size n. Thus, when standardizing probabilities relating to $\bar{X}$, divide by $\sigma/\sqrt{n}$, *not* σ. Also, note that, in accordance with the qualification "approximately" in Definition 2 we have used "$\doteq$" rather than "=" when going from probabilities relating to $\bar{X}$ to probabilities relating to Z.

TABLE 2

Procedures for computing probabilities relating to $\bar{X}$ using the CLT[a]

(a) $P(\bar{X} < A) \doteq P\left(Z < \frac{A - \mu}{\sigma/\sqrt{n}}\right)$

(b) $P(\bar{X} > A) \doteq P\left(Z > \frac{A - \mu}{\sigma/\sqrt{n}}\right) = 1 - P\left(Z < \frac{A - \mu}{\sigma/\sqrt{n}}\right)$

(c) $P(A < \bar{X} < B) \doteq P\left(\frac{A - \mu}{\sigma/\sqrt{n}} < Z < \frac{B - \mu}{\sigma/\sqrt{n}}\right)$

$= P\left(Z < \frac{B - \mu}{\sigma/\sqrt{n}}\right) - P\left(Z < \frac{A - \mu}{\sigma/\sqrt{n}}\right)$

[a] A and B are constants.

EXAMPLE 7 The mean number of arrivals per hour at the emergency room in a hospital is 6.5 with a standard deviation $\sigma = 2.4$. If the number of arrivals for 30 randomly selected hours is to be recorded, what is the probability that the mean number of arrivals will exceed 7.0?

Solution In this case, from the CLT, $\bar{X}$ is approximately $N[6.5, 2.4/\sqrt{30}]$ or $N[6.5, .438]$. Following the procedure in Table 2(b) with $A = 7$, we obtain

$$P(\bar{X} > 7.0) \doteq P\left(Z > \frac{7.0 - 6.5}{.438}\right) = 1 - P(Z < 1.14)$$

$$= 1 - .8729 = .1271$$

This probability is shown in Figure 5 as the area under the $N[6.5, .438]$ distribution of $\bar{X}$ to the right of 7.0.

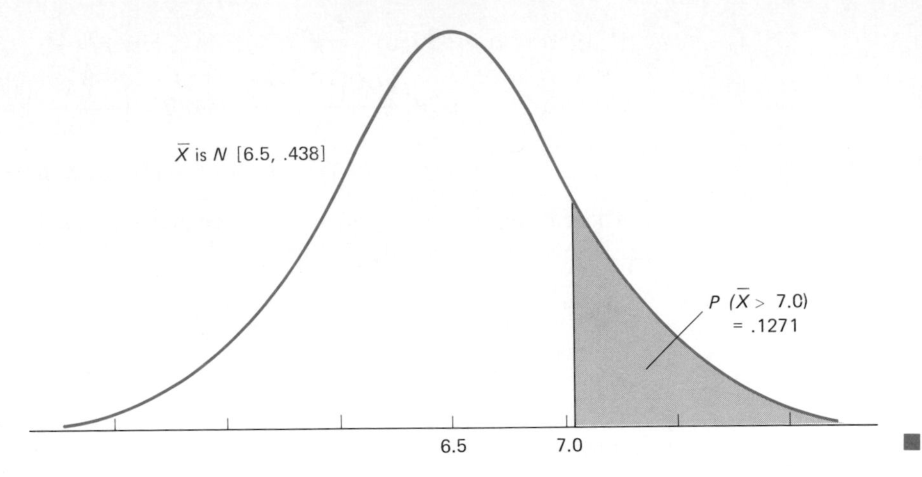

FIGURE 5

$P(\overline{X} > 7.0)$ in Example 7

THE CENTRAL LIMIT THEOREM FOR PROPORTIONS

When the population consists of only ones and zeros (corresponding to successes and failures), the population mean is $\mu = \pi$, the proportion of successes in the population, and the variance is $\sigma^2 = \pi(1 - \pi)$. The mean of the zeros and ones *in the sample* is, in turn, $\overline{X} = \overline{P}$. In this case the CLT in Definition 2 can be expressed as follows:

> DEFINITION 4
>
> Call $\overline{P}$, the random variable, the **proportion** of successes in a random sample from a population in which the proportion of successes is π. For sufficiently large values of n, the sampling distribution of $\overline{P}$ will be approximately normal with mean π and standard deviation $\sqrt{\pi(1 - \pi)}/\sqrt{n} = \sqrt{\pi(1 - \pi)/n}$. In other words, $\overline{P}$ is approximately $N[\pi, \sqrt{\pi(1 - \pi)/n}]$. Standardizing the random variable $\overline{P}$, we obtain $(\overline{P} - \pi)/\sqrt{\pi(1 - \pi)/n}$, which is approximately $N[0, 1]$.

This version of the CLT enables us to compute probabilities relating to the sample proportion $\overline{P}$. The details are set out in Table 3.

TABLE 3

Computing probabilities about $\overline{P}$[a]

(a) $P(\overline{P} < A) \doteq P\left(Z < \dfrac{A - \pi}{\sqrt{\pi(1 - \pi)/n}}\right)$

(b) $P(\overline{P} > A) \doteq 1 - P\left(Z < \dfrac{A - \pi}{\sqrt{\pi(1 - \pi)/n}}\right)$

(c) $P(A < \overline{P} < B) \doteq P\left(Z < \dfrac{B - \pi}{\sqrt{\pi(1 - \pi)/n}}\right) - P\left(Z < \dfrac{A - \pi}{\sqrt{\pi(1 - \pi)/n}}\right)$

[a] A and B are constants.

EXAMPLE 8 Twenty percent of registered voters in a city are Hispanic. If a sample of 80 voters is to be selected, what is the probability that the proportion of Hispanic voters in the sample is (a) greater than .25; (b) less than .2?

Solution In this case, π, the proportion of successes in the population, is .2 and $n = 80$. Call $\bar{P}$ the proportion of the sample that are Hispanic. From Definition 4, the sampling distribution of $\bar{P}$ is $N[.2, \sqrt{(.2)(.8)/80}]$ or $N[.2, .045]$. This distribution, together with the probabilities $P(\bar{P} > .25)$ and $P(\bar{P} < .2)$, is shown in Figure 6.

(a) From procedure (b) in Table 3 with $A = .25$,

$$P(\bar{P} > .25) \doteq P\left(Z > \frac{.25 - .2}{.045}\right) = P(Z > 1.11)$$

$$= 1 - P(Z < 1.11) = 1 - .8665 = .1335$$

(b) From procedure (a) in Table 3, with $A = .2$,

$$P(\bar{P} < .2) \doteq P\left(Z < \frac{.2 - .2}{.045}\right) = P(Z < 0) = .5$$

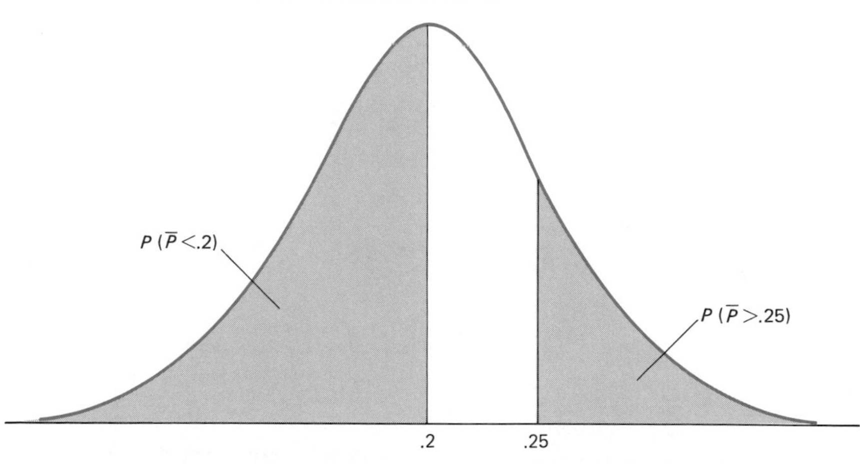

FIGURE 6

Probabilities $P(\bar{P} > .25)$ and $P(\bar{P} < .2)$ in Example 8

The CLT for proportions (Definition 4) has a different sample size requirement than that for means in Definition 2. *We may assume that $\bar{P}$ has a normal distribution as long as both $n\pi$ and $n(1 - \pi)$ are greater than or equal to 5.*

In Example 8 the values for n and π were $n = 80$ and $\pi = .2$, so that both $n\pi = 80(.2) = 16$ and $n(1 - \pi) = 80(.8) = 64$ exceed 5.

Definition 4 enables us to find binomial probabilities which are beyond the range of the binomial tables, Tables A1 and A2. For instance, in Example 8 call Y the number of successes (Hispanic voters) in the sample. Now $\bar{P} = Y/n = Y/80$. $P(\bar{P} > .25)$ can be expressed in terms of Y, as follows.

$P(\bar{P} > .25) = P(Y/80 > .25) = P(Y > .25(80)) = P(Y > 20)$. Thus had the original question been to find $P(Y > 20)$, we could proceed by reversing the above process and changing Y to a proportion as $P(Y/80 > 20/80) = P(\bar{P} > .25)$, which we can then compute as we did in Example 8.

We stress this point further in the next example.

EXAMPLE 9 The city fire chief believes that 30% of the department calls are false alarms or, conversely, that 70% are genuine. If the chief is correct and the department receives 105 calls in a week, what is the probability that (a) more than 70 are genuine; (b) fewer than 80 are genuine; (c) between 60 and 70 are genuine?

Solution We regard the 105 calls as a sample from the population of all calls, 70% of which are genuine. This sample of calls, therefore, constitutes a binomial experiment with $n = 105$ and $\pi = .7$. We require $P(Y > 70)$, $P(Y < 80)$, and $P(60 < Y < 70)$, where Y is the number of genuine calls. These probabilities are found by translating them into probabilities relating to $\bar{P}$ and using the procedures in Table 3. First we note that these procedures can be applied since both $n\pi = 105(.7) = 73.5$ and $n(1 - \pi) = 105(.3) = 31.5$ exceed 5. From Definition 4, then $\bar{P}$ is approximately $N[.7, \sqrt{(.7)(.3)/105}]$ or $N[.7, .0447]$.

$$\text{(a) } P(Y > 70) = P\left(\frac{Y}{105} > \frac{70}{105}\right) = P(\bar{P} > .67)$$

Applying procedure (b) in Table 3,

$$P(\bar{P} > .67) \doteq P\left(Z > \frac{.67 - .7}{.0447}\right) = P(Z > -.67) = 1 - P(Z < -.67)$$
$$= 1 - .2514 = .7486$$

$$\text{(b) } P(Y < 80) = P\left(\frac{Y}{105} < \frac{80}{105}\right) = P(\bar{P} < .76)$$

Applying procedure (a) in Table 3,

$$P(\bar{P} < .76) \doteq P\left(Z < \frac{.76 - .7}{.0447}\right) = P(Z < 1.34) = .9099$$

$$\text{(c) } P(60 < Y < 70) = P\left(\frac{60}{105} < \frac{Y}{105} < \frac{70}{105}\right) = P(.57 < \bar{P} < .67)$$

Applying procedure (c) in Table 3,

$$P(.57 < \bar{P} < .67) \doteq \left(\frac{.57 - .7}{.0447} < Z < \frac{.67 - .7}{.0447}\right)$$
$$= P(-2.91 < Z < -.67)$$
$$= P(Z < -.67) - P(Z < -2.91) = .2514 - .0018$$
$$= .2496$$

These three probabilities are shown in Figure 7.

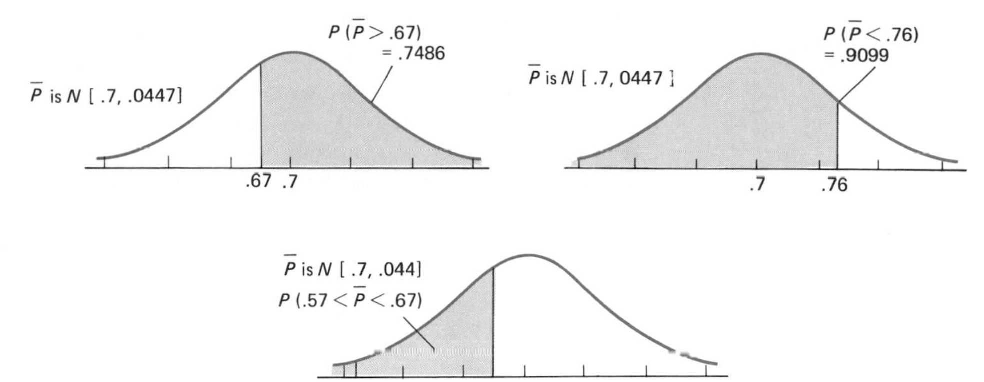

FIGURE 7

Probabilities in Example 9

To summarize we can compute binomial probabilities relating to Y by translating them into probabilities relating to $\overline{P}$ and using the procedures in Table 3.

THE DISTRIBUTION OF $\overline{X}_1 - \overline{X}_2$

In future chapters we shall be interested in the behavior of (1) $\overline{X}_1 - \overline{X}_2$, the *difference* between two sample means, and (2) $\overline{P}_1 - \overline{P}_2$, the difference between two sample proportions. First, extending the CLT to the distribution of $\overline{X}_1 - \overline{X}_2$ in large samples, we obtain:

> **DEFINITION 5**
>
> Suppose that large, random samples of size n_1 and n_2 (n_1 and $n_2 \geq 30$) are to be selected independently from populations with means μ_1 and μ_2 and standard deviations σ_1 and σ_2. In this case $\overline{X}_1 - \overline{X}_2$ is approximately $N[\mu_1 - \mu_2, \sqrt{\sigma_1^2/n_1 + \sigma_2^2/n_2}]$. Standardizing the random variable $\overline{X}_1 - \overline{X}_2$, we obtain
>
> $$\frac{\overline{X}_1 - \overline{X}_2 - (\mu_1 - \mu_2)}{\sqrt{\sigma_1^2/n_1 + \sigma_2^2/n_2}}$$
>
> which is approximately $N[0, 1]$.

For example, suppose that samples of size $n_1 = 50$ and $n_2 = 60$ are to be taken from populations with respective means $\mu_1 = 100$ and $\mu_2 = 100$ and standard deviations $\sigma_1 = 15$ and $\sigma_2 = 18$. In this case the distribution of the difference in sample means, $\overline{X}_1 - \overline{X}_2$, is approximately normal with mean $\mu_1 - \mu_2 = 100 - 100 = 0$ and standard deviation of $\sqrt{15^2/50 + 18^2/60} = 3.146$.

In the case of the distribution of $\bar{P}_1 - \bar{P}_2$ in large samples, we obtain:

DEFINITION 6

Suppose that large, random samples of size n_1 and n_2 (n_1 and $n_2 \geq 30$) are to be selected independently from populations with respective proportion of successes π_1 and π_2. In this case $\bar{P}_1 - \bar{P}_2$ is approximately $N[\pi_1 - \pi_2, \sqrt{\pi_1(1 - \pi_1)/n_1 + \pi_2(1 - \pi_2)/n_2}]$. Standardizing the random variable $\bar{P}_1 - \bar{P}_2$, we obtain

$$\frac{\bar{P}_1 - \bar{P}_2 - (\pi_1 - \pi_2)}{\sqrt{\pi_1(1 - \pi_1)/n_1 + \pi_2(1 - \pi_2)/n_2}}$$

which is approximately $N[0, 1]$.

For example, suppose that samples of size $n_1 = n_2 = 100$ are to be selected from populations in which the respective proportion of successes are $\pi_1 = .6$ and $\pi_2 = .5$. In this case the distribution of $\bar{P}_1 - \bar{P}_2$ is approximately normal with a mean of $\pi_1 - \pi_2 = .6 - .5 = .1$ and a standard deviation of

$$\sqrt{\frac{(.6)(1 - .6)}{100} + \frac{(.5)(1 - .5)}{100}} = .07$$

Notice that the sample size requirement in Definition 6 is similar to that in Definition 5 rather than to that in Definition 4.

Definitions 5 and 6 will be applied to a variety of problems in later chapters. They are introduced here as they are generalizations of the CLT.

PROBLEMS 7.2

9. A sample of 60 marriages recorded in California in 1970 are to be selected and $\bar{P}$, the proportion that have ended in divorce, computed. If π, the proportion of all such marriages that have ended in divorce, is .3, compute (a) $P(\bar{P} < .25)$; (b) the probability that $\bar{P}$ will be within .03 of π.
10. In a certain jurisdiction the mean number of years served in prison by persons convicted of second-degree murder is 12.4 with a standard deviation of 2.8 years. If a sample of 50 such persons is to be selected, what is the probability that the sample mean time spent in prison (a) exceeds 13.0 years; (b) lies between 12.0 and 12.6 years; (c) equals 12.0 years?
11. A random sample of 100 households is to be selected in a large county and $\bar{X}$, the average distance from the nearest medical facility recorded. Compute (a) $P(\bar{X} < 5.0)$; (b) $P(\bar{X} > 5.8)$; (c) the probability that $\bar{X}$ lies within .2 mile of μ. Use the fact that $\mu = 4.8$ miles and $\sigma = 3.2$ miles.
12. The mean number of passengers per run of a particular bus route in a year is $\mu = 37.5$. The standard deviation is $\sigma = 18.1$. A random sample of 75 runs is to be selected, and the number of passengers on each run is recorded. If $\bar{X}$ is the sample mean, compute (a) $P(35 < \bar{X} < 40)$; (b) $P(\bar{X} < 30)$; (c) $P(\bar{X} > 37.5)$.

13. The distribution of the heights of adult males is normal with a mean of 68 inches and a standard deviation of 6 inches. Call $\bar{X}$ the mean height in a random sample of n adult males. Find $P(\bar{X} > 69)$ if (a) $n = 9$; (b) $n = 36$; (c) $n = 100$. Draw the distribution in each case showing the required area.
14. The mean monthly telephone bill for students with telephones in their dormitory rooms is \$24.50. The standard deviation is $\sigma = \$10.40$. If a random sample of 60 monthly bills for such students is to be selected and the mean $\bar{X}$ obtained, compute (a) $P(\bar{X} > 23)$; (b) the 75th percentile of the distribution of $\bar{X}$.
15. A brand of contact lenses is advertised as lasting an average of 20 months. A consumer group doubts that the mean is that high and intends to randomly sample 25 persons who wear this brand and record the value for $\bar{X}$, the mean time that the lenses last. Assuming that μ *is* 20 and $\sigma = 2.5$ months, compute $P(\bar{X} < 18.9)$. If, in fact, the sample mean *is* found to be less than 18.9 months, what does this suggest about the advertised claim? (Assume that the lifetime of these lenses is normally distributed.)
16. A sample of 100 of the recorded births at a large hospital last year will be selected. Assuming that each birth is equally likely to be a boy or girl, compute the probability of (a) more than 55 boys; (b) between 45 and 55 boys; (c) less than 48 girls.
17. A random sample of 30 of the restaurants in a city will be selected and Y, the number that have a nonsmoking area, will be recorded. If the proportion of all restaurants in the city that have such an area is $\pi = .65$, compute (a) $P(Y < 15)$; (b) $P(13 < Y < 16)$; (c) $P(Y > E(Y))$.
18. A survey of 1200 voters is to be selected. Call $\bar{P}$ the proportion in the sample who feel that the economy is improving. If the corresponding population proportion is $\pi = .2$, compute (a) $P(\bar{P} < .22)$; (b) $P(\bar{P} > .195)$; (c) the probability that $\bar{P}$ will lie within .01 of π.
19. Suppose that 70% of graduating biology majors get jobs related to biology. If a random sample of 80 such graduates is to be selected, what is the probability that $\bar{P}$, the sample proportion who get such jobs, is (a) between .67 and .73 (b) less than .67?
20. Suppose that for the average player the probability of winning at solitaire is .1. For an expert the probability of winning is presumably higher. A person claiming to be an expert will play the game 150 times. If the player wins more than 22 games, the judges will declare the person an expert. What is the probability that the player will be declared an expert if:
    (a) She is in fact just an average player?
    (b) She is an expert with a probability of .2 of winning each game?
    (c) She is a superstar with a probability of .25 of winning each game?

Solution to Problem 9

In this case $\pi = .3$ and $n = 60$.

(a) Following procedure (a) in Table 3 with $A = .25$, we can write

$$P(\bar{P} < .25) \doteq P\left(Z < \frac{.25 - .3}{\sqrt{(.3)(.7)/60}}\right)$$

$$= P(Z < -.85) = .1977$$

(b) the probability that $\bar{P}$ is within .03 of π (.3) can be written $P(.27 < \bar{P} < .33)$.

Following procedure (c) in Table 3 with $A = .27$ and $B = .33$, we can write

$$\begin{aligned} P(.27 < \bar{P} < .33) &\doteq P\left(Z < \frac{.33 - .3}{\sqrt{(.3)(.7)/60}}\right) - P\left(Z < \frac{.27 - .3}{\sqrt{(.3)(.7)/60}}\right) \\ &= P(Z < .51) - P(Z < -.51) \\ &= .695 - .305 \\ &= .39 \end{aligned}$$

SECTION 7.3
REASONING IN PROBABILITY AND STATISTICS

At this point it will be helpful to distinguish between quantities based on information on the entire population or populations and those based on the sample or samples. Specifically:

> DEFINITION 7
>
> Any quantity based only on the data in a sample or samples is a **statistic.**

Although the word "statistics" refers to the entire field of study, as in the title of this text, it also has this limited meaning in the present context. Examples of statistics, in this sense, include the sample mean, $\bar{x}$, standard deviation, s, and the difference in sample means, $\bar{x}_1 - \bar{x}_2$.

> DEFINITION 8
>
> A **parameter** is any quantity based on the data in the entire population(s).

The population mean, μ, the standard deviation, σ, and the difference between the mean number (or the proportion of successes) in two populations, $\mu_1 - \mu_2$ and $\pi_1 - \pi_2$ are examples of parameters.

In Table 4 we summarize the important statistics and parameters that you will need in subsequent work in statistics.

TABLE 4

Statistics and parameters

QUANTITY	STATISTIC	PARAMETER
Mean	$\bar{x}$	μ
Variance	s^2	σ^2
Standard deviation	s	σ
Proportion of successes	$\bar{p}$	π
Difference between two means	$\bar{x}_1 - \bar{x}_2$	$\mu_1 - \mu_2$
Difference between two proportions	$\bar{p}_1 - \bar{p}_2$	$\pi_1 - \pi_2$

In Chapters 3 to 6 we have assumed that the values for the relevant parameters were known and we computed the probabilities associated with statistics such as $\bar{x}$, $\bar{p}$, or $\bar{x}_1 - \bar{x}_2$. In Example 8, for instance, we assumed that the proportion of Hispanic voters in a population was $\pi = .2$ and then computed the probability that the proportion of Hispanics in a sample of $n = 80$ voters would exceed .25, that is, $P(\bar{P} > .25)$. In Example 6 we assumed that $\mu = 5000$ and $\sigma = 500$ and then computed $P(4950 < \bar{x} < 5050)$.

In reality it is extremely unlikely that the value of the population proportion π or the population mean μ or other parameters will be known. Indeed, in reality, sample(s) are generally taken with the express purpose of using the sample information to make judgments about some aspect (i.e., parameter) of the population. In Example 8, for instance, the researcher might want to use the sample proportion, say $\bar{p} = .25$, to make a judgment about π, in reality the unknown proportion of Hispanics among all voters. Similarly, in Example 6 the actual mean lifetime in a sample of electronic flashes, say $\bar{x} = 4928$, for instance, might be used as an estimate for the population mean μ. Reasoning in this way, from the sample back to the population, is the essence of the statistical method.

When studying probability the values of the relevant parameters were specified and we sought the probability of obtaining various outcomes associated with a statistic. By contrast, in statistical inference we will reason in the opposite way. The outcome "the value of a statistic" provides the basis for a judgment or inference about the appropriate (unknown) parameter. In Table 5 we summarize the differences between the two types of reasoning just described.

In Chapter 8 we gradually move into matters statistical by discussing some of the practical issues involved in organizing a sample or samples so that we can make valid judgments about the population(s).

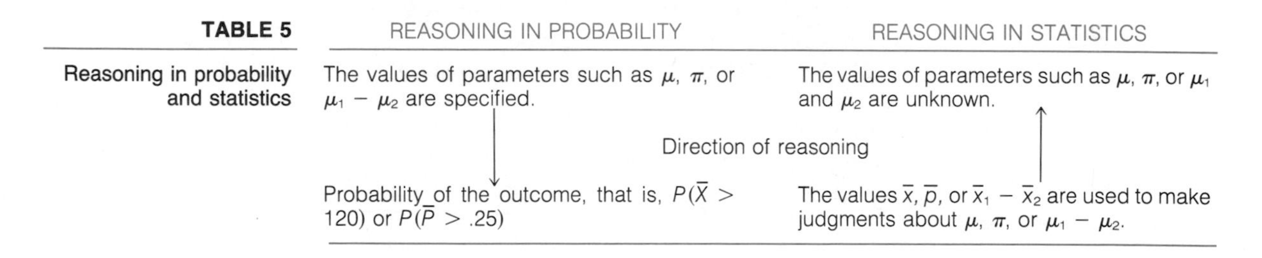

TABLE 5

Reasoning in probability and statistics

REASONING IN PROBABILITY		REASONING IN STATISTICS
The values of parameters such as μ, π, or $\mu_1 - \mu_2$ are specified.		The values of parameters such as μ, π, or μ_1 and μ_2 are unknown.
↓	Direction of reasoning	↑
Probability of the outcome, that is, $P(\bar{X} > 120)$ or $P(\bar{P} > .25)$		The values $\bar{x}$, $\bar{p}$, or $\bar{x}_1 - \bar{x}_2$ are used to make judgments about μ, π, or $\mu_1 - \mu_2$.

SECTION 7.4
SUMMARY AND REVIEW

In this chapter we examined the behavior of sample means, and of the difference between two sample means, over repeated samples.

Beginning with a population of numbers having mean μ and standard deviation σ, we select a sample of size n from this population. We then denote the numbers in the sample as the random variables $X_1, X_2, \ldots, X_n$, and the random variable, the sample mean as

$$\bar{X} = \frac{X_1 + X_2 + \cdots + X_n}{n}$$

Once the sample has been selected, we obtain *observed* values for $X_1, X_2, \ldots, X_n$ and $\bar{X}$. We write these as $x_1, x_2, \ldots, x_n$ and $\bar{x}$.

If we take repeated samples of size n, then for *each* successive sample the random variable $\bar{X}$ will take values $\bar{x}_1, \bar{x}_2, \ldots$. Over all possible samples these values (means) will themselves have a mean, variance, and standard deviation given by

$$\mu_{\bar{X}} = E(\bar{X}) = \mu \qquad \text{(Eq. 7.1)}$$

$$\sigma_{\bar{X}}^2 = \text{Var}(\bar{X}) = \frac{\sigma^2}{n} \qquad \text{(Eq. 7.2)}$$

$$\sigma_{\bar{X}} = \text{SD}(\bar{X}) = \frac{\sigma}{\sqrt{n}} \qquad \text{(Eq. 7.3)}$$

A special case of these formulas occurs where the population consists of only ones (for successes) and zeros (for failures). The population can then be represented by the probability distribution

x	0	1
$P(x)$	$1 - \pi$	π

where π is the proportion of ones in the population. The mean of these zeros and ones in the population is $\mu = \pi$ and the variance is $\sigma^2 = \pi(1 - \pi)$. In this special case we replace $\bar{X}$ in Equations 7.1, 7.2, and 7.3 by $\bar{P}$, μ by π and σ^2 by $\pi(1 - \pi)$. These equations thus become

$$\mu_{\bar{P}} = E(\bar{P}) = \pi \qquad \text{(Eq. 7.4)}$$

$$\sigma_{\bar{P}}^2 = \text{Var}(\bar{P}) = \frac{\pi(1 - \pi)}{n} \qquad \text{(Eq. 7.5)}$$

$$\sigma_{\bar{P}} = \text{SD}(\bar{P}) = \sqrt{\frac{\pi(1 - \pi)}{n}} \qquad \text{(Eq. 7.6)}$$

If the population is *normal,* the *distribution* of $\bar{X}$ is also normal with mean μ and standard deviation $\sigma/\sqrt{n}$. That is, $\bar{X}$ is $N[\mu, \sigma/\sqrt{n}]$. This is true regardless of the size of the sample n.

It is a remarkable fact that even if the population is *not* normal provided that $n \geq 30$, the distribution of the sample mean $\bar{X}$ is approximately normal, again with mean μ and standard deviation $\sigma/\sqrt{n}$. This result is called the **central limit theorem.**

Procedures for computing probabilities relating to $\overline{X}$ are given in Table 2. When using these procedures, it is important to remember that the standard deviation of $\overline{X}$ is $\sigma/\sqrt{n}$, *not* σ.

The CLT can also be applied when the population consists of zeros and ones. In this case $\overline{X} = \overline{P}$, $\mu = \pi$, and $\text{SD}(\overline{P}) = \sqrt{\pi(1 - \pi)/n}$, where π is the proportion of successes in the population. In this case, the CLT provides that $\overline{P}$ is approximately $N[\pi, \sqrt{\pi(1 - \pi)/n}]$. This is true provided that $n\pi$ and $n(1 - \pi)$ are greater than or equal to 5.

Table 3 contains procedures for computing probabilities about $\overline{P}$. These procedures enable us to compute binomial probabilities when the values for n are beyond the range of the binomial tables. This is done by changing the number of successes into a proportion by dividing it by n.

The CLT can be extended to the *difference* between two sample means $(\overline{X}_1 - \overline{X}_2)$ and the difference between two sample proportions $(\overline{P}_1 - \overline{P}_2)$. The approximately normal distributions of these two quantities are referred to in Definitions 5 and 6.

The last section of this chapter represents a transition from the methods of probability to those of statistics. A **statistic** is any quantity based only on the data in the sample(s) (Definition 7). A **parameter** is any quantity based on the data in the entire population (Definition 8). A list of the important statistics and parameters can be found in Table 4.

In probability, values for the relevant parameters are given and we seek the probability of a particular value or range of values for a statistic. In statistics we argue in precisely the opposite way. The sample outcome in the form of the value of a statistic is known and provides the basis for a judgment or inference about the appropriate unknown parameter.

REVIEW PROBLEMS

GENERAL

1. Which of the following statements about the central limit theorem (CLT) are true? Where you think a statement is false, explain why.
   (a) The CLT assumes that the population being sampled is normal.
   (b) The CLT becomes more applicable the larger the sample size.
   (c) The CLT is concerned with the mean of a single sample.
   (d) The CLT can be used whether the population is continuous or discrete.
2. Suppose that of all the single records released by U.S. record companies last year only 25% made the top-40 hit parade. If a random sample of 100 singles are selected, what is the probability that more than 30% made the top-40 hit parade?
3. The mean and standard deviation of all contributions to a major charity in a year are μ = \$69 and σ = \$48. If a random sample of 100 contributions is selected, what is the probability that the sample mean will (a) exceed \$80; (b) lie within \$10 of μ?
4. In Review Problem 3, a proportion $\pi = .33$ of contributors gave \$100 or more. If a sample of 100 contributions is selected, what is the probability that the proportion that are \$100 or over (a) exceeds .4; (b) lies within .025 of π?

5. The mean weight of 11-year-old boys in a school district is 63 pounds, the standard deviation, 8 pounds. A random sample of n boys is to be selected and their weights recorded.

   (a) Compute $P(62 < \overline{X} < 64)$ if (i) $n = 30$; (ii) $n = 60$; (iii) $n = 100$.

   (b) Explain why this probability increases as n increases.

6. A population has a mean μ and standard deviation σ. If a random sample of 100 values is randomly selected, what is the probability that the sample mean will lie within $\sigma/10$ of μ?

7. In Texas the mean size of farms is $\mu = 770$ acres with a standard deviation of 480 acres.

   (a) If a random sample of 40 farms is selected, what is the probability that the mean acreage in the sample will (i) exceed 900; (ii) lie between 700 and 800?

   (b) Suppose that five independently selected random samples, each of 40 farms, is selected. What is the probability that the sample mean will exceed 900 acres in at least one of the five samples?

8. A random sample of 50 U.S. Navy recruits is to be selected. Members of the sample will be asked whether or not they intend to reenlist. Suppose that the proportion of all such recruits intending to reenlist is $\pi = .65$. What is the probability that the proportion in the sample intending to reenlist is (a) less than .6; (b) less than .65; (c) between .7 and .75?

9. A normal population has mean -120 and standard deviation 20. If a random sample of size 12 is to be selected from this population, what is the probability that the sample mean will (a) exceed -112; (b) not exceed -100.

10. The mean number of job offers to graduates of a business school is $\mu = 4.5$. The standard deviation is 2.8. If a random sample of 30 graduates are selected, what is the probability that $\overline{X}$, the mean number of job offers in the sample, (a) lies between 4 and 5; (b) lies between 4.2 and 4.8?

11. In Review Problem 10, how many graduates should be selected in order that $P(4.2 < \overline{X} < 4.8) = .6$?

12. Suppose that over the past 10 years the distribution of the number of copies of first novels sold is approximately $N[4700, 2600]$. If a random sample of 20 first novels is to be selected, what is the probability that the mean number of copies sold will exceed 5000?

13. The U.S. Army orders 10,000 advanced four-wheel drive vehicles from Brash, Inc. On the average the vehicles are to travel 2500 miles before a breakdown. The Army agrees to accept the entire order if the mean mileage to breakdown ($\overline{X}$) in a random sample of 35 vehicles exceeds 2250. Call μ the mileage to breakdown over all 10,000 vehicles.

    (a) Assuming that the standard deviation of mileage to breakdown is $\sigma = 850$, compute (i) $P(\overline{X} < 2250)$ if $\mu = 2500$ as required; (ii) $P(\overline{X} > 2250)$ if μ is only 2100.

    (b) What mistake will occur (i) if $\overline{X} < 2250$ when $\mu = 2500$; (ii) if $\overline{X} > 2250$ when $\mu = 2100$?

14. The probability that a basketball player will foul out of a game is .3. Whether he fouls out of one game is independent of whether he fouls out of any other. He plays in 80 games in a season and will get a bonus if he fouls out in fewer than 18 games. What is the probability that he will get the bonus?

HEALTH SCIENCES

15. The mean time that it takes a particular anesthetic drug to produce unconsciousness is 11.5 minutes with a standard deviation of 1.8 minutes.

    (a) If a random sample of 40 patients is to be selected and $\overline{X}$, the average time for the drug to work computed, what is the probability that (i) $\bar{X}$ will exceed 12.0 minutes; (ii) $\bar{X}$ will lie between 11.5 and 12.0 minutes?

(b) Suppose that two such samples of 40 patients is selected. What is the probability that (i) neither sample mean will exceed 12.0 minutes; (ii) one of the sample means exceeds 12.0?

16. In 15% of cases a particular medication produces serious side effects. If a random sample of 200 users of this medication is selected, what is the probability that the proportion in the sample who experience serious side effects (a) exceeds .2; (b) lies between .12 and .18?

17. Referring to Review Problem 16, suppose that three studies each involving 200 randomly selected users of this medication are carried out. What is the probability that in only one of the three studies the sample proportion of patients experiencing the side effects exceeds .2?

18. A new method of performing a particular medical procedure is to be tested on a random sample of 100 patients. It will be adopted for general use if it is successful on at least 85% of the sample. Call π the probability that the new method will be successful on a patient.

(a) What is the probability that the new method will be adopted for general use if $\pi = .8$?

(b) What is the probability that the new method will not be adopted for general use if $\pi = .9$?

19. The probability that a newly hatched tadpole lives to maturity is .2. What is the probability that among 200 newly hatched tadpoles the number living to maturity (a) is greater than 50; (b) lies between 40 and 50?

SOCIOLOGY AND GOVERNMENT

20. The mean caseload over all social workers employed by a state is $\mu = 22.2$. The standard deviation is $\sigma = 4.1$. If a random sample of 42 such social workers is selected, what is the probability that the average caseload among those sampled exceeds 23?

21. Seventy percent of voters in a state favor an increase in the size of the state legislature. If a random sample of 400 voters is selected:

(a) What is the probability that the sample proportion favoring this move will (i) exceed .75; (ii) lie between .68 and .72?

(b) How large a value of n would be necessary so that $P(.68 < \bar{P} < .72) = .9$?

22. Each of the voters in a random sample of 1000 is to be asked whether or not they think of themselves as being associated with a political party. Suppose, in fact, that a proportion $\pi = .70$ of all voters think of themselves as being so associated. If $\bar{P}$ is the corresponding proportion in the sample, compute $P(.68 < \bar{P} < .72)$. Interpret your answer.

23. Sixty-four percent of residents of Hendon oppose a town ordinance that would permit the building of multiple-family dwellings in the town. Call $\bar{P}$ the proportion in a random sample of 200 residents who oppose the ordinance.

(a) What is $P(\bar{P} < .6)$?

(b) What is the probability that (i) more than 140 of those sampled will oppose the ordinance; (ii) more than 65 of those sampled are not opposed to the ordinance?

ECONOMICS AND MANAGEMENT

24. Among all families earning over \$20,000 a year 18% have opened Individual Retirement Accounts (IRAs). If a random sample of 600 families earning over \$20,000 a year is selected, what is the probability that the proportion in the sample having an IRA (a) exceeds .2; (b) lies between .17 and .19?

25. The average number of passengers per journey carried between two rural towns by a major intercity bus company is $\mu = 6.4$. The standard deviation is $\sigma = 2.6$. If a random sample of 30 journeys is selected, what is the probability that the sample average number of passengers (a) lies between 6 and 7; (b) exceeds 7.5?

26. Last year the average mortgage approved by the Third National Bank of Commerce was \$74,150. The standard deviation was \$8120. A random sample of the records of 16 such loans is to be selected. Assume that the distribution of the mortgage amount is normal.

    (a) What is the probability that the mean mortgage in the sample is (i) less than \$74,000; (ii) lies between \$74,000 and \$75,000?

    (b) Why was the assumption of normality necessary in this case?

27. The proportion of adults in Albion county that are unemployed is $\pi = .16$. A random sample of 140 adults residing in the county is to be randomly selected. What is the probability that the unemployment rate in the sample exceeds .20?

EDUCATION AND PSYCHOLOGY

28. In a test of pattern recognition a group of 40 subjects are asked to pick out 10 circles hidden in a picture. The time taken by each subject is recorded. If the mean and standard deviation of the time taken over all potential participants is $\mu = 14$ minutes and $\sigma = 6$ minutes, find the probability that the sample mean time is (a) greater than 15 minutes; (b) between 16 and 17 minutes.

29. At a large university 65% of students would like to see a change in the class scheduling grid.

    (a) If a sample of 60 students is randomly selected, what is the probability that (i) more than 70% of those sampled, and (ii) between 70 and 80% of those sampled favor a change in the grid?

    (b) The registrar agrees to set up a committee to consider alternatives to the grid if more than 40 students in the sample favor a change. What is the probability that such a committee will be set up?

30. A psychologist has constructed a test which she believes will accurately reflect a person's degree of competitiveness. The higher the score on the test, the more competitive the person is assumed to be. The test is given to a random sample of 50 female and 50 male sophomores. Call $\overline{X}_1$ and $\overline{X}_2$ the sample mean scores for the men and women, respectively. Suppose that the mean and standard deviation of scores for all potential male participants are $\mu_1 = 76$ and $\sigma_1 = 18$ and that the corresponding figures for women are $\mu_2 = 69$ and $\sigma_2 = 20$. What is the probability that (a) $\overline{X}_1 > 80$; (b) $\overline{X}_2 < 70$; (c) both sample means exceed 70? [In part (c) assume the independence of the sample means.]

SCIENCE AND TECHNOLOGY

31. A machine manufactures printed circuits for computers. Approximately 8% of them are defective and have to be discarded. If a sample of 120 circuits is randomly selected:

    (a) What is the probability that the proportion that are defective is (i) greater than .1; (ii) less than .05?

    (b) What is the probability that the number that are defective lies between 10 and 15?

32. A company's computer periodically "crashes" (breaks down). The number of hours between 30 consecutive crashes are to be carefully recorded. If the mean time between these 30 crashes, $\overline{X}$, is less than a week, the computer will be replaced by the manufacturer. What is the probability that the computer will be replaced if, in fact, the potential mean time between all crashes is $\mu = 190$ hours and the standard deviation is $\sigma = 58$ hours?

SUGGESTED READING

AYER, A. J. "Chance"; WEAVER, W. "Probability"; and KAC, M. "Probability." Readings in *Mathematics in the Modern World*. San Francisco: Foreman, 1968.

MOORE, D. S. *Statistics: Concepts and Controversies*. San Francisco: Freeman, 1979, pp. 237–268.

STANCL, D. L. and STANCL, M. L. *Applications of College Mathematics*. Lexington, Mass.: D. C. Heath, 1983, pp. 323–414.

ROSS, S. *A First Course in Probability*. New York: Macmillan, 1976.

8 SURVEYS AND EXPERIMENTS

LUNG CANCER AND SMOKING

One of the warnings that is to appear on the side of cigarette packs reads:

> Smoking causes lung cancer, heart disease, emphysema and may complicate pregnancy.

Is this warning correct? To what extent is it possible to determine a *causal link* between cigarette smoking and such diseases as lung cancer and emphysema?

This is one of the issues explored in Section 8.3.

SECTION 8.1

THE DESIGN OF STATISTICAL STUDIES

It is helpful to classify statistical studies into two broad groups: surveys and experiments.

The term **survey** is often used simply as a nontechnical alternative to the word "sample." Generally, the information in the sample is used to draw conclusions about a single variable or a number of separate variables. The following are examples of studies that can be classified as surveys.

EXAMPLE 1 Opinion polls that report the percentage of prospective voters favoring one candidate or one issue are essentially surveys.

■

EXAMPLE 2 In a poll consisting of a sample of 1050 members of a large labor union, 370, or 35%, voted to go on strike on the issue of company-paid medical insurance.

■

EXAMPLE 3 The federal government's Current Population Survey collects monthly information on a sample of households. The type of information collected varies from month to month but usually includes economic data such as income; employment status (i.e., employed or not); and demographic data such as age, number of children, and size of household.

■

In contrast to surveys, **experiments** attempt to determine the extent of a *cause-and-effect relationship* between two or more variables. The following are examples of experiments.

EXAMPLE 4 Numerous studies both in this country and abroad have attempted to link smoking, particularly cigarette smoking, with lung cancer and other health problems.

■

EXAMPLE 5 One of the largest and most costly experiments of all times was the 1954 Salk vaccine experiment. This was designed to investigate whether the Salk vaccine offered people any protection from the effects of poliomyelitis, more commonly known as polio. The positive results led to a program that has almost completely eliminated polio in most of the world.

■

EXAMPLE 6 Many studies have attempted to determine whether seeding clouds with silver iodide pellets increases the amount of rain.

■

EXAMPLE 7 An experiment was recently completed at Simmons College to determine the effect of different techniques for memorizing information on the amount of material recalled. The 90 participants were divided into three groups. Two of the groups were trained to use special techniques to memorize the material while the third group received no training. At the end of the training period all 90 persons were given a test to determine the amount of information recalled. ■

To summarize, a primary purpose of the survey is descriptive, whereas that of the "experiment" is determining the extent to which a change in one variable may *cause* a change in a second variable. This does not imply, however, that all statistical studies can be neatly pigeonholed into one of these two categories. When conducting a survey, for instance, we often compute the correlation coefficient as well to determine the *strength* of the relationship between two variables. Recall, however, that a large value for r does not necessarily indicate a *causal* relationship between the two variables involved (see Chapter 2).

In order to continue our discussion of surveys and experiments, we require two additional concepts: **precision** and **bias.** The meanings of both of these terms can be illustrated by the following example.

EXAMPLE 8 As an advertising gimmick an automobile dealer fills a car with 55,555 pennies. He then places a sign in his window inviting customers to "guesstimate" the number of pennies he has placed in the car. Customers may take up to five guesses. After a month the person with the guess closest to 55,555 wins the car. The five guesses by customers A, B, C, D, and E are as follows:

A	62,238	79,321	84,381	67,001	74,209
B	31,709	48,870	44,108	23,870	26,751
C	41,098	42,379	40,007	41,382	41,994
D	48,631	57,421	53,241	62,340	55,321
E	53,948	56,091	55,632	54,921	57,100

These five sets of guesses are shown as dot diagrams in Figure 1. The vertical line corresponds to the true number of pennies (i.e., 55, 555) on each horizontal line.

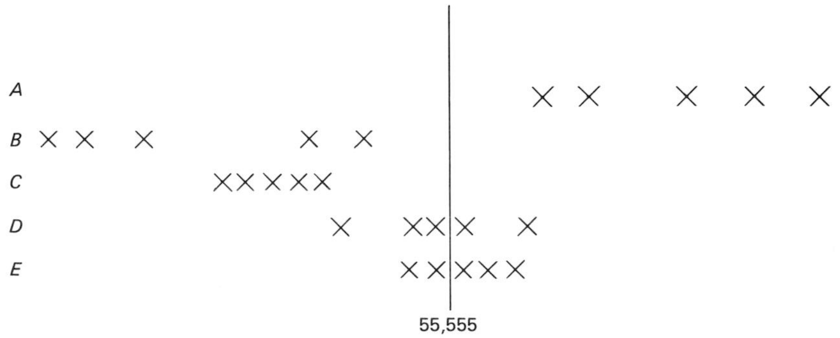

A
B
C
D
E
55,555

FIGURE 1

Dot diagrams for the five sets of guesses in Example 8

The five guesses made by A are both widely scattered *and* systematically higher than the true figure. We describe these guesses as *imprecise* and *biased* (upward). The numbers for B are also widely scattered but are systematically lower than the true figure. These are imprecise and biased (downward). The numbers for C vary little but are systematically below 55,555. We describe them as *precise but biased* (downward). Although the guesses by D are imprecise, they are unbiased in the sense that they seem to average out close to the true figure (in fact, $\overline{X}_D = 55{,}390.8$). Person E's guesses represent a statistician's delight. The numbers are both *precise* and *unbiased*.

As you can see, **precision** can be viewed as indicating how close *together* a sequence of values are (*note:* not how close to the true value). Indeed, the standard deviation is a good measure of precision. For example, $S_D = 5067.9$ but $S_E = 1189.9$.

■

We shall consider the issue of precision in statistical studies in detail in Chapters 9 to 13.

Bias is a new and subtle concept which we will explain as we proceed through this chapter. There are two forms of bias in statistical studies: (1) **external bias,** which poses a particular problem in surveys; and (2) **internal bias,** which is primarily a problem in experiments. We begin with a discussion of external bias in the next section and consider internal bias in the one following.

SECTION 8.2 SURVEYS, SAMPLING, AND EXTERNAL BIAS

EXTERNAL BIAS

External bias occurs where a sample systematically differs from the population in important respects. Consider, for instance, the following example.

EXAMPLE 9

Assume that an investigation is being conducted to determine the proportion of voters in a certain precinct who favor a change in the city charter. The change will permit renovation of neighborhoods of historic interest. A sample of 100 voters is to be selected and interviewed.

It might be convenient to select 100 voters who live close together, or the first 100 voters on the voting list, or even to interview the first 100 voters that one meets in the precinct.

Since the change will affect some neighborhoods more than others, voters who live close together in the same neighborhood may have similar views on the change. But these views may well not reflect those of the entire precinct. Similarly, voters listed together alphabetically may belong to just a few families whose views are quite atypical of those of the entire precinct. And interviewing the first 100 voters one meets on the street biases the results against the kind of voters, office workers, or aged residents, for instance, who are unlikely to be at that location at that time.

■

When sampling is conducted as described in Example 9, it is said to be *biased* as the following definition indicates.

> DEFINITION 1
>
> A sampling procedure is **biased** if it tends to favor the selection of certain numbers of the population.

There is a way to avoid the type of bias we have been concerned with above, and we turn to this next.

RANDOM SAMPLING

We have frequently used the terms "taken at random" or "a random sample" with little explanation. The word "random" conveys the impression of haphazardness, but the statistician uses the term "**random sample**" in a well-defined way as follows:

> DEFINITION 2
>
> A **random sample** of size n is one selected by a process that gives each and every group of n members of the population the same chance of forming the sample.

Such a process exists and can be illustrated with a simple example.

EXAMPLE 10

Assume that there were 4873 voters in the precinct referred to in Example 9. We are to select 100 of them at random. We begin by assigning to each voter a number between 1 and 4873 inclusive. This may be done in any way that is convenient. Two examples are the use of alphabetical order or street address. When the numbers have been assigned, a table of random numbers such as that in Table A10 may be used. The numbers in Table A10 were computer generated in such a way that any position is equally likely to be occupied by any of the digits 0, 1, 2, . . . , 9. For convenience the digits have been arranged in groups of five.

We now proceed as follows. A random starting point is selected. One technique is to take a birthdate. (10-13-48 is the birthday of the wife of one of the authors.) Add the digits in the birthdate: $1 + 0 + 1 + 3 + 4 + 8 = 17$ and take the 17th five-digit entry in the first column as your starting point. Since 4873 is a four-digit number, we need examine only the first four of the five digits of the entry. If the number represented by these four digits is less than or equal to 4873, the voter with that number is the first one selected. If the number exceeds 4873, move to the number below and continue in this way until a number that is 4873 or lower is obtained. A glance at Table A10 indicates that the first such number is 4419, so the voter corresponding to that number is selected. We could as easily move along each row of numbers or diagonally across the page or indeed use any system that we might prefer as long as (1) the system is decided on before beginning the selection, and (2) the same system is used consistently throughout the process.

Using this procedure, we select as many four-digit numbers between 1 and 4873 as required. For instance, to select a sample of 100 voters, we move from number to number as described above until we have found 100 numbers 4873 or lower. The voters associated with these numbers will constitute the sample.

Before beginning the selection process, it is necessary to decide whether sampling is to be with or without replacement. When sampling is with replacement, a particular voter may be included in the sample more than once. In this case, if a four-digit number is obtained which has previously been selected, the voter associated with that number is included again. When sampling is without replacement (the more usual practice), a repeated number is skipped.

Using Table A10, as described a sample of 10 voters selected *without* replacement would consist of those corresponding to the numbers

4419 2132 328 2279 2060

4189 3544 4804 4188 1551

The sampling procedure above is the simplest, but not the most efficient, method for selecting a sample of 100 voters. A more efficient procedure would be to expand the random numbers used to include not only 1 through 4873 but also 5000 through 9873, subtracting 5000 from any number selected that is over 5000. Thus, if the number 5089 were picked, the voter assigned the number 89 is selected. Had the numbers been limited to 1 through 4873, the number 5089 would have been disregarded and it would have taken longer to select the sample.

Finally, we emphasize that it is not the contents of the sample that makes it random; it is the *method by* which it is selected.

■

Even randomly selected samples may be subject to bias if, for example, a population list is used that is out of date, incomplete, or in some other way differs from the intended population—the *target* population.

A classic and spectacular example of this kind of bias was the sample selected by the now defunct magazine *the Literary Digest*, [see Bryson, (1976)]. In an attempt to forecast the result of the 1936 presidential election, the magazine mailed out approximately 10 million ballots to people whose names had been taken largely from telephone directories, magazine subscription lists, and registers of automobile owners. Several million of the cards were returned and just over 40% favored Franklin D. Roosevelt. A few weeks later in the actual election, Roosevelt received more than 60% of the votes. The discrepancy between the two results was due, at least in part, to the fact that the people in the sampled population, those in the telephone directories, and so on, differed significantly from the target population—all voters—in that it underrepresented less-well-off voters, many of whom did not have a telephone or own a car. These people voted overwhelmingly for F.D.R.

Today, when practically all adults in the United States have telephones, populations consisting of the holders of all telephone numbers in an area are frequently used in opinion polling and marketing research. Telephone directories are, however, generally *not* used because they omit recent arrivals and those with unlisted numbers. Instead, the (last four digits of) numbers are usually selected at random.

The *Literary Digest* poll illustrates another form of bias that is easy to overlook, namely *non-response* bias. Only one-fourth of those who were sent ballots returned them. This would not have been a problem if those who returned the ballots had similar voting intentions to those who did not. But as the election results suggested, a far greater proportion of Landon supporters than of Roosevelt supporters bothered to respond. Nowadays professional pollsters take great pains to obtain information from at least some of the nonrespondents.

Even if a random sample is selected from the target population and there is a 100% response, there is still no guarantee that the composition of a sample will mirror that in the population. Referring to our earlier example of randomly selecting 100 voters, we might, for instance, obtain a sample which, *by chance*, underrepresents (or overrepresents) those favoring the proposed change in the city charter.

For reasons to be described in Chapter 9, however, a randomly selected sample (and only such a sample) enables us to use probability theory to measure the extent to which the percentage in the sample (favoring the change, in this case) is likely to deviate from the corresponding percentage in the population.

PROBLEMS 8.2

1. The admissions office at a large college is interested in the proportion of freshmen for whom this college was their first choice. It will ask the relevant question of a sample of 50 students. Why should the admissions office not take as its sample the first 50 students who apply and are accepted?
2. A store wishes to select a random sample of 15 from the 87 names on its mailing list. Explain how such a sample could be selected and use Table A10 to select a sample (without replacement).
3. A college registrar wishes to select a sample of 25 from the 696-person freshman class. No student will be allowed to appear more than once in the sample. Explain how such a sample could be selected.
4. A population consists of the eight numbers 0, 1, 2, 3, 4, 5, 6, and 7. Use Table A10 to obtain a sample of 15 (with replacement, of course) from this population.
5. A newspaper invites readers to cut out and fill in a ballot contained in the paper. The choice is between candidates A and B. One week later the newspaper publishes the results of this poll as a "forecast" of the election result. On what grounds could you criticize the validity of this "sample"?
6. Eighty percent of a senator's home-state mail on the subject of gun control legislation is opposed to such legislation. How might you counter the argument that this "sample" of reactions reflects the views of all voters in the state?
7. A state legislative committee wishes to obtain the views of a sample of 50 senior citizens on proposed legislation dealing with medical assistance payments to the

elderly. Why should the 50 senior citizens *not* be selected (a) by choosing 50 of the residents of the nursing home nearest to the capital; (b) by choosing the first 50 senior citizens encountered in the street?

8. The Department of Labor selects a random sample of 1400 persons receiving unemployment benefits. Telephone interviews are completed with 684 of those selected. Of these, 30% reported that they had been unemployed for more than a year. Give two reasons why it would be risky to conclude that 30% of all presently unemployed persons have been unemployed for at least a year.

Solution to Problem 1

The problem is that those students who are the first to apply may differ systematically from all accepted students in their attitude toward the school. Such a sample for instance, will probably be biased in favor of those who are more likely than all students to have viewed the college as their first choice.

SECTION 8.3
EXPERIMENTS AND INTERNAL BIAS

In Section 8.1 we described the primary purpose of an experiment as "exploring the extent to which a change in one variable may *cause* a change in another." We begin our discussion of experiments with two examples.

EXAMPLE 11 A psychologist is interested in the impact of day care on the sociability of toddlers. She identifies 20 children who spend almost all week at a day-care center and another group of children who live in the same area but stay home with their mothers. Each of the 40 children is observed in a variety of situations and given a sociability rating which runs from 0 to 50. The higher the score, the more sociable the child. The mean score for the day-care children is 34.8 and for those staying at home 27.7. Does the difference in mean score $34.8 - 27.7 = 7.1$ reflect the positive effect of day care on the sociability of children?

■

EXAMPLE 12 A doctor is interested in the effect of sleep deprivation on the reasoning ability of students. Sixty student volunteers are randomly divided into three groups, *A*, *B*, and *C*. Members of group *A* are allowed three hours of sleep and members of group *B* five hours of sleep, while members of group *C* are instructed to obtain their customary night's sleep. In each case the prescribed sleep pattern is continued for four consecutive nights under controlled conditions. At the end of this time the students are asked to read a long article and then are given a 30-question test based on the article. The mean scores for the three groups are *A* (18.7), *B* (19.3), and *C* (18.5). Do these results suggest that lack of sleep has little effect on the scores?

■

In Example 11 day care is assumed to have some effect on sociability, and in Example 12 sleep is deprivation is assumed to have some effect on reasoning ability. With these relationships in mind we distinguish between two types of variables used in experiments.

DEFINITION 3

The variable, the change in whose behavior we wish to study, is called the **dependent variable** or the **response variable**.

In Example 11 the dependent variable was sociability rating, and in Example 12 it was reasoning ability as measured by the ability to comprehend written materials.

DEFINITION 4

The variable which may explain or "cause" changes in the dependent variable is called the **independent variable**.

In Example 11 the (qualitative) independent variable was "child care." In Example 12 the (quantitative) independent variable was "amount of sleep."

The independent variable, child care, has two levels, "home care" and "day care." These two levels are referred to as **treatments**. The independent variable, amount of sleep, takes values corresponding to the different amounts of sleep permitted the three groups. These values are also referred to as treatments.

To summarize:

DEFINITION 5

The various levels of the independent variable (if it is qualitative) or the values (if it is quantitative) are referred to as **treatments**.

We illustrate these ideas with a further example.

EXAMPLE 13 A chemist is interested in comparing the mean sulfur content of four different chemicals, the standard one, A, and three new ones, B, C, and D. Twelve samples of each of A, B, C, and D are made available to the chemist. In this example the dependent variable is "percentage of sulfur" and the independent variable is "type of chemical." There are four treatments, A, B, C, and D. ■

In Examples 11 and 12 the experiments were performed on human beings. In Example 13 the experiment was performed on inanimate objects (chemical samples). Generally:

> DEFINITION 6
>
> Each individual object on which an experiment is performed is called an **experimental unit** or unit of study. When the units are human, they are referred to as **subjects.**

Subjects (or experimental units) in experiments generally can be divided or divide themselves into groups which are defined as follows:

> DEFINITION 7
>
> One group receives the normal or standard treatment. This is called the **control group.**

> DEFINITION 8
>
> The group(s) receiving the other, usually new or innovative, treatment(s) are called the **experimental group(s).**

In Example 11 the control group was composed of the children who stayed home and the experimental group was composed of the day-care children. In Example 12 the control group (C) was composed of those students who enjoyed their customary night's sleep and the experimental groups (A and B), of those students who were allowed restricted amounts of sleep.

The control group provides a baseline against which to compare the effectiveness of the treatment(s). In Example 11, for instance, assume that the mean sociability for the day-care children was 34.8. In and of itself this figure tells us very little. But suppose we add that the mean score for home-care children was 27.7. If we compare these two scores, it now *appears* that day care improves sociability.

CONFOUNDING VARIABLES AND INTERNAL BIAS

We stressed the word "appears" above because it is possible that the difference in mean scores may *not* be due to type of care but to a third kind of variable, called a **confounding variable.**

It is known, for instance, that the presence of siblings affects a child's sociability. Now assume that 12 of the day-care children, but only six of the home-care children, had at least one sibling. Perhaps it was this difference that was responsible for the higher mean score in the day-care group, rather than the day-care environment. The presence or absence of siblings is an example of a confounding variable.

In the same way variables such as natural ability or normal sleeping habits would be potentially confounding variables in Example 12. If the members of the three groups differed greatly on these two variables, this might, at least in part, be responsible for the lack of significant differences in test scores.

More generally:

> DEFINITION 9
>
> A **confounding variable** is one which (a) is distributed differently among the members of the control and experimental group(s), and (b) has an effect on the dependent variable.

When the effect (or lack of effect) of an independent variable is disguised to some extent by a confounding variable, the experiment is said to contain **internal bias.** More formally:

> DEFINITION 10
>
> **Internal bias** occurs in experiments when differences in the dependent variable between the groups are (a) attributed to the independent variable, but (b) are at least partly due to one or more confounding variables.

The next two examples further illustrate internal bias.

EXAMPLE 14 Forty-eight of the students taking a statistics course agree to participate in an experiment to compare the standard lecture method with a self-paced plan. The independent variable is "type of teaching method" and the dependent variable is "the final exam score."

It is found that the average score of students in the lecture section is 4.5 points higher than that in the self-paced section. It appears that the self-paced method is inferior. But it is also reported that the students were allowed to *select* the treatment they would receive. It is, therefore, quite possible that the self-paced section ended up with a much greater proportion of students weak in math skills. As a result, some of the 4.5-point difference was due to the resulting imbalance between the members of the two classes. The concentration of the weaker math students in the self-paced class would then have biased the experiment against this type of instruction, and math skills would be the confounding variable responsible for the bias.

There may also be a substantially higher proportion of blue-eyed students in the self-paced section than in the lecture section. But eye color has little or no effect on test scores and is, therefore, not a confounding variable.

■

EXAMPLE 15 Numerous studies have attempted to link cigarette smoking and lung cancer. In all cases the death rate for this kind of cancer is substantially higher for smokers than for nonsmokers. Further, the more cigarettes smoked, the greater a person's chance of contracting lung cancer. The following figures, for males, are typical:

GROUP	MALE SMOKERS	MALE NONSMOKERS
Death rate from lung cancer per 100,000 persons	90	10

Male smokers are nine times as likely to die from lung cancer as are nonsmokers. But do these figures prove that smoking causes lung cancer? Let's see.

People choose whether or not to smoke, and there is a possibility that there may be some physiological (and confounding) variable, call it "factor X," which is responsible for both lung cancer and a desire for nicotine. An argument can be made that it is the greater presence of factor X in smokers than in nonsmokers which is the cause of *both* lung cancer and increased smoking.

■

The internal bias caused by confounding variables can be controlled by randomized assignment.

RANDOMIZED ASSIGNMENT

In Section 8.2 we indicated how external bias in surveys can be controlled by random *sampling*. In much the same way internal bias in experiments can be controlled by the random *assignment* of subjects (or units) to treatments. When this is done, the experiment is referred to as a **randomized experiment.**

Random assignment ensures that the distribution of each potentially confounding variable will be the same for each group. In Example 12, for instance, a potentially confounding variable was the number of hours of sleep to which a student is accustomed. The fact that students were assigned to the three sleep regimens at random helped ensure that this and *all* other variables would have approximately the same distribution in each of the three groups. Similarly, in Example 14, if the students were randomly assigned to the two sections, 24 to one section and 24 to the other, this would help to ensure that both sections were approximately comparable with regard to math competency and all other potentially confounding variables.

There are a number of different ways to assign subjects randomly to treatments. The simplest method involves assigning each subject directly to one of the treatments, using a table of random numbers or some other appropriate device. The result of this procedure is referred to as a **completely randomized design.** This type of design for Example 14 is shown graphically in Figure 2.

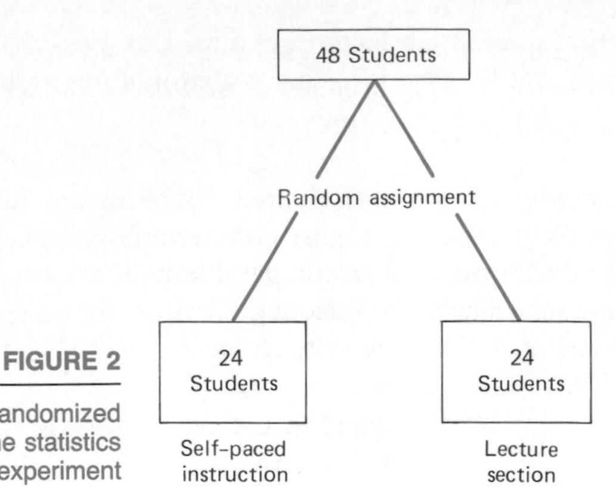

FIGURE 2

Completely randomized design for the statistics instruction experiment

EXAMPLE 16 One way to settle whether or not smoking *causes* lung cancer is to construct a completely randomized design. An ideal but repugnant method of doing this would be to randomly select perhaps 4000 10-year-olds and randomly divide them into four groups of 1000 each, one group smoking 60 cigarettes a day, the second 30 a day, the third 10, and the fourth none at all. The random assignment of children would ensure that factor X—if it exists—would then be distributed almost evenly over the four groups. Suppose that the death rates from lung cancer were then as follows:

GROUP	*A* (60 PER DAY)	*B* (30 PER DAY)	*C* (10 PER DAY)	*D* (NONE)
Death rate from lung cancer per 100,000 persons	243	92	60	12

Because of the randomization we could conclude that cigarette smoking does cause lung cancer or, more precisely, that the more you smoke, the greater are your chances of contracting lung cancer.

■

Recall from Section 8.2 that random sampling for a survey results in a sample that is representative of the population but does not guarantee that the sample so obtained will be a mirror image of the population. Similarly, in an experiment a completely randomized design results in groups that are approximately similar but not exactly comparable on all potentially confounding variables.

RANDOMIZED BLOCK DESIGN

On occasion the researcher may recognize a confounding variable which is believed to have a very strong effect on the dependent variable. In such cases it may be desirable to ensure that the groups are more exactly comparable with respect to this variable than might be obtained by simple randomization. This is most usually done by the use of what is called a **randomized block design.** The following is an example of such a design.

EXAMPLE 17 Referring back to the statistics course experiment (Example 14), it is known that a student's performance in statistics is highly correlated with math competency. Therefore, the two sections should be as comparable as possible with respect to this variable. This can be achieved by selecting the 48 students so that 16 have high scores (H) on a math competency test, 16 have moderate (M) scores, and 16 low (L) scores. These three categories are called **blocks.** Next, students within each block are randomly assigned to the two teaching methods (treatments), as illustrated in Figure 3 on the next page. In effect, this procedure "blocks out" the effect of the differences in math competency.

In general, this design is used to ensure that the various groups are comparable with respect to a potentially confounding variable. Members of each block should be as similar to each other as possible and also as different as possible from the members of other blocks—in each case on the basis of this variable.

As Figure 3 illustrates, the various groups are comparable with respect to the confounding variable (in this case math competency), and this variable therefore has the same effect on them all.

■

A randomized block design would also be helpful in the sleep experiment discussed in Example 12. In this case we might block on the basis of verbal SAT scores. First, we would divide the 60 students into four blocks of 15 each. Block I would consist of those students with the highest 15 scores, block II would consist of those 15 with the next highest scores, and so on. Then within each block five randomly chosen students would be assigned to each treatment (amount of sleep).

The next example is a special kind of randomized block design.

EXAMPLE 18 A new form of therapy for patients suffering from agoraphobia is to be compared with the standard therapy that has been in use at a hospital. Twenty pairs of patients are carefully selected so that each pair is matched with regard to sex, race, approximate age, and approximate severity of the disorder. One member of each pair will then be randomly selected to receive the new therapy. The other member of the pair will receive the standard treatment. In this case the independent variable is the type of therapy. The dependent variable is a subjective measure of the improvement in each patient after six months of treatment, as judged by an outside panel of consultants. This design is shown in Figure 4.

This experimental plan is called a **matched pairs design** and, in this case, is a special case of a randomized block design with 20 blocks and two subjects per block (i.e., each pair of patients constitute a block).

■

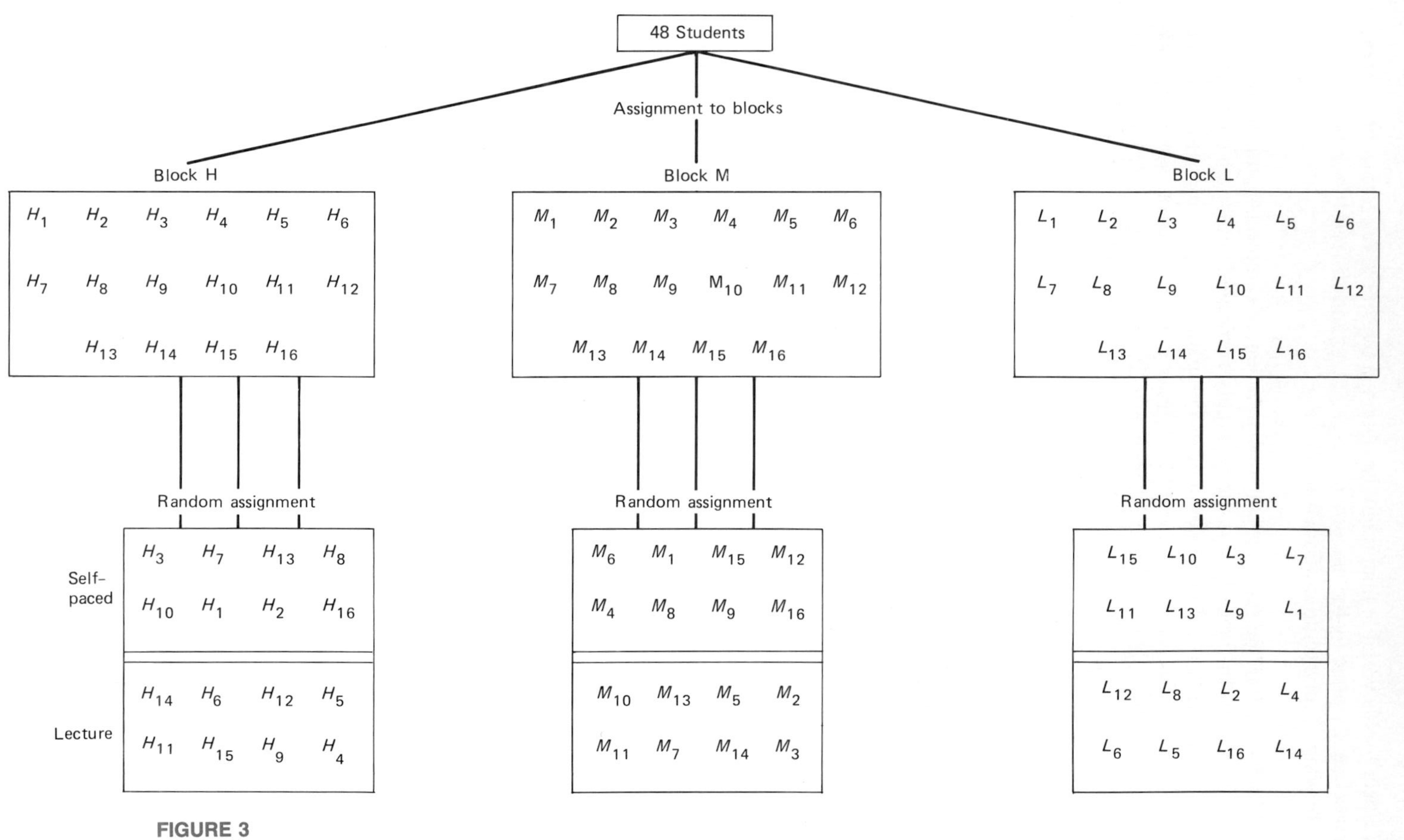

FIGURE 3

Randomized block design for the teaching statistics experiment

FIGURE 4

Matched pairs design in Example 18

Pair	Experimental therapy	Standard therapy
1	Patient E1	Patient S1
2	Patient E2	Patient S2
.	.	.
.	.	.
.	.	.
20	Patient E20	Patient S20

More generally, a matched pairs design involves the use of only two equal-sized groups: a control and an experimental group. Each member of the experimental group is matched with a member of the control group. The intention is that each pair should be as alike as possible *except* for the independent variable. The next example illustrates a special case of a matched pair design.

EXAMPLE 19 Twenty-five persons enroll in a weight-reducing program to test its effectiveness. Each person's weight will be recorded when he or she enrolls and again after one month. This design is shown in Figure 5. In this case each person in effect serves as his or her own matched pair, that is, as his or her own control. The advantage of using measurements taken on the same person at different times is the exact comparability of the experimental and control groups on all potentially confounding variables, such as age, initial weight, sex, and so on.

FIGURE 5

Matched pairs design in Example 19

Subject	Weight before (Control)	Weight after (Experimental)
1	$W_{1,C}$	$W_{1,E}$
2	$W_{2,C}$	$W_{2,E}$
3	$W_{3,C}$	$W_{3,E}$
.	.	.
.	.	.
.	.	.
25	$W_{25,C}$	$W_{25,E}$

■

QUASI-EXPERIMENTS

There are numerous situations, usually involving people, in which random assignment is not possible. Practically all such situations fall into three categories:

(a) Those in which treatments are assigned by nature rather than the experimenter. Examples include experiments in which the independent variable is sex, race, nationality, and the presence or absence of a disease or an inherited ability or other trait.

(b) Those in which researchers are prevented from assigning subjects to treatments at random because of ethical, moral, or political considerations. In such cases subjects consciously or unconsciously select themselves into the treatment groups. The day-care example and the smoking example illustrate this type of problem.

(c) Those in which treatments are assigned nonrandomly by an agency frequently without regard to the needs of experimentation. For example, applicants for a new housing project may be selected by a housing authority on the basis of need rather than at random. Researchers interested in comparing those accepted with those rejected—with regard to health in later life for instance—will have to take this method of assignment into account. (See Review Problem 17.)

Studies such as these are referred to as **quasi-experiments** or **observational studies.** In quasi-experiments the disturbing effects of confounding variables cannot be removed by randomization. A prominent statistician, William Cochran, divides confounding variables in quasi-experiments into three types based on the experimenter's perception of the importance of their effect on the independent variable. These three types and methods for controlling for them, where possible, are described below.

Type I are those confounding variables that are considered so important that some means for controlling their confounding effects must be employed, preferably in the design of the study. The primary means for removing the effect of variables of this type is to select the participants so that each confounding variable has at least approximately the same distribution for all groups. Procedures to accomplish this are called **matching** and are described below.

As a simple example of matching, the confounding effect of the variable "sex" can be eliminated by ensuring that each group has the same proportion of males and females as every other group. Similarly, if age is viewed as a potentially confounding variable, an attempt can be made to match each subject in one group with a person of approximately the same age in each of the other groups.

Matching is similar to blocking in that the purpose in both cases is the control of bias, and this is achieved by the careful selection of subjects from a pool. In blocking, subjects are selected *before* they are assigned to treatments (e.g., self-paced or classroom instruction). In matching, on the other hand, this selection is made *after* potential subjects in the pool have assigned themselves (or been assigned) to treatments (e.g., smokers or nonsmokers).

Given a large enough pool of potential subjects, it is possible to match experimental and control groups on any number of possible confounding variables. In practice, it is rare to match on more than three or four of the most important of such variables. We emphasize that if two groups are matched on a particular variable, this variable cannot then be responsible for differences between groups on the dependent variable.

Type II, the second type of confounding variables, are those for which adjustments are made only *after* the data suggest that they are necessary. This would occur, for instance, where it appears the results might, in part, be attributed to a confounding variable and the groups are found to differ substantially on this variable. Methods for deciding whether groups vary enough with respect to possibly confounding variables of this type will be discussed in later chapters. Methods for making such adjustments are given in Anderson et al.

Type III, the third type of confounding variables, are those for which no adjustments will be made because (1) they are considered sufficiently unrelated to the dependent variable, (2) they are too difficult to measure, or (3) they are unknown to the experimenter.

Those type III variables which the experimenter is unable to measure, or is unaware of, are particularly troublesome. In Example 11, for instance, a variable that would be difficult to measure is the disposition on the part of parents to give their children an opportunity to mix with other children of different types or backgrounds. It might be the greater disposition on the part of the day-care parents to do this rather than the day care itself which is responsible for the higher sociability scores (in the day-care group). Similarly, as indicated in Example 15, in the cigarette–lung cancer controversy, factor X is a type III variable which cannot be identified by the researcher.

The next example illustrates that matched pair designs occur in the context of quasi-experiments as well as randomized experiments.

EXAMPLE 20 A member of the education department at a university is interested in comparing the performance of boys and girls on a city-wide mathematics test taken by junior high school students. Fifty boys and 50 girls are carefully selected from the junior high. Each boy is matched with a girl so that they are the same age, have the same number of math courses, and have the same family background. Finally, the math scores for each pair are obtained. In this case the independent variable is sex, so no randomized assignment is possible. ■

EXTERNAL BIAS IN EXPERIMENTS

In this section we have emphasized the major issue in designing experiments as the *assignment* of participants to treatments in order to avoid internal bias. This does not mean, however, that the random *selection* of participants need not be considered. In fact, the comments in the preceding section concerning selection procedures for surveys also apply to experiments. The results of experiments can be generalized with greater validity if the participants are randomly selected from the target population. Unfortunately, this is rarely possible in practice, so that experimenters need to be cautious in generalizing the results of their experiments.

In Example 18, for instance, suppose that the new therapy is found to result in a considerable improvement over the standard treatment. The extent to which this finding can be generalized will depend on how typical the hospital's staff, size, and patients are.

PROBLEMS 8.3

9. A bakery is interested in the effect of oven temperature on the moistness of angel cake. Twenty identical batters are made up in the morning and randomly divided into four groups of five each. The first five are cooked at 300°F, then the next five are cooked at 350°F, the next five at 400°F, and the last five at 450°F, all for 40 minutes. Thirty minutes after being removed from the oven, each cake is given a moistness rating on a scale of 0 to 10 (the moister the cake, the higher the rating).
   (a) What are the independent and the dependent variables?
   (b) What are the experimental units?

(c) Is this a randomized experiment or a quasi-experiment? Explain your choice.

(d) Does the order in which the batters are tested suggest a potentially confounding variable?

10. Briefly, outline an experiment that you are familiar with (or one that you would like to perform). State the independent, dependent and potentially confounding variables. Was randomization used? If not why not?

11. (a) Explain why it would be helpful to block by SAT verbal scores in Example 12.

(b) Draw a picture similar to that of Figure 3, showing a possible randomized block design for this experiment.

12. The effects of two experimental pain-killing drugs A and B are evaluated in an experiment. Seventy-two patients have volunteered to participate. Drugs A and B are each given to 24 randomly selected patients. The remaining 24 patients receive the standard (control) drug (C). The dependent variable is a measure of the seriousness of the side effects of the drugs.

(a) What is the independent variable?

(b) What is the experimental unit?

(c) It is suspected that the seriousness of the side effects may be related to the sex and race (black or white) of the patient. Assuming that you can control the sex and racial composition of the 72 patients, explain how you would modify the design to create a randomized block design with four blocks.

(d) Why would it be inappropriate to allow the patients to select which drug to receive?

(e) Why would it be inappropriate to select the 48 most seriously ill patients and assign drug A or B to them?

13. A commuter has just moved to a new home and has a choice of using two routes (A and B) to his job. He is interested in discovering the one that is, on average, the faster. After flipping a coin he begins with A and travels each route on alternate workdays until he has used each route for 15 days. Each day he records the time to get to work.

(a) What are the independent and dependent variables?

(b) What are the units of study?

(c) Is this a randomized experiment or a quasi-experiment? Explain your answer.

(d) Can you think of any potentially confounding variables?

14. A consumer testing agency wants to test a tire manufacturer's claim that its recently developed (and expensive) Z tire significantly reduces braking distance as compared with its standard Y tire. The agency uses 20 cars in an experiment. These include two identical cars of 10 different models. One of each model is selected on a coin toss and is fitted with Z tires. The other car is fitted with Y tires. All 20 cars are given the same series of braking tests and the total stopping distance is recorded for each. The distances for the 10 cars with Z tires are to be compared to those for the 10 cars with Y tires.

(a) What are the independent and dependent variables?

(b) What are the experimental units?

(c) What kind of experiment is this?

(d) What potential confounding variables has this experiment controlled for?

15. A researcher interviews 30 women whom she judges to be successful. She would like to use the information obtained to isolate those variables which are good indicators of a woman's being successful. [In this example the researcher has not identified the independent variable(s) beforehand.]

(a) What is the dependent variable in this case?

(b) What are the experimental units?

(c) What essential element is missing in this study?

(d) Explain why the researcher's efforts will be fruitless without this element.

(e) What steps could be taken to correct the problem?

16. As part of a study investigating the effect of smoking on infant birthweight a physician examines the records of 40 nonsmoking mothers, 40 light-smoking, and 40 heavy-smoking mothers. The mean infant birthweights (in pounds) for the three groups are

	NONSMOKERS	LIGHT SMOKERS	HEAVY SMOKERS
MEAN INFANT BIRTHWEIGHTS	7.56	7.25	7.08

(a) What are the independent and dependent variables?

(b) What are the experimental units?

(c) Is this a randomized experiment or a quasi-experiment? Explain your choice.

(d) What are potential confounding variables in this case?

(e) Explain how you would eliminate the effects of at least some of the variables in part (d).

17. In a particular state the speed limit on sections of highway other than interstate highways has been lowered from 55 mph to 50 mph. To examine the effect of this move the state's secretary of transportation wants to compare the accident rates at 20 randomly selected sites six months before and six months after the lowering of the speed limit.

(a) Why should the secretary wait six months and then examine accident rates at that time? Why go back six months before the change and examine accident rates at that time?

(b) What are the independent and dependent variables?

(c) What are the experimental units?

(d) What are potentially confounding variables in this case?

18. Morris (1979) describes the Leboyer method of childbirth. With this method the child is born into a warm, quiet atmosphere and is gently massaged by both doctor and mother for 4 or 5 minutes. This is in contrast to the harsh lights, the activity, and the quick slap on the bottom found in the normal delivery room. Morris comments:

> The preliminary results of one study of 50 Leboyer babies have shown them to be more content and more socially adaptive than other babies. They have fewer problems with eating and sleeping and have excellent relationships with their parents (Rapoport, cited in Englund, 1974). This may give some substance to Leboyer's argument that babies should be treated like "persons," not "mere squalling digestive tracts."

(a) Why do you think the study described here was a quasi-experiment?

(b) Criticize the validity of the conclusions expressed here, concentrating on the effects of potentially confounding variables.

19. In Example 14 the mean score in the self-paced section was 4.5 points lower than that in the lecture section. We pointed out though that it was possible the experiment was biased because "the self-paced section ended up with a much greater

proportion of those students weak in math skills. As a result, some of the 4.5-point difference was due to the resulting imbalance."

Suppose now that the mean score in the self-paced section was 4.5 points *higher* than that in the lecture section. Using the same assumptions as in Example 14, explain why the bias indicated above would be far less important in this case.

Solution to Problem 9

(a) The independent variable is temperature; the dependent variable is the moistness rating.

(b) The experimental units are the batters.

(c) This is a randomized experiment: a completely randomized design, to be precise. Each of the batters are randomly assigned to one of the four temperatures (treatments).

(d) One can imagine that the longer the batters are left waiting to be baked, the dryer they will become. To avoid any bias caused by some groups being baked later than others, all the cakes should be baked at the same time.

SECTION 8.4
SUMMARY AND REVIEW

Statistical studies may be divided into two broad groups: surveys and experiments.

In a **survey,** sample data are used to draw conclusions about a single variable or, more usually, a number of separate variables.

In an **experiment**, data are used to determine the extent of a cause-and-effect relationship between two or more variables.

Two concepts, those of **precision** and **bias,** are important to the understanding of surveys and experiments. Bias was discussed in this chapter and precision will be discussed in later chapters.

There are two forms of bias: external and internal.

External bias occurs when a sample systematically differs from the population in important respects. This is especially relevant to surveys. A sampling procedure is biased if it tends to favor the selection of certain members of the population.

External bias can be avoided by sampling randomly. One method for doing this is to use a table of random numbers such as that in Table A10.

A randomly selected sample may still be biased if a population list is used that is (1) out of date, (2) incomplete, or (3) in some way differs from the target population; (4) there is considerable nonresponse.

Even with random selection, an appropriate list, and 100% response, a random sample may still not mirror the population. But with a randomly selected sample (and only such a sample) probability theory enables us to measure the extent to which the sample is likely to differ from the population.

In experiments it is necessary to distinguish between three types of variables: (1) independent variables (2) dependent variables, and (3) confounding variables.

The **dependent** variable is the variable whose behavior we wish to study. The **independent** variable is the variable that we believe may explain or "cause" changes in the dependent variable. A **confounding** variable is one that (1) is distributed differently among members of the control and experimental groups and (2) affects the dependent variable.

The various levels (if the independent variable is qualitative) or the values (if it is quantitative) are referred to as **treatments.**

Internal bias occurs when differences between groups on the dependent variable are attributed to the independent variable but are at least partly due to one or more confounding variables.

Internal bias can be controlled by the random assignment of subjects or units to treatments. When such an assignment is made directly to treatments, the experiment is referred to as a **completely randomized design.**

When one or more confounding variables have a strong effect on the dependent variable, a **randomized block design** is used. This involves the assignment of subjects (or units) to blocks so that the confounding variable(s) are similarly distributed among the members in each block. Subjects are then randomly assigned from blocks to treatments.

There are situations, usually involving people, in which random assignment is not possible. Such studies are referred to as **quasi-experiments** or **observational studies** and include situations where:

(a) Subjects are assigned to treatments by nature rather than the experimenter. This occurs, for example, when the independent variable is sex, race, or age.
(b) The researcher is prevented from assigning subjects to treatments by ethical, moral, or political reasons. In such cases the subjects usually select themselves into treatment groups, either consciously or unconsciously.
(c) treatments were assigned nonrandomly by an agency frequently without regard to the needs of experimentation.

The primary means for removing the effects of confounding variables in quasi-experiments is **matching.** Matching is similar to blocking in that the purpose, in both cases, is the control of bias and this is achieved by the careful selection of subjects from a pool. In blocking, subjects are selected *before* they are assigned to treatments. In matching, on the other hand, this selection is made *after* the subjects in the pool have assigned themselves (or been assigned) to treatments.

A **matched pairs design** occurs when each member of the control group is "matched" with a member of the experimental group. This design can be used in either a randomized experiment or a quasi-experiment.

REVIEW PROBLEMS

GENERAL

1. Explain in your own words the rationale for selecting a random rather than a nonrandom sample.

2. Summarize in your own words the rationale for performing a randomized experiment rather than a quasi-experiment whenever possible.
3. Why is it impossible to compute probabilities such as $P(\bar{X} > 368)$ if the sample is *not* randomly selected? (Assume that the population mean and standard deviation are known.)
4. The host of a children's television show invites viewers at home to call the program and record "yes" if they like the program and "no" if they do not. In all, 406 called with a "yes" and 38 with a "no." The host claims that these figures indicate how popular the program is. Comment on the host's claim.
5. One of the two newspapers in a large city invites its readers to fill out and mail a coupon indicating whether or not they approve of a downtown development scheme. Of the 5988 coupons returned, 4733 indicated opposition to the scheme. The newspaper claims that these figures indicate that residents of the city are overwhelmingly opposed to the scheme. Comment on this claim.
6. A hockey player will experiment with a new aluminum hockey stick. He intends to use the new stick in the first half and a regular stick in the second half of each of the next 16 games. At the end of this period he will compare the number of points scored with the two sticks.
   (a) Criticize this experimental plan.
   (b) What would be a better plan?
7. Restaurants frequently leave suggestion cards on the table inviting diners to make suggestions or comments about the meal. Why are these cards likely to give the restaurants a biased view of their diners' impressions?
8. An opinion pollster has been hired to select a sample of 1000 adults in the state, to visit those selected in the evening, and to help them fill out a questionnaire on various aspects of their leisure activities. The pollster finally reports that only 540 of those sampled could be found at home. Explain why, in addition to reducing the sample size, this nonresponse will probably bias the results obtained.
9. Eighty students agree to participate in an experiment to investigate the effects of various amounts of alcohol on reaction time to simulated automobile incidents. The students are first divided at random into four groups of 20. The average reaction time of each student is measured *before* drinking alcohol and then again afterward. The amounts of alcohol consumed are group A, six beers; group B, four beers; and group C, two beers. Students in group D were given a glass of water. The dependent variable is the final reaction time minus the initial time.
   (a) What is the independent variable in this case?
   (b) Is this a quasi-experiment or a randomized experiment?
   (c) What is the experimental unit?
   (d) What is the purpose of including one group (D) which is given no alcohol?
   (e) Explain how you would have created a randomized block design with blocks based on sex.
10. Referring to Review Problem 9, explain what problems would arise if groups C and D were tested in the middle of the day, and groups A and B in the early evening?

HEALTH SCIENCES

11. A pediatrician is interested in learning about the many ways in which prospective parents prepare for the birth of their child. Her first reaction is to sample the views of couples enrolled in childbirth classes. Why would this result in a biased sample?
12. An insurance company is interested in investigating the effect on medical bills of providing coverage that will pay for outpatient care as well as full hospitalization.

They can afford to provide such coverage on a trial basis for 600 of their clients in one city.

(a) Indicate a procedure by which an experimental and control group of clients could be compiled.

(b) What are the independent and dependent variables?

13. In many medical experiments today a double-blind procedure is used. With this procedure neither the patient (subject) nor the person responsible for evaluating the patient's progress is aware of which patients have received which treatments. Why do you think this procedure is used?

14. In 1945 the water supply for Newburgh, New York, was treated with sodium fluoride. Kingston, New York, a nearby town of roughly the same size, which did not fluoridate its water supply, was chosen as a control town. After 10 years it was found that children in Newburgh age 6 to 9 years had 58% less tooth decay than children of the same age in Kingston. Considerable efforts were made to determine that children in the two towns did not differ significantly with resepct to growth, bone structure, vision, and a host of other variables.

(a) Is this a randomized experiment or a quasi-experiment?

(b) What is the experimental unit?

(c) What are the independent and dependent variables?

(d) Why was it necessary to include a control town?

(e) Why did the researchers check that the two towns were comparable with regard to so many variables?

(f) Can we conclude that fluoridated water produces less tooth decay than unfluoridated water?

15. The Lanarkshire Milk Experiment (Gosset, 1971) in Lanarkshire, Scotland, in 1930 was designed to measure the benefits, if any, of giving young children free milk in school. For a period of four months 5000 children received raw milk, 5000 pasteurized milk, and 10,000 children—the control group—no milk. The 20,000 children were selected from 67 schools. For the most part up to 200 children in each school were selected at random to receive milk and another 200 were selected at random to act as controls. However, "in any particular school when there was any group to which these [random] methods had given an undue proportion of well-fed or ill-nourished children, others were substituted in order to obtain a more level selection." After four months the weight gain and gain in height were recorded for each of the 20,000 children. On each variable the average gain for the milk children was greater than for the control children.

(a) What is the experimental unit in this experiment?

(b) What are the independent and dependent variables?

(c) What is wrong with allowing the schools the discretion indicated in quotation marks above?

SOCIOLOGY AND GOVERNMENT

16. Immediately after American military forces invaded a Caribbean island, telephone calls to the White House numbered three to one in favor of the invasion. The President referred to these figures as evidence of the support in the country for his policy. Comment on the president's view.

17. In the 1950s Johns Hopkins University investigators began a study of the physical and social benefits of low-cost public housing for the poor (Freedman and others,1978). The experimental group consisted of a random sample of 300 families from approximately 800 that had applied for a new public housing project in Baltimore and had been approved by the Baltimore Housing Authority. The control group consisted of 300 families that had applied for the same project but had

been rejected by the housing authority. All 600 families in the study *had* lived in roughly the same slum area. After three years the experimental group was found to be both happier and healthier than the control group.

(a) Is this a randomized experiment or a quasi-experiment?

(b) What are the independent and the dependent variables?

(c) Can we conclude that public housing produces happier and healthier people?

18. As part of a study of the needs of prisoners when released from prison a group of 60 prisoners about to be released were randomly divided into two groups of 30 each. Those in group *A* received special assistance in finding a job, finding housing, and reuniting with their families. Group *B* received only the normal (minimal) assistance. After two years the 60 persons were traced and it was recorded whether or not each of them had a job; if so, how long they had held the job; and whether or not they had had any further brushes with the law. It was found that those who had received the special assistance had lived more stable lives and had considerably fewer brushes with the law than the other group.

(a) Is this a randomized experiment or a quasi-experiment?

(b) What is the experimental unit?

(c) What are the independent and the dependent variables?

(d) Can we conclude that this kind of special assistance is responsible for the more stable and law-abiding lives?

ECONOMICS AND MANAGEMENT

19. A company owning a chain of women's clothing stores is interested in sampling its customers to obtain information about their shopping habits. The company is advised that its credit-card holders form a convenient list from which to sample. Explain why the use of this list would likely result in a biased sample.

20. As part of an advertising campaign an automobile manufacturer invites a group of 50 persons to be driven blindfolded over a course, first in the company's sedan and then in a luxury automobile. Each rider is asked to rate and compare each of the rides. In what way would using the luxury automobile last bias the experiment? What would be a better system?

21. A company is interested in determining which of three types of packaging, *A*, *B*, or *C*, appeals the most to potential customers. A random sample of 120 retail outlets is selected as part of an experiment. These outlets are then randomly divided into three groups, each consisting of 40 stores. The stores in group I will receive the company's items packaged in *A*. Those in group II will receive packaging *B*, and group III, packaging *C*. The number of items sold in a 12-week period is recorded for each of the 120 stores and the sales per store compared.

(a) Is this a randomized experiment or a quasi-experiment?

(b) What are the independent and the dependent variables?

(c) What is the experimental unit?

(d) How has the company avoided problems of external bias?

22. Refer to Review Problem 21. The company statistician suggests that bias in the experiment would be more effectively controlled if the 120 stores were blocked on the basis of the number of items normally sold (fewer than 10 per week, between 10 and 25, between 25 and 40 and more than 40 per week).

(a) Explain how the 120 stores could be selected and blocked in this way.

(b) Why would such blocking be likely to create a less biased experiment?

EDUCATION AND PSYCHOLOGY

23. A high school principal is convinced of the benefit to children of studying Latin. She selects a random sample of those seniors who have had at least a year of Latin and a random sample of those seniors who have not. The mean verbal SAT score for the Latin scholars is 584. The corresponding mean for those without Latin is 524. The principal points out to her teachers that taking Latin produces on average a 60-point increase in verbal SAT scores. Why is the principal at fault in her reasoning?

24. At a liberal arts college, students needing remedial math may elect to take either a standard introductory mathematics course or a new "mind over math" clinic. In either case these students have to retake the school's math competency test. Those taking the introductory mathematics course score an average of 27 on this test. Graduates of the clinic score an average of 34. Advocates of the clinic claim that this difference demonstrates the superiority of their approach. Comment on this claim.

25. An important question in psychology that has not been fully resolved is the extent to which IQ is determined by heredity and by environment. Perhaps the best evidence on the issue is obtained in the rare instances when identical twins are separated at or close to birth and brought up in different environments. When such twins have been located and their IQs recorded, the scores for each twin have been very close.

(a) Explain why identical twins separated at birth and raised in different environments represent the best method for deciding this issue.

(b) Why is it impossible from a practical viewpoint to conduct a randomized experiment to answer this question?

(c) If you could manipulate people at will, how would you set up a randomized experiment to answer the questions?

(d) What conclusions seem appropriate given that the IQ scores for such twins are very close?

26. A famous experiment (Rosenthal and Jacobson, 1968) involved the effect of teacher expectations on students' academic performance. All the children in an elementary school were given an IQ test. The teachers were told that it was a test designed to discover the extent to which children can be expected to improve in the future. Approximately 20% of the children in the school were selected at random and their teachers informed that the test had predicted that these children would make considerable progress in the coming year. At the end of the year all the children were given another IQ test. When the increases in test scores were studied, it was found that the children "expected" to improve had in fact attained a greater average increase than the other students.

(a) Is this a randomized experiment or a quasi-experiment? Explain.

(b) What are the independent and the dependent variables?

(c) Can we conclude that positive teacher expectations do have an effect on student performance? (*Note:* The results obtained were later questioned for other methodological reasons.)

SCIENCE AND TECHNOLOGY

27. Why is it so much easier to perform randomized experiments in the natural sciences than in the social sciences?

28. In a certain area of the country the average amount of rainfall in September over the past three years was .15 inch. This September clouds in the region are "seeded" with silver iodide pellets in an effort to increase rainfall. The rainfall this September was .22 inch. Do these data "prove" that seeding increases rainfall?

REFERENCES

BRYSON, M. C. "The Literary Digest Poll: Making of a Statistical Myth," *The American Statistician,* Vol. 30, (Nov. 1976), pp. 184–185.

ENGLUND, S, Birth without violence. New York Times Magazine, December 8, 1974, pp. 113–120.

FREEDMAN, D., PISANI, R., and PURVES, R. *Statistics*. New York: W. W. Norton, 1978, pp. 11–12.

GOSSET, W. S. ("Student"). "The Lanarkshire Milk Experiment," in Steger, J.A., ed. *Readings in Statistics for the Behavioral Scientist*. New York: Holt, Rinehart and Winston, 1971, pp. 159–168.

MORRIS, C. G. *Psychology: An Introduction,* 3rd ed. Englewood Cliffs, N.J.: Prentice-Hall, 1979, p. 77.

ROSENTHAL, R. and JACOBSON, L. *Pygmalion in the Classroom. Teacher Expectations and Pupils' Intellectual Development*. New York: Holt, Rinehart and Winston, 1968.

SUGGESTED READING

More complete introductions to sampling practice and theory can be found in:

KISH, L. *Survey Sampling*. New York: Wiley, 1965.

SLONIN, M. J. *Sampling in a Nutshell*. New York: Simon and Schuster, 1973.

STUART, A. *Basic Ideas of Scientific Sampling*. New York: Hafner, 1962.

Randomized experiments are treated thoroughly in:

KIRK, R. E. *Experimental Design: Procedures for the Behavioral Sciences:* Belmont, California: Brooks/Cole, 1968.

Two new excellent books discuss bias and other problems in quasi-experiments:

ANDERSON, S., et al. *Statistical Methods for Comparative Studies*. New York: Wiley, 1980.

COCHRAN, W. G. *Planning and Analysis of Observational Studies*. New York: Wiley, 1983.

An early overview of this important topic can be found in:

KISH, L. "Some Statistical Problems in Research Designs," *American Sociological Review,* Vol. 24, (1959), pp. 328–338.

9 ESTIMATION IN LARGE SAMPLES

REDUCING MEDICAL COSTS

The *New York Times* recently reported on the results of a health survey of 1530 randomly selected adults. Fifty-eight percent of those surveyed were willing to have routine illnesses treated by a nurse or a doctor's assistant rather than by a doctor if this would reduce the cost of health care. Anticipating the question of the reliability of figures based on so small a fraction of *all* U.S. adults the *Times* printed the following explanation: "In theory it can be said that in 95 cases out of 100 the results based on the entire sample differ no more than 3 percentage points in either direction from what would have been obtained by interviewing all adult Americans."

You are asked to interpret this explanation in Review Problem 20 on page 322.

SECTION 9.1
INTRODUCTION TO ESTIMATION

ESTIMATION

In Chapter 7 we said that statistical inference involved making judgements about aspects of a population based on information (data) in a sample from the population. It is convenient to think of statistical inference as falling into one of two categories: (1) estimation and (2) hypothesis testing. We discuss hypothesis testing in Chapter 10. We begin our discussion of estimation in this chapter with an example.

EXAMPLE 1

Of considerable interest to colleges and universities is the unknown proportion of high school seniors who intend to go on to college. This is designated π. An educational consultant selects a random sample of 500 seniors of whom 190 report that they intend to go on to college. The consultant might then (1) use the proportion of the sample intending to go on to college ($\bar{p} = 190/500 = .38$) as an estimate of the population proportion, π; (2) ask how close this estimate is to π; and (3) wonder whether a more accurate estimate for π might be obtained by selecting a larger sample of students. These are some of the issues that arise in estimation.

■

More generally, as the word suggests, in estimation we are interested in using the data in the sample(s) to estimate the unknown value of parameters such as π, μ, $\pi_1 - \pi_2$, or $\mu_1 - \mu_2$. In this chapter we discuss estimation using samples which are large enough for the central limit theorem to apply (i.e., $n \geq 30$ or n_1 and $n_2 \geq 30$). These are referred to as large samples. First, however, we shall need to make reference to a new Z table.

CENTRAL Z VALUES

Table A5 provides values for the standardized normal random variable Z such that different areas, denoted by γ (the lowercase Greek letter gamma), lie between $-z$ and z. For instance, the area between -1.96 and 1.96 is $P(-1.96 < Z < 1.96) = .95$. Similarly, the area between -1.282 and 1.282 is .8, that is, $P(-1.282 < Z < 1.282) = .8$. We illustrate this last value in Figure 1.

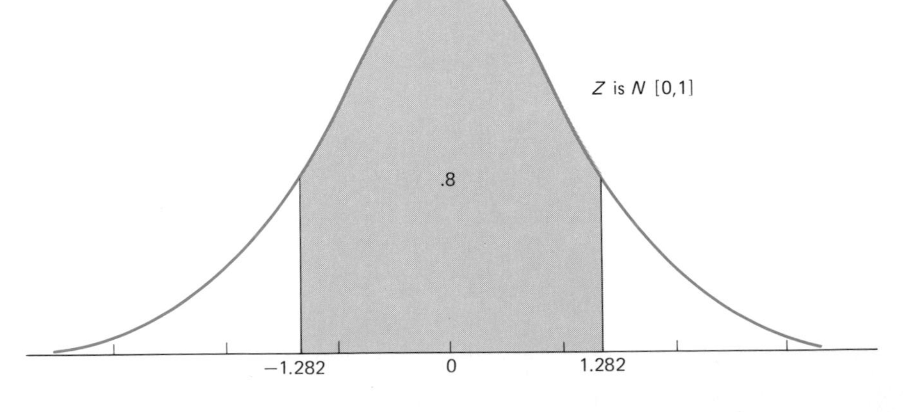

FIGURE 1

The area under the $N[0, 1]$ distribution between -1.282 and 1.282 is .8

PROBLEMS 9.1

1. Find the value z such that the area under the $N[0, 1]$ distribution between $-z$ and z is $\gamma = .99$.
2. Find the value z such that the area under the $N[0, 1]$ distribution between $-z$ and z is (a) $\gamma = .75$; (b) $\gamma = .90$.

Solution to Problem 1

From Table A5 with $\gamma = .99$, the values of $-z$ and z are -2.576 and 2.576.

SECTION 9.2

CONFIDENCE INTERVAL FOR THE POPULATION MEAN

An economist is interested in μ, the mean monthly expenditure for food per family in a town. Not every family can be interviewed, so the economist selects a random sample of families and uses $\bar{x}$, the mean expenditure in the sample, as an estimate for μ. Suppose, for instance, that $\bar{x} = \$129.40$. We refer to this number as a **point estimate** for μ since this single value is used to estimate μ. What is missing in the point estimate, however, is any indication of how *close* it is to the unknown μ. In Chapter 7 we noted that provided $n \geq 30$, the distribution of the random variable $\bar{X}$ is approximately normal with a mean of μ and standard deviation, $SD(\bar{X}) = \sigma/\sqrt{n}$. Recall that $\sigma/\sqrt{n}$ measures the extent to which $\bar{X}$ varies around μ in repeated samples. Suppose, for example, that σ is known to be approximately \$25 and $n = 100$ families. Then $SD(\bar{X}) = 25/\sqrt{100} = \$2.50$. Since $\bar{X}$ is approximately normally distributed, a specific value $\bar{x}$ is unlikely to be farther than 2 standard deviations—or in this case $2(2.50) = \$5.00$—from μ. We can now formalize this approach by constructing what is known as a "confidence interval" for the unknown μ.

A **confidence interval** for the population mean μ is an interval constructed around the observed sample mean $\bar{x}$ in which we can be reasonably sure that μ lies. The interval is usually presented in the form (L, U), where L is the lower bound and U is the upper bound of the interval. We can construct a confidence interval which we may be 99%, 95%, 80%, and so on, sure includes μ. We begin by constructing a 95% confidence interval.

A 95% CONFIDENCE INTERVAL FOR μ

Recall from pages 251 and 252 that if $n \geq 30$ we can regard the random variable $(\bar{X} - \mu)/(\sigma/\sqrt{n}) = Z$ as $N[0, 1]$. From Table A5 the area between $Z = -1.96$ and $Z = 1.96$, that is, $P(-1.96 < Z < 1.96)$ is .95. We can thus write

$$P\left(-1.96 < \frac{\bar{X} - \mu}{\sigma/\sqrt{n}} < 1.96\right) = .95$$

We can rearrange the inequality,

$$-1.96 < \frac{\bar{X} - \mu}{\sigma/\sqrt{n}} < 1.96$$

which is inside the parentheses, so that only μ lies between the inequality signs. Multiplying each of the three components by $\sigma/\sqrt{n}$, we obtain

$$-1.96 \frac{\sigma}{\sqrt{n}} < \bar{X} - \mu < 1.96 \frac{\sigma}{\sqrt{n}}$$

Multiplying through by -1 requires that we change the direction of the inequality signs. The result is

$$1.96 \frac{\sigma}{\sqrt{n}} > \mu - \bar{X} > -1.96 \frac{\sigma}{\sqrt{n}}$$

(note that $-(\bar{X} - \mu) = \mu - \bar{X}$)

Adding $\bar{X}$ to each of the three components, we obtain

$$\bar{X} + 1.96 \frac{\sigma}{\sqrt{n}} > \mu > \bar{X} - 1.96 \frac{\sigma}{\sqrt{n}}$$

or, equivalently,

$$\bar{X} - 1.96 \frac{\sigma}{\sqrt{n}} < \mu < \bar{X} + 1.96 \frac{\sigma}{\sqrt{n}}$$

The ends of the interval $\bar{X} - 1.96\ \sigma/\sqrt{n}$ to $\bar{X} + 1.96\ \sigma/\sqrt{n}$ are determined by its midpoint $\bar{X}$. Here $\bar{X}$ is a random variable. This interval is therefore referred to as a *random interval*.

It will be more convenient for computation to express the interval in the form

$$\bar{X} \mp 1.96 \frac{\sigma}{\sqrt{n}}$$

Notice that the minus sign is placed above the plus sign. This is to indicate that the interval runs from

$$\bar{X} - 1.96 \frac{\sigma}{\sqrt{n}} \quad \text{to} \quad \bar{X} + 1.96 \frac{\sigma}{\sqrt{n}}$$

Once the sample has been selected we obtain an observed sample mean, $\bar{x}$. Replacing $\bar{X}$ by $\bar{x}$, we obtain the result defined as follows:

DEFINITION 1

$$\bar{x} - 1.96 \frac{\sigma}{\sqrt{n}} \quad \text{to} \quad \bar{x} + 1.96 \frac{\sigma}{\sqrt{n}}$$

is a 95% confidence interval for μ.

In our earlier notation, $L = \bar{x} - 1.96\ \sigma/\sqrt{n}$ and $U = \bar{x} + 1.96\ \sigma/\sqrt{n}$. Applying Definition 1 to the expenditure data with $\bar{x} = \$129.40$, $n = 100$, and $\sigma = \$25$, we obtain a 95% confidence interval as $129.40 \mp 1.96(25/\sqrt{100})$, which is 129.40 ∓ 4.90 or \$124.50 to \$134.30.

It is tempting to say that there is a .95 probability that μ lies in the interval above, but this is incorrect. We can say there is a .95 probability that μ *will* lie in the interval but only *before* we select the sample. Once the sample is selected the value of $\bar{x}$ (and therefore of the interval) is fixed. In the "expenditure" case this interval is \$124.50 to \$134.30 and μ is either in *this* interval or it is not. Therefore we can only say that we are *95% confident* that μ lies in the internal \$124.40 to \$134.30.

Finally, we emphasize that it is much more important that you understand the interpretation of the interval in Definition 1 than its derivation.

THE CONFIDENCE INTERVAL FOR μ AS A PERCENT

We noted from Table A5 that the area between -1.96 and 1.96 is $\gamma = .95$. In general we speak of a confidence interval in terms of a percentage. To do so, we multiply γ by 100. Thus, if $\gamma = .95$, we get a $[100(.95)]\% = 95\%$ confidence interval.

Thus in general:

> DEFINITION 2
>
> If a random sample of size $n \geq 30$ is selected, a $(100\gamma)\%$ confidence interval for μ is $\bar{x} - Z_{[\gamma]}(\sigma/\sqrt{n})$ to $\bar{x} + Z_{[\gamma]}(\sigma/\sqrt{n})$. Here $Z_{[\gamma]}$ is the *value* for Z such that the area between $-Z_{[\gamma]}$ and $Z_{[\gamma]}$ is γ.

Now suppose that our economist preferred an 80% confidence interval for μ. Referring to Table A5 with $\gamma = .8$, the corresponding value for $Z_{[.8]}$ is 1.282. With $\bar{x} = 129.40$, $\sigma = \$25$, and $n = 100$, as before, an 80% confidence interval for μ is

$$129.40 \mp 1.282\left(\frac{25}{\sqrt{100}}\right) = 129.40 \mp 3.205$$

$$= \$126.20 \text{ to } \$132.60$$

We may be 80% confident that the mean expenditure per family, μ, lies between \$126.20 and \$132.60.

Notice that the 80% confidence interval is narraower than the 95% confidence interval and thus we are estimating μ more accurately. We pay for this greater accuracy, however, by being less confident that the interval contains μ.

As the entries in Table A5 suggest, the larger the value for γ, that is, the more confidence we wish to have in our inference, the larger $Z_{[\gamma]}$ and the wider the resulting interval. In short, when constructing confidence intervals, keep in mind the trade-off between accuracy and the degree of confidence in that accuracy.

EXAMPLE 2 A sociologist wishes to estimate the mean number of hours, μ, that children in a city watch TV during a particular week. A random sample of n children is selected and the number of hours spent watching TV recorded. Assume that the standard deviation of the time spent watching TV over all children in the city is $\sigma = 2.5$ hours. If the sample mean is $\bar{x} = 14.7$ hours, find a 90% confidence interval for μ if the sample size is (a) $n = 64$; (b) $n = 100$; (c) $n = 400$.

Solution From Table A5, $Z_{[.9]} = 1.645$. Using Definition 2, the required interval is

$$14.7 \mp 1.645\left(\frac{2.5}{\sqrt{n}}\right)$$

Successively substituting $n = 64$, $n = 100$, and $n = 400$ into this result, we obtain the following confidence intervals:

n	*Confidence interval*
(a) 64	14.19 to 15.21
(b) 100	14.29 to 15.11
(c) 400	14.49 to 14.91

(Do verify these intervals.)

We interpret the interval in (a) as follows. We can be 90% confident that the mean time that children watch TV in this week is between 14.19 and 15.21 hours.

Notice that for fixed values for γ and σ, the width of the interval decreases as n increases. For a given degree of confidence, the *larger* the *sample* we select, the *more accurately* we can estimate μ.

■

Although we cannot interpret a computed confidence interval probabilistically, there is a long-run relative frequency interpretation. Consider, for instance, an 80% confidence interval for μ. Were we repeatedly to select similar samples of the same size from a population and compute an 80% confidence interval for μ with each sample, approximately 80% of such intervals would contain μ.

We illustrate this interpretation graphically in Figure 2. There we show a (skewed) population with mean $\mu = 40$ and standard deviation $\sigma = \sqrt{80}$. Fifty random samples each of size $n = 50$ were selected from this population by a computer and for each sample an 80% confidence interval was constructed and represented by a line at the right. Intervals that do not contain $\mu = 40$ are indicated. Of the 50 intervals, 44 or 88% contain 40. This is a little higher proportion than expected but is not unreasonable in only 50 samples.

In general, then, we can also interpret a (100γ)% confidence interval for μ by stating that (100γ)% of similarly obtained intervals would contain μ.

Definition 2 for a confidence interval for μ contains the population standard deviation σ. In reality, σ is rarely known. If the sample size $n \geq 30$,

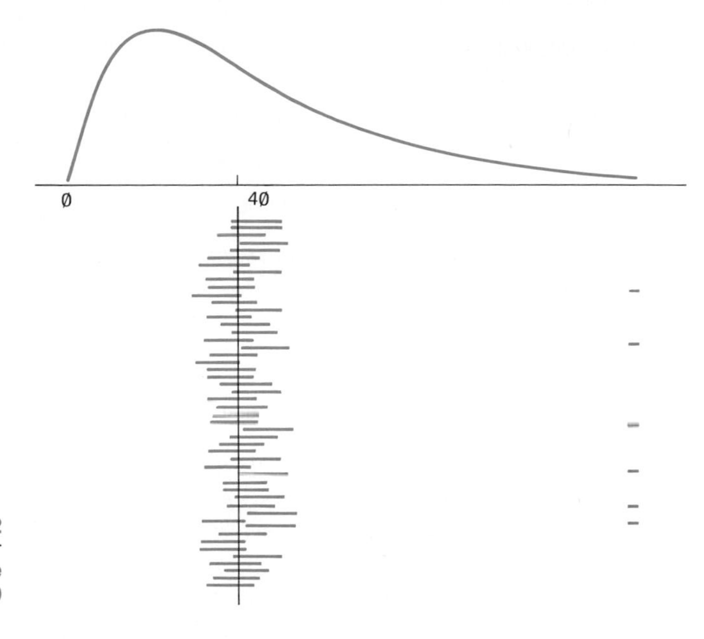

FIGURE 2

Fifty 80% confidence intervals for $\mu = 40$

however, the sample standard deviation

$$s = \sqrt{\frac{\sum (x_i - \bar{x})^2}{n - 1}} = \sqrt{\frac{n \sum x_i^2 - (\sum x_i)^2}{n(n - 1)}}$$

is usually a satisfactory point estimate for σ.

Replacing σ by s in Definition 2 we obtain an *approximate* confidence interval, or more formally:

> DEFINITION 3
>
> If a random sample of size $n \geq 30$ is selected, an approximate $(100\gamma)\%$ confidence interval for μ is
>
> $$\bar{x} - Z_{[\gamma]} \frac{s}{\sqrt{n}} \quad \text{to} \quad \bar{x} + Z_{[\gamma]} \frac{s}{\sqrt{n}}$$

EXAMPLE 3 An author has written a book containing 520 pages. By selecting a random sample of 40 pages with replacement, she wishes to estimate μ, the mean number of words per page *and* T, the total number of words in the book. The following summary data were obtained for the 40 pages:

$$\sum^{40} x_i = 7324 \text{ words} \qquad \sum^{40} x_i^2 = 1{,}350{,}760$$

Find a 90% confidence interval for both μ and the total number of words in the book.

Solution The sample mean is $\bar{x} = \Sigma x_i/40 = 7324/40 = 183.1$ words per page and the sample standard deviation is:

$$s = \sqrt{\frac{40(1{,}350{,}760) - 7324^2}{40(39)}} = \sqrt{249.6308}$$

$$= 15.80 \text{ words per page}$$

Using Definition 3, an approximate 90% confidence interval for the mean number of words per page, μ, is

$$183.1 \mp 1.645 \frac{15.8}{\sqrt{40}} = 183.1 \mp 4.1$$

$$= 179.0 \text{ to } 187.2$$

If the total number of words in the book is T, then μ, the mean number of words per page, is $\mu = T/520$. So $T = 520\mu$. Since $\bar{x} = 183.1$ is a point estimate for μ, a point estimate for $T = 520\mu$ is $520\bar{x} = 520(183.1) = 95{,}212$ words. We can also obtain an approximate 90% confidence interval for T by multiplying both bounds of the confidence interval for μ by 520 to obtain

$$520(179.0) \text{ to } 520(187.2) \quad \text{or} \quad 93{,}080 \text{ to } 97{,}344$$

We may be approximately 90% confident that the total number of words in the book is between 94,080 and 97,344.

■

In the problem set below we will not distinguish between exact and approximate confidence intervals for μ.

PROBLEMS 9.2

3. In a random sample of 40 private four-year colleges the mean student/faculty ratio was 12.9.

   (a) Assuming that $\sigma = 1.85$, compute a 99% confidence interval for μ, the mean student/faculty ratio over all private four-year colleges.

   (b) How would the interval change if σ was not known but the sample variance were $s^2 = 2.86$?

4. In a random sample of water bills paid by 50 homeowners in a town, the mean bill was \$48.53. Find an 80% confidence interval for μ, the mean bill for all homeowners in the town. Assume that (a) $\sigma = \$60$; (b) $\sigma = \$25$; (c) $\sigma = \$10$. Comment on the relationship between the width of the confidence interval and the magnitude of the population standard deviation.

5. Use the speed data in Problem 8 of Section 1.2 (page 17) to find an 85% confidence interval for μ, the mean speed of cars on the Pennsylvania Turnpike. Use the fact that $\Sigma x_i = 2407$ and $\Sigma x_i^2 = 146{,}343$.

6. The IRS wishes to estimate μ, the mean number of dependents claimed by families earning under \$8000 a year. A random sample of 400 such families is selected and the number of dependents claimed by each is recorded. Find and interpret a 95% confidence interval for μ. Use the fact that $\Sigma x_i = 1060$ and $\Sigma x_i^2 = 3283$.

7. The mean number of mathematics courses that had been taken by a random sample of 30 second-semester seniors at a college is $\bar{x} = 1.86$. The standard deviation is 2.04.

(a) Compute a 90% confidence interval for the mean number of mathematics courses over all such seniors.

(b) Interpret your interval.

(c) Suppose that there are 730 members of the senior class. Find a 90% confidence interval for the *total* number of mathematics courses taken by the entire class.

8. Consider again the 60 numbers in Example 1 of Chapter 1 (page 9). Each number is the time in hours to complete a brain operation using a new procedure. Regarding these times as a sample from the potential population of times, find a 95% confidence interval for μ, the potential mean time. Use the fact that $\Sigma x_i = 605.30$ hours and $\Sigma x_i^2 = 6259.69$.

9. An apple grower wishes to estimate the mean number of apples per tree in his 3800-tree orchard. He selects a random sample of 50 trees and counts the number of apples on each. The sample mean and variance of the number of apples per tree are $\bar{x} = 698.5$ and $s^2 = 6889.3$, respectively.

(a) Obtain an 80% confidence interval for (i) the mean number of apples per tree; (ii) the total number of apples in the orchard.

(b) Interpret both of these intervals.

10. In a random sample of 35 members of the congregation of a synagogue, the mean age was found to be 45.4 years with a standard deviation of 9.6 years. Find a 90% confidence interval for the mean age over all members of the congregation.

11. Each week for a year the quality control experts at an automoble manufacturing plant intensively test 50 cars and record the number of defects found in each. They then compute an 80% confidence interval for the mean number of defects per car. In how many of the weeks would you expect the resulting confidence intervals to contain the true mean number of defects per car?

Solution to Problem 3

(a) Using Definition 2 with $\bar{x} = 12.9$, $\sigma = 1.85$, and $n = 40$, a 99% confidence interval for μ is

$$12.9 \mp Z_{[.99]} = \frac{1.85}{\sqrt{40}}$$

From Table A5, $Z_{[.99]} = 2.576$ and our interval becomes

$$12.9 \mp 2.576\left(\frac{1.85}{\sqrt{40}}\right) = 12.9 \mp .75$$

$$= 12.15 \text{ to } 13.65$$

(b) If the value for σ is not known, we replace it by the sample standard deviation $s = \sqrt{s^2} = \sqrt{2.86} = 1.691$.

The resulting confidence interval is, using Definition 3,

$$12.9 \mp 2.576\left(\frac{1.691}{\sqrt{40}}\right) = 12.9 \mp .69$$

$$= 12.21 \text{ to } 13.59$$

SECTION 9.3

CONFIDENCE INTERVAL FOR THE POPULATION PROPORTION OF SUCCESSES

The results of Section 9.2 can be used to obtain a confidence interval for π, the unknown proportion of successes in a population. As in Section 7.2, we view the population as consisting of ones (for successes) and zeros (for failures). By replacing μ by π, σ by $\sqrt{\pi(1 - \pi)}$, and $\bar{x}$ by $\bar{p}$ (the proportion of successes in the sample) in Definition 2, the confidence interval for π becomes $\bar{p} \mp Z_{[\gamma]}\sqrt{\pi(1 - \pi)/n}$ or

$$\bar{p} - Z_{[\gamma]}\sqrt{\frac{\pi(1 - \pi)}{n}} \quad \text{to} \quad \bar{p} + Z_{[\gamma]}\sqrt{\frac{\pi(1 - \pi)}{n}}$$

As you can see, the confidence interval for π depends on the value for π itself. But we do not know this. In fact, we select the sample in order to estimate π. We therefore replace π with the corresponding sample proportion $\bar{p}$ and thus find an approximate confidence interval for π; specifically

DEFINITION 4

If in a random sample of size n both $n\bar{p}$ and $n(1 - \bar{p})$ are ≥ 5, then an approximate $(100\gamma)\%$ confidence interval for π is

$$\bar{p} - Z_{[\gamma]}\sqrt{\frac{\bar{p}(1 - \bar{p})}{n}} \quad \text{to} \quad \bar{p} + Z_{[\gamma]}\sqrt{\frac{\bar{p}(1 - \bar{p})}{n}}$$

EXAMPLE 4 A random sample of 80 employees of a large automobile manufacturing company is selected. Of these, 25 reported owning an imported automobile. Find a 90% confidence interval for π, the proportion of all employees who own an imported car.

Solution The proportion in the sample who own a foreign automobile is $\bar{p} = 25/80 = .3125$. This number is a point estimate for the unknown π. From Definition 4 a 90% confidence interval for π is

$$.3125 \mp 1.645\sqrt{\frac{(.3125)(1 - .3125)}{80}} = .3125 \mp .08525$$

$$= .227 \text{ to } .398$$

We interpret this interval in much the same way as we did those in Section 9.2. We can be 90% confident that the proportion of all employees owning an imported car is between .227 and .398. Alternatively, if similar samples of 80 employees were selected and a 90% confidence interval for π computed in each case, approximately 90% of the intervals would contain π.

■

We can obtain a confidence interval for the *percentage* of successes in the population by multiplying the two bounds of the confidence interval for π by 100. For instance, in Example 4 a 90% confidence interval for the percentage (100π) of employees owning an imported car is

$$100(.227) \text{ to } 100(.398) \quad \text{or} \quad 22.7 \text{ to } 39.8\%$$

(Note that there are two percentages, the percent of employees and the percent of confidence.)

Notice from Definition 4 that, as before, all else remaining constant (1) the larger the sample size, n, the narrower the interval and the more accurately we estimate π; and (2) the more confident we wish to be, the larger γ needs to be, the larger $Z_{[\gamma]}$ becomes, and the wider the resulting interval will be.

PROBLEMS 9.3

12. A company that conducts opinion polls selects a random sample of 1200 from the list of registered voters in a state. Each voter is asked whom they intend to vote for in the forthcoming presidential primary election. Of these, 579 intend to vote for the incumbent. Find a 95% confidence interval for the *percentage* of all voters in the state who intend to vote for the incumbent. Interpret your interval.
13. Forty-eight of 150 adults interviewed reported that baseball was their favorite sport. Find a 90% confidence interval for the proportion of all adults for whom baseball is their favorite sport.
14. The subscription manager for a magazine selects a random sample of 50 subscribers, of whom only 15 are women. Find an 85% confidence interval for the percentage of all subscribers who are women.
15. A random sample of 2000 voters in the United States were asked for their views on a wide range of political and social issues. Only 482 felt very optimistic about the future, but 1899 felt that the United States was still the best country in the world in which to live. Find 90% confidence intervals for the proportion of all voters holding each of these opinions.
16. A random sample of 60 children were interviewed as they left the zoo. Twelve of the children reported that the elephant was their favorite animal. Find a 90% confidence interval for the proportion of all children visiting the zoo who prefer the elephant.
17. A random sample of 1500 adults agreed to record which programs they watched on TV for a period of time. All 1500 did so and 494 of them reported watching a program on the effects of nuclear war. Find an 80% confidence interval for the proportion of all adults who watched this program. Interpret this interval.
18. In 200 at-bats a professional baseball player gets 52 hits. If we regard these 200 at-bats as a random sample from the player's career, find a 95% confidence interval for the player's career batting "average" (i.e., proportion of at-bats that were hits).
19. The polling company Arbitron's ratings report estimates of the number of people listening to different radio stations in a city. In a random sample of 2000 adults in a city, 268 reported listening to WRRO at least once in the last week.

    (a) Find an 80% confidence interval for the proportion, π, of all adults in the city who listened to this station at least once in the last week.

    (b) If there are 3,500,000 adults in the city, find a point estimate for T, the total number of adults in the city who listened to WRRO.

(c) Find an 80% confidence interval for T.

20. In Example 1 an educational consultant selected a random sample of 500 high school seniors, of whom 190 reported that they intended to go on to college. Find a 90% confidence interval for the proportion of all high school seniors who intend to go on to college.

Solution to Problem 12

The proportion in the sample favoring the incumbent is $\bar{p} = 579/1200 = .4825$. Using Definition 4, a 95% confidence interval for the corresponding proportion in the population is

$$.4825 \mp 1.96\sqrt{\frac{(.4825)(1 - .4825)}{1200}} = .4825 \mp .0283$$

A 95% confidence interval for the *percentage* in the population favoring the incumbent is then

$$48.25 \mp 2.83 = 45.4 \text{ to } 51.1$$

We can be 95% confident that the true percentage favoring the incumbent (at least at the time the poll was taken) is between 45.5% and 51.1%. Instead of reporting the results of an opinion poll in this way, numbers such as 2.83% (or more likely, 3%) are frequently reported in newspapers as the "margin of error." The result above might be reported as "48% plus or minus a margin of error of 3%." Implicit (although usually unstated) in such reports is the degree of confidence (95%) with which the margin of error is valid.

SECTION 9.4

CONFIDENCE INTERVALS FOR THE DIFFERENCE BETWEEN TWO POPULATION MEANS AND TWO POPULATION PROPORTIONS

A CONFIDENCE INTERVAL FOR $\mu_1 - \mu_2$

It is quite straightforward to extend the results of Sections 9.2 and 9.3 to obtain confidence intervals for the *difference* between two population means, $\mu_1 - \mu_2$, and between two population proportions, $\pi_1 - \pi_2$. We begin with confidence intervals for $\mu_1 - \mu_2$.

We assume that a sample of size n_1 is selected from population 1 and, independently, a sample of size n_2 is selected from population 2. The purpose of the samples is to estimate $\mu_1 - \mu_2$. We noted in Definition 5 in Chapter 7 that if the sample sizes n_1 and n_2 are both ≥ 30, the difference in sample means, $\bar{X}_1 - \bar{X}_2$, is approximately normal with a mean $\mu_1 - \mu_2$ and standard deviation $\sqrt{\sigma_1^2/n_1 + \sigma_2^2/n_2}$. This suggests that a reasonable point estimate for $\mu_1 - \mu_2$ is the corresponding difference in the observed sample means, $\bar{x}_1 - \bar{x}_2$. In this case, as in the one-sample case, the standard deviations in the two populations, σ_1 and σ_2, are rarely known. However, as with single large samples, we can substitute the corresponding sample standard deviations, s_1 and s_2.

A confidence interval for $\mu_1 - \mu_2$ can therefore be obtained by substituting $\bar{x}_1 - \bar{x}_2$ for $\bar{x}$ and $\sqrt{s_1^2/n_1 + s_2^2/n_2}$ for $s/\sqrt{n}$ in Definition 3. More formally:

DEFINITION 5

If independent and randomly selected samples of size n_1 and n_2 (n_1 and $n_2 \geq 30$) are selected, an approximate $(100\gamma)\%$ confidence interval for $\mu_1 - \mu_2$ is

$$\bar{x}_1 - \bar{x}_2 - Z_{[\gamma]}\sqrt{\frac{s_1^2}{n_1} + \frac{s_2^2}{n_2}} \quad \text{to} \quad \bar{x}_1 + Z_{[\gamma]}\sqrt{\frac{s_1^2}{n_1} + \frac{s_2^2}{n_2}}$$

We illustrate the use of Definition 5 in the following example.

EXAMPLE 5 Using radar, state police record the speed of a random sample of 80 cars selected just before the imposition of the 55-mph speed limit. At the same location another sample consisting of the speed of 60 cars is taken two weeks after the new limit is imposed. (a) Use the results below to find a 90% confidence interval for $\mu_1 - \mu_2$, the difference in the true mean speed before and after the new speed limit. (b) Does the confidence interval suggest that the true mean speeds are equal (i.e., $\mu_1 = \mu_2$)?

BEFORE	AFTER
$n_1 = 80$	$n_2 = 60$
$\bar{x}_1 = 62.5$	$\bar{x}_2 = 56.9$
$s_1 = 7.2$	$s_2 = 6.7$

Solution (a) Substituting these values into Definition 5, a 90% confidence interval for $\mu_1 - \mu_2$ is

$$(62.5 - 56.9) \mp 1.645\sqrt{\frac{7.2^2}{80} + \frac{6.7^2}{60}} = 5.6 \mp 1.94$$

$$= 3.66 \text{ to } 7.54 \text{ mph}$$

We can be 90% confident that the *difference* (decrease) in mean speed lies between 3.66 and 7.54 mph.

The difference $\bar{x}_1 - \bar{x}_2 = 62.5 - 56.9 = 5.6$ mph is a point estimate for the unknown $\mu_1 - \mu_2$. By computing the interval above, we obtain an assessment of the accuracy of this point estimate.

(b) The interval 3.66 to 7.54 can be regarded as providing a range of plausible values for $\mu_1 - \mu_2$. If the population means were equal, $\mu_1 - \mu_2$ would equal 0. Since the value $\mu_1 - \mu_2 = 0$ is not even close to being in the interval, we regard it as implausible, given the data. The interval does suggest that the mean speed has, in fact, declined by an amount somewhere between 3.66 and 7.54 mph.

■

A CONFIDENCE INTERVAL FOR $\pi_1 - \pi_2$

Assume now that we wish to estimate $\pi_1 - \pi_2$, the difference in the proportion of successes in two populations. To do this, a sample of size n_1 is selected from population 1 and, independently, a sample of size n_2 is selected from population 2.

In Definition 5 we noted that provided that n_1 and n_2 are ≥ 30, the distribution of the difference in sample proportions $\bar{P}_1 - \bar{P}_2$ is approximately normal with a mean of $\pi_1 - \pi_2$ and standard deviation

$$\sqrt{\frac{\pi_1(1 - \pi_1)}{n_1} + \frac{\pi_2(1 - \pi_2)}{n_2}}$$

This suggests that a reasonable point estimate for the unknown difference in the population proportions, $\pi_1 - \pi_2$, is $\bar{p}_1 - \bar{p}_2$, the difference in the observed sample proportions. Since π_1 and π_2 are unknown, we replace π_1 by $\bar{p}_1$ and π_2 by $\bar{p}_2$. A confidence interval for $\pi_1 - \pi_2$ is then as given in Definition 6.

> DEFINITION 6
>
> If independent and randomly selected samples of size n_1 and n_2 (n_1 and $n_2 \geq 30$) are selected, then an approximate $(100\gamma)\%$ confidence interval for $\pi_1 - \pi_2$ is
>
> $$\bar{p}_1 - \bar{p}_2 - Z_{[\gamma]}\sqrt{\frac{\bar{p}_1(1 - \bar{p}_1)}{n_1} + \frac{\bar{p}_2(1 - \bar{p}_2)}{n_2}} \quad \text{to}$$
>
> $$\bar{p}_1 - \bar{p}_2 + Z_{[\gamma]}\sqrt{\frac{\bar{p}_1(1 - \bar{p}_1)}{n_1} + \frac{\bar{p}_2(1 - \bar{p}_2)}{n_2}}$$

EXAMPLE 6 In a 1972 random sample of 100 families in a town, 40% reported owning their homes. Ten years later in a sample of 200 families from the town, 44% reported owning their homes. Compute an approximate 80% confidence interval for the change in the proportion owning their homes within the 10-year period.

Solution Call π_1 the unknown proportion of all families owning their homes in 1982 and π_2 the corresponding unknown proportion in 1972. The sample proportions are $\bar{p}_1 = .44$ and $\bar{p}_2 = .40$. Here $n_1 = 200$ and $n_2 = 100$. An approximate 80% confidence interval for $\pi_1 - \pi_2$ is

$$.44 - .40 \mp 1.282\sqrt{\frac{(.44)(.56)}{200} + \frac{(.4)(.6)}{100}} = .04 \mp .077$$

$$= -.037 \text{ to } .117$$

We can be 80% confident that the 10-year change in the proportion owning their homes lies between $-.037$ and $.117$. The fact that this interval includes the value $\pi_1 - \pi_2 = 0$ suggests that the change in proportion could be negative as well as positive, although the bounds suggests that it is more likely to be positive.

An 80% confidence interval for the change in the *percentage* of families owning their homes is $100(-.037)$ to $100(.117)$, or -3.7 to 11.7. We can say with 80% confidence that the *change* in percentage was anywhere from a *decline* of 3.7% to an *increase* of 11.7%.

■

The value for a single proportion π must lie between zero and 1, but $\pi_1 - \pi_2$ can be negative. Accordingly, a confidence interval for $\pi_1 - \pi_2$ may well contain negative values, as in Example 6.

Thus far we have assumed that we are comparing two quite separate and independent samples. But frequently we are interested in comparing subgroups of a single sample. For example, suppose that a random sample of 85 college student records have been selected. Assume that 50 of the students have taken the language placement test in Spanish, the remaining 35 in French. We might want to compare the mean scores for the two languages by computing a confidence interval for the difference in mean scores for *all* those students taking the Spanish and *all* those taking the French test (see Problem 28). It can be shown that the procedures for independent samples can be applied in this situation with the two sample sizes here being $n_1 = 50$ and $n_2 = 35$.

PROBLEMS 9.4

21. In a particular school 70 boys and 60 girls took a nationwide test of math computation.
    (a) The mean and standard deviation for the boys were $\bar{x}_1 = 80.4$ and $s_1 = 14.6$. The corresponding figures for the girls were $\bar{x}_2 = 78.3$ and $s_2 = 16.9$. Treating these students as a random sample of all students taking the test, compute a 90% confidence interval for $\mu_1 - \mu_2$, the nationwide difference in the mean scores of the boys and girls.
    (b) Suppose, in addition, that 43 boys and 38 girls obtained scores of 75 or more. Use this information to find a 90% confidence interval for $\pi_1 - \pi_2$, the difference in the proportion of all boys and girls obtaining scores of 75 or more.
22. Prior to the final draft of a book, the number of errors on each of a random sample of 40 pages was recorded. The mean number of errors per page was 1.1 with a standard deviation of .5. An independently selected sample of 40 pages was selected from the final draft. The mean number of errors in this sample was .4 per page with a standard deviation of .3. Compute a 98% confidence interval for the *change* (decrease) in the mean number of errors per page over all pages. Does this interval reassure you that the mean number of errors has declined?
23. Three months prior to a presidential election exactly two-thirds of a random sample of 900 voters favored the incumbent president. In an independent sample of 1050 voters selected one week before the election, 588 reported favoring the president.

Compute an 85% confidence interval for the decline in the *percentage* of all voters favoring the president.

24. Random samples of adults are drawn from communities A and B and the following information on annual income obtained:

	SAMPLE A	SAMPLE B
n	90	90
$\bar{x}$	\$18,472	\$17,994
s	\$2,605	\$2,417

Compute and interpret a 90% confidence interval for $\mu_A - \mu_B$ the difference in mean income per adult for the two communities.

25. In a random sample of 50 private colleges, 17 reported offering at least one doctoral program. In a random sample of 50 public colleges the corresponding figure was 11.
(a) Find a 92% confidence interval for $\pi_1 - \pi_2$, the difference between the proportion of all private colleges (I) and all public colleges (II) offering at least one doctoral program.
(b) Interpret this interval. How would you interpret the fact that the interval contains zero?

26. The administrator of a health plan selects a random sample of 50 members who live in the town where the health plan's clinic is located. The mean number of visits to the clinic in the past year by these 50 members was 1.63 with a standard deviation of 1.59. Forty of these members made at least one visit. In a random sample of 50 members who live in other towns, the mean number of visits to the clinic was 1.50 with a standard deviation of 1.54. Thirty-six of the 50 made at least one visit. Call μ_1 and μ_2 the mean number of visits per member over all members living in the town where the clinic is located and those living in other towns, respectively.
(a) Find a 90% confidence interval for $\mu_1 - \mu_2$.
(b) Call π_1 and π_2 the proportions of all members who made at least one visit among those living in the same town as the clinic and among those living in other towns, respectively. Find a 90% confidence interval for $\pi_1 - \pi_2$.

27. A random sample of 380 automobile accident reports were classified by size of automobile (large or small) and whether or not any of the occupants were killed.

	SIZE OF AUTOMOBILE	
	SMALL	LARGE
FATAL	39	21
NOT FATAL	181	139
TOTAL	220	160

Find a 90% confidence interval for the difference in fatality rates for small and large cars.

28. Fifty of the students in a sample of 85 students had taken the language placement

test in Spanish; the remainder had taken the test in French. The mean and standard deviation of scores for the two groups are summarized as follows:

	SPANISH	FRENCH
$\overline{x}$	38.4	35.2
s	9.4	10.2

Use these data to find an 80% confidence interval for the difference in mean scores for all students taking the Spanish test and all students taking the French test.

Solution to Problem 21

(a) From Definition 5 a 90% confidence interval for $\mu_1 - \mu_2$ is

$$(80.4 - 78.3) \mp 1.645\sqrt{\frac{14.6^2}{70} + \frac{16.9^2}{60}} = 2.1 \mp 4.6$$

$$= -2.5 \quad \text{to} \quad 6.7$$

(b) In this case $\overline{p}_1$ = proportion of boys scoring 75 or more points = $43/70 = .614$. For the girls, $\overline{p}_2 = 38/60 = .633$. A 90% confidence interval for $\pi_1 - \pi_2$ is

$$(.614 - .633) \mp 1.645\sqrt{\frac{(.614)(.386)}{70} + \frac{(.633)(.367)}{60}} = -.019 \mp .140$$

$$= -.159 \quad \text{to} \quad .121$$

SECTION 9.5
SAMPLE SIZE FOR ESTIMATION

SAMPLE SIZE FOR ESTIMATING A SINGLE MEAN

In Example 2 we illustrated the point that keeping the value for γ fixed, the width of the confidence interval decreases (i.e., the accuracy with which we estimate μ improves) as the sample size n is increased.

In this section we consider how to find the sample size necessary to achieve a desired accuracy (i.e., an interval of specified width) together with a specified degree of confidence.

We begin with an example.

EXAMPLE 7 A researcher plans to select a sample of first-grade girls in order to estimate μ, the mean height of all such girls in the state. The sample should be large enough so that the researcher will be 95% confident that the mean height in the sample ($\overline{x}$) will be within .5 inch of μ. Previous studies suggest that the standard deviation of the height of female first graders is $\sigma = 2.8$ inches. How many such children (n) should be selected?

Solution From Definition 1 a 95% confidence interval for μ is

$$\bar{x} \mp 1.96 \frac{\sigma}{\sqrt{n}}$$

But we want an interval wide enough so that the length of that part of the interval on each side of $\bar{x}$ has a value .5 inch. Symbolically, this can be written

$$1.96 \frac{\sigma}{\sqrt{n}} = .5$$

Rearranging this formula to express n in terms of the other components, we first multiply both sides by $\sqrt{n}$ to obtain

$$1.96\sigma = .5\sqrt{n}$$

or, equivalently,

$$\sqrt{n} = \frac{1.96\sigma}{.5}$$

Squaring both sides, we find that

$$n = \frac{1.96^2\sigma^2}{(.5)^2}$$

Substituting $\sigma = 2.8$ inches, we have

$$n = \frac{1.96^2(2.8^2)}{.5^2} = 120.47 \text{ girls}$$

We cannot, of course, select .47 of a girl. By custom when computing sample sizes we always round *up* to the next whole number regardless of the digits following the decimal place. We therefore conclude that the researcher should select 121 female first graders.

■

In this example the acceptable or required difference between the sample mean and the population mean was .5 inch. In general, we refer to this difference as the **required accuracy** and denote it by the letter A. Notice that since A represents how close the sample mean is to be to the population mean, the *smaller* A is, the more accurate the estimate, and vice versa.

Generalizing, we now obtain the sample size (n) necessary to achieve a required accuracy (A) with a specified confidence coefficient (γ). From Definition 2, a $(100\gamma)\%$ confidence interval for μ is

$$\bar{x} \mp Z_{[\gamma]} \frac{\sigma}{\sqrt{n}}$$

The value for n should be such that $Z_{[\gamma]}(\sigma/\sqrt{n}) = A$. Rearranging this result as we did in Example 7, we obtain

$$n = \frac{Z_{[\gamma]}^2 \sigma^2}{A^2} \qquad \text{(Eq. 9.1)}$$

The values for A and γ are usually specified by the researcher and, as always, $Z_{[\gamma]}$ is found from Table A5.

EXAMPLE 8 Suppose that in Example 7 the researcher would be satisfied if she could be 80% (instead of 95%) confident that $\bar{x}$ is within .5 inch of μ. In this case the required sample size would be

$$n = \frac{Z_{[.8]}^2(2.8^2)}{.5^2} = \frac{(1.282)^2(2.8)^2}{(.5)^2} = 51.54$$

A sample of only 52 girls would be required.

■

"GUESSTIMATING" THE POPULATION STANDARD DEVIATION

The major difficulty in applying Equation 9.1 is that the population standard deviation, σ, is rarely known. Since the determination of sample size must *precede* the selection of the sample, we do not even have a sample standard deviation, s, to use as an estimate for σ. In practice, there are three ways of obtaining a replacement value for σ in Equation 9.1.

1. A "pilot" survey might be conducted. The standard deviation, s, in this small sample may be substituted for σ.
2. There might be earlier studies of this or similar populations from which estimates for σ may be found. We assumed this in Examples 7 and 8.
3. In some cases, although the researcher may not be willing or able to "guesstimate" σ, it is possible to guesstimate the range of values (R) within which most of the population lies. In such a case the crude approximation $\sigma \doteq R/4$ can be used. For instance, it might be reasonable to assume that nearly all female first graders are between 24 and 36 inches tall. The range of heights would then be $R = 12$ inches and a guesstimate for σ would be $12/4 = 3$ inches. The result, $\sigma \doteq R/4$, is derived from the fact that for a normal population approximately 95% ("most") of the values lie within two standard deviations of the mean (i.e., within a range of 4 standard deviations). Therefore, $R \doteq 4\sigma$ or $\sigma \doteq R/4$. This estimate for σ is thus more accurate the closer the population is to normal.

EXAMPLE 9 A town selectman wishes to estimate the mean monthly rent per unit in the town. She wishes to select a large enough sample of rental units to be 90% confident that the mean rent in the sample ($\bar{x}$) will be within (a) \$30 and (b) \$15 of the mean (μ) of all rentals in the town. She knows that most units in the town rent for between \$180 and \$660 a month. How many units should she select?

Solution The range of monthly rents is $R = 660 - 180 = \$480$. An estimate for σ is thus $R/4 = 480/4 = \$120$.

(a) Using Eq. 9.1 with $A = \$30$, we have

$$n = \frac{Z^2_{[.90]}\sigma^2}{A^2} = \frac{(1.645)^2(120)^2}{30^2} = 43.30 \text{ or } 44 \text{ units}$$

(b) Using Eq. 9.1 with $A = \$15$, we have

$$n = \frac{Z^2_{[.90]}\sigma^2}{A^2} = \frac{(1.645)^2(120)^2}{15^2} = 173.19 \text{ or } 174 \text{ units}$$

Notice that when A is halved, the value for n is quadrupled.

■

THE EFFECT OF CHANGES IN γ, σ, AND A ON n

It is worth examining how changes in the three components $Z_{[\gamma]}$, σ, and A in Equation 9.1 affect the value for n:

First, given a specific value for σ and a desired accuracy, A, the more confident we wish to be (i.e., the larger γ and $Z_{[\gamma]}$), the larger the sample n must be.

Similarly, if the accuracy A and the level of confidence (γ) are specified, then the more varied the population (i.e., the larger σ), the larger the sample n must be.

The relationship between A and n is somewhat different. First note that increasing the accuracy means reducing the *value for* A (i.e., we want $\bar{x}$ to be closer to μ). Now referring to Equation 9.1, we see that for given values of σ and γ, the *smaller* the value for A, the larger n must be. More precisely, we say that n varies inversely with the square of A. For instance, as A decreases by a factor of 2, the value for n will increase by a factor of 4.

Similarly, if A increases, n will decrease, as illustrated in Example 9.

SAMPLE SIZE FOR ESTIMATING A SINGLE PROPORTION

For situations where we want to estimate π, there is a formula for n analogous to Equation 9.1. Specifically,

$$n = \frac{Z^2_{[\gamma]}\pi(1 - \pi)}{A^2} \qquad \text{(Eq. 9.2)}$$

Instead of σ, Equation 9.2 for n depends on the unknown value for π. Since, as before, the sample has not yet been taken, we do not have a sample proportion, $\bar{p}$, to substitute for π. A substitute for π in Equation 9.2 can be obtained from the proportion of successes in either a comparable study or in a pilot survey, but there is a special feature of the product $\pi(1 - \pi)$ which enables us to find a conservative value for n directly.

Notice first that the larger the value of the product $\pi(1 - \pi)$ in Equation 9.2, the larger n will have to be. There is, however, an upper limit on the value of $\pi(1 - \pi)$. To see this, consider Figure 3, where we have plotted a graph of $\pi(1 - \pi)$. The values for $\pi(1 - \pi)$ are on the vertical axis and values for π on the horizontal axis. The same scale is used for both axes.

The value π lies between zero and 1. Notice that the largest value for $\pi(1 - \pi)$ is .25, which occurs when $\pi = .5$. The farther π is from .5, in either

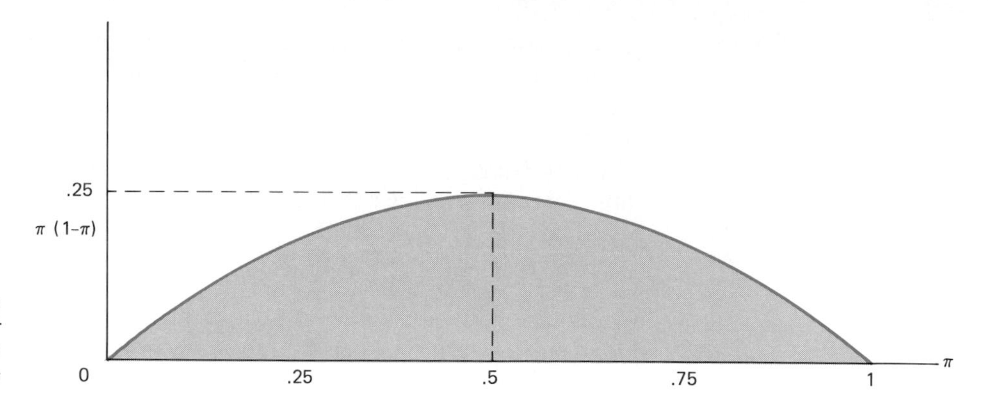

FIGURE 3

Plotting $\pi(1 - \pi)$ against π

direction, the lower the product $\pi(1 - \pi)$. Thus, if we replace π by .5 in Equation 9.2, the value for $\pi(1 - \pi)$ will be as high as it *can* be and the sample size will then be as large as it *need* be. In this case Equation 9.2 becomes

$$n = \frac{Z^2_{[\gamma]}(1/2)(1 - 1/2)}{A^2}$$

or

$$n = \frac{Z^2_{[\gamma]}}{4A^2} \qquad \text{(Eq. 9.3)}$$

EXAMPLE 10 A polling organization wishes to estimate π, the proportion of voters in a state who intend to vote for a particular candidate in the forthcoming election. The sample should be large enough so they may be 80% confident that the sample proportion $\overline{p}$ is within .025 of π. Use Equation 9.3 to obtain the required sample size.

Solution Substituting $Z_{[.8]} = 1.282$ and $A = .025$ in Equation 9.3, we obtain

$$n = \frac{1.282^2}{4(.025)^2} = 657.41 \text{ or } 658 \text{ voters}$$

■

In Example 10 it is reasonable to expect that under ordinary circumstances the population proportion, π, will not be terribly far from .5 (it might be .4 or .65 but not .15 or .92, for instance). In such a case it makes sense to use Equation 9.3. But if it is known or strongly suspected that π is very small or very large (i.e., far from .5), then using Equation 9.3, would result in taking a sample that is unnecessarily large. An example will illustrate this point.

EXAMPLE 11 A drug company has developed a new vaccine and is concerned with the proportion, π, of potential users of the drug who will develop the desired immunity. The company intends to estimate π by testing the drug on a group of subjects and computing the proportion, $\bar{p}$, who develop the immunity. The company wants to be 95% confident that $\bar{p}$ will be within .04 of π. They are quite sure that π is greater than .85. How many subjects should the vaccine be tested on?

Solution In this case since it is known that π is so far above .5, it makes no sense to use Equation 9.3 for n. Instead, we replace π in Equation 9.2 by .85. Substituting $Z_{[.95]} = 1.96$ and $A = .04$, we obtain

$$n = \frac{1.96^2(.85)(1 - .85)}{(.04)^2} = 306.128 \text{ or } 307 \text{ subjects}$$

Had we used Equation 9.3, which replaces π by .5, we would have obtained

$$n = \frac{1.96^2}{4(.04)^2} = 600.25 \text{ or } 601 \text{ subjects}$$

Knowing that π is at least .85 has saved the company the expense of testing approximately 300 subjects.

Suppose that the company was interested in the proportion, π, of potential users who will *not* develop immunity. Knowing that π was at most .15, they would replace π in Equation 9.2 by .15, $1 - \pi$ by .85, and would have obtained the same sample size, $n = 307$.

■

The lessons of examples 10 and 11 can be summarized as follows:

1. If the researcher has no knowledge of the value for π *or* if it is suspected that π is close to .5, use Equation 9.3.
2. If it is known that π is a long way from .5, replace π in Equation 9.2 by the closest value to .5 that π could realistically take (.85 in Example 11).

In the relevant problems to follow this section, we shall provide guidance as to the appropriate approach.

It is interesting to note that the sample size necessary to obtain a specified accuracy does *not* appear to depend on the size of the population. Suppose, for instance, that a pollster wanted to estimate the proportion of voters who feel that unemployment is the country's number 1 economic problem. A random sample of 1500 voters selected from the entire United States will provide the same accuracy as a similar sized sample for the state of California or the city of Santa Monica, California. The explanation for this lies in the fact that, provided the sample is a tiny fraction of the population—as we have generally been assuming—the distribution of $\bar{P}$ or $\bar{X}$ does not depend on the population size. (See Cochran, 1977 for further details.)

MULTIPLE ESTIMATES OF n

In practice, when a sample is selected, many variables are recorded, both quantitative and qualitative, and for each variable a suitable size of sample may be obtained using the methods of this section. But these sample sizes are likely to differ. Ideally, the largest sample size obtained among the four or five most important variables would be the one to choose. Frequently, however, financial considerations necessitate some compromise among the four or five sample sizes. This compromise may be the median value for n, or more usually simply the maximum that financial restrictions will permit.

PROBLEMS 9.5

29. A random sample of students at a state university is to be selected by the registrar's office in order to estimate the proportion π of all students favoring the new teaching schedule. The registrar wants an accuracy of .075 with 90% confidence. The value for π may well be close to .5. How many students should be selected?
30. We wish to estimate, with an accuracy of .04, the proportion of voters in Massachusetts favoring a tax cut. How large a random sample should be selected? Assume that $\gamma = .80$ and that the population proportion is close to .5.
31. A company wishes to set up a quality control program. Each week they would like to estimate the proportion of defective items with an accuracy of .05, with probability .95. How many items should be selected for weekly inspection if it is known that the defective rate would never exceed .1?
32. By tossing a bent coin n times we intend to estimate the probability π that the coin will come up heads. We wish to be 95% confident that the proportion of heads, $\bar{p}$, in the n tosses will be within .02 of π. What value for n is necessary? Assume that π may still be close to .5.
33. A school board is interested in estimating the proportion of high school students (π) who walk to school. A sample of high school students will be asked whether they walk to school. The school board wishes to be 90% confident that the sample proportion will be within .05 of π. They are sure that π is at least .8. How large a sample of students should be selected?
34. We wish to be 95% confident that the mean of a random sample is within 10 of the population mean. Most members of the population lie between 800 and 1000. How large a sample should be selected? Would the size of the sample be affected if most members of the population lie between 2200 and 2400?
35. A researcher is interested in estimating the mean time (μ) it takes adults to learn a certain task. He wants to be 80% confident that the mean time for a sample of adults is within 5 minutes of μ. In a small pilot study the standard deviation of learning times was $s = 33$ minutes. How large a sample of adults should be tested?
36. A health plan wishes to estimate μ, the mean number of visits per plan participant in a year. How large a sample of participants' records should be selected to estimate μ with an accuracy of .25 visit with probability .9? A similar study at a similar organization in California suggests that the standard deviation of the number of visits per person is approximately 2.1.
37. A sociologist is interested in the age at which women marry in the county. She intends to estimate the mean age (μ) by reviewing the ages of the women in a sample of marriage records. An accuracy of .4 year is required with probability .9. She feels that in most marriages in the county the woman is between 19 and 42 years of age. How many records should she select?

38. A sociologist wants to estimate the mean time, μ, between date of indictment and the trial date for persons charged in New York State. How many records should be selected so that the sample mean time will lie within 4 days of μ with 95% confidence? Assume that most people wait between 320 and 400 days.
39. The mean weight of adult males in country I is μ_I and in country II, μ_{II}. A researcher intends to select a sample of n adult males in each country and record the weight of each. He would like to be 90% confident that the difference in sample mean weights, $\bar{x}_1 - \bar{x}_2$, will be within three kilograms of $\mu_I - \mu_{II}$. Assuming that $\sigma_1 = 12.5$ and $\sigma_2 = 11.8$ kilograms, compute the number of adults, n, that must be selected in each country.

Solution to Problem 29

Since it is suspected that π is close to .5, we use Equation 9.3 with $A = .075$ and $Z_{[.9]} = 1.645$ to obtain

$$n = \frac{1.645^2}{4(.075)^2} = 120.17 \text{ or } 121 \text{ students}$$

SECTION 9.6
SUMMARY AND REVIEW

CONFIDENCE INTERVALS

As the word suggests, estimation is concerned with the use of sample data to obtain an estimate of the unknown value of a parameter. These parameters include μ, π, $\mu_1 - \mu_2$, and $\pi_1 - \pi_2$.

We regard the sample mean $\bar{x}$ as a **point estimate** for μ. Going further, when $n \geq 30$, a $(100\gamma)\%$ confidence interval for μ is

$$\bar{x} - Z_{[\gamma]}\frac{\sigma}{\sqrt{n}} \quad \text{to} \quad \bar{x} + Z_{[\gamma]}\frac{\sigma}{\sqrt{n}} \qquad \text{(Def. 2)}$$

where $Z_{[\gamma]}$ is the value for Z such that the area between $-Z_{[\gamma]}$ and $Z_{[\gamma]}$ is γ and is found from Table A5. The interval may be interpreted as follows: We may be $(100\gamma)\%$ confident that the interval contains μ. Alternatively, we can state that $(100\gamma)\%$ of similarly obtained confidence intervals will contain μ.

A convenient form of the confidence interval for computation is $\bar{x} \mp Z_{[\gamma]}(\sigma/\sqrt{n})$. An approximate $(100\gamma)\%$ confidence interval for μ is found by substituting the sample standard deviation s for σ in Definition 2. The result is

$$\bar{x} - Z_{[\gamma]}\frac{s}{\sqrt{n}} \quad \text{to} \quad \bar{x} + Z_{[\gamma]}\frac{s}{\sqrt{n}} \qquad \text{(Def. 3)}$$

Similarly, $\bar{p}$, the proportion of successes in a sample, is a point estimate for π, the unknown proportion of successes in the population. If both $n\bar{p}$ and $n(1 - \bar{p})$ are ≥ 5, an approximate $(100\gamma)\%$ confidence interval for π is

$$\bar{p} - Z_{[\gamma]}\sqrt{\frac{\bar{p}(1 - \bar{p})}{n}} \quad \text{to} \quad \bar{p} + Z_{[\gamma]}\sqrt{\frac{\bar{p}(1 - \bar{p})}{n}} \qquad \text{(Def. 4)}$$

A confidence interval for the *percentage* $(100\pi)\%$ of successes in the population is obtained by multiplying the bounds of the confidence interval for π by 100.

If n_1 and n_2 are both ≥ 30, an approximate $(100\gamma)\%$ confidence interval for the unknown difference between two population means $(\mu_1 - \mu_2)$ is

$$\bar{x}_1 - \bar{x}_2 - Z_{[\gamma]}\sqrt{\frac{s_1^2}{n_1} + \frac{s_2^2}{n_2}} \quad \text{to} \quad \bar{x}_1 - \bar{x}_2 + Z_{[\gamma]}\sqrt{\frac{s_1^2}{n_1} + \frac{s_2^2}{n_2}} \qquad \text{(Def. 5)}$$

If n_1 and n_2 are both ≥ 30, an approximate $(100\gamma)\%$ confidence interval for the difference between two population proportions $\pi_1 - \pi_2$ is

$$\bar{p}_1 - \bar{p}_2 - Z_{[\gamma]}\sqrt{\frac{\bar{p}_1(1 - \bar{p}_1)}{n_1} + \frac{\bar{p}_2(1 - \bar{p}_2)}{n_2}} \quad \text{to}$$

$$\bar{p}_1 - \bar{p}_2 - Z_{[\gamma]}\sqrt{\frac{\bar{p}_1(1 - \bar{p}_1)}{n_1} + \frac{\bar{p}_2(1 - \bar{p}_2)}{n_2}} \qquad \text{(Def. 6)}$$

ESTIMATING SAMPLE SIZE

If we wish to be $(100\gamma)\%$ confident that $\bar{x}$ is within A of μ, the required sample size is

$$n = \frac{Z_{[\gamma]}^2 \sigma^2}{A^2} \qquad \text{(Eq. 9.1)}$$

Since the value for σ is rarely known, we replace it by a "guesstimate."

Similarly, in order that we may be $(100\gamma)\%$ confident that $\bar{p}$ is within A of π, we require a sample size of

$$n = \frac{Z_{[\gamma]}^2 \pi(1 - \pi)}{A^2} \qquad \text{(Eq. 9.2)}$$

If nothing is known about the value for π or if it is known to be close to .5, replace π by .5 in Equation 9.2 to obtain

$$n = \frac{Z_{[\gamma]}^2}{4A^2} \qquad \text{(Eq. 9.3)}$$

If it is known that π is a long way from .5, replace π in equation 9.2 by the closest value to .5 that π could conceivably take.

REVIEW PROBLEMS

GENERAL

1. A random sample of size $n = 70$ is selected from a population with (unknown) mean μ and standard deviation 8.
   (a) Find a 92% confidence interval for μ if (i) $\bar{x} = 78$; (ii) $\bar{x} = 578$.
   (b) How does the value for $\bar{x}$ affect the accuracy with which we can estimate μ?

2. Thirty of the observations in a random sample of size $n = 80$ are successes. Compute an 80% confidence interval for the proportion of successes in the population.
3. A 90% confidence interval for a population mean, μ is computed as (6.37 to 7.14). Why is it incorrect to interpret this interval as $P(6.37 < \mu < 7.14) = .9$?
4. A coin is suspected of being biased in favor of heads. In 80 tosses of this coin heads came up 55 times. Compute a 75% confidence interval for the probability that the coin comes up heads on any throw.
5. A survey is taken of 60 medium-sized city libraries over the United States to determine, among other things, the annual book expenditure and the number of annual book thefts. The resulting data on these two questions are as follows:

	BOOK EXPENDITURE (THOUSANDS OF DOLLARS)	THEFTS OF BOOKS
$\sum^{60} x_i$	720	19,440
$\sum^{60} x_i^2$	11,136	6,650,191

Compute 95% confidence intervals for (a) the mean annual expenditure (on books); (b) the annual theft average for all such libraries.

6. In a random sample of 120 adult residents of a town the distribution of the number of visits to the movies in the past month is

NUMBER OF VISITS	0	1	2	3	4	5
NUMBER OF RESIDENTS	45	29	20	15	7	4

(a) Compute (i) the mean; (ii) the standard deviation of the number of visits; (iii) a 90% confidence interval for the mean number of visits per adult in the town.

(b) If there are 19,200 adults in the town, find a 90% confidence interval for the total number of visits to the movies made by all adults in the town in the past month.

7. Referring to Review Problem 6, find a 90% confidence interval for (a) the proportion of adults in the town who have *not* been to a movie in the past month; (b) the total number of adults in the town who have not been to a movie in the past month.
8. Sometimes interval estimates of a population mean, μ, are expressed as $(\bar{x} - s/\sqrt{n}$ to $\bar{x} + s/\sqrt{n})$. With what degree of confidence in computing a confidence interval does this correspond?
9. In a random sample of 35 rush hours the mean number of illegal left turns per hour at an intersection is $\bar{x}_1 = 5.9$. The standard deviation is $s_1 = 2.4$. Sometime after new and larger signs are erected, the number of left turns is recorded in each of another sample of 35 rush hours. In this case $\bar{x}_2 = 4.2$ and $s_2 = 1.7$. Find an 85% confidence interval for $\mu_1 - \mu_2$, the true decline in the mean number of illegal left turns.
10. In a survey of 40 journal articles in psychology, 24 made some use of the statistical methods in this text. In a similar survey of 50 journal articles in sociology, 23 made a similar use of statistics.

(a) Find an 80% confidence interval for the proportion, π_1, of all journal articles in psychology that use statistics.

(b) With the obvious notation find an 80% confidence interval for $\pi_1 - \pi_2$.

11. A random sample of size 80 is to be selected from a certain population. It is decided that the sample mean should be three times as accurate in estimating μ as originally planned. How many additional observations should be taken to achieve this new level of accuracy?

12. Two polling agencies are to select random samples from the same voting population and compute 90% confidence intervals for the proportion, π, favoring a particular candidate. What is the probability that (a) both intervals will contain π; (b) only one of the intervals will contain π?

HEALTH SCIENCES

13. A hospital dietician wishes to estimate μ, the mean birthweight of infants born to mothers who have received dietary counseling during pregnancy. She wishes to estimate μ with an accuracy of 50 grams with probability .9. Similar studies on such mothers suggest a standard deviation of birthweight of 400 grams.

(a) How many maternity records should she randomly sample for the study?

(b) How many records should she sample if an accuracy of 100 grams is considered adequate?

14. A community health plan wishes to estimate μ, the mean number of visits per plan participant in a year. How large a random sample of participants' records should be selected in order to estimate μ with an accuracy of .20 visit with probability .8? A similar study at a health maintenance organization in California suggests that the standard deviation of the number of visits per person is approximately 2.1.

15. A professional standards review organization wishes to estimate μ, the mean number of days spent in the hospital for women having a particular operation in hospitals within its jurisdiction. The records of a sample of 80 such women were sampled and analyzed with the following results:

$$\sum x_i = 609.6 \text{ days} \qquad \sum x_i^2 = 4993.31$$

Obtain a 95% confidence interval for μ.

16. An official of a health center is interested in estimating the proportion of patients (π) who travel to the center by public transportation. A random sample of 60 patients is randomly selected and each is asked how they traveled to the center. Suppose that 20 used public transportation. Compute a 90% confidence interval for π.

17. Eighty students from rural communities have a mean absolute visual threshold of 2.01 (in appropriate units) with standard deviation .31, whereas 120 students from urban communities have a mean absolute visual threshold of 1.89 with a standard deviation .29. Find an 85% confidence interval for the difference ($\mu_R - \mu_U$) in mean absolute visual threshold.

18. A group of 200 volunteers participate in a medical experiment. The group is randomly divided into two groups. The first group, I, is provided with a diet high in vitamin C. The second group, II, is allowed to eat and drink as usual. The mean and standard deviation of the number of colds in a year were recorded for each group.

GROUP I	$\bar{x}_1 = .87$	$s_1 = .32$	$n_1 = 100$
GROUP II	$\bar{x}_2 = 1.37$	$s_2 = .51$	$n_2 = 100$

(a) Compute a 90% confidence interval for $\mu_1 - \mu_2$. [Here μ_1 is the (potential) mean number of colds per year for all people using this diet. Similarly, μ_2 is the mean for all people not using the diet.]

(b) Does there appear to be any real difference in the mean number of colds in the two groups?

19. A researcher wishes to estimate the difference in proportions, $\pi_1 - \pi_2$, having blood type A in two communities I and II. A sample of size 100 is selected from each community. Thirty-seven persons in sample I and 42 in sample II have blood type A. Compute and interpret a 90% confidence interval for $\pi_1 - \pi_2$.

20. Refer to the case study at the beginning of the chapter. Interpret the explanation in quotes for a person unfamiliar with statistics.

SOCIOLOGY AND GOVERNMENT

21. A researcher wishes to estimate μ, the mean number of hours that the television sets are on each day in a particular community. A sample of 200 homes is selected at random. Among these homes the TV was on an average of 5 hours and 22 minutes with a standard deviation s = 1 hour and 35 minutes. Construct a 90% confidence interval for μ.

22. A political scientist wishes to select a sample of business leaders in order to estimate the proportion, π, that favor government aid to ailing industries. If an accuracy of .04 is required, how large a sample should be selected? Use $\gamma = .8$.

23. A sociologist is interested in estimating the proportion π_1 of all college students at the university who believe that marijuana use should be decriminalized. In a random sample of 150 students, 119 favor such decriminalization. Find a 75% confidence interval for π_1.

24. The sociologist referred to in Review Problem 23 is also interested in comparing the attitude of college students toward decriminalizing marijuana use with that of people of the same age in the town who are not in college. In a random sample of 80 such persons, 53 favored such decriminalization. If π_2 is the corresponding proportion over all such young people in the town (excluding the college students), find a 75% confidence interval for $\pi_1 - \pi_2$.

25. A newspaper wishes to select a random sample of voters in order to estimate the proportion, π, that favor a proposition that would ban smoking in all public buildings in the state.

(a) How large a sample should the paper select so as to be 95% sure that the sample proportion is within .03 of π? Assume that π is close to .5.

(b) Suppose that this number of voters are interviewed and 640 favor the smoking ban. Find a 95% confidence interval for π.

(c) Interpret this interval.

(d) Does the width of this interval reflect the accuracy specified in part (a)?

26. The Social Security Administration wishes to estimate the mean monthly payment made to the disabled in a particular state. A random sample of 300 such persons is selected with the followng results:

$$\bar{x} = \$229.37 \qquad s^2 = 2836.63$$

Compute an 80% confidence interval for the mean of such payments in the state.

27. In a presidential election 12 polling organizations intend to sample the same population of voters over the same period of time. Each intends to compute a 90% confidence interval for π, the proportion of all voters favoring the incumbent. What is the probability that (a) all 12 confidence intervals contain π; (b) two of the confidence intervals do not contain π?

ECONOMICS AND MANAGEMENT

28. A news magazine wishes to estimate π, the proportion of its subscribers who own their homes. A sample of 1100 subscribers reveals that 493 own their homes.
(a) Obtain a 98% confidence interval for π.
(b) Obtain a 98% confidence interval for the percentage of subscribers who own their homes.

29. A company is considering the introduction of an in-house health service for its employees and their families. To get some sense of the magnitude of the task, the personnel officer wishes to estimate μ, the mean number of persons per employee family (including the employee). A random sample of 50 employees is selected and interviewed. The mean number of persons per family is 3.72 and the standard deviation 1.8. (Assume that only one person per family works for the company.)
(a) Compute a 90% confidence interval for μ.
(b) If the company employs 14,500 persons, how would you estimate T, the total number of persons that the health service would have to serve?
(c) Find a 90% confidence interval for T.

30. A large company wishes to estimate the average monthly pension that it pays to former employees. A random sample of 90 former employees is selected and the following results obtained:

$$\bar{x} = 494 \qquad \sum (x_i - \bar{x})^2 = 936{,}360$$

(a) Compute a 95% confidence interval for the true mean monthly pension.
(b) Suppose that the company is paying pensions to 17,400 former employees. How might you use the sample data to estimate T, the total monthly expenditure on pensions?
(c) Find a 95% confidence interval for T.

31. In Review Problem 30, the researcher responsible for the sample gave the following interpretation of the confidence interval in part (a): "The probability that this interval contains our true mean monthly expenditure is .95." Why is this interpretation incorrect?

32. A company wishes to set up a quality control program. Each week they would like to estimate the percentage of defective items with an accuracy of 1.5% with probability .95. How many items should be selected weekly for inspection if it is known that the defective rate would never exceed .15?

33. Periodically, a questionnaire is sent by the Department of Commerce to random samples of 200 company presidents. The figures below are the numbers of presidents indicating approval of the government's monetary policy (i) two years ago and (ii) last month. Find a 90% confidence interval for the change, $\pi_1 - \pi_2$, in the proportion favoring this policy.

	TWO YEARS AGO (I)	LAST MONTH (II)
SAMPLE SIZE	200	200
NUMBER INDICATING APPROVAL	155	121

34. In the two samples referred to in Review Problem 33, company presidents were also asked to record the number of new employees they had acquired over the past

year. The mean and standard deviations of the number of new employees are as follows:

	TWO YEARS AGO (I)	LAST MONTH (II)
SAMPLE SIZE	200	200
MEAN NUMBER OF NEW EMPLOYEES	185.4	149.9
STANDARD DEVIATION OF THE NUMBER OF NEW EMPLOYEES	70.9	80.6

Find a 90% confidence interval for the change $\mu_1 - \mu_2$ in the population mean number of new employees per company.

EDUCATION AND PSYCHOLOGY

35. The time that it takes each of a random sample of 60 students to complete a professional examination is recorded. The average time in the sample is 2.86 hours. The standard deviation is .20 hour. Find and interpret an 80% confidence interval for the mean time it takes all students to complete this examination.
36. In a random sample of 40 classrooms in a school district on one day, seven were being taught by substitute teachers. Find a 90% confidence interval for the proportion of all classes on that day that were being taught by a substitute teacher.
37. A psychology student wishes to estimate μ, the mean time it takes undergraduates to perform a certain memory test. She is willing to pay subjects \$2.50 each and is advised that she should use sufficient subjects so as to estimate μ with an accuracy of 2 minutes with probability .8. Her preliminary research suggests that most subjects complete the test in between 30 and 90 minutes. Approximately how much money should be set aside to pay subjects in this experiment?
38. Creativity tests are given to groups of Anglo-Americans and Mexican-Americans with the following results:

	ANGLO-AMERICANS (I)	MEXICAN-AMERICANS (II)
SAMPLE SIZE	100	110
SAMPLE MEAN	68.4	70.2
SAMPLE STANDARD DEVIATION	10.9	14.4

(a) Compute an 80% confidence interval for $(\mu_1 - \mu_2)$.

(b) What does the confidence interval tell you about the extent to which one group is more creative than the other?

39. A guidance counselor wants to compare the mean length of time it takes to complete two different checklists (I and II). Fifty people were randomly selected and asked to complete checklist I, while 50 others completed checklist II. The following results were obtained:

$$\bar{x}_1 = 52.3 \qquad \bar{x}_2 = 49.1$$

$$s_1^2 = 16.0 \qquad s_2^2 = 12.2$$

(a) Find a 90% confidence interval for $\mu_1 - \mu_2$.

(b) Does there appear to be any real difference in the mean times involved?

SCIENCE AND TECHNOLOGY

40. A quality control engineer wishes to estimate μ, the mean length of life of a particular appliance. Testing a random sample of 200 such appliances reveals the following data (x is in hours):

$$\sum x_i = 23{,}905 \qquad \sum x_i^2 = 2{,}907{,}034$$

Find and interpret a 98% confidence interval for μ.

41. The Department of Defense intends to select one of two radar systems, I or II. In independent tests of the two systems, system I detected 95 of 108 aircraft, while system II detected 89 of 104 aircraft.

(a) If π_1 is the probability of system I detecting an aircraft, find a 95% confidence interval for π_1.

(b) With the obvious notation find a 95% confidence interval for $\pi_1 - \pi_2$.

(c) Does your interval in part (b) suggest any important difference between π_1 and π_2?

42. A scientist is interested in estimating μ, the mean temperature inside a chemical reactor. She will obtain temperature readings at a sample of locations inside the reactor. How large a random sample of readings should she take to be 80% confident that the sample mean temperature lies within 8 degrees Celsius of μ? (Assume $\sigma = 48$ degrees Celsius)

SUGGESTED READING

Some aspects of estimation that we have not covered can be found in the following books:

COCHRAN, W.G. *Sampling Techniques*. 3rd Ed, New York: Wiley, 1977

LEVIN, R.I. and RUBIN, P.H. *Applied Elementary Statistics*. Englewood Cliffs, N.J.: Prentice-Hall, 1980.

IMAN, R.L. and CONOVER, W.J. *A Modern Approach to Statistics*. New York: Wiley, 1983.

10 HYPOTHESIS TESTING IN LARGE SAMPLES

REDUCING THE LENGTH OF PRISON SENTENCES

A judge recommends that judges under her jurisdiction hand down shorter sentences than before to persons convicted of minor crimes. Her argument is that for such persons, short sentences are as effective a deterrent as longer ones and are less of a burden on the prison service.

Complete data for the previous five years indicate that the average sentence for these persons was 540 days. A law reform group is interested in determining whether the mean length of sentence (μ) for such criminals *since* the judge's recommendation is substantially less than 540 days. A random sample of 40 such cases is selected and the average length of sentence $\bar{x}$ = 518.3 days computed. Since 518.3 *is* less than 540, it would seem that there has been a decline in mean sentence length. But is it not possible to obtain a sample mean as low as 518.3 days even if in fact the judge's suggestion has been ignored, and μ is still 540 days? How should we determine whether the sample data are consistent with (1) there having been no decline or (2) there having been a significant decline in the mean length of sentence?

This is a hypothesis-testing problem. See Review Problem 6 on page 365 for further details.

SECTION 10.1

THE P VALUE

Chapter 9 was devoted to methods for estimating parameters such as μ, π, $\mu_1 - \mu_2$, or $\pi_1 - \pi_2$. In this chapter we introduce the second branch of statistical inference—hypothesis testing. A dictionary defines an hypothesis as "an unproved or unverified assumption that can be . . . accepted as probable in the light of established facts." Frequently, a researcher assumes that a parameter, μ, for instance, has a particular value and then determines whether this value is consistent with the sample data (the "established facts"). As the title suggests, this chapter is concerned with methods for doing just this when the sample or samples are large ($n \geq 30$ or n_1 and $n_2 \geq 30$).

EXAMPLE 1 A tire manufacturer claims that with normal driving a new brand of radial ply tire will last an average of 48,000 miles. A consumers group feels that the company is overestimating the tire life and asks the Consumer Protection Agency (CPA) to look into the question.

The agency first establishes the ground rules that it will follow. Call the mean life span of *all* such tires, μ. It is the value for μ that is in dispute. The CPA poses two hypotheses:

1. The manufacturer is correct and the mean life span of all such tires is 48,000 miles. This is expressed as $H_0: \mu = 48{,}000$ and is called the **null hypothesis**.
2. The consumer group is correct and the mean life span is *less* than 48,000 miles. This is expressed as H_1: $\mu < 48{,}000$ and is called the **alternative hypothesis.**

The decision as to which hypothesis is more appropriate will be based on the results of testing a random sample of 50 tires.

Assume that the mean life span of the 50 sampled tires is $\bar{x} = 45{,}286$ miles and the standard deviation, $s = 6012.60$ miles. Since $\bar{x} = 45{,}286$ is less than 48,000, it seems as though the data support H_1. However, the CPA will ask: Even if μ were 48,000 miles, is it not possible to obtain a sample mean as small as 45,286 by chance? Accepting that this is possible, the question, then, is *how* possible? Before answering this question, we review briefly some aspects of the behavior of the random variable $\bar{X}$, the sample mean.

In this case since $n = 50$ exceeds 30, the distribution of $\bar{X}$ is approximately $N[\mu, \sigma/\sqrt{n}]$. Using s in place of σ as we did in Chapter 9, we can write $s/\sqrt{n} = 6012.60/\sqrt{50} = 850.31$. Thus, *if* H_0 is true, $\bar{X}$ is approximately $N[48{,}000, 850.31]$ or, equivalently, $Z = (\bar{X} - 48{,}000)/850.31$ is $N[0, 1]$. When we substitute the observed value $\bar{x} = 45{,}286$, we obtain the standardized value $z = (45{,}286 - 48{,}000)/850.31 = -3.19$. This value indicates that $\bar{x} = 45{,}286$ is more than three standard deviations below 48,000.

The value -3.19 is shown in Figure 1. We can now answer the question posed above. i.e., "Even if μ were 48,000 miles, is it not possible to obtain a sample mean as small as 45,286 by chance?" Since Z is continuous we cannot find $P(Z = -3.19)$ but we can find $P(Z < -3.19)$, which is the probability of

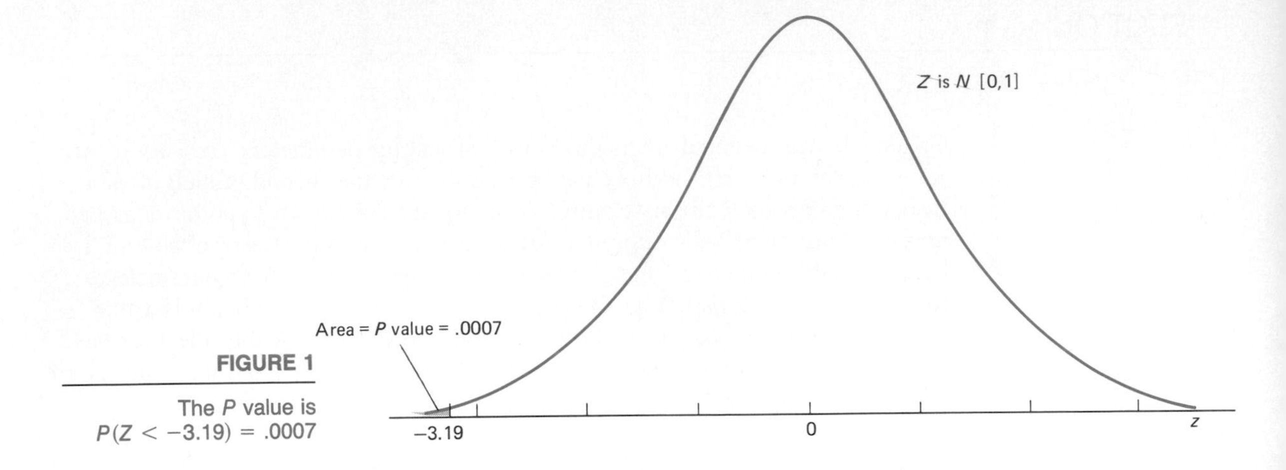

FIGURE 1

The P value is $P(Z < -3.19) = .0007$

getting a value for $\bar{X}$ which is as far (or further than) 3.19 standard deviations below the value for the hypothesized population mean. From Table A3, $P(Z < -3.19)$ is .0007. This probability is the tiny area to the left of -3.19 in Figure 1. We call this probability the P value. The fact that the P value is so small suggests that if H_0 were true, there would be only a minuscule chance (7 out of 10,000) of obtaining a sample mean (45,286) so far below 48,000. The CPA should, therefore, reject H_0 in favor of H_1 and report that the data support the consumers group's claim. Notice we have said only that "the data support" or perhaps "the data suggest." We shall have more to say about this point in a moment.

You may wonder why we began by assuming that H_0 was true rather than focusing on the claim of the consumer group (H_1) directly since that is what prompted the study in the first place. The problem is that as the problem was posed, the consumers did not specify how far below 48,000 μ was: 200 miles; 1000; 2000? Therefore, there is no alternative value against which to judge $\bar{x} = 45{,}286$.

To continue, suppose that the sample mean tire life was $\bar{x} = 47{,}691$ miles. Should H_0 be rejected in this case? Following the same procedure as before, we compute

$$z = \frac{\bar{x} - 48{,}000}{850.31} = \frac{47{,}691 - 48{,}000}{850.31} = -.36$$

This value indicates that 47,691 is just over one-third of a standard deviation below the hypothesized population mean. The P value in this case is $P(Z < -.36)$, the probability of getting a sample mean (47,691) at least .36 standard deviation below 48,000 miles. From Table A3 the P value is .3594. This is shown in Figure 2 as the substantial area to the left of $Z = -.36$. We interpret this P value as follows: If H_0 is true, the chance of getting a sample mean at least .36 standard deviation below 48,000 miles is almost .36 (i.e., 36 out of 100). In other words, if H_0 is true, a sample mean $\bar{x}$ somewhere in this

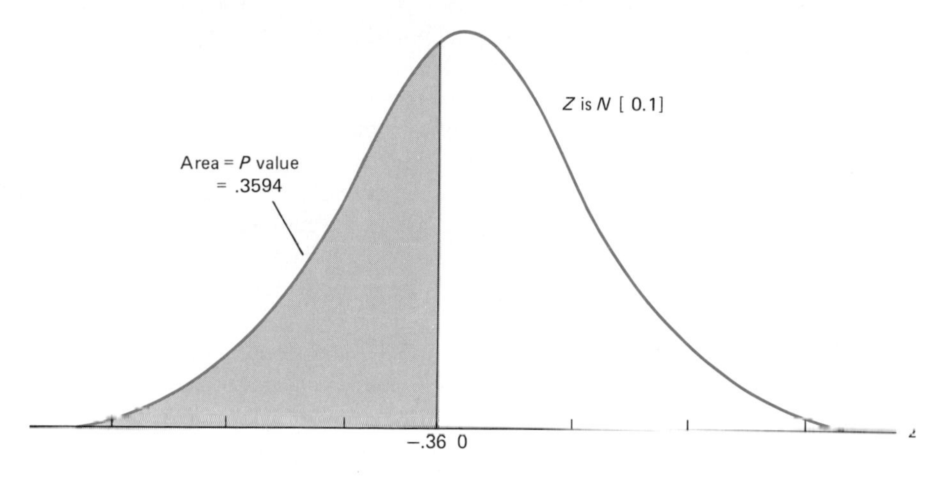

FIGURE 2

The P value is $P(Z < -.36) = .3594$

range is quite reasonable. Accordingly, the CPA should *not reject* H_0 but report that the data are consistent with the manufacturer's claim.

Notice that we used the phrase "should not reject H_0" rather than "accept H_0." The difference may seem trivial, but it is not. Accepting H_0 would imply that μ *was* 48,000 miles. But since we cannot test *all* such tires to be produced, we cannot prove this. We can only measure the extent to which the data *support* H_0. In the same way, when we rejected H_0 above, this only meant that the data *suggested* (but did not prove) that H_0 was incorrect and therefore that H_1 was correct.

Finally, before leaving this example, suppose that $\bar{x}$ was 49,200 miles. Since this value for $\bar{x}$ *exceeds* 48,000, it would provide no support for H_1, the hypothesis that μ is *less* than 48,000. In this case there is no need to compute a P value. The CPA should merely note that the result of its tests were inconsistent with H_1, and therefore it should not reject H_0.

■

Generalizing from this example, the P value provides a measure of the extent to which the data support or do not support H_0. More specifically if the P value is *large*, the sample value, $\bar{x}$, is consistent with H_0 and we should *not* reject H_0. If the P value is *small*, the sample value, $\bar{x}$, is inconsistent with H_0 and we should *reject* H_0. The question remains how small is small? We shall address this a bit later.

We can now define the P value as follows:

> DEFINITION 1
>
> The ***P* value** is the probability, if H_0 is true, of getting a sample result at least as extreme as the one obtained.

In some cases (as in Example 1) "at least as extreme" means "less than"; in others it means "more than"; and in still others, "less than or equal to."

EXAMPLE 2 A medical review board approves a mean stay in the hospital for patients having a particular operation as 6.0 days. The board claims that the average for Medicare patients (i.e., those for whose bills the government is responsible) has been substantially longer than 6.0 days. To examine this claim a sample of 100 Medicare patients who have had this operation in the past year is selected. In this case $\bar{x}$ and s are, respectively, the sample mean and standard deviation of the length of stay. (a) State the null and alternative hypotheses. (b) What decision should be made if $s = 1.62$ days and (i) $\bar{x} = 6.32$ days; (ii) $\bar{x} = 6.03$ days; (iii) $\bar{x} = 5.86$ days; (iv) $\bar{x} = 6.18$ days?

Solution (a) Call the unknown mean length of hospitalization for all such Medicare patients having this operation μ. The review board has indicated that μ should be 6.0 days but are concerned that it may be significantly higher. The choice for this problem is therefore between $H_0: \mu = 6.0$ and $H_1: \mu > 6.0$.

(b) Now, if $H_0: \mu = 6.0$ is true,

$$\frac{\bar{X} - 6.0}{s/\sqrt{n}} = \frac{\bar{X} - 6.0}{1.62/\sqrt{100}} = \frac{\bar{X} - 6.0}{.162} = Z$$

has the $N[0, 1]$ distribution.

(i) $\bar{x} = 6.32$ days. Notice that this value for $\bar{x}$ is consistent with $H_1: \mu > 6.0$ in that 6.32 is greater than 6.0. The value for Z corresponding to $\bar{x} = 6.32$ is $z = (6.32 - 6.0)/.162 = 1.98$. Continuing, the P value is $P(Z > 1.98)$, the probability of getting a sample mean at least 1.98 standard deviations greater than 6.0. From Table A3 this probability is .0239 or a bit over 2 out of 100. If H_0 is true, there is only a small chance of getting a sample mean as large or larger than 6.32. In other words, if $\mu = 6.0$, such values for $\bar{x}$ would be so unusual that the board should reject H_0, concluding that the data support H_1 (i.e., the 100 patients were selected from a population with a mean significantly greater than 6.0 days). This P value is shown in Figure 3(i).

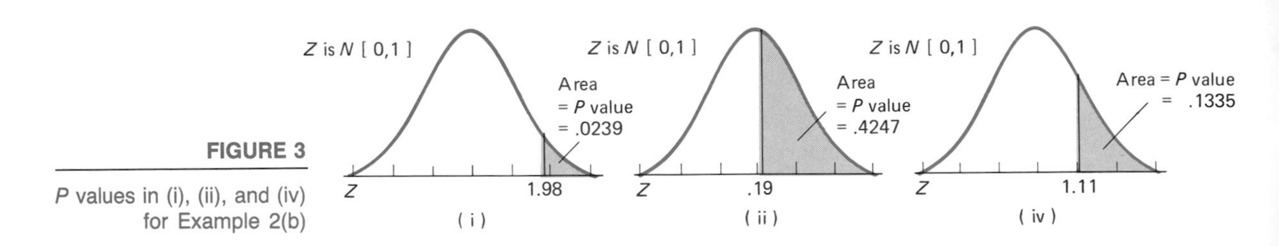

FIGURE 3

P values in (i), (ii), and (iv) for Example 2(b)

(ii) $\bar{x} = 6.03$ days. Again, this value for $\bar{x}$ is consistent with H_1, but how consistent? The Z value is $z = (6.03 - 6.0)/.162 = .19$, and the corresponding P value is $P(Z > .19) = .4247$. If H_0 is true, there is a considerable chance (more than 4 out of 10) of getting a value as far from 6.0 as 6.03. Therefore,

the board should not reject H_0. The data suggest that the mean stay is not significantly above 6.0. The P value, .4247, is shown in Figure 3(ii).
(iii) $\bar{x} = 5.86$ days. This value for $\bar{x}$ is on the "wrong" side of the null hypothesis mean in relation to H_1 and therefore offers no support for H_1. There is no reason to compute a P value in this case and the board should simply not reject H_0.
(iv) $\bar{x} = 6.18$ days. This value is consistent with H_1 but again, how consistent? The value for Z corresponding to $\bar{x} = 6.18$ is $z = (6.18 - 6.0)/.162 = 1.11$. The corresponding P value is $P(Z > 1.11) = .1335$. If H_0 is true, there is a 13% chance of getting a sample mean at least as extreme as (i.e., larger than) 6.18 days. This P value is shown in Figure 3(iv). Should the board reject H_0 or not on the basis of this P value? The value .1335 is not very small as in (i), or reasonably large as in (ii). Is an event that would occur 13 times in 100 trials unusual enough to reject H_0? Clearly, some measure is needed which indicates when an outcome is sufficiently unusual to warrant rejecting H_0. We shall return to this point in the next section.

■

We can now generalize the procedure in Examples 1 and 2. We began by distinguishing between two types of hypotheses:

1. The **null hypothesis:** This assumes a specific value, μ_0 (e.g., 48,000 miles or 6.0 days), for the unknown population mean, μ, and is of the form $H_0 : \mu = \mu_0$.
2. The **alternative hypothesis:**
   (a) The population mean is less than μ_0, written $H_1 : \mu < \mu_0$, or
   (b) The population mean is greater than μ_0, written $H_1 : \mu > \mu_0$.

It is important to keep in mind that we are often more interested in showing that the alternative hypothesis is correct. The problem is, as we said earlier, that H_1 does not provide a specific value for μ against which to judge the value for $\bar{x}$. Rather we must decide whether the data are or are not consistent with the null hypothesis (H_0) value for μ. If it is *inconsistent,* we reject the null hypothesis in favor of H_1. Otherwise, we do not reject H_0.

Now in large random samples if $H_0 : \mu = \mu_0$ is true the random variable $\bar{X}$ will be approximately $N[\mu_0, \sigma/\sqrt{n}]$ or, equivalently $Z = \dfrac{\bar{X} - \mu_0}{\sigma/\sqrt{n}}$ will have the $N[0, 1]$ distribution. Substituting the observed value $\bar{x}$ for $\bar{X}$, we obtain the standardized value $z = \dfrac{\bar{x} - \mu_0}{\sigma/\sqrt{n}}$. This is called the **test statistic** and indicates how many standard deviations $\bar{x}$ is from μ_0. Because we have detailed tables giving probabilities for the $N[0, 1]$ distribution, we can go one step further and compute the P value. In this context the P value is the probability of getting a value for $\bar{x}$ at least z standard deviations from μ_0. The step-by-step procedure for deciding whether or not to reject H_0 is given in Table 1.

TABLE 1 Procedure for deciding whether or not to reject $H_0: \mu = \mu_0$ with a large random sample ($n \geq 30$)

STEP	ALTERNATIVE HYPOTHESIS (a) $H_1: \mu > \mu_0$	(b) $H_1: \mu < \mu_0$
1. Do not reject H_0 if $\bar{x}$ is inconsistent with H_1, i.e., if:	$\bar{x} < \mu_0$	$\bar{x} > \mu_0$
Perform the test if:	$\bar{x} \geq \mu_0$	$\bar{x} \leq \mu_0$
2. Compute:	$z = \dfrac{\bar{x} - \mu_0}{s/\sqrt{n}}$	
3. The P value is:	$P(Z > z)$	$P(Z < z)$
4. Check how small the P value is.		
5. Conclusion: Reject H_0 if the P value is very small. Otherwise, do not reject H_0.		
Assumption: A large ($n \geq 30$) random sample		

EXAMPLE 3 A machine produces bolts which should have a mean length of 7.5 centimeters (cm). The operator suspects that bolts are being produced which are on the average smaller than 7.5 cm. In a sample of 60 bolts the following summary data were obtained:

$$\sum^{60} x_i = 449.76 \text{ cm} \qquad \sum^{60} x_i^2 = 3371.5814$$

(a) What is μ in this case? (b) What are the null and alternative hypotheses? (c) What decision should be made on the basis of the data above? (d) How would your decision change if $\Sigma\, x_i = 450.12$ and $\Sigma\, x_i^2 = 3376.9346$?

Solution (a) In this case μ is the true (but unknown) mean length of all bolts produced by the machine.

(b) $H_0: \mu = 7.5$, the intended length
$H_1: \mu < 7.5$, the suspected length

(c) We compute $\bar{x} = \Sigma\, x_i/60 = 449.76/60 = 7.496$ and

$$s = \sqrt{\frac{60(3371.5814) - 449.76^2}{60(59)}} = \sqrt{.0030583} = .0553$$

Following the steps in Table 1 with $\mu_0 = 7.5$, $n = 60$, and $\bar{x} = 7.496$:

1. Since $\bar{x} = 7.496 < 7.5$, we perform the test as follows:
2. $z = \dfrac{7.496 - 7.5}{.0553/\sqrt{60}} = -.56$
3. The P value $= P(Z < -.56) = .2877$.

4/5. Since the P value is substantial, we conclude that $\bar{x} = 7.496$ is consistent with H_0. The data suggest that the mean bolt length is not significantly less than 7.5 cm.

(d) In this case $\bar{x} = 450.12/60 = 7.502$, which is *greater* than, that is, on the "wrong" side of, 7.5. Do not reject H_0.

■

PROBLEMS 10.1

1. We wish to test whether μ, the mean time to complete a surgical operation, is significantly greater than 200 minutes. In a random sample of 60 such operations, the standard deviation of the time taken is 12.0 minutes. What conclusion can be reached if (a) $\bar{x} = 190.6$; (b) $\bar{x} = 206.5$?
2. An educational planner suspects there has been a significant decline in the mean age at which students begin college. Complete figures for two years ago indicate a mean age of 18.20 years. A random sample of 40 student records are randomly selected from this year's freshman class at the state university. The mean age at orientation is 18.17 years and the standard deviation .39 year.
   (a) Define μ in this case.
   (b) State H_0 and H_1
   (c) What decision should be made?
   (d) Obtain an 80% confidence interval for μ.
3. An effort was recently made to clean up the river flowing through a city. Prior to this effort, a study indicated a concentration of pollutants in the river of 74 milligrams per liter. One year after the cleanup attempt, random samples of 1 liter each were taken from 35 sites along the river. The mean and standard deviation of the amount of pollutants over the 35 sites were computed with the following results:

   $$\bar{x} = 48.2 \text{ milligrams} \qquad s = 16.1 \text{ milligrams}$$

   (a) Define μ in this case.
   (b) State H_0 and H_1.
   (c) What decision should be made?
4. A town's school committee selects a random sample of 75 students who have recently taken the SAT. Their scores are analyzed with the result that $\bar{x} = 509.8$ and $s = 125.6$. The committee wishes to know whether students in the town score, on the average, higher than 500?
   (a) Define μ in this case.
   (b) State H_0 and H_1.
   (c) What conclusion should be reached?
5. In Example 2 of Chapter 1, we recorded the number of previous attempts at an examination by each of 66 students. The mean and standard deviation of the number of attempts were $\bar{x} = 1.2$ (from Example 9 of Chapter 1) and $s = 1.3$ (from Example 14 of Chapter 1).
   (a) Regarding these 66 students as a random sample from the population of all students taking the examination, test whether the population mean number of previous attempts is less than 1.7.
   (b) Find a 90% confidence interval for μ.
6. In a random sample of 400 high school teachers in a state the mean income is $20,054 and the standard deviation, $3750. Do these data suggest that the mean salary over all such teachers in the state is greater than $20,000?
7. Problem 8 of Section 1.2 contains the speed of each of 40 cars traveling on the Pennsylvania Turnpike. Use these data to test whether the mean speed of all cars on the turnpike exceeds the 55-mph speed limit. (In Problem 33 of Section 1.4 we gave the values $\Sigma x_i = 2407$ and $\Sigma x_i^2 = 146,343$.)
8. A psychological test given by the U.S. Air Force has only recently been given to

women recruits. The mean score for men is 22.5. A researcher is interested in finding whether on the average women will score higher than men.

(a) Specify the hypotheses that are appropriate here.

(b) In a sample of scores by 32 women recruits the mean score was 24.8 and the standard deviation 5.6. What conclusion do these data suggest?

(c) What conclusion would be appropriate if $\bar{x} = 22.1$ and $s = 5.6$?

(d) Obtain a 95% confidence interval for μ.

9. Table 1 lists the procedure for computing P values for testing $H_0 : \mu = \mu_0$. Why are these procedures applicable only when $n \geq 30$?
10. Explain why the P value as defined in Table 1 cannot exceed .5.
11. A brand of contact lenses is advertised as lasting an average of 20 months. A consumer group doubts that the mean is that high and randomly samples 40 persons who use this brand. In this sample the average length of life is 18.9 months and the standard deviation 2.5 months. What do these data suggest about the advertised claim? (Compare this problem with Problem 15 of Section 7.2.)
12. Draw a picture showing the area corresponding to the P value in Example 3(c).

Solution to Problem 1

We are to test $H_0 : \mu = 200$ against $H_1 : \mu > 200$. Following the procedure in Table 1 with $\mu_0 = 200$:

(a) The value $\bar{x} = 190.6$ is less than 200, so we simply do not reject H_0. The data suggest that μ is not significantly greater than 200.

(b) (1) Since $\bar{x} = 206.5$ we perform the test.

(2) We compute $z = \dfrac{206.5 - 200}{12.0/\sqrt{60}} = 4.20$.

(3) The P value is $P(Z > 4.20) \doteq 0$.

(4/5) Since the P value is so small, we reject H_0. The data suggest that μ is significantly greater than 200.

SECTION 10.2
THE LEVEL OF SIGNIFICANCE

In Section 10.1 we rejected H_0 in favor of H_1 when the P value was very small. But just how small a P value justifies rejecting H_0? In Example 2(iii), for instance, the P value was .1335. Is this sufficiently small to warrant rejecting H_0?

The usual approach to this problem is to set up a criterion or standard against which we can compare the P value. It is important that this be done *before* seeing the data. The next example indicates how such a criterion is used.

EXAMPLE 4 As part of the presampling ground rules in Example 1, the CPA agrees with both parties to reject the manufacturer's claim ($H_0 : \mu = 48{,}000$) in favor of the consumers' claim ($H_1 : \mu < 48{,}000$) if the P value is less than or equal to .02. What decision should be made (a) if $\bar{x} = 45{,}286$ miles; (b) if $\bar{x} = 47{,}691$ miles?

Solution (a) From Example 1 with $\bar{x} = 45{,}286$, the P value was $P(Z < -3.19) = .0007$. Since this probability is much less than .02, the CPA should reject H_0.

(b) Now, what if $\bar{x} = 47{,}691$? In this case the corresponding P value was $P(Z < -.36) = .3594$. This time, since the P value is *more* than .02, the CPA should not reject H_0.

In Figure 4 these two P values are shown as the shaded areas, together with the tail area .02, illustrating graphically that in part (a) the P value is less than .02, but in part (b) the P value greatly exceeds .02.

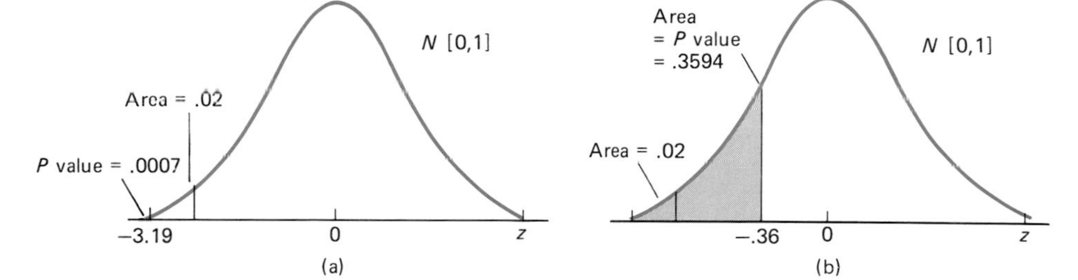

FIGURE 4

P values in Example 4 as areas compared to the tail area .02

In general, the value of the criterion against which we compare the P value is designated α (the Greek lowercase letter alpha) and is called the **level of significance**. Thus in Example 4(a) we speak of "rejecting H_0 at the .02 level of significance."

The rule for rejecting H_0 can be stated as:

> DEFINITION 2
>
> Reject H_0 if the P value is $\leq \alpha$.

When reporting the result of a test, it is important to note the value of α that was used. Suppose, for instance, that the P value was .027. Had $\alpha = .05$ been selected, we would *reject H_0 at the .05 level of significance*. Had we selected $\alpha = .01$, however, we would *not reject* H_0 at the .01 level of significance.

To understand Definition 2 it is helpful to think of the P value as measuring the extent to which the data do support H_0. The smaller the P value, the less support the data provide for H_0. We view α as an indicator of just how little support the data must provide for H_0 before we are willing to reject H_0 in favor of H_1.

EXAMPLE 5 Using the information in Example 2 and $\alpha = .05$, what would you decide if (i) $\bar{x} = 6.32$ days; (ii) $\bar{x} = 6.03$ days; (iii) $\bar{x} = 5.86$ days; (iv) $\bar{x} = 6.18$ days?

Solution (i) The P value associated with $\bar{x} = 6.32$ is $P(Z > 1.98) = .0239$. This is less than .05, and we reject H_0 at the .05 level of significance. The data suggest that the mean time spent in the hospital by Medicare patients is significantly greater than 6.0 days.

(ii) When $\bar{x} = 6.03$ days, the P value is $P(Z > .19) = .4247$. Since this value is much greater than $\alpha = .05$, we cannot reject H_0 at the .05 level of significance. The data suggest that the mean time spent in the hospital is *not* significantly greater than 6.0 days.

(iii) Recall from Example 2 that when $\bar{x} = 5.86$ (i.e., *below* 6.0), there is no need to compute the P value. We merely state that we cannot reject H_0 on the basis of these data.

(iv) When $\bar{x} = 6.18$, the P value is $P(Z > 1.11) = .1335$, which exceeds $\alpha = .05$ and again we do not reject H_0 at the .05 level of significance. The conclusion is the same as in (ii).

■

It is important to interpret the result of a test in the context in which it was performed. For example, in addition to the statement "we reject H_0 at the .05 level of significance," it will be helpful to add the interpretation "the data suggest that the mean stay in the hospital is significantly greater than 6.0 days." More hospital administrators are likely to understand the second statement than the first.

When interpreting the results of a statistical test, keep in mind the distinction between *statistical* significance and *practical* significance. In Example 5(i), for instance, when $\bar{x} = 6.32$ days we rejected $H_0: \mu = 6.0$, concluding that "the data suggest that the mean time . . . is significantly greater than 6.0 days." This is what we mean by statistical significance. It occurs when the probability of so extreme a sample result if H_0 is true is sufficiently small. But the medical review board may only be interested in whether μ appears to be of the order of 7.0 or 7.5 days and may regard $\bar{x} = 6.32$ as not sufficiently greater than 6.0 days to have any practical significance. This is what we mean when we refer to practical significance. We shall introduce methods for addressing this distinction in Chapter 11.

You may wonder how a value for α is chosen. Before we answer this, notice that the lower the value for α, the more unusual the sample results have to be before we reject H_0. For instance, with $\alpha = .1$, we reject H_0 only if as extreme a sample mean would occur less than once in 10 times, *were* H_0 *true*. If $\alpha = .01$, we reject H_0 only if as extreme a sample mean could occur less than one time in 100, *were* H_0 *true*. A low value for α is appropriate where a decision to reject H_0 has important consequences. For instance, the tire manufacturer in Examples 1 and 4 may be required to withdraw advertising and/or pay a fine if the claim (that the mean lifetime of the tire is 48,000 miles) is rejected by the CPA. In this case the CPA would want to be cautious and use a small value for α, perhaps $\alpha = .02$ as we suggested, or even .01 or .005.

Although values from .005 to .15 are occasionally used, two values, $\alpha = .05$ and $\alpha = .01$, are the most widely used in statistical work today. These values, although arbitrary, have become so standard that with some exceptions, we shall use them in the remainder of the text. The next example illustrates how we use both $\alpha = .05$ and $\alpha = .01$.

EXAMPLE 6 A district school board in a large city is interested in learning whether the mean sixth-grade score on a city-wide test *for its* district is significantly higher than the city-wide average of 75 points. A sample of 40 sixth graders from this district is selected, their scores recorded, and the sample mean and standard deviation of the scores computed. What should the board conclude if $s = 13.4$ points and (a) $\bar{x} = 81.07$; (b) $\bar{x} = 76.61$; (c) $\bar{x} = 78.79$?

Solution Call μ the (unknown) mean test score for this school district. We will test $H_0: \mu = 75$ against $H_1: \mu > 75$.

In each case $s/\sqrt{n} = 13.4/\sqrt{40} = 2.119$. Also notice that following the procedures in Table 1, in each case, (a), (b), and (c), the value for $\bar{x}$ is greater than 75 and is therefore consistent with H_1.

(a) When $\bar{x} = 81.07$, $z = (81.07 - 75)/2.119 = 2.86$ and the P value is $P(Z > 2.86) = .0021$.

Since the P value $< .01$, we reject H_0 at the .01 level of significance. The data *strongly* suggest that the mean test score for this district is significantly greater than 75.

(b) When $\bar{x} = 76.61$, $z = (76.61 - 75)/2.119 = .76$ and the P value is $P(Z > .76) = .2236$.

Since the P value exceeds .05, we cannot reject H_0 at the .05 level of significance. The data suggest that the mean test score for this district is *not* significantly greater than 75.

(c) When $\bar{x} = 78.79$, $z = (78.79 - 75)/2.119 = 1.79$ and the P value is $P(Z > 1.79) = .0367$.

Since the P value is greater than .01, we cannot reject H_0 at the .01 level of significance. However, since .0367 is less than .05, we can reject H_0 at the .05 level of significance. This data also suggest that the mean test score for the district is significantly greater than 75. However, since H_0 is rejected only at the .05, not at the .01 level, we now omit the word "strongly."

Each of these three P values is shown in Figure 5 as tail areas under the $N[0, 1]$ distribution. For comparison purposes the tail areas .05 and .01 are also shown.

FIGURE 5

Three P values in Example 6 compared to $\alpha = .01$ and $\alpha = .05$

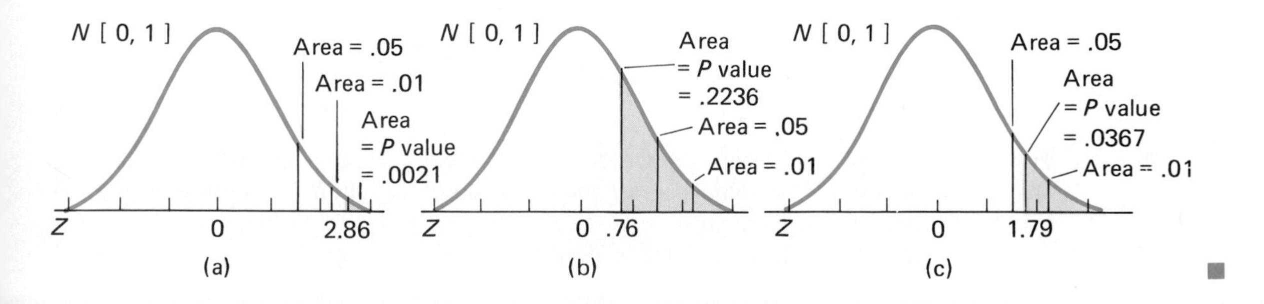

■

The rules for applying both $\alpha = .05$ and $\alpha = .01$ are summarized graphically in Figure 6.

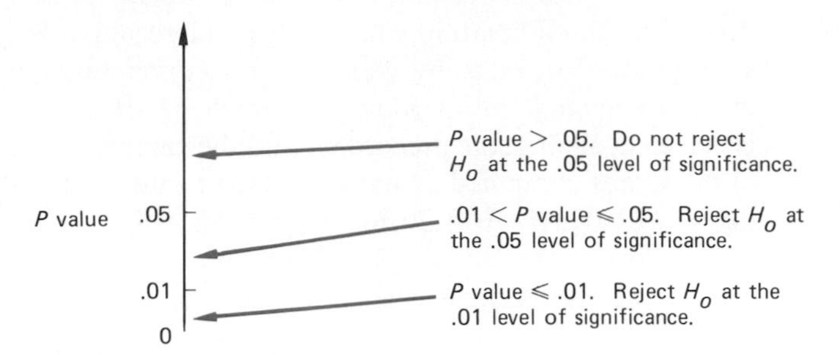

FIGURE 6

Graphic summary of the rules for rejecting H_0, using $\alpha = .05$ and $\alpha = .01$

Let us apply the rules in Figure 6 to the following P values: .247, .007, and .043. The conclusions are:

1. If the P value = .247, do not reject H_0 at the .05 level of significance since $.247 > .05$.
2. If the P value = .007, reject H_0 at the .01 level of significance because $.007 < .01$.
3. If the P value = .043, reject H_0 at the .05 level of significance since $.043 < .05$ but $.043 > .01$.

In Table 2 we expand the procedure for testing H_0 in Table 1 to include the use of $\alpha = .01$ and $\alpha = .05$.

TABLE 2

Procedure for deciding whether or not to reject $H_0: \mu = \mu_0$ at the α level of significance with a large random sample ($n \geq 30$)

STEP	ALTERNATIVE HYPOTHESIS (a) $H_1: \mu > \mu_0$	ALTERNATIVE HYPOTHESIS (b) $H_1: \mu < \mu_0$
1. Do not reject H_0 if $\bar{x}$ is inconsistent with H_1, i.e., if:	$\bar{x} < \mu_0$	$\bar{x} > \mu_0$
Perform the test if:	$\bar{x} \geq \mu_0$	$\bar{x} \leq \mu_0$
2. Compute:	$z = \dfrac{\bar{x} - \mu_0}{s/\sqrt{n}}$	
3. The P value is:	$P(Z > z)$	$P(Z < z)$
4. Using $\alpha = .05$ or .01, reject H_0 if:	P value $\leq \alpha$	

Assumption: A large ($n \geq 30$) random sample

EXAMPLE 7 A sample of 32 new applicants for automobile insurance are asked to fill out a new form. The sample mean and the standard deviation of the time needed to complete the form are $\bar{x} = 11.12$ minutes and $s = 3.68$ minutes. Do these data suggest that the mean time to complete the form is significantly greater than 10 minutes?

Solution Call μ the mean time to complete the form over all potential users. We will test $H_0: \mu = 10.0$ against $H_1: \mu > 10.0$. Following the procedure in Table 2, column a, with $n = 32$ and $\mu_0 = 10.0$:

1. The value $\bar{x} = 11.12 > 10$ and so is consistent with H_1.
2. $z = \dfrac{11.12 - 10}{3.68/\sqrt{32}} = 1.72.$
3. The P value is $P(Z > 1.72) = .0427$.

4/5. Since $.01 < P$ value $< .05$, we reject H_0 at the .05 level of significance. The data suggest that the mean time to complete the form is significantly greater than 10.0 minutes.

■

Most research reports today include the P value, although many statistics textbooks develop methods for hypothesis testing which do not include it.

The reason the P value is now included in reports is best explained in terms of past practices. For many years researchers commonly reported that H_0 was or was not rejected at only a *single* level of significance, frequently .05. Without the P value, a reader interested in evaluating the research with a different level of significance had no way to do so. Therefore, in addition to reporting whether H_0 is (or is not) rejected at one of the more commonly used levels of significance—.05 or .01—we strongly recommend that you include the P value whenever possible.

ALTERNATIVE CRITERION FOR REJECTING H_0

Comparing the P value with α is the preferable but not the only method for testing hypotheses about μ. Another criterion is based on the value for z. Consider an example in which we wish to test $H_0: \mu = \mu_0$ against $H_1: \mu > \mu_0$ using a single level of significance, $\alpha = .05$. We will reject H_0 if the P value $\leq .05$. We know from Table A4 that the value for z which has an area of .05 to the right is $Z_{95} = 1.645$. This value is shown in Figure 7. If the value for z is less than 1.645, the corresponding P value will exceed .05. This is shown in Figure 7(a). If the value for z exceeds 1.645, the P value will be less than .05 and we will reject H_0. This is shown in Figure 7(b). This indicates that an alternative method for testing $H_0: \mu = \mu_0$ against $H_1: \mu > \mu_0$ with $\alpha = .05$ would be to reject H_0 if $z = (\bar{x} - \mu_0)/(s/\sqrt{n}) \geq 1.645$. You should verify that if we used $\alpha = .1$ instead of .05 we would reject H_0 if $z > 1.282$.

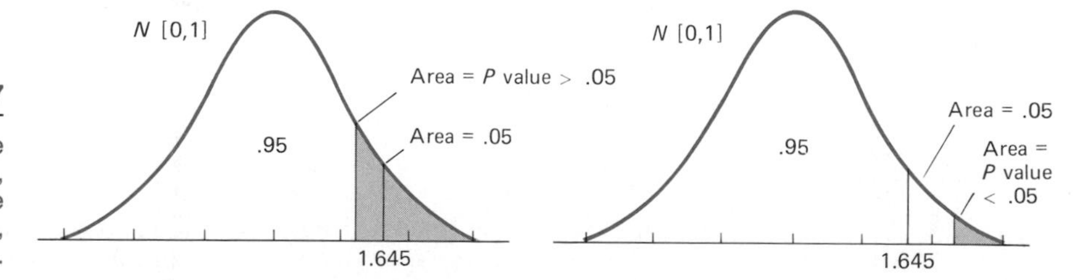

FIGURE 7

Showing that if the value for z is less than 1.645, the P value $> .05$. If the value for z exceeds 1.645, the P value $< .05$.

We will follow this approach to testing hypotheses in later chapters when it is not easy to compute the P value.

THE IMPORTANCE OF RANDOM SAMPLING

In Chapter 9 we emphasized that a confidence interval can be misleading if the sample on which it is based is biased in some way. The same is true for hypothesis testing. In Example 7, for instance, suppose that consciously or unconsciously the person selecting the sample tended to favor those applicants who are likely to have trouble filling out the form. In this case the large sample mean time ($\bar{x} = 11.12$) may well be due to this bias, and the suggestion that the population mean greatly exceeds 10 minutes may, therefore, be incorrect. In other words, the results of the test may be misleading.

Strictly speaking, the P value—which is just a probability—should be computed only when the sample is random.

PROBLEMS 10.2

In the following problems, unless a value for α is specified, use $\alpha = .05$ and $.01$. In each case sketch the P value.

13. Several years ago graduating students at a state university were getting an average of 2.7 job offers per student. The placement office suspects that the mean for the last class is substantially less than 2.7. In a random sample of 40 graduates of the last class the mean and standard deviation of the number of job offers were $\bar{x} = 2.18$ and $s = 1.74$.
    (a) Define μ in this case.
    (b) What are the appropriate null and alternative hypotheses?
    (c) What conclusion do the data suggest?
14. An anthropologist is interested in determining whether the mean gestation period for pregnant women in a South American tribe is significantly longer than 270 days. She assumes that the gestation period for the next 60 pregnancies in the tribe can be considered a random sample of such periods.
    (a) Define μ in this case.
    (b) State H_0 and H_1.
    (c) What decision should be made if the mean and standard deviation of the gestation period for the 60 women are:
    (i) $\bar{x} = 275.28$ days and $s = 11.51$ days?
    (ii) $\bar{x} = 272.85$ days and $s = 11.51$ days?
    (iii) $\bar{x} = 267.39$ days and $s = 11.51$ days?
    (iv) $\bar{x} = 280.65$ days and $s = 11.51$ days?
15. The data in Problem 6 of Section 1.2 (page 16) are the temperatures for the last 70 years at midday on July 4 in a small New England town. Regarding these 70 temperatures as a random sample from all past and future July fourths, test whether the mean temperature for this date and time is significantly less than 80°F. (Use the results $\Sigma\, x_i = 5586$ and $\Sigma\, x_i^2 = 448{,}578$.)
16. Referring to Problem 15, what decision would you come to if you used (a) $\alpha = .1$; (b) $\alpha = .15$; (c) $\alpha = .2$?
17. In Example 1 in Section 1.2 we provided the time, in hours, needed to perform 60 brain operations using a new neurosurgical procedure. Use these data to test whether the potential mean time for such brain operations is significantly greater than 10 hours. (In Example 12 in Chapter 1, it was found that $\Sigma\, x_i = 605.3$ and $s = 1.61$ hours.)

18. In Problem 17, what decision would you come to if (a) $\alpha = .1$; (b) $\alpha = .2$? How large would the level of significance, α, have to be before you would reject H_0?
19. The owner of a chicken farm is interested in using a new chicken feed which it is claimed will significantly improve the gain in weight of chicks. Before investing heavily in the new and more expensive feed, she plans to test it out on 80 two-week-old chicks. She knows that the traditional feed produces a mean gain of 19 grams with a standard deviation of 2.8 grams between the second and third weeks of life.

    (a) State the appropriate null and alternative hypothesis, defining μ.

    (b) What would you advise the owner if the mean gain in weight for the 80 chicks during the week of test is (i) $\bar{x} = 18.29$ grams; (ii) $\bar{x} = 20.15$ grams; (iii) $\bar{x} = 19.57$ grams; (iv) $\bar{x} = 19.02$ grams?
20. Problem 5 of Section 1.2 contains the amount spent by the first 36 persons going through the checkout line in a supermarket. Regarding these data as a random sample from all supermarket customers, test whether the mean amount spent by customers is significantly greater than \$15. (In Problem 32 in Section 1.4 we provided the summary results $\Sigma\, x_i = 702.47$ and $\Sigma\, x_i^2 = 20{,}980.601$.)
21. Until fairly recently successful mortgage applicants spent an average 25% of income on their mortgage payments. It has been assumed that this percentage has increased in recent years with the increase in interest rates. In a random sample of 45 recent successful mortgage applicants the average percentage of income spent on the mortgage payment was 32.1%. The standard deviation was 7.9%.

    (a) Do these data suggest that over *all* successful mortgage applicants the mean percentage is now significantly greater than 25%?

    (b) How would your conclusion above change if the sample mean had been 22.1%?
22. In Problem 21(a), would your conclusion change if you used (a) $\alpha = .005$; (b) $\alpha = .001$?
23. Draw a sketch indicating the P value in Example 7 compared to $\alpha = .05$ and $\alpha = .01$.

Solution to Problem 13

(a) In this case μ is the mean number of job offers for the *last* graduating class.

(b) $H_0: \mu = 2.7$; $H_1: \mu < 2.7$.

(c) Following the procedure in Table 2, $z = (2.18 - 2.7)/(1.74/\sqrt{40}) = -1.89$ and the P value $= P(Z < -1.89) = .0294$.
Since this value lies between .01 and .05, we reject H_0 at the .05 level of significance. The data suggest that the mean number of job offers made to the last graduating class is significantly less than 2.7.

SECTION 10.3

TESTING HYPOTHESES ABOUT PROPORTIONS

The procedures for testing hypotheses are similar whatever the context. In this section we apply the procedures developed in Sections 10.1 and 10.2 to testing hypotheses about a population proportion.

Assume that a population has an (unknown) proportion of successes; call it π. Recall that selecting a random sample of size n from this population is a binomial experiment if (1) sampling is with replacement or (more usually) (2) the population is large compared to n. We call the number of successes in such a sample y and the proportion of successes $\overline{p} = y/n$.

From Section 7.2 if $n\pi$ and $n(1 - \pi)$ are both ≥ 5, the random variable $\overline{P}$ is approximately $N[\pi, \sqrt{\pi(1 - \pi)/n}]$. Equivalently, the random variable

$$\frac{\overline{P} - \pi}{\sqrt{\pi(1 - \pi)/n}} = Z$$

is approximately $N[0, 1]$.

An example will illustrate how the procedure for testing hypotheses about π parallels that in Table 2.

EXAMPLE 8 A company produces computer chips. With the present process, only 85% are perfect. As a test, a total of 1200 chips are produced with a new process that it is hoped will have a higher "perfect" rate. What decision should be made if the number of perfect chips is (a) $y = 1038$; (b) $y = 1010$; (c) $y = 1049$?

Solution Call π the (potential) proportion of "perfect" chips among all the chips made with the *new* process. We wish to test $H_0: \pi = .85$ against $H_1: \pi > .85$. If H_0 is true, $\pi = .85$ and

$$Z = \frac{\overline{P} - 85}{\sqrt{(.85)(.15)/1200}} = \frac{\overline{P} - .85}{.0103}$$

is approximately $N[0, 1]$. In each of (a), (b), and (c) we will follow the procedure in Table 2 but replace μ by π, $\overline{x}$ by $\overline{p}$, and $s/\sqrt{n}$ by $\sqrt{(.85)(.05)/1200} = .0103$.

(a) If $y = 1038$, $\overline{p} = 1038/1200 = .865$, which is consistent with H_1 since .865 exceeds .85. In this case $z = (.865 - .85)/.0103 = 1.46$ and the P value is $P(Z > 1.46) = .0721$.

Since the P value exceeds .05, we cannot reject H_0 at the .05 level of significance. The data suggest that the new process is not significantly better than the old.

(b) If $y = 1010$, $\overline{p} = 1010/1200 = .842$, which is inconsistent with H_1 (i.e., is *below* .85). Therefore, we simply do not reject H_0.

(c) If $y = 1049$, $\overline{p} = 1049/1200 = .874$, which is greater than .85 and therefore consistent with H_1. In this case $z = (.874 - .85)/.0103 = 2.33$ and the P value is $P(Z > 2.33) = .0099$.

Since this P value is less than .01, we may reject H_0 at the .01 level of significance. The data strongly suggest that the "perfect" rate for the new process is significantly greater than .85.

The P values in (a) and (c) are shown in Figure 8(a) and (b), respectively.

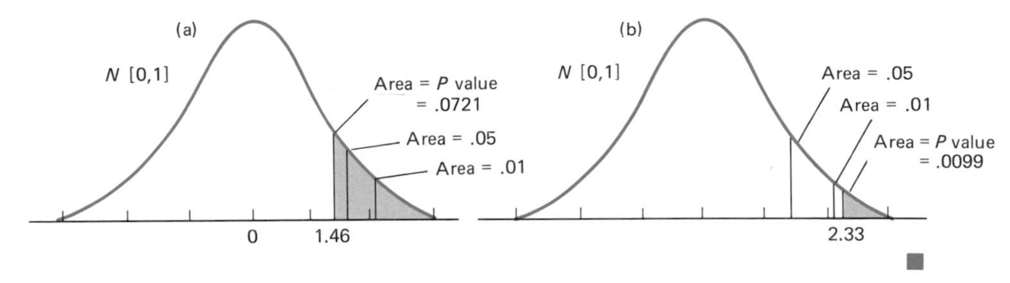

FIGURE 8

P values in Example 8(a) and (c)

In the last example we wrote $\mathrm{SD}(\bar{P}) = \sqrt{(.85)(.15)/1200}$ using the information that the null hypothesis value for π, was .85 and $n = 1200$. This illustrates the point that when we test hypotheses about π, H_0 not only specifies a value for π, it also determines the value for $\mathrm{SD}(\bar{P}) = \sqrt{\pi(1 - \pi)/n}$.

The conclusion in Example 8(c), "the data strongly suggest that the 'perfect' rate for the new process was significantly greater than .85," again illustrates the distinction between *statistical* significance and *practical* significance. Statistically the "perfect" rate for the new process was significantly higher than for the old. But is the apparent improvement (from .85 to approximately .874) of practical use? It is quite possible that such an improvement would not justify the expense and inconvenience of installing the new process. As we indicated earlier, this distinction will be discussed in detail in Chapter 11.

In Table 3 we have summarized the procedure for testing $H_0: \pi = \pi_0$ against either $H_1: \pi > \pi_0$ or $H_1: \pi < \pi_0$, where π_0 is the assumed population proportion.

TABLE 3

Procedure for deciding whether or not to reject $H_0: \pi = \pi_0$ at the α level of significance with a large random sample

STEP	ALTERNATIVE HYPOTHESIS (a) $H_1: \pi > \pi_0$	ALTERNATIVE HYPOTHESIS (b) $H_1: \pi < \pi_0$
1. Do not reject H_0 if p is consistent with H_1, i.e., if:	$\bar{p} < \pi_0$	$\bar{p} > \pi_0$
Perform the test if:	$\bar{p} \geq \pi_0$	$\bar{p} \leq \pi_0$
2. Compute:	$z = \dfrac{\bar{p} - \pi_0}{\sqrt{\pi_0(1 - \pi_0)/n}}$	
3. The P value is:	$P(Z > z)$	$P(Z < z)$
4. Using $\alpha = .05$ or .01, reject H_0 if:	P value $\leq \alpha$	

Assumption: A large random sample [$n\pi_0$ and $n(1 - \pi_0) \geq 5$]

We illustrate the procedure in Table 3 in the next two examples.

EXAMPLE 9 Supporters of a bill that would reintroduce the death penalty in a state claim that 55% of the voters support the move. Opponents feel that this figure is too high and commission a study involving a random sample of 200 of the voters in the state. Of the 200 interviewed, 97 favor the bill. Define π and state both hypotheses. What conclusion do the data suggest?

Solution Call π the (unknown) proportion of all voters in the state favoring the bill. We wish to test $H_0: \pi = .55$ (the supporters' claim) against $H_1: \pi < .55$ (the opponents' belief).

Following the procedure in Table 3 with $\pi_0 = .55$, $n = 200$, and

$$\sqrt{\frac{\pi_0(1 - \pi_0)}{200}} = \sqrt{\frac{(.55)(.45)}{200}} = .0352$$

1. Since $y = 97$, $\bar{p} = 97/200 = .485$, which is less than .55 and therefore consistent with H_1.
2. $z = \dfrac{.485 - .55}{.0352} = -1.85$, and thus
3. The P value $= P(Z < -1.85) = .0322$
4. Since .0322 lies between .01 and .05, we reject H_0 at the .05 level of significance. The data suggest that the proportion of voters favoring the bill is significantly less than .55.

■

EXAMPLE 10 Earlier this century the eminent British statistician R. A. Fisher proposed a slight variation of the following experiment as a good example of hypothesis testing. A woman claims that after tasting a cup of tea with milk, she can, in most cases, detect which of the two ingredients were poured first. To test this claim, she is given 40 cups of tea at regular intervals. In each case the order in which the ingredients were poured into the cup was determined by tossing a fair coin. The woman named the correct order in 29 of the 40 cases. Do these data suggest that (1) the woman was guessing or (2) her claim is valid?

Solution Call π the probability at each tasting that the woman knows the correct order. The two hypotheses are:

$H_0: \pi = .5$ in effect the woman simply guesses at each cup

$H_1: \pi > .5$ she does have some discriminating skill

Since $40(.5)$ and $40(1 - .5)$ both exceed 5, we may apply the procedures in Table 3 with $\pi_0 = .5$, $n = 40$, and

$$\sqrt{\frac{\pi_0(1 - \pi_0)}{40}} = \sqrt{\frac{(.5)(.5)}{40}} = .0791$$

1. Here $y = 29$, $\bar{p} = 29/40 = .725$, which is greater than .5 and so is consistent with H_1.
2. In this case $z = \dfrac{.725 - .5}{.0791} = 2.84$.
3. The P value $= P(Z > 2.84) = .0023$.
4. Since the P value $< .01$, we reject H_0 at the .01 level of significance. The data strongly suggest that the woman's claim is true.

■

PROBLEMS 10.3

24. Over the past 10 years 50% of Americans applying for passports listed Western Europe as their destination. Passport officials sense that this proportion is declining. In a random sample of 100 passport applications for the past two months, Western Europe was reported as the destination in 41.
    (a) Define π in this case.
    (b) State H_0 and H_1.
    (c) What conclusion do these data suggest?
25. Under federal law some people over 65 years of age are eligible for an aid program. Thirty-five percent of those eligible for the program were expected to enroll. Officials believe that the actual percentage enrolled is below this figure. In a pilot survey of 1000 eligible persons, only 294 were enrolled.
    (a) Define π in this case.
    (b) What are H_0 and H_1?
    (c) What conclusion do these data suggest?
    (d) Obtain an 80% confidence interval for π.
26. A student takes a 30-question multiple-choice test. There are four choices for each question. The student gets nine correct answers. The teacher is interested in determining whether this result suggests that the student simply guessed at the correct answer in each case.
    (a) What is π in this case?
    (b) State H_0 and H_1.
    (c) What should be the teacher's conclusion?
27. Last year twenty percent of all sales of homes in a state were made to first-time buyers. In a random sample of 200 sales completed thus far this year, only 36 were to first-time buyers. Do these data suggest that the present proportion of purchasers who are first-time buyers is significantly less than .20?
28. A company auditor selects a random sample of 50 accounts. The mean amount outstanding in these accounts is $\bar{x} = 40.34$ dollars with a standard deviation of $s = 15.84$ dollars. In 33 of the selected accounts the amount outstanding exceeds 50 dollars. Does this sample data suggest that (a) the mean amount outstanding over all accounts (μ) is significantly below 50 dollars; (b) the proportion π of all accounts in which more than 50 dollars is outstanding is significantly greater than .5?
29. Referring to the data in Problem 28:
    (a) Compute a 90% confidence interval for μ.
    (b) Compute a 90% confidence interval for π.
30. In a sample of 80 seniors graduating with a major in biology, a proportion $\bar{p} = .525$ get jobs related to the field. Does this result suggest that the proportion of all graduating biology majors who get such jobs is substantially below .6?
31. A college has long held the view that about 40% of its undergraduates represent the first generation of their family to attend college. In a sample of 80 present undergraduates, 24 report that they are the first generation to attend college. Do these data suggest that the corresponding proportion among all students is now below .4?
32. Draw a picture showing the P value in Example 9.

Solution to Problem 24

(a) Here π is the proportion of applications *in the past two months* listing Western Europe as their destination.

(b) $H_0: \pi = .5$; $H_1: \pi < .5$.

(c) The sample proportion is $\bar{p} = 41/100 = .41$.

Following the procedure in Table 3, we have

$$z = \frac{.41 - .5}{\sqrt{(.5)(1 - .5)/100}} = -1.8$$

and the corresponding P value is $P(Z < -1.8) = .0359$. We reject H_0 at the .05 level of significance. The data suggest that π is significantly less than .5.

SECTION 10.4

TESTING HYPOTHESES ABOUT THE DIFFERENCE BETWEEN TWO POPULATION MEANS AND PROPORTIONS

In Section 7.3 we presented the probability distribution of the difference between (1) two sample means $\bar{X}_1 - \bar{X}_2$ and (2) two sample proportions $\bar{P}_1 - \bar{P}_2$. In Chapter 9 we used these results to obtain confidence intervals for $\mu_1 - \mu_2$ and $\pi_1 - \pi_2$, respectively. In this chapter we use these results to test hypotheses about $\mu_1 - \mu_2$ and $\pi_1 - \pi_2$.

TESTING HYPOTHESES ABOUT $\mu_1 - \mu_2$

On many occasions, especially in experiments, the researcher wants to test whether one population mean is larger than another, given only the corresponding sample means $\bar{x}_1$ and $\bar{x}_2$. Using the null hypothesis, $H_0: \mu_1 - \mu_2 = 0$, provides a standard (i.e., zero) against which to compare the difference in sample means $(\bar{x}_1 - \bar{x}_2)$.

In Chapter 9 we used the fact that in two large independently selected samples the random variable

$$Z = \frac{\bar{X}_1 - \bar{X}_2 - (\mu_1 - \mu_2)}{\sqrt{s_1^2/n_1 + s_2^2/n_2}}$$

has approximately the $N[0, 1]$ distribution. We now consider how to use this result to test $H_0: \mu_1 - \mu_2 = 0$ against $H_1: \mu_1 - \mu_2 > 0$. The procedure is briefly as follows. The data will be consistent with H_1 if $\bar{x}_1 - \bar{x}_2 > 0$. If $H_0: \mu_1 - \mu_2 = 0$ is true, the value for Z corresponding to the observed difference $\bar{x}_1 - \bar{x}_2$ will be

$$z = \frac{\bar{x}_1 - \bar{x}_2 - 0}{\sqrt{s_1^2/n_1 + s_2^2/n_2}}.$$

This is the test statistic. Here z is the number of standard deviations that $\bar{x}_1 - \bar{x}_2$ is from 0 (the null hypothesis value for $\mu_1 - \mu_2$).

We illustrate this procedure with an example and then set out the general rules.

EXAMPLE 11 A high priority of the new school committee in a town is a reduction in school absenteeism. They have approved a class award program based on attendance figures, and after a year they wish to investigate whether absenteeism has declined. A random sample of 60 students is selected for the year prior to initiation of the board's program, and the number of days of school missed in the year is recorded for each student. A similar random sample of 60 students' records is tabulated one year after the new program is instituted. Let us call the mean number of days lost over all students in the years before and after the program μ_1 and μ_2, respectively. (a) What hypotheses are appropriate here? (b) What conclusion is appropriate if $\bar{x}_1 = 17.32$ days and $s_1 = 6.92$ days, and if $\bar{x}_2 = 13.72$ days and $s_2 = 8.02$ days. (c) Viewing this study as a quasi-experiment, what confounding variables would concern you?

Solution (a) We should test $H_0: \mu_1 - \mu_2 = 0$ (or equivalently, $\mu_1 = \mu_2$) against $H_1: \mu_1 - \mu_2 > 0$ (or equivalently, $\mu_1 > \mu_2$).

(b) Since $\bar{x}_1 - \bar{x}_2 = 17.32 - 13.72 = 3.60$, which is >0, the data are consistent with $H_1: \mu_1 - \mu_2 > 0$.

The value of Z corresponding to $\bar{x}_1 - \bar{x}_2 = 3.60$ is

$$z = \frac{\bar{x}_1 - \bar{x}_2}{\sqrt{s_1^2/n_1 + s_2^2/n_2}} = \frac{3.60}{\sqrt{6.92^2/60 + 8.02^2/60}} = \frac{3.6}{1.3675} = 2.63$$

The P value is then $P(Z > 2.63) = .0043$. Since this value is less than .01, we reject H_0 at the .01 level of significance. The data strongly suggest that the mean number of days lost per student has declined significantly.

The P value $= .0043$ is shown in Figure 9 as the small area under the $N[0, 1]$ distribution to the right of 2.63.

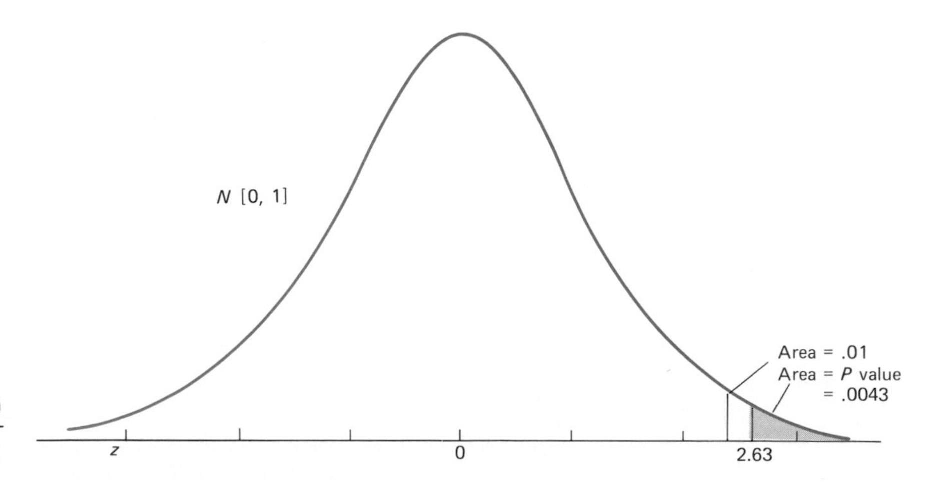

FIGURE 9

P value for Example 11

(c) In this quasi-experiment a variable (1) whose distribution differs in the two years and (2) which affects absenteeism will be a confounding variable. Such a variable may be in part responsible for the difference in sample means, thus rendering the test result meaningless. Examples of such variables might include a flu epidemic in the first year but not in the second, or the daily temperature over two years (it might have been colder on average in the first year than in the second).

■

When we test whether or not one population mean is significantly larger than another, we need only consider alternative hypotheses of the form $H_1: \mu_1 - \mu_2 > 0$ (and not $H_1: \mu_1 - \mu_2 < 0$). The reason is that we can always designate the mean that we suspect may be larger as μ_1 and the other mean as μ_2. If the sample mean $\bar{x}_1$ then turns out to be smaller than $\bar{x}_2$, this will be inconsistent with the hypothesis $H_1: \mu_1 - \mu_2 > 0$ and we will not reject H_0. There is no need to go any further. Before illustrating these points, we have outlined the general procedure for testing $H_0: \mu_1 - \mu_2 = 0$ against $H_1: \mu_1 - \mu_2 > 0$ in Table 4.

TABLE 4

Procedure for testing $H_0: \mu_1 - \mu_2 = 0$ in large, random, and independent samples

STEP	$H_1: \mu_1 - \mu_2 > 0$ (μ_1 ASSUMED LARGER)
1. Do not reject H_0 if $\bar{x}_1 - \bar{x}_2$ is inconsistent with H_1, i.e., if:	$\bar{x}_1 - \bar{x}_2 < 0$
Perform the test if:	$\bar{x}_1 - \bar{x}_2 > 0$
2. Compute:	$z = \dfrac{\bar{x}_1 - \bar{x}_2}{\sqrt{s_1^2/n_1 + s_2^2/n_2}}$
3. The P value is:	$P(Z > z)$
4. Using $\alpha = .05$ or $.01$, reject H_0 if:	P value $\leq \alpha$

Assumptions: Two large (n_1 and $n_2 \geq 30$), random, and independently selected samples

Another example will further clarify this two-sample test.

EXAMPLE 12 Do boys watch more TV than girls? Random samples of 75 boys and 70 girls were selected from the elementary grades in a large school district, and the number of hours each child spent watching TV during a specific week was recorded. The results were (in hours) as follows:

	n	$\bar{x}$	s
BOYS	75	20.21	6.90
GIRLS	70	18.85	5.40

State (a) the null and alternative hypotheses; (b) your conclusions; (c) your conclusions had $\bar{x}$ for the boys been 18.75 and $\bar{x}$ for the girls 20.5.

Solution In view of the question at the beginning of the example, we use the subscript 1 for boys and 2 for girls. Thus μ_1 is the (unknown) mean time spent watching TV for all elementary-grade boys in the district in that week. We define μ_2 similarly for girls.

(a) The hypotheses are $H_0: \mu_1 - \mu_2 = 0$ and $H_1: \mu_1 - \mu_2 > 0$.

(b) We follow the steps in Table 4:

1. The sample difference $\bar{x}_1 - \bar{x}_2 = 20.21 - 18.85 = 1.36$ is > 0 and is therefore consistent with H_1.
2. In this case,

$$z = \frac{1.36}{\sqrt{6.9^2/75 + 5.4^2/70}} = \frac{1.36}{1.0254} = 1.33$$

3. The P value is $P(Z > 1.33) = .0918$.
4. Since .0918 exceeds .05, we do not reject H_0 at the .05 level of significance. The data suggest that the mean time that boys watch TV is *not* significantly greater than that for girls. Although the sample mean for the boys exceeds that for the girls, the difference is not statistically significant.

The P value is shown as the right tail area in Figure 10.

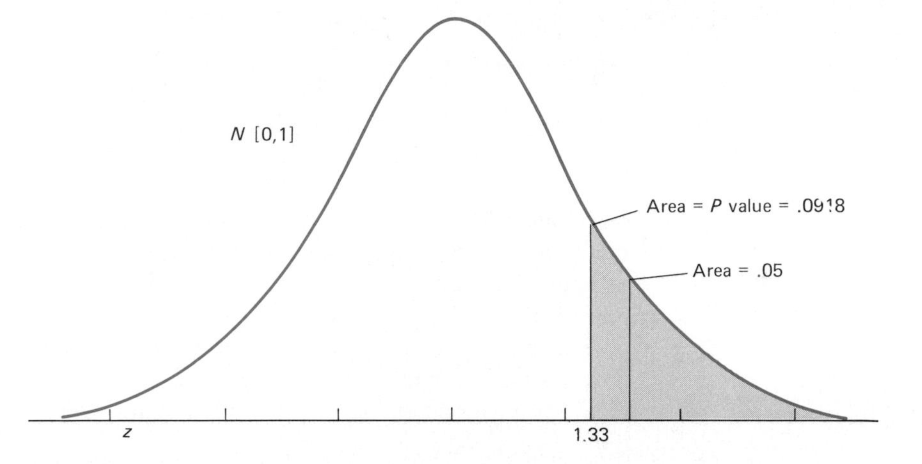

FIGURE 10

P value for Example 12

(c) Had $\bar{x}$ for the boys been 18.75 and $\bar{x}$ for the girls 20.5, we would still have labeled the boys 1 and the girls 2 since, before selecting the sample, we believed that the boys' time was larger. In this case, however, since $\bar{x}_1 - \bar{x}_2 < 0$ $(18.75 - 20.5 = -1.15)$, the sample data would be inconsistent with $H_1: \mu_1 - \mu_2 > 0$ and we would not reject H_0.

■

We are generally interested in testing the null hypothesis referred to in Table 4 (i.e., $H_0: \mu_1 - \mu_2 = 0$), but occasionally we wish to use a null hypothesis value for $\mu_1 - \mu_2$ other than zero. For instance, in Example 12 we might want to test the claim that on average boys watch TV at least one hour more per week than girls. In this case the hypotheses are $H_0: \mu_1 - \mu_2 = 1$ and $H_1: \mu_1 - \mu_2 > 1$. Using the data in Example 12(b), we compute

$$z = \frac{\bar{x}_1 - \bar{x}_2 - 1}{\sqrt{6.9^2/75 + 5.4^2/70}} = \frac{1.36 - 1}{1.0254} = .35$$

and the P value is $P(Z > .35) = .3632$. Since $.3632 > .05$ we would not reject H_0 at the .05 level of significance.

TESTING HYPOTHESES ABOUT $\pi_1 - \pi_2$

We turn now to testing hypotheses about $\pi_1 - \pi_2$ where π_1 and π_2 are the unknown proportion of successes in each of two populations. Samples of size n_1 and n_2 (n_1 and $n_2 \geq 30$), respectively, are to be selected from these populations. If $\bar{P}_1$ and $\bar{P}_2$ are the random variables, the respective proportion of successes in the two samples, then from Chapter 7,

$$Z = \frac{\bar{P}_1 - \bar{P}_2 - (\pi_1 - \pi_2)}{\text{SD}(\bar{P}_1 - \bar{P}_2)} = \frac{\bar{P}_1 - \bar{P}_2 - (\pi_1 - \pi_2)}{\sqrt{\pi_1(1 - \pi_1)/n_1 + \pi_2(1 - \pi_2)/n_2}}$$

is $N[0, 1]$. The *observed* proportions of successes are $\bar{p}_1 = y_1/n_1$ and $\bar{p}_2 = y_2/n_2$, respectively.

We shall test $H_0: \pi_1 - \pi_2 = 0$ against $H_1: \pi_1 - \pi_2 > 0$. *If* H_0 is true, π_1 and π_2 are equal (though unknown). In this case it will be helpful to call their common value π, so that $\pi_1 = \pi_2 = \pi$. In this case $\text{SD}(\bar{P}_1 - \bar{P}_2)$ can be written as

$$\text{SD}(\bar{P}_1 - \bar{P}_2) = \sqrt{\frac{\pi(1 - \pi)}{n_1} + \frac{\pi(1 - \pi)}{n_2}} = \sqrt{\pi(1 - \pi)\left(\frac{1}{n_1} + \frac{1}{n_2}\right)}$$

Since $\bar{p}_1 = y_1/n_1$ is a point estimate for π_1 and $\bar{p}_2 = y_2/n_2$, a point estimate for π_2 both of these sample proportions will be point estimates for π. To estimate π it therefore makes sense to combine the two sample proportions into one. The combined sample size would then be $n_1 + n_2$, of which a total of $y_1 + y_2$ are successes. Using the subscript c for the word "combine," we can now use the combined sample proportion, $\bar{p}_c = (y_1 + y_2)/(n_1 + n_2)$ as an estimate for π. Replacing π by $\bar{p}_c$ in this last formula, we obtain

$$\text{SD}(\bar{P}_1 - \bar{P}_2) = \sqrt{\bar{P}_c(1 - \bar{P}_c)\left(\frac{1}{n_1} + \frac{1}{n_2}\right)}$$

The test statistic is thus

$$z = \frac{\bar{p}_1 - \bar{p}_2}{\sqrt{\bar{p}_c(1 - \bar{p}_c)\left(\frac{1}{n_1} + \frac{1}{n_2}\right)}}$$

In the next example we illustrate how z is computed and the test performed.

EXAMPLE 13 A drug company conducts a test comparing a new experimental insecticide with their standard brand. Under controlled conditions the standard brand kills 425 of 500 mosquitos within 1 minute, whereas the experimental spray kills 459 of 500 mosquitos in the same time. Does this suggest that the experimental insecticide is significantly more effective than the standard brand?

Solution Call π_1 the proportion of all mosquitos that would be killed within a minute with the experimental spray. Define π_2 similarly for the standard spray. The null and alternative hypotheses are

$H_0: \pi_1 - \pi_2 = 0$ there is no difference in the death rates for the two sprays

$H_1: \pi_1 - \pi_2 > 0$ the experimental spray has a higher death rate

In this case $n_1 = n_2 = 500$, $y_1 = 459$, and $y_2 = 425$. Accordingly,

(i) $\overline{p}_1 = y_1/n_1 = 459/500 = .918$
(ii) $\overline{p}_2 = y_2/n_2 = 425/500 = .850$
(iii) $\overline{p}_c = (y_1 + y_2)/(n_1 + n_2) = (459 + 425)/(500 + 500) = .884$

and finally,

$$\sqrt{\overline{p}_c(1 - \overline{p}_c)\left(\frac{1}{n_1} + \frac{1}{n_2}\right)} = \sqrt{(.884)(1 - .884)\left(\frac{1}{500} + \frac{1}{500}\right)} = .0203$$

Since the difference in sample proportions, $\overline{p}_1 - \overline{p}_2 = .918 - .850 = .068 > 0$, the data are consistent with H_1. In this case

$$z = \frac{\overline{p}_1 - \overline{p}_2}{\sqrt{\overline{p}_c(1 - \overline{p}_c)(1/n_1 + 1/n_2)}} = \frac{.068}{.0203} = 3.35$$

and the P value is $P(Z > 3.35) = .0004$. This P value is much less than .01, so we reject H_0 at the .01 level of significance. The data strongly suggest that the experimental spray is significantly more effective than the standard spray. ■

In Example 13 we used (i), (ii), and (iii) to emphasize that we need to compute three sample proportions, $\overline{p}_1$, $\overline{p}_2$, *and* the combined sample proportion $\overline{p}_c$. Note that $\overline{p}_c = (y_1 + y_2)/(n_1 + n_2)$ is equal to $(\overline{p}_1 + \overline{p}_2)/2$ *only* if $n_1 = n_2$ (see Problem 42).

As with $\mu_1 - \mu_2$, when we test hypotheses about the difference $\pi_1 - \pi_2$, we need only consider the alternative hypothesis of the form $H_1: \pi_1 - \pi_2 > 0$, choosing subscripts as before. For instance, in Example 13 we used the subscript 1 for the experimental spray because it was thought that the associated death rate would be higher than that for the standard spray. This enables us to always use the one-sided alternative $H_1: \pi_1 - \pi_2 > 0$.

The procedure for testing $H_0: \pi_1 - \pi_2 = 0$ is summarized in Table 5.

TABLE 5

Procedure for testing $H_0: \pi_1 - \pi_2 = 0$ in large, random, and independent samples

STEP	$H_1: \pi_1 - \pi_2 > 0$ (π_1 ASSUMED LARGER)
1. Do not reject H_0 if $\bar{p}_1 - \bar{p}_2$ is inconsistent with H_1, i.e., if:	$\bar{p}_1 - \bar{p}_2 < 0$
Perform the test if:	$\bar{p}_1 - \bar{p}_2 > 0$
2. Compute:	$\bar{p}_c = \dfrac{y_1 + y_2}{n_1 + n_2}$ and $z = \dfrac{\bar{p}_1 - \bar{p}_2}{\sqrt{\bar{p}_c(1 - \bar{p}_c)\left(\frac{1}{n_1} + \frac{1}{n_2}\right)}}$
3. The P value is:	$P(Z > z)$
4. Using $\alpha = .05$ or $.01$, reject H_0 if:	P value $\leq \alpha$

Assumptions: Two large (n_1 and $n_2 \geq 30$), random, and independently selected samples

Now let us apply this procedure to another problem.

EXAMPLE 14 In 1974, 40 families in a random sample of 100 (in a certain town) reported owning their own homes. Ten years later a survey was conducted to find whether there had been a significant increase in the proportion of families in the same town owning their own homes. A random sample of 200 families was selected, and 88 of them owned their homes. What conclusion can be drawn from these data?

Solution Call π_1 and π_2 the (unknown) proportions of all families owning their homes in 1984 and 1974, respectively. The two hypotheses are $H_0: \pi_1 - \pi_2 = 0$ and $H_1: \pi_1 - \pi_2 > 0$.

In this case $n_1 = 200$, $y_1 = 88$, $n_2 = 100$, and $y_2 = 40$. As a result, $\bar{p}_1 = 88/200 = .44$ and $\bar{p}_2 = 40/100 = .4$. We follow the procedure in Table 5:

1. In this case $\bar{p}_1 - \bar{p}_2 = .44 - .40 = .04$, which is > 0 and therefore consistent with H_1. Thus we compute

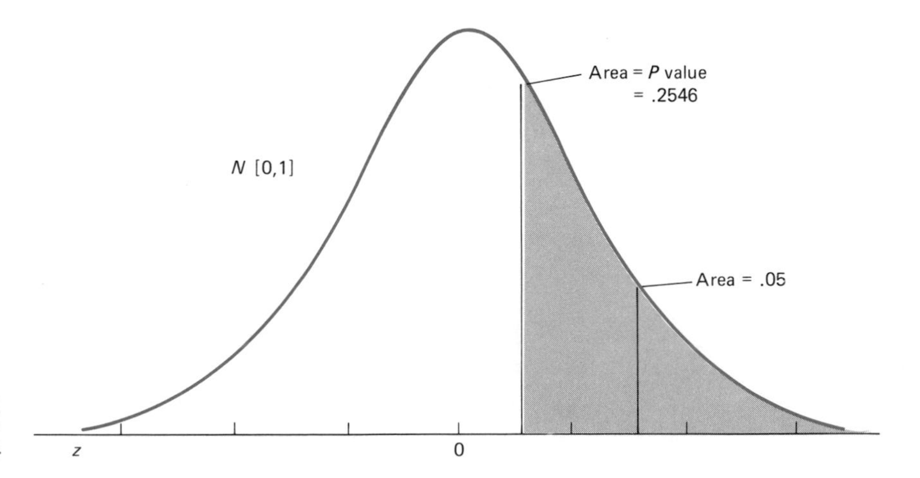

FIGURE 11

P value for Example 14

2. $\bar{p}_c = (88 + 40)/(200 + 100) = 128/300 = .4267$, and

$$\sqrt{\bar{p}_c(1 - \bar{p}_c)\left(\frac{1}{n_1} + \frac{1}{n_2}\right)} = \sqrt{(.4267)(.5733)\left(\frac{1}{200} + \frac{1}{100}\right)} = .0606$$

Thus: $z = .04/.0606 = .66$ and

3. The P value $= (Z > .66) = .2546$.
4. Since the P value is larger than .05, we cannot reject H_0 at the .05 level of significance. Although on the surface there appeared to be an increase in the proportion of families owning their homes, the data suggest that the increase was not a significant one. The P value, .2546, is shown in Figure 11.

■

PROBLEMS 10.4

33. In a recent study separate random samples of 100 young and 100 middle-aged persons were asked various questions relating to smoking. Only 35 of the young persons smoked, while 43 of the middle-aged group smoked. Do the data suggest that a significantly lower proportion of younger people smoke than middle-aged persons?
34. Are women more physically dexterous than men? Random samples of 120 men and 120 women participate in a test to examine this question (among others). Sixty-seven men and 75 women pass the test.
    (a) Define π_1 and π_2 in this case.
    (b) State H_0 and H_1.
    (c) Perform the appropriate test and state your conclusion.
35. In the study referred to in Problem 34 the average score for the women was 153.6 with a standard deviation of 20.9. The corresponding statistics for the men were 134.3 and 18.7.
    (a) Define μ_1 and μ_2 in this case.
    (b) Do these data suggest that on the average women score significantly higher than men on this test?
36. Two species of beetle appear to differ only in their overall length, species A appearing to be on average larger than B. In a random sample of 36 members of species A, the average length is 2.19 centimeters and the standard deviation .17 centimeter. In a random sample of 42 members of species B, the corresponding figures are 2.14 centimeters and .22 centimeter.
    (a) Define μ_1 and μ_2 in this case.
    (b) State H_0 and H_1.
    (c) Carry out the appropriate test and state your conclusion.
37. Do doctors in general practice have longer work weeks than those of doctors who specialize? To investigate this and other questions a random sample of 80 general practitioners and 80 specialists are asked to record their work hours for a specific week. The results, in hours, are as follows:

	n	$\bar{x}$	s
GENERAL PRACTITIONERS	80	55.9	5.4
SPECIALISTS	80	53.7	4.

(a) Define μ_1 and μ_2 in this case.
(b) State H_0 and H_1.
(c) Carry out the appropriate test and state your conclusion.

38. Random samples of 150 cars were selected prior to and after a campaign to encourage car-pooling among commuters. In the precampaign sample, 70 of the cars contained only the driver. In the postcampaign sample, the corresponding number was 52.
(a) Define π_1 and π_2 in this case.
(b) State H_0 and H_1.
(c) Carry out the appropriate test and state your conclusion.

39. Refer to the samples in Problem 38. The data on the number of occupants per car in both samples are as follows:

	n	$\overline{x}$	s
PRECAMPAIGN	150	1.88	.89
POSTCAMPAIGN	150	2.19	1.04

Do these data suggest a significant increase in the number of occupants per car after the campaign?

40. Are professional baseball players older on average than professional hockey players? Suppose that in random samples of 40 major league players from both sports, the following data were obtained:

	BASEBALL	HOCKEY
MEAN AGE	28.6	27.0
STANDARD DEVIATION OF AGE	6.3	4.8

What conclusion can be drawn from these data?

41. A large corporation is initiating a program of subsidized bus passes for its employees. Two months prior to adopting the program a random sample of 75 employees was selected. Of the 75, 24 used the bus service. In an independent random sample of 75 employees taken two months after the program was initiated, 34 reported using the bus service.
(a) Do these data suggest that the proportion of all employees using the bus service has increased significantly?
(b) If it has, is it necessarily true that the pass program was responsible?

42. Show that if the two sample sizes n_1 and n_2 are equal, then $\overline{p}_c$, the combined proportion of successes, is simply $(\overline{p}_1 + \overline{p}_2)/2$.

Solution to Problem 33

Call π_1 and π_2 the proportions of young and middle-aged persons, respectively, who smoke. We wish to test $H_0: \pi_1 - \pi_2 = 0$ against $H_1: \pi_1 - \pi_2 > 0$.

Following the procedure in Table 5, the difference in sample proportions is $\overline{p}_1 - \overline{p}_2 = .43 - .35 = .08$, which is greater than 0 and therefore consistent with H_1. The combined proportion is $\overline{p}_c = (43 + 35)/(100 + 100) = .39$, so

$$\sqrt{\overline{p}_c(1 - \overline{p}_c)\left(\frac{1}{n_1} + \frac{1}{n_2}\right)} = \sqrt{(.39)(.61)\left(\frac{1}{100} + \frac{1}{100}\right)} = .069$$

and $z = .08/.069 = 1.16$. The P value $= P(Z > 1.16) = .123$, which exceeds .05, so we cannot reject H_0 at the .05 level of significance. The data suggest no significant difference for the two age groups as to the proportion who smoke.

SECTION 10.5 TWO-SIDED TESTS

In many hypothesis-testing problems the researcher is not able or willing to specify that either μ or π may be greater than or less than the null hypothesis value, μ_0 or π_0, or that a difference between two means or proportions is $>$ or $<$ 0. For example, a sociologist may have no idea whether boys watch more television than girls, or vice versa. In this case, the sociologist would express the alternative hypothesis as $H_1: \mu_1 - \mu_2 \neq 0$. Alternative hypotheses of this form might also include $H_1: \pi_1 - \pi_2 \neq 0$, $H_1: \mu \neq 100$, or $H_1: \pi \neq .7$. Tests involving such alternative hypotheses are called *two-sided* or *two-tailed tests*, as contrasted with those involving $H_1: \mu > \mu_0$, $H_1: \pi < \pi_0$, and so on, which are called *one-sided* or *one-tail tests*.

The procedure for conducting two-sided tests is similar to that used for one-sided tests, differing only in the following two respects:

1. The data are *always* consistent with H_1. As a result, we always compute the P value.
2. The words "in either direction" are added to the definition of the P value for one-sided tests (Definition 1) so that it now reads:

> **DEFINITION 3**
>
> The P value for a two-sided test is the probability, *if H_0 is true*, of getting a sample result at least as extreme in either direction as the one obtained.

The importance of these differences will be seen in the following example.

EXAMPLE 15 The mean salary for workers in an industry is believed to be \$19,800. The personnel officer of a large corporation in the industry wants to find out if μ, the mean salary for all workers in the corporation, is significantly different from the industry figure, but is unwilling to speculate whether it is greater or less than \$19,800. The answer can be found by using a two-sided test in which the null and alternative hypotheses are $H_0: \mu = 19{,}800$ and $H_1: \mu \neq 19{,}800$.

A random sample of 100 employees is selected and the mean salary $\bar{x} = \$20,362$ and the standard deviation $s = \$4000.20$ computed. From Section 10.1 the test statistic is

$$z = \frac{\bar{x} - \mu_0}{s/\sqrt{n}} = \frac{20,362 - 19,800}{4000.2/\sqrt{100}} = 1.40$$

Had H_1 been $H_1: \mu > 19,800$, we would have computed the P value as $P(Z > 1.40)$. However, Definition 3 suggests that we should compute the probability of getting a sample mean 1.40 standard deviations from μ_0 *in either direction*. Because the $N[0, 1]$ distribution is symmetric around 0, we compute the P value for this two-sided test as

$$\begin{aligned} P \text{ value} &= P(Z > 1.40) + P(Z < -1.40) \\ &= 2P(Z > 1.40) \\ &= 2(.0808) = .1616 \end{aligned}$$

This P value is shown in Figure 12.

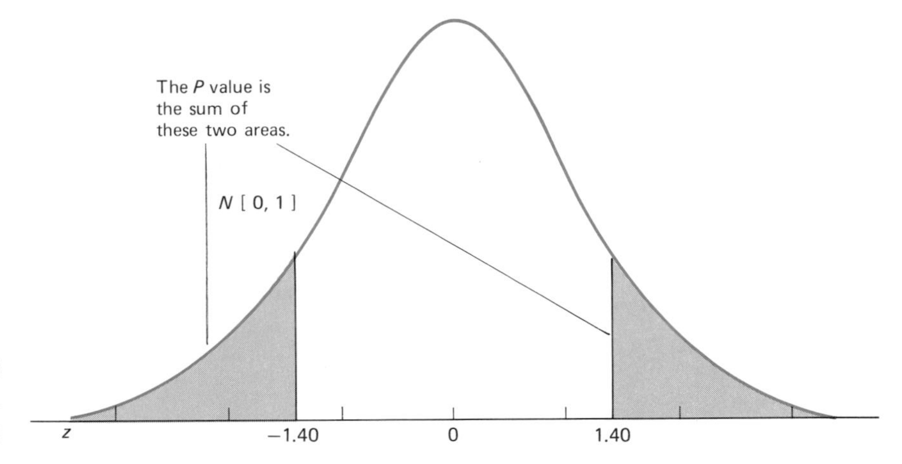

FIGURE 12

Two-sided P value for Example 15

As with a one-sided test, we compare the P value with .05 and if necessary with .01. In this case since the P value exceeds .05, we do not reject H_0 at the .05 level of significance. The data suggest that the mean salary for workers in the corporation is not significantly different from \$19,800.

■

Notice that in this last example the P value was simply twice the P value for $H_1: \mu > 19,800$. This suggests that to obtain the P value for a two-sided test, we should double the P value that would have been obtained for the appropriate one-sided test, and in fact, this is the procedure that is generally followed.

The details associated with this procedure are spelled out in Table 6.

TABLE 6

Procedure for performing a large-sample two-sided test of $H_0: \mu = \mu_0$, $H_0: \pi = \pi_0$, $H_0: \mu_1 - \mu_2 = 0$, or $H_0: \pi_1 - \pi_2 = 0$

1. Compute the appropriate test statistic (z) as indicated in Table 2, 3, 4, or 5.
2. (a) If $z > 0$, the P value for the two-sided test is $2P(Z > z)$.
   (b) If $z < 0$, the P value for the two-sided test is $2P(Z < z)$.
3. Using $\alpha = .05$ or $.01$, reject H_0 if the P value $\leq \alpha$.

Suppose, for instance, that in Example 15, $\bar{x}$ were \$18,707 instead of \$20,362. We follow the steps in Table 6:

1. The corresponding value for z is

$$z = \frac{18{,}707 - 19{,}800}{4000.2/\sqrt{100}} = -2.73$$

2(b). Since this is less than zero, the P value is $2P(Z < -2.73) = 2(.0032) = .0064$.

3. Since this is less than .01, we reject H_0 at the .01 level of significance. The data suggest that the mean salary for the corporation is significantly less than \$19,800.

EXAMPLE 16 We wish to check whether a particular coin is biased. There is no reason to suspect a bias in favor of heads or tails, so we adopt the null and alternative hypotheses

$$H_0: \pi = .5 \quad \text{and} \quad H_1: \pi \neq .5$$

where π is the probability of throwing a head at each toss. What decision should be made if when the coin is tossed 100 times, 55 heads are obtained?

Solution We regard the 100 tosses as a sample from the population of all possible tosses of this coin. The proportion of heads in the sample is $\bar{p} = 55/100 = .55$.

Following the procedure in Table 6:

1. From Table 3 the value for the test statistic is

$$z = \frac{\bar{p} - \pi_0}{\sqrt{\pi_0(1 - \pi_0)/n}} = \frac{.55 - .5}{\sqrt{(.5)(1 - .5)/100}} = 1.0$$

2(a). Since this value exceeds 0, the P value is

$$2P(Z > 1.0) = 2(.1587) = .3174$$

3. This value exceeds .05, so we do not reject H_0 at the .05 level of significance. The data suggest that the probability of a head is not significantly different from .5 and therefore that the coin is not biased.

■

EXAMPLE 17 Five years ago 407 of a sample of 1000 voters in a certain town opposed handgun legislation. Last year 575 voters in a sample of 1500 voters in the town opposed the legislation. Do these data suggest any significant change in the proportion of all voters opposed to this legislation?

Solution Designate the proportion of all voters who opposed the legislation five years ago and last year as π_1 and π_2, respectively. The hypotheses to be tested are:

$$H_0: \pi_1 - \pi_2 = 0 \text{ (or } \pi_1 = \pi_2)$$
$$H_1: \pi_1 - \pi_2 \neq 0 \text{ (or } \pi_1 \neq \pi_2)$$

From Table 5 in Section 10.4, we compute the sample proportions (i) $\bar{p}_1 = 407/1000 = .407$; (ii) $\bar{p}_2 = 575/1500 = .383$, and (iii) the combined proportion $\bar{p}_c = (407 + 575)/(1000 + 1500) = 982/2500 = .393$; and finally

$$\sqrt{\bar{p}_c(1 - \bar{p}_c)\left(\frac{1}{n_1} + \frac{1}{n_2}\right)} = \sqrt{(.393)(1 - .393)\left(\frac{1}{1000} + \frac{1}{1500}\right)} = .020$$

The difference in sample proportions is $\bar{p}_1 - \bar{p}_2 = .407 - .383 = .024$, so the test statistic is $z = .024/.02 = 1.2$ and the P value is

$$2P(Z > 1.2) = 2(.1151) = .2302$$

Since the P value exceeds .05, we cannot reject H_0 at the .05 level of significance. Although there has been a decline in the sample proportion opposing the legislation, the difference is not statistically significant.

Note that had the alternative hypothesis been $H_1: \pi_1 - \pi_2 > 0$ (instead of $\pi_1 - \pi_2 \neq 0$), we would still have not rejected H_0 at the .05 level of significance because for the one-sided test the P value would have been .1151, which also exceeds .05.

The P value, .2302, is shown in Figure 13 as the total of the two shaded areas under the $N[0, 1]$ distribution.

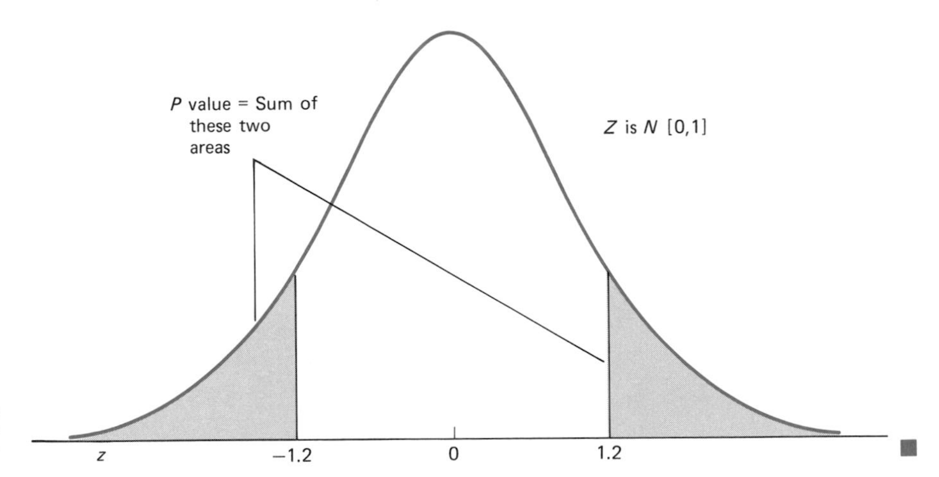

FIGURE 13

P value for Example 17

Example 17 illustrates a point that we have mentioned but not stressed before. *Hypotheses should be formulated before seeing the data*. It is not appropriate to formulate or modify hypotheses *after* seeing the data. To do this would bias the test in the direction indicated by the data. In example 17 we assumed that prior to seeing the results of both samples, the researcher was unwilling to specify whether the change in the proportion opposing the legislation (i.e., $\pi_1 - \pi_2$) would be positive or negative. Hence the alternative hypothesis $H_1: \pi_1 - \pi_2 \neq 0$. Having seen that $\bar{p}_1 - \bar{p}_2$ is positive, the researcher might be tempted to change H_1 to $H_1: \pi_1 - \pi_2 > 0$ but this modification would have biased the test in favor of rejecting H_0.

PROBLEMS 10.5

43. A die is to be rolled 600 times to test whether π, the probability of a 4, is significantly different from 1/6. What conclusion can be drawn if a 4 appears 76 times?
44. In a random sample of 38 weeks during the last ten years, the mean number of industrial accidents per week at a manufacturing plant was $\bar{x} = 2.7$ with a standard deviation of $s = 2.4$. Use these data to test whether the mean weekly accident rate at the plant over the past ten years is significantly different from 2.2.
45. Draw a picture showing the P value (= .3172) in Example 16.
46. A test has traditionally been taken only by male Air Force recruits. Only 70% of such recruits passed the test. The same test is now to be given on a trial basis to a random sample of 240 recent female recruits. Use the fact that 186 of the women passed the test to check whether the (potential) proportion of all female recruits passing the test is significantly different from .7.
47. In Problem 46, assume that the mean score for all males was 64.5. Among the 240 females taking the test, the mean and standard deviation of scores were $\bar{x} = 66.79$ and $s = 17.70$. Do these data suggest that the (potential) mean score for all females is significantly different from 64.5?
48. (a) Using the information in Problem 46, compute a 95% confidence interval for the proportion of all females who will pass the test.

    (b) Using the information in Problem 47, compute a 95% confidence interval for the mean score for all females taking the test.
49. The average length in a random sample of 32 of last year's baseball games is 2.37 hours with a standard deviation of $s = .43$ hour. Does this suggest that the mean length of all baseball games last year is significantly different from 2.5 hours? Would your conclusion change if you used $\alpha = .1$ (not .01)?
50. A random sample of 80 manufacturing companies who use steel in their products is selected. Twenty-six of the companies reported using imported steel. Does this suggest that the proportion of all such companies using imported steel is significantly different from .3?
51. Using the data in Problem 50, compute an 80% confidence interval for the proportion of all steel-using companies who use imported steel.
52. We wish to compare the mean number of visits to a library made by persons in populations A and B last year. A random sample of 100 persons from each population resulted in the following summary data:

	MEAN NUMBER OF VISITS	STANDARD DEVIATION
SAMPLE *A*	2.32	1.94
SAMPLE *B*	2.09	2.33

Do these data suggest any significant differences in mean number of visits between the two populations?

53. In the two samples referred to in Problem 52, 30 of the persons in sample A and 26 of those in sample B did not make any visits to the library. Do these data suggest any significant difference in the proportion of all persons in the two populations not making any visits to a library?

54. Do men and women differ in their likelihood of being color-blind? To investigate this question, a random sample of 1750 persons was selected. Nineteen of the 902 women were color-blind, as were 10 of the men. Do these data suggest any significant difference in the proportion of each sex who are color-blind?

55. Two friends, one living in London and the other in New York, mail each other 40 postcards at the same (local) time on the same day. The mean and standard deviation of the number of days each set of postcards took to reach their destination are as follows:

ROUTE	$\overline{x}$	s
London to New York	7.4	2.1
New York to London	8.3	2.1

Do these data suggest any significant difference in the mean time it takes mail to travel between the two cities?

Solution to Problem 43

We will test $H_0: \pi = 1/6 = .1667$ against $H_1: \pi \neq .1667$. The proportion of 4's in the sample is $\overline{p} = 76/600 = .1267$. Following the steps in Table 6, we compute the test statistic

$$z = \frac{\overline{p} - \pi_0}{\sqrt{\pi_0(1 - \pi_0)/n}} = \frac{.1267 - .1667}{\sqrt{(.1667)(.8333)/600}} = \frac{-.04}{.0152} = -2.63$$

Since this value is less than 0, the P value is

$$2P(Z < -2.63) = 2(.0043) = .0086$$

Since this value is less than .01, we reject H_0 at the .01 level of significance. The data suggest that for this die, the probability of getting a 4 is significantly *less* than 1/6.

SECTION 10.6

AN ANALOGY AND A DISTINCTION

The analogy between a statistical test and a criminal trial provides considerable insight into hypothesis testing. In a criminal trial the jury decides whether the defendant is innocent or guilty of the charges. In a statistical test the researcher—and the readers of the research report—are the jury. The null hypothesis (H_0) and the alternative hypothesis (H_1) correspond to the innocence and guilt of the defendant, respectively. Similarly, the researcher does not know which is true, H_0 or H_1, just as the jury does not know whether the defendant is innocent or guilty—otherwise, why have the trial?

The jury must decide the question of guilt or innocence on the basis of evidence presented to them in court. The researcher correspondingly must choose between H_0 and H_1 based on the evidence in the sample or samples.

In hypothesis testing we begin with the assumption that H_0 is correct and only reject that hypothesis—in favor of H_1—if the P value is small. This procedure corresponds to the convention, at least in Anglo-Saxon law, that the defendant is presumed innocent unless "proven" guilty beyond a reasonable doubt. The "reasonable doubt" in criminal trials refers, of course, to the degree of uncertainty that is acceptable before the assumption of innocence is rejected. In hypothesis testing the level of significance, α is "reasonable doubt." But "beyond a reasonable doubt" still leaves the possibility that the defendant might be found guilty when, in fact, innocent.

In hypothesis testing the error that is analogous to convicting an innocent person is rejecting the null hypothesis when, in fact, H_0 is true. We call such an error a **type I error.** The possibility of falsely convicting an innocent person cannot be quantified, but the corresponding probability of rejecting H_0 when it is true and thus committing a type I error can—as follows: We reject H_0 if the probability of so extreme a sample value, if H_0 *were* true (i.e., the P value), is $\leq \alpha$. Since the probability α determines when we will reject H_0 it is the probability of a type I error.

Any legal requirements which benefit the defendant, such as restricting the types of evidence that the prosecution may present, further reduces the possibility of convicting an innocent person wrongly. This corresponds somewhat to lowering the level of significance, α, in a statistical test.

While it is important to conduct criminal trials so that the possibility of wrongly convicting an innocent person is small, it is also important to avoid another kind of error—that of not convicting a guilty person. The problem is that as you *lower* the chance of creating the first mistake, you will often *raise* the chance of creating the second kind of mistake.

In hypothesis testing the error analogous to not convicting a guilty person is not rejecting H_0 when H_1 is true. This is called a **type II error.**

The next example illustrates how type I and type II errors may occur.

EXAMPLE 18 A farmer claims that on the average his crates contain 100 pounds of oranges. The wholesaler believes that this is too high a figure and as a test weighs a random sample of 35 crates, with the result that $\bar{x} = 98.77$ pounds and $s = 4.23$ pounds.

In this case we call μ the mean weight of all the crates shipped by the farmer and test $H_0: \mu = 100$ against $H_1: \mu < 100$. The value $\bar{x} = 98.77$ is consistent with H_1, and the Z value is

$$z = \frac{\bar{x} - \mu_0}{s/\sqrt{n}} = \frac{98.77 - 100}{4.23/\sqrt{35}} = -1.72$$

The corresponding P value is $P(Z < -1.72) = .0427$. Since this is less than .05, we reject H_0 at the .05 level of significance. The data suggest that μ is significantly below 100 pounds.

Now suppose that $H_0: \mu = 100$ *is* actually true but by chance the sample overrepresented lighter crates, so that the sample mean (98.77) was small enough to reject H_0. We would then have made a type I error (i.e., H_0 was incorrectly rejected). Of course, in reality we can never know whether a type I error has occurred. We can only increase or decrease its likelihood of occurrence by varying the value of α.

By contrast, suppose that $\bar{x} = 99.42$ and $s = 4.23$ pounds, so that the z value is $(99.42 - 100)/(4.23/\sqrt{35}) = -.81$ and the P value is $P(Z < -.81) = .209$.

This exceeds .05, so we do not reject H_0 at the .05 level of significance. The data suggest that μ is not significantly below 100 pounds. But if μ were, in fact only 98 pounds, we would have made a type II error in *not* rejecting H_0 when H_1 was true (i.e., μ was less than 100).

■

At this point we return to our analogy with a criminal trial. Such a trial would be viewed as seriously flawed if there were large probability that a guilty person would be found innocent. Yet such a trial could be held, and a verdict reached. Similarly, as Sections 1 through 5 demonstrate, it is quite simple to perform a statistical test without ever considering the probability of a type II error.

The two types of errors possible in a statistical test are summarized in Table 7.

TABLE 7

The two types of errors in hypothesis testing

ACTION	FACT	ERROR
Reject H_0 if:	H_0 true	Type I
Do not reject H_0 if:	H_0 false	Type II

Sections 10.1 to 10.5 have been concerned with the procedures for performing a test and have focused on α, the probability of a type I error. By contrast, in Chapter 11 we focus on methods for computing the probability of a type II error.

SECTION 10.7
SUMMARY AND REVIEW

TESTING HYPOTHESES ABOUT μ

When testing hypotheses about the unknown population mean, μ, we assume a value for μ and designate it μ_0. We then set up two hypotheses:

1. That μ_0 is the true mean. This is called the **null hypothesis** and is written as H_0: $\mu = \mu_0$.
2. That μ_0 is *not* the true mean. This is called the **alternative hypothesis** and can be one of three types: $H_0: \mu > \mu_0$ or $H_0: \mu < \mu_0$, which are referred to as one-sided alternatives, and $H_0: \mu \neq \mu_0$, which is two-sided.

Our decision as to which of H_0 and H_1 is the more plausible is based on the observed sample mean $\bar{x}$ in a large ($n \geq 30$) sample. For one-sided tests we first check whether $\bar{x}$ is on the appropriate side of μ_0. For example, for H_0: $\mu = 500$ and H_1: $\mu > 500$ if $\bar{x}$ were 493.6, there is no doubt that these data favor H_0 rather than H_1 and we simply do not reject H_0. If the value of $\bar{x}$ *is* on the appropriate side of μ_0, we compute the **test statistic** $z = (\bar{x} - \mu_0)/(s/\sqrt{n})$. The ***P* value,** $P(Z > z)$ or $P(Z < z)$, is the probability of getting a value for $\bar{x}$ at least z standard deviations from μ_0. Procedures for computing P values are given in Table A3.

A small P value means that there is little chance of getting a sample mean so extreme as the one obtained if $\mu = \mu_0$, and hence H_0 should be rejected in favor of H_1. The larger the P value, the more consistent the data are with H_0, and if the P value is sufficiently large, H_0 should *not* be rejected.

We speak of not rejecting H_0 rather than accepting H_0 because the latter suggests that we believe that μ is exactly μ_0. In fact, neither H_0 nor H_1 can be *proved* correct.

It should be noted that the researcher is often more interested in testing H_1 (Is the mean life time of these tires significantly less than 48,000 miles?) rather than H_0. However, since only H_0 provides a specific value against which to compare the observed sample value, we begin by assuming that H_0 is true.

THE LEVEL OF SIGNIFICANCE

Before seeing the data the decision must be made as to how small the P value need be before H_0 will be rejected. The value chosen is called the **level of significance** and designated α. We reject H_0 if the P value is $\leq \alpha$ (Definition 2). The smaller the value of α, the more unusual the sample mean has to be before H_0 is rejected.

In practice, the values $\alpha = .05$ and $\alpha = .01$ are the ones most widely used. With these two values the following decisions are possible.

P VALUE	DECISION
P value $> .05$	Do not reject H_0 at the .05 level of significance.
$.01 < P$ value $\leq .05$	Reject H_0 at the .05 level of significance.
P value $\leq .01$	Reject H_0 at the .01 level of significance.

Procedures for performing one-sided tests of $H_0: \mu = \mu_0$ are provided in Table 2.

When reporting the results of hypothesis tests, it is important to:

1. Specify the P value.
2. Specify the level of significance at which H_0 is or is not rejected.
3. Interpret the results in the appropriate context.

TESTS ABOUT THE PROPORTION OF SUCCESSES

We are frequently interested in using the observed *proportion* of successes in the sample, $\bar{p} = y/n$, to test a hypothesis about π, the proportion of successes in the population. The null hypothesis will be of the form H_0: $\pi = \pi_0$ and the one-sided alternative either $H_1: \pi > \pi_0$ or $H_1: \pi < \pi_0$. Detailed procedures for testing H_0 are given in Table 3. They are used, however, only if $n\pi_0$ and $n(1 - \pi_0)$ both exceed 5.

TESTING HYPOTHESES ABOUT THE DIFFERENCE BETWEEN TWO POPULATION MEANS AND PROPORTIONS

Hypotheses concerning the differences between two means or two proportions are stated as:

$$H_0: \mu_1 - \mu_2 = 0 \qquad \text{and} \qquad H_0: \pi_1 - \pi_2 = 0$$
$$H_1: \mu_1 - \mu_2 > 0 \qquad \qquad H_1: \pi_1 - \pi_2 > 0$$

These hypotheses are tested on the basis of the corresponding differences in sample means $(\bar{x}_1 - \bar{x}_2)$ and sample proportions $(\bar{p}_1 - \bar{p}_2)$ respectively. Note that there is no need to test $H_1: \mu_1 - \mu_2 < 0$ or $H_1: \pi_1 - \pi_2 < 0$ since we can always designate the mean or proportion we believe to be larger as μ_1 or π. Procedures for testing hypotheses about $\mu_1 - \mu_2$ are given in Table 4 and hypotheses about $\pi_1 - \pi_2$ in Table 5.

TWO-SIDED TESTS

When the researcher is unwilling to specify whether μ, π, $\mu_1 - \mu_2$, or $\pi_1 - \pi_2$ is greater than or less than the null-hypothesis value, a two-sided test is appropriate. In this case the alternative hypothesis will be of the form $H_1: \mu \neq \mu_0$, $H_1: \pi_1 - \pi_2 \neq 0$, and so on.

The P value for a two-sided test is the probability, if H_0 is true, of getting a sample result at least as extreme *in either direction* as the one obtained (Definition 3). The details of how this definition is used in practice are given in Table 6.

In Section 10.6 we discussed the analogy between hypothesis testing and criminal trials. In both cases two kinds of errors are possible. In hypothesis testing these errors are:

A type I error: rejecting H_0 when H_0 is true.
A type II error: *not* rejecting H_0 when H_1 is true.

The probability of a type I error is α.

REVIEW PROBLEMS

GENERAL

1. Explain in your own words (a) your understanding of the P value; (b) the rationale for rejecting H_0 if the P value $\leq \alpha$.
2. We wish to test $H_0: \mu = 500$ against $H_1: \mu < 500$, where the population standard deviation is $\sigma = 57$. The sample size is $n = 60$.
   (a) Compute the P value if the observed sample mean is (i) $\bar{x} = 463.3$; (ii) $\bar{x} = 497.3$. In each case state your conclusion.
   (b) What decision should you come to if $\bar{x} = 510.8$?

3. A random sample of size 100 is to be taken from a population with $\sigma = 5$ in order to test $H_0: \mu = 40$ against $H_1: \mu > 40$.
   (a) Compute the P value and state your conclusion if $\bar{x} = 40.3$.
   (b) How would your decision be affected if σ were .5 instead of 5?
4. We wish to test $H_0: \mu = 75$ against $H_1: \mu \neq 75$. Using a random sample of size $n = 45$, what conclusion should you come to if (a) $\bar{x} = 80.6$ and $s = 14.8$; (b) $\bar{x} = 80.6$ and $s = 29.6$; (c) $\bar{x} = 71.7$ and $s = 6.2$?
5. We wish to test $H_0: \pi = .5$ against $H_1: \pi < .5$ using a random sample of size $n = 50$. What conclusion should be reached if (a) $y = 21$; (b) $y = 16$; (c) $y = 27$?
6. Refer to the case study on page 326. (a) State H_0 and H_1. (b) What conclusion is appropriate if the standard deviation in the sample is $s = 70.0$ days?
7. Over the past five years the mean age at death in the United States has been 70.5 years. In a random sample of the biographies of 50 "famous persons" who died in this period, the mean age at death was 72.0 years with a standard deviation of $s = 6.3$ years. Do these data suggest that famous people on the average live longer than the rest of us?
8. We wish to compare the mean heights of two varieties of wheat, A and B. B is an experimental variety. A sample of 50 stalks of each variety is selected with the following results (in centimeters):

VARIETY	$\bar{x}$	s
A	94	16
B	100	15

   Do these data suggest that the mean height for B is significantly greater than that for A?
9. In a survey of 500 voters in Virginia, 235 thought the president was doing a good job. In a similar survey of 800 voters in Connecticut, 479 thought the same. Do these data suggest that the proportion of all voters favoring the president is different in the two states?
10. In a random sample of 50 spectators at a hockey game, the mean age was 36.4 years with a standard deviation of 12.4 years. Additionally, 38 of the 50 were males. The next night at a basketball game at the same arena the mean age in a random sample of 50 spectators was 32.5 with a standard deviation of 15.2 years and 27 of the spectators were males. Do these data suggest that (a) on the average hockey tends to attract older persons than basketball; (b) basketball attracts a significantly greater proportion of women than does hockey?
11. A tobacco manufacturer claims that the average amount of tar in his cigarettes is only 9.8 milligrams. A consumer group believes that this figure underestimates the true mean amount of tar. A random sample of n cigarettes of this brand are selected and the amount of tar per cigarette is determined. The sample mean and the standard deviation of the amount of tar are $\bar{x} = 9.905$ and $s = .543$ milligram.
    (a) Define μ in this case.
    (b) State H_0 and H_1.
    (c) What conclusion should be reached if the data above were based on (i) $n = 40$; (ii) $n = 80$; (iii) $n = 200$ cigarettes?
    (d) Explain how your conclusion depends on the sample size, n.

12. Refer to the data in Review Problem 9 of Chapter 9 (page 320). Use these data to test whether there has been a significant decline in the mean number of illegal left turns since the new signs were erected.
13. Refer to the data in Review Problem 10 of Chapter 9 (page 320). Do these data suggest that there is a significant difference between the proportion of psychology and sociology articles that use statistics?

HEALTH SCIENCES

14. It is hoped that the administration of a new drug will significantly reduce the average recovery time for a particular operation. Until now the mean recovery time has been 36 hours. The next 40 patients having this operation at a large teaching hospital are treated with the new drug.

    (a) State H_0 and H_1, defining μ.

    (b) Suppose that the standard deviation of recovery time among the 40 patients is 3.82 hours. Compute the P value and state your conclusion if (i) $\bar{x} = 36.09$ hours; (ii) $\bar{x} = 35.95$ hours; (iii) $\bar{x} = 33.95$ hours.
15. A study is concerned with the survival time, in months, for persons who have a cancer which has reached a critical stage of development. To date, with the standard treatment the mean length of survival is 18.8 months. A new treatment which it is hoped will prolong life is tested on 38 patients. Their mean survival time is $\bar{x} = 23.1$ months with a standard deviation of $s = 8.1$ months. Do these data suggest that the new treatment significantly increases the mean survival time?
16. In Review Problem 15 assume that with the standard treatment only 50% of patients survive beyond one year (after the critical stage). With the new treatment 28 of the 38 patients survive beyond one year. Do these data suggest that the new treatment significantly increases the proportion of all such patients surviving beyond one year?
17. In the past when elementary school children in a large city were screened, 8% were found to have auditory defects. In a recent study of 400 children in the city who had been screened within the past month, 35 were found to have such defects.

    (a) Do these data suggest a significant increase in the proportion of children having auditory defects?

    (b) How would your conclusion change if the number of children with auditory defects was not 35 but (i) 30; (ii) 44?
18. A study followed a group of 500 smokers and 500 nonsmokers until all had died. Of the smokers, 185 died of lung cancer. Of the nonsmokers, 28 died of lung cancer.

    (a) Do these data suggest that the probability of dying from lung cancer is significantly greater for smokers than for nonsmokers?

    (b) Why is a statistical test not really necessary in this case?
19. As a part of an experiment in a community counseling center, 80 new clients are assigned at random to be interviewed by either a student or a staff clinician. The variable of interest is the number of days between the initial visit and the follow-up visit. The resulting data are summarized as follows:

	INTERVIEWER	
	STUDENT	STAFF
n	40	40
$\bar{x}$	51.3	42.7
s	22.6	22.2

Do these data suggest that the mean interval between visits is significantly greater when the interviewer is a student than when the interviewer is a staff member?

20. Refer to Review Problem 15 of Chapter 9 (page 321). The PSRO standards indicate a mean stay in the hospital of 7.5 days for this operation. Use the data to test whether the actual mean stay in hospitals within the PSRO's jurisdiction is significantly greater than 7.5 days.
21. Use the data in Review Problem 17 of Chapter 9 (page 321) to test whether there is a significant difference in mean absolute visual threshold for students in rural and urban areas.
22. Use the data in Review Problem 19 of Chapter 9 (page 322) to test whether there is any significant difference in the proportion with blood type A for the two communities.

SOCIOLOGY AND GOVERNMENT

23. A social worker wants to demonstrate that welfare recipients in her region spend, on the average, more on heating than the $80 per month allowed by the government.

    (a) State the null and alternative hypothesis, defining μ.

    (b) A random sample of 50 welfare recipients spend an average of $88.91 per month with standard deviation $21.05. Compute the *P* value for these data and state your conclusion.
24. A complete analysis of all criminal records in a state for last year showed that the mean waiting time between the date of charge and date of trial was 142 days. Certain court reforms designed to lower waiting times were undertaken at the end of the year. A random sample of 200 cases were selected from the records of all persons charged *this* year. The mean waiting time in the sample was 134.7 days and the standard deviation was 38.4 days. Do these data suggest a significant decline in the mean waiting time for trial?
25. A review of last year's court records indicated that 30% of all criminal convictions were given to first-time offenders. In a random sample of 60 convictions for this year, only 12 were to first-time offenders. Do these data suggest that for this year, the proportion of all conviction that are given first offenders is significantly below .3?
26. Federal audits of a state's welfare caseload indicate that the mean number of errors per case over the past five years was 1.2. This year welfare workers are using a greatly simplified guide which it is hoped will significantly reduce the mean number of errors. A random sample of 90 cases is selected to investigate whether such a decline has occurred. The number of errors per case is summarized as follows:

NUMBER OF ERRORS	0	1	2	4
NUMBER OF CASES	42	30	16	2

    (a) State H_0 and H_1, defining μ.

    (b) Compute $\bar{x}$ and s.

    (c) Test H_0 and state your conclusion.
27. As part of a study of the sociability of young teenagers a group of seventy-five 13-year-old boys and sixty 13-year-old girls were each asked for their number of close friends. The results are summarized as follows:

	BOYS	GIRLS
n	75	60
$\bar{x}$	1.86	2.24
s	.62	.71

Do these data suggest that, on the average, girls have significantly more close friends than boys?

28. A pollster wishes to test the attitudes toward communist nations of Korean War veterans and Vietnam War veterans. A random sample of 100 Korean and 200 Vietnam War veterans are questioned, with the following results:

	TOUGH POLICY YES	TOUGH POLICY NO
KOREAN WAR VETERANS	65	35
VIETNAM WAR VETERANS	110	90

Do these data suggest that the proportion of Korean War veterans favoring a tough policy is significantly greater than that for Vietnam War veterans?

29. Use the data in Review Problems 23 and 24 of Chapter 9 (page 322) to test whether a greater proportion of college students than comparably aged noncollege people favor decriminalizing marijuana use.

ECONOMICS AND MANAGEMENT

30. In a random sample of 120 telephone company accounts, 35 were more than a month overdue. Does this suggest that significantly more than 25% of all accounts are more than a month overdue?

31. It has been suggested that wealthy Americans save an average of 12% of their income. Doubting whether the true figure is that high, an economist selects a random sample of 100 families with incomes greater than $50,000 a year. Among these families the mean percentage of income that was saved last year is 11.2 with a standard deviation of 3.5. Do these data suggest that the mean percentage of income that is saved in such families is significantly less than 12%?

32. A statistician at the Department of Labor is asked to investigate whether there has been any significant increase in the mean salary paid to CETA workers compared to last year's mean figure of $8530. (Most CETA workers are paid by the government while they learn a skill on the job.) The statistician selects a random sample of 100 such workers and computes the following data: $\bar{x} = \$8777$; $s = \$1320$.
    (a) What hypotheses are appropriate here? Define μ.
    (b) What conclusion should the statistician draw?
    (c) What conclusion should the statistician draw if $\bar{x} = \$8507$?

33. A department store will advertise in a local newspaper if the results for a random sample of 60 customers suggest that the newspaper is read by 40% of all the store's customers. In fact, 22 of the interviewed customers read the newspaper. Do these data suggest that significantly fewer than 40% of the customers read the newspaper?

34. It has been widely believed that until recently homeowners lived in their homes for an average of 8.4 years before selling them. A national realty company believes there has been a trend toward homeowners keeping their homes for a longer period. To test whether this is so, a random sample of 60 families that have recently sold

their homes is selected. The mean time that these families had lived in their homes was 8.94 years. The standard deviation was 3.49 years. What can be concluded from these data?

35. A consumer group believes that supermarkets in poor sections of a large city charge more than do those in other sections. The same "basket" of goods is bought in 40 supermarkets located in the poor section and in 40 supermarkets located in other sections. The results are as follows:

	POOR SECTION	OTHER SECTIONS
n	40	40
$\bar{x}$	\$60.98	\$59.09
s	\$4.17	\$4.33

What can be concluded from these data?

36. A company is planning a large advertising campaign in a city. Prior to the campaign a market research survey finds that of 200 homeowners randomly selected, only 42 had heard of the company. In a random sample of 200 homeowners selected three months after the end of the campaign, 75 reported having heard of the company. Do these data suggest that the proportion of home owners who are aware of the company's name has increased?

    The company was hoping to increase the proportion of homeowners knowing of the company by at least .1. Can they conclude that the increase in the proportion is significantly greater than .1?

37. An automobile manufacturer claims that its cars (model A) require fewer warranty-covered repairs during the first two years than do those of its competitor (model B). An automobile club tests this claim by asking 50 randomly selected recent purchasers of each model to record the number of warranty-covered repairs made during the first two years. The summary figures are:

	MODEL A	MODEL B
n	50	50
$\bar{x}$	2.95	3.11
s	.92	1.14

Do these data suggest that the claim of the manufacturer of model A is correct?

38. Use the data in Review Problem 33 of Chapter 9 (page 323) to test whether there has been a significant decline in the proportion of company presidents favoring the U.S. president's monetary policy over the past two years.

39. Use the data in Review Problem 34 of Chapter 9 (page 323) to test whether there has been a significant change in the mean number of new employees per company over the past two years.

EDUCATION AND PSYCHOLOGY

40. Fourth-grade reading scores in a city had averaged 107 points on a certain test. A new experimental program designed to improve the reading skills of fourth graders is adopted for one year. To test the effectiveness of the program, at the end of the year a sample of 60 fourth graders is randomly selected and their reading scores obtained.

(a) Define μ in this case.

(b) What are H_0 and H_1?

(c) What conclusion should be reached if $\bar{x} = 118.7$ and $s = 36.4$?

41. In a random sample of 40 private colleges, 14 offer a Master of Arts in Education degree. Do these data suggest that significantly less than 45% of all private colleges offer this degree?

42. All the fourth-grade students in a city have taken the same mathematics courses and are required to take a national test of math skills. A random sample of 100 fourth-grade boys and 100 fourth-grade girls is selected and their scores on this test recorded. The results are summarized as follows:

	GIRLS	BOYS
NUMBER	100	100
MEAN SCORE	78.4	74.3
STANDARD DEVIATION OF SCORE	27.4	29.8
NUMBER SCORING ABOVE THE 80TH PERCENTILE (FOR THE NATION)	22	25

(a) Do these data suggest any significant difference in the proportion of both sexes who score above the 80th percentile?

(b) Do these data suggest any significant difference in the mean scores for the two sexes?

43. A self-assessment test was given to 100 college students in the United States and Greece. The mean and standard deviation of the scores for each country are as follows:

	U.S.	GREECE
MEAN	30	34
STANDARD DEVIATION	10	8

Do these data suggest that on the average Greeks score significantly higher than Americans?

44. In an experiment 100 volunteer college students are randomly divided into two groups, A and B. Both groups take the same test but in two different environments, "pressured" for those in A and "relaxed" for those in B. The resulting scores are summarized as follows:

	A	B
n	50	50
$\bar{x}$	65.7	69.8
s	21.4	17.6

Do these data suggest that, on the average, students in a relaxed environment score higher than do those in a pressured environment?

45. Use the data in Review Problem 36 of Chapter 9 (page 324) to test whether significantly more than 15% of classrooms are being taught by substitute teachers.
46. Use the data in Review Problem 38 of Chapter 9 (page 324) to test whether the mean score for Anglo-Americans is significantly greater than that for Mexican-Americans.
47. Use the data in Review Problem 39 of Chapter 9 (page 324) to test whether the mean time to complete checklist I is significantly different from that for checklist II.

SCIENCE AND TECHNOLOGY

48. A company selects 60 random samples each of 1 kilogram of pig iron. For each sample the amount of silicon (in grams) is recorded. The company wishes to test whether the mean amount of silicon per kilogram associated with the production process is significantly less than 10. What conclusion is appropriate if $\bar{x} = 9.84$ grams and $s = 1.91$ grams?
49. It is known that the mean time between malfunctions for a certain type of electronic equipment is 510 hours. A new model is to be introduced for which the mean time between malfunctions may be less than or greater than 510. A number of these models are put into service and give rise to 61 malfunctions with a mean time between them of 554 hours and a standard deviation of 182 hours. What can be concluded from these data?
50. In a cloud-seeding experiment 50 clouds were seeded with silver nitrate pellets in the hope that this would increase the chances of their producing rain. A control group of 50 similar clouds was not seeded. The mean and standard deviation (in inches) of rainfall produced by the two groups of clouds are:

	SEEDED	UNSEEDED
$\bar{x}$	1.21	1.06
s	.83	.92

Do these data suggest that cloud seeding significantly improves the mean yield of rain per cloud?
51. In the cloud-seeding experiment referred to in Review Problem 50, 42 of the seeded and 36 of the unseeded clouds produced a measurable amount of rain. Do these data suggest that clouds that are seeded are more likely to produce rain than are unseeded ones?
52. We wish to compare the strength of two alloys, A_1 and A_2. Forty samples of each alloy are tested with the following summary of breaking strengths in pounds per square inch:

	A_1	A_2
n	40	40
$\bar{x}$	24,980	25,410
s	1,072	994

Do these data suggest any significant difference in the mean breaking strength for the two alloys?

53. Use the data in Review Problem 41 of Chapter 9 (page 325) to test whether there is any significant difference in the probability of detecting an aircraft for the two radar systems.

SUGGESTED READINGS

FEINBERG, W. E. "Teaching the type I and type II errors: the judicial process," *The American Statistician*, Vol. 25, (June 1971), pp. 30–32.

GIBBONS, J. D. and PRATT, J. W. "P values: Interpretation and Methodology," *The American Statistician*, Vol. 29, (February 1975), pp. 20–25.

There is an excellent series of readable articles covering many aspects of hypothesis testing in:

STEGER, J. A., ed. *Readings in Statistics for the Behavioral Scientist*. New York: Holt, Rinehart and Winston, 1971, pp. 199–311.

11 THE SENSITIVITY OF A TEST (OPTIONAL)

THE ARMY VERSUS BRASH—AGAIN

The U.S. Army orders 10,000 advanced four-wheel-drive vehicles from Brash, Inc. The contract specifies that, on average, the vehicles are to travel at least 2500 miles before a breakdown. The Army will test drive a sample of 35 of the vehicles to be delivered by Brash before accepting delivery of the remainder. The Army will *not* accept the remainder if the mean mileage before breakdown is found to be significantly less than 2500.

This time we are interested in finding how likely it is that the Army will be required to accept delivery if, in fact, the mean mileage to breakdown over the 10,000 vehicles is only 2100 miles? See Review Problem 12 on page 394.

SECTION 11.1

THE REJECTION REGION

As we illustrated in Section 10.2, a P value $\leq \alpha$ is not the only possible criterion for rejecting H_0. In this section we indicate how to obtain a criterion for rejecting H_0 *in terms of* $\bar{x}$.

Assume that we are to test $H_0: \mu = \mu_0$ against $H_1: \mu > \mu_0$ based on a random sample of size n. For purposes of illustration we shall use a single value $\alpha = .05$ for the desired level of significance. Following the procedure in Table 2 of Section 10.2 (page 338) if $\bar{x}$ is $> \mu_0$, we would compute $z = (\bar{x} - \mu_0)/(\sigma/\sqrt{n})$ and the P value as $P(Z > z)$. We are to reject H_0 if this P value $\leq .05$. At the bottom of Figure 1 we indicate (from Table A4) that the value of Z with an area of .05 to the right is 1.645—the 95th percentile for Z. Now if the value for $z = (\bar{x} - \mu_0)/(\sigma/\sqrt{n})$ exceeds 1.645, the P value will be less than .05. If the value for z is less than 1.645, the P value will exceed .05. Thus a criterion equivalent to rejecting H_0 if the P value $\leq .05$ is

$$z \left(\text{or } \frac{\bar{x} - \mu_0}{\sigma/\sqrt{n}} \right) \geq 1.645$$

This is equivalent to rejecting H_0 if

$$\bar{x} \geq \mu_0 + \frac{\sigma}{\sqrt{n}}(1.645)$$

We show this equivalence in Figure 1, where we denote the value $\mu_0 + (\sigma/\sqrt{n})(1.645)$ as C.

The value $C = \mu_0 + (\sigma/\sqrt{n})(1.645)$ is called the **critical point** or **critical** value for $\bar{x}$. The *range* of values of $\bar{x}$ *above* C is called the **rejection region** or **critical region** since these are the values for $\bar{x}$ for which we will reject H_0. The range of values of $\bar{x}$ below C is the **region of nonrejection**, which, for

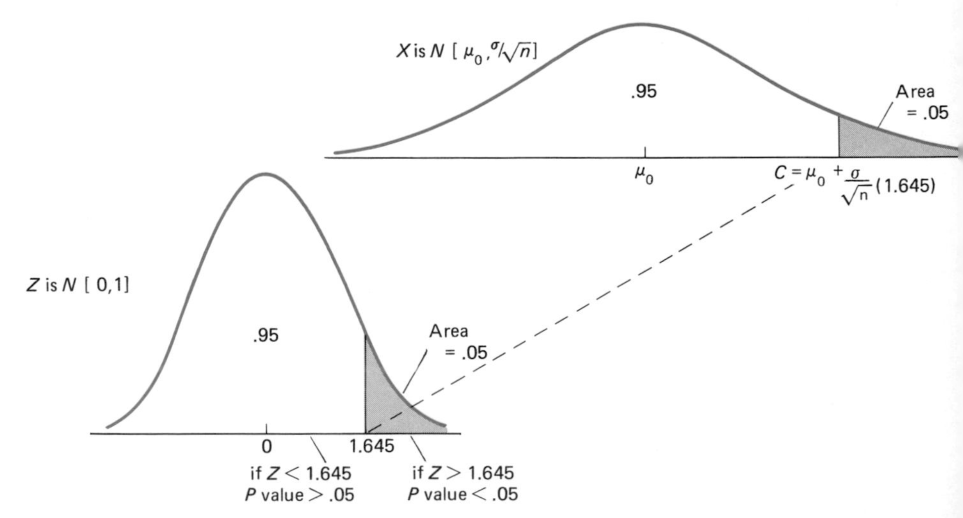

FIGURE 1

Rejecting H_0 if $z = (\bar{x} - \mu_0)/(\sigma/\sqrt{n}) \geq 1.645$ is equivalent to rejecting H_0 if $\bar{x} \geq \mu_0 + (\sigma/\sqrt{n})(1.645)$

convenience, we call the **acceptance region**. We show these regions in Figure 2.

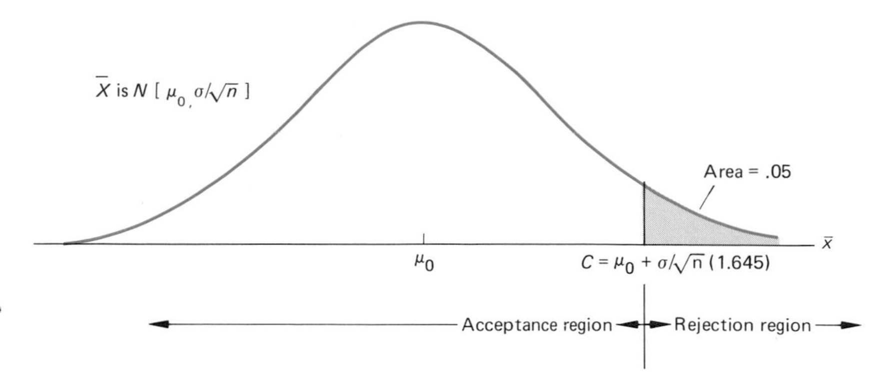

FIGURE 2

Acceptance and rejection regions

Now suppose that we were to use $\alpha = .01$ rather than .05. In this case the formula for C will be the same as before except that $Z_{95} = 1.645$ is replaced by $Z_{99} = 2.326$. More generally, for $H_1: \mu > \mu_0$ and arbitrary α the critical point will be

$$C = \mu_0 + \frac{\sigma}{\sqrt{n}} (Z_{100(1-\alpha)})$$

You should verify that if $\alpha = .05$, $Z_{100(1-.05)} = Z_{95} = 1.645$, and that if $\alpha = .01$, $Z_{100(1-.01)} = Z_{99} = 2.326$.

Had we been interested in $H_1: \mu < \mu_0$ rather than $H_1: \mu > \mu_0$, the critical point and the rejection region would be in the *left* tail of the $N[\mu_0, \sigma/\sqrt{n}]$ distribution. With $\alpha = .05$, for instance, the critical point would be $\mu_0 + (\sigma/\sqrt{n})(-1.645)$, where -1.645 is, from Table A4, the 5th percentile of Z (i.e. Z_5). More generally, for $H_1: \mu < \mu_0$, and any α the critical point is

$$C = \mu_0 + \frac{\sigma}{\sqrt{n}} (Z_{100\alpha})$$

The method for determining the value of C and the location of the acceptance and rejection regions for one-sided tests of $H_0: \mu = \mu_0$ are given in Table 1. These regions are shown in Figure 3. Notice that the rejection region is *above* μ_0 for $H_1: \mu > \mu_0$ and *below* μ_0 for $H_1: \mu < \mu_0$. In each case if $\bar{x}$ is not in the rejection region, it will be in the acceptance region and we will *not* reject H_0.

TABLE 1

Rejection and acceptance regions for one-sided tests of $H_0: \mu = \mu_0$

NULL HYPOTHESIS $H_0: \mu = \mu_0$

ALTERNATIVE HYPOTHESIS	(a) $H_1: \mu > \mu_0$	(b) $H_1: \mu < \mu_0$
Value of C	$C = \mu_0 + \frac{\sigma}{\sqrt{n}} (Z_{100(1-\alpha)})$	$C = \mu_0 + \frac{\sigma}{\sqrt{n}} (Z_{100\alpha})$
Rejection region	$\bar{x} \geq C$	$\bar{x} \leq C$
Acceptance region	$\bar{x} < C$	$\bar{x} > C$

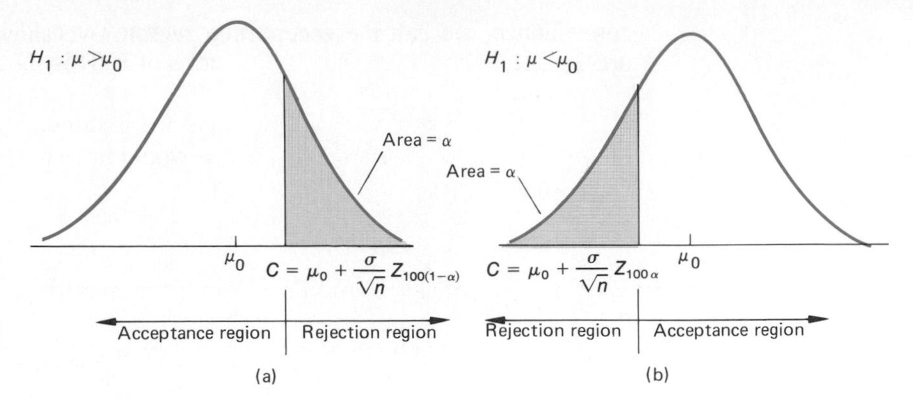

FIGURE 3

Rejection Region for α corresponding to (a) $H_1: \mu > \mu_0$ and (b) $H_1: \mu < \mu_0$

Rejection regions for the two-sided test of $H_0: \mu \neq \mu_0$ can be obtained. This will not be done in this book, however.

Although when performing tests in Chapter 10 we generally used $\alpha = .01$ and $\alpha = .05$, in this chapter we use a variety of values for α to illustrate the points that we wish to make.

EXAMPLE 1

Until recently the mean waiting time to see a doctor at a public clinic was 17.5 minutes with a standard deviation of 15.8 minutes. A new appointment system is set up which it is hoped will significantly reduce the waiting time.

After the system is introduced, a sample will be selected consisting of the waiting times for 40 patients. Call $\bar{x}$ the sample mean waiting time. Below what value for $\bar{x}$ can the clinic conclude that the (potential) mean waiting time (μ) for all patients using the new system is significantly below 17.5 minutes. Use $\alpha = .025$. You may assume that for the new system $\sigma = 15.8$ the same as for the old system.

Solution

The sample is to be used to test $H_0: \mu = 17.5$ against $H_1: \mu < 17.5$. Following the rule in Table 1(b) with $\mu_0 = 17.5$, $\sigma/\sqrt{n} = 15.8/\sqrt{40} = 2.5$ and $\alpha = .025$, we will reject H_0 if $\bar{x} \leq C = 17.5 + 2.5(Z_{2.5}) = 17.5 + 2.5(-1.96) = 12.6$. We will reject $H_0: \mu = 17.5$ at the .025 level of significance if $\bar{x} \leq 12.6$ minutes. We will *not* reject H_0 if $\bar{x} > 12.6$ minutes. These rejection and acceptance regions are shown in Figure 4.

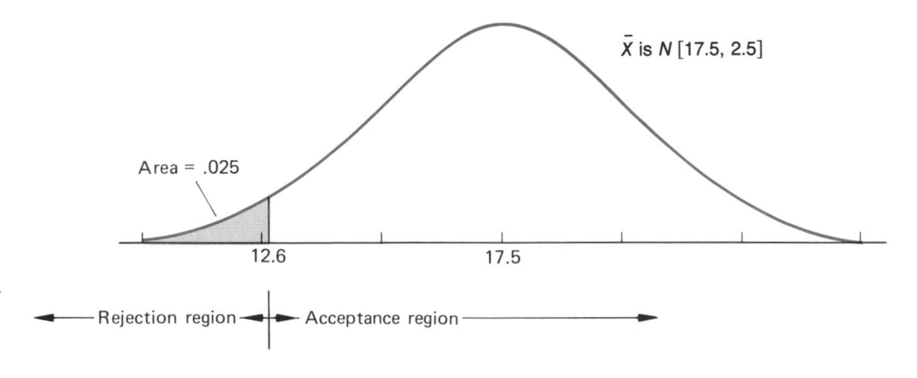

FIGURE 4

Rejection and Acceptance Regions for Example 1.

Did you notice that we did not *perform* any tests in this section? We merely set up the criteria for rejecting H_0 in terms of $\bar{x}$. This should, of course, be done before obtaining the data.

In this and the next two sections we assume, for simplicity, that σ is known. In reality a "guesstimate" for σ would have to be obtained using one of the methods suggested in Section 9.5.

PROBLEMS 11.1

For each problem draw a picture showing the rejection and acceptance regions.

1. A random sample of 400 is to be selected in order to test $H_0: \mu = 2100$ against $H_1: \mu > 2100$. Use Table 1 to find the rejection and acceptance regions corresponding to $\alpha = .03$. Assume that $\sigma = 450$.
2. We wish to test $H_0: \mu = .5$ against $H_1: \mu < .5$. Use Table 1 to find the rejection and acceptance regions corresponding to $\alpha = .1$. Use $\sigma = .008$ and assume a sample size of $n = 50$.
3. A government official claims that workers in public transit systems earn an average of $28,500 a year. Union officials feel that this figure is too high and intend to record the salaries of a random sample of 80 such workers.
   (a) What is μ in this case?
   (b) State H_0 and H_1.
   (c) Find the rejection and acceptance regions corresponding to $\alpha = .05$. Assume that $\sigma = \$4000$.
4. A mechanical engineer wants to test whether the tensile strength of an alloy is significantly greater than 2.3 tons per square inch. A total of 35 samples are available for testing. Assume that $\sigma = .095$ tons. Find the rejection and acceptance regions corresponding to (a) $\alpha = .15$; (b) $\alpha = .1$; (c) $\alpha = .05$.
5. At a private college a random sample of n student records is to be selected to test whether the mean GPA is significantly greater than 2.5. Assuming that $\sigma = .5$ and $\alpha = .1$, find the rejection and acceptance regions corresponding to (a) $n = 40$ students; (b) $n = 100$ students; (c) $n = 256$ students.
6. An independent evaluator has been selected to test whether the mean time it takes a police department to respond to an emergency call is significantly less than 2.5 minutes. The evaluator will record the time taken to respond to a random sample of 50 calls. Assume that $\sigma = 1.4$ minutes.
   (a) State H_0 and H_1.
   (b) Find the rejection and the acceptance regions corresponding to $\alpha = .1$.

Solution to Problem 1

From Table 1 the critical point is

$$C = 2100 + \frac{450}{\sqrt{400}} Z_{100(1-.03)} = 2100 + 22.5(Z_{97}) = 2100 + 22.5(1.881)$$

$$= 2142.32$$

The rejection region is $\bar{x} \geq 2142.32$; the acceptance region, $\bar{x} < 2142.32$.

SECTION 11.2
DETERMINING THE SENSITIVITY OF A TEST

In Section 10.6 we referred to rejecting H_0 when it is true as a type I error and *not* rejecting H_0 when it is false (i.e., H_1 is true) as a type II error. The probability of a type I error is α. In this section we are concerned with the probability of a type II error.

We begin by reexamining Example 1, which was concerned with μ, the potential mean waiting time for patients under a new appointment system. The two hypotheses were $H_0: \mu = 17.5$ and $H_1: \mu < 17.5$. The standard deviation of waiting time was assumed to be 15.8 minutes. The sample consists of the waiting times of 40 patients. Thus $\sigma/\sqrt{n} = 15.8/\sqrt{40} = 2.5$. From Example 1 we will reject H_0 at the .025 level of significance if the P value $\leq .025$ or, equivalently, if $\bar{x} \leq 12.6$ minutes. On the other hand, we will *not* reject H_0 if $\bar{x} > 12.6$ minutes.

If H_0 is true, the random variable $\bar{X}$ is approximately $N[17.5, 2.5]$. This distribution, together with the rejection region $\bar{x} \leq 12.6$ and the acceptance region $\bar{x} > 12.6$, is shown in Figure 5.

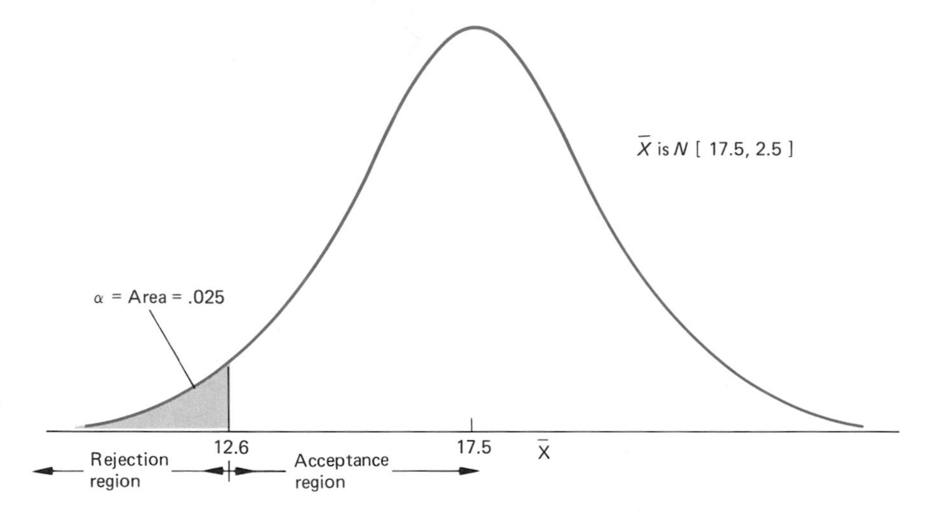

FIGURE 5

Rejection and Acceptance Regions for the appointments example.

Even if $H_0: \mu = 17.5$ is true, it is still possible—by chance—to select a sample with a mean $\bar{x}$ less than 12.6 minutes. Then H_0 would be rejected when it should *not* be, and a type I error would occur. The probability of this happening is just the probability of $\bar{x} < 12.6$ when $\mu = 17.5$, which is $\alpha = .025$.

Now suppose that the sample mean $\bar{x}$ were above 12.6. Then we would not reject H_0. If μ were in fact 17.5, this would be the correct decision. But suppose that μ were not 17.5. Say, for example, that it were 14 minutes. Then we would *not* reject $H_0: \mu = 17.5$ when, in fact, it was false and a type II error would occur. What is the probability of this happening? It is the probability that

$\bar{x} > 12.6$ if $\mu = 14$. This is written $P(\overline{X} > 12.6 \mid \mu = 14)$ and is called $\beta(14)$ and read as "beta 14." (β is the Greek lowercase letter beta.) But if the new mean waiting time, μ, is in reality 14, then the distribution of $\overline{X}$ is approximately $N[14, 2.5]$ and $\beta(14)$ is the area to the right of 12.6 under *this* distribution, not under the $N[17.5, 2.5]$ distribution. The area corresponding to $\beta(14)$ is shaded in Figure 6, where for comparative purposes we also show the $N[17.5, 2.5]$ distribution. Notice that the acceptance region (i.e., values for $\bar{x}$ above 12.6) is defined the same way in both cases.

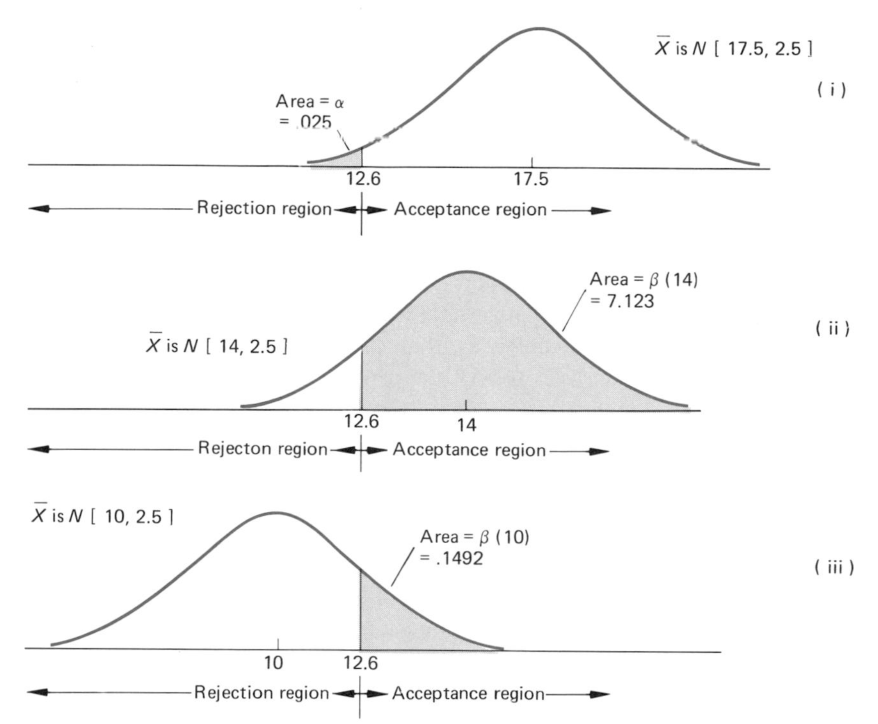

FIGURE 6

The shaded areas to the right of 12.6 in (ii) and (iii) are $\beta(14)$ and $\beta(10)$, respectively.

We compute the probability $\beta(14)$ by standardizing, just as we computed probabilities about $\overline{X}$ in Section 7.2. Thus

$$\begin{aligned}\beta(14) &= P(\overline{X} > 12.6 \mid \mu = 14)^1 \\ &= P\left(Z > \frac{12.6 - 14}{2.5}\right) \\ &= P(Z > -.56) = .7123\end{aligned}$$

Before interpreting this value, let us look at a slightly different case. Assume that instead of μ being 14, it were 10 minutes. Again, if the sample

[1] $P\,(\overline{X} > 12.61\mu = 14)$ is not a conditional probability. We merely use the information $\mu = 14$ to compute $P(\overline{X} > 12.6)$.

mean were above 12.6 minutes, we would not reject $H_0: \mu = 17.5$ when, in fact, it was false, and a type II error would occur. What is the probability of this?

If $\mu = 10$, the distribution of $\bar{X}$ will be approximately $N[10, 2.5]$ and not $N[14, 2.5]$ or $N[17.5, 2.5]$. The $N[10, 2.5]$ distribution is shown at the bottom of Figure 6. If $\mu = 10$, the probability of a type II error is $P(\bar{X} \geq 12.6 \mid \mu = 10)$ and is denoted $\beta(10)$. Proceeding as we did before, we have

$$\beta(10) = P(\bar{X} > 12.6 \mid \mu = 10) = P\left(Z > \frac{12.6 - 10}{2.5}\right)$$

$$= P(Z > 1.04) = .1492$$

The value $\beta(10) = .1492$ is the shaded area to the right of 12.6 under the $N[10, 2.5]$ distribution in Figure 6. How shall we interpret the two values $\beta(14) = .7123$ and $\beta(10) = .1492$?

Assume that the administrator were interested in rejecting H_0 if an improvement of at least 3.5 minutes (from 17.5 to 14) had been achieved. Even if this did happen and μ *were* in fact 14.0, there is a large (.7123) probability of obtaining a sample mean above 12.6 and thus (falsely) concluding that μ is not significantly less than 17.5 minutes. Although α, the probability of a type I error, is small, $\beta(14)$, the probability of a type II error, is large—and there is a good chance of reaching the wrong decision.

On the other hand, suppose that the administrator were interested in rejecting H_0 only if an improvement from 17.5 minutes to at least 10.0 minutes had occurred. If μ were 10, $\beta(10)$, the probability of not rejecting H_0, would only be .1492. In this case the probabilities of making *both* types of error are small.

Notice that there is not a single value for β. In fact, in the present example since $H_1: \mu < 17.5$, we could compute $\beta(\mu)$ for any μ less than 17.5. Typically, the researcher will select μ_1, a *specific* alternative value for μ (such as 14 or 10 above) and compute $\beta(\mu_1)$—the probability of not rejecting H_0 if $\mu = \mu_1$. The lower the value for $\beta(\mu_1)$ (i.e., the less the likelihood of a Type II error), the more *sensitive* the test is said to be for the specific alternative $\mu = \mu_1$.

All other factors being equal, the greater the difference between the values μ_0 and μ_1, the smaller the value of $\beta(\mu_1)$ and the more sensitive the test is said to be. You can see this in Figure 6 by comparing the areas corresponding to $\beta(14)$ and $\beta(10)$.

We now generalize the procedures for computing $\beta(\mu_1)$ as follows:

1. Find the critical point C for $H_0: \mu = \mu_0$ and α.
2. Specify the corresponding acceptance region. This will be the range of values for $\bar{x} \geq C$ or $\bar{x} \leq C$ depending on the alternative hypothesis (see Table 2).
3. Compute $\beta(\mu_1)$, the probability of the sample mean falling in the acceptance region, for the $N[\mu_1, \sigma/\sqrt{n}]$ distribution (i.e., when $\mu = \mu_1$).

This procedure is summarized in Table 2.

TABLE 2 Procedure for computing $\beta(\mu_1)$ for one-sided tests of $H_0: \mu = \mu_0$

STEP	ALTERNATIVE HYPOTHESIS (a) $H_1: \mu > \mu_0$	(b) $H_1: \mu < \mu_0$
1. The critical point is:	$C = \mu_0 + \frac{\sigma}{\sqrt{n}} Z_{100(1-\alpha)}$	$C = \mu_0 + \frac{\sigma}{\sqrt{n}} Z_{100\alpha}$
2. Specify the acceptance region:	$\bar{x} < C$	$\bar{x} > C$
3. $\beta(\mu_1)$ is:	$P(\bar{X} < C \mid \mu = \mu_1) = P\left(Z < \frac{C - \mu_1}{\sigma/\sqrt{n}}\right)$	$P(\bar{X} > C \mid \mu = \mu_1) = P\left(Z > \frac{C - \mu_1}{\sigma/\sqrt{n}}\right)$

THE RELATIONSHIP BETWEEN α AND $\beta(\mu_1)$

In the following example we apply the procedure in Table 2 in order to examine the relationship between β and α.

EXAMPLE 2 A large urban school district is interested in determining whether their students have, on the average, higher SAT verbal scores than the national mean (450 that year with a standard deviation of 80). A random sample of $n = 100$ students will be selected in order to test $H_0: \mu = 450$ against $H_1: \mu > 450$, where μ is the (unknown) district mean score. The district office would like to reject H_0 if $\mu = 465$ or higher. Compute $\beta(465)$ if (a) $\alpha = .1$; (b) $\alpha = .01$.

Solution Assume first that σ, the standard deviation of scores in the district, is also 80 points.

(a) $\alpha = .1$. Following the procedure in column (a), Table 2, the critical value for $\bar{x}$ corresponding to $\alpha = .1$ is

$$C = \mu_0 + \frac{\sigma}{\sqrt{n}} Z_{100(1-\alpha)}$$

$$= 450 + \frac{80}{\sqrt{100}} Z_{100(1-.1)} = 450 + 8(Z_{90}) = 450 + 8(1.282)$$

$$= 460.256$$

The acceptance region thus consists of values of $\bar{x} < 460.256$. In other words, the researcher will accept (i.e., not reject) $H_0: \mu = 450$ if $\bar{x} < 460.256$. The probability of this occurring (i.e., of a type II error) if μ is really 465 is

$$\beta(465) = P(\bar{X} < 460.256 \mid \mu = 465)$$

$$= P\left(Z < \frac{460.256 - 465}{8}\right) = P(Z < -.59) = .2776$$

The critical value, 460.256, is shown as the 90th percentile of the $N[450, 8]$ distribution of $\bar{X}$ at the top of Figure 7. The critical value is also shown in relation to the $N[465, 8]$ distribution of $\bar{X}$. The probability $\beta(465) = .2776$ is the area to the left of 460.256 under this last distribution. Notice that $\beta(\mu_1)$

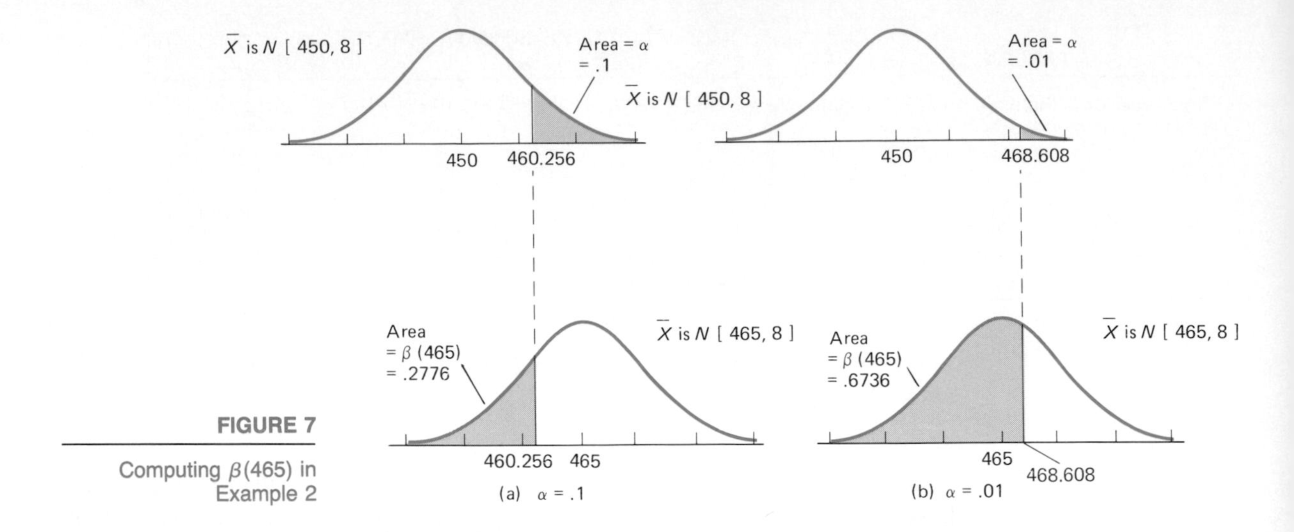

FIGURE 7

Computing $\beta(465)$ in Example 2

is to the *left* of C as contrasted with Figure 6 since, in the present case $H_1: \mu > \mu_1$.

(b) $\alpha = .01$. The critical value corresponding to $\alpha = .01$ is $C = 450 + 8(Z_{99}) = 450 + 8(2.326) = 468.608$. The corresponding acceptance region is $\bar{x} < 468.608$, so that $\beta(465)$ is

$$\begin{aligned}\beta(465) &= P(\overline{X} < 468.608 \mid \mu = 465) \\ &= P\left(Z < \frac{468.608 - 465}{8}\right) = P(Z < .45) \\ &= .6736\end{aligned}$$

The null hypothesis $H_0: \mu = 450$ will be accepted (i.e., not rejected) if $\bar{x} < 468.608$. The probability that this decision is wrong if μ is really 465 is .67.

The critical value $C = 468.608$ and the probability $\beta(465) = .67$ corresponding to $\alpha = .01$ are shown in Figure 7(b).

Our results are summarized below:

$$\alpha = .1 \qquad \beta(465) = .2776$$

$$\alpha = .01 \qquad \beta(465) = .6736$$

These figures suggest that the smaller we make α, the larger $\beta(465)$ will be, and vice versa.The reason for this can be seen in Figure 7.

As α is made smaller, the value of C is farther from 450, resulting in a larger acceptance region and hence a larger value for $\beta(\mu_1)$, in the case above—$\beta(465)$. This inverse relationship between α and $\beta(\mu_1)$ is perfectly general. The smaller the value for α, the larger any $\beta(\mu_1)$ will be.

■

We emphasize that α is a probability associated with H_0 and is therefore an area under the distribution of $\overline{X}$ if $H_0: \mu = \mu_0$ is true. By contrast, $\beta(\mu_1)$ is a probability associated with the specific alternative $\mu = \mu_1$ and is therefore an area under the distribution of $\overline{X}$ if $\mu = \mu_1$.

The various decisions that can be made in a statistical test are summarized in Table 3. However, since the value for μ is not known, we cannot *know* whether our eventual decision is correct or not.

TABLE 3

Decisions in hypothesis testing

	BASED ON THE SAMPLE WE: REJECT H_0	WITH PROBABILITY	DO NOT REJECT H_0	WITH PROBABILITY
$H_0: \mu = \mu_0$ IS TRUE	Type I Error	α	Correct Decision	$1 - \alpha$
H_0 IS FALSE ($H_1: \mu = \mu_1$ IS TRUE)	Correct Decision	$1 - \beta(\mu_1)$	Type II Error	$\beta(\mu_1)$

Two new probabilities, $1 - \alpha$ and $1 - \beta(\mu_1)$, are shown in Table 3. The value $1 - \alpha$ is the probability of *correctly* accepting (i.e., not rejecting) H_0 when H_0 is true. In Example 2 if $\alpha = .1$, $1 - \alpha = .9$ and is the area under the $N[450, 8]$ distribution above the acceptance region, $\bar{x} < 460.256$ in Figure 7. The probability $1 - \beta(\mu_1)$ is the probability of *correctly* rejecting H_0 if $\mu = \mu_1$ and is known as the **power** of the test if $\mu = \mu_1$. In Example 2 if $\alpha = .1$, $1 - \beta(465) = 1 - .2776 = .7224$ and is the area under the $N[465, 8]$ distribution above the *rejection* region $\bar{x} \geq 460.256$. This large value for $1 - \beta(465)$ indicates that the test is quite powerful for the alternative $\mu = 465$. On the other hand, note that if $\alpha = .01$ instead of .1, $1 - \beta(465) = 1 - .6736 = .3264$ and in this case the test is not very powerful for $\mu = 465$.

THE RELATIONSHIP BETWEEN n AND $\beta(\mu_1)$

There is an important relationship between $\beta(\mu_1)$ and the sample size, n. This is illustrated in the following example.

EXAMPLE 3 A company claims that the mean weight of its cans of peaches is 370 grams with a standard deviation of $\sigma = 20$ grams. A consumer testing agency, feeling that this figure is too high, plans to select and weigh a random sample of n cans. The two hypotheses in this case are $H_0: \mu = 370$ and $H_1: \mu < 370$, where μ is the unknown true mean weight of the cans. To assure that there is a small chance of falsely rejecting the company's claim, the agency will use $\alpha = .025$. Compute $\beta(365)$ if (a) $n = 50$; (b) $n = 150$.

Solution (a) $n = 50$. Following the procedure in column (b) of Table 2, the critical value for $\bar{x}$ is $C = 370 + (20/\sqrt{50})Z_{2.5} = 370 + 2.828(-1.96) = 364.46$. The corresponding acceptance region is $\bar{x} > 364.46$, and therefore

$$\beta(365) = P(\overline{X} > 364.46 \mid \mu = 365)$$

$$= P\left(Z > \frac{364.46 - 365}{2.828}\right) = P(Z > -.19) = .5753$$

This value for $\beta(365)$ is the area to the right of the critical point 364.46 under the $N[365, 2.828]$ distribution of $\bar{X}$ in Figure 8. With a sample of 50 cans there is a large probability (.5753) of making a type II error and of not rejecting $H_0: \mu = 370$ when in fact $\mu = 365$.

(b) If $n = 150$, the critical value for $\bar{x}$ becomes $C = 370 + (20/\sqrt{150})(-1.96) = 370 + (1.633)(-1.96) = 366.80$. The acceptance region is $\bar{x} > 366.80$; consequently,

$$\begin{aligned}\beta(365) &= P(\bar{X} > 366.80 \mid \mu = 365)\\ &= P\left(Z > \frac{366.80 - 365}{1.633}\right) = P(Z > 1.1)\\ &= .1357\end{aligned}$$

This probability is the area to the right of 366.80 under the $N[365, 1.633]$ distribution of $\bar{X}$ in Figure 8.

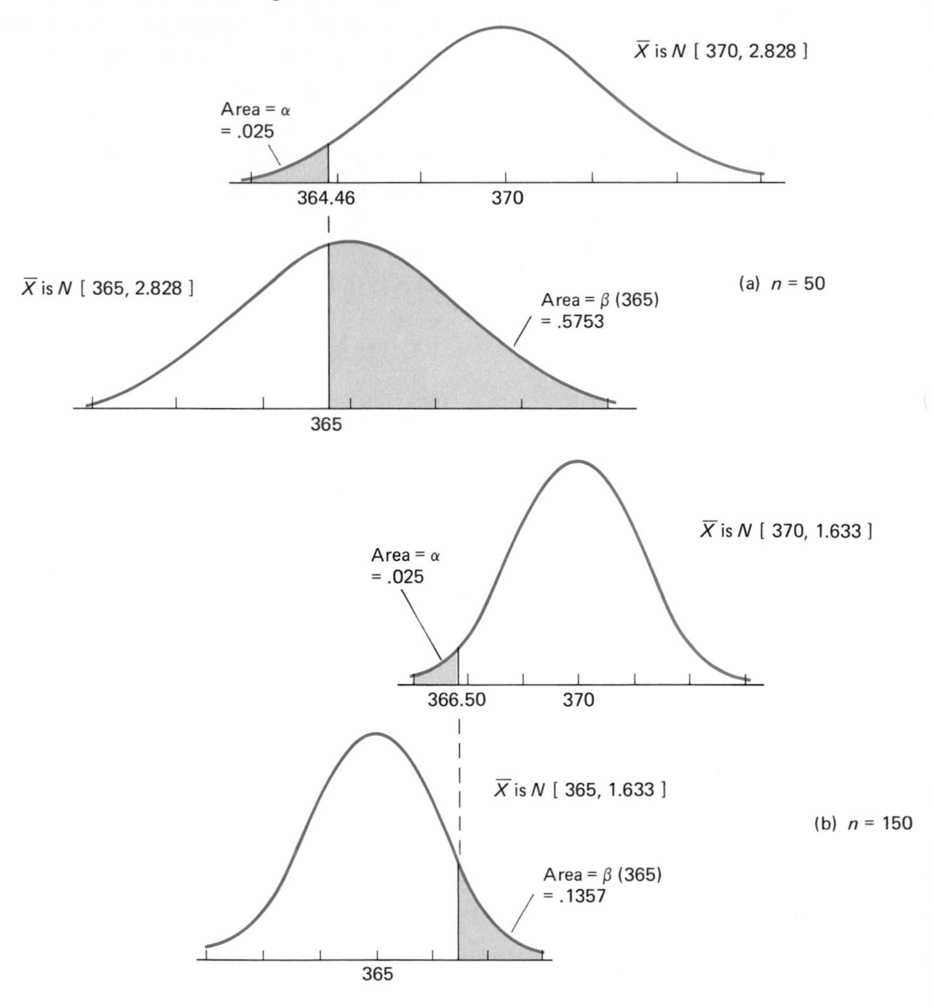

FIGURE 8

$\beta(365) = .5753$ when $n = 50$ and .1357 when $n = 150$

With a sample of 50 cans the probability of not rejecting H_0 if μ is, in fact, 365 grams (and not 370) is .5753, but with a 150-can sample this probability is .1357. The test is far more sensitive when $n = 150$ than when $n = 50$. This can be explained as follows: As n increases, the value of $\mathrm{SD}(\overline{X}) = 20/\sqrt{n}$ gets smaller and the distribution of $\overline{X}$ thus narrows (as shown in Figure 8) and C, the critical point, gets closer and closer to 370, resulting in a smaller acceptance region. As this region becomes smaller, $\beta(365)$, the area above this region under the $N[365, 20/\sqrt{n}]$ distribution, also gets smaller.

■

There are two messages we would leave before completing this section.

First: The specification of a particular alternative value for μ (i.e., μ_1) is essentially a nonstatistical decision. In our earlier example (Example 1) the clinic administrator should decide the magnitude of the decline in mean waiting time (from 17.5 to μ_1) that it is worthwhile to detect. Similarly, a medical researcher interested in learning whether the mean recovery time for an operation has been improved with a new procedure should decide on the magnitude of increase worth detecting. In this, as in all other cases, the decision should be made before the study is begun.

Second: If the proper decisions are made initially, one may conduct a test so that the probability of *not* discriminating between H_0 and H_1 is minimized. In Example 3, for instance, the agency wishes to detect a difference of 5 grams from the company's claim. Had they arbitrarily decided on $n = 50$, however, there would be a 58% chance of not rejecting H_0 when, in fact, such a difference did exist. An analysis similar to that in Example 3 would have alerted the agency to the fact that a sample larger than 50 was necessary.

In the next section we present procedures for determining just how large a sample should be selected.

PROBLEMS 11.2

7. An educational group is interested in the percentage of private college income that is derived from tuition and fees. Call μ the (unknown) mean of this percentage over all private colleges. The group intends to select a random sample of $n = 50$ such colleges in order to test $H_0 : \mu = 35$ against $H_1 : \mu < 35$. Assuming that $\sigma = 5\%$ and using $\alpha = .075$, compute (a) $\beta(33)$; (b) the power of the test if $\mu = 33$.
8. In a particular problem the level of significance is .02. Which of the following statements are true? If a statement is not true, explain what is wrong.
   (a) The probability of a type II error is .02.
   (b) The probability of not rejecting H_0 if H_0 is true is .98.
   (c) The probability of rejecting H_0 if H_1 is true is .98.
   (d) The probability of a type I error is .02.
   (e) The probability of rejecting H_0 if H_0 is true is .02.
9. In a statistical test $\alpha = \beta(40) = .05$. Which of the following statements are true? If a statement is not true, explain what is wrong.

(a) The probability of not rejecting H_0 when $\mu = 40$ is .05.

(b) The probability of not rejecting H_0 when H_0 is true is equal to the probability of rejecting H_0 when $\mu = 40$.

(c) The probability of falsely rejecting H_0 is .95.

(d) The probability of not rejecting H_0 when $\mu = 40$ is .95.

(e) The probability of not rejecting H_0 when H_0 is true is equal to the probability of not rejecting H_0 when $\mu = 40$.

(f) The probability of not rejecting H_0 when H_0 is true is .95.

10. We wish to test $H_0: \mu = 45$ against $H_1: \mu > 45$ and we compute $\beta(47) = .092$.
    (a) Interpret the value .092.
    (b) Is this test sensitive or insensitive against the alternative $\mu = 47$?

11. We wish to test whether the mean IQ in a population is significantly greater than 115. A random sample of 40 persons will be selected and tested. Assume that $\sigma = 10$.
    (a) Define μ in this case.
    (b) State the two hypotheses.
    (c) If $\alpha = .05$, compute $\beta(\mu_1)$ for $\mu_1 = 122, 120, 118$, and 116.
    (d) Describe the relationship between μ_1 and $\beta(\mu_1)$.

12. A pharmaceutical company advertises that the mean time required for its brand of aspirin to dissolve is 9 minutes with a standard deviation of 2.5 minutes. The Food and Drug Administration doubts that the company's aspirins dissolve this fast and intends to record the time to dissolve for $n = 75$ tablets.
    (a) Define μ in this case.
    (b) State the null and the alternative hypotheses.
    (c) Compute $\beta(10)$ if (i) $\alpha = .005$; (ii) $\alpha = .05$.

13. A magazine editor is interested in learning whether the mean years of education of the magazine's subscribers is significantly less than 18.5 years. A sample of n subscribers will be selected.
    (a) Define μ in this case and state H_0 and H_1.
    (b) If σ is assumed to be 2.3 years and $\alpha = .05$ is used, compute $\beta(18)$ if (i) $n = 40$; (ii) $n = 100$; (iii) $n = 400$.
    (c) Interpret your value for $\beta(18)$ if $n = 400$.

14. A sociologist is interested in testing whether the mean age at which men marry in the county is significantly greater than 28.5 years. She intends to record the man's age at marriage for a random sample of 200 recent marriage records and use a level of significance of $\alpha = .025$.
    (a) State H_0 and H_1.
    (b) Assuming that $\sigma = 6$ years, (i) compute $\beta(30)$; (ii) interpret this value.
    (c) How powerful is the test if $\mu = 30$?

Solution to Problem 7

(a) Following the procedure in Table 2 (b) with $\mu_0 = 35$, $n = 50$, $\alpha = .075$, and $\sigma = 5$, the critical point is $C = 35 + (5/\sqrt{50})Z_{7.5} = 35 + (5/\sqrt{50})(-1.44) = 33.982$. The acceptance region is $\bar{x} > 33.982$ and

$$\beta(33) = P(\bar{X} > 33.982 \mid \mu = 33) = P\left(Z > \frac{33.982 - 33}{5/\sqrt{50}}\right) = P(Z > 1.39) = .0823$$

(b) The power of the test if $\mu = 33$ is $1 - \beta(33) = 1 - .0823 = .9177$. This is a powerful test for the alternative $\mu_1 = 33$.

SECTION 11.3
SAMPLE SIZE FOR HYPOTHESIS TESTING

In Section 11.2 you saw that if all other factors are equal, as the sample size n is *increased*, the value of $\beta(\mu_1)$ will *decrease*. This suggests that we may be able to find the sample size necessary for a desired value of both α *and* $\beta(\mu_1)$. The next example will illustrate how this can be done.

EXAMPLE 4 In Example 3 we tested $H_0: \mu = 370$ grams against $H_1: \mu < 370$ grams. The standard deviation, σ, was assumed to be 20 grams and we used the value $\alpha = .025$. We found that if $n = 50$ cans were selected, $\beta(365)$, the probability of not rejecting H_0 if, in fact, $\mu = 365$ grams $= .5753$, but if $n = 150$, $\beta(365) = .1357$.

Now, how many cans have to be sampled to guarantee the values $\alpha = .025$ and $\beta(365) = .05$? This situation is represented graphically in Figure 9.

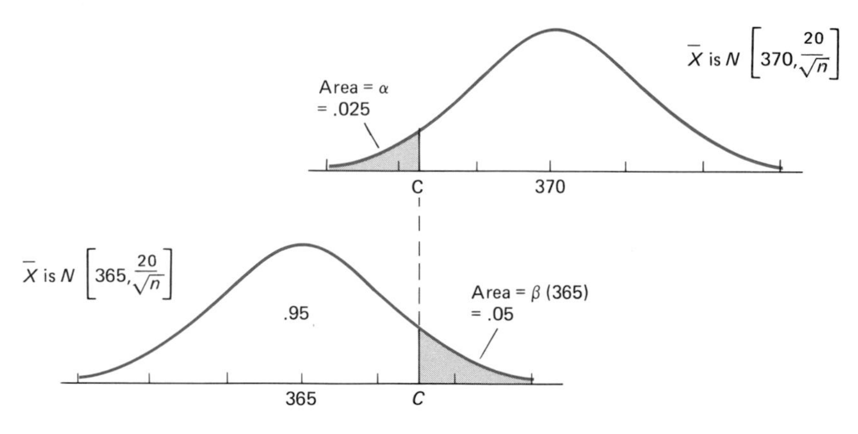

FIGURE 9

Fixing $\alpha = .025$ and $\beta(365) = .05$ in Example 4

If H_0 is true, $\mu = 370$ and $\bar{X}$ is approximately $N[370, 20/\sqrt{n}]$. If H_1 is true, $\mu = 365$ and $\bar{X}$ is approximately $N[365, 20/\sqrt{n}]$. You can see from Figure 9 that the critical point C must satisfy two conditions. It must be both (1) the 2.5th percentile of the $N[370, 20/\sqrt{n}]$ distribution and (2) the 95th percentile of the $N[365, 20/\sqrt{n}]$ distribution. Symbolically, we may write (using Equation 6.1 with $\sigma/\sqrt{n}$ in place of σ)

(1) $$C = 370 + \frac{20}{\sqrt{n}}(-1.96) = 370 - \frac{20}{\sqrt{n}}(1.96)$$

(2) $$C = 365 + \frac{20}{\sqrt{n}}(1.645)$$

Since the right-hand side of each of these equations is equal to C, they are also equal to each other. Thus

$$370 - \frac{20}{\sqrt{n}}(1.96) = 365 + \frac{20}{\sqrt{n}}(1.645)$$

Isolating elements associated with n, we obtain

$$\frac{20}{\sqrt{n}}(1.96 + 1.645) = 370 - 365$$

or

$$\sqrt{n} = \frac{20(1.96 + 1.645)}{(370 - 365)}$$

and

$$n = \frac{20^2(1.96 + 1.645)^2}{(370 - 365)^2}$$

$$= 207.94$$

Recall that in Section 9.5 when we estimated the sample size necessary to obtain a confidence interval of specified width, we always rounded up the value for n. We do the same here and conclude that $n = 208$ cans should be selected to guarantee that $\alpha = .025$ and $\beta(365) = .05$.

■

More generally, assume a one-sided test of $H_0: \mu = \mu_0$ and a level of significance α. What sample size (n) is necessary to guarantee that the probability of a type II error is some specified β if, in fact, $\mu = \mu_1$?

Without going into detail, generalizing the procedure in Example 4 produces the formula

$$n = \frac{\sigma^2[Z_{100(1-\alpha)} + Z_{100(1-\beta)}]^2}{(\mu_0 - \mu_1)^2} \qquad \text{(Eq. 11.1)}$$

This formula is appropriate for either $H_1: \mu > \mu_0$ or $H_1: \mu < \mu_0$.

EXAMPLE 5 Referring to the details of Example 2, we were to test $H_0: \mu = 450$ against $H_1: \mu > 450$, where μ was the unknown mean SAT verbal score in a large urban school district. The standard deviation of such scores was $\sigma = 80$. How large a sample would be necessary to guarantee that $\alpha = .05$ and $\beta(465) = .1$?

Solution In this case

$$Z_{100(1-\alpha)} = Z_{100(1-.05)} = Z_{95} = 1.645$$

$$Z_{100(1-\beta)} = Z_{100(1-.1)} = Z_{90} = 1.282$$

Inserting these values and $\mu_0 = 450$, $\mu_1 = 465$, and $\sigma = 80$ into Equation 11.1, we obtain

$$n = \frac{80^2(1.645 + 1.282)^2}{(450 - 465)^2}$$

$$= 243.69 \quad \text{or} \quad 244 \text{ students}$$

■

THE EFFECT OF α, β, μ_0, μ_1, AND σ ON n

It is worth examining the effect of each of the components of Equation 11.1 on the magnitude of n when other components are held constant.

First, the *smaller* we make α and β, the *larger* the corresponding percentiles in Equation 11.1 [i.e., $Z_{100(1-\alpha)}$ and $Z_{100(1-\beta)}$] will be and hence the larger n will have to be. For instance, if α and β are both .1, both percentiles will be $Z_{90} = 1.282$. If, however, α and β are both .01, both percentiles will be $Z_{99} = 2.326$ (see Problem 23).

The closer μ_1, the particular alternative, is to μ_0, the smaller the denominator in Equation 11.1 and the larger n will be.

Finally, the more varied the population being sampled (i.e., the larger σ), the larger the n has to be.

CAN THE SAMPLE SIZE BE TOO LARGE?

At the end of Section 11.2 we noted that with too small a sample size, the test may be insensitive to a desired alternative. This means that although we would like to reject H_0 if $\mu = \mu_1$, the probability of *not* doing this [i.e., $\beta(\mu_1)$] is high.

Sometimes too *large* a sample size in hypothesis testing can also cause misunderstanding. For instance, in Example 3 the consumer agency wanted to test $H_0: \mu = 370$ against $H_1: \mu < 370$ with $\sigma = 20$ and $\alpha = .025$. Suppose that a sample of $n = 10{,}000$ cans are to be selected. Consider the probability of rejecting H_0 if the true mean were only .5 below 370 (that is, 369.5). It is not hard to show that in this case $\beta(369.5) = .2946$ and if $\mu = 369.5$ grams the power of the test is $1 - \beta(369.5) = .7054$. This last figure indicates that there is greater than a 70% chance of rejecting the producer's claim even if the mean weight of the cans is only $370 - 369.5 = .5$ gram below that claimed.

This is an example of a case where the test may well result in a *statistically* significant difference from the manufacturer's claim, but the difference is of no *practical* significance. The consumer agency would be quite unlikely to have an interest in rejecting H_0 if μ were so close to 370. The test is *too* sensitive for its purpose.

Where a very large sample is selected or available, many statisticians prefer to compute a confidence interval for the unknown μ rather than perform a test with no practical significance. For instance, in the example above suppose that the sample of 10,000 cans resulted in the values $\bar{x} = 369.6$ and $s = 19.4$. If we perform the test, then from Table 2 of Chapter 10 (page 338) we would compute

$$z = \frac{369.6 - 370}{19.4/\sqrt{10{,}000}} = -2.06$$

As a result, the P value would be

$$P(Z < -2.06) = .0197$$

We reject H_0 at the .025 level of significance. The data suggest that the mean weight of the cans is significantly less than 370.

Using these same results, a 90% confidence interval for μ is, using Definition 3 of Chapter 9 (page 301),

$$\bar{x} \mp 1.645 \frac{s}{\sqrt{n}} = 369.6 \mp 1.645\left(\frac{19.4}{\sqrt{10{,}000}}\right)$$

$$= 369.6 \mp .32$$

$$= 369.28 \text{ to } 369.92$$

In this example it is more useful to know that the mean weight lies between 369.28 and 369.92 grams than that μ is statistically (but clearly not practically) less than 370 grams.

PROBLEMS 11.3

15. Suppose that over all deliveries the mean infant birthweight is 3400 grams with a standard deviation of 680 grams. Are the infants born to heavy-smoking mothers significantly smaller than average? A random sample of the records of such mothers is to be selected to test this. How many such records should be selected in order that $\alpha = .025$ and $\beta(3200) = .2$?
16. Department of Transportation research indicated that, prior to a new campaign to encourage carpooling, the mean number of riders per car on an expressway was 1.7 with a standard deviation 1.4. After the campaign the research group would like to select a random sample of cars on the expressway in order to test whether or not the post-campaign mean number of riders per car (μ) is significantly greater than 1.7.
    (a) How many cars should be selected so that $\alpha = \beta(2.0) = .05$? Assume that σ is still 1.4.
    (b) Interpret the value $\beta(2.0) = .05$ here.
17. In Problem 16, how would n change if it were required that $\alpha = \beta(2.2) = .05$?
18. It is known that the mean lifespan of a species of beetle is 6.5 months. It is suspected that a subspecies has a somewhat shorter mean lifespan. A random sample of members of this subspecies will be observed.
    (a) State H_0 and H_1.
    (b) Define μ.
    (c) Assuming that the standard deviation of the length of life in the subspecies is $\sigma = 1.4$ months, find the sample size necessary to guarantee that $\alpha = .1$ and $\beta(6.0) = .2$.
19. In Problem 11 in Section 11.2, we wanted to test whether the mean IQ in a population is significantly greater than 115. Assuming that $\sigma = 10$, find the sample size necessary so that $\alpha = \beta(118) = .025$.
20. With the sample size suggested in Problem 19, what decision should be made if the mean and standard deviation of IQ in the sample are $\bar{x} = 116.54$ and $s = 12.4$?

21. In Problem 12 in Section 11.2, the Food and Drug Administration intended to test whether it takes significantly longer than 9 minutes for a brand of aspirin to dissolve. They originally intended to dissolve a sample of $n = 75$ aspirins. How many *additional* aspirins would have to be tested in order to guarantee that $\alpha = .005$ and $\beta(10) = .05$? (Assume that $\sigma = 2.5$ minutes.)
22. With the sample size suggested in Problem 21, what decision should be made if the mean dissolving time in the sample is $\bar{x} = 10.05$ minutes and $s = 2.56$ minutes?
23. A student using Equation 11.1 with $\alpha = \beta(11.2) = .1$ obtains the value $n = 100$. The student's advisor suggests (perhaps overcautiously) using $\alpha = \beta(11.2) = .01$. How many *additional* observations would need to be selected if the advisor's advice were followed?

Solution to Problem 15

Following Equation 11.1 with $\alpha = .025$, $\beta = .2$, $\mu_0 = 3400$, $\mu_1 = 3200$, and $\sigma = 680$:

$$n = \frac{680^2(Z_{97.5} + Z_{80})^2}{(3400 - 3200)^2} = \frac{680^2(1.96 + .842)^2}{200^2}$$

$$= 90.76 \text{ or, rounding up, 91 records}$$

SECTION 11.4

SUMMARY AND REVIEW

In this chapter we focused on computing the probability of a type II error (i.e., of not rejecting H_0 when H_1—and not H_0—is true). For simplicity we concentrated only on one-sided tests of H_0: $\mu = \mu_0$.

THE CRITICAL VALUE C AND THE ACCEPTANCE AND REJECTION REGIONS

We first find a value of $\bar{x}$ such that the P value $= \alpha$. This value is designated C and is called the **critical point** or **critical value** of $\bar{x}$. This value, in turn, determines the **rejection** and **acceptance regions.** If $\bar{x}$ falls in the rejection region, we reject H_0 in favor of H_1; if $\bar{x}$ falls in the acceptance region, we accept (i.e., do not reject) H_0.

The method for determining the value of C and the location of the rejection and acceptance regions for the two alternative hypotheses, $H_1: \mu > \mu_0$ and $H_1: \mu < \mu_0$, are outlined in Table 1.

THE SENSITIVITY OF A TEST

A type I error occurs if H_0 is rejected when it is actually true. The probability of making a type I error is α. The researcher can control the chance of making this type of error by deciding on the value for α.

A type II error is *not* rejecting H_0 when, in fact, it is false [i.e., $\mu = \mu_1$ (and not μ_0)]. *Prior* to obtaining the sample it is possible to compute the probability of a type II error. We denote this probability $\beta(\mu_1)$, which should be read as "beta mu one." Of course, we can always compute the probability of a type II error after the data have been obtained, but by then the "damage" (a large probability of a type II error) may already have been done.

As the notation $\beta(\mu_1)$ suggests, there is not a single value for β. For $H_1: \mu > \mu_0$, there is a value of $\beta(\mu_1)$ for every value of $\mu_1 > \mu_0$. Similarly, for $H_1: \mu < \mu_0$, there is a value of $\beta(\mu_1)$ for every value of $\mu_1 < \mu_0$.

The researcher will usually have a specific value of μ_1 in mind when setting up the test. He would like to reject H_0 if, in fact, $\mu = \mu_1$ (and not μ_0).

In general, the smaller the value of $\beta(\mu_1)$, the more sensitive the test for the specific alternative $\mu = \mu_1$.

The general procedure for computing $\beta(\mu_1)$ involves three steps (see Table 2).

All other factors remaining constant,

1. The *farther* μ_1 is from μ_0 the *smaller* the value for $\beta(\mu_1)$ (see Figure 6).
2. The *smaller* we make α the *larger* $\beta(\mu_1)$ (see Figure 7).
3. The *larger* the sample size, n, the *smaller* $\beta(\mu_1)$ (see Figure 8).

Two probabilities related to α and $\beta(\mu_1)$ are:

1. $1 - \alpha$ is the probability of *correctly* not rejecting H_0 when $H_0: \mu = \mu_0$ is true.
2. $1 - \beta(\mu_1)$ is the probability of *correctly* rejecting H_0 when $\mu = \mu_1$.

The probability $1 - \beta(\mu_1)$ is called the **power** of the test. The larger the value of $1 - \beta(\mu_1)$, the more powerful the test is for the alternative $\mu = \mu_1$ (that is, the greater the probability of correctly rejecting H_0 if $\mu = \mu_1$).

The various decisions that can be made in a statistical test are summarized in Table 3.

Since the value for μ is unknown, we cannot know whether the eventual decision is correct or not. What we can do, however, is work toward a test in which both α and $\beta(\mu_1)$ in relation to a particular alternative are small and known. This can be done by determining the sample size necessary to obtain specific values for α and $\beta(\mu_1)$.

For one-sided tests of $H_0: \mu = \mu_0$ with a level of significance α the size of sample necessary to guarantee a specific value $\beta(\mu_1)$ is

$$n = \frac{\sigma^2[Z_{100(1-\alpha)} + Z_{100(1-\beta)}]^2}{(\mu_0 - \mu_1)^2} \qquad \text{(Eq. 11.1)}$$

If too large a sample size n is selected, the test may result in a difference between $\bar{x}$ and μ_0 that is statistically significant but of little *practical* significance. When a very large sample is selected, it may therefore be more useful to compute a confidence interval for the unknown μ.

REVIEW PROBLEMS

GENERAL

1. Given the information $H_0: \mu = 18$, $H_1: \mu > 18$, $\sigma = .07$, and $n = 35$, find (a) the rejection region corresponding to $\alpha = .025$; (b) $\beta(18.02)$.
2. Given the information $H_0: \mu = 400$, $H_1: \mu < 400$, $\sigma = 80$, and $n = 60$, find

$\beta(380)$ if (a) $\alpha = .2$; (b) $\alpha = .05$. Comment on the relationship between α and β in the example.

3. Why does it make more sense to compute $\beta(\mu_1)$ *before* rather than after selecting the sample?
4. In a particular problem the probability of not rejecting H_0 when $H_0: \mu = .75$ is true is .9 and the probability of not rejecting H_0 when $\mu = .80$ is .075. What is (a) α; (b) $1 - \beta(.80)$?
5. We wish to test $H_0: \mu = K$ against $H_1: \mu > K$ using a random sample of size $n = 75$. (a) Find the rejection region (in terms of σ and K) corresponding to $\alpha = .05$.

   (b) Complete $\beta(K + \sigma/10)$, the probability of not rejecting H_0 if in fact μ is one-tenth of a standard deviation greater than K.
6. In Review Problem 5, how large does n have to be in order that $\alpha = .05$ and $\beta(K + \sigma/10) = .025$?
7. There are M students in an elementary statistics class. For an assignment on hypothesis testing the instructor uses a computer to generate, independently for each student, a random sample of 100 values from a population with mean 20. Students are asked to use the data to test $H_0: \mu = 20$ against $H_1: \mu > 20$ using a level of significance of α (of course, for each of the M tests H_0 is exactly true).

   (a) Show that the probability (p) of at least one of the M students making a type I error is $p = 1 - (1 - \alpha)^M$.

   (b) Using the value $\alpha = .05$, compute the value of p for $M = 5, 10, 50$, and 100.

   (c) Comment on how the probability of at least one type I error varies with the number of tests.
8. Suppose that we are interested in testing $H_0: \mu_1 - \mu_2 = 0$ against $H_1: \mu_1 - \mu_2 > 0$ with a level of significance α. The variances in the two populations are σ_1^2 and σ_2^2. It can be shown that the (equal) sample size, n, that must be taken from each population in order to guarantee that $\beta(T)$, the probability of not rejecting H_0 if in fact $\mu_1 - \mu_2 = T$, is β is obtained from the formula

$$n = \frac{(\sigma_1^2 + \sigma_2^2)[Z_{100(1-\alpha)} + Z_{100(1-\beta)}]^2}{T^2}$$

   Use this result to solve the following problem. We wish to test whether men in a specific profession earn significantly more than women. This will be done by comparing the mean annual salary in random samples of n men and women in the profession. How large does n have to be in order that $\alpha = \beta(500) = .05$? Assume that the standard deviations of annual income for men and women are \$2500 and \$2000, respectively.

HEALTH SCIENCES

9. Among a very large population the average number of days of work lost per person due to colds is 2.1 per year with a standard deviation of .7 day. A random sample of 50 persons from this population agree to participate in an experiment. They are prescribed a diet high in ascorbic acid (vitamin C) in the hope of reducing the number of days lost to colds. At the end of the year the mean number of days lost to colds will be computed for these 50 persons.

   (a) State the null and alternative hypotheses, carefully defining μ.

   (b) Find the rejection region corresponding to $\alpha = .05$.

   (c) Compute $\beta(1.9)$.
10. The *Boston Globe* (January 23, 1984) reported on research which indicated that among adult Americans the current blood cholesterol level is 210 milligrams per deciliter (mg/dl) with a standard deviation of approximately 40 mg/dl. There is

considerable evidence that as people gradually adjust to a diet lower in cholesterol, this mean value will decline. In two years time how large a sample of adults would be required to test whether the mean cholesterol level had declined to 205 mg/dl? Assume no change in variability and use $\alpha = \beta(205) = .1$.

11. Refer to Review Problem 10. Suppose that next year a researcher plans to use blood cholesterol levels obtained in a study of 20,000 adult Americans in order to test whether μ has declined from 210 to 205. Explain why the researcher would be better advised to find a confidence interval for μ rather than test this hypothesis.

SOCIOLOGY AND GOVERNMENT

12. The Army orders 10,000 advanced four-wheel-drive vehicles from Brash, Inc. The contract specifies that, on average, the vehicles are able to travel at least 2500 miles before a breakdown. The Army will test drive a sample of 35 of the vehicles to be delivered by Brash before accepting delivery of the remainder. The Army will *not* accept the remainder if the mean mileage before breakdown is found to be significantly less than 2500. What is the probability that the Army will be required to accept delivery if, in fact, the mean mileage breakdown over the 10,000 vehicles is only 2100 miles? (Use $\alpha = .05$ and assume $\sigma = 850$ miles)

13. A criminologist wants to know if recent legislation has resulted in a significant decline in the mean length of sentence of defendants imprisoned for marijuana violations in the state. Official reports indicate that last year, prior to the legislation, the mean length of sentence for such violations was 38.6 months and the standard deviation was 9.2 months. The criminologist intends to randomly sample the records of 60 persons convicted of such crimes since the legislation went into effect.

    (a) What is μ in this case?

    (b) What are H_0 and H_1?

    (c) How powerful is this test if in fact $\mu = 35$ months? (Use $\alpha = .05$ and assume that σ is still 9.2 months.)

    (d) Interpret your answer in part (c).

14. Refer to Review Problem 13.

    (a) What decision should the criminologist make if the sample results were $\bar{x} = 37.4$ and $s = 10$ months?

    (b) Would the decision be correct if μ were in fact 35 months?

15. Two years ago a city disbursed an average of \$655 per recipient to the aged and disabled to assist in paying increased heating costs. The standard deviation was \$282. There is some feeling at city hall that, although complete figures are not available, the mean amount disbursed last year was significantly larger than \$655. To investigate this, a random sample of 100 recipient records from the past year will be selected and the amount disbursed will be recorded.

    (a) State the appropriate hypotheses.

    (b) Find the rejection region corresponding to $\alpha = .1$.

    (c) Compute (i) $\beta(670)$; (ii) $\beta(685)$; (iii) $\beta(700)$; (iv) $\beta(715)$.

    (d) City authorities would like to know if the mean amount disbursed went up to \$670 (i.e., there was an increase of \$15 over the previous year). The consulting statistician points out that it is more realistic to expect to detect an increase of \$45 or more. Use your answers in part (c) to explain the statistician's point.

16. Refer to Review Problem 15. How large a sample of recipient records should be taken to obtain the values $\alpha = .1$ and $1 - \beta(670) = .9$?

ECONOMICS AND MANAGEMENT

17. A department store is concerned that credit-card customers are taking longer to pay their bills than in the past. Some years ago a special audit of accounts showed that the mean interval between mailing the bill and receipt of the payment was 12.5 days and the standard deviation was 7.2 days. The company is particularly interested in finding whether the mean interval has increased by 2.5 days. Rather than undertake a complete audit of accounts, the company statistician recommends investigating a random sample of 72 accounts on the grounds that this number will be sufficient to perform a test with the properties that $\alpha = .05$ and $\beta(15) = .1$.
   (a) Verify the statistician's calculations.
   (b) Put yourself in the statistician's shoes and explain to the vice-president what the values $\alpha = .05$ and $\beta(15) = .1$ mean in this context.
18. A financial analyst with a company that deals extensively with municipal bonds is interested in testing whether the mean return (μ) for tax-exempt bonds issued by small cities in the United States is significantly less than 8.2%. The analyst intends to base the test on a random sample of 40 such city bonds. How powerful is the test if, in fact, μ is 8.0%? (Assume that $\sigma = .52\%$ and use $\alpha = .025$.)
19. How many cities would the analyst referred to in Review Problem 18 have needed to select to obtain a test with power .95 when $\mu = 8.0\%$?

EDUCATION AND PSYCHOLOGY

20. A psychologist has developed a written test designed to measure the "independence" of college students. The scale runs from 0 to 30, the higher the score, the more independent the student. The mean score and the standard deviation of scores are expected to be close to 18 and 4.5, respectively. However, initial trials with the new test suggest that the mean score (μ) might be significantly less than 18. To investigate this point, the psychologist intends to ask a random sample of 70 students to take the test.
   (a) Find the rejection region corresponding to the test of $H_0: \mu = 18$ and $H_1: \mu < 18$ using $\alpha = .05$.
   (b) Compute $\beta(17)$.
   (c) Interpret your answer in part (b).
   (d) What decision should be reached if the 70 tests result in the values $\bar{x} = 17.29$ and $s = 4.46$?
   (e) Would the decision in part (d) be correct if in fact $\mu = 17$? Explain.
21. Refer to Review Problem 20. How many students would the psychologist need to test in order to make $\beta(17)$ as low as .25?
22. In this problem you are asked to adapt the techniques of this chapter to tests involving π rather than μ. A college alumnae office wishes to test whether the proportion (π) of all alumni favoring a new graduate program is significantly greater than .6. A random sample of $n = 200$ alumni will be selected and the proportion ($\bar{p}$) favoring the program ascertained.
   (a) Using $\alpha = .05$, find the rejection region for $\bar{p}$ for testing $H_0: \pi = .6$ against $H_1: \pi > .6$. (*Hint:* Use the procedure in Table 1 with $\pi_0 = .6$ in place of μ_0 and $\sqrt{\pi_0(1 - \pi_0)}$ in place of σ.)
   (b) Compute $\beta(.7)$. (Follow the procedure in Table 2 with $\pi_1 = .7$ in place of μ, and $\sqrt{\pi_1(1 - \pi_1)}$ in place of σ.)
23. Refer to Review Problem 22.
   (a) What decision would be appropriate if after the sample was selected it was found that $\bar{p} = .67$?
   (b) Would your decision be correct if in fact $\pi = .7$?

24. An educational foundation is interested in the percentage of college income that is derived from tuition and fees. Call μ the mean of this percentage over all private colleges. The foundation intends to select a sample of $n = 50$ private colleges in order to test $H_0: \mu = 35$ against $H_1: \mu < 35$. Assuming that $\sigma = 5\%$ and using $\alpha = .075$, compute (a) $\beta(33)$; (b) the power of the test if $\mu = 33$.

SCIENCE AND TECHNOLOGY

25. A company producing pig iron is concerned with the mean amount, μ, of silicon (in grams) per kilogram of pig iron produced. Specifically, they wish to test $H_0: \mu = 10$ against $H_1: \mu < 10$ using a random sample of $n = 60$ separate portions of 1 kilogram of pig iron.
    (a) Compute $\beta(9.8)$ assuming that $\alpha = .05$ and $\sigma = 2.0$ grams.
    (b) Comment on the sensitivity of the test for the alternative $\mu_1 = 9.8$.
26. Suppose that in Review Problem 25 the sample results were $\bar{x} = 9.84$ and $s = 1.91$ grams. Use these results to test $H_0: \mu = 10.0$ against $H_1: \mu < 10$.
27. The melting point of a new alloy is required to be 1200°C. Forty samples of the alloy are to be produced with the intention of testing whether the true melting point is significantly less than 1200°C. The standard deviation of the melting point is assumed to be approximately 25°C. If μ is, in fact, only 1185°C, what is the probability that H_0 will be rejected? Use $\alpha = .1$.

12 INFERENCE IN SMALL SAMPLES FROM NORMAL POPULATIONS

SELLING YOUR HOME OR CAR

Home owners planning to sell their houses must decide whether to try to do so themselves or to use the services of a real estate broker. A factor in this decision is the extent to which brokers provide comparable estimates of home values. In one town there are two real estate brokers, Allen and Baldwin. A group of 10 home owners ask each broker for an appropriate asking price for their houses. The results (in thousands of dollars) are as follows:

HOUSE	*A*	*B*
1	82.5	78.9
2	112.0	110.0
3	105.5	108.8
4	122.0	116.2
5	99.5	106.2
6	126.8	118.5
7	87.5	86.5
8	142.5	135.6
9	104.9	110.0
10	96.5	92.0

Do these data suggest that, on average, the estimates made by these two brokers are significantly different? The same approach would apply to selecting a used-car dealer. (Notice that each of the groups of data above contain only 10 values—nowhere near the 30 required in Chapter 10.) See Review Problem 25 on page 465.

SECTION 12.1

THE t DISTRIBUTION

The inferences we made about μ in chapters 9 and 10 were based on the fact that in large samples ($n \geq 30$) the standardized form for the sample mean $\bar{X}$, $(\bar{X} - \mu)/(s/\sqrt{n})$, has approximately the standardized normal $N[0, 1]$ or Z distribution. This is true regardless of the shape of the population being sampled.

Unfortunately, in the case of small samples, $(\bar{X} - \mu)/(s/\sqrt{n})$ does not generally have the $N[0, 1]$ distribution. There are two reasons for this.

1. The central limit theorem frequently does not hold.
2. We can no longer regard the sample standard deviation s as an accurate estimate for σ.

We can resolve this difficulty but only when the *population itself is normal*. In this case the distribution of $(\bar{X} - \mu)/(s/\sqrt{n})$ has a known form called the ***t* distribution.** Unlike the Z distribution, there is no single t distribution but rather a family of distributions. Members of the t family differ in accordance with what is called the "number of degrees of freedom," designated "df" or ν. For example, the t distribution with 17 degrees of freedom is referred to as the t_{17} distribution. For the present we note that the "number of degrees of freedom" is a whole number greater than or equal to 1.

The t distributions were first studied by the statistician W. S. Gosset in a paper published in 1908 under the pseudonym "Student." As a result the t distribution is sometimes referred to as "Student's t distribution." Gosset worked for the Guinness Brewery in Dublin and used the pseudonym "Student" because the brewery objected to its employees publishing articles on their research. Guinness was presumably concerned that other breweries would learn of their work in quality control.

Gosset verified his theoretical derivation by (1) assuming that the heights and middle-finger lengths of 3000 Dublin criminals were each normally distributed and (2) investigating the behavior of $(\bar{X} - \mu)/(s/\sqrt{n})$ for both height and finger length over 750 samples (each of size 4) selected from the 3000 subjects.

The t_3, the t_{20}, and, for purposes of comparison, the standardized normal (Z) distributions are each shown in Figure 1.

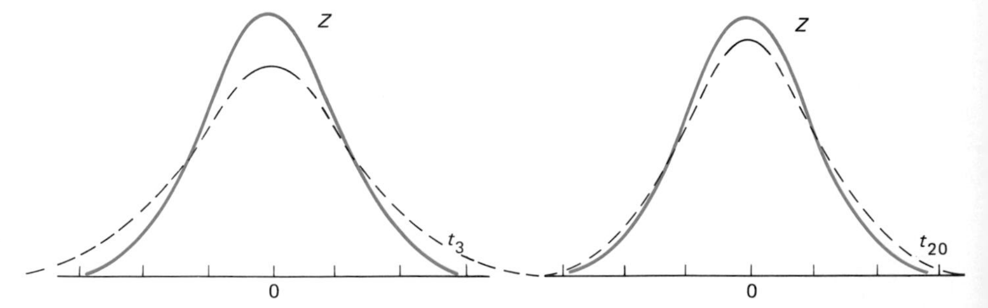

FIGURE 1

t_3, t_{20}, and Z distributions compared

As you can see, t distributions are, like the Z distribution, continuous and symmetric around 0. However, they have a greater variability (that is, are more spread out) than the Z distribution. As you can see in Figure 1, the larger the number of degree of freedom, the more the corresponding t distribution looks like the Z distribution.

As we just indicated if a sample is selected from a normal population $(\bar{X} - \mu)/(s/\sqrt{n})$ will have the t distribution. More specifically, it has the t distribution with $v = n - 1$ degrees of freedom. For example, over repeated random samples of size $n = 15$ from an $N[\mu, \sigma]$ population $(\bar{X} - \mu)/(s/\sqrt{15})$, has the t_{14} distribution. Under similar conditions, if $n = 25$, $(\bar{X} - \mu)/(s/\sqrt{n})$ has the t_{24} distribution. More generally:

DEFINITION 1

In random samples of size n drawn from an $N[\mu, \sigma]$ distribution, $(\bar{X} - \mu)/(s/\sqrt{n})$ has the t_{n-1} distribution.

Notice that Definition 1 is true for *any n*, but we will only need to use this result when $n < 30$. Though there is not complete agreement as to the dividing line between large and small samples, in this and other texts large samples refer to those with $n \geq 30$ and small to those with $n < 30$. As we show in Figure 1, when n is large the t_{n-1} and the $N[0, 1]$ distribution are almost indistinguishable and so, in fact, for $n \geq 30$ we obtain much the same results whichever we use. However, recall from Chapter 7 that for $n \geq 30$ we assumed that $(\bar{X} - \mu)/(s/\sqrt{n})$ is approximately $N[0, 1]$ *regardless of the shape of the population*. Thus, in large samples we did not have to make the restrictive assumption of a normal population as we do in this chapter.

USING THE t TABLES

In this chapter we use the t distributions in much the same way as that in which we used the Z distribution in Chapters 9 and 10. However, because there are many t distributions, we cannot tabulate the area under each of them as thoroughly as we did the Z distribution.

Table A6 contains values associated with a large number of t distributions. A section of the table is shown in Figure 2. The left hand column, designated ν, lists the different degrees of freedom. Each entry in the table serves two different purposes. For instance, consider the value 1.415 in the row corresponding to the t_7 distribution. When we test hypotheses we shall refer to 1.415 as $t_{7,.1}$, indicating that the area to the right of 1.415 (or to the left of -1.415) is .1. When we construct confidence intervals it will then be consistent with our notation in Chapter 9 to refer to 1.415 as $t_{7(.8)}$ (note the parentheses in this case), indicating that the area *between* -1.415 and 1.415 is .8. These relationships are shown in the sketch of the t_7 distribution in Figure 2(a). The sketch of the t_7 distribution in Figure 2(b) shows that the area to the right of $2.365 = t_7$ (or to the left of

FIGURE 2

Section from Table A6 showing values for the t_7 distribution

	$t\nu, .25$	$t\nu, .2$	$t\nu, .15$	$t\nu, .1$	$t\nu, .05$	$t\nu, .025$	$t\nu, .01$	$t\nu, .005$
	$t\nu_{(.5)}$	$t\nu\,(.6)$	$t\nu\,(.7)$	$t\nu\,(.8)$	$t\nu\,(.9)$	$t\nu\,(.95)$	$t\nu\,(.98)$	$t\nu\,(.99)$
1	1.000	1.376	1.963	3.078	6.314	12.706	31.821	63.657
2	.817	1.061	1.386	1.886	2.920	4.303	6.965	9.925
3	.765	.978	1.250	1.638	2.353	3.183	4.541	5.841
4	.741	.941	1.190	1.533	2.132	2.776	3.747	4.604
5	.727	.920	1.156	1.476	2.015	2.571	3.365	4.032
6	.718	.906	1.134	1.440	1.943	2.447	3.143	3.707
7	.711	.896	1.119	1.415	1.895	2.365	2.998	3.500
8	.706	.889	1.100	1.387	1.860	2.306	2.896	3.355
9	.703	.883	1.100	1.383	1.833	2.262	2.821	3.250
10	.700	.879	1.093	1.372	1.813	2.228	2.764	3.169

Area = α = .1 γ = .8   Area = α = .1 t_7 −1.415   0   1.415

(a)

Area = α = .025 γ = .95   Area = α = .025 t_7 −2.365   0   2.365

(b)

$-2.365 = -t_7$) is .025 and that the area between -2.365 and 2.365 is .95. We therefore label 2.365 as both $t_{7,.025}$ and $t_{7(.95)}$.

More generally, we shall describe a value in the table as (1) $t_{\nu,\alpha}$, the value of t_ν having an area α to the right, or (2) as $t_{\nu(\gamma)}$, the value of t_ν such that the area between $-t_{\nu(\gamma)}$ and $t_{\nu(\gamma)}$ is γ.

EXAMPLE 1 Use Table A6 to find the value of (a) $t_{12(.9)}$; (b) $t_{12,.025}$; (c) $t_{18,.1}$; (d) $t_{18(.99)}$.

Solution From Table A6:

(a) $t_{12(.9)} = 1.782$; (b) $t_{12,.025} = 2.179$; (c) $t_{18,.1} = 1.33$; (d) $t_{18(.99)} = 2.878$. ■

PROBLEMS 12.1

1. Using Table A6, find (a) $t_{17,.01}$; (b) $t_{9,.05}$; (c) $t_{24(.9)}$; (d) $t_{12(.99)}$; (e) $t_{14,.1}$; (f) $t_{30(.8)}$.
2. For the following t distributions find the values of t corresponding to A, B, C, and D.

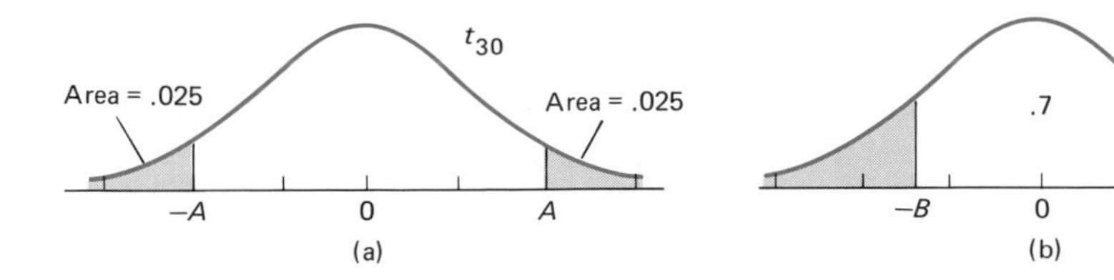

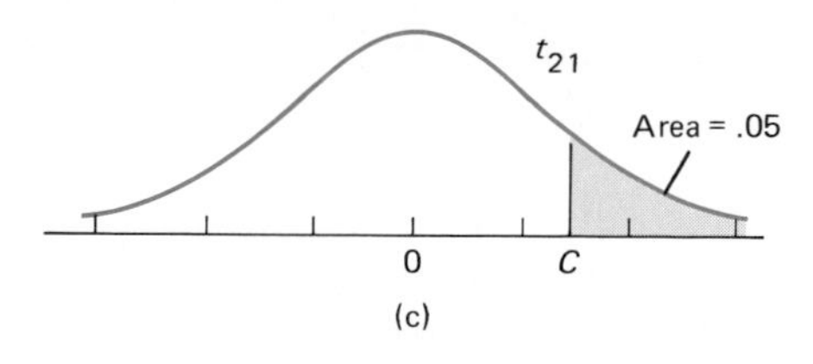

(c)

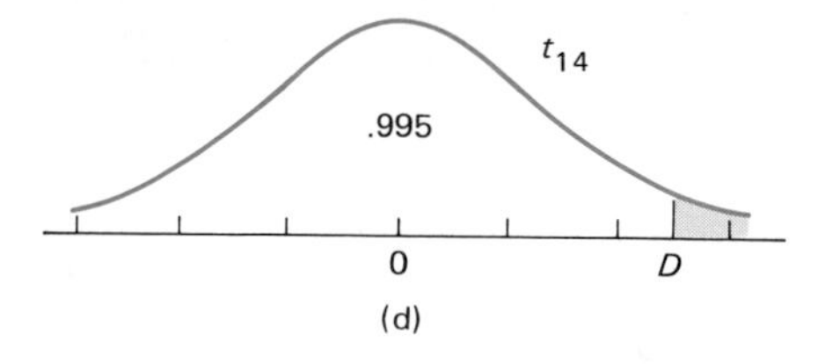

(d)

3. Find (a) $P(t_{17} > 1.333)$; (b) $P(t_{17} > -1.333)$; (c) $P(t_{24} < 1.059)$; (d) $P(-.685 < t_{24} < .685)$.
4. Find A, B, C, D, E, F, and G such that:
   (a) $P(t_4 > A) = .05$ (b) $P(t_4 > B) = .01$
   (c) $P(t_{19} < C) = .01$ (d) $P(t_{19} < D) = .05$
   (e) $P(-E < t_{12} < E) = .8$ (f) $P(t_5 < F) = .975$
   (g) $P(0 < t_{28} < G) = .35$.

Solution to Problem 1

(a) With $\alpha = .01$, $t_{17,.01} = 2.567$.
(b) With $\alpha = .05$, $t_{9,.05} = 1.833$.
(c) With $\gamma = .9$, $t_{24(.9)} = 1.711$.
(d) With $\gamma = .99$, $t_{12(.99)} = 3.055$.
(e) With $\alpha = .1$, $t_{14,.1} = 1.345$.
(f) With $\gamma = .8$, $t_{30(.8)} = 1.310$.

SECTION 12.2

CONFIDENCE INTERVALS FOR THE POPULATION MEAN WITH SMALL SAMPLES FROM A NORMAL POPULATION

In Section 9.2 we noted that in large ($n \geq 30$) samples the sample mean is a point estimate for the unknown μ. But we went one step further and constructed a confidence interval for μ around $\bar{x}$. Since such confidence intervals were based on the central limit theorem, we did not need to make any assumptions about the population.

In small samples $\bar{x}$ is again a point estimate for μ. We can also construct a confidence interval for μ, this time, however, using the t distribution. But we can do this only *if the population is normal.*

The derivation of the confidence interval is similar to that provided for large samples at the beginning of Section 9.2. The result is summarized below.

> DEFINITION 2
>
> If a random sample of size n is selected from a $N[\mu, \sigma]$ population where both μ and σ are unknown, then a $(100\gamma)\%$ confidence interval for μ is
>
> $$\bar{x} - t_{n-1(\gamma)}\frac{s}{\sqrt{n}} \quad \text{to} \quad \bar{x} + t_{n-1(\gamma)}\frac{s}{\sqrt{n}}$$

The form of this confidence interval is similar to the large sample interval except that the value $Z_{[\gamma]}$ is replaced by the value $t_{n-1(\gamma)}$.

EXAMPLE 2 Twelve patients who undergo a particular operation are anesthetized with a new drug. Their recovery times, in hours, are 3.9, 4.2, 4.2, 3.8, 3.6, 4.4, 4.1, 3.4, 4.0, 3.5, 3.7, and 4.2. Assume that (1) these 12 patients are thought of as a random sample from all potential patients given the drug, and (2) that recovery time is normally distributed. Compute and interpret an 80% confidence interval for μ, the mean recovery time for this drug.

Solution Following the results in Chapter 1 with these data, we compute $\Sigma x_i = 47$ and $\Sigma x_i^2 = 185.2$ and thus

$$\bar{x} = \frac{\Sigma x_i}{12} = \frac{47}{12} = 3.917 \text{ hours}$$

and

$$s = \sqrt{\frac{12(185.2) - 47^2}{12(11)}} = \sqrt{.10152} = .3186$$

In this case we need, from Table A6, $t_{n-1(\gamma)} = t_{11(.8)} = 1.363$. From Definition 2 an 80% confidence interval for μ is

$$3.917 \mp 1.363\left(\frac{.3186}{\sqrt{12}}\right) = 3.917 \mp .1254$$

$$= 3.79 \text{ to } 4.04$$

We can be 80% confident that the mean recovery time over all potential users of the drug lies between 3.79 and 4.04 hours.

■

Notice that in Example 2 we rounded off only the final answer. This is in accordance with our rules for rounding in Section 1.3. Also, in accordance with these rules, the bounds were rounded to one more decimal place than the original data.

EXAMPLE 3 A truck rental company selects a random sample of 25 completed rental agreements from last month. The mean and standard deviation of the number of miles traveled are $\bar{x} = 173.41$ and $s = 64.32$, respectively. Compute a 95% confidence interval for μ, the mean mileage per agreement for the month. What assumptions did you make?

Solution From Table A6, $t_{n-1(\gamma)} = t_{24(.95)} = 2.064$. A 95% confidence interval for μ is therefore

$$173.41 \mp 2.064\left(\frac{64.32}{\sqrt{25}}\right) = 173.41 \mp 26.551$$

$$= 146.86 \text{ to } 199.96$$

We can be 95% confident that μ is in this interval. This conclusion is valid only if the distribution of mileage per agreement last month is normal. ■

In practice, the population need only be *approximately* normal for the confidence interval in Definition 2, and hence that in Example 3, to be appropriate. We take up this point in detail later in this chapter.

Provided the population is (near) normal, the confidence interval in Definition 2 can be used for *any* n. In practice, however, when $n \geq 30$ the large sample confidence interval for μ,

$$\bar{x} - z_{[\gamma]}\frac{s}{\sqrt{n}} \quad \text{to} \quad \bar{x} + z_{[\gamma]}\frac{s}{\sqrt{n}}$$

is used. This requires no assumptions about the shape of the population.

PROBLEMS 12.2

5. In a random sample of size n from a normal population the sample mean is 70.4 and the sample standard deviation is 6.9. Compute an 80% confidence interval for μ if (a) $n = 5$; (b) $n = 15$; (c) $n = 25$.
6. The following is a random sample from a normal population. Compute a 95% confidence interval for the population mean.

17.3	18.9	17.4	16.1	18.2	18.5	
16.6	18.2	17.6	17.2	18.4	18.1	$\sum x_i = 316.9$
17.0	16.5	16.8	17.0	18.0	19.1	$\sum x_i^2 = 5591.83$

7. In a random sample of the records of 22 house sales in a city in one month the mean sale price was $85,360 with a standard deviation of $24,105.
   (a) Compute a 90% confidence interval for μ, the mean sale price in the city for that month.
   (b) What assumptions did you make in computing this confidence interval?
   (c) Interpret the interval.
8. An author selects a random sample of 12 pages from his new book. A count of the number of words per page resulted in the following values:

365	402	369	379	413	444	
359	411	399	420	481	377	$\sum x_i^2 = 1{,}949{,}229$

   (a) Compute a 90% confidence interval for the mean number of words per page.
   (b) What assumptions were necessary in computing this interval?
   (c) If the book contains 575 pages, compute a 90% confidence interval for the total number of words in the entire book.
9. The following distribution represents the number of visits to the movies in the last two months as reported by a random sample of 28 teenagers in a neighborhood.

NUMBER OF VISITS	0	1	2	3	4	5	6	7	8	9
NUMBER OF TEENAGERS	2	4	5	7	5	2	0	2	0	1

Compute the mean and standard deviation of the number of visists and a 90% confidence interval for the mean number of visits per teenager in the neighborhood.

10. In a sample of 17 recent Ph.D.s in English the mean time interval between graduation from college and graduate school is 5.2 years with a standard deviation of 1.4 years.

    (a) Compute a 90% confidence interval for the mean time interval over all such Ph.D.s.

    (b) Answer the same question if the mean and standard deviation were based on 70 graduates.

11. The number of visitors to a science museum is recorded on each of a random sample of six weekdays. The mean and standard deviation of the number of visitors are $\bar{x} = 2012.4$ and $s = 363.9$.

    (a) Find an 80% confidence interval for the mean number of visitors per weekday during the entire year.

    (b) What assumption did you make?

12. A random sample of 18 students at a college are asked to record the approximate number of hours of sleep they receive on a particular night. The results are summarized as follows (in hours):

$$\sum x_i = 130.2 \qquad \sum x_i^2 = 948.53$$

Find a 90% confidence interval for the mean number of hours of sleep for all students at the college.

13. A researcher is interested in the mean length, μ, of marriages that end in divorce in a California county. In a random sample of 12 divorce records the following numbers of years of marriage were recorded:

3.8 7.1 6.2 4.9 8.1 1.9 31.8 3.8 2.6 2.8 23.6 5.2

Why do these data suggest that the normality assumption would not be appropriate in this case?

Solution to Problem 5

Using Definition 2, an 80% confidence interval for μ is

$$70.4 \mp t_{n-1(.8)}\left(\frac{6.9}{\sqrt{n}}\right)$$

(a) With $n = 5$, $t_{4(.8)} = 1.533$ and the interval is

$$70.4 \mp 1.533\left(\frac{6.9}{\sqrt{5}}\right) = 70.4 \mp 4.73$$

$$= 65.7 \text{ to } 75.1$$

(b) With $n = 15$, $t_{14(.8)} = 1.345$ and the interval is

$$70.4 \mp 1.345\left(\frac{6.9}{\sqrt{15}}\right) = 70.4 \mp 2.40$$

$$= 68.0 \text{ to } 72.8$$

(c) With $n = 25$, $t_{24(.8)} = 1.318$ and the interval is

$$70.4 \mp 1.318\left(\frac{6.9}{\sqrt{25}}\right) = 70.4 \mp 1.82$$

$$= 68.6 \text{ to } 72.2$$

SECTION 12.3

TESTING HYPOTHESES ABOUT THE MEAN IN SMALL SAMPLES FROM A NORMAL POPULATION

ONE-SIDED TESTS In Chapter 10 we tested hypotheses about μ in large samples by computing the P value and comparing it with $\alpha = .01$ and $\alpha = .05$. At the end of Section 10.2 we also described an equivalent procedure based on the value for $z = (\bar{x} - \mu)/(s/\sqrt{n})$ rather than the P value.

Our approach to testing hypotheses about μ in small samples parallels this second approach but is based on the t rather than the Z distribution and is only valid if the population itself is at least approximately normal.

As with large samples we begin with one-sided tests of the form $H_0: \mu = \mu_0$ and either $H_1: \mu > \mu_0$ or $H_1: \mu < \mu_0$. From Definition 1, if the sample is selected from a normal population, then if H_0 is true, $\mu = \mu_0$ and $t = (\bar{X} - \mu_0)/(s/\sqrt{n})$ has the t_{n-1} distribution. Evidence in favor of H_0 occurs where the value for $\bar{x}$ is close to μ_0, so that the value for $t = (\bar{x} - \mu_0)/(s/\sqrt{n})$ will be close to 0. We will reject H_0 in favor of $H_1: \mu > \mu_0$ if we get a value for $\bar{x}$ unusually far *above* μ_0 or equivalently a value of $t = (\bar{x} - \mu_0)/(s/\sqrt{n})$ unusually greater than 0. Similarly, we will reject H_0 in favor of $H_1: \mu < \mu_0$ if we get a value for $\bar{x}$ unusually far *below* μ_0 or, equivalently, a value for $t = (\bar{x} - \mu_0)/(s/\sqrt{n})$ unusually less than 0.

An example will show how this procedure, known as the "t test," works.

EXAMPLE 4 The following data were the recovery times for the 12 patients in Example 2:

3.9 4.2 4.2 3.8 3.6 4.4 4.1 3.4 4.0 3.5 3.7 4.2

Assume that prior to the use of this new drug the mean recovery time was 3.6 hours. One of the doctors involved in the experiment claims that although the new drug has several advantages, it also appears to increase the length of time for recovery. Calling μ the (potential) mean recovery time for patients receiving the new drug, we test $H_0: \mu = 3.6$ hours against $H_1: \mu > 3.6$ hours. The first hypothesis, H_0, states there is no increase in mean recovery time with the new drug, while H_1 states that there is an increase.

Solution As we indicated above, we will reject H_0 if

$$t = \frac{\bar{x} - \mu_0}{s/\sqrt{n}} = \frac{\bar{x} - 3.6}{s/\sqrt{n}}$$

is sufficiently greater than 0. But how much greater? To be consistent with large-sample methods we will (1) reject H_0 at the .01 level of significance if $t = (\bar{x} - 3.6)/(s/\sqrt{n})$ exceeds $t_{11,.01} = 2.718$, (2) reject H_0 at the .05 level of significance if $t = (\bar{x} - 3.6)/(s/\sqrt{n})$ lies between $t_{11,.05} = 1.796$ and $t_{11,.01} = 2.718$, and (3) will *not* reject H_0 at the .05 level of significance if $t = (\bar{x} - 3.6)/(s/\sqrt{n})$ is *less* than $t_{11,.05} = 1.796$.

In Example 2 we computed $\bar{x} = 3.917$, $s = .3186$, and $s/\sqrt{n} = .3186/\sqrt{12} = .092$, so that $t = (3.917 - 3.6)/.092 = 3.45$. This observed value for t together with $t_{11,.05}$ and $t_{11,.01}$ are shown in Figure 3.

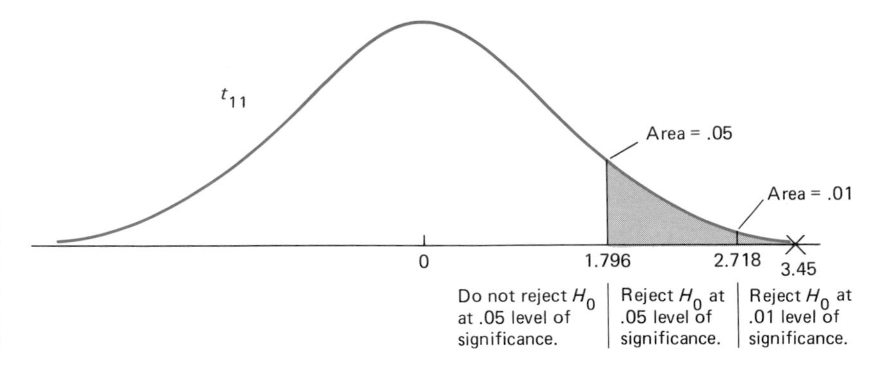

FIGURE 3

Observed value $t = 3.45$ shown relative to $t_{11,.05} = 1.796$ and $t_{11,.01} = 2.718$

Since $t = 3.45$ exceeds $t_{11,.01} = 2.718$, we reject H_0 at the .01 level of significance. The data strongly suggest that the true mean recovery for the new drug is significantly greater than 3.6 hours. We emphasize that this conclusion is valid only if the distribution of recovery time is at least approximately normal.

Incidentally, the P value corresponding to $t = 3.45$ is $P(t_{11} > 3.45)$, which is the tiny area to the right of 3.45 under the t_{11} distribution in Figure 3. Because of the lack of detail in Table A6, we cannot find this probability exactly. However, since the area to the right of 2.718 is .01, the area to the right of 3.45 is certainly smaller than .01, so were we to use the P value we would, again, reject H_0 at the .01 level of significance.

■

This example illustrates that rejecting H_0 at the .01 level of significance if $t \geq 2.718$ is exactly equivalent to rejecting H_0 if the P value $\leq .01$.

We do not compute the P value in small samples because Table A6 for the t distribution lacks the necessary detail. (The tabulated values for the t_{17} distribution, for example, occupy one line in Table A6, compared to the two pages of values for Z in Table A3.)

In Table 1 we summarize and generalize the procedure for testing

$H_0: \mu = \mu_0$ against either $H_1: \mu > \mu_0$ or $H_1: \mu < \mu_0$. We emphasize the importance of the normality assumption by stating it at the foot of the table.

TABLE 1

Procedure for performing one-sided tests of $H_0: \mu = \mu_0$ in small samples from a normal population

STEP	ALTERNATIVE HYPOTHESIS (a) $H_1: \mu > \mu_0$	(b) $H_1: \mu < \mu_0$
1. Compute:	$t = \dfrac{\bar{x} - \mu_0}{s/\sqrt{n}}$	
2. Find $t_{n-1,.05}$ and $t_{n-1,.01}$ from Table A6.		
3. *Conclusion:* Using $\alpha = .05$ or $.01$, reject H_0 at the α level of significance if:	$t \geq t_{n-1,\alpha}$	$t \leq t_{n-1,\alpha}$
Assumption: The sample is drawn at random from a normal population		

In Example 4 we followed the procedure in Table 1 with $\mu_0 = 3.6$, $n = 12$, $\bar{x} = 3.917$ and $s = .3186$.

EXAMPLE 5 An industry standards laboratory wants to test whether the true mean breaking strength of sections of cable is significantly *less* than 1600 pounds. The laboratory chooses a random sample of 24 sections and finds the breaking strength of each section. What can they conclude if $s = 177.83$ pounds and (a) $\bar{x} = 1570.81$ pounds; (b) $\bar{x} = 1491.22$ pounds; (c) $\bar{x} = 1532.79$ pounds? What assumptions are necessary in this case?

Solution The null and alternative hypotheses are $H_0: \mu = 1600$ and $H_1: \mu < 1600$, where μ is the true mean breaking strength per cable section. We follow the procedure in Table 1 with $n = 24$, $\mu_0 = 1600$, $s = 177.83$, and $s/\sqrt{n} = 177.83/\sqrt{24} = 36.30$.

(a) If $\bar{x} = 1570.81$, $t = (1570.81 - 1600)/36.3 = -.804$. From Table A6, $t_{23,.05} = 1.714$ and $t_{23,.01} = 2.500$. Notice that we are working with negative values of t and therefore since $-.804$ is *greater* than $t_{23,.05} = -1.714$, we do not reject H_0 at the .05 level of significance. The data suggest that the mean breaking strength is not significantly less than 1600 pounds. The value $-.804$ is shown in relation to -1.714 in Figure 4.

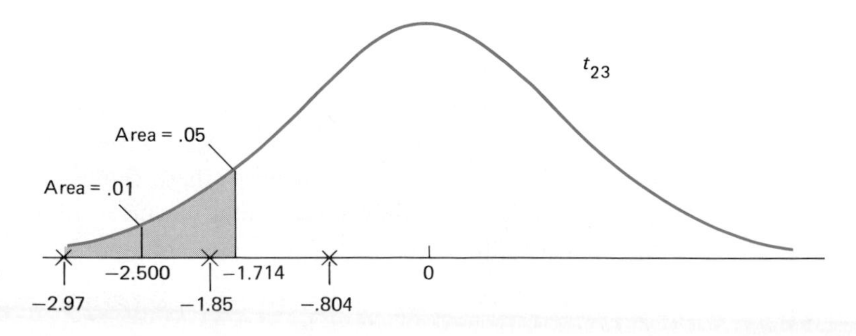

FIGURE 4

t values in Example 5

(b) If $\bar{x} = 1491.22$, $t = (1491.22 - 1600)/36.3 = -3.00$. Since this value is less than -2.500, we reject H_0 at the .01 level of significance. The data strongly suggest that the mean breaking strength is significantly below 1600 pounds.

(c) If $\bar{x} = 1532.79$, $t = (1532.79 - 1600)/36.3 = -1.85$. Since this value lies between -1.714 and -2.500, we reject H_0 at the .05 level of significance. The data again suggest that the mean breaking strength is significantly below 1600 pounds.

The values -3.00 and -1.85 are both shown in relation to -2.500 and -1.714 in Figure 4.

The tests above are valid *only* if the distribution of breaking strength is approximately normal.

■

TWO-SIDED TESTS

There is also a two-sided t test of $H_0: \mu = \mu_0$. In Table 2 we provide the procedure for testing whether to reject $H_0: \mu = \mu_0$ in favor of $H_1: \mu \neq \mu_1$, using $\alpha = .05$ or $\alpha = .01$.

TABLE 2

Procedure for testing $H_0: \mu = \mu_0$ against $H_1: \mu \neq \mu_0$ in small samples from a normal population

STEP	$H_1: \mu \neq \mu_0$
1. Compute:	$t = \dfrac{\bar{x} - \mu_0}{s/\sqrt{n}}$
2. Find $t_{n-1,.025}$ and $t_{n-1,.005}$ from Table A6.	
3. *Conclusion:* Using $\alpha = .05$ or .01, reject H_0 at the α level of significance if:	$\lvert t \rvert \geq t_{n-1,\alpha/2}$
Assumption: The sample is drawn at random from a normal population	

The next example illustrates how the procedure in Table 2 is applied.

EXAMPLE 6

A passport office claims to take an average of 17.5 days to process an application for a first passport. An internal investigation at the passport office involves sampling 16 applications and recording the number of days from application to the date on which each passport is available to the applicant. Do the following data suggest that the mean time to process a passport application is significantly different from 17.5 days: (a) $\bar{x} = 19.875$, $s = 3.934$; (b) $\bar{x} = 16.9375$, $s = 3.934$?

Solution

We test $H_0: \mu = 17.5$ against $H_1: \mu \neq 17.5$, where μ is the true mean time to process a passport application. In both cases, $s/\sqrt{n} = 3.934/\sqrt{16} = .9835$.

(a) $\bar{x} = 19.875$. Following the procedure in Table 2, $t = (19.875 - 17.5)/.9835 = 2.415$, so that $\lvert t \rvert = 2.415$. Also, from Table A6, $t_{15,.025} = 2.132$ and $t_{15,.005} = 2.947$. Since $\lvert t \rvert$ lies *between* these values, we reject H_0 at the .05 level of significance. The data suggest that the mean time to process a passport is significantly greater than 17.5 days.

(b) $\bar{x} = 16.9375$. In this case $t = (16.9375 - 17.5)/.9835 = -.57$. Thus

$|t| = .57$. This value is much less than 2.132, so we cannot reject H_0 at the .05 level of significance. The data suggest that the mean processing time is not significantly different from 17.5 days.

Notice that for $\alpha = .05$ we use $t_{n-1,.025}$ and for $\alpha = .01$ we use $t_{n-1,.005}$. Note also that had $|t|$ been 4.132, we would have rejected H_0 at the .01 level of significance since this value is greater than 2.947.

As we emphasized in Table 2, we assume that the distribution of processing time is normal.

■

t TESTS WITH A SINGLE VALUE FOR α

The procedures in Tables 1 and 2 were based on $\alpha = .05$ and $\alpha = .01$, as were those in Chapter 10. These procedures are readily generalized to any value for α. We provide the details in Definition 3.

> DEFINITION 3
>
> Assume a random sample of size n from a normal population. As before, $t = (\bar{x} - \mu_0)/(s/\sqrt{n})$. We are to test $H_0: \mu = \mu_0$ at the α level of significance.
>
> (a) For $H_1: \mu > \mu_0$, reject H_0 if $t \geq t_{n-1,\alpha}$.
>
> (b) For $H_1: \mu < \mu_0$, reject H_0 if $t \leq -t_{n-1,\alpha}$.
>
> (c) For $H_1: \mu \neq \mu_0$, reject H_0 if $|t| \geq t_{n-1,\alpha/2}$.

Suppose in Example 5(c) that the researcher used $\alpha = .025$ for the test of $H_0: \mu = 1600$ against $H_1: \mu < 1600$. In this case, $t = -1.85$. From Table A6, $-t_{23,.025} = -2.069$. Since -1.85 is *not* < -2.069, we do not reject H_0 at the .025 level of significance.

Finally, note that the t distribution is not appropriate when making inferences about proportions. This is true because a 0/1 population *cannot* be even approximately normal.

PROBLEMS 12.3

In the following problems, whether you are explicitly asked to do so or not, it would be helpful to (i) define μ, (ii) state H_0 and H_1, and (iii) interpret the results of your test in terms of the context of the problem.

14. The complaints of bank customers about the number of people waiting in line to see an officer suggest that on the average there are six people in line. In an attempt to reduce this figure the bank introduces a scheme which involves a more flexible use of staff. Two weeks after beginning the scheme the bank records the number of people in line at 14 randomly chosen times. The results are

    2 4 7 9 4 6 7 6 4 3 3 4 5 6

    (a) Define μ in this case.

(b) What conclusions can be drawn from these data?

(c) What assumptions did you make in performing this test?

15. Answer Problem 14 if the value $\alpha = .025$ had been used.

16. A telephone company executive is interested in learning whether the mean monthly charges for long-distance calls for private subscribers is much greater than \$25. In a random sample of 25 such customers' bills the long-distance charges were recorded. The summary results are:

$$\sum x_i = 763.61$$

$$\sum x_i^2 = 26{,}653.7470$$

(a) Compute the mean and the standard deviation of charges.

(b) What are the appropriate null and alternative hypotheses?

(c) Use the results in part (a) to test H_0.

(d) What assumption did you make in part (c)?

17. Last year an intensive study indicated that the mean level of carbon monoxide for a city was 9.4 parts per million. Since then the city has taken steps which it is hoped will reduce this figure. In a pilot sample of 18 randomly selected sites in the city the following carbon monoxide readings (in parts per million) were obtained:

8.6	6.4	7.2	10.5	8.7	10.7	
5.4	5.7	3.9	4.5	3.6	7.6	
6.8	10.9	10.2	7.9	9.4	7.9	$\sum x_i^2 = 1117.29$

(a) On the basis of the sample mean alone, do these data suggest a significant decline in the mean carbon monoxide level in the city?

(b) Use a t test to check your conclusion. What assumptions did you make in performing the test?

18. The mean pulse rate for adult males is approximately 70 beats per minute. In a sample of 12 males classified as "20% overweight" the mean pulse rate was 76.8 with a standard deviation of $s = 4.2$ beats. Do these findings suggest that the mean pulse rate for all such overweight males is significantly greater than 70? What assumption is implicit in your test?

19. Would your conclusion in Problem 18 have been any different had we used the value $\alpha = .005$?

20. It is claimed that cartons of a particular cereal weigh an average of 510 grams. A family wonders if this figure is an overstatement. When they weigh seven unopened cartons of this brand the results are (in grams)

503 512 510 505 508 507 506

$$\sum x_i^2 = 1{,}801{,}427$$

(a) Use these data to test whether the true mean weight of such cartons is significantly less than 510 grams.

(b) What assumption did you have to make in performing this test?

21. A psychologist has constructed a test designed to measure emotional maturity. She expects the mean score on the test to be 32. A random sample of 10 adults take

the test, with the following results:

25 26 19 28 37 29 28 31 26 22

(a) Use these data to perform a two-sided test to determine whether the (potential) mean score for the test is significantly different from 32. (b) What assumption is implicit in your test?

$$\sum x_i^2 = 7561$$

22. A pharmaceutical company has been marketing a brand of antacid which, they are satisfied, relieves heartburn in 15 minutes on the average. They are now developing a "super" version of this brand which they hope will reduce this mean time. When the new product is tested on 24 subjects, the average time to relieve heartburn is 13.8 minutes. The standard deviation was only $s = 2.3$ minutes. Do these results suggest that the mean time to relieve heartburn with the new product is significantly less than 15 minutes?
23. The class that graduated from a college four years ago took an average of 4.9 science courses. Since then a new distribution requirement has been introduced. The registrar is interested in the effect this requirement has had on the enrollment in science courses. In a pilot sample of the records of 25 seniors about to graduate the mean number of science courses was $\bar{x} = 6.4$. The standard deviation was $s = 3.9$. Do these results suggest that the mean number of science courses taken by the present graduating class is significantly different from 4.9?
24. In Problem 23, how would your conclusion change if the value $\alpha = .1$ had been used?
25. As part of a larger study on the effect of sleep on student performance, a researcher is interested in finding whether the students at a college sleep significantly less than 8.0 hours per night on the average. A random sample of 18 students are asked to record the approximate number of hours they sleep on a particular night. The results are summarized as follows:

$$\sum x_i = 130.2 \qquad \sum x_i^2 = 948.53$$

What can the researcher conclude from these data?

26. A sociologist claims that teenagers in a town visit the movies an average of four times every two months. The following distribution represents the number of visits to the movies in the last two months as reported by a random sample of 28 teenagers in a particular town.

NUMBER OF VISITS	0	1	2	3	4	5	6	7	8	9
NUMBER OF TEENAGERS	2	4	5	7	5	2	0	2	0	1

Perform a two-sided test on these data to test the sociologist's claim.

Solution to Problem 14

(a) Here μ is the mean number of people in line *after* the scheme. We test $H_0: \mu = 6$ against $H_1: \mu < 6$.

(b) Following the procedures in Table 1, we compute

$$\bar{x} = 5.0 \quad \text{and} \quad s = 1.9215 \text{ persons and hence}$$

$$t = \frac{\bar{x} - \mu_0}{s/\sqrt{n}} = \frac{5.0 - 6.0}{1.9215/\sqrt{14}} = -1.947$$

From Table A6, $t_{13,.05} = 1.771$ and $t_{13,.01} = 2.650$. Since -1.947 lies between -1.771 and -2.650, we reject H_0 at the .05 level of significance. The data suggest that the mean number of people in line *after* the scheme is significantly *less* than 6.0.

(c) We assume that the distribution of the number of people in line after the scheme is approximately normal. (The number of people in line cannot be exactly normally distributed since it is a discrete, not a continuous random variable.)

SECTION 12.4

INFERENCES ABOUT $\mu_1 - \mu_2$ BASED ON TWO SMALL, INDEPENDENT SAMPLES FROM NORMAL POPULATIONS

ONE-SIDED TESTS

In Sections 9.4 and 10.4 we indicated how to obtain confidence intervals for, and test hypotheses about, the differences between two population means $\mu_1 - \mu_2$ with large samples ($n_1, n_2 \geq 30$). But what if only small samples ($n_1, n_2 < 30$) are available? Assume, for example, that a consumer organization suspects an unusual difference in the recorded use of electricity by homeowners in two different cities. The organization has limited funds, so it obtains data on the number of kilowatthours for $n = 12$ homeowners in city I and $n_2 = 8$ in city II. The results are:

CITY I	32.9	24.4	23.8	32.1	38.9	35.6	27.4	44.8	37.5	40.9	17.6	40.3
CITY II	30.4	28.1	34.1	19.6	23.4	35.7	40.2	26.4				

Do these data suggest that homeowners in city I used more electricity on the average than those in city II?

Had the sample sizes been $n_1 = 120$ and $n_2 = 80$, then from Section 10.4 we could have used the fact that

$$\frac{(\bar{X}_1 - \bar{X}_2) - (\mu_1 - \mu_2)}{\sqrt{s_1^2/n_1 + s_2^2/n_2}} \quad \text{is approximately } N[0, 1]$$

to test whether μ_1 was significantly greater than μ_2.

You might assume that where the *two populations were normal,* then, as in the one-sample case, the standardized form above would have a t distribution. Unfortunately, in this case statistical life is not so simple and the requirements for a t distribution in the two-sample case are in fact more severe. In addition to both populations being normal, they must also have *equal standard deviations* (i.e., $\sigma_1 = \sigma_2$). Where this is true we denote their common value σ, so that

$\sigma_1 = \sigma_2 = \sigma$. Since σ is rarely known, an estimate is needed. Keeping these ideas in mind, the details of the two-sample t distribution follow:

DEFINITION 4

A random sample of $n_1 < 30$ is taken from an $N[\mu_1, \sigma]$ population. A random sample of size $n_2 < 30$ is taken independently from an $N[\mu_2, \sigma]$ population. Assume that all the parameters μ_1, μ_2, and σ are unknown. Then the standardized random variable

$$\frac{(\bar{X}_1 - \bar{X}_2) - (\mu_1 - \mu_2)}{s_p\sqrt{1/n_1 + 1/n_2}}$$

has the t distribution with $n_1 + n_2 - 2$ degrees of freedom.

In this new random variable s_1^2 and s_2^2 are gone and a new term, s_p, appears. We define the *square* of this term, s_p^2, as follows:

$$s_p^2 = \frac{(n_1 - 1)s_1^2 + (n_2 - 1)s_2^2}{n_1 + n_2 - 2} \qquad \text{(Eq. 12.1)}$$

s_p^2 is called the **pooled variance**

Where did s_p in Definition 4 come from? First note that since we assumed that the standard deviations and hence the variances in the two populations were equal, the variances of the two sample means are, respectively, $\text{Var}(\bar{X}_1) = \sigma^2/n_1$ and $\text{Var}(\bar{X}_2) = \sigma^2/n_2$. The variance of the difference in sample means is then

$$\text{Var}(\bar{X}_1 - \bar{X}_2) = \frac{\sigma^2}{n_1} + \frac{\sigma^2}{n_2} \qquad \text{(see Section 7.2)}$$

$$= \sigma^2\left(\frac{1}{n_1} + \frac{1}{n_2}\right)$$

and

$$\text{SD}(\bar{X}_1 - \bar{X}_2) = \sigma\sqrt{\frac{1}{n_1} + \frac{1}{n_2}}$$

We could use either of the two sample variances s_1^2 or s_2^2 as an estimate of the unknown σ^2. However, to base the estimate on as many observations as possible, it makes sense to combine or pool the two estimates. On the other hand, since one of the estimates s_1^2 or s_2^2 may be based on more observations than the other, we give more weight to the one based on the larger number of observations. The result is s_p^2 in Equation 12.1. For instance, if $n_1 = 5$ and $n_2 = 25$,

$$s_p^2 = \frac{4s_1^2 + 24s_2^2}{28}$$

(We followed much the same reasoning when we computed the combined proportion $\bar{p}_c$ in Section 10.4.)

DEFINITION 5

A test of the null hypothesis that there is no difference between the population means (i.e., $H_0: \mu_1 - \mu_2 = 0$) is based on the fact that if H_0 is true, then from Definition 4,

$$t = \frac{\bar{X}_1 - \bar{X}_2 - 0}{s_p\sqrt{1/n_1 + 1/n_2}}$$

has the $t_{n_1+n_2-2}$ distribution.

If both populations are normal with equal standard deviations, sampling is independent, and H_0 is true, we would expect $\bar{X}_1 - \bar{X}_2$ and hence t in Definition 5 to be close to zero. For $H_1: \mu_1 - \mu_2 > 0$ we reject H_0 if t is unusually large, and for $H_1: \mu_1 - \mu_2 \neq 0$ we reject H_0 if $|t|$ is unusually large. Recall from Section 10.4 that by designating μ_1 as the mean thought to be the larger of the two, we need only consider alternative hypotheses of the form $H_1: \mu_1 - \mu_2 > 0$ and *not* $H_1: \mu_1 - \mu_2 < 0$. The details for testing $H_0: \mu_1 - \mu_2 = 0$ against $H_1: \mu_1 - \mu_2 > 0$ are given in Table 3.

TABLE 3 Procedure for testing $H_0: \mu_1 - \mu_2 = 0$ against $H_1: \mu_1 - \mu_2 > 0$ in small samples from normal populations

STEP	$H_1: \mu_1 - \mu_2 > 0$
1. Compute:	(a) $\bar{x}_1$, $\bar{x}_2$, s_1^2, and s_2^2
	(b) $s_p^2 = \dfrac{(n_1 - 1)s_1^2 + (n_2 - 1)s_2^2}{n_1 + n_2 - 2}$
and hence:	(c) $t = \dfrac{\bar{x}_1 - \bar{x}_2 - 0}{s_p\sqrt{1/n_1 + 1/n_2}}$
2. Find $t_{n_1+n_2-2,.05}$ and $t_{n_1+n_2-2,.01}$ from Table A6.	
3. *Conclusion:* Using $\alpha = .05$ or .01, reject H_0 at the α level of significance if:	$t \geq t_{n_1+n_2-2,\alpha}$

Assumption: Random samples are drawn independently from two normal populations having the same standard deviation

We apply this procedure in the next example.

EXAMPLE 7 Use the electricity data to test whether the mean use in city I is significantly greater than that in city II.

Solution The null and alternative hypotheses are $H_0: \mu_1 - \mu_2 = 0$ and $H_1: \mu_1 - \mu_2 > 0$, where μ_1 and μ_2 are the mean number of kilowatt hours per household in the two cities. We assume that the use of electricity in both cities is normally distributed with equal standard deviations.

The summary data based on the two samples are computed as follows:

	Σx_i	Σx_i^2	n
CITY I	396.2	13,825.90	12
CITY II	237.9	7395.79	8

Thus $\bar{x}_1 = 396.2/12 = 33.017$ and $\bar{x}_2 = 237.9/8 = 29.738$, and

$$s_1^2 = \frac{12(13{,}825.90) - 396.2^2}{12(11)} = 67.6997$$

$$s_2^2 = \frac{8(7395.79) - 237.9^2}{8(7)} = 45.8912$$

From Equation 12.1 the pooled variance is

$$s_p^2 = \frac{11(67.6997) + 7(45.8912)}{18} = 59.2186$$

and thus $s_p = 7.695$. Continuing with the procedure in Table 3, we have

$$t = \frac{33.017 - 29.738}{7.695\sqrt{1/12 + 1/8}} = \frac{3.279}{3.512} = .934$$

From Table A6, with $\nu = n_1 + n_2 - 2 = 18$, $t_{18,.05} = 1.734$ and $t_{18,.01} = 2.552$. Since $t = .934 < 1.734$, we cannot reject H_0 at the .05 level of significance. The data suggest that the mean electricity use in city I is not significantly greater than that in city II.

The value $t = .934$ is shown in relation to $t_{18,.05} = 1.734$ and $t_{18,.01} = 2.552$ in Figure 5. In this case the P value corresponding to $t = .934$ is $P(t_{18} > .934)$, which is the area to the right of .934 under the t_{18} distribution in Figure 5. Again, because the t table is not as complete as the Z table, we cannot find the appropriate P value to compare with α. However, since .934 is less than $t_{18,.05}$, the P value (the area to the right of .934) will exceed .05.

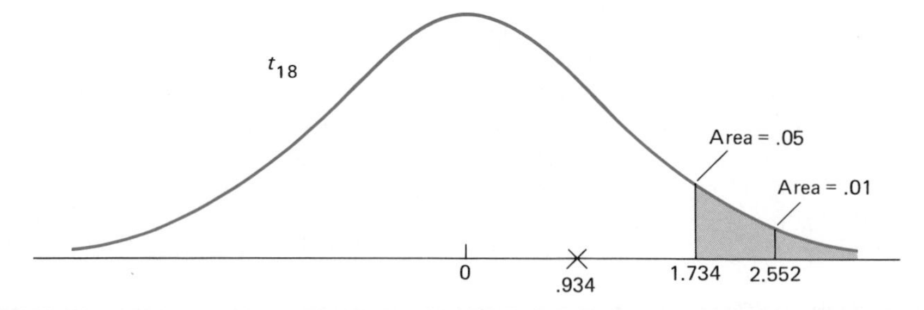

FIGURE 5

Value $t = .934$ in Example 7 shown in relation to $t_{18,.05}$ and $t_{18,.01}$

■

Notice how closely the details in Table 3 correspond to those in Table 1 for testing $H_0: \mu = \mu_0$ against $H_1: \mu > \mu_0$. The only differences are that in Table 3

$$t = \frac{\overline{x}_1 - \overline{x}_2 - 0}{s_p\sqrt{1/n_1 + 1/n_2}} \quad \text{instead of} \quad \frac{\overline{x} - \mu_0}{s/\sqrt{n}}$$

and the corresponding number of degrees of freedom are $\nu = n_1 + n_2 - 2$ instead of $\nu = n - 1$.

TWO-SIDED TESTS

In Table 4 we indicate how to perform a two-sided test of $H_0: \mu_1 - \mu_2 = 0$ against $H_1: \mu_1 - \mu_2 \neq 0$. The differences between this table and Table 2, for testing $H_0: \mu = \mu_0$ against $H_1: \mu \neq \mu_0$, are precisely those listed above.

TABLE 4

Procedure for testing $H_0: \mu_1 - \mu_2 = 0$ against $H_1: \mu_1 - \mu_2 \neq 0$ in small samples from normal populations

STEP	$H_1: \mu_1 - \mu_2 \neq 0$
1. Compute:	(a) $\overline{x}_1, \overline{x}_2, s_1^2$, and s_2^2
	(b) $s_p^2 = \dfrac{(n_1 - 1)s_1^2 + (n_2 - 1)s_2^2}{n_1 + n_2 - 2}$
and hence:	(c) $t = \dfrac{\overline{x}_1 - \overline{x}_2 - 0}{s_p\sqrt{1/n_1 + 1/n_2}}$
2. Find $t_{n_1+n_2-2,.025}$ and $t_{n_1+n_2-2,.005}$ from Table A6.	
3. *Conclusion:* Using $\alpha = .05$ or $.01$, reject H_0 at the α level of significance if:	$\|t\| \geq t_{n_1+n_2-2,\alpha/2}$

Assumption: Random samples are drawn from normal populations having the same standard deviation

The next example illustrates how to apply this procedure.

EXAMPLE 8

An experiment is to be conducted to compare two methods for teaching typing. Forty students with no previous experience are divided randomly into two groups, one of which receives method I and the other method II. After three weeks of training, each student is given a typing test. The resulting scores in words per minute are recorded and summarized as follows:

$$\overline{x}_1 = 51.82 \qquad \overline{x}_2 = 46.71 \qquad n_1 = n_2 = 20$$

$$s_1 = 5.91 \qquad s_2 = 8.42$$

If μ_1 and μ_2 are the true mean scores over all potential users of the two methods, test $H_0: \mu_1 - \mu_2 = 0$ against $H_1: \mu_1 - \mu_2 \neq 0$. What assumptions do we need to make?

Solution

Answering the last question first, we assume that the distribution of the number of words per minute (for all potential students) for the two methods are approximately normal and their standard deviations are equal.

Following the procedures in Table 4, we have

$$s_p^2 = \frac{19(5.91)^2 + 19(8.42)^2}{38}$$

$$= \frac{19(5.91^2 + 8.42^2)}{38} = \frac{5.91^2 + 8.42^2}{2}$$

$$= 52.912$$

and thus $s_p = \sqrt{52.912} = 7.274$. Notice when the sample sizes are equal, s_p^2 is simply the average of the two sample variances, that is, $(s_1^2 + s_2^2)/2$. Continuing, we have

$$t = \frac{51.82 - 46.71}{7.274\sqrt{1/20 + 1/20}} = \frac{5.11}{2.30} = 2.22$$

From Table A6, $t_{38,.025} = 2.024$ and $t_{38,.005} = 2.712$. Since the value $|t| = 2.22$ lies between 2.024 and 2.712, we reject H_0 at the .05 level of significance. The data suggest that the mean number of words per minute for method I is significantly greater than for method II. The value $t = 2.22$ is shown in Figure 6 in relation to $t_{38,.025}$ and $t_{38,.005}$.

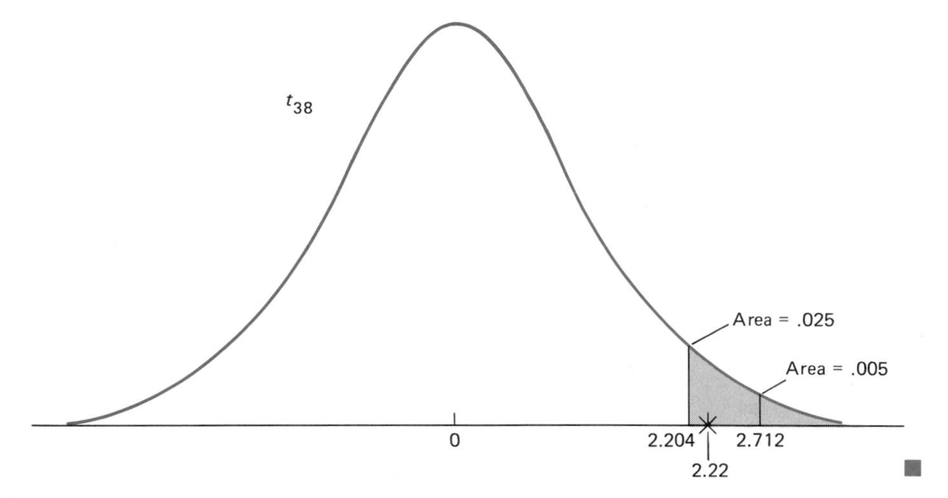

FIGURE 6

Value $t = 2.22$ shown in relation to $t_{38,.025} = 2.024$ and $t_{38,.005} = 2.712$ ■

We can generalize the procedures in Tables 3 and 4 to take into account the use of any single value for α. The details are provided below.

> DEFINITION 6
>
> Assume that random samples of size n_1 and n_2 are taken independently from two normal populations with the same standard deviations. As before,
>
> $$t = \frac{\overline{x}_1 - \overline{x}_2 - 0}{s_p\sqrt{1/n_1 + 1/n_2}}$$
>
> A level of significance α is to be used to test $H_0: \mu_1 - \mu_2 = 0$.
> (a) For $H_1: \mu_1 - \mu_2 > 0$; reject H_0 if $t > t_{n_1+n_2-2,\alpha}$.
> (b) For $H_1: \mu_1 - \mu_2 \neq 0$; reject H_0 if $|t| > t_{n_1+n_2-2,\alpha/2}$.

In Example 8, for instance, had we used only $\alpha = .01$, we would have rejected H_0 only if $|t| > t_{38,.005}$. Since in fact $|t| = 2.22$ is *not* $> t_{38,.005} = 2.712$, we cannot reject H_0 at the .01 level of significance.

CONFIDENCE INTERVALS FOR $\mu_1 - \mu_2$ IN SMALL SAMPLES

In addition to using Definition 4 for hypothesis testing, it can also be used to set confidence intervals for $\mu_1 - \mu_2$ in small samples. The details follow.

> DEFINITION 7
>
> A random sample of size n_1 is taken from a $N[\mu_1, \sigma]$ population. Another random sample of size n_2 is independently taken from a $N[\mu_2, \sigma]$ population. A $(100\gamma)\%$ confidence interval for $\mu_1 - \mu_2$ is
>
> $$(\bar{x}_1 - \bar{x}_2) - t_{n_1+n_2-2(\gamma)} s_p \sqrt{\frac{1}{n_1} + \frac{1}{n_2}} \quad \text{to} \quad (\bar{x}_1 - \bar{x}_2) + t_{n_1+n_2-2(\gamma)} s_p \sqrt{\frac{1}{n_1} + \frac{1}{n_2}}$$

EXAMPLE 9 Use the data in Example 7 to find a 90% confidence interval for $\mu_1 - \mu_2$, the difference in the mean use of electricity per household in the two cities.

Solution From Example 7, $\bar{x}_1 - \bar{x}_2 = 3.279$ and $s_p\sqrt{1/n_1 + 1/n_2} = 3.512$. From Table A6, $t_{n_1+n_2-2(\gamma)} = t_{18(.9)} = 1.734$. Thus a 90% confidence interval for $\mu_1 - \mu_2$ is

$$3.279 \mp 1.734(3.512) = -2.811 \text{ to } 9.369$$

We may be 90% confident that the true *difference* in mean usage lies between -2.811 and 9.369. This suggests that μ_1 may be less than μ_2 by as much as 2.8 kilowatts or may exceed μ_2 by as much as 9.4 kilowatts. By contrast, had these means and standard deviations occurred with sample sizes of 120 and 80 rather than 12 and 8, a 90% confidence interval for $\mu_1 - \mu_2$ would be (from Section 9.4 and Example 7)

$$(33.017 - 29.738) \mp 1.645\sqrt{\frac{67.6997}{120} + \frac{45.8912}{80}} = 1.524 \text{ to } 5.034$$

The interval -2.811 to 9.369 is an example of the lack of precision that often accompanies results based on small samples. ■

PROBLEMS 12.4

In the problems involving hypothesis testing we again urge you to (i) define μ_1 and μ_2, (ii) state H_0 and H_1, (iii) interpret the results of your test in terms consistent with the context of the problem, and (iv) outline the assumptions you make in performing the test.

27. Twenty cigarettes of brand 1 and 20 of brand 2 were tested for nicotine content (in milligrams). The following summary data were obtained:

BRAND 1	$\bar{x}_1 = 1.36$	$s_1 = .22$
BRAND 2	$\bar{x}_2 = 1.27$	$s_2 = .15$

We wish to know whether, on the average, brand 1 contains significantly more nicotine than brand 2.

(a) Define μ_1 and μ_2 in this case.

(b) State the two hypotheses.

(c) What is the appropriate number of degrees of freedom?

(d) Conduct the appropriate test and state your conclusion.

(e) What assumptions did you make in part (d)?

28. Forty similar-sized pork roasts are randomly divided into two groups. Group I contains 18 roasts and group II, 22 roasts. All those in I are cooked at 400°F and those in II at 300°F. the amount of thiamine (in micrograms per gram) in each roast after cooking was recorded, with the following results:

	n	$\bar{x}$	s
I	18	15.72	.549
II	22	15.41	.477

(a) Define μ_1 and μ_2 in this case.

(b) Do these results suggest any difference between the mean amount of thiamine retained when roasts are cooked at these two temperatures?

29. Use the data in Problem 28 to obtain an 80% confidence interval for $\mu_1 - \mu_2$. Interpret your interval.

30. Samples of size $n_1 = 15$ and $n_2 = 12$ are independently selected from two normal populations, I and II. The sample means and standard deviations are:

$$\bar{x}_1 = 55.4 \qquad s_1 = 1.67 \qquad \bar{x}_2 = 52.8 \qquad s_2 = 1.94$$

(a) Test $H_0: \mu_1 - \mu_2 = 0$ against $H_1: \mu_1 - \mu_2 > 0$.

(b) Test $H_0: \mu_1 - \mu_2 = 0$ against $H_1: \mu_1 - \mu_2 \neq 0$.

31. Horticulturists recorded the time in days it took each of 30 marigold seeds, 15 of variety *A* and 15 of variety *B*, to grow to full bloom. The results are summarized as follows:

	n	$\bar{x}$	s
VARIETY *A*	15	90.4	5.2
VARIETY *B*	15	86.6	5.6

(a) Do these data suggest any significant difference in the mean time for the two varieties to reach full bloom?

(b) What assumptions were necessary to perform the test?

32. Answer Problem 31 if the researcher wished to use $\alpha = .1$.

33. In response to customer complaints about the length of the line to see a bank officer, the bank intends to introduce a scheme involving a more flexible use of

staff. Prior to the introduction of the scheme the bank records the number of people in line at 14 randomly selected times. Two weeks after the scheme is introduced the number of people in line is again recorded at 14 randomly selected times. The data are as follows:

AFTER	2	4	7	9	4	6	7	6	4	3	3	4	5	6
BEFORE	7	4	5	6	6	7	11	5	6	7	8	3	4	6

Do these data suggest that the mean length of the line to see an officer has been significantly reduced? This is a more realistic version of Problem 14.

34. Use the data in Problem 33 to obtain a 90% confidence interval for the improvement in the mean length of line.

35. The ages of eight men and eight women randomly sampled from the list of registered members of a political party are as follows:

MEN	35	42	62	29	44	39	51	47	$\Sigma x_i^2 = 15{,}941$
WOMEN	50	29	42	39	29	56	38	39	$\Sigma x_i^2 = 13{,}568$

(a) Do those data suggest that the mean age of men in the party significantly exceeds that of women?
(b) What assumptions do you need to make in this case?

36. Use the data in Problem 35 to find a 95% confidence interval for the difference between the mean age of men and women in the party. Interpret your interval.

37. A sample of seven people is selected from the head of a large line waiting for 400 tickets for a major sporting event. Each person is asked to estimate the number of people in line, with the following results:

$$427 \quad 432 \quad 401 \quad 520 \quad 450 \quad 480 \quad 460 \qquad \sum x_i^2 = 1{,}444{,}654$$

Another six people selected from the end of the line are asked to make the same estimate, with the following results:

$$442 \quad 421 \quad 364 \quad 389 \quad 394 \quad 452 \qquad \Sigma x_i^2 = 1{,}015{,}962$$

Do these data suggest that on the average people at the head of the line make higher estimates of the length of the line than those at the end of the line?

38. A pharmaceutical company wishes to compare their standard blend of antacid with a new "improved version" which it is hoped will speed up relief from heartburn. A group of 48 patients are randomly divided into two groups of 24. Members of group A are given the standard blend to relieve heartburn. Members of group B are given the improved blend. The time to obtain relief is recorded for each person. The summary results (in minutes) are as follows:

	STANDARD BLEND	IMPROVED BLEND
$\bar{x}$	15.2	13.0
s	3.4	4.3

Do these results suggest that the improved blend relieves heartburn significantly faster than the standard blend? (*Note:* In this case $\nu = n_1 + n_2 + 2 = 48 - 2 = 46$ is not in Table A6. Use the corresponding values for $\nu = 40$ instead.)

39. Answer Problem 38 if the value $\alpha = .025$ is used.

Solution to Problem 27

(a) The values μ_1 and μ_2 are the mean nicotine content over all cigarettes of brand I and brand II, repectively.

(b) $H_0: \mu_1 - \mu_2 = 0$; $H_1: \mu_1 - \mu_2 > 0$ (or $\mu_1 > \mu_2$)

(c) In this case $\nu = n_1 + n_2 - 2 = 20 + 20 - 2 = 38$.

(d) Following the procedure in Table 3, we have

$$\bar{x}_1 = 1.36,\ \bar{x}_2 = 1.27,\ s_1^2 = .22^2 = .0484 \text{ and } s_2^2 = .15^2 = .0225$$

$$s_p^2 = \frac{19(.0484) + 19(.0225)}{38} = .03545$$

and thus $s_p = .1883$, and

$$t = \frac{1.36 - 1.27}{.1883\sqrt{1/20 + 1/20}} = 1.511$$

From Table A6, $t_{38,\,.05} = 1.686$ and $t_{38,\,.01} = 2.428$. Since $1.511 < 1.686$, we cannot reject H_0 at the .05 level of significance. The data suggest that the mean nicotine content for brand I is not significantly greater than that for brand II.

(e) We assume that for both brands the distribution of nicotine content per cigarette is approximately normal with the same variance.

SECTION 12.5
ANALYSIS OF PAIRED DATA

In this section we reapply the methods of Section 12.3 to a new type of problem. In the process we reintroduce some of the experimental design issues discussed in Chapter 8. The following is an example of this type of problem.

EXAMPLE 10 Twenty students participate in a study of the effect of alcohol on the response time to a simulated automobile incident. After several practice runs each student's reaction time is recorded twice, once *before* drinking three beers and again a half-hour *after* drinking the beers. The resulting reaction times (in seconds) are shown in Table 5. Do these data suggest that drinking beer increases reaction time?

At first glance it is tempting to view these results as consisting of two sets of data, the 20 "before beer" measurements and the 20 "after beer" measurements, and thus perform the two-sample test of the preceding section. Notice, though, that one of the assumptions of the two-sample test—namely,

TABLE 5 Difference in reaction times for 20 students before and after drinking beer

STUDENT	BEFORE BEER	AFTER BEER	DIFFERENCE (AFTER) − (BEFORE)
1	3.4	3.9	.5
2	4.1	4.2	.1
3	3.6	3.5	−.1
4	2.9	2.9	0
5	3.7	4.1	.4
6	5.0	5.2	.2
7	4.2	4.6	.4
8	3.9	4.6	.7
9	3.6	3.7	.1
10	2.8	3.2	.4
11	4.0	4.3	.3
12	3.3	3.1	−.2
13	3.7	4.0	.3
14	3.2	3.5	.3
15	4.1	4.4	.3
16	4.1	4.1	0
17	3.9	4.2	.3
18	4.0	4.2	.2
19	2.7	2.9	.2
20	3.5	3.8	.3

independence of the samples—is violated here. The "after" measurements were intentionally recorded on the same persons for whom a "before" measurement was taken. This very *dependence*, however, suggests the appropriate approach. We can compute the individual *differences* in reaction times, regarding the data as a single sample of differences and hence apply the one-sample test of Section 12.3.

To emphasize that we are working with individual differences, we use the notation $d_1, d_2, \ldots, d_{20}$ instead of $x_1, x_2, \ldots, x_{20}$ and $\bar{d}$ and s_d instead of $\bar{x}$ and s to denote the mean and standard deviation of the differences. If we call μ_d the mean difference over all potential participants in the experiment, we can write the null and alternative hypotheses as

$H_0: \mu_d = 0$ the three beers have no effect on reaction time

$H_1: \mu_d > 0$ the three beers significantly increase the reaction time

Applying the usual formulas for the sample mean and standard deviation to the 20 differences in Table 5, we obtain

$$\bar{d} = \frac{\Sigma\, d_i}{20} = \frac{.5 + .1 + \cdots + .3}{20} = \frac{4.7}{20} = .235$$

$$s_d^2 = \frac{20\, \Sigma\, d_i^2 - (\Sigma\, d_i)^2}{20(19)}$$

$$= \frac{20(.5^2 + .1^2 + \cdots + .3^2) - 4.7^2}{20(19)}$$

$$= \frac{20(1.95) - 22.09}{380} = .0445$$

and $s_d = \sqrt{.0445} = .21095$.

Following the rules in Table 1 for testing hypotheses about μ in small samples (with $\bar{d}$ in place of $\bar{x}$ and $\mu_0 = 0$), we have

$$t = \frac{\bar{d} - 0}{s_d/\sqrt{n}} = \frac{.235}{.21095/\sqrt{20}} = \frac{.235}{.04717} = 4.98$$

From Table A6, $t_{19,.05} = 1.729$ and $t_{19,.01} = 2.539$.

Since $t = 4.98 > 2.539$, we reject H_0 at the .01 level of significance. The data strongly suggest that the persons tested have, on the average, longer reaction times after drinking beer than before.

■

A CONFIDENCE INTERVAL FOR μ_d

To compute a confidence interval for μ_d we need only apply Definition 2 on page 401. For example, using the beer data, a 90% confidence interval for μ_d is

$$\bar{d} \mp t_{19(.9)} \frac{s_d}{\sqrt{n}}$$

or

$$.235 \mp 1.729\left(\frac{.21095}{\sqrt{20}}\right) = .153 \text{ to } .317 \text{ seconds}$$

We can be 90% confident that the mean change (increase) in reaction time lies between .153 and .317 second.

In using the confidence interval above, we assume that d_i, the differences in reaction times, were drawn from a population that is at least approximately normal.

MATCHED PAIRS DESIGNS AGAIN

The reaction-time data above provide an example of what is frequently referred to as *paired* or *matched* data. Readers familiar with the material in Chapter 8 may recognize the experiment in Example 10 as a **matched pairs design.** We introduced this procedure earlier as a means of controlling for *bias* when planning an experiment. Using this design in the example above, by taking both measurements on the same person, we eliminate differences (and hence bias) between the control group (the 20 people measured before the beer) and the experimental group (the same 20 people measured after the beer).

An important instance of a matched pairs design occurs when the same variable is measured on each sample member both before and after some event. In such a case we are interested in the *change* in the value of the variable which, we hope, reflects the effect of the intervening event. Example 10 is an example of one such context. Others include:

A new drug designed to reduce blood pressure is administered to a group of patients suffering from high blood pressure. The blood pressure of each patient is recorded before and after the drug is given, and the change in blood pressure recorded.

A group of people are weighed before and after two weeks participation in a weight-loss program. The weight loss is recorded for each person.

A company runs a large advertising campaign in each of 20 cities. The number of items sold before and after the campaign and the difference in these numbers is recorded for each city.

The number and proportion of students receiving financial aid at a group of 15 private colleges is recorded in the year before and the year after a government cutback in aid to college students.

A somewhat different type of situation involving paired data is illustrated in the following example.

EXAMPLE 11 An experiment is conducted involving 20 children divided into 10 pairs. The members of each pair are carefully matched so that they are similar with respect to sex, age, IQ, and previous math training. They have not yet studied multiplication. One member of each pair is randomly chosen and taught multiplication using method A. The other member is taught multiplication using method B. Test scores for the 20 children after the learning experience are:

PAIR	1	2	3	4	5	6	7	8	9	10
Method A	47	42	26	50	20	29	37	35	40	23
Method B	38	44	31	46	21	21	37	41	30	28
$d = A - B$	9	−2	−5	4	−1	8	0	−6	10	−5

Do these data suggest that the mean scores for the two methods are significantly different?

Solution Call the potential mean difference in the population of interest here, μ_d. We wish to test $H_0: \mu_d = 0$ against $H_1: \mu_d \neq 0$. A two-sided alternative is used here since we have no preconceived idea of which method is better.

From the 10 observed differences, $A - B$, we compute $\bar{d} = 1.2$ and $s_d = 6.125$. Following the procedure in Table 2 on page 408 (again, with $\bar{d}$ instead of $\bar{x}$ and $\mu_0 = 0$), we compute

$$t = \frac{\bar{d} - 0}{s_d/\sqrt{n}} = \frac{1.2}{6.125/\sqrt{10}} = .62$$

From Table A6, $t_{9,.025} = 2.262$ and $t_{9,.005} = 3.25$. Since $|.62|$ is less than 2.262, we cannot reject H_0 at the .05 level of significance. Although the children in the study receiving method A do better than those receiving method B, the difference is not statistically significant.

■

We have reintroduced matched pairs designs here rather than in Chapter 10, since they usually occur with small samples. However, if such a design involves 30 or more pairs, we may apply the large-sample methods of Chapter 10. Suppose, for instance, the values $\bar{d} = 1.2$ and $s_d = 6.125$ above had been obtained with 110 instead of 10 pairs of observations. In that case we would have performed the two-sided test of $H_0: \mu_d = 0$ against $H_1: \mu_d \neq 0$ following the procedures in Section 10.5. We compute

$$z = \frac{\bar{d} - 0}{s_d/\sqrt{n}} = \frac{1.2}{6.125/\sqrt{110}} = 2.05$$

The P value is then

$$2P(Z > 2.05) = 2(.0202) = .0404$$

Since the P value $< .05$, we reject H_0 at the .05 level of significance. The data (with $n = 110$) suggest that students studying with method A score significantly better than those studying with method B.

An increase in n results in a *decrease* in the value of $s/\sqrt{n}$ and hence to a larger value for $t = (\bar{d} - \mu_0)/s/\sqrt{n})$ [or $z = (\bar{d} - \mu_0)/(s/\sqrt{n})$ in the large sample example above], and hence to a more sensitive test.

It is highly instructive to compare the design and analysis in Example 10 to the design and analysis of the two-independent-sample problem in the following example.

EXAMPLE 12 Forty students participate in a study of the effect of alcohol on the time needed to respond to a simulated automobile incident. The students are given several practice runs on the simulator and are then randomly divided into two groups of 20. The time to react for the 20 students in group B is recorded. The 20 students in group A are given three beers and a half-hour later their reaction times are recorded. The times (in seconds) for the two groups are:

GROUP A (BEER)				GROUP B (NO BEER)			
4.4	3.0	4.1	4.4	3.9	3.4	3.6	3.3
4.5	4.0	4.3	4.1	4.0	4.2	3.8	4.1
3.0	3.7	3.1	3.8	2.6	3.6	3.7	3.8
4.5	3.7	5.1	4.2	4.3	4.9	3.8	2.8
2.8	3.6	4.1	4.0	3.7	4.0	3.5	2.7

Do these data suggest that drinking beer increases reaction time?

Solution Apart from the fact that these data represent independent samples, they seem quite similar to the data in Example 10. For instance, the mean times for the two groups are $\bar{x}_A = 3.920$ and $\bar{x}_B = 3.685$. The difference in mean times $\bar{x}_A - \bar{x}_B = 3.920 - 3.685 = .235$ is exactly the same as the mean difference in times, $\bar{d} = .235$ in Example 10.

To get a sense of the spread of times in the two groups, we computed the range of times as $R_A = 5.1 - 2.8 = 2.3$ and $R_B = 4.9 - 2.6 = 2.3$ seconds. It is simple to show that the range of "before beer" times and "after beer" times in Example 10 are also both 2.3 seconds.

Having established the superficial similarity of the two data sets, we now analyze the two independent samples in this example. We test $H_0: \mu_1 - \mu_2 = 0$ against $H_1: \mu_1 - \mu_2 > 0$, where 1 refers to group A and 2 to group B. The summary data are:

	$\overline{x}$	s
GROUP A (I)	3.920	.5935
GROUP B (II)	3.685	.5518

Following the procedure in Table 3, we have

$$s_p^2 = \frac{19(.5935)^2 + 19(.5518)^2}{38} = .3284$$

$$s_p = \sqrt{.3284} = .5730$$

and hence

$$t = \frac{\overline{x}_1 - \overline{x}_2 - 0}{s_p\sqrt{1/n_1 + 1/n_2}} = \frac{3.920 - 3.685}{.5730\sqrt{1/20 + 1/20}}$$

$$= \frac{.235}{.1812} = 1.30$$

From Table A6, $t_{38,\,.05} = 1.686$ and $t_{38,\,.01} = 2.428$.

Since 1.30 is less than 1.686, we cannot reject H_0 at the .05 level of significance. The data suggest that beer drinkers do not, on the average, have a significantly longer reaction time than nondrinkers.

Compare this conclusion with that in Example 10, where, because the value $t = (\overline{d} - 0)/(s_d/\sqrt{n}) = 4.98$ was so large, we rejected H_0 at the .01 level of significance. By contrast, in the two-sample case,

$$t = \frac{\overline{x}_1 - \overline{x}_2 - 0}{s_p\sqrt{1/n_1 + 1/n_2}}$$

is only 1.30. This in spite of the fact that both data sets are quite similar. How are we to account for the very different t values? As we mentioned earlier, the difference in the means $\overline{x}_1 - \overline{x}_2 = .235$ in Example 12 is exactly the same as $\overline{d} = .235$ in Example 10. Thus the numerators of both t values are the same. Therefore, the difference in the two decisions is due solely to the fact that in the paired t test, $\text{SD}(\overline{d}) = S_d/\sqrt{n} = .04717$, whereas in the two-sample t test $\text{SD}(\overline{X}_1 - \overline{X}_2) = .1812$, almost four times the value of $\text{SD}(\overline{d})$. Why this large difference, and what does it mean?

We can identify two sources of variation among the 40 reaction times in Example 12:

1. Variation due to the fact that 20 of the subjects drank beer while the other 20 did not; in the language of Chapter 8, this is variation due to the *independent* variable.
2. Variation due to differences among individuals.

Even among the 20 persons who had no beer, reaction times will vary due to age, weight, sex, drinking habits, and a host of other factors. The same will be true for the 20 beer drinkers.

By contrast, when we compare the "before" and "after" measurements for the *same individuals,* as in Example 10, we are eliminating this second, between-individual, source of variation. As a result, the $SD(\overline{d})$ reflects only the variation in reaction time due to drinking three beers, and we can determine the effect of the beer drinking much more precisely.

More generally, in addition to providing a means of minimizing bias, the matched pair design usually provides a more *precise* experiment than one in which the experimental and control groups are *not* matched. This is true whether the matched pair design involves taking both measurements on the same individual or matching individuals in two groups.

■

STATISTICAL TESTS AND EXPERIMENTS

It is often easy to overlook the extent to which the interpretation of a statistical test depends on the experimental design. For instance, consider again the experiment in Example 12 where the reaction time of 40 students were recorded. Twenty of the students had three beers. The other 20 did not. Suppose that the mean reaction time for the beer drinkers was found to be significantly greater than that for the nonbeer drinkers. If, as we suggested in Example 12, students were assigned to the two groups at random, it is reasonable to attribute the significant differences to the effects of the beer. But suppose that students had been allowed to choose whether or not to join the beer group. In this case the significant difference *may* be due to the effect of the beer, but on the other hand, it may be due to one or more other variables, such as normal drinking habits for example.

Without randomized assignment the situation would be similar to the smoking and lung cancer example in Chapter 8. In all studies the proportion of smokers getting lung cancer is invariably statistically significantly greater than the corresponding proportion of nonsmokers. But this does not *prove* that smoking causes lung cancer. The difference in proportions may be due to a "confounding" variable (such as anxiety) which may be found in much greater concentration in the smoking group.

A statistical test can only tell whether chance alone could have been responsible for the observed results. Rejecting H_0 suggests that some factor other than chance is operating, but only with a randomized experiment can we identify the independent variable as the "factor" responsible.

PROBLEMS 12.5

40. For the following paired data test, decide whether the A measurements are significantly greater than the B measurements.

A	10.3	7.5	9.7	11.3	15.1	10.3	9.3	17.3
B	9.2	7.2	8.6	11.5	14.8	9.8	8.9	16.4

A	12.2	4.9	8.3	12.6	12.9	7.4	13.4
B	12.0	4.3	8.0	11.5	12.1	6.5	12.5

41. A student performs eight calculations on each of two calculators and records in seconds the time it takes to complete each calculation. She uses a VS45 to compute the first four calculations and a PL67 to compute the second four. She then reverses the order for the next eight calculations. The recorded times are as follows:

CALCULATIONS	1	2	3	4	5	6	7	8
VS45	15.8	10.3	20.6	9.9	12.4	14.1	13.1	16.4
PL67	17.7	9.5	24.2	12.6	14.1	14.0	12.4	20.3

Compute the eight differences between VS45 and PL67 times; then test for any significant difference between the mean times for the two calculators.

42. The grade-point average (GPA) of 10 randomly selected college students is recorded at the end of the fall and spring semesters of their senior year, as follows:

STUDENT	1	2	3	4	5	6	7	8	9	10
Fall Semester	3.2	2.7	2.2	2.6	3.3	3.4	2.9	2.0	2.6	3.0
Spring Semester	2.8	2.4	1.9	3.0	3.3	3.2	3.1	1.8	2.3	2.9

Do these data suggest that on the average, students' GPA's decline significantly in their final semester of college?

43. Use the data in Problem 42 to obtain an 80% confidence interval for the mean decline in GPA.

44. The weights upon enrollment in a weight-loss program and the weights after two weeks in the program are recorded for 25 males. The results are as follows:

PERSON	1	2	3	4	5	6	7	8	9
Original Weight	245	190	310	261	224	197	308	325	242
Two-week Weight	236	180	302	251	210	183	291	319	231

PERSON	10	11	12	13	14	15	16	17	18
Original Weight	292	231	258	248	186	242	340	266	210
Two-week Weight	280	218	245	243	171	222	328	250	204

PERSON	19	20	21	22	23	24	25
Original Weight	244	222	380	229	316	282	259
Two-week Weight	223	216	370	218	303	276	245

The program claims that participants can expect to lose considerably more than 10 pounds in the first two weeks.

(a) What is μ_d in this case?

(b) Do these data support the program's claim?

45. (a) Use the data in Problem 44 to find a 90% confidence interval for the mean two-week weight loss for this program.

(b) Interpret your interval.

46. Arrange the 25 differences in Problem 44 in a frequency histogram and draw the corresponding relative frequency histogram. Does your histogram suggest that weight loss is approximately normally distributed?

47. In a recent study 120 randomly selected voters were asked to grade the governor's performance in office on a scale of 1 to 10. The higher the score, the better the perceived performance. A year later the same voters were again asked to grade the governor on a similar scale. The differences between the original scores and the second scores were found and the following summaries computed.

$$\bar{d} = .45 \qquad s_d = 2.08$$

Test whether these data suggest a significant decline in the perception of the governor's performance. (Remember to use the large-sample test.)

Solution to Problem 40

The 15 differences $d = A - B$ are

1.1 .3 1.1 −.2 .3 .5 .4 .9 .2 .6 .3 1.1 .8 .9 .9

We test $H_0: \mu_d = 0$ against $H_1: \mu_d > 0$ using the procedures in Table 1 with $\bar{d}$ in place of $\bar{x}$ and $\mu_0 = 0$. In this case $\bar{d} = .613$, $s_d = .3944$, and

$$t = \frac{\bar{d} - 0}{s_d/\sqrt{n}} = \frac{.613}{.3944/\sqrt{15}} = 6.02$$

From Table A6, $t_{14,.05} = 1.761$ and $t_{14,.01} = 2.624$. Since $t = 6.02$ is considerably larger than 2.624, we reject H_0 at the .01 level of significance.

SECTION 12.6

THE CHI-SQUARE AND *F* DISTRIBUTIONS

THE CHI-SQUARE DISTRIBUTION

In this section we introduce two new continuous distributions, the **chi-square** and **F distributions.**

The chi-square distribution derives its name from the Greek lowercase letter chi, pronounced "*ky*" as in "sky" without the s, and is written χ. Although we refer to *the* chi-square distribution, as in the case of the *t* distribution there is not one but a family of distributions. Each chi-square (χ^2) distribution also has associated with it a number of degrees of freedom. Thus the notation χ^2_ν denotes the chi-square distribution with ν degrees of freedom. If a random variable Y has a χ^2_ν distribution, it is called a **chi-square random variable.**

In Figure 7 we show the χ^2_1, χ^2_3, χ^2_5, and χ^2_{10} distributions.

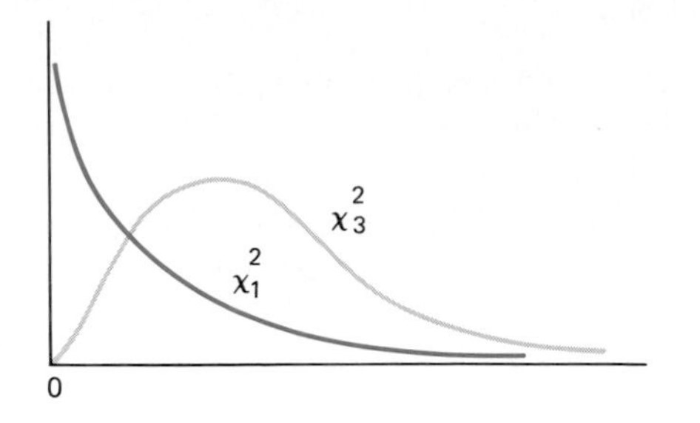

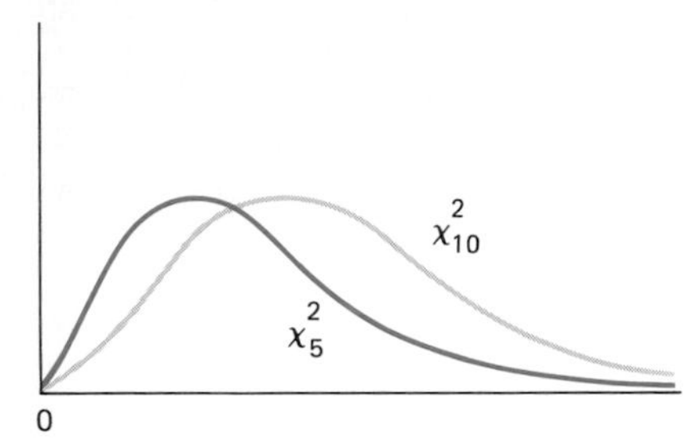

FIGURE 7

Various χ^2 distributions

All χ^2 distributions are asymmetric (i.e., not symmetric), and with the exception of the χ^2_1 distribution, have roughly the same shape. However, as the number of degrees of freedom increases, the corresponding distributions are closer to being symmetric. Each chi-square random variable takes only positive values, with the mean value in each case equal to the number of degrees of freedom, ν. Thus the mean of the χ^2_5 distribution $= \nu = 5$ and the mean of the χ^2_{16} distribution $= 16$.

Values of χ^2 with various specified areas to their *right* are tabulated in Table A7 for a variety of χ^2_ν distributions. As the sketch at the top of that table shows, a typical entry, denoted $\chi^2_{\nu,\alpha}$ represents the value of χ^2_ν which has an area α to the right. Entries are provided for $\alpha = .005, .01, .025, .05, .1, .9, .95, .975, .99$, and $.995$.

EXAMPLE 13 In Figure 8 we show a section of Table A7 for the distribution together with two sketches of that distribution which illustrate that $\chi^2_{5,.975} = .83$ and that $\chi^2_{5,.05} = 11.07$. Also notice that since the area to the right of .83 is .975, the area to the left of .83 is $1 - .975 = .025$.

FIGURE 8

The values $\chi^2_{5,.05} = 11.07$ and $\chi^2_{5,.975} = .83$

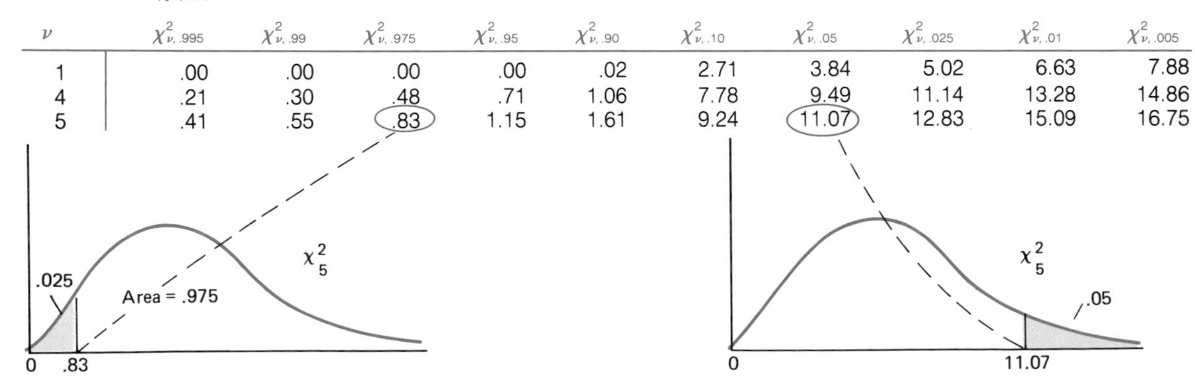

ν	$\chi^2_{\nu,.995}$	$\chi^2_{\nu,.99}$	$\chi^2_{\nu,.975}$	$\chi^2_{\nu,.95}$	$\chi^2_{\nu,.90}$	$\chi^2_{\nu,.10}$	$\chi^2_{\nu,.05}$	$\chi^2_{\nu,.025}$	$\chi^2_{\nu,.01}$	$\chi^2_{\nu,.005}$
1	.00	.00	.00	.00	.02	2.71	3.84	5.02	6.63	7.88
4	.21	.30	.48	.71	1.06	7.78	9.49	11.14	13.28	14.86
5	.41	.55	.83	1.15	1.61	9.24	11.07	12.83	15.09	16.75

EXAMPLE 14 Use Table A7 to find (a) $\chi^2_{3,.95}$ and $\chi^2_{3,.05}$; (b) the area between $\chi^2_3 = .35$ and 7.81; (c) $\chi^2_{15,.99}$ and $\chi^2_{15,.01}$; (d) the area between $\chi^2_{15} = 5.23$ and 30.58.

Solution (a) From Table A7, $\chi^2_{3,.95} = .35$. This is the value of χ^2_3 with an area $= .95$ to the right. Similarly, $\chi^2_{3,.05} = 7.81$ is the value of χ^2_3 with an area $= .05$ to the right.

(b) As you can see in Figure 9, the area to the right of the value .35 is .95 and includes the area between the values .35 and 7.81 plus the area .05 to the right of 7.81. The area between .35 and 7.81 is therefore $.95 - .05 = .9$.

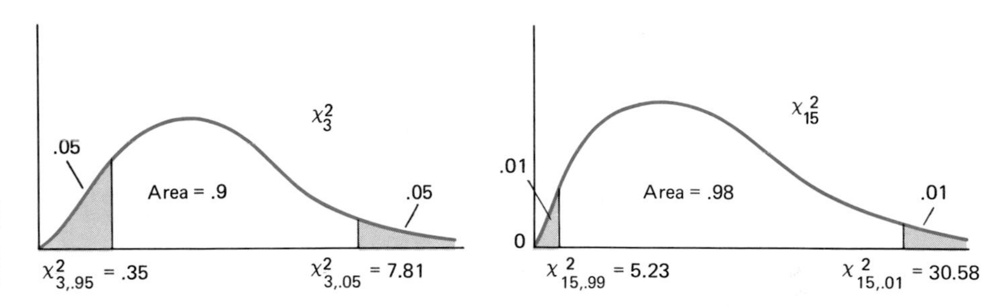

FIGURE 9

Chi-square values obtained in Example 14

(c) From Table A7, $\chi^2_{15,.99} = 5.23$ and $\chi^2_{15,.01} = 30.58$.

(d) As shown in Figure 9, the area .99 to the right of the value 5.23 includes the area between 5.23 and 30.58 and the area .01 to the right of 30.58. The area between 5.23 and 30.58 is therefore $.99 - .01 = .98$.

■

THE *F* DISTRIBUTION

Another important distribution is the *F* distribution which received its name from Ronald Fisher, who was the first to make use of it. The *F*, like the χ^2_ν and *t* distributions, is a family of distributions. Unlike the others, however, each *F* distribution is identified by *two* numbers of degrees of freedom, ν_1 and ν_2. There is a different *F* distribution for every combination of values for ν_1 and ν_2. For example, the $F_{4,6}$ and the $F_{2,10}$ distributions are shown in Figure 10.

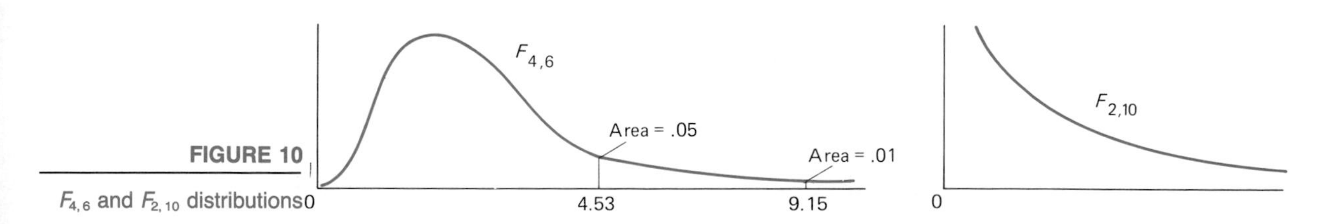

FIGURE 10

$F_{4,6}$ and $F_{2,10}$ distributions

F distributions are continuous and are similar to chi-square distributions in that they take only positive values and are asymmetric. The notation $F_{\nu_1,\nu_2,\alpha}$ is used to refer to the value of F_{ν_1,ν_2} having an area α to the *right*. For example, in Figure 10 we show that $F_{4,6,.05} = 4.53$ is the value for $F_{4,6}$ having an area $\alpha = .05$ to the right. The value for $F_{4,6,.01} = 9.10$ is also shown in Figure 10.

Values of $F_{\nu_1, \nu_2, \alpha}$ for a large number of combinations of values of ν_1 and ν_2 and for $\alpha = .25, .1, .05, .025, .01$, and $.005$ are provided in Table A8. A section of this table for the $F_{4,6}$ distribution is shown in Figure 11. The six values $F_{4,6,.005}, F_{4,6,.01}, \ldots F_{4,6,.25}$ can be found where the column $\nu_1 = 4$ and the row $\nu_2 = 6$ intersect. Note that $F_{4,6,.25} = 1.79$, $F_{4,6,.1} = 3.18$, and so on.

		ν_1									
ν_2	α	1	2	3	4	5	6	7	8	9	10
(6)	.25	1.62	1.76	1.78	1.79	1.79	1.78	1.78	1.78	1.77	1.77
	.1	3.78	3.46	3.29	3.18	3.11	3.05	3.01	2.98	2.96	2.94
	.05	5.99	5.14	4.76	4.53	4.39	4.28	4.21	4.15	4.10	4.06
	.025	8.81	7.26	6.60	6.23	5.99	5.82	5.70	5.60	5.52	5.46
	.01	13.7	10.9	9.78	9.15	8.75	8.47	8.26	8.10	7.98	7.87
	.005	18.6	14.5	12.9	12.0	11.5	11.1	10.8	10.6	10.4	10.2

FIGURE 11

Tabulated values for the $F_{4,6}$ distribution

■

EXAMPLE 15 Use Table A8 to find (a) $F_{3,6,.1}$, (b) $F_{3,6,.025}$, and (c) $F_{4,9,.01}$. Mark each of these values on the appropriate distribution.

Solution From Table A8 with $\nu_1 = 3$ and $\nu_2 = 6$, (a) $F_{3,6,.1} = 3.29$, (b) $F_{3,6,.025} = 6.60$, and (c) with $\nu_1 = 4$ and $\nu_2 = 9$, $F_{4,9,.01} = 6.42$. These three values are shown in Figure 12.

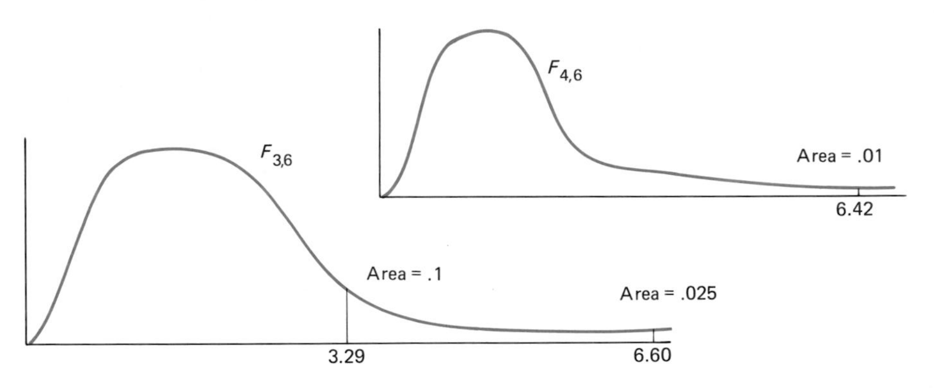

FIGURE 12

F values found in Example 15

Table A8 contains only upper-tail values for each F distribution. There is no need to provide lower-tail values since these can be obtained from upper-tail values and the definition below.

DEFINITION 8

The value of F_{ν_1, ν_2} which has an area $1 - \alpha$ to the right is the reciprocal of the value of F_{ν_2, ν_1} which has an area α to the right. Symbolically,

$$F_{\nu_1, \nu_2, 1-\alpha} = \frac{1}{F_{\nu_2, \nu_1, \alpha}}$$

Note that the F_{ν_1, ν_2} and the F_{ν_2, ν_1} are different distributions.

Applying Definition 8, assume that we need $F_{4, 10, .95}$, the value of $F_{4, 10}$ with an area .95 to the right. This is the reciprocal of $F_{10, 4, 1-.95} = F_{10, 4, .05}$.

From Table A8, $F_{10, 4, .05} = 5.96$, so that $F_{4, 10, .95} = 1/5.96 = .168$. The relationship between $F_{4, 10, .95}$ and $F_{10, 4, .05}$ is shown in Figure 13.

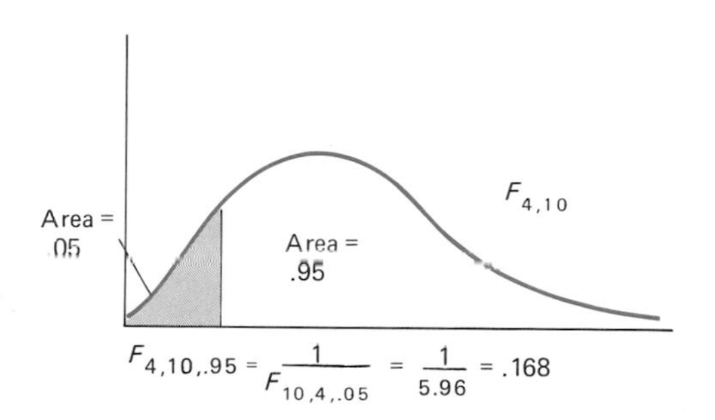

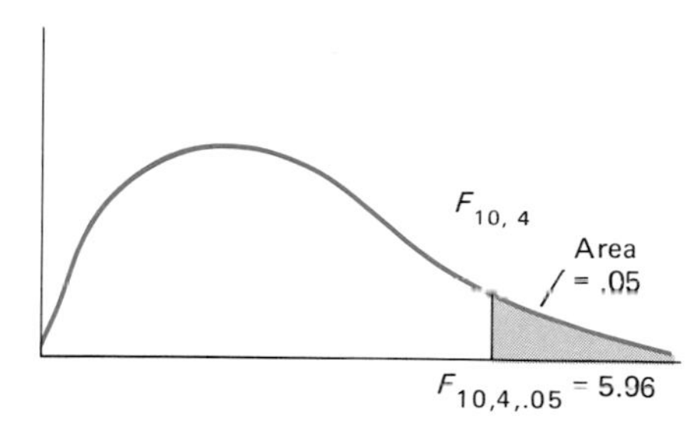

FIGURE 13

Relationship between $F_{4, 10, .95}$ and $F_{10, 4, .05}$

EXAMPLE 16 Use Definition 8 and Table A8 to obtain (a) $F_{3, 7, .995}$; (b) $F_{8, 8, .9}$.

Solution (a) $F_{3, 7, .995} = 1/F_{7, 3, .005} = 1/44.4 = .0225$.
(b) $F_{8, 8, .9} = 1/F_{8, 8, .1} = 1/2.59 = .386$. ■

PROBLEMS 12.6

For each of the following problems, sketch the appropriate distribution(s), marking the appropriate values and areas.

48. (a) Find (i) $\chi^2_{6, .025}$; (ii) $\chi^2_{6, .005}$; (iii) $\chi^2_{6, .99}$; (iv) $\chi^2_{6, .9}$.
    (b) If the random variable Y is χ^2_6, what is μ_Y?
49. (a) Find (i) $\chi^2_{10, .1}$; (ii) $\chi^2_{10, .025}$; (iii) $\chi^2_{10, .975}$; (iv) $\chi^2_{10, .95}$.
    (b) If the random variable Y is χ^2_{10}, what is μ_Y?
50. If the random variable Y is χ^2_7 find (a) $P(Y > 16.01)$; (b) $P(Y > 18.48)$; (c) $P(Y > 2.17)$; (d) $P(Y < 2.83)$; (e) $P(Y < 2.17)$; (f) $P(Y < 16.01)$.
51. If the random variable Y is χ^2_8, find (a) $P(\chi^2_{8, .95} < Y < \chi^2_{8, .05})$ (this probability is the area between $\chi^2_{8, .95}$ and $\chi^2_{8, .05}$); (b) $P(\chi^2_{8, .9} < Y < \chi^2_{8, .1})$.
52. Find (a) $F_{4, 13, .01}$; (b) $F_{2, 8, .025}$; (c) $F_{12, 17, .05}$; (d) $F_{12, 17, .005}$.
53. Use Definition 8 to find (a) $F_{9, 30, .95}$; (b) $F_{30, 9, .99}$; (c) $F_{30, 9, .95}$; (d) $F_{6, 6, .975}$, (e) $F_{4, 12, .995}$.
54. If the random variable F has the $F_{12, 10}$ distribution, find (a) $P(F > F_{12, 10, .05})$; (b) $P(F > F_{12, 10, .005})$; (c) $P(F < F_{12, 10, .025})$. (You will not need Table A8.)
55. If the random variable F has the $F_{20, 12}$ distribution, find (a) $P(F_{20, 12, .975} < F < F_{20, 12, .025})$; (b) $P(F_{20, 12, .995} < F < F_{20, 12, .005})$. (You will not need Table A8.)

Solution to Problem 48

(a) From Table A7, (i) $\chi^2_{6,.025} = 14.45$; (ii) $\chi^2_{6,.005} = 18.55$; (iii) $\chi^2_{6,.99} = .87$; (iv) $\chi^2_{6,.9} = 2.20$.

(b) Since the mean of a chi-square random variable is the corresponding number of degrees of freedom, $\mu_Y = 6$.

SECTION 12.7

INFERENCES ABOUT σ^2 OR σ

At this point we consider how to make inferences about the unknown population variance, σ^2, or standard deviation, σ. As we indicated earlier, the sample standard deviation,

$$s = \sqrt{\frac{\Sigma\,(x_i - \bar{x})^2}{n - 1}}$$

can be regarded as a point estimate for σ (and s^2 as a point estimate for σ^2). This is true regardless of the shape of the population being sampled. On the other hand we can find a confidence interval for σ^2 or test a hypothesis about σ^2 only if the sample is drawn from a *normal population*. When this is the case, we use the following definition to make inferences about σ^2 or σ.

> DEFINITION 9
>
> If a random sample of size n is to be selected from a $N[\mu, \sigma]$ population, then $y = (n - 1)s^2/\sigma^2$ has the χ^2_{n-1} distribution.

We first use this definition to find a confidence interval for σ^2 or σ.

CONFIDENCE INTERVALS FOR σ^2

Suppose that we intend to take a random sample of size $n = 12$ from a normal population in order to obtain a 95% confidence interval for σ^2. We assume that the value for μ is also unknown. Since the sample size is $n = 12$, we shall refer to the χ^2_{11} distribution. From Table A7 the value of χ^2_{11} that has an area of .025 to the right is $\chi^2_{11,.025} = 21.92$. Further, the value of χ^2_{11} that has an area of .025 to the left is $\chi^2_{11,.975} = 3.82$. Thus the area between 3.82 and 21.92 is $1 - .025 - .025 = .95$. This is shown in Figure 14.

From Definition 9, $(n - 1)s^2/\sigma^2$ has the χ^2_{11} distribution and thus the probability is .95 that $3.82 < (n - 1)s^2/\sigma^2 < 21.92$.

We obtain a 95% confidence interval for σ^2 by rearranging this inequality, isolating σ^2 in the middle. To do this, we first take the reciprocal of each of the three components. To anticipate the results of this step, notice that for example, $3 < 4$, but taking reciprocals we get the inequality $1/4 < 1/3$. On taking the

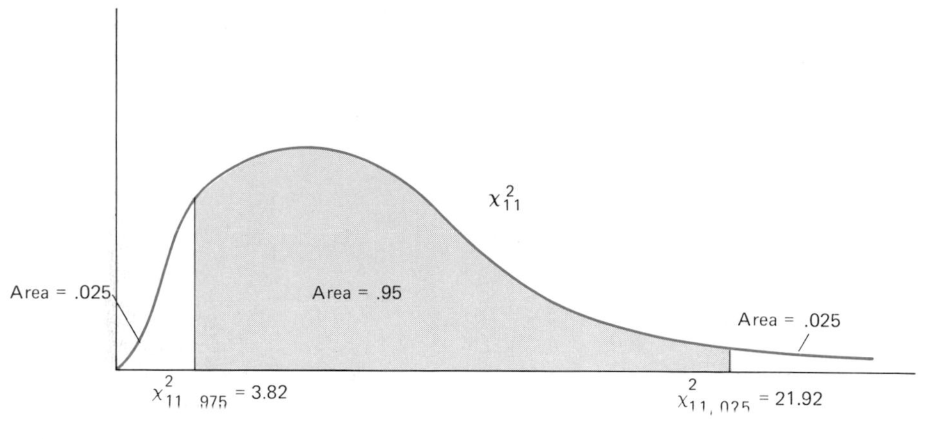

FIGURE 14

Showing that $P(3.82 < \chi^2_{11} < 21.92) = .095$

reciprocals of our original components, we get

$$\frac{1}{21.92} < \frac{\sigma^2}{(n-1)s^2} < \frac{1}{3.82}$$

All we need do now is multiply each of these three components by $(n-1)s^2$ to get

$$\frac{(n-1)s^2}{21.92} < \sigma^2 < \frac{(n-1)s^2}{3.82}$$

Prior to selecting the sample, we can say that the probability that the random interval $(n-1)s^2/21.92$ to $(n-1)s^2/3.82$ will contain σ^2 is .95. However, once the sample is selected and the value for s^2 computed, this interval is fixed and we say that we are 95% confident that σ^2 is in the interval. For example, with $n = 12$ and $s = 13.42$, we can be 95% confident that σ^2 lies in the interval $11(13.42)^2/21.92$ to $11(13.42)^2/3.82$, or 90.38 to 518.60. More generally:

> **DEFINITION 10**
>
> If a random sample of size n is selected from a $N[\mu, \sigma]$ population, a $(100\gamma)\%$ confidence interval for σ^2 is
>
> $$\frac{(n-1)s^2}{\chi^2_{n-1,(1-\gamma)/2}} \quad \text{to} \quad \frac{(n-1)s^2}{\chi^2_{n-1,(1+\gamma)/2}}$$
>
> A $(100\gamma)\%$ confidence interval for σ is
>
> $$\sqrt{\frac{(n-1)s^2}{\chi^2_{n-1,(1-\gamma)/2}}} \quad \text{to} \quad \sqrt{\frac{(n-1)s^2}{\chi^2_{n-1,(1+\gamma)/2}}}$$

The values $\chi^2_{n-1,(1-\gamma)/2}$ and $\chi^2_{n-1,(1+\gamma)/2}$ are shown in Figure 15. We also indicate in the figure that these values contain the central $(100\gamma)\%$ of the χ^2_{n-1}

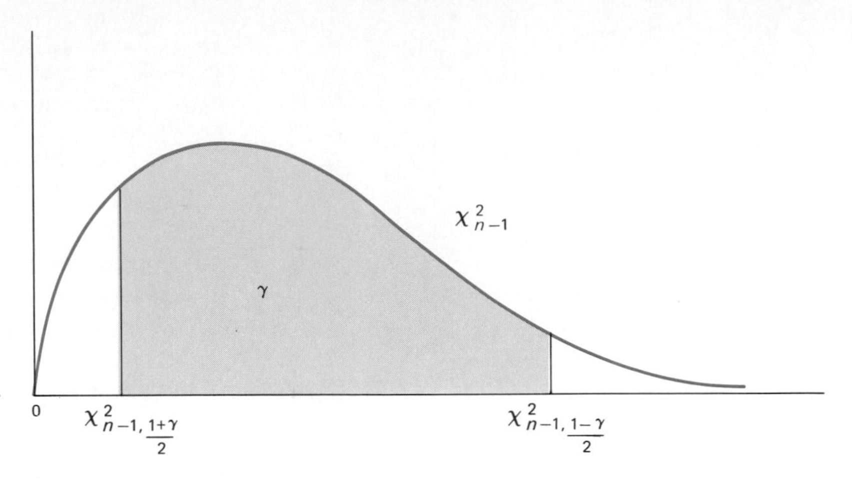

FIGURE 15

Showing the values $\chi^2_{n-1,1-\gamma/2}$ and $\chi^2_{n-1,1+\gamma/2}$

distribution. For instance, with $\gamma = .95$, as above, these values are $\chi^2_{n-1,(1+\gamma)/2} = \chi^2_{11,(1+.95)/2} = \chi^2_{11,.975} = 3.82$ and $\chi^2_{n-1,(1-\gamma)/2} = \chi^2_{11,(1-.95)/2} = \chi^2_{11,.025} = 21.92$.

■

EXAMPLE 17 The amount of rainfall in a month is recorded at a random sample of 15 locations around the state. The readings are (in inches) 3.5, 2.9, 2.7, 3.7, 3.4, 2.7, 2.9, 3.3, 3.0, 3.0, 3.1, 2.8, 3.4, 3.3, and 3.4. Assuming the distribution of rainfall is normal find an 80% confidence interval for σ, the standard deviation of the amount of rainfall in the state.

Solution We first compute $\Sigma\, x_i = 47.1$ and $\Sigma\, x_i^2 = 149.25$. The sample variance is then

$$s^2 = \frac{n\,\Sigma\, x_i^2 - (\Sigma\, x_i)^2}{n(n-1)} = \frac{(15)(149.25) - (47.1)^2}{(15)(14)} = .09686$$

From Definition 10 we require $\chi^2_{n-1,(1+\gamma)/2} = \chi^2_{14,(1+.8)/2} = \chi^2_{14,.9} = 7.79$ and $\chi^2_{n-1,(1-\gamma)/2} = \chi^2_{14,(1-.8)/2} = \chi^2_{14,.1} = 21.06$. An 80% confidence interval for σ is then $\sqrt{(14)(.09686)/21.06}$ to $\sqrt{(14)(.09686)/7.79}$, or .254 to .417. We can be 80% confident that σ lies in this interval.

■

Chi-square distributions are not symmetric and are not centered at 0 as are the Z and the t distributions. As a result, confidence intervals for σ^2 or σ do not have the nice $a \mp b$ form that previous confidence intervals have had.

TESTING HYPOTHESES ABOUT σ^2 OR σ

Definition 9 can also be used to test hypotheses about the unknown σ^2 or σ. Suppose, for instance, that we wish to test whether the variance of a normal population is significantly greater than a specific value σ^2. The null and alternative hypotheses are $H_0: \sigma^2 = \sigma_0^2$ and $H_1: \sigma^2 > \sigma_0^2$. Now, if H_0 is true and $\sigma^2 = \sigma_0^2$ then, from Definition 9, $y = (n-1)s^2/\sigma_0^2$ will have the χ^2_{n-1} distribution. From Section 12.6 we know that the mean of the distribution is ν, so that if H_0 is true, the mean value of y is $n - 1$. Evidence favoring $H_0: \sigma^2 = \sigma_0^2$ will be a value for s^2 close to σ_0^2 or, equivalently, since s^2/σ_0^2 will then be close to 1, a value for y close to $n - 1$. Conversely, evidence for $H_1: \sigma^2 > \sigma_0^2$ will be a value for s^2 much greater than σ_0^2 or a value for y much greater than $n - 1$. Thus we should reject H_0 if the value for the test statistic $y = (n - 1)s^2/\sigma^2$ is unusually far above $n - 1$. Just how far above we determine from the chi-square table. Similarly, if we were testing $H_0: \sigma^2 = \sigma_0^2$ against $H_1: \sigma^2 < \sigma_0^2$, we would reject H_0 if the value for $y = (n - 1)s^2/\sigma_0^2$ were unusually far *below* $n - 1$. For the two-sided test of $H_0: \sigma^2 = \sigma_0^2$ against $H_1: \sigma^2 \neq \sigma_0^2$, we reject H_0 if y is sufficiently far above *or* below $n - 1$. The exact procedures for testing H_0 are given in Table 6.

TABLE 6

Procedures for testing $H_0: \sigma^2 = \sigma_0^2$ in random samples from a normal population

1. Compute s and $y = (n - 1)s^2/\sigma_0^2$.

(a) $H_1: \sigma^2 > \sigma_0^2$
2. Find $\chi^2_{n-1,.05}$ and $\chi^2_{n-1,.01}$ from Table A7.
3. *Conclusion:* Using $\alpha = .05$ and .01, reject H_0 at the α level of significance if $y \geq \chi^2_{n-1,\alpha}$.

(b) $H_1: \sigma^2 < \sigma_0^2$.
2. Find $\chi^2_{n-1,.95}$ and $\chi^2_{n-1,.99}$ from Table A7.
3. *Conclusion:* Using $\alpha = .05$ and .01, reject H_0 at the α level of significance if $y \leq \chi^2_{n-1,1-\alpha}$.

(c) $H_1: \sigma^2 \neq \sigma_0^2$.
2. Find $\chi^2_{n-1,.025}$, $\chi^2_{n-1,.005}$, $\chi^2_{n-1,.975}$, and $\chi^2_{n-1,.995}$ from Table A7.
3. *Conclusion:* Using $\alpha = .05$ and .01, reject H_0 at the α level of significance if either $y \geq \chi^2_{n-1,\alpha/2}$ or if $y \leq \chi^2_{n-1,1-\alpha/2}$.

Assumption: That a random sample is drawn from a normal population

We illustrate the one-sided test first.

EXAMPLE 18 A certain make of coffee-dispensing machine needs to be reset when there is too much variability in the amount of coffee being dispensed. The manufacturer recommends to its service staff that a machine be reset when there is reason to believe that the standard deviation of the amount of coffee dispensed exceeds .2 ounce. Samples of 10 cups are selected from each of the three machines on a college campus. The sample standard deviations of the amount dispensed are given below. Which, if any, of the machines need to be reset?

MACHINE	*A*	*B*	*C*
s	.284	.236	.314

Solution For each machine we will test $H_0: \sigma = .2$ against $H_1: \sigma > .2$ or, equivalently, $H_0: \sigma^2 = .04$ against $H_1: \sigma^2 > .04$, where σ^2 is the present variance in the amount of coffee dispensed by a machine. We apply the procedure in Table 6 to the data for each machine in turn.

(a) Since $s = .284$, $s^2 = .284^2 = .080656$ and $y = (n - 1)s^2/.04 = 9(.080656)/.04 = 18.15$. From Table A7, $\chi^2_{9,.05} = 16.92$ and $\chi^2_{9,.01} = 21.67$.

Since $y = 18.15$ is greater than 16.92 but is not greater than 21.67, we reject H_0 at the .05 level of significance. Machine A should be reset.

(b) Since $s = .236$, $s^2 = .236^2 = .055696$ and $y = 9(.055696)/.04 = 12.532$. Using the same χ_9^2 values as in (a), note that 12.532 is less than 16.92, so we cannot reject H_0 at the .05 level of significance. Machine B need not be reset.

(c) Since $s = .314$, $s^2 = .098596$ and $y = 9(.098596)/.04 = 22.184$. As this value exceeds 21.67, we reject H_0 at the .01 level of significance. Machine C should be reset.

The value for y for each machine is shown in Figure 16.

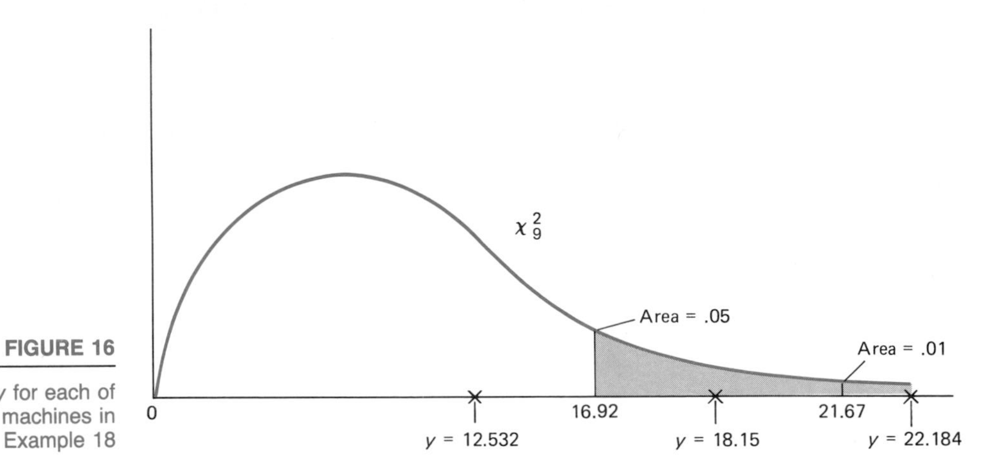

FIGURE 16

Values for y for each of the machines in Example 18

■

EXAMPLE 19 The variance of reading scores for all seventh graders in a particular state is 192. Although mean scores are available for each school district, variances are not. An official at one school district is interested in learning whether the variability of scores for the district differs from that for the entire state. The reading scores for a random sample of 20 seventh graders in the district are obtained and the following summary data computed:

$$\sum x_i = 1484 \qquad \sum x_i^2 = 111{,}937$$

Do these data suggest that the variability in scores for the district differ significantly from 192?

Solution We follow the procedure for the two-sided test in Table 6. Call σ^2 the variance of scores in the district. The hypotheses are $H_0: \sigma^2 = 192$ and $H_1: \sigma^2 \neq 192$. The sample variance is:

$$s^2 = \frac{n \sum x_i^2 - (\sum x_i)^2}{n(n - 1)} = \frac{(20)(111{,}937) - 1484}{20(19)} = 96.0105$$

Therefore,

$$y = \frac{(n-1)s^2}{\sigma_0^2} = \frac{19(96.0105)}{192} = 9.5$$

From Table A7 we obtain the values $\chi^2_{19,.005} = 38.58$, $\chi^2_{19,.025} = 32.85$, $\chi^2_{19,.975} = 8.91$, and $\chi^2_{19,.995} = 6.84$. Since $y = 9.5$ is not less than 8.91, and not greater than 32.85 we cannot reject H_0 at the .05 level of significance. The data suggest that the variance of scores for the district is not significantly different from that for the entire state. We indicate the value $y = 9.5$ on the χ^2_{19} distribution in Figure 17. Notice from this figure that had y been 7.76, for example, we would have rejected H_0 at the .05 level of significance. Similarly, had y been 39.02, we would have rejected H_0 at the .01 level of significance.

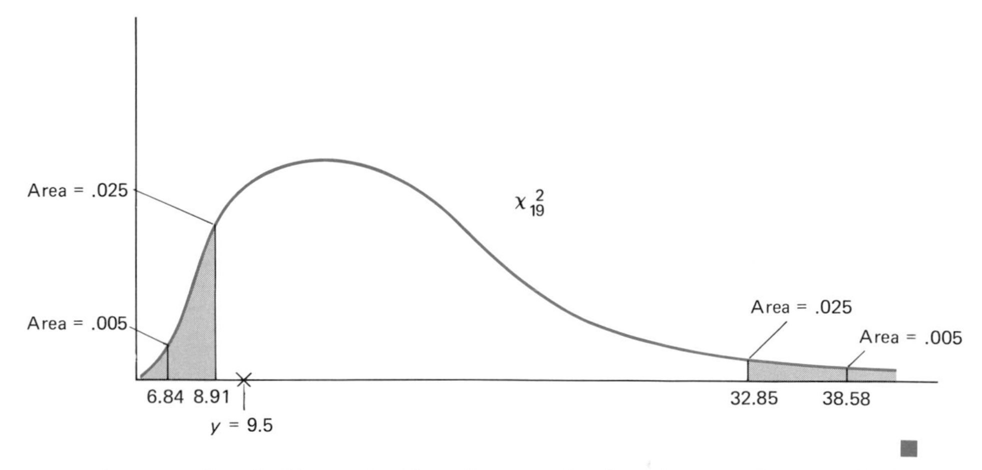

FIGURE 17

The value $y = 9.5$ in Example 19

■

Suppose that in Example 19, prior to selecting the sample the official suspected that the variability of scores for the district was significantly *less* than 192. In this case the appropriate hypotheses are $H_0 : \sigma^2 = 192$ and $H_1 : \sigma^2 < 192$. As before, $y = 9.5$. Following the procedure in Table 6, we find from Table A7 that $\chi^2_{19,.95} = 10.12$ and $\chi^2_{19,.99} = 7.63$. Since 9.5 is less than 10.12 but is not less than 7.63, we reject H_0 at the .05 level of significance.

PROBLEMS 12.7

56. A sample of size $n = 20$ is to be selected from a normal population with unknown mean and standard deviation. How small must s, the sample standard deviation, be in order to reject $H_0 : \sigma^2 = 100$ in favor of $H_1 : \sigma^2 < 100$? (Use only $\alpha = .05$.)
57. An inspector for a match manufacturing company selects a sample of 27 matches. The length of time (in seconds) that each match stays lit when held in a specific position is recorded. The following summary data are computed:

$$\sum x_i = 710.01 \qquad \sum x_i^2 = 18{,}801.50$$

(a) Find a 90% confidence interval for the standard deviation of time lit for this brand of matches.

(b) What assumptions did you make?

58. A random sample of 15 mentally impaired patients are given a test of manual dexterity. The mean time to complete the test is 12.68 minutes. The standard deviation is 3.9 minutes.

(a) Find a 95% confidence interval for the (i) mean time and (ii) standard deviation of the time it takes all such patients to complete the task.

(b) What assumptions were necessary in part (a)?

59. In Problem 58, assume that the standard deviation of the time taken to complete the task among the general population is 2.3 minutes. Do the data in Problem 58 suggest that the standard deviation of time among mentally impaired persons is significantly different from that of the general population?

60. Employees of an insurance company who wish to move up to a higher job category must take a written examination. The mean and standard deviation of scores in this test were previously 24.6 and 6.5, respectively. This year a new type of test is to be given. The mean score for the new test is expected to be roughly the same as the old, but there is considerable interest in learning whether the variability of scores for the new test is significantly different from that for the old. The scores of a sample of five employees taking the new test are 16, 26, 38, 10, and 37. Do these data suggest that the variability of scores on the new test is significantly different from that on the old?

61. Using the data in Problem 60, obtain an 80% confidence interval for the standard deviation of scores for the new test.

62. A manufacturer of refrigerators receives supplies of metal plates supposedly of uniform thickness. Shipments from the supplier are accepted only if the standard deviation of thickness does not appear to be significantly greater than .002 centimeter. In a sample of 12 plates randomly taken from a large shipment, the standard deviation of thickness is $s = .0026$.

(a) Should this shipment be accepted?

(b) What if s were .0036?

63. In Problem 62, if $s = .0036$:

(a) Find an 80% confidence interval for σ^2.

(b) Interpret your interval.

64. Of concern to teachers at a nursery school are children who arrive either much earlier or much later than the official starting time. One of the teachers believes that the standard deviation of the *difference* between the time of arrival and the official starting time is 10 minutes. In an effort to reduce this variability teachers mail out a request to parents during the Christmas vacation asking them to bring their children to school closer to the starting time. On the fifth day after the vacation the differences from the starting time are -6, 11, 8, 9, -9, -3, -4, 4, 2, 7, -5, -11, 0, 10, and -6. (For instance, -6 indicates that a child arrived 6 minutes early.)

(a) Do these data suggest that there has been a significant decline in the variability in arrival times?

(b) What assumptions were necessary?

65. A sample of size $n = 12$ is to be selected from a normal population with unknown mean and variance. For what range of values of s, the sample standard deviation, will the null hypothesis $H_0: \sigma^2 = .02$ *not* be rejected in favor of $H_1: \sigma^2 \neq .02$? (Use only $\alpha = .05$.)

66. Why is the phrase "prior to selecting the sample" in the last paragraph of Section 12.7 necessary?

Solution to Problem 56

From Table 6 with $n = 20$, $\sigma_0^2 = 100$, and $\alpha = .05$, we will reject H_0 if

$$y = \frac{(n-1)s^2}{\sigma_0^2} = \frac{19s^2}{100} < \chi^2_{19,\,1-.05}$$

Since $\chi^2_{19,\,1-.05} = \chi^2_{19,\,.95} = 10.12$, we reject H_0 if $.19s^2 \leq 10.12$ or if $s^2 < 10.12/.19 = 53.263$. Equivalently, we can reject H_0 if $s < \sqrt{53.263} = 7.298$. We will reject $H_0\!:\sigma^2 = 100$ in favor of $H_1\!:\sigma^2 < 100$ if the sample standard deviation $s < 7.298$.

SECTION 12.8

INFERENCES ABOUT σ_1^2/σ_2^2

As we have noted, inferences about two population means focus on the difference between them (i.e., $\mu_1 - \mu_2$). However, this approach is not possible with the variances of two populations. In this case we make inferences about the *ratio* of one variance to another (σ_1^2/σ_2^2) or one standard deviation to another (σ_1/σ_2). Specifically, comparisons between σ_1^2 and σ_2^2—whether in the form of confidence intervals or tests—are based on the following.

DEFINITION 11

A random sample of size n_1 is taken from an $N[\mu_1, \sigma_1]$ population. Another random sample of size n_2 is taken independently from a second $N[\mu_2, \sigma_2]$ population. Assume that the parameters μ_1, μ_2, σ_1 and σ_2 are unknown. In this case $F = (s_2^2/s_1^2)(\sigma_1^2/\sigma_2^2)$ has the $F_{n_2-1,\,n_1-1}$ distribution.

First we consider how this result can be used to find a confidence interval for σ_1^2/σ_2^2.

CONFIDENCE INTERVALS FOR σ_1^2/σ_2^2

Suppose that samples of size $n_1 = 21$ and $n_2 = 9$ are to be selected independently from two normal populations. We want a 90% confidence interval for the ratio σ_1^2/σ_2^2. The appropriate F distribution is, from Definition 11, $F_{8,20}$. From Table A8, $F_{8,20,.05}$, the value for $F_{8,20}$ which has an area of .05 to the right, is 2.45. From Definition 8 the value for $F_{8,20}$, which has an area of .05 to the *left*, is $F_{8,20,.95} = 1/F_{20,8,.05} = 1/3.15 = .317$. These two F values are shown in Figure 18. Notice that the area between .317 and 2.45 is $1 - .05 - .05 = .9$. There is thus a .9 probability that an $F_{8,20}$ random variable will lie between .317 and 2.45. But from Definition 11, $F = (s_2^2/s_1^2)(\sigma_1^2/\sigma_2^2)$ has the $F_{8,20}$ distribution.

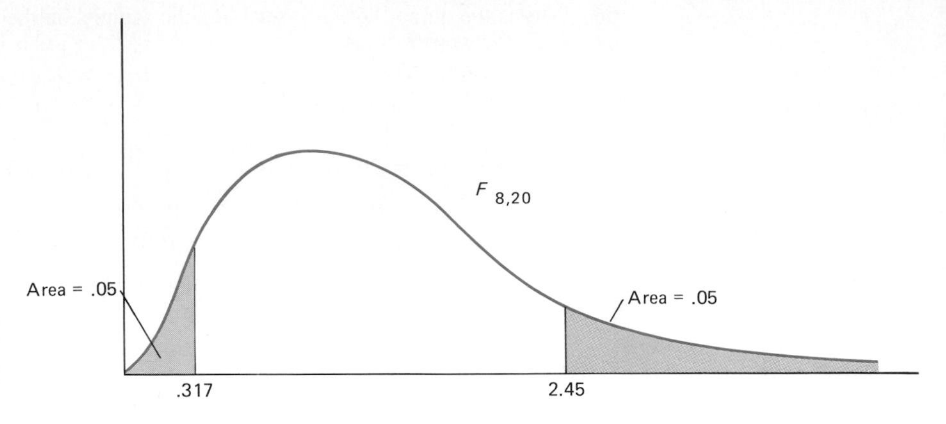

FIGURE 18

The values $F_{8, 20, .05} = 2.45$ and $F_{8, 20, .95} = .317$

Therefore, we can say that prior to selecting the samples there is a .9 probability that

$$.317 < \frac{s_2^2}{s_1^2} \cdot \frac{\sigma_1^2}{\sigma_2^2} < 2.45$$

By multiplying all three of these components by s_1^2/s_2^2, we can isolate σ_1^2/σ_2^2 in the middle, obtaining

$$.317 \frac{s_1^2}{s_2^2} < \frac{\sigma_1^2}{\sigma_2^2} < 2.45 \frac{s_1^2}{s_2^2}$$

Once the samples are selected and values for s_1 and s_2 computed, the interval is fixed and we say that we can be 90% confident that σ_1^2/σ_2^2 lies in this interval. Suppose, for example, that $s_1^2 = 84.9$ and $s_2^2 = 62.7$. We can be 90% confident that σ_1^2/σ_2^2 lies in the interval .317(84.9)/(62.7) to 2.45(84.9)/(62.7), or .429 to 3.32. This confidence interval for σ_1^2/σ_2^2 suggests that σ_1^2 may be as small as 43% of σ_2^2 to as large as 3.3 times σ_2^2. A 90% confidence interval for σ_1/σ_2 is then $\sqrt{.429}$ to $\sqrt{3.32}$ or .655 to 1.82. These results can be generalized as follows:

DEFINITION 12

A random sample of size n_1 is taken from an $N[\mu_1, \sigma_1]$ population. Another random sample of size n_2 is taken independently from a second $N[\mu_2, \sigma_2]$ population. A $(100\gamma)\%$ confidence interval for σ_1^2/σ_2^2 is

$$\frac{1}{F_{n_1-1,\, n_2-1,\, (1-\gamma)/2}} \frac{s_1^2}{s_2^2} \quad \text{to} \quad F_{n_2-1,\, n_1-1,\, (1-\gamma)/2} \frac{s_1^2}{s_2^2}$$

and a $(100\gamma)\%$ confidence interval for σ_1/σ_2 is

$$\sqrt{\frac{1}{F_{n_1-1,\, n_2-1,\, (1-\gamma)/2}}} \frac{s_1}{s_2} \quad \text{to} \quad \sqrt{F_{n_2-1,\, n_1-1,\, (1-\gamma)/2}} \frac{s_1}{s_2}$$

EXAMPLE 20 An archeologist is interested in comparing the variability of cranial widths (the widths of skulls) for skulls found at two different sites, I and II. Use the following data to find a 95% confidence interval for σ_1/σ_2 the ratio of the standard deviations of cranial widths (the data are in inches). What assumptions are necessary?

SITE	n	s
I	8	.388
II	5	.555

Solution From Definition 12 with $\gamma = .95$, a 95% confidence interval for σ_1/σ_2 is

$$\sqrt{\frac{1}{F_{7,4,.025}}}\frac{s_1}{s_2} \quad \text{to} \quad \sqrt{F_{4,7,.025}}\frac{s_1}{s_2}$$

From Table A8, $F_{7,4,.025} = 9.07$ and $F_{4,7,.025} = 5.52$. Thus the required interval is

$$\sqrt{\frac{1}{9.07}}\left(\frac{.388}{.555}\right) \quad \text{to} \quad \sqrt{5.52}\left(\frac{.388}{.555}\right)$$

or .232 to 1.643. We can be 95% confident that σ_1/σ_2 lies in this interval.

We assume (1) that the skulls represent random samples of skulls found at the two sites, and (2) the distribution of cranial widths for skulls located at each of the sites is normal.

■

We can use a slight variant of Definition 11 to test whether the variances, or equivalently, the standard deviations, of two normal populations are the same. Specifically, we can test $H_0: \sigma_1^2 = \sigma_2^2$ against either $H_1: \sigma_1^2 > \sigma_2^2$ or $H_1: \sigma_1^2 \neq \sigma_2^2$.

The result we need to do this is summarized in Definition 13.

DEFINITION 13

With the normality assumptions of Definition 11, $F = \dfrac{s_1^2}{s_2^2} \cdot \dfrac{\sigma_2^2}{\sigma_1^2}$ has the F_{n_1-1,n_2-1} distribution.

Now, tests of $H_0: \sigma_1^2 = \sigma_2^2$ are based on the fact that if H_0 is true, then $\sigma_2^2/\sigma_1^2 = 1$ and the test statistic $F = s_1^2/s_2^2$ has the F_{n_1-1,n_2-1} distribution. Evidence favoring H_0 occurs when s_1^2 is close in value to s_2^2 or, in other words, when the ratio $F = s_1^2/s_2^2$ is close to 1. Evidence for $H_1: \sigma_1^2 > \sigma_2^2$ occurs when s_1^2 is unusually greater than s_2^2 or equivalently, if $F = s_1^2/s_2^2$ is unusually greater than 1. Evidence for the two-sided alternative $H_1: \sigma_1^2 \neq \sigma_2^2$ occurs when $F = s_1^2/s_2^2$ is either unusually far above or unusually far below 1. The exact procedures for testing $H_0: \sigma_1^2 = \sigma_2^2$ are given in Table 7.

TABLE 7

Procedures for testing $H_0: \sigma_1^2 = \sigma_2^2$

1. Compute $F = s_1^2/s_2^2$.

(a) $H_1: \sigma_1^2 > \sigma_2^2$

2. Find $F_{n_1-1, n_2-1, .05}$ and $F_{n_1-1, n_2-1, .01}$ from Table A8.
3. *Conclusion:* Using $\alpha = .05$ and $.01$, reject H_0 at the α level of significance if $F \geq F_{n_1-1, n_2-1, \alpha}$.

(b) $H_1: \sigma_1^2 \neq \sigma_2^2$

(A)

2. If $F \geq 1$, find $F_{n_1-1, n_2-1, .025}$ and $F_{n_1-1, n_2-1, .005}$ from Table A8.
3. *Conclusion:* Using $\alpha = .05$ and $.01$ reject H_0 at the α level of significance if $F \geq F_{n_1-1, n_2-1, \alpha/2}$.

(B)

2. If $F < 1$ find $\dfrac{1}{F_{n_2-1, n_1-1, .025}}$ and $\dfrac{1}{F_{n_2-1, n_1-1, .005}}$ from Table A8.
3. *Conclusion:* Using $\alpha = .05$ and $.01$ reject H_0 at the α level of significance if $F \leq \dfrac{1}{F_{n_2-1, n_1-1, \alpha/2}}$.

Assumption: Random samples are drawn independently from normal populations.

The next two examples illustrate these procedures.

EXAMPLE 21 An economist living in Detroit is interested in whether there has been an increase in the variability of rental prices of apartments in that city since 1978. In that year the standard deviation of rental prices in a random sample of 25 apartments was \$87. In a recent sample of 21 apartments the standard deviation of monthly rent was \$162. Do these data suggest that there has been a significant increase in the variability of rental prices?

Solution We test $H_0: \sigma_1^2 = \sigma_2^2$ against $H_1: \sigma_1^2 > \sigma_2^2$, where σ_1^2 is the recent variance of rentals in all apartments and σ_2^2 the variance of all such rentals in 1978. The sample results we shall need are $n_1 = 21$, $n_2 = 25$, $s_1^2 = 162^2 = 26{,}244$ and $s_2^2 = 87^2 = 7569$.

We now follow the procedure in Table 7.

1. $F = s_1^2/s_2^2 = 26{,}244/7569 = 3.467$.
2. From Table A8 with $\nu_1 = n_1 - 1 = 20$ and $\nu_2 = n_2 - 1 = 24$ we find $F_{20, 24, .05} = 2.03$ and $F_{20, 24, .01} = 2.74$.
3. Since $F = 3.467$ exceeds even 2.74 we reject H_0 at the .01 level of significance. The data suggest that the variability in recent rental prices is significantly greater than that for 1978.

The tabulated values, $F_{20, 24, .05} = 2.03$, $F_{20, 24, .01} = 2.74$ and the observed value $F = 3.467$, are shown in Figure 19.

In this example if the economist had no reason to expect that one variance was larger than the other, the two-sided alternative $H_1: \sigma_1^2 \neq \sigma_2^2$ would be appropriate. In this case, following the procedure in Table 7 and since

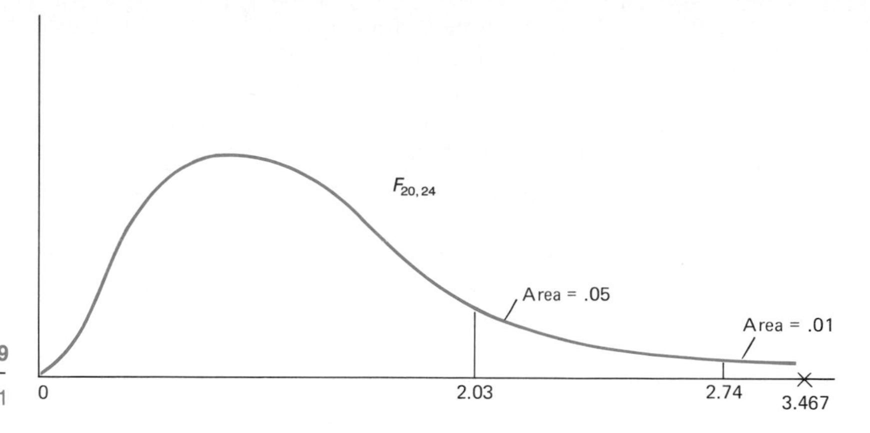

FIGURE 19

F values for Example 21

$F = 3.467 > 1$ we compare this value to $F_{20, 24, .025} = 2.23$ and $F_{20, 24, .005} = 3.06$. Since 3.467 exceeds even 3.06 we would again reject H_0 at the .01 level of significance.

■

It frequently happens that the required F_{ν_1, ν_2} distribution does not appear in Table A8. In this case we use the tabulated F distribution with the largest numbers of degrees of freedom smaller than ν_1 and ν_2. Suppose in Example 21, for instance, the respective sample sizes had been $n_1 = 26$ and $n_2 = 33$. The appropriate $F_{25, 32}$ distribution does not appear in Table A8. Therefore, we would refer instead to values for the $F_{24, 30}$ distribution. Similarly, we would use the $F_{17, 20}$ distribution in place of the $F_{17, 23}$ distribution.

In Section 12.4 we pointed out that confidence intervals for, or tests about, $\mu_1 - \mu_2$ in small samples require that the two populations be normal with equal variances. An application of the F test in Table 7 is testing whether the assumption of equal variances is violated. The next example illustrates this type of application.

EXAMPLE 22 In Example 8 on page 416 we assumed that the variability of typing scores for the two teaching methods σ_1^2 and σ_2^2 were equal. Is this assumption justified?

Solution With no indication of which variance might be larger, we perform a two-sided test of $H_0: \sigma_1^2 = \sigma_2^2$ against $H_1: \sigma_1^2 \neq \sigma_2^2$ following the appropriate procedure in Table 7.

1. Using the data in Example 8, we compute $F = s_1^2/s_2^2 = 5.91^2/8.42^2 = .493$. (B). Since $F < 1$, we
2. Find $1/F_{19, 19, .025}$ and $1/F_{19, 19, .005}$ from Table A8. Since the $F_{19, 19}$ distribution does not appear there, we use instead the values $1/F_{15, 19, .025} = 1/2.62 = .382$ and $1/F_{15, 19, .005} = 1/3.59 = .279$. Since $F = .493$ is not smaller than even .382, we cannot reject $H_0: \sigma_1^2 = \sigma_2^2$ at the .05 level of significance. The data suggest that the variance of typing scores, although perhaps not exactly equal, are not significantly different.

■

We shall have a lot more to say about this use of the F test in the next section when we discuss assumptions.

The inferences about σ and σ_1/σ_2 that we have discussed in this section and Section 12.7 are valid for any n. The reason they are included in this chapter on *small* sample inferences is because, like small sample inferences about μ, they require the assumption that the population or populations be normal.

PROBLEMS 12.8

67. Use the following data to test whether there is a significant difference in the population variances:

n	s^2
9	2917.4
16	1842.2

68. A criminologist is interested in comparing the consistency of length of sentence given to persons convicted of drug offences in two jurisdictions. The length of sentence for each of a random sample of 16 such offenders in jurisdiction I is obtained and the standard deviation $s = 1.28$ years is recorded. Similarly, in a random sample of 21 such offenders from jurisdiction II, the standard deviation is $s = 2.46$ years.

    (a) Do these data suggest that the variability of length of sentence is significantly different for the two jurisdictions?

    (b) What assumptions were necessary?

69. Refer to the data in Problem 68. Find a 90% confidence interval for σ_1^2/σ_2^2. Interpret your interval.

70. In each of cases (a), (b), (c), and (d), samples are drawn independently from two normal populations. In each case, test whether the population variances are significantly different.

	(a)		(b)		(c)		(d)	
n	12	18	27	13	12	12	5	25
s^2	76.4	61.1	.007	.041	10.3	39.4	1.13	1.72

71. Over the years a statistics professor has used two forms of a final exam, A and B. She is reasonably sure that, on the average, students score equally well on the two tests but is interested in finding whether the *variability* of scores is different for the two forms. She selects a random sample of 20 students who were given form A and, independently, a random sample of 20 who were given form B. Examining their final scores, she obtains the following data:

	EXAM A	EXAM B
Σx_i	1518	1528
Σx_i^2	124,749.64	118,152.8

Use these results to test whether the variability of scores for the two tests differ significantly.

72. Use the data in Problem 27 on page 418 to test for a significant difference in the variability of nicotine content for the two brands.
73. Refer to the electricity use data summarized in Example 7 on page 414. Test whether the variability of use in city I is significantly greater than that in city II.
74. A major concern of a labor union in negotiating with a company is the belief that there is greater variability in the hours of overtime in the Barnet plant relative to the Ackley plant. To investigate this issue, an independent consulting company selects a sample of 20 union members from each plant. The average number of hours of overtime per week for the past three months was recorded for each person. The following summary data were obtained (one of the Ackley members had left, reducing the sample size for that plant):

	n	Σx_i	Σx_i^2
Ackley	19	179.4	1744.85
Barnet	20	208.0	2284.61

(a) What hypotheses seem appropriate here?
(b) Use the data to test H_0.
(c) What assumptions were necessary?

75. In Problem 74, test whether there is any significant difference in the mean amount of overtime for union members in the two plants.

Solution to Problem 67

Call $n_1 = 9$, $n_2 = 16$, $s_1^2 = 2{,}917.4$, and $s_2^2 = 1{,}842.2$. We test $H_0: \sigma_1^2 = \sigma_2^2$ against $H_1: \sigma_1^2 \neq \sigma_2^2$ using the procedure in Table 7.

1. $F = s_1^2/s_2^2 = 2917.4/1842.2 = 1.584$.
   Since $F > 1$;
2. From Table A8 we find $F_{n_1-1, n_2-1, .025} = F_{8, 15, .025} = 3.20$ and $F_{n_1-1, n_2-1, .005} = F_{8, 15, .005} = 4.67$.
3. Since F is not greater than even 3.20, we cannot reject H_0 at the .05 level of significance. The data suggest that there is no significant difference between the two population variances.

SECTION 12.9

ASSUMPTIONS IN SMALL-SAMPLE INFERENCES

ASSUMPTIONS IN INFERENCES BASED ON ONE SMALL SAMPLE

The procedures for making inferences about μ and σ using a single sample assume that (1) the sample is randomly drawn and (2) the population is normal. We discussed the importance of random sampling in Chapter 8. Frequently, inferences are made using data from samples that are not randomly selected. In such cases the degree of faith that one has in the inference must be a matter of judgment.

There are two simple graphic ways to check the assumption of normality. We can construct a dot diagram if the sample size is less than 20 or a relative frequency histogram if n is ≥ 20. Note especially that a distribution need not be exactly symmetric to validate the assumptions of normality. Rather, we look for evidence of a population that is far from normal in shape in order to invalidate the assumption.

Two dot diagrams and two relative frequency histograms are portrayed in Figure 20. They are all consistent with the conclusion that the data came from a normal population. The nearly symmetric dot diagram (i) for example, is almost exactly what we would expect for a small sample from a normal population.

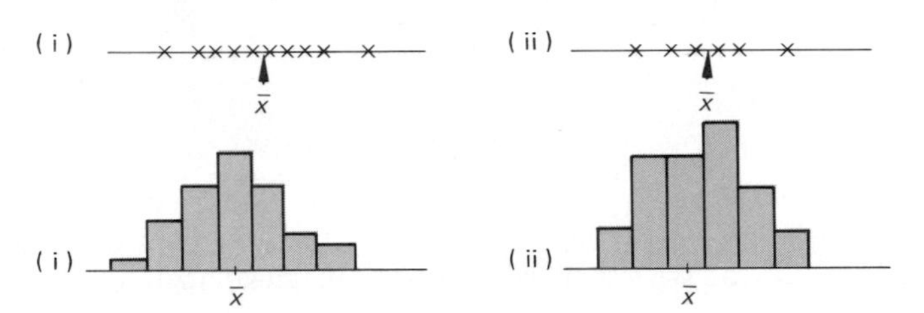

FIGURE 20

Two dot diagrams and two histograms of samples consistent with the conclusion that they came from normal populations

By contrast, in Figure 21 we show two dot diagrams and two relative frequency histograms all of which suggest that the populations being sampled are far from normal. The shape of the population suggested by each dot diagram is shown beneath it. If a test or confidence interval were based on the data represented in any of the last four diagrams (Figure 20), it is likely that it would be invalid and perhaps lead to erroneous interpretations.

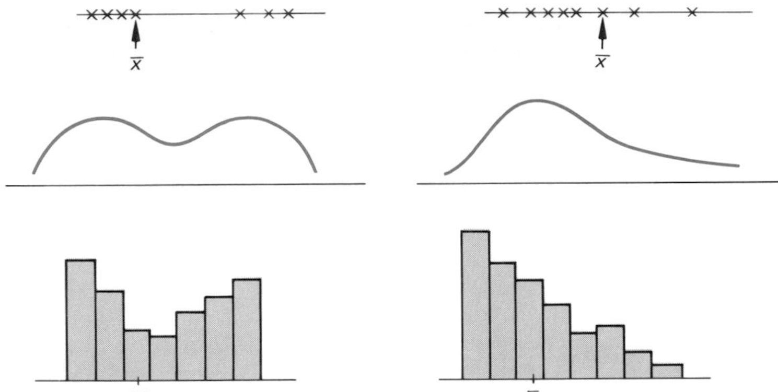

FIGURE 21

Two dot diagrams and two histograms which appear inconsistent with having come from a normal population

EXAMPLE 23 In Example 10 on page 421 we assumed that the distribution of the differences, d_i, in reaction times were approximately normal. A frequency (and a relative frequency) distribution for these differences are shown in Figure 22, together with the relative frequency histogram for these data.

FIGURE 22

Frequency and relative frequency distributions and the relative frequency histogram for the 20 differences from Example 10

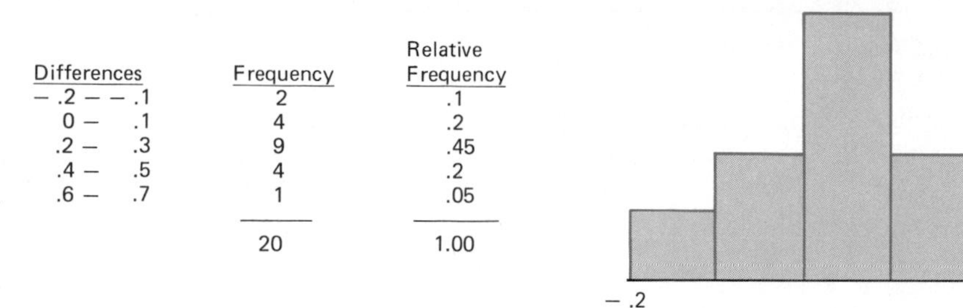

Differences	Frequency	Relative Frequency
−.2 − −.1	2	.1
0 − .1	4	.2
.2 − .3	9	.45
.4 − .5	4	.2
.6 − .7	1	.05
	20	1.00

The histogram in Figure 22 certainly suggests that the distribution of differences is not far from normal.

■

ASSUMPTIONS IN INFERENCES BASED ON TWO SMALL SAMPLES

In Section 12.4 we emphasized that inferences about $\mu_1 - \mu_2$ in small samples require the assumption that the two samples be drawn independently from approximately normal populations with equal standard deviations (and hence, variances).

Two samples can be considered independent if the selection of either of them is unaffected by the contents of the other. In experiments, independence between groups is obtained if the groups are randomly formed. The assumption that each population is approximately normal can be examined as we did above with a sample from a single population.

In Example 21 we illustrated how the assumption of equal variances in the two-independent-sample t test can be examined using the F test in Section 12.8. If $H_0: \sigma_1^2 = \sigma_2^2$ is rejected in favor of $H_1: \sigma_1^2 > \sigma_2^2$ or $H_1: \sigma_1^2 \neq \sigma_2^2$, the t test should not be used. We should emphasize though that a nonsignificant difference between σ_1^2 and σ_2^2 does not prove that they are equal. Recent evidence does suggest however that the t test works well even with moderate departures from both the equal variance and the normality assumptions.

The F test, like the t test, requires that the two populations be normal. But the F test is disturbed far more easily by departures from normality than is the t test. Particularly if the sample sizes are not equal, even slight departures from normality can invalidate conclusions based on the F test. As a consequence, when examining the normality assumption for the F test, our graphical method should be reinforced by theoretical or empirical evidence favoring a normal population.

Because of its sensitivity to departures from normality, the F test of the equality of two variances may be inappropriately used even when the t case can safely be used. For this reason the F test of the assumption of equal variances in a two-sample t test should be used only if there is considerable evidence that the population is very close to normal.

If the one- or two-sample t tests are not appropriate, we recommend the nonparametric techniques discussed in Chapter 16.

PROBLEMS 12.9

76. Each of the pairs of histograms shown represents two independent samples. In each case, do the histograms suggest that the samples were selected from normal populations? If not, speculate on how a population might differ from the normal.

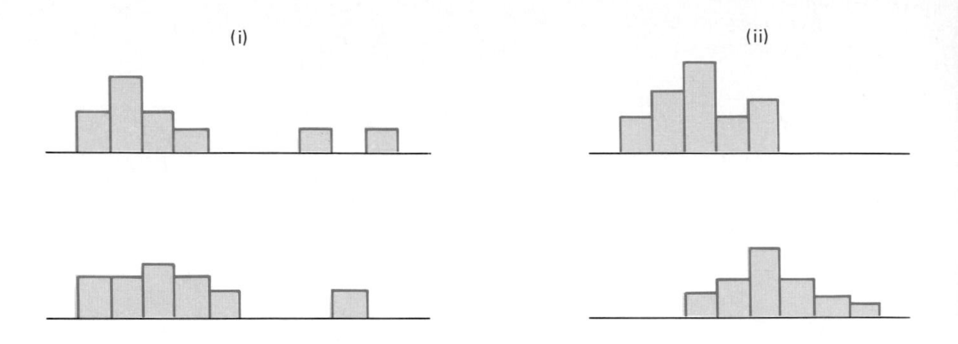

77. Each of the pairs of dot diagrams shown represents two independent samples. In each case, do the diagrams suggest that the samples were drawn from normal populations? If not, speculate on how a population might differ from the normal.

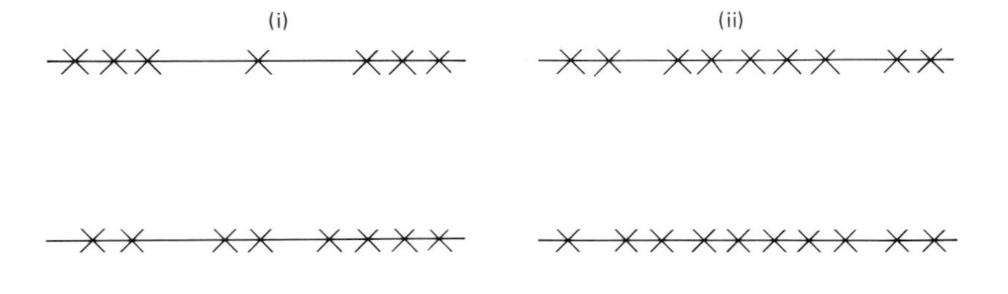

78. Construct a dot diagram for the following data and decide whether they appear to come from a normal population.

$$9.0 \quad -2.2 \quad -5.2 \quad 3.8 \quad -1.1 \quad 7.9 \quad -.1 \quad -5.7 \quad 10.3 \quad -4.9$$

79. Construct a dot diagram for each of the samples of kilowatt hours on page 412. Do these data appear to have come from normal populations?

Solution to Problem 76

The two histograms in (i) come from populations that do not appear to be normal. The histograms suggest populations each of which has a much longer right tail than left tail. Neither histogram in (ii) is far from symmetric. Both are therefore consistent with having come from normal populations.

SECTION 12.10

USING COMPUTER OUTPUT: INFERENCES ABOUT μ AND $\mu_1 - \mu_2$ USING MINITAB

In this section we indicate how Minitab can be used to obtain confidence intervals for or test hypotheses about μ or $\mu_1 - \mu_2$. We shall not deal with the less important topic of inferences about σ or σ_1/σ_2.

INFERENCES IN LARGE SAMPLES

In this text we used methods involving the Z distribution for large samples ($n \geq 30$) and the t distribution for small samples ($n < 30$). When the t distribution is used, Minitab will compute the values of s, the sample mean. When the Z distribution is used, however, Minitab requires that the value for σ be specified. In Chapters 9 and 10, when the population standard deviation, σ, was not known (which is usually the case), we used s as an estimate of σ. Similarly, when using Minitab with large samples, if σ is not known, the value of s is used in its place. An example will illustrate this point.

EXAMPLE 24 Referring to the nutrition data in Section 1.6, we wish to obtain a 90% confidence interval for both the mean birthweight and mean birth length of infants born to low-income mothers. Since the sample size $n = 68$ is > 30, we use the large-sample confidence interval in Definition 3 of Chapter 9 (page 301). The appropriate commands are shown in Table 8. Notice that we first ask for the standard deviation of the variable and only then for the appropriate confidence interval. In the case of birthweight, for instance, the command is "Z INTERVAL 90 PERCENT CONFIDENCE, SIGMA = 450.83, DATA IN C7." Notice that we used the value $s = 450.83$ for sigma (σ). The output consists of the value for $\bar{x}$ (3234), s (450), $\text{SD}(\bar{X}) = s/\sqrt{n}$ (55), and the corresponding interval (3144 to 3324).

```
MTB > STANDARD DEVIATION OF C7
   ST.DEV. =       450.83
MTB > ZINTERVAL 90 PERCENT CONFIDENCE, SIGMA=450.83, DATA IN C7

THE ASSUMED SIGMA = 451

               N      MEAN    STDEV   SE MEAN     90.0 PERCENT C.I.
C7            68      3234      451        55  (   3144,     3324)
```

TABLE 8

Using Minitab to find a confidence interval for the mean birthweight and birth length

■

The commands necessary to test hypotheses involving a large sample are similarly constructed.

EXAMPLE 25 Again referring to the nutrition data, it was of interest to test whether the mean weight gain (μ) during pregnancy for all low-income women was significantly different from 31.5 pounds. In the notation of Chapter 10, we test $H_0 : \mu = 31.5$ against $H_1 : \mu \neq 31.5$. The 68 weight gains are stored in C10. The necessary commands are shown in Table 9. As in Example 24, we first ask for the standard deviation of weight gain (13.074) and then for the desired test with the command "Z TEST OF MU = 31.5 VS THE ALTERNATE = 0, SIGMA = 13.074, DATA IN C10." The value 0 is used here to indicate the two-sided alternative $H_1 : \mu \neq \mu_0$. The values -1 and $+1$ are used to indicate the one-sided alternatives $H_1 : \mu < \mu_0$ and $H_1 : \mu > \mu_0$, respectively. The output in this case consists of a statement of the hypotheses, summary data including the sample mean weight gain $\bar{x} = 29.8$, and the corresponding P value, which is from Section 10.5, $2P(Z < -1.09) = .28$. Since this value greatly exceeds .05, we cannot reject H_0 at the .05 level of significance. The data suggest that the mean weight gain for these women is not significantly different from 31.5 pounds.

```
MTB > STANDARD DEVIATION OF C10
    ST.DEV. =        13.074
MTB > ZTEST OF MU=31.5 VS THE ALTERNATE=0, SIGMA=13.074, DATA IN C10

TEST OF MU = 31.5 VS MU N.E. 31.5
THE ASSUMED SIGMA = 13.1

              N      MEAN     STDEV    SE MEAN        Z      P VALUE
C10          68      29.8      13.1        1.6    -1.09         0.28
```

TABLE 9

Using Minitab to perform a Z test

■

Minitab does not perform Z tests based on two large independent samples of the kind discussed in Section 10.4.

INFERENCES IN SMALL SAMPLES The Minitab commands for small-sample inferences about μ_1 and $\mu_1 - \mu_2$ using the t distribution are also straightforward. We begin with an example of a one-sample confidence interval and t test.

EXAMPLE 26 A new pesticide is being tested. The data entered in Table 10 are the number of larvae of a certain pest in each of 18 test plots after spraying with the pesticide. We first obtain a 95% confidence interval for μ, the potential mean number of larvae per test plot. To do so, we use the command "TINTERVAL 95 PERCENT CONFIDENCE, DATA IN C1." We also wish to test whether μ is

```
MTB > SET DATA INTO C1
DATA> 10,19,5,22,16,14,6,23,21,0,9,12,17,2,17,22,6,9
DATA> TINTERVAL 95 PERCENT CONFIDENCE, DATA IN C1

              N       MEAN     STDEV  SE MEAN    95.0 PERCENT C.I.
C1           18      12.78      7.26      1.7  (     9.2,     16.4)

MTB > TTEST OF MU=18 VS THE ALTERNATE=-1,DATA IN C1

TEST OF MU = 18.0 VS MU L.T. 18.0

              N       MEAN     STDEV   SE MEAN        T      P VALUE
C1           18      12.78      7.26       1.7    -3.05       0.0036
```

TABLE 10

Small-sample confidence interval and t test

significantly less than 18. This time we ask for a t test with the command "T TEST OF MU = 18 VS THE ALTERNATE = −1, DATA IN C1." The −1 indicates the alternative hypothesis $H_1: \mu < 18$. Notice that Minitab computes not only the value for t (−3.05) but also the corresponding P value, which in this case is $P(t_{17} < -3.05) = .0036$.

■

We refer again to the nutrition data to illustrate a t test involving two independent samples.

EXAMPLE 27 Of considerable interest is the effect on infant birthweight (BTWT) of the mother's smoking habits. We will compare the birthweights of infants born to nonsmoking mothers with those born to smoking mothers. The values for BTWT were given in C3. Nonsmoking mothers were indicated by a 1 in C3, light and heavy smokers by 2 and 3, respectively. We first have to isolate the "nonsmoking weights" and the "smoking weights" in separate columns. The commands to achieve this are given in Table 11. The command "CHOOSE ROWS WITH 1 IN C3, CORR. ROWS OF C7 INTO C11, C20" takes the "nonsmoking" weights and puts them in column C20. Minitab also requires that the 1's from C3 be placed in a new column, in this case C11. The next command similarly places the "light smoking" weights into C21 and the "heavy smoking" weights into C22. Finally, the command "JOIN C21 TO C22, PUT INTO C23" combines these last two sets of infant birthweights and puts them in C23. We wish to test $H_0: \mu_1 - \mu_2 = 0$ against $H_1: \mu_1 - \mu_2 > 0$, where μ_1 is the mean BTWT for low-income nonsmoking mothers and μ_2 is the corresponding mean for smoking mothers. The two-sample t test is requested with the command "POOLED T, ALTERNATE = +1, DATA IN C20, C23." In addition to the test results, the output includes the summary data for the two samples (there are

```
MTB > CHOOSE ROWS WITH 1 IN C3 CORR. ROWS OF C7 INTO C11,C20
MTB > CHOOSE ROWS WITH 2 IN C3 CORR. ROWS OF C7 INTO C12,C21
MTB > CHOOSE ROWS WITH 3 IN C3 CORR. ROWS OF C7 INTO C13,C22
MTB > JOIN C21 TO C22, PUT INTO C23
MTB > POOLEDT, ALTERNATE=+1, DATA IN C20,C23
```

```
TWOSAMPLE T FOR C20 VS C23
       N       MEAN      STDEV    SE MEAN
C20   49       3297        472         67
C23   19       3071        353         81

95 PCT CI FOR MU C20 - MU C23: (-13, 465)
TTEST MU C20 = MU C23 (VS GT): T=1.89 P=0.032 DF=66.0
```

TABLE 11

Separating the weights of infants born to smoking and nonsmoking mothers and performing a two-sample *t* test

49 nonsmokers and 19 smokers) and a 95% confidence interval for $\mu_1 - \mu_2$ (-13 to 465). The value for t (with 66 degrees of freedom) is 1.89 and the corresponding P value is .032.

■

Finally, we present an example of a paired t test (i.e., a t test involving paired data). An example of a paired Z test was described in Example 25, where the weight gain in C10 was the difference between a mother's prepregnancy weight and weight at delivery. The hypothesis to be tested was $H_0: \mu = 31.5$, where μ was the mean weight gain for all low-income women.

EXAMPLE 28 A math teacher is interested in finding whether students score better on multiple-choice tests than on tests with open-ended questions. As an experiment, the 22 students in the class were given a two-hour test, one half of which contained open-ended questions and the other half, multiple-choice questions. The students were able to spend an hour on each part. The scores on the open-ended part of the test (in C1) and on the multiple-choice part (in C2) are entered in Table 12. The appropriate null and alternative hypotheses are $H_0: \mu_d = 0$ and $H_1: \mu_d > 0$, where μ_d is the potential mean difference, "score on the multiple-choice section—score on open-ended section." To test this hypothesis, Minitab is asked to place the 22 differences in C3 and then to perform the appropriate test.

```
MTB > READ DATA INTO C1,C2
DATA> 33,28
DATA> 38,38
DATA> 30,35
DATA> 21,32
DATA> 33,45
DATA> 25,31
DATA> 32,42
DATA> 34,34
DATA> 40,41
DATA> 32,35
DATA> 34,38
DATA> 32,31
DATA> 38,39
DATA> 18,21
DATA> 43,46
DATA> 35,37
DATA> 34,34
DATA> 30,36
DATA> 26,23
DATA> 21,28
DATA> 18,24
DATA> 38,32
DATA> SUBTRACT C1 FROM C2, PUT INTO C3
     22 ROWS READ
MTB > TTEST OF MU=0 VS THE ALTERNATE=+1,DATA IN C3
```

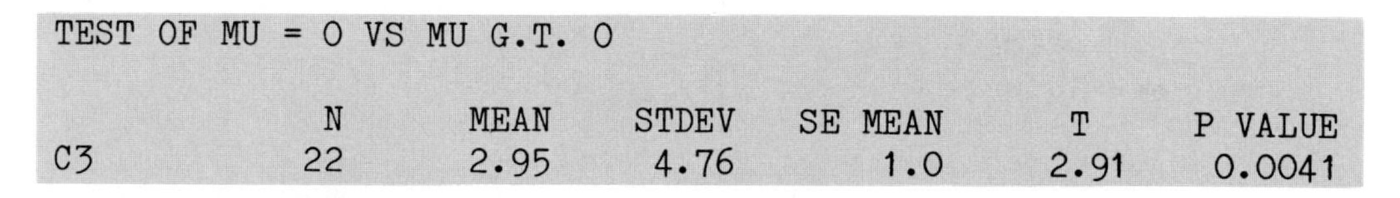

TEST OF MU = 0 VS MU G.T. 0

	N	MEAN	STDEV	SE MEAN	T	P VALUE
C3	22	2.95	4.76	1.0	2.91	0.0041

TABLE 12

t test for the paired test data

PROBLEMS 12.10

80. Interpret the confidence interval (a) for μ in Example 24; (b) for $\mu_1 - \mu_2$ which is given in addition to the test results in Example 27.
81. (a) Use the summary data in the output in Table 10 and the procedure in Table 1 to verify the value $t = -3.05$.
    (b) Interpret the results of this test.
82. (a) Use the summary data in the output in Table 11 and the procedure in Table 3 to verify the value $t = 1.89$.

(b) Intepret the results of this test.

83. The following data are the longest hitless streaks for each of a random sample of 35 major league baseball players.

```
MTB > SET DATA INTO C1
DATA> 25,18,40,13,17,27,12,24,19,16,28,32,19,24,13,20,16,19,16
DATA> 15,19,18,14,21,21,15,16,26,24
DATA> 19,18,12,17,17,20
DATA> STANDARD DEVIATION OF C1
   ST.DEV. =       5.9040
MTB > ZINTERVAL 80 PERCENT CONFIDENCE, S=5.9040, DATA IN C1

THE ASSUMED SIGMA = 5.90

              N       MEAN      STDEV   SE MEAN    80.0 PERCENT C.I.
C1           35      19.71       5.90       1.0  (    18.4,     21.0)
```

(a) Verify the computations of the confidence interval.

(b) What is it a confidence interval for?

(c) Interpret the interval.

84. Referring to the nutrition data in Table 8 of Chapter 1 (page 48) it was feared that on the average women were spending less than the suggested 90 minutes with the nutritionist. To test $H_0: \mu = 90$ against $H_1: \mu < 90$ using the times in C9, we asked Minitab to perform a Z test as follows:

```
MTB > STANDARD DEVIATION OF C9
   ST.DEV. =       19.364
MTB > ZTEST OF MU=90 VS THE ALTERNATE=-1,SIGMA=19.364,DATA IN C9

TEST OF MU = 90.0 VS MU L.T. 90.0
THE ASSUMED SIGMA = 19.4

             N      MEAN     STDEV   SE MEAN        Z     P VALUE
C9          68      86.2      19.4       2.3    -1.63       0.052
```

Interpret the resulting output.

85. A golfer complains that he does not seem to be scoring as well with his new clubs as he did with his old ones. With the old clubs he averaged 86.5. The nine scores that he obtained with his new clubs are entered in C1 as follows:

```
MTB > SET DATA INTO C1
DATA> 87,90,91,90,88,92,85,88,90
DATA> TTEST OF MU=86.5 VS THE ALTERNATE=+1, DATA IN C1

TEST OF MU = 86.5 VS MU G.T. 86.5

             N      MEAN     STDEV   SE MEAN        T     P VALUE
C1           9     89.00      2.18      0.73     3.44      0.0044
```

(a) What are the appropriate hypotheses?

(b) Interpret the test results.

86. Refer to the ambition scores in Problem 49 of Chapter 1 (page 54). We wish to test for a significant difference in the mean ambition scores for sixth-grade boys and girls. The appropriate output is as follows:

```
MTB > SET DATA INTO C1
DATA> 16,18,20,21,23,23,26,27,29,29,30,31,33,33,35,37,38,40,45,49
DATA> SET DATA INTO C2
DATA> 16,14,20,22,22,25,26,27,27,28,32,34,35,36,38,40,42,46
DATA> POOLEDT,ALTERNATE=0,DATA IN C1,C2
```

```
TWOSAMPLE T FOR C1 VS C2
     N      MEAN      STDEV    SE MEAN
C1  20     30.15       8.85        2.0
C2  18     29.44       9.02        2.1

95 PCT CI FOR MU C1 - MU C2: (-5.2, 6.6)
TTEST MU C1 = MU C2 (VS NE): T=0.24 P=0.81 DF=36.0
```

(a) Use the summary data to verify the value $t = .24$.

(b) Interpret this result.

87. In the following printout we have separated the times spent with the nutritionist (C9) into those for mothers who breastfed their infants (C6 = 1) and those for mothers who did not (C6 = 0).

```
MTB > CHOOSE ROWS WITH 0 IN C6 CORR. ROWS OF C9 INTO C11,C15
MTB > CHOOSE ROWS WITH 1 IN C6 CORR. ROWS OF C9 INTO C12,C16
MTB > POOLEDT, ALTERNATE=0, DATA IN C15,C16
```

```
TWOSAMPLE T FOR C15 VS C16
      N      MEAN      STDEV    SE MEAN
C15  32      79.9       20.0        3.5
C16  36      91.7       17.2        2.9

95 PCT CI FOR MU C15 - MU C16: (-20.9, -2.8)
TTEST MU C15 = MU C16 (VS NE): T=-2.63 P=0.011 DF=66.0
```

(a) Indicate the nature of the test we performed.

(b) Interpret the results.

SECTION 12.11
SUMMARY AND REVIEW

In the case of small ($n < 30$) samples the standardized form of $\overline{X}$, $(\overline{X} - \mu)/(s/\sqrt{n})$ does not generally have the $N[0, 1]$ distribution (as it does with large samples). There are two reasons for this: (1) the central limit theorem, which states that in large samples $\bar{X}$ will be approximately normal, frequently does not hold for small samples; and (2) in small samples we cannot regard s as an accurate estimate for σ. If the population itself is approximately normally distributed, however, then $(\bar{X} - \mu)/(s/\sqrt{n})$ will have the t_{n-1} distribution.

THE t DISTRIBUTION

Although we speak of *the* t distribution, there is a family of t distributions, a specific member being designated t_ν. Here ν is called the **number of degrees of freedom** (df) and is a whole number ≥ 1. Like the $N[0, 1]$, each t_ν distribution is symmetric around 0. The larger the value for ν, the more the corresponding t_ν distribution looks like the $N[0, 1]$ distribution.

Table A6 contains values of t_ν for different values of ν. Each entry can be designated in two ways: (1) $t_{\nu,\alpha}$, the value of t_ν having an area of α to the right, or (2) $t_{\nu(\gamma)}$, the value of t_ν such that the area between $-t_{\nu(\gamma)}$ and $t_{\nu(\gamma)}$ is γ.

CONFIDENCE INTERVALS FOR μ

A sample is selected from a $N[\mu, \sigma]$ population where both μ and σ are unknown. When $n < 30$, a $(100\gamma)\%$ confidence interval for μ is then

$$\bar{x} - t_{n-1(\gamma)} \frac{s}{\sqrt{n}} \quad \text{to} \quad \bar{x} + t_{n-1(\gamma)} \frac{s}{\sqrt{n}} \qquad \text{(Def. 2)}$$

Definition 2 can also be used when $n \geq 30$. However since, in this case the assumption of normality is not required, the large sample interval for μ is usually used with the value $Z_{[\gamma]}$ replacing $t_{n-1(\gamma)}$. The interpretation is exactly the same.

TESTING HYPOTHESES ABOUT μ

Since tables of the t distribution lack the necessary detail, we do not compute the P value when making tests of hypotheses about μ involving small samples from a normal population. Instead, we use the value of $t = (\bar{x} - \mu_0)/(s/\sqrt{n})$, where, as before, μ_0 is the null hypothesis value for μ.

The basic idea behind t tests of $H_0: \mu = \mu_0$ is to reject H_0 if the test statistic $t = (\bar{x} - \mu_0)/(s/\sqrt{n})$ is unusually *larger* than 0 for $H_1: \mu > \mu_0$, unusually *smaller* than 0 for $H_1: \mu < \mu_0$, or unusually *different* from 0 for $H_1: \mu \neq \mu_0$. The procedures for performing one-sided tests of $H_0: \mu = \mu_0$ are given in Table 1, those for two-sided tests in Table 2. In each case we use $\alpha = .05$ and .01. For any single value of α we follow the procedures in Definition 3.

Inferences about $\mu_1 - \mu_2$ in small samples are based on samples of size n_1 and $n_2 < 30$ selected from a $N[\mu_1, \sigma]$ and a $N[\mu_2, \sigma]$ population respectively. (Note the assumption of *equal* standard deviations.)

Tests of $H_0: \mu_1 - \mu_2 = 0$ involve computing the test statistic

$$t = \frac{\bar{x}_1 - \bar{x}_2 - 0}{s_p\sqrt{1/n_1 + 1/n_2}}$$

and rejecting H_0 if this is unusually large in the case of $H_1: \mu_1 - \mu_2 > 0$ or unusually different from 0 in the case of $H_1: \mu_1 - \mu_2 \neq 0$. The details for $\alpha = .05$ and $\alpha = .01$ are given in Tables 3 and 4.

In the formula for t,

$$s_p^2 = \frac{(n_1 - 1)s_1^2 + (n_2 - 1)s_2^2}{n_1 + n_2 - 2} \qquad \text{(Eq. 12.1)}$$

s_p^2 is called the pooled variance and is a *weighted* average of the two sample variances s_1^2 and s_2^2.

With the same assumptions as above, we define a $(100\gamma)\%$ confidence interval for $\mu_1 - \mu_2$ as

$$\bar{x}_1 - \bar{x}_2 - t_{n_1+n_2-2(\gamma)} s_p \sqrt{\frac{1}{n_1} + \frac{1}{n_2}} \quad \text{to} \quad \bar{x}_1 - \bar{x}_2 + t_{n_1+n_2-2(\gamma)} s_p \sqrt{\frac{1}{n_1} + \frac{1}{n_2}}$$

(Def. 7)

ANALYSIS OF PAIRED DATA

Data obtained in one of the following ways is called **paired** or **matched data:**

1. The same variable (e.g., reaction time) is measured for each member twice, frequently before and after some intervening event. In such cases we want to see if there has been any change—hopefully reflecting the effect of this event.
2. In other cases two groups are set up, one member of each group being carefully matched with one of the other. Some difference between the two groups following a differential treatment is then evaluated. For example, they may be treated with two procedures and the differences between the resulting scores for each matched pair calculated.

Such data arise from the matched pairs design discussed in Chapter 8.

To emphasize that we are analyzing differences, the differences themselves are designated $d_1, d_2, \ldots, d_n$, their mean $\bar{d}$, and their standard deviation, s_d. We designate the mean difference in the population of differences (assumed to be normal) as μ_d. We generally test $H_0: \mu_d = 0$ against either $H_1: \mu_d > 0$ or $H_1: \mu_d \neq 0$ using the one-sample methods in Table 1 or 2. If the sample size is $n \geq 30$, use the methods of Chapter 10.

It is of particular interest to compare the matched pairs design with the two independent samples design. With two independent samples two sources of variability are present: (1) that due to the effect of the event or procedure under study (i.e., the independent variable; in Example 10, some of the group received beer, others did not); and (2) that due to differences between individuals—some people simply react faster than others. In a matched pairs design this second source of variability is eliminated, making for a more precise analysis.

THE CHI-SQUARE AND F DISTRIBUTIONS

The χ_ν^2 distributions take only positive values and are asymmetric with mean ν, the number of degrees of freedom. The notation $\chi_{\nu,\alpha}^2$ refers to the value of the χ_ν^2 distribution having an area α to the *right*. Values of $\chi_{\nu,\alpha}^2$ for a large number of values for ν are given in Table A7.

We refer to the F_{ν_1,ν_2} distribution, where ν_1 and ν_2 are degrees of freedom ($\nu_1, \nu_2 \geq 1$). The F_{ν_1,ν_2} distributions take only positive values and are asymmetric. The notation $F_{\nu_1,\nu_2,\alpha}$ refers to the value of the F_{ν_1,ν_2} distribution having an area α to the *right*. Values of $F_{\nu_1,\nu_2,\alpha}$ for a large number of combinations of values for ν_1 and ν_2 and for $\alpha = .25, .1, .05, .025, .01$, and $.005$ are given in Table A8.

INFERENCES ABOUT σ OR σ^2

Inferences about σ^2 are based on the fact that if a sample of size n is to be selected from a $N[\mu, \sigma]$ population then $y = (n - 1)s^2/\sigma^2$ has the χ^2_{n-1} distribution. As a consequence, a $(100\gamma)\%$ confidence interval for σ^2 is

$$\frac{(n-1)s^2}{\chi^2_{n-1,(1-\alpha)/2}} \quad \text{to} \quad \frac{(n-1)s^2}{\chi^2_{n-1,(1+\alpha)/2}}$$

and a $(100\gamma)\%$ confidence interval for σ is

$$\sqrt{\frac{(n-1)s^2}{\chi^2_{n-1,(1-\gamma)/2}}} \quad \text{to} \quad \sqrt{\frac{(n-1)s^2}{\chi^2_{n-1,(1+\gamma)/2}}} \qquad \text{(Def.10)}$$

The procedure for testing $H_0: \sigma^2 = \sigma_0^2$ is based on the fact that if H_0 is true, $y = (n - 1)s^2/\sigma_0^2$ has the χ^2_{n-1} distribution. The details of the test are given in Table 6.

INFERENCES ABOUT σ_1^2/σ_2^2

Confidence intervals for σ_1^2/σ_2^2 or σ_1/σ_2 are based on the fact that, if independent samples are selected from normal populations, $F = (s_2^2/s_1^2)(\sigma_1^2/\sigma_2^2)$ has the $F_{n_2-1,n,-1}$ distribution. As a consequence, a $(100\gamma)\%$ confidence interval for σ_1^2/σ_1^2 is

$$\frac{1}{F_{n_1-1,n_2-1,(1-\gamma)/2}}\frac{s_1^2}{s_2^2} \quad \text{to} \quad F_{n_2-1,n_1-1,(1-\gamma)/2}\frac{s_1^2}{s_2^2}$$

and a $(100\gamma)\%$ confidence interval for σ_1/σ_2 is

$$\sqrt{\frac{1}{F_{n_1-1,n_2-1,(1-\gamma)/2}}}\frac{s_1}{s_2} \quad \text{to} \quad \sqrt{F_{n_2-1,n_1-1,(1-\gamma)/2}}\frac{s_1}{s_2}$$

The procedure for testing $H_0: \sigma_1^2 = \sigma_2^2$ is based on the fact that $F = (s_1^2/s_2^2) \cdot (\sigma_2^2/\sigma_1^2)$ has the F_{n_1-1,n_2-1} distribution. Thus, if H_0 is true, $F = s_1^2/s_2^2$ alone has the F_{n_1-1,n_2-1} distribution. The exact test procedures are set out in Table 7.

EXAMINING ASSUMPTIONS

It is possible to check for *normality* of a population by using the sample data to construct a dot diagram if $n < 20$ or a relative frequency histogram if $n \geq 20$. We look for evidence of a population that is far from normal. This occurs where the histogram, for example, is very asymmetric.

REVIEW PROBLEMS

GENERAL

1. A zoologist is interested in aspects of the behavior of Antarctic seals. As part of the research the length of time spent by such a seal on successive dives is recorded for a sample of 20 dives. The following times (in seconds) were recorded:

30	40	24	47	55	67	44	33	49	53
65	44	25	19	47	52	61	57	42	28

(a) Obtain a 95% confidence interval for the true mean time that this seal spends in dives.

(b) What assumptions were necessary in this case?

2. A landlord is considering the purchase of a device which the gas company claims will help the furnace work more efficiently and hence more cheaply than before. Before making a decision, the landlord asks six colleagues who have the device to provide her with their gas bills for the January before and the January after the device was installed. The results are as follows (in dollars):

COLLEAGUE	1	2	3	4	5	6
Before	182	110	215	129	117	170
After	136	112	184	100	102	145

(a) Do these data suggest that on average the device significantly lowers the gas bill?

(b) What assumptions did you have to make in part (a)?

3. Two friends of one of the authors agreed to record their bowling scores for a month, with the following results:

BOWLER	NUMBER OF GAMES	SCORE MEAN	SCORE MEDIAN	STANDARD DEVIATION
L	17	167.2	166.0	6.2
J	22	174.5	172.0	12.9

(a) Do these data suggest that the variability in scores for the more tempermental bowler *J* is significantly greater than that for *L*?

(b) What can you deduce about the symmetry of *L*'s scores from the fact that the mean and median are so close?

4. Refer to Review Problem 3. Calling bowler *J*, 1, and bowler *L*, 2, find a 90% confidence interval for σ_1/σ_2. Interpret your interval.
5. Explain why the normality assumption is not necessary in large-sample inferences.
6. Why can't the t distribution be used to make inferences about the population proportion, π, in small samples?
7. Explain in your own words why, in general, a matched pairs experiment is preferable to one involving two independent samples or groups.
8. The average number of points per game scored by the leading shooter in the National Basketball Association for the years 1947–1953 and the years 1976–1982 are

1947	23.2	1976	31.1
1948	21.0	1977	31.1
1949	28.3	1978	27.2
1950	27.4	1979	29.6
1951	28.4	1980	33.1
1952	25.4	1981	30.7
1953	22.3	1982	32.3

Do these data suggest that there has been a significant increase in the average number of points per game by such leading shooters?

9. Refer to Review Problem 8. Test whether there has been any significant change in the variability of the leading scorer's averages.

10. Gould (1981) reported the following early data on the cranial capacity (in cubic inches) of various English and Anglo-American persons:

NATIONALITY	n	$\bar{x}$	s[a]
English	5	96	3.50
Anglo-American	7	90	3.75

[a] The standard deviations are estimates.

Do these data suggest any significant difference in the mean cranial capacity for the two nationalities?

11. Assume at a set location on a highway, state police stop a random sample of motorists and measure the alcohol content of their blood. The same measurement is taken on those motorists that are stopped for speeding. The following data are the percent of alcohol in the blood for all persons stopped on a Friday evening who have at least a trace of alcohol in their blood.

MOTORISTS STOPPED AT RANDOM					MOTORISTS STOPPED FOR SPEEDING				
.062	.025	.030	.051	.015	.062	.029	.012	.059	.150
.048	.057	.069	.069	.074	.034	.049	.042	.058	.060
.035	.039	.052	.042	.091	.058	.126	.079	.032	.068
.041	.018	.027	.046	.041	.038	.049	.052	.066	
.036									
$\Sigma x_i = .968$		$\Sigma x_i^2 = .05207$			$\Sigma x_i = 1.123$		$\Sigma x_i^2 = .08513$		

Do these data suggest that the mean percent alcohol among all those motorists not speeding (the population from which presumably those chosen at random are selected) is significantly below that for all speeding motorists?

12. Refer to Review Problem 11. Multiply the values in each group by 1000 (to eliminate the decimal points). Arrange each of the samples into a stem-and-leaf display. Do either of the displays give the impression that the data are significantly nonnormal?

13. A sample of size $n_1 = 13$ is taken from a $N[\mu_1, 50]$ population. Independently, a sample of size $n_2 = 17$ is taken from a $N[\mu_2, 25]$ population. The results are as follows:

$$\bar{x}_1 = 373.9 \qquad \bar{x}_2 = 362.6$$

$$s_1 = 44.2 \qquad s_2 = 31.9$$

(a) Use these data to test $H_0: \sigma_1^2 = \sigma_2^2$ against $H_1: \sigma_1^2 > \sigma_2^2$.

(b) Did you make the correct decision in part (a)?

HEALTH SCIENCES

14. The concentration of a particular amino acid in the brain fluid of a sample of (I) 25 healthy adult women and (II) 25 emotionally disturbed adult women are recorded and the following summary data obtained:

	n	Σx_i	Σx_i^2
HEALTHY WOMEN (I)	25	105.4	451.14
EMOTIONALLY DISTURBED WOMEN (II)	25	125.2	630.76

(a) Do these data suggest that the mean concentration of this amino acid in brain fluid is on the average higher in emotionally disturbed women than in healthy women?

(b) With the obvious notation, obtain a 90% confidence interval for $\mu_2 - \mu_1$.

(c) Perform a two-sided test of $H_0: \sigma_1 = \sigma_2$.

(d) What assumptions were necessary in parts (a), (b), and (c)?

15. Pfeffer et al. (1983) assigned an "assaultiveness" score to each of 103 children receiving psychiatric help. The mean and standard deviation of such scores was computed for children (I) exhibiting and (II) not exhibiting mental retardation, as follows:

		$\bar{x}$	s	n
MENTAL	PRESENT (I)	4.0	1.2	12
RETARDATION	ABSENT (II)	3.1	1.6	91

Do these data suggest a significant difference in the mean assaultiveness score for the two groups? (Use a t test.)

16. In an experiment involving 12 undergraduate women, a dynamometer was used to record the isometric strength of the hamstring when the knee is flexed at (i) 5°, and (ii) 45° to the horizontal. The force produced in each case (in pounds) was as follows:

SUBJECT	FORCE PRODUCED[a] 5°	45°
1	22.2	18.5
2	24.0	16.7
3	29.2	25.5
4	39.2	34.3
5	31.3	23.3
6	34.8	32.0
7	31.7	30.0
8	35.5	29.2
9	35.8	28.0
10	34.3	29.3
11	37.7	35.3
12	34.7	31.8

[a] These data were provided by Deanna DiTommaso.

Call μ_d the mean difference in force produced at 5° and 45°. Test whether μ_d is significantly greater than 3 pounds.

17. Referring to Review Problem 16, sketch the 12 differences in force on a dot diagram. Does the diagram suggest a significantly nonnormal population of differences?

SOCIOLOGY AND GOVERNMENT

18. A reporter for a leisure magazine has devised a rule for obtaining a composite index of how culturally attractive a community is. He would like to estimate the mean value of this index for all small cities (those with populations of less than 25,000) in the United States. To do this he selects a random sample of eight such cities and computes a cultural index for each one. The results are 35, 29, 14, 33, 24, 20, 30, and 26. Use these data to obtain an 80% confidence interval for the mean value of the index over all such cities.
19. According to the *Sourcebook of Criminal Justice Statistics,* 1982, the mean length of sentence for defendants imprisoned for marijuana violations in U.S. District Courts in 1980 was 37.2 months. A criminologist is interested in determining whether the corresponding mean figure for this year, μ, is significantly higher than this. Suppose that in a random sample of 22 such defendants this year, the average sentence was 42.2 months and the standard deviation was 8.4 months.
    (a) What conclusion can the criminologist draw from these data?
    (b) What assumptions were necessary in this case?
20. Answer Review Problem 19 if the sample data had been based on 62 rather than 22 observations.
21. The office of the state insurance commissioner is interested in comparing the service records of the two major insurance companies in the state. The following data are based on the time (in days) it takes to process random samples of 24 burglary claims made to each company:

	$\bar{x}$	s
MUTUAL INSURANCE	29.4	7.4
STATE INSURANCE	26.6	12.1

    (a) Test whether the variance of service time in the two companies is significantly different.
    (b) Test for a significant difference in the mean service times for the two companies.
    (c) What assumptions were necessary in parts (a) and (b)?
    (d) Answer part (b) if the sample sizes were 124 rather than 24.
    (e) What assumptions were necessary in part (d)?

ECONOMICS AND MANAGEMENT

22. Two airlines, I and II, fly between New York and Los Angeles. For each of 12 consecutive months the percentage of flights between these two cities that arrived within 15 minutes of the scheduled arrival time was recorded for each airline.

MONTH	1	2	3	4	5	6	7	8	9	10	11	12
AIRLINE I	75	74	66	49	54	64	73	81	88	72	75	72
AIRLINE II	79	82	62	55	56	64	81	86	84	80	81	76

    (a) Do these data suggest any significant difference in the punctuality of the two airlines?
    (b) What assumptions did you make in part (a)?

23. According to the Council for Financial Aid to Education, corporations contributing to nonprofit organizations give an average of 1.77% of their pretax net income. An economist is interested in finding whether, on the average, banks give a significantly smaller percentage than this. In a random sample of eight banks that make some contribution, the percentages of income that are given are

1.16 .98 1.70 1.95 1.32 1.27 1.84 1.23

(a) What is μ in this case?
(b) What are H_0 and H_1?
(c) What conclusions do these data suggest?
(d) What assumptions were necessary in this case?

24. Refer to the data in Review Problem 23. Obtain and interpret a 90% confidence interval for σ.

25. In one town there are two real estate brokers, Allen and Baldwin. Each of 10 home owners asks each broker for an approximate asking price for his or her house. The results (in thousands of dollars) are as follows:

HOUSE	*A*	*B*
1	82.5	78.9
2	112.0	110.0
3	105.5	108.8
4	122.0	116.2
5	99.5	106.2
6	126.8	118.5
7	87.5	86.5
8	142.5	135.6
9	104.9	110.0
10	96.5	92.0

Do these data suggest that, on average, these two brokers give significantly different estimates?

26. Several years ago an economist selected a random sample of small companies in the state and recorded the anticipated first-quarter percentage profits for each of them. He is interested in comparing the results in that survey with the expected profits of similar companies for the forthcoming quarter. He is uncertain whether to obtain the current figures (i) from the same companies as before, or (ii) from another sample of companies selected independently of the first. He is leaning toward option (ii), having learned in a statistics course of the importance of independence when comparing two samples. Advise him on this point.

EDUCATION AND PSYCHOLOGY

27. In a particular school district the teachers contract calls for a student/teacher ratio of 23.5. In a random sample of 20 classrooms on one day the average student/teacher ratio was $\bar{x} = 25.3$. The standard deviation was $s = 5.9$. Do these sample results suggest that the district-wide mean student/teacher ratio is significantly greater than 23.5?

28. Herman (1967) reported the mean and standard deviation of the Minnesota Teacher Attitude Inventory (MTAI) for 14 athletes and 28 nonathletes. The results were as follows:

	n	$\overline{x}$	s
ATHLETES	14	116.00	31.11
NONATHLETES	28	119.54	32.41

Do these data suggest that, on average, athletes score lower than nonathletes on the MTAI?

29. A psychologist is interested in learning how quickly six-year-olds complete a task as compared to five-year olds. He is sure that on average the older children will take less time than the younger but is interested in comparing the variability of the times for the two groups. A group of 12 five-year-olds and 20 six-year-olds were asked to complete the task, with the following results (in minutes):

	n	$\overline{x}$	s
FIVE-YEAR-OLDS I	12	15.8	3.9
SIX-YEAR-OLDS II	20	10.2	2.8

With the obvious notation, (a) compute a 90% confidence interval for $\mu_1 - \mu_2$; (b) test for a significant difference between σ_1 and σ_2.

SCIENCE AND TECHNOLOGY

30. A chemist is concerned with the mean percentage, μ, of impurities in a certain chemical product. In a random sample of 16 cases of the product the percentages of impurities were:

6.4	5.5	6.0	6.2	6.7	6.2	5.3	5.8
6.1	6.5	6.6	5.5	5.9	6.1	6.9	5.9

(a) Use these data to obtain a 95% confidence interval for μ.

(b) Interpret your interval.

(c) Plot these data on a dot diagram and comment on its degree of symmetry.

(d) Answer part (a) if the mean and standard deviation calculated in the sample above had been based on 60 rather than 16 cases.

31. Refer to the data in Review Problem 30. Find a 90% confidence interval for σ, the standard deviation of the percentage of impurities per case.

32. A sample of 12 cases of oranges imported from a South American country are examined for the presence of the pesticide ethylene dibromide (EDB). The mean level of EDB per case is $\overline{x} = 1455$ parts per billion (ppb) with a standard deviation of $s = 480$ ppb.

(a) Use these results to find an 80% confidence interval for the mean level of EDB in cases of oranges from this country.

(b) Test whether this mean is significantly greater than 1300 ppb.

33. The National Security Agency arranges an experiment to compare the precision of a satellite-based and a land-based navigation system. Each system is used to estimate the location of 15 randomly selected points around the world whose locations are known precisely. For each point the error (in meters) in each system's estimate is recorded. The results are as follows:

LOCATION	LAND SYSTEM	SATELLITE SYSTEM	LOCATION	LAND SYSTEM	SATELLITE SYSTEM
1	52	47	9	34	29
2	80	68	10	58	38
3	27	29	11	54	40
4	54	51	12	20	18
5	47	22	13	32	39
6	34	37	14	38	24
7	36	31	15	54	35
8	40	44			

(a) Find (i) a 90% confidence interval for the mean error for the satellite-based system; (ii) a 90% confidence interval for the mean difference between the land-based and the satellite based error.

(b) Test whether the mean difference referred to in part (a, ii) is significantly greater then 0.

REFERENCES

Gould, S. J. *The Mismeasure of Man,* New York: Norton, 1981, p. 55.

Herman, W. L. "Teaching attitudes as related to academic grades and athletic ability of prospective physical education teachers," *Journal of Educational Research,* 61 (1967), pp. 40–42.

Pfeffer, C. R. et al. "Predictors of Assaultiveness in Latency Age Children," American Journal of Psychiatry, Vol. 14, #1, Jan. 1983, pp. 31–35.

SUGGESTED READING

Boneau, C. A. "The Effects of Violations of Assumptions Underlying the t Test," *Psychological Bulletin,* 57 (1960), pp. 49–64.

Steger, J. A., ed. *Readings in Statistics for the Behavioral Scientist.* New York: Holt, Rinehart and Winston, 1971, pp. 199–311.

13 ANALYSIS OF VARIANCE

WHAT IS THE FASTEST ROUTE?

A bus company plans to begin service between two cities. Four routes, *A*, *B*, *C*, and *D*, are being considered. To assess differences in the mean time for the four routes, a bus makes the trip between the cities 32 times, taking each route eight times. The times (in hours) for each trip are as follows:

A	*B*	*C*	*D*
6.30	6.50	6.81	6.27
6.45	6.66	6.72	6.00
6.18	6.44	6.93	6.30
6.33	6.37	6.83	6.37
5.95	6.30	6.60	6.15
6.07	6.55	6.53	6.18
6.25	6.18	6.60	6.09
6.13	6.27	6.44	6.29

How can the bus company use these data to determine whether there are any significant differences among the mean times for the four routes? See Review Problem 18 on page 521.

SECTION 13.1

ONE-WAY ANALYSIS OF VARIANCE

In Section 12.4 we used the two-sample t test to test for a significant difference between the means of two normal populations. In this chapter we introduce a test for differences among *three or more* such means. The following example is typical of this type of problem.

EXAMPLE 1 An experiment is conducted at a teaching hospital to measure the effect three different anesthetic drugs I, II, and III have on the recovery time of the next 24 patients to undergo a specific operation. These patients are randomly assigned to three groups of eight. One group receives drug I, another drug II, and the last drug III. The recovery times in hours for the 24 patients are as follows:

DRUG I	DRUG II	DRUG III
5.7	4.3	6.2
4.8	5.2	3.8
6.0	5.4	3.8
5.2	4.2	6.4
6.3	5.5	6.2
5.0	3.8	6.0
6.0	5.6	4.0
5.0	4.4	3.6

Do these data suggest that the true mean recovery times μ_{I}, μ_{II}, and μ_{III} differ significantly? We will return to this example in a moment.

■

The technique for analyzing data such as that in Example 1 is called the **analysis of variance** and was developed in the 1920s by the statistician Ronald Fisher (of F-test fame) as a way to evaluate agricultural experiments. We introduce this new technique with another example.

EXAMPLE 2 Data set B in Table 1 represents the recovery times for the three drugs from Example 1. Data sets A and C represent similar recovery times but were prepared only for purposes of comparison.

TABLE 1

Three data sets, each consisting of three samples

	A			B			C		
	I	II	III	I	II	III	I	II	III
	5.5	4.8	5.0	5.7	4.3	6.2	4.2	3.0	7.3
	5.5	4.8	5.0	4.8	5.2	3.8	1.8	6.4	6.4
	5.5	4.8	5.0	6.0	5.4	3.8	8.3	5.8	.9
	5.5	4.8	5.0	5.2	4.2	6.4	5.7	3.6	2.1
	5.5	4.8	5.0	6.3	5.5	6.2	3.9	4.0	8.6
	5.5	4.8	5.0	5.0	3.8	6.0	5.4	4.2	7.0
	5.5	4.8	5.0	6.0	5.6	4.0	2.8	6.0	5.1
	5.5	4.8	5.0	5.0	4.4	3.6	7.9	7.0	2.6
Total	44.0	38.4	40.0	44.0	38.4	40.0	40.0	40.0	40.0
$\bar{x}$	5.5	4.8	5.0	5.5	4.8	5.0	5.0	5.0	5.0

In data set A the three sample means differ, yet all observations *within* each sample have the same value. In this case, we say that there is variability *between*[1] sample means but no variability *within* each sample. By contrast, in data set C there is considerable variability *within* each sample, but since the sample means are identical, there is no variability *between* sample means. Sets A and B are extreme examples. More realistically, in data set B there is variation among the values (observations) *within* each sample and also variation *between* sample means.

■

Before formalizing these ideas, we introduce the term "sums of squares," abbreviated SS. Sums of squares are almost variances but do not have the divisor that one finds in a variance. For instance, $\Sigma(x_i - \bar{x})^2/(n - 1)$ is a variance, whereas $\Sigma(x_1 - \bar{x})^2$ is a sum of squares.

Now suppose that we independently select r samples from r different populations. Call the respective sample sizes $n_1, n_2, \ldots, n_r$. We next record the value for each member of each sample on the same variable. The means of the r samples are denoted $\bar{x}_1, \bar{x}_2, \ldots, \bar{x}_r$ and the sample variances $s_1^2, s_2^2, \ldots, s_r^2$. If we regard all the data as a single large sample, it will contain $n_1 + n_2 + \cdots + n_r$ values. We shall refer to this sum as N. We call the mean of this large sample, obtained by adding all of the N values and dividing by N, the "grand" mean and denote it $\bar{x}.$ (note the dot following $\bar{x}$). We now introduce a fundamental equality.

$$\underbrace{\sum^{N} (x_i - \bar{x}.)^2}_{SS_{\text{total}}} = \underbrace{\sum^{r} n_i(\bar{x}_i - \bar{x}.)^2}_{SS_{\text{between}}} + \underbrace{\sum^{r} (n_i - 1)s_i^2}_{SS_{\text{within}}} \qquad \text{(Eq. 13.1)}$$

The SS_{total} in Equation 13.1 reflects the extent to which all the individual N values vary around the grand mean, $\bar{x}.$. The SS_{between} measures the extent to which the r sample means $\bar{x}_1, \bar{x}_2, \ldots, \bar{x}_r$ vary around the grand mean, $\bar{x}.$. Finally, the SS_{within} summarizes the extent to which the values (x_i) within each sample vary around that sample mean $(\bar{x}_i)$. In Table 2 we compute these three sums of squares for the recovery times data in Example 1. (In this case $n_1 = n_2 = n_3 = 8$ and $N = 8 + 8 + 8 = 24$.)

Notice that $2.080 + 17.360 = 19.440$. Notice also that of the *total* variability about the grand mean (19.440), a large amount (17.360) is due to variability in recovery times *within* drug treatments. We shall return to this point in the next section.

It is interesting to apply Equation 13.1 to data sets A and C in Table 1. In data set A there is *no* variability within each sample, so that $s_1^2 = s_2^2 = s_3^2 = 0$, $SS_{\text{within}} = 0$, so that $SS_{\text{total}} = SS_{\text{between}}$. By contrast, in data set C, $\bar{x}. = (40 + 40 + 40)/24 = 5.0$, which is the same as $\bar{x}_1$, $\bar{x}_2$, and $\bar{x}_3$. As a consequence, the value of $\bar{x}_i - \bar{x}.$ is $5 - 5 = 0$ for each i in SS_{between} and $SS_{\text{between}} = \sum^{3} 8(\bar{x}_i - \bar{x}.)^2 = 0$ and $SS_{\text{total}} = SS_{\text{within}}$.

[1]Although not grammatically precise, the word "between" is used for historical reasons.

TABLE 2

Computation of the sums of squares for the recovery-time data in Example 1 ($r = 3$)

DRUG I ($n_1 = 8$) x	x^2	DRUG II ($n_2 = 8$) x	x^2	DRUG III ($n_3 = 8$) x	x^2
5.7	32.49	4.3	18.49	6.2	38.44
4.8	23.04	5.2	27.04	3.8	14.44
6.0	36.00	5.4	29.16	3.8	14.44
5.2	27.04	4.2	17.64	6.4	40.96
6.3	39.69	5.5	30.25	6.2	38.44
5.0	25.00	3.8	14.44	6.0	36.00
6.0	36.00	5.6	31.36	4.0	16.00
5.0	25.00	4.4	29.16	3.6	12.96
44.0	244.26	38.4	187.74	40.0	211.68

$$\bar{x}_1 = \frac{44}{8} = 5.5 \qquad \bar{x}_2 = \frac{38.4}{8} = 4.8 \qquad \bar{x}_3 = \frac{40}{8} = 5.0$$

$$s_1^2 = \frac{8(244.26) - 44^2}{8(7)} = .32286 \qquad s_2^2 = \frac{8(187.74) - 38.4^2}{8(7)} = .48857 \qquad s_3^2 = \frac{8(211.68) - 40^2}{8(7)} = 1.6686$$

$$\bar{x}. = \frac{5.7 + 4.8 + \cdots + 5.5 + \cdots + 4.0 + 3.6}{24} = \frac{122.4}{24} = 5.1$$

$$SS_{\text{between}} = \sum^{3} n_i(\bar{x}_i - \bar{x}.)^2$$
$$= 8(5.5 - 5.1)^2 + 8(4.8 - 5.1)^2 + 8(5.0 - 5.1)^2 = 2.08$$

$$SS_{\text{within}} = \sum^{3} (n_i - 1)s_i^2 = 7(.32286) + 7(.48857) + 7(1.6686) = 17.360$$

$$SS_{\text{total}} = \sum^{24} (x_i - \bar{x}.)^2 = (5.7 - 5.1)^2 + (4.8 - 5.1)^2 + \cdots + (3.6 - 5.1)^2 = 19.440$$

Associated with any sum of squares is a number of degrees of freedom. For the recovery time data the SS_{between} has $r - 1$ degrees of freedom or $3 - 1 = 2$; the SS_{within} has $N - r$ or $24 - 3 = 21$ degrees of freedom; and finally, the SS_{total} has $N - 1 = 24 - 1 = 23$ degrees of freedom.

When we divide a sum of squares by its associated number of degrees of freedom, we obtain a variance-like quantity which is referred to as a "mean square" and denoted by MS.

For the recovery time data, for instance,

$$MS_{\text{between}} = \frac{SS_{\text{between}}}{r - 1} = \frac{2.08}{2} = 1.04$$

$$MS_{\text{within}} = \frac{SS_{\text{within}}}{N - r} = \frac{17.360}{21} = .8267$$

$$MS_{\text{total}} = \frac{SS_{\text{total}}}{N - 1} = \frac{19.440}{23} = .8452$$

These three sums of squares and mean squares are conveniently displayed in an **analysis of variance table** (or **ANOVA** table for short), as in Table 3.

TABLE 3

Analysis of variance (ANOVA) table for the recovery-time data

SOURCE OF VARIABILITY	SS	df	MS
Between drug means	2.080	$r - 1 = 2$	1.04
Within drugs	17.360	$N - r = 21$	.8267
Total	19.440	$N - 1 = 23$	

Notice that just as the SS_{between} and SS_{within} add to SS_{total}, so the number of degreees of freedom associated with SS_{between} and SS_{within} add to the number of degrees of freedom for SS_{total}. Thus in Table 3, $2 + 21 = 23$ and, more generally, $(r - 1) + (N - r) = N - 1$.

We did not include MS_{total} in Table 3. This quality is not required when testing for differences among the r population means.

The form of the sums of squares in Equation 13.1 is not the most efficient computationally. A preferable form is outlined below.

1. We call the N values $x_1, x_2, \ldots, x_N$.
2. T_1 is the sum of the values in sample 1. $T_2, T_3, \ldots, T_r$ are similarly defined. The sum of all the T_i's is designated T (i.e., $T = T_1 + T_2 + \cdots + T_r$).
3. $\sum^{N} x_i^2$ is the sum of all the N squared values.

The computational formulas are then

$$SS_{\text{between}} = \sum^{r} \frac{T_i^2}{n_i} - \frac{T^2}{N} \qquad \text{(Eq. 13.2)}$$

$$SS_{\text{within}} = \sum^{N} x_i^2 - \sum^{r} \frac{T_i^2}{n_i} \qquad \text{(Eq. 13.3)}$$

$$SS_{\text{total}} = \sum^{N} x_i^2 - \frac{T^2}{N} \qquad \text{(Eq. 13.4)}$$

Thus all three sums of squares can be computed from the three quantities

$$(1) \sum^{N} x_i^2 \qquad (2) \sum^{r} \frac{T_i^2}{n_i} \qquad (3) \frac{T^2}{N}$$

Since SS_{between} and SS_{within} add to SS_{total}, we need compute only two of these three sums of squares. We recommend computing SS_{total} and SS_{within} and then finding SS_{between} as $SS_{\text{total}} - SS_{\text{within}}$. Applying Equations 13.3 and 13.4 to the recovery-time data we first compute

$$(1) \sum^{N} x_i^2 = \sum^{24} x_i^2 = 5.7^2 + 4.8^2 + \cdots + 4.0^2 + 3.6^2$$
$$= 643.68$$

$$(2) \sum^{r} \frac{T_i^2}{n_i} = \frac{44^2}{8} + \frac{38.4^2}{8} + \frac{40^2}{8} = 626.32$$

$$(3) \frac{T^2}{N} = \frac{(44 + 38.4 + 40)^2}{24} = \frac{122.4^2}{24} = 624.24$$

From Equations 13.3 and 13.4,

$$SS_{\text{total}} = \sum^{24} x_i^2 - \frac{T^2}{N} = 643.68 - 624.24 = 19.44$$

$$SS_{\text{within}} = \sum^{24} x_i^2 - \sum^{3} \frac{T_i^2}{n_i} = 643.68 - 626.32 = 17.36$$

$$SS_{\text{between}} = 19.44 - 17.36 = 2.08$$

These results correspond to those in Table 2.

The following example should further clarify how an ANOVA table is constructed.

EXAMPLE 3 Three technicians working in a photographic laboratory have the following numbers of successfully developed pictures on each of six randomly selected days. Three counts for technician III are lost.

I			II			III		
182	165	196	190	179	208	161	178	165
157	164	180	186	178	211			

(a) Construct the ANOVA table for these data. (b) Do there appear to be any significant differences among the three means?

Solution (a) In this case $n_1 = n_2 = 6$, $n_3 = 3$, and $N = n_1 + n_2 + n_3 = 15$; and $T_1 = 1044$, $T_2 = 1152$, $T_3 = 504$, and $T = 2700$. We shall need

$$\sum^{N} x_i^2 = \sum^{15} x_i^2 = 182^2 + 165^2 + \cdots + 178^2 + 165^2 = 489{,}746$$

$$\sum^{3} \frac{T_i^2}{n_i} = \frac{1044^2}{6} + \frac{1152^2}{6} + \frac{504^2}{3} = 487{,}512$$

$$\frac{T^2}{N} = \frac{2700^2}{15} = 486{,}000$$

From Equations 13.3 and 13.4,

$$SS_{\text{total}} = \sum^{15} x_i^2 - \frac{T^2}{N} = 489{,}746 - 486{,}000 = 3746$$

$$SS_{\text{within}} = \sum^{15} x_i^2 - \sum^{3} \frac{T_i^2}{n_i} = 489{,}746 - 487{,}512 = 2234$$

and by subtraction, $SS_{\text{between}} = 3746 - 2234 = 1512$.

The number of degrees of freedom are $r - 1 = 3 - 1 = 2$ for SS_{between}, $N - r = 15 - 3 = 12$ for SS_{within}, and $N - 1 = 15 - 1 = 14$ for SS_{total}. The ANOVA table is as follows:

SOURCE OF VARIABILITY	SS	df	MS
Between technician means	1512	2	756
Within technicians	2234	12	186.167
Total	3746	14	

(b) It is interesting to look a little further at the means for the three samples in this problem. The means are $\bar{x}_1 = T_1/6 = 1044/6 = 174$, $\bar{x}_2 = T_2/6 = 1152/6 = 192$, and $\bar{x}_3 = T_3/3 = 504/3 = 168$. In Figure 1 we display these mean numbers of successfully developed pictures for the three technicians. In these samples technician II developed more pictures successfully on average than the other two, but it is difficult to tell by inspection whether there are any statistically significant differences among the three technicians. We take up this issue in the following section.

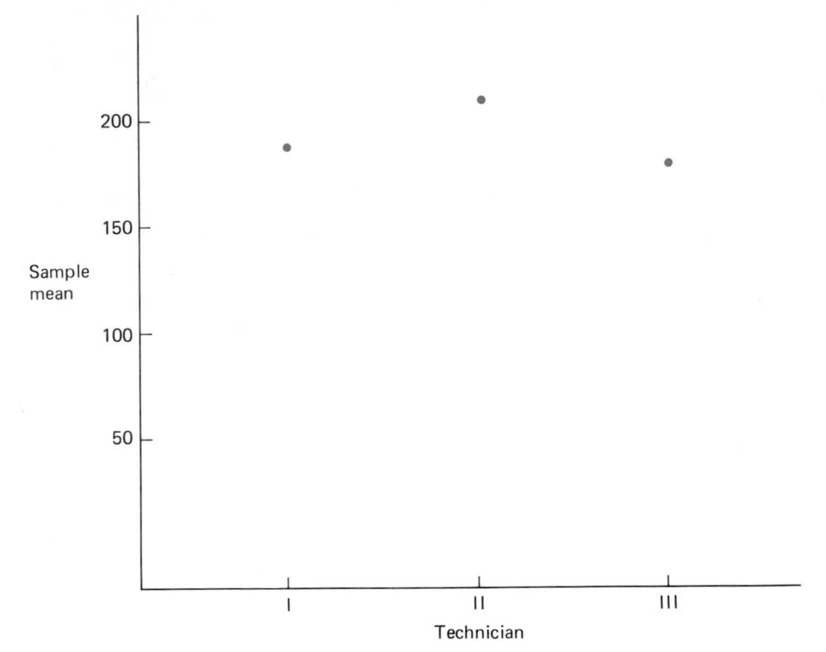

FIGURE 1

Graph showing the three sample means in Example 3

■

PROBLEMS 13.1

Save the ANOVA tables you construct. You will need them in Section 13.2.

1. Assume there is a question as to whether a certain literary work was written by either A, B, or C. One method that has been used to "determine" authorship in such cases is to search for key words which are used to different extents in known works of the three authors. Suppose, for example, it was suspected that "when" was such a word. Various segments, each of 10,000 words, are then selected from the known works of A, B, and C and the number of "whens" is recorded for each segment. Possible results are as follows:

AUTHOR A			AUTHOR B			AUTHOR C		
28	18	25	30	34	36	32	32	28
23	23	36	33	26	24	32	34	36
29	42	28	25	28		32	31	31
30	17	21				31		
$\Sigma x_i^2 = 9106$			$\Sigma x_i^2 = 7102$			$\Sigma x_i^2 = 10{,}215$		

Use these data to construct the appropriate ANOVA table.

2. Random samples of size $n_1 = 14$, $n_2 = 8$, and $n_3 = 10$ are selected from three populations. Complete the following ANOVA table.

SOURCE OF VARIABILITY	SS	df	MS
Between-sample means			
Within-samples	20.73		
Total	24.82		

3. Random samples of size $n = 12$ are selected from each of five populations. Complete the ANOVA table.

SOURCE OF VARIABILITY	SS	df	MS
Between sample means	1081	4	
Within samples			
Total	4382		

4. The number of kilowatts of electricity used in a month by random samples of six households in city I, city II, and city III are as follows:

CITY I			CITY II			CITY III		
32.9	24.4	23.8	30.8	28.1	34.1	38.9	26.2	42.0
32.1	38.9	35.6	19.6	23.4	35.7	38.4	38.2	28.3
$\Sigma x_i^2 = 6055.19$			$\Sigma x_i^2 = 5107.27$			$\Sigma x_i^2 = 7698.34$		

Use these data to construct the appropriate ANOVA table.

5. A company has three machines of varying ages producing Twickets. On each of nine days samples of 200 Twickets are randomly selected from each machine for inspection. The number that are defective from each machine for each day is recorded. Construct the ANOVA table for these data.

MACHINE I	II	III
9	13	8
13	16	12
19	14	14
20	11	13
18	11	6
11	15	12
16	14	15
14	10	11
16	11	11

$$\sum^{27} x_i^2 = 4889$$

6. The birthweight (in pounds) of each of a group of 60 newborns were classified by the smoking habits of the mother. The results were as follows:

I NONSMOKERS			II LIGHT SMOKERS	III MODERATE SMOKERS		IV HEAVY SMOKERS
8.4	5.5	8.4	7.6	6.1	5.6	7.0
10.2	8.7	8.8	5.1	3.9	6.3	6.0
7.0	6.7	6.7	6.7	5.9	5.7	3.9
8.6	8.0	8.2	8.4	5.9	6.3	6.9
6.6	10.4	8.9	5.4	6.5	5.5	7.6
7.7	4.6	8.8	7.1	7.2		
5.8	5.4	6.2	9.0	8.3		
7.2	8.3	5.8	8.5	7.7		
7.1	6.4	8.1	6.9	5.9		
9.3	8.9	8.1	7.8	6.0		
$\Sigma x_i^2 = 1805.24$			$\Sigma x_i^2 = 540.49$	$\Sigma x_i^2 = 588.60$		$\Sigma x_i^2 = 205.58$

Construct the ANOVA table for these data.

7. A random sample of 20 students is selected from each undergraduate class in a college and the grade-point average (GPA) for each student is recorded. Use the following summary data to construct the appropriate ANOVA table.

	Σx_i^2	T_i
FRESHMAN	147.191	52.16
SOPHOMORE	155.133	54.13
JUNIOR	183.059	59.73
SENIOR	188.623	60.86

8. The bills for medical services for the last 10 patients to undergo an appendectomy without complications at each of four major hospitals are as follows (in thousands of dollars):

I	3.07	3.12	2.72	3.03	3.71	3.16	3.11	2.78	3.80	2.99
II	2.97	2.99	2.98	2.35	3.10	3.25	2.90	2.27	2.18	2.92
III	2.71	2.68	3.18	2.53	2.97	2.82	2.16	2.30	2.74	3.14
IV	2.20	1.42	2.56	1.68	2.32	3.06	1.85	2.17	1.87	2.04

$$\sum^{40} x_i^2 = 301.3512 \qquad T = 107.80$$

(a) Construct the ANOVA table for these data.

(b) Compute the four sample means and plot them as we did the means in Figure 1.

(c) Which hospital seems the most expensive? The least expensive?

9. A sample of eight counties is selected from each of New Jersey, New York, and Pennsylvania. For each county the median family income was recorded (in thousands of dollars). The following summary data were recorded:

	Σx_i^2	T_i
NEW JERSEY	3444.39	165.51
NEW YORK	2561.99	142.40
PENNSYLVANIA	2516.82	141.78

Construct the ANOVA table for these data.

Solution to Problem 1

Here $N = n_1 + n_2 + n_3 = 12 + 8 + 10 = 30$. To construct the ANOVA table, we need the following results:

$$\sum x_i^2 = 9106 + 7102 + 10{,}215 = 26{,}423$$

$$T_1 = 320 \qquad T_2 = 236 \qquad T_3 = 319 \qquad T = 875$$

Then

$$SS_{\text{total}} = \sum x_i^2 - \frac{T^2}{N} = 26{,}423 - \frac{875^2}{30} = 902.167$$

$$SS_{\text{within}} = \sum x_i^2 - \left(\frac{T_1^2}{12} + \frac{T_2^2}{8} + \frac{T_3^2}{10}\right)$$

$$= 26{,}423 - \left(\frac{320^2}{12} + \frac{236^2}{8} + \frac{319^2}{10}\right) = 751.567$$

$$SS_{\text{between}} = 902.167 - 751.567 = 150.6$$

The mean squares are

$$MS_{within} = \frac{751.567}{27} = 27.836$$

$$MS_{between} = \frac{150.6}{2} = 75.3$$

The ANOVA table is:

SOURCE OF VARIABILITY	SS	df	MS
Between-author means	150.6	2	75.3
Within-author means	751.567	27	27.836
Total	902.167	29	

SECTION 13.2

THE *F* TEST IN ANALYSIS OF VARIANCE

The ANOVA table can be used to test whether or not the r samples were drawn from populations with equal means. Before going any further, however, you may wish to review the discussion on sources of variability at the end of Section 12.5 and the F distribution in Section 12.6.

AN INTUITIVE EXAMPLE

Suppose that we select a sample of eight observations from each of three populations I, II and III. Three possibilities, A, B, and C, are illustrated in Figure 2. In each case members of sample I are indicated by a circle (○), members of sample II by a cross (×), and members of sample III by a triangle (△).

In situation A the differences between the sample means are large when compared to the differences among individual values within each sample. In the language of this chapter the variability within samples is small compared to that between sample means.

By contrast, in situation B, although the variability between sample means is the same as in A, the variability within samples is so large that it appears to swamp differences between sample means. The 24 values overlap so much that they look as though they could have all come from populations with the same mean, the differences in the sample means being due only to chance.

In situation C the variability within each sample is the same as that in the corresponding sample in B. However, in this case the sample means are much farther apart than in either A or B. The result is now similar to the case in A, where the within-sample variability is small compared to that among all the sample means and, in consequence, the sample means appear to be very different.

The message here is that, in order to investigate the equality or inequality of population means, it is important to judge differences between sample

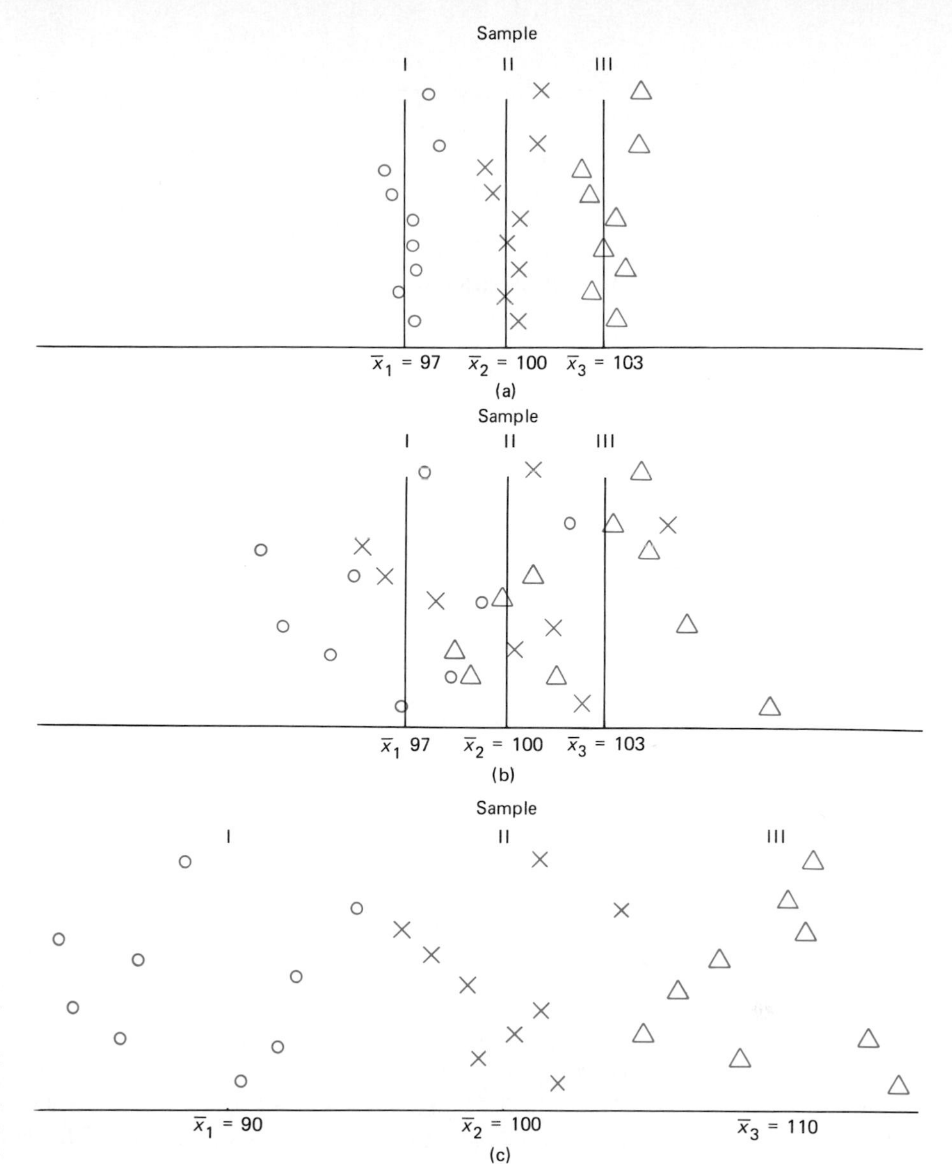

FIGURE 2

Three situations *A*, *B*, and *C*, illustrating that differences between sample means have to be judged in relation to variability within samples

means only as they are related to the variation that exists *within samples*. The larger the variability between sample means *relative* to the varability within samples, the more significant the difference between sample means will appear to be. The term "significant" is used here in an intuitive rather than a statistical sense. We shall now formalize this idea.

THE *F* TEST IN ANALYSIS OF VARIANCE

Assume that independent samples are drawn from r normal populations with respective means $\mu_1, \mu_2, \ldots, \mu_r$ all having the *same* variance $\sigma_1^2 = \sigma_2^2 = \cdots = \sigma_r^2 = \sigma^2$. We are to test $H_0: \mu_1 = \mu_2 = \mu_3 = \cdots = \mu_r$ against H_1: these means are *not* all equal.

If the assumptions above are valid and if H_0 is true, the ratio $F = \text{MS}_{\text{between}}/\text{MS}_{\text{within}}$ will have the $F_{r-1,N-r}$ distribution. The first number of degrees of freedom in F, $r - 1$, is associated with $\text{SS}_{\text{between}}$ and the second, $N - r$, is associated with $\text{SS}_{\text{within}}$. It can be shown that evidence in favor of H_0 will be a value for F close to 1. Evidence in favor of H_1 will be a value of F unusually far above 1. This will occur when $F \geq F_{r-1,N-r,.05}$. Before formalizing this F test, we illustrate it using the recovery-time data from Example 1.

EXAMPLE 4 Use the ANOVA table corresponding to the recovery-time data in Example 1 to perform the F test to decide whether there is any significant difference in the true mean recovery times for the three anesthetic drugs.

Solution The appropriate ANOVA table (from Table 3) is reproduced in Table 4, with an additional column labeled "F." In this column we have inserted the value of the ratio

$$F = \frac{\text{MS}_{\text{between}}}{\text{MS}_{\text{within}}} = \frac{1.04}{.8267} = 1.258$$

TABLE 4

ANOVA table for the recovery-time data, including the F ratio

SOURCE OF VARIABILITY	SS	df	MS	F
Between drug means	2.080	2	1.04	1.258
Within drugs	17.360	21	.8267	
Total	19.440	23		

Calling μ_1, μ_2, and μ_3 the unknown mean recovery times for all patients receiving the respective drugs, we now test $H_0: \mu_1 = \mu_2 = \mu_3$ against the alternative H_1: the three means are not all equal.

From our earlier discussion, the appropriate number of degrees of freedom for F in this case are $r - 1 = 2$ and $N - r = 21$. Since the $F_{2,21}$ distribution is not tabulated in Table A8, we use the $F_{2,20}$ distribution instead. From Table A8, $F_{2,20,.05} = 3.49$ and $F_{2,20.01} = 5.85$. Since $F = 1.258$ is far below 3.49, we cannot reject H_0 at the .05 level of significance. The data suggest no significant differences among the mean recovery times for the three drugs. The values $F = 1.258$, $F_{2,20,.05} = 3.49$, and $F_{2,20,.01} = 5.85$ are shown in Figure 3.

Generalizing the procedure above, we begin with r normal populations with respective means $\mu_1, \mu_2, \ldots, \mu_r$ and common variance σ^2. Independently drawn random samples are taken from each population. In experiments we treat the experimental groups as though they were randomly selected samples, although they are often not. (This issue was discussed in Chapter 8.)

It may be helpful at this point to identify the populations and the dependent and independent variables in Examples 1 and 3. In Example 1 we assume three populations: the recovery times for all patients who might receive Drug 1, and the same for Drug II and Drug III. The independent variable is then "type of drug" (with 3 levels) and the dependent variable "recovery time."

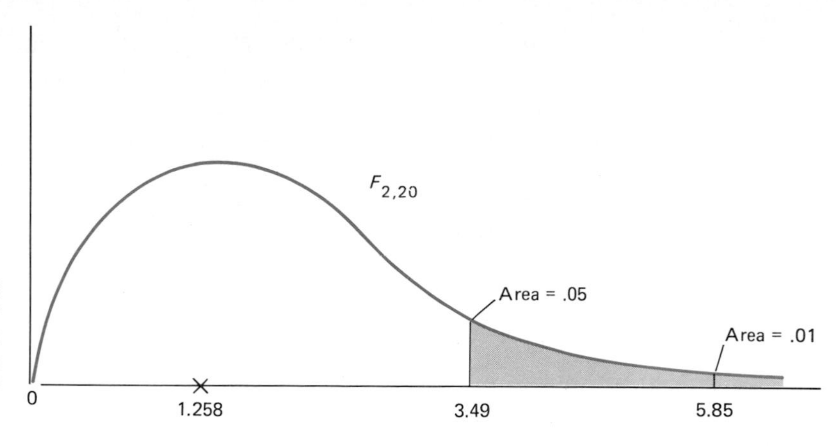

FIGURE 3

Value $F = 1.258$ shown on the $F_{2,20}$ distribution in relation to $F_{2,20,.05} = 3.49$ and $F_{2,20,.01} = 5.85$.

In Example 3 we again assume three populations: the number of pictures developed by the first technician for all potential days, the same for the second and the third technician. The independent variable is then "Technician" (three levels) and the dependent variable "Number of pictures developed."

We test $H_0: \mu_1 = \mu_2 = \cdots = \mu_r$ against $H_1: \mu_1, \mu_2, \ldots, \mu_r$ are *not* all equal.

If H_0 is true, the test statistic $F = \text{MS}_{\text{between}}/\text{MS}_{\text{within}}$ has the $F_{r-1,N-r}$ distribution. We reject H_0 if the observed value for F is unusually large. The details are given in Table 5.

TABLE 5

Procedure for performing the F test of $H_0: \mu_1 = \mu_2 = \cdots \mu_r$ against $H_1: \mu_1, \mu_2, \ldots, \mu_r$ are not all equal

1. Compute $F = \text{MS}_{\text{between}}/\text{MS}_{\text{within}}$.
2. Find $F_{r-1,N-r,.05}$ and $F_{r-1,N-r,.01}$ from Table A8.
3. *Conclusion:* Using $\alpha = .05$ or .01, reject H_0 if $F \geq F_{r-1,N-1,\alpha}$.

Assumption: Random samples are drawn independently from r normal populations, all having the same standard deviations

EXAMPLE 5 We reproduce below the ANOVA table for the photographic data in Example 3 on page 473. The last column contains the value

$$F = \frac{\text{MS}_{\text{between}}}{\text{MS}_{\text{within}}} = \frac{756}{186.167} = 4.06$$

SOURCE OF VARIABILITY	SS	df	MS	F
Between technician means	1512	2	756	4.06
Within technicians	2234	12	186.167	
Total	3746	14		

We shall test $H_0: \mu_1 = \mu_2 = \mu_3$ against $H_1: \mu_1, \mu_2, \mu_3$ are not all equal, where the μ_i are the mean number of successful pictures per day for the three technicians. From Table A8, $F_{2,12,.05} = 3.89$ and $F_{2,12,.01} = 6.93$. Since $F = 4.06$ exceeds 3.89 but not 6.93, we reject H_0 at the .05 level of significance. The data suggest that the mean number of successful pictures per technician are *not* the same. These various F values are shown in Figure 4.

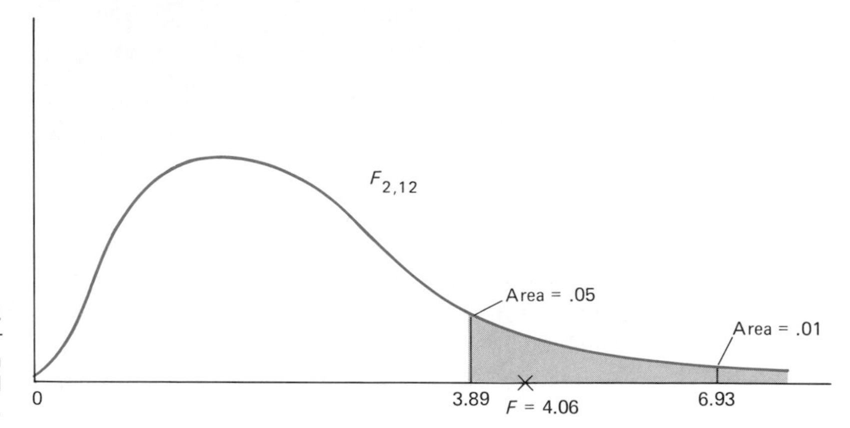

FIGURE 4

Value $F = 4.06$ in relation to $F_{2,12,.05} = 3.89$ and $F_{2,12,.01} = 6.93$.

Incidentally, the P value corresponding to the test statistic $F = 4.06$ is $P(F_{2,12} > 4.06)$, which is the area to the right of 4.06 under the $F_{2,12}$ distribution. The area will lie between .05 and .01, so that we would also have rejected H_0 at the .05 level of significance had we used the P value.

■

Notice how closely the F test conforms to our intuitive introduction at the beginning of the section. We reject H_0 only if the ratio of the *between-group mean square* to the *within-group mean square* is sufficiently large.

If the F test is significant, it is appropriate to account for the apparent difference among $\mu_1, \mu_2, \ldots, \mu_r$ by examining the corresponding sample means. There are statistical methods for examining differences among the sample means following a significant F test. In Example 5, for instance, we might ask if μ_2 is significantly different from μ_1 and μ_3. This is done by testing whether there is any significant difference between μ_2 and the average of μ_1 and μ_3. This question arose from an examination of the means in Figure 1 on page 474. Methods for performing such tests come under the general heading of **multiple comparison methods.** (See Ott, 1984)

Together, the process of constructing the ANOVA table and performing the F test is known as a **one-way analysis of variance.** The "one" refers to the fact that the data are classified on the basis of only one independent variable. As you saw this was "type of drug" and "technician" in the examples above.

PROBLEMS 13.2

10. Refer to the data in Problem 1 on page 474.
    (a) Do these data suggest a significant difference per 10,000 words in the mean frequency of use of the word "when" among the three authors?
    (b) Would you conclude that checking the freqeuncy of use of this word is a good way to discriminate among the works of the three authors?
11. As part of a larger experiment, 30 three-month-old infants are randomly divided into five groups of six infants each. The infants in a group are repeatedly shown one of five multicolored designs A, B, C, D, and E. The average time in seconds spent watching the design is recorded for each infant with the following results:

	A	B	C	D	E
	9.8	9.8	8.5	7.9	7.6
	10.3	12.3	9.6	6.9	10.6
	13.6	11.1	9.5	6.6	5.6
	10.5	10.6	7.4	7.6	10.1
	8.6	11.6	7.6	8.9	10.5
	11.1	10.9	9.9	9.1	8.6
Σx_i^2	694.51	736.27	465.19	373.36	487.70

    (a) Construct the ANOVA table for these data.
    (b) State the null and alternative hypotheses, defining the μ's.
    (c) Perform the F test and state your conclusions.
12. The accounts department of a company records the length of telephone calls (in minutes) between its East and West Coast offices for a sample of eight calls from each of three major divisions. The resulting data are as follows:

	PURCHASING	SALES	MARKETING
	23.5	25.6	31.2
	20.0	32.7	15.0
	9.0	11.1	20.6
	12.6	10.7	20.5
	22.7	15.9	23.2
	10.2	28.2	20.9
	19.6	41.8	19.3
	29.3	20.3	20.0
Σx_i^2	3053.99	5169.73	3790.59

    (a) Construct the ANOVA table for these data.
    (b) Use the F test to test for significant differences in the mean length of calls for the three divisions.
13. Researchers have constructed an index of home safety for the elderly. The higher the score, the more safe the home. Samples of 10 homes from low-, medium-, and high-income elderly communities are graded, with the following results:

											Σx_i^2
HIGH INCOME	79	67	70	58	60	78	63	74	65	57	45,597
MODERATE INCOME	53	46	42	59	61	58	63	51	62	52	30,373
LOW INCOME	42	53	34	46	44	43	33	35	42	42	17,472

(a) Construct the ANOVA table for these data.

(b) Use the F test to test for significant differences in the mean score for the three communities.

(c) Summarize the differences between the mean scores for the three groups.

14. Refer to the smoking data in Problem 6 on page 476. Test for significant differences in the mean birthweight for children of mothers with these smoking habits.

15. A total of 29 students participated in a study of the effect of alcohol on reaction time to a simulated automobile accident. The students were randomly assigned to one of four groups. The reaction times of the students were measured twice at intervals of an hour. Three of the four groups drank varying amounts of beer half an hour before the second test. The *difference* between reaction times on the second and first tests was recorded for all students, with the following results (in seconds):

	NO BEER	TWO BEERS	FOUR BEERS	SIX BEERS
	−.21	.20	.45	.02
	−.02	.36	.22	.42
	−.17	.17	.12	.43
	.03	.24	.34	.42
	.03	.18	.35	.37
	−.17	.08	.30	.27
		.34	.24	.55
		.06	.12	
Σx_i^2	.1041	.4141	.6654	1.0504

(a) Define μ_1, μ_2, μ_3, and μ_4 in this case.

(b) Test for significant differences in the mean change in reaction time for the four groups.

(c) Can we conclude that more beer tends to lengthen reaction times to the simulated automobile accident?

(d) Would your conclusion in part (c) change if students had been allowed to choose which group to join? Explain.

16. Use the data in Problem 4 on page 475 to test for significant differences in the mean use of electricity between the three cities.

17. Use the data in Problem 7 on page 476 to test for significant differences in GPA between the four classes.

18. Use the data in Problem 5 on page 475 to test for significant differences in the mean number of defectives per 200 Twickets between machines.

19. Use the data in Problem 8 on page 476 to test for significant differences in mean medical costs for the four hospitals.

20. The following data are the number of days between breakdowns for three machines:

MACHINE A			MACHINE B			MACHINE C		
2.4	24.3	99.7	286.1	90.7	155.0	118.4	432.0	69.7
322.5	9.8	121.3	192.1	66.9	26.3	166.2	289.0	47.1
167.5	28.6	139.2	82.1	92.2	97.4	272.6	11.4	297.4
54.7	52.6	42.7	33.8	437.2	143.5	237.4	8.5	290.7
223.9	243.7	99.4	300.2	132.0	76.4	198.7	8.7	88.5
385.6	307.3	85.9	13.6	250.7	124.4	209.1	22.8	142.9
353.2	432.7		407.9	406.4		112.8	75.6	

Group each sample into five intervals of the form 0–99.9, 100–199.9 and so on. Plot the relative frequency histogram for each sample. Why would an F test be inappropriate in this case?

21. The authors used a computer to generate four samples each of size 5 but all selected from a $N[10, 2]$ population:

	A	B	C	D
	13.2	9.2	13.0	6.9
	12.5	7.2	8.6	8.9
	11.9	9.3	7.6	8.9
	9.7	12.4	8.3	11.3
	6.4	9.3	7.7	12.3
Σx_i^2	607.15	463.22	428.90	485.01

(a) Perform the F test for these data.

(b) Did your test suggest the correct answer? Explain.

22. For each of 25 families the average number of years of education for the father and mother are as follows, classified by the number of children in the family:

NUMBER OF CHILDREN				
0	1	2	3	4
14.5	16.5	10.5	10.5	12.0
13.0	12.0	11.0	12.0	13.0
	12.0	15.3	11.0	10.0
	14.5	13.5	15.5	
	14.0	16.0	14.0	
		12.0	11.5	
		13.0		
		12.0		
		13.5		

Do these data suggest any significant differences in the mean education level for the five groups?

Solution to Problem 10

From the solution to Problem 1 on page 477, we obtain the following table (we have added the value $F = 75.3/27.836 = 2.705$):

SOURCE OF VARIABILITY	SS	df	MS	F
Between author means	150.6	2	75.3	2.705
Within authors	751.567	27	27.836	
Total	902.167	29		

(a) Since Table A8 does not contain the $F_{2,27}$ distribution, we refer to the $F_{2,24}$. Note that $F_{2,24,.05} = 3.40$ and $F_{2,24,.01} = 5.61$. Since $F = 2.705$ is less than 3.40, we cannot reject H_0 at the .05 level of significance. The data suggest no significant differences among the three mean frequencies.

(b) Since there appear to be no significant differences in the frequency with which "when" is used, this word is *not* a good discriminator.

SECTION 13.3

ESTIMATION IN ANOVA AND THE RELATIONSHIP BETWEEN THE F AND t TESTS

In this section we discuss two special topics related to the one-way analysis of variance. The first involves *estimating* μ_i and $\mu_i - \mu_j$.

ESTIMATING μ_i and $\mu_i - \mu_j$

It is frequently of interest to use a confidence interval to estimate one of the r population means, μ_i, or the difference between two of the r population means, $\mu_i - \mu_j$.

In the case of one mean, we proceed as follows:

DEFINITION 1

If the assumptions in Table 5 apply, then a $(100\gamma)\%$ confidence interval for μ_i is

$$\bar{x}_i - t_{N-r(\gamma)}\frac{s_*}{\sqrt{n_i}} \quad \text{to} \quad \bar{x}_i + t_{N-r(\gamma)}\frac{s_*}{\sqrt{n_i}}$$

where n_i is the number of values, $\bar{x}_i$ is the mean in the ith group (or sample), and $s_* = \sqrt{s_*^2} = \sqrt{\text{MS}_{\text{within}}}$ is found from the ANOVA table.

The number of degrees of freedom for t is then that for $\text{MS}_{\text{within}}$.

EXAMPLE 6 Using the data in Example 3, find a 90% confidence interval for the mean daily output of technician II (μ_2).

Solution From Example 3, $T_2 = 1152$, $n_2 = 6$, so $\bar{x}_2 = 1152/6 = 192$. From the ANOVA table, $s_* = \sqrt{\text{MS}_{\text{within}}} = \sqrt{186.167} = 13.6443$, and from Table A6, $t_{N-r(.9)} = t_{12(.9)} = 1.782$.

A 90% confidence interval for μ_2 is

$$192 \mp 1.782\left(\frac{13.6443}{\sqrt{12}}\right) = 184.98 \quad \text{to} \quad 199.02$$

We can be 90% confident that μ_2 lies between 184.98 and 199.02 photographs per day.

■

Suppose now that instead of estimating one population mean, we want to estimate $\mu_i - \mu_j$, the difference between two of the r population means. We use the following definition:

> DEFINITION 2
>
> If the assumptions in Table 5 apply, then a $(100\gamma)\%$ confidence interval for $\mu_i - \mu_j$ is
>
> $$(\bar{x}_i - \bar{x}_j) - t_{N-r(\gamma)}s_*\sqrt{\frac{1}{n_i} + \frac{1}{n_j}} \quad \text{to} \quad (\bar{x}_i - \bar{x}_j) + t_{N-r(\gamma)}s_*\sqrt{\frac{1}{n_i} + \frac{1}{n_j}}$$
>
> where as before $s_* = \sqrt{\text{MS}_{\text{within}}}$. Here $\bar{x}_i - \bar{x}_j$ is the difference between the ith and jth sample means, which are based on n_i and n_j values, respectively.

EXAMPLE 7 Using the data in Example 3, find a 90% confidence interval for the difference, $\mu_2 - \mu_1$, in the mean daily output for technicians II and I.

Solution From Example 3, $\bar{x}_2 = 192$ and $\bar{x}_1 = 174$, $\bar{x}_2 - \bar{x}_1 = 18$, $n_2 = n_1 = 6$, $s_* = \sqrt{\text{MS}_{\text{within}}} = \sqrt{186.167} = 13.6443$, and from Table A6, $t_{N-r(.9)} = t_{12(.9)} = 1.782$.

From Definition 2, a 90% confidence interval for $\mu_2 - \mu_1$ is

$$18 \mp (1.782)(13.6443)\sqrt{\frac{1}{6} + \frac{1}{6}} = 18 \mp 14.038$$

$$= 3.96 \text{ to } 32.04$$

We can be 90% confident that the *difference* in daily output between technicians II and I lies between 3.96 and 32.04 photographs per day (or 4 to 32).

■

The confidence intervals in Definitions 1 and 2 are similar to those for μ alone and for $\mu_1 - \mu_2$ in Chapter 12 (Definitions 2 and 7). The differences are:

1. We now use $s_* = \sqrt{\text{MS}_{\text{within}}}$ instead of s or of s_p, the pooled standard deviation in the case of $\mu_1 - \mu_2$.
2. We now use $N - r$ degrees of freedom for t instead of $(n - 1)$ or $(n_1 + n_2 - 2)$.

To explain these differences we first recall that $SD(\overline{X}) = \sigma/\sqrt{n}$, but in the confidence interval for μ we used s as an estimate of σ, the unknown population standard deviation. Similarly, in the confidence interval for $\mu_1 - \mu_2$, we used the square root of the pooled sample variance, $\sqrt{s_p^2} = s_p$, as an estimate of σ, the standard deviation common to the two populations.

In this chapter we have assumed that the r population variances σ_1^2, $\sigma_2^2, \ldots, \sigma_r^2$ are all equal (to σ^2). Since this time we are interested in obtaining a confidence interval for the mean of one population, or the difference between the means of two of *several* populations it makes sense to obtain a weighted average of *all* the r sample variances: $s_1^2, s_2^2, \ldots$, and s_r^2. $MS_{within} = s_*^2$ is just such a weighted average. To see this, recall that in Equation 13.1 we defined SS_{within} as $\sum^r (n_i - 1)s_i^2$, and dividing by $n - r$ degrees of freedom, we obtained

$$MS_{within} = \frac{(n_1 - 1)s_1^2 + (n_2 - 1)s_2^2 + \cdots + (n_2 - 1)s_r^2}{N - r}$$

$$= \frac{(n_1 - 1)s_1^2 + (n_2 - 1)s_2^2 + \cdots + (n_r - 1)s_r^2}{(n_1 - 1) + (n_2 - 1) + \cdots + (n_r - 1)}$$

which is a weighted average of the sample variances.

If $r = 2, N = n_1 + n_2$, MS_{within} becomes s_p^2 in Equation 12.1 on page 413 and in this case the confidence intervals in Definition 2 above and Definition 7 in Chapter 12 (page 418) are identical.

Definition 2 for the confidence interval for $\mu_i - \mu_j$ applies *only* when the groups to be compared, the ith and jth, are selected *before* seeing the data: for instance, if, after seeing that $\overline{x}_3$ was the largest sample mean and $\overline{x}_1$ was the smallest, it would *not* be appropriate to compute a confidence interval for $\mu_3 - \mu_1$. The reason is the implicit assumption in Definition 2 that the r samples are selected at random and independently. This assumption is violated if we select the groups with the largest and smallest sample means for comparison.

THE F TEST AND THE TWO-SAMPLE t TEST

In view of our findings above, it may not be surprising that when $r = 2$, the F test is exactly equivalent to the two-sample t test described in Section 12.4.

When $r = 2$ the null hypothesis for the F test is $H_0: \mu_1 = \mu_2$ (or $\mu_1 - \mu_2 = 0$). The alternative is $H_1: \mu_1, \mu_2$ are not equal, which is written as $H_1: \mu_1 \neq \mu_2$ or $H_1: \mu_1 - \mu_2 \neq 0$. As always, we assume that both populations are (approximately) normal with equal variances. But this situation is precisely that associated with the two-sided, two-sample t test in Section 12.4. When $r = 2$ the numbers of degrees of freedom for F are $r - 1 = 1$ and $N - r = n_1 + n_2 - r = n_1 + n_2 - 2$. In the next example we demonstrate that $F_{1,n_1+n_2-2} = t_{n_1+n_2-2}^2$ and that the two tests are indeed equivalent.

EXAMPLE 8 A class of 20 students are randomly divided into two groups of 10 each. Group I is taught using a lecture format. Group II is taught the same topic in small-group sessions. The scores on the final test are summarized as follows:

	T_i	Σx_i^2	s_i	n_i
LECTURES (I)	462.4	21,619.152	5.14	10
SMALL-GROUP SESSIONS (II)	496.1	24,803.621	4.62	10

Using these data, we obtain the following ANOVA table:

SOURCE OF VARIABILITY	SS	df	MS	F
Between method means	56.784	1	56.784	2.378
Within methods	429.876	18	23.882	
Total	486.660	19		

The appropriate hypotheses are $H_0: \mu_1 = \mu_2$ and $H_1: \mu_1 \neq \mu_2$. From Table A8, $F_{1, 18, .05} = 4.41$ and $F_{1, 18, .01} = 8.29$. Since $F = 2.378$ is *less* than 4.41, we cannot reject H_0 at the .05 level of significance. The data suggest no significant difference in mean scores for the two teaching methods.

Using the same data, we may perform a two-sided two-sample t test. The hypotheses are also $H_0: \mu_1 = \mu_2$ and $H_1: \mu_1 \neq \mu_2$. Following the procedure in Section 12.4, the appropriate number of degrees of freedom $= n_1 + n_2 - 2 = 18$. The sample means are $\bar{x}_1 = T_1/10 = 462.4/10 = 46.24$ and $\bar{x}_2 = T_2/10 = 49.61$ and the difference $\bar{x}_1 - \bar{x}_2 = -3.37$.

The pooled variance is

$$s_p^2 = \frac{9(5.14)^2 + 9(4.62)^2}{18} = 23.882 \qquad \text{and} \qquad s_p = 4.887$$

(Notice that $s_p^2 = \text{MS}_{\text{within}}$, as we indicated earlier.)

The value for $|t|$ is

$$\left| \frac{-3.37}{4.887\sqrt{1/10 + 1/10}} \right| = 1.542$$

From Table A6, $t_{18, .025} = 2.101$ and $t_{18, .005} = 2.878$. Since 1.542 is less than 2.101, we do not reject H_0 at the .05 level of significance. The conclusion is the same as for the F test.

Notice that the square of the computed value for t, $t^2 = 1.542^2 = 2.378$, is the computed value for F obtained above. Furthermore, the square of the critical values $t_{18, .025}^2 = 2.101^2 = 4.41$ and $t_{18, .005}^2 = 2.878^2 = 8.29$ are precisely the critical values for the F test above: 4.41 and 8.29, respectively. Thus the two test procedures are exactly equivalent. Rejecting H_0 at the .01 level of significance if $F > 8.29$ is equivalent to rejecting H_0 at the .01 level if $|t| > \sqrt{8.29} = 2.878$.

■

The equivalence of the two tests does not mean that when comparing two sample means either test is appropriate. The alternative hypothesis for the F test when $r = 2$ is always $H_1: \mu_1 \neq \mu_2$ or, equivalently, $H_1: \mu_1 - \mu_2 \neq 0$. On the other hand, with the t test we have the option of performing a one- or a two-sided test. If our alternative is two-sided (i.e., $H_1: \mu_1 \neq \mu_2$), either test may be used. But if our alternative is one-sided ($H_1: \mu_1 > \mu_2$, for instance), then only the t test can be used. In Example 8, for instance, if the mean score for the small-group method was expected to be higher than that for the lecture method, H_1 would be $H_1: \mu_2 - \mu_1 > 0$ and the appropriate test would be a one-sided t test.

PROBLEMS 13.3

23. Refer to the drug data at the beginning of Section 13.1. Suppose that drug II is a new experimental drug. Use the results in Tables 2 and 3 both appearing on page 471, respectively, to find an 80% confidence interval for (a) the mean recovery time, μ_2, for all patients using drug II; (b) the difference $\mu_1 - \mu_2$ in mean recovery times for all patients using drugs I and II.
24. A college admissions office selects two random samples each containing 25 students. The first consists of students who are entering the college as freshmen and the second of students who were accepted by the college but who elected to go elsewhere. The combined SAT verbal and math scores for all 50 students were recorded. The summary data are as follows:

	ATTENDING COLLEGE (A)	ATTENDING ANOTHER COLLEGE (B)
n	25	25
Σx_i	25,392	26,532
Σx_i^2	25,971,400	28,263,200

    (a) Use the methods in Section 12.4 to perform a t test for a significant difference in the means for the two groups.
    (b) Perform an F test for a significant difference in the mean for the two groups.
    (c) Show that s_p^2 in part (a) is the same as $\text{MS}_{\text{within}}$ in part (b).
    (d) Show that the square of the calculated t value in part (a) is the calculated F value in part (b).
25. Refer to Problem 24.
    (a) Use Definition 7 of Chapter 12 (page 418) to find a 95% confidence interval for $\mu_B - \mu_A$.
    (b) Show that you get the same answer if you use Definiton 2.
26. Refer to Problem 15 on page 484. Find an 80% confidence interval for the mean difference in reaction time for those who had no beer and those who had two beers.
27. Refer to Problem 6 on page 476. Find and interpret a 90% confidence interval for (a) the mean birthweight of babies born to heavy-smoking mothers; (b) the difference between the mean birthweight of babies born to light-smoking and heavy-smoking mothers.
28. As used in elementary school, mastery learning includes students working together on one level of a subject regardless of grade level. Students leave the group only when they have completely mastered the material. In one school district some of the children were learning in this way, whereas others were learning in the more

traditional classroom. The reading scores of a random sample of "mastery learning" fifth graders and a random sample of traditional learning fifth graders were recorded. The summary results are as follows:

	MASTERY LEARNING	TRADITIONAL LEARNING
n	24	20
Σx_i	1,684	1,205
Σx_i^2	122,322	75,127

(a) Use both the F and the t test to check for evidence of a significant difference in mean reading scores for the two learning techniques. Show that both tests will result in the same conclusion.

(b) The school board would like to know whether these data prove that students who study with mastery learning will score higher than other children. Answer this question if (i) the children were randomly assigned to the two teaching methods; (ii) the students' parents decided which method the children would receive.

Solution to Problem 23

(a) From Tables 2 and 3 we obtain the values $\bar{x}_2 = 4.8$ hours, $n_2 = 8$ and $s_* = \sqrt{\text{MS}_{\text{within}}} = \sqrt{.8267} = .9092$. Finally, from Table A6, $T_{N-r(.8)} = t_{21(.8)} = 1.323$. Substituting these values into Definition 1, an 80% confidence interval for μ_2 is

$$4.8 \mp (1.323)\left(\frac{.9092}{\sqrt{8}}\right) = 4.8 \mp .425$$

$$= 4.38 \text{ to } 5.22$$

We can be 80% confident that the mean recovery time for drug II is between 4.38 and 5.22 hours.

(b) In addition to the numbers found in part (a), we need, from Table 2, the values $\bar{x}_1 = 5.5$ and $n_1 = 8$. Substituting in Definiton 2, an 80% confidence interval for $\mu_1 - \mu_2$ is

$$(5.5 - 4.8) \mp (1.323)(.9092)\sqrt{\frac{1}{8} + \frac{1}{8}} = .7 \mp .601$$

$$= .099 \text{ to } 1.301 \text{ or more conveniently } .1 \text{ to } 1.3$$

We can be 80% confident that the *difference* in mean recovery time ($\mu_1 - \mu_2$) is between .1 and 1.3 hours.

SECTION 13.4

TWO-WAY ANALYSIS OF VARIANCE

In the one way analysis of variance we compare the effect of different levels of *one* independent variable on a dependent variable. In this section we indicate the methodology for, and the advantages of, determining the joint effect of *two* independent variables on the dependent variable.

Suppose, for example, that the statistics department at Academic U. is interested in the effects on student scores given in the introductory statistics course of (1) three different textbooks, those of Brown, Green, and White, and (2) two teaching plans, the usual lecture plan and a self-paced plan.

Sixty of the students enrolled in the course are selected at random to participate in an experiment. The 60 students are randomly assigned to six sections, each with 10 students. Each section, called an **experimental combination,** is to be taught with a combination of one text and one teaching method (see Figure 5).

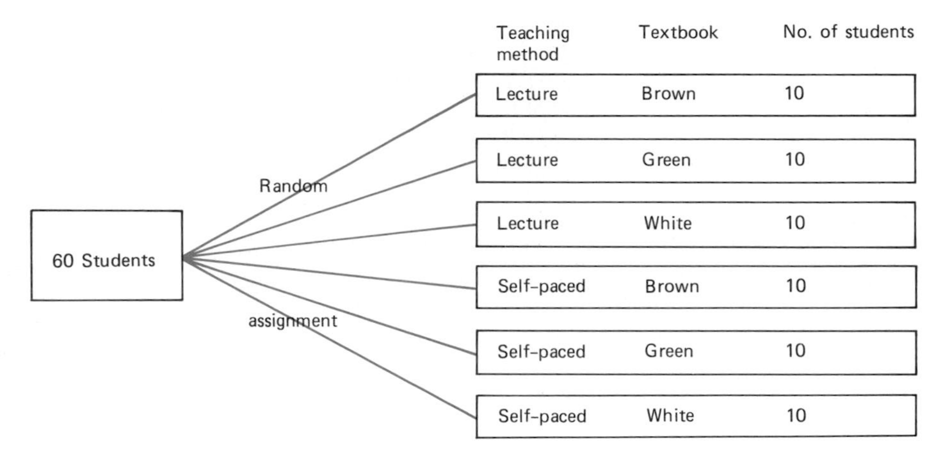

FIGURE 5

Assignment of 60 students to six experimental combinations

The values for the dependent variable, students' final examination course scores, are shown in Table 6.

The scores fall into six **cells,** one for each experimental combination. In Table 6 we have included cell, method, and textbook totals and means as well as the sum of all 60 *squared* scores (370,027).

"MOST" OF A TWO-WAY ANOVA

In this case the researchers will first want to test for significant differences (1) between the mean scores for the two teaching methods, and (2) among the mean scores for the three textbooks. To do this, we need to construct most of a *two-way* analysis of variance table. The components of the table that are immediately necessary can be obtained by analogy with the corresponding components of the one-way ANOVA table.

First, we need a between teaching method sum of squares. By analogy with the form of $SS_{between}$ in Equation 13.2, (page 472) using the data in Table 6 we compute this as

$$SS_{methods} = \left(\frac{2254^2}{30} + \frac{2403^2}{30}\right) - \frac{4657^2}{60} = 370.0167$$

TABLE 6

Final scores for the 60 students

TEACHING METHOD	TEXTBOOK BROWN	 GREEN	 WHITE	ROW TOTAL (MEAN)
Lecture	83	73	61	
	73	74	56	
	55	71	63	
	82	60	87	
	82	73	61	
	98	75	87	
	77	83	91	
	95	99	49	
	76	62	78	
	84	80	66	
Total (mean)	805(80.5)	750(75.0)	699(69.9)	2254(75.13)
Self-paced	75	81	96	
	68	[illegible]	78	
	99	76	96	
	59	84	85	
	83	79	79	
	69	83	90	
	72	79	95	
	83	84	80	
	70	78	49	
	87	89	75	
Total (mean)	765(76.5)	815(81.5)	823(82.3)	2403(80.10)
Column Total (mean)	1570(78.50)	1565(78.25) $\Sigma x_i^2 - 370{,}027$	1522(76.10)	4657(77.62)

Similarly, a between textbook sum of squares will be

$$SS_{\text{textbook}} = \left(\frac{1570^2}{20} + \frac{1565^2}{20} + \frac{1522^2}{20}\right) - \frac{4657^2}{60} = 69.6333$$

We shall also need a SS_{within}, which in this case measures the variability *within cells*. It is computed by analogy to its counterpart in Equation 13.3 as

$$SS_{\text{within}} = 370{,}027 - \left(\frac{805^2}{10} + \frac{750^2}{10} + \frac{699^2}{10} + \frac{765^2}{10} + \frac{815^2}{10} + \frac{823^2}{10}\right)$$
$$= 7436.5$$

Finally (this time by analogy with Equation 13.4), the SS_{total} is found as

$$SS_{\text{total}} = 370{,}027 - \frac{4657^2}{60} = 8566.1833$$

These four sums of squares with the corresponding numbers of degrees of freedom and mean squares are shown in Table 7. You can check that, as indicated in Table 7, the three sums of squares and the three df's do not add to the total. As you can see, this is due to a missing entry. We will return to this entry later.

TABLE 7 Partially complete two-way ANOVA table for the statistics example

SOURCE OF VARIABILITY	SS	df	MS	F
Between method means	370.0167	1	370.0167	2.687
Between textbook means	69.6333	2	34.8167	.2528
* * *	*	*	*	*
Within cells	7436.50	54	137.7130	
Total	8566.1833	59		

The first step in testing for significant differences between the method means is to compute the ratio $F = \text{MS}_{\text{methods}}/\text{MS}_{\text{within}} = 2.687$. We should compare this figure with $F_{1,54,.05}$ and $F_{1,54,.01}$ but since the $F_{1,54}$ distribution is not listed in Table A8, we compare 2.687 to $F_{1,40,.05} = 4.08$ and $F_{1,40,.01} = 7.31$. Since 2.687 is less than 4.08, we conclude that there is no significant difference in the mean scores for the two teaching methods.

Similarly, we test for significant differences among the textbook means by comparing the ratio $F = \text{MS}_{\text{textbook}}/\text{MS}_{\text{within}} = .2528$ to $F_{2,40,.05} = 3.23$ and $F_{2,40,.01} = 5.18$. Since .2538 is considerably less than 3.23, we conclude that there are no significant differences among the mean scores for the three texts.

INTERACTION

In Table 8 we have abstracted from Table 6 the cell, text, and method means.

TABLE 8 Cell, text, and method means for the statistics data

		TEXTBOOK			
		BROWN	GREEN	WHITE	MEAN
TEACHING METHOD	LECTURE	80.5	75.0	69.9	75.13
	SELF-PACED	76.5	81.5	82.3	80.10
	MEAN	78.50	78.25	76.10	

Notice that on the average, the Brown text produced the highest mean score (78.50) of the three texts. At the same time the self-paced method scored higher (80.10) than the lecture method. This suggests that the optimum combination is self-paced instruction with Brown. The fact is, however, that this *combination* produced only the fourth best mean (76.5) of the six. This apparent contradiction is due to the **interaction** between textbook and teaching method. We define this new term as follows:

DEFINITION 3

Two independent variables, A and B, are said to **interact** if the effect of A on the dependent variable is different for different levels of B (and vice versa).

In this example the effect of the teaching method on statistics scores differs by textbook. The self-paced method is better with Green and White, but the lecture method is better with Brown.

To better see the nature of interaction, assume that the results of this experiment were those in Table 9.

TABLE 9

Table of means showing no interaction

		TEXTBOOK			
		BROWN	GREEN	WHITE	MEAN
TEACHING METHOD	LECTURE	81	74	70	75
	SELF-PACED	86	79	75	80
	MEAN	83.5	76.5	72.5	

In this table there is *no* interaction between the two variables. On the average self-paced learning is better than lectures by 5 points, *and this is true for each text*. Also, *for each teaching method*, Brown is better than Green by 7 points and better than White by 11 points.

In Figure 6(a) we have represented each of the six cell means from Table 9 by a point. With no interaction, the lines connecting the means for the different methods are parallel. By contrast, look at the corresponding lines for the original cell means from Table 8 in Figure 6(b). The extent to which the lines are not parallel indicates the extent of the interaction.

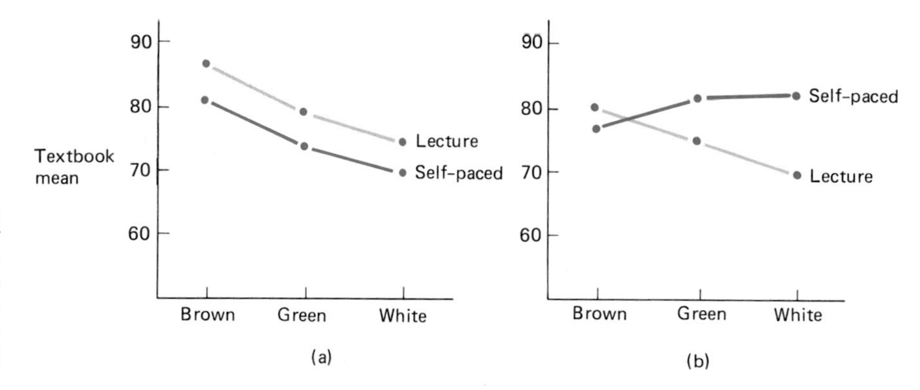

FIGURE 6

Line plots of the cell means, indicating (a) no interaction from Table 9 and (b) interaction from Table 8

We can also test for the significance of the interaction between the two variables. As a first step we introduce the general case of the two-way analysis of variance. We shall need the following notation (the terms or values are from Table 6). When quantitative data are classified on the basis of two independent variables, we treat one as the "row" variable (teaching method) and the other as the "column" variable (textbook). In this case the row variable has r levels (r = 2 methods) and the column variable c levels (c = 3 texts). There will thus be rc experimental combinations and cells. The number of values within each cell is n(n = 10 scores). (We shall only consider the case with equal numbers of values per cell.) There will thus be rcn (= 60) values in all. We shall also need the following terms:

M is the sum of the rcn squared values (M = 370,027).
$C_1, C_2, \ldots, C_C$ are the column totals (C_1 = 1570, C_2 = 1565, C_3 = 1522).
$R_1, R_2, \ldots, R_r$ are the row totals (R_1 = 2254, R_2 = 2403).
$T_1, T_2, \ldots, T_{rc}$ are the *cell* totals. (Here T_1 = 805, T_2 = 750, T_3 = 699, T_4 = 765, T_5 = 815, T_6 = 823.)
G is the grand total (G = 4657).
The grand mean is G/rcn.

Corresponding to Equation 13.1 is the following equality for a two way analysis of variance.

$$SS_{total} = SS_{rows} + SS_{columns} + SS_{interaction} + SS_{within} \qquad \text{(Eq. 13.5)}$$

SS_{total} measures the extent to which all rcn values vary around the grand mean. SS_{rows} measures the extent to which the row means, and $SS_{columns}$, the extent to which the column means vary around the grand mean. The $SS_{interaction}$, which is the entry missing in Table 7, is that portion of the total variability that is due to the effect of *combinations* of levels of the independent variables over and above their separate effects. Finally, we note that the SS_{within} is that portion of the total variability due to variables *other* than the two independent variables (and their combinations).

The formulas for these various sums of squares are:

$$SS_{total} = M - \frac{G^2}{rcn} \qquad \text{(Eq. 13.6)}$$

$$SS_{rows} = \frac{R_1^2 + R_2^2 + \cdots + R_r^2}{cn} - \frac{G^2}{rcn} = \frac{1}{cn}\sum^{r} R_i^2 - \frac{G^2}{rcn} \qquad \text{(Eq. 13.7)}$$

$$SS_{columns} = \frac{C_1^2 + C_2^2 + \cdots + C_c^2}{rn} - \frac{G^2}{rcn} = \frac{1}{rn}\sum^{c} C_i^2 - \frac{G^2}{rcn} \qquad \text{(Eq. 13.8)}$$

$$\mathrm{SS}_{\text{within}} = M - \frac{T_1^2 + T_2^2 + \cdots + T_{rn}^2}{n} = M - \frac{1}{n}\sum^{rc} T_i^2$$

(Eq. 13.9)

(Do check that these formulas give precisely the values we found earlier for the statistics course data.)

Finally, by subtraction, we have

$$\mathrm{SS}_{\text{interaction}} = \mathrm{SS}_{\text{total}} - \mathrm{SS}_{\text{rows}} - \mathrm{SS}_{\text{columns}} - \mathrm{SS}_{\text{within}}$$

(Eq. 13.10)

From Table 7 for the statistics data,

$$\mathrm{SS}_{\text{interaction}} = 8566.1833 - 370.0167 - 69.6333 - 7436.50$$
$$= 690.0333$$

The $\mathrm{SS}_{\text{total}}$ has $rcn - 1$ degrees of freedom; $\mathrm{SS}_{\text{rows}}$, $r - 1$; $\mathrm{SS}_{\text{columns}}$, $c - 1$; $\mathrm{SS}_{\text{interaction}}$, $(r - 1)(c - 1)$; and $\mathrm{SS}_{\text{within}}$, $rcn - rc = rc(n - 1)$. The complete ANOVA table for the statistics course data is shown in Table 10.

TABLE 10

Two-way ANOVA table for the statistics example

SOURCE OF VARIABILITY	SS	df	MS	F
Between method means	370.0167	$r - 1 = 1$	370.0167	$F_R = 2.687$
Between textbook means	69.6333	$c - 1 = 2$	34.8167	$F_C = .2528$
Interaction	690.0333	$(r - 1)(c - 1) = 2$	345.0167	$F_I = 2.505$
Within cells	7436.50	$rc(n - 1) = 54$	137.7130	
Total	8566.1833	$rcn - 1 = 59$		

We can now formalize the test for significant differences between row and column means that we performed earlier and introduce the F test for interaction.

F TESTS IN THE TWO-WAY ANOVA

Call the true means for the r levels of the row variable $\mu_1, \mu_2, \ldots, \mu_r$, and those for the c levels of the column variable $U_1, U_2, \ldots, U_c$. The appropriate null and alternative hypotheses are:

$$H_0: \mu_1 = \mu_2 = \cdots = \mu_r \qquad H_1: \mu_1, \mu_2, \ldots, \mu_r \text{ are } \textit{not} \text{ all equal}$$

$$H_i: U_1 = U_2 = \cdots = U_c \qquad H_1: U_1, U_2, \ldots, U_c \text{ are } \textit{not} \text{ all equal}$$

The third pair of hypotheses are:

H_0: there is no interaction between the two independent variables

H_1: there is some significant interaction

The assumptions needed to test these hypotheses are set out below.

> DEFINITION 4
>
> We assume that the values in each cell are a random sample from a normal population and that all rc populations have the same standard deviation.

If the assumptions apply and all three null hypotheses are true, then

$$F_R = \text{MS}_{\text{rows}}/\text{MS}_{\text{within}} \text{ has the } F_{r-1,\,rc(n-1)} \text{ distribution}$$

$$F_C = \text{MS}_{\text{columns}}/\text{MS}_{\text{within}} \text{ has the } F_{c-1,\,rc(n-1)} \text{ distribution}$$

$$F_I = \text{MS}_{\text{interaction}}/\text{MS}_{\text{within}} \text{ has the } F_{(r-1)(c-1),\ rc(n-1)} \text{ distribution}$$

The procedures for performing the F test of the three hypotheses are set out in Table 11.

TABLE 11

Procedure for performing the F test for significant differences among (a) row means, (b) column means, and (c) the F test for interaction

1. Compute F_R, F_C, and F_I from the ANOVA table.
2. Find $F_{w,rc(n-1)}$, .05 and $F_{w,rc(n-1)}$, .01 from Table A8, where $w = r - 1$, $c - 1$, and $(r - 1)(c - 1)$ for the respective tests.
3. *Conclusion:* Using $\alpha = .05$ or .01, reject H_0 if $F_R, F_C, F_I \geq F_{w,rc(n-1),\alpha}$.

Assumptions: As in Definition 4

We have already found no significant differences between teaching method (row) means and text (column) means. To test for significant interaction, we compute $F_I = 345.0167/137.7130 = 2.505$. Since $rc(n - 1) = 54$, and the $F_{2,54}$ distribution is not in Table A8 we find instead $F_{2,40,.05} = 3.23$ and $F_{2,40,.01} = 5.18$. Since 2.505 is less than 3.23, we cannot reject H_0 at the .05 level of significance. The data suggest no significant interaction between textbook and teaching method.

ANALYSIS OF VARIANCE IN A RANDOMIZED BLOCK DESIGN

We pointed out in Chapter 8 that even in a completely randomized design, if the number of values in each group is small, the groups may still differ considerably with regard to a potentially confounding variable. Suppose, for instance, that in Example 1 recovery time for men is generally longer than that for women. It would be important that the groups receiving the different drugs have the same proportion of men. This can be done by forming two groups (called "blocks"), one consisting of 12 male and the other of 12 female patients and then assigning four members of each block to each of the three drugs. Such a **randomized block design** is shown in Figure 7.

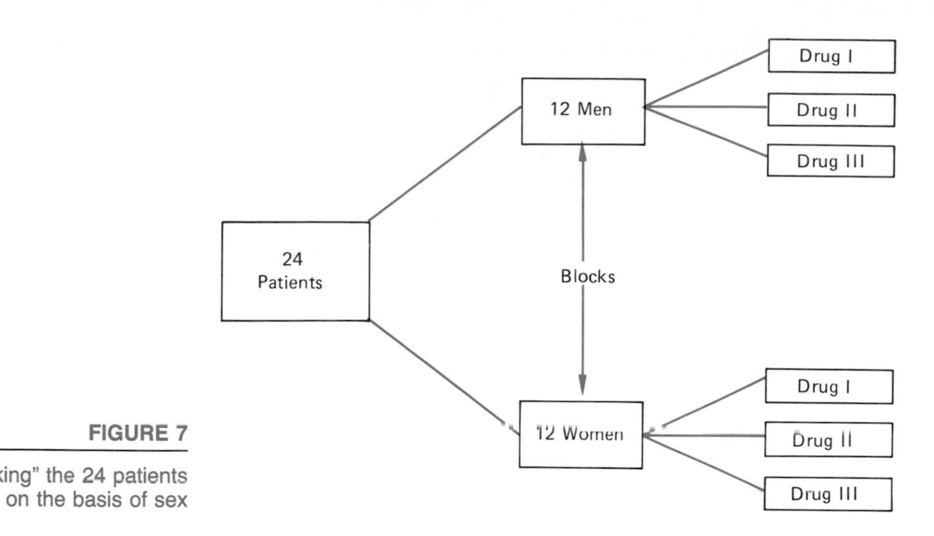

FIGURE 7

"Blocking" the 24 patients on the basis of sex

In Table 12(a) we reproduce the 24 recovery times from Example 1, and in Table 12(b) we indicate how they might look in a randomized block design.

TABLE 12

24 recovery times from Example 1 (a) and in a possible randomized block format (b)

(a) COMPLETELY RANDOMIZED DESIGN

DRUG I	DRUG II	DRUG III	TOTAL
5.7	4.3	6.2	
4.8	5.2	3.8	
6.0	5.4	3.8	
5.2	4.2	6.4	
6.3	5.5	6.2	
5.0	3.8	6.0	
6.0	5.6	4.0	
5.0	4.4	3.6	
44.0	38.4	40.0	122.4

$\Sigma x_i^2 = 643.68$

(b) RANDOMIZED BLOCK DESIGN

	DRUG I	DRUG II	DRUG III	TOTAL
Males	5.7	5.6	6.2	
	6.0	5.4	6.4	
	6.0	5.2	6.2	
	6.3	5.5	6.0	
	24.0	21.7	24.8	70.5
Females	4.8	4.4	3.8	
	5.2	4.2	3.8	
	5.0	4.3	4.0	
	5.0	3.8	3.6	
	20.0	16.7	15.2	51.9
	44.0	38.4	40.0	122.4

$\Sigma x_i^2 = 643.68$

In Table 12(b) The longer recovery times for each drug belong to the men. Notice that the column (drug) totals, the grand total, and the sum of the square values are the same for both arrangements.

We can now construct a two-way analysis of variance table for these data. First, from Table 12(b), $M = \sum^{24} x_i^2 = 643.68$, $C_1 = 44.0$, $C_2 = 38.4$, $C_3 = 40.0$, $R_1 = 70.5$, $R_2 = 51.9$, $T_1 = 24.0$, $T_2 = 21.7$, $T_3 = 24.8$, $T_4 = 20.0$, $T_5 = 16.7$, $T_6 = 15.2$, and $G = 122.4$.

From Equations 13.6 to 13.10 with $n = 4$, $c = 3$, and $r = 2$:

$$\text{SS}_{\text{total}} = M - \frac{G^2}{rcn} = 643.68 - \frac{122.4^2}{24} = 19.440$$

$$\text{SS}_{\text{rows}} = \text{SS}_{\text{sex}} = \frac{1}{cn}\sum^{r} R_i^2 - \frac{G^2}{rcn} = \frac{70.5^2 + 51.9^2}{12} - \frac{122.4^2}{24} = 14.415$$

$$\text{SS}_{\text{column}} = \text{SS}_{\text{drugs}} = \frac{1}{rn}\sum^{c} C_i^2 - \frac{G^2}{rcn}$$

$$= \frac{1}{8}(44.0^2 + 38.4^2 + 40.0^2) - \frac{122.4^2}{24} = 2.08$$

$$\text{SS}_{\text{within cells}} = M - \frac{1}{n}\sum^{rc} T_i^2$$

$$= 643.68 - \frac{1}{4}(24^2 + 21.7^2 + 24.8^2 + 20.0^2 + 16.7^2 + 15.2^2)$$

$$= .715$$

Finally, $\text{SS}_{\text{interaction}} = 19.440 - 14.415 - 2.08 - .715 = 2.230$.

The resulting ANOVA table is shown in Table 13. We have put $\text{SS}_{\text{columns}}$ before SS_{rows} to provide an easy contrast to the one-way ANOVA in Table 4 on page 480.

TABLE 13 Two way ANOVA table for the recovery time data

SOURCE OF VARIABILITY	SS	df	MS	F
Between drug means (columns)	2.080	2	1.04	$F_C = 26.196$
Between sex means (rows)	14.415	1	14.415	$F_R = 363.10$
Sex, drug interaction	2.230	2	1.115	$F_I = 28.09$
Within cells	.715	18	.0397	
Total	19.440	23		

The hypotheses of greatest interest here are $H_0: U_1 = U_2 = U_3$ and $H_1: U_1, U_2, U_3$ are not all equal, where U_1, U_2, and U_3 are the mean recovery times over all potential users of drugs I, II, and III, respectively. These were the hypotheses we tested in Example 4. Following the procedure in Table 11 and using Table A8, $F_{2, 18, .05} = 3.55$ and $F_{2.18, .01} = 6.01$. Since, from Table 13, $F_C = 26.196$ is far greater than 6.01, we reject H_0 at the .01 level of significance. The data strongly suggest significant differences among the mean recovery times for the three drugs.

We can also test two other pairs of hypotheses:

1. $H_0: \mu_1 = \mu_2$ against $H_1: \mu_1 \neq \mu_2$, where μ_1 and μ_2 are the potential mean recovery times for each sex. Following the procedure in Table 11 and using Table A8, $F_{1, 18, .05} = 4.41$ and $F_{1, 18, .01} = 8.29$. Since, from Table 13, $F_R = 363.1$ is far greater than 8.29, we reject H_0 at the .01 level of significance. The data strongly suggest a difference between the mean recovery times for men and women.

2. H_0: there is no interaction between sex and drugs; H_1: there is some significant interaction. Following the procedure in Table 11 and using Table A8, we have $F_{2,18,.05} = 3.55$ and $F_{2,18,.01} = 6.01$. Since from Table 13, $F_I = 28.09$ is much greater than 6.01, we reject H_0 at the .01 level of significance. The data suggest a very significant interaction between sex and these drugs.

There is a difference between the randomized block design above and the teaching example earlier. In the latter case the department was interested in discovering the effect of *both* (independent) variables, text and teaching method, on statistics scores. In the example above, the researcher is concerned primarily with the effect of only one of the independent variables, type of drug. Indeed, the researcher probably blocked by sex precisely because it was *known* that recovery times vary substantially by sex.

Blocking by sex eliminates one source of potential bias. It also, however, enables us to eliminate a great deal of unwanted variability in recovery times. To see this recall from Example 4 on page 480 that the F ratio of interest was $F = \text{MS}_{\text{drugs}}/\text{MS}_{\text{within}} = 1.04/.8267 = 1.258$, leading us to conclude that there were *no* significant differences among the recovery times. By contrast, in Table 13 the F value of interest was $F_C = \text{MS}_{\text{drugs}}/\text{MS}_{\text{within}} = 1.04/.0397 = 26.196$, leading to the conclusion that there *were* significant differences in the mean recovery times for the three drugs. Both of these conclusions were based on the same data. How can we account for the difference?

Quite simply, the difference is that $\text{MS}_{\text{within}}$ is only .0397 in the randomized block design compared to .8267 in the completely randomized design earlier. To understand this difference, notice that in Example 4 the variability in recovery time due to variables *other* than drugs, $\text{SS}_{\text{within}} = 17.360$, was so large as to outweigh or obscure the sum of squares between drug means, $\text{SS}_{\text{drugs}} = 2.08$. By contrast, in Table 13 the value 17.360 was broken down into SS_{sex} (14.415), $\text{SS}_{\text{interaction}}$ (2.230), and $\text{SS}_{\text{within}}$ (.715). The variability in recovery times due to sex alone and that due to the interaction of sex and drugs have been split off and their effects can therefore be removed and considered separately, as described above. What remains, $\text{SS}_{\text{within}} = .715$, is random or chance variability in recovery times due to variables *other* than drugs and sex. We can now more properly compare this with the variability due to drugs. As a consequence, $\text{MS}_{\text{within}}$ in Table 13 is only $.715/18 = .0397$, compared to $\text{MS}_{\text{within}} = 17.360/21 = .8267$ earlier. Eliminating the variability due to sex results in a more *precise* test of the significance of the difference in drug means.

PROBLEMS 13.4

29. (a) Construct the two-way ANOVA table for the following data:

		VARIABLE A			
		A_1		A_2	
VARIABLE B	B_1	−1	−5	−1	5
	B_2	0	2	6	10
	B_3	5	1	0	4

(b) Test for significant (i) differences between the means for variable B; (ii) differences between the means for variable A; (iii) interaction between A and B.

30. Refer to the data in Problem 29.

    (a) Compute the mean score for each of the six cells.

    (b) For each level of variable B, plot the means for each level of variable A. Connect the points.

    (c) Do these variables appear to interact? Explain.

31. Does the season of the year make any difference to the mileage a car is capable of getting? A consumer organization undertook to answer this question. During winter, spring, summer, and fall six cars of a particular model were selected, three of which were manual and three automatic transmissions. In each case the car was driven over an identical course and the mileage recorded. The following results were obtained:

	WINTER	SPRING	SUMMER	FALL
MANUAL	20.9 21.3 21.7	22.4 22.9 22.2	22.0 22.3 21.7	23.6 23.0 23.0
AUTOMATIC	19.5 20.3 20.8	22.0 21.5 21.3	21.4 20.7 20.9	21.9 22.6 21.5

Perform a two-way ANOVA for these data, testing for significant (a) differences between seasons; (b) differences between types of transmissions; (c) interaction.

32. Refer to Problem 31. Obtain the mean mileage for each of the six cells. For each type of transmission plot the season means and connect the four points.

    (a) On average, in which season is mileage the highest? The lowest?

    (b) Does there seem to be any interaction between season and type of transmission?

    (c) Is your answer in part (b) consistent with that in part (c) of Problem 31?

33. A company plans to test a new training program for its trainees. Eighteen men and 10 women trainees participate in an experiment. The men and the women are each randomly divided into equal-sized groups of size 9 and 5, respectively. In each case one of the groups (A) receives the traditional training program, whereas the other (B) receives the new training program. At the end of the programs all 28 participants are given an objective test. The mean scores for the test are as follows:

	PROGRAM A	PROGRAM B	MEAN
MEN	76.0	67.0	71.5
WOMEN	63.0	81.0	72.0
MEAN	71.4	72.0	

(a) For both men and women plot the mean scores for the two programs and connect the points.

(b) Does there appear to be any interaction between sex and type of program? If so, explain its nature.

34. Thirty pilots take part in an experiment to compare three different control panel designs, A, B, and C, on a flight simulator. Six mock-ups were built so that each control panel could be tried on two types of aircraft (I and II).

The 30 pilots are randomly divided into six groups of five and then each group is assigned to a panel/aircraft combination. Each pilot is asked to react to a number of simulated incidents. The following data are the reaction times (in seconds) to respond fully to a simulated stall:

		PANEL	
	A	*B*	*C*
AIRCRAFT I	7.3 7.9 8.2	9.5 8.9 10.0	7.7 8.3 8.0
	6.9 8.7	8.7 8.9	7.6 8.4
TYPE II	8.4 8.7 8.1	9.5 9.6 8.9	8.7 8.7 8.7
	9.1 7.7	8.7 9.8	8.0 7.9

(a) Construct the two-way ANOVA table for these data.

(b) Test for significant (i) differences between aircraft types; (ii) differences between panels; (iii) interaction between these two variables.

35. Refer to Problem 34.

(a) Compute the mean reaction time for each cell. For each aircraft type, plot the panel means and connect the points.

(b) On average, which panel has the smallest mean reaction time?

(c) Does there appear to be any interaction between aircraft type and panel design?

(d) Is your answer in part (c) consistent with your answer to part (b, iii) in Problem 34?

36. An agronomist has 18 plots of ground available for an experiment. Nine of the plots are irrigated and the others are not. Each set of nine plots forms a block on which three fertilizers, A, B, and C, are to be tested. The following data are the yield of soybeans (in bushels per acre) for each plot:

	FERTILIZER *A*			FERTILIZER *B*			FERTILIZER *C*		
WITH IRRIGATION	21.9	22.8	22.7	24.2	24.2	23.4	20.6	21.0	21.3
WITHOUT IRRIGATION	19.6	19.0	20.2	20.4	21.6	20.9	21.6	20.4	21.0

(a) Ignoring whether the plot is irrigated or not, perform a one-way ANOVA with six observations per group.

(b) Perform a two-way ANOVA that takes into account whether or not the plot is irrigated.

(c) Does your decision about differences between fertilizers in part (b) differ from that in part (a)? Explain.

37. Refer to the data in Problem 36. Determine graphically whether there appears to be any interaction between the two variables.

Solution to Problem 29

(a) In this case $r = 3$, $c = 2$ and $n = 2$

$$M = (-1)^2 + (-5)^2 + 0^2 + \cdots + 4^2 = 234$$

$$C_1 = 2,\ C_2 = 24$$

$$R_1 = -2,\ R_2 = 18,\ R_3 = 10$$

$$T_1 = -6,\ T_2 = 4,\ T_3 = 2,\ T_4 = 16,\ T_5 = 6,\ T_6 = 4, \text{ and } G = 26$$

From Equations 13.6 to 13.9 we obtain

$$SS_{total} = 234 - 26^2/12 = 177.667$$

$$SS_{rows} = \frac{(-2)^2 + 18^2 + 10^2}{4} - \frac{26^2}{12} = 50.667$$

$$SS_{columns} = \frac{2^2 + 24^2}{6} - \frac{26^2}{12} = 40.333$$

$$SS_{within} = 234 - \frac{(-6)^2 + 4^2 + 2^2 + 16^2 + 6^2 + 4^2}{2} = 52$$

Finally,

$$SS_{interaction} = 177.667 - 50.667 - 40.333 - 52 = 34.667$$

The ANOVA table is then

SOURCE OF VARIABILITY	SS	df	MS	*F*
Between rows (levels of *B*)	50.667	2	25.333	2.92
Between columns (levels of *A*)	40.333	1	40.333	4.65
A, *B* Interaction	34.667	2	17.333	2.00
Within cells	52.000	6	8.667	
Total	177.667	11		

(b) (i) $H_0: \mu_{B_1} = \mu_{B_2} = \mu_{B_3}$ $H_1: \mu_{B_1}, \mu_{B_2}, \mu_{B_3}$ are not all equal
From Table A8 $F_{2,6,.05} = 5.14$ and $F_{2,6,.01} = 10.9$. Since $F = 2.92 < 5.14$ we do not reject H_0 at the .05 level of significance. The data suggest that the means for variable *B* are not significantly different.
(ii) $H_0: U_{A_1} = U_{A_2}$ $H_1: U_{A_1} \neq U_{A_2}$
From Table A8 $F_{1,6,.05} = 5.99$ and $F_{1,6,.01} = 13.7$. Since $F = 4.65 < 5.99$ we do not reject H_0 at the .05 level of significance. The data suggest that the means for variable *A* are not significantly different.
(iii) H_0: there is no interaction between *A* and *B*
H_1: there is significant interaction between *A* and *B*
From Table A8 $F_{2,6,.05} = 5.14$ and $F_{2,6,.01} = 10.9$. Since $F = 2.00 < 5.14$ we do not reject H_0 at the .05 level of significance. The data suggest no significant interaction between *A* and *B*.

SECTION 13.5

USING COMPUTER OUTPUT: ANALYSIS OF VARIANCE WITH MINITAB

ONE-WAY ANOVA

There are a number of ways to request a one-way analysis of variance in Minitab depending on the format of the data. We illustrate one of the techniques below.

EXAMPLE 9 A company purchases 16 trucks at the same time, five from one manufacturer, four from another, and seven from a third. All the trucks are to be used in roughly the same way and to the same extent. The costs (in dollars) of maintaining the 16 trucks for two years are as follows:

A	B	C
2031	2074	2483
2224	1943	2309
2505	2059	2531
2196	2184	2444
2104		2319
		2476
		2229

Regarding these three groups of trucks as random samples of the respective manufacturers' trucks, do they suggest significant differences among the mean costs for the three populations of trucks?

The null and alternative hypotheses in this case are $H_0: \mu_1 = \mu_2 = \mu_3$ and $H_1: \mu_1, \mu_2, \mu_3$ are *not* all equal, where μ_1, μ_2, and μ_3 are the mean two-year maintainance costs for each make of truck.

The three groups of data are entered into columns C1, C2, and C4, respectively, using the "SET DATA. . ." command. (The author entering these data meant to enter the third set of data into C3 but hit the "4" button by mistake.) The printout is shown in Table 14. A complete analysis of variance table is obtained with the command "AOVONEWAY ON DATA IN C1, C2, C4."

The first part of the subsequent output consists of the ANOVA table. We still have to check how unusually large the value $F = 8.22$ is. From Table A8 with $\nu_1 = 2$ and $\nu_2 = 13$, $F_{2,13,.05} = 3.81$ and $F_{2,13,.01} = 6.70$. Since 8.22 is greater than 6.70, we reject H_0 at the .01 level of significance. The data suggest significant differences among the three mean costs.

The remaining output, shown in Table 14, is useful when, as in this case, the population means are found to differ significantly. Minitab prints out the means and standard deviations for each sample, but more usefully, it computes and illustrates a 95% confidence interval for each population mean. These intervals are based on Definition 1 on page 486. The fact that the confidence interval for μ_3 barely overlaps that for μ_1 and does not overlap at all with that

```
MTB > SET DATA INTO C1
DATA> 2031,2224,2505,2196,2104
DATA> SET DATA INTO C2
DATA> 2074,1943,2059,2184
DATA> SET DATA INTO C4
DATA> 2483,2309,2531,2444,2319,2476,2229
DATA> AOVONEWAY ON DATA IN C1,C2,C4
```

```
ANALYSIS OF VARIANCE
SOURCE      DF        SS         MS         F
FACTOR       2    298162     149081      8.22
ERROR       13    235669      18128
TOTAL       15    533831
                                     INDIVIDUAL 95 PCT CI'S FOR MEAN
                                     BASED ON POOLED STDEV
LEVEL        N      MEAN      STDEV  -+---------+---------+---------+-----
C1           5      2212        181            (-------*-------)
C2           4      2065         99   (--------*--------)
C4           7      2399        112                        (------*------)
                                     -+---------+---------+---------+-----
POOLED STDEV =     135             1920      2080      2240      2400
```

TABLE 14

Analysis of variance for the "truck" data

for μ_2 suggests that μ_3 is substantially greater than the other two means. This explains the significant differences among the three means.

The output in Table 14 also includes the value for the pooled standard deviation, $s_* = \sqrt{\text{MS}_{\text{error}}} = 135$ $(=\sqrt{18{,}128})$.

■

In the next example we illustrate an analysis of variance using the nutrition data in Table 8 of Chapter 1 (page 48).

EXAMPLE 10 It was of interest to determine whether the mean times spent with the nutritionist differed for the three ethnic groups. The null and alternative hypotheses are $H_0: \mu_1 = \mu_2 = \mu_3$ and $H_1: \mu_1, \mu_2, \mu_3$ are not all equal, where μ_1, μ_2, and μ_3 are the potential mean times that low-income white, black, and Hispanic women, respectively, would spend with the nutritionist.

In this case the required times for the three groups are not in separate columns as they were in Example 9, but are all in C9. The levels of the variable "ethnic group" are indicated by the values 1, 2, or 3 in C1. The command that is required here is "ONEWAY, DATA IN C10, LEVELS IN C1." This command and the resulting output are shown in Table 15.

```
MTB > ONEWAY, DATA IN C10, LEVELS IN C1
```

```
ANALYSIS OF VARIANCE ON C10
SOURCE      DF        SS        MS        F
C1           2       442       221     1.30
ERROR       65     11010       169
TOTAL       67     11452
                                    INDIVIDUAL 95 PCT CI'S FOR MEAN
                                    BASED ON POOLED STDEV
LEVEL        N      MEAN     STDEV  ---------+---------+---------+-------
    1       31     32.52     16.24                  (--------*--------)
    2       25     27.04      9.04    (---------*---------)
    3       12     28.42     10.17  (     ----------*--------------)
                                    ---------+---------+---------+-------
POOLED STDEV = 13.01                     25.0      30.0      35.0
```

TABLE 15

Analysis of variance of the "times with nutritionist" data

To check the significance of the value $F = 1.30$, we note from Table A8 that $F_{2,60,.05} = 3.15$ and $F_{2,60,.01} = 4.98$. Since 1.30 is less than 3.15, we cannot reject H_0 at the .05 level of significance. The data suggest no significant differences among the mean times for the three ethnic groups. The confidence intervals shown below indicate that the mean for white women (C1 = 1) may be somewhat larger than that for black women (C1 = 2) and Hispanic women (C1 = 3) but from the F test we know that the differences are not statistically significant.

■

Minitab can also perform a two-way analysis of variance, but the data should be in a format close to that described in Example 10. We illustrate the required format and the appropriate commands in the next example.

EXAMPLE 11 Thirty-six people participate in a study of depth perception under different lighting conditions. The subjects were first divided into three "blocks" on the basis of age: young (1), middle-aged (2), and older (3). Each of the 12 members of a block were randomly assigned to one of three treatment groups, A, B, or C. All 36 subjects were asked to judge how far they were from a number of different points. An average "error" in judgment (in feet) was recorded for each subject. Group A were shown the points in bright sunshine, group B under cloudy conditions, and group C at twilight. The 36 average "errors" are as follows:

	A (1)	B (2)	C (3)
YOUNG (1)	5.2	7.1	8.9
	4.3	6.4	10.4
	5.8	7.9	8.4
	6.0	5.6	7.9
MIDDLE AGED (2)	5.2	7.0	9.5
	4.2	7.1	9.0
	6.0	8.2	7.6
	5.4	6.0	10.7
OLD (3)	6.2	8.0	10.6
	7.1	6.7	12.1
	6.7	9.1	13.1
	6.5	9.2	12.4

These data have to be entered as follows: The actual error values (5.2, 4.3, . . . , 12.4) go into one column (C1 in Table 16). The corresponding level of the row variable, age (1, 2, or 3), is entered into a second column (C2 in Table 16) and the corresponding level of the column variable, light, into a third column (C3 in Table 16). The 36 triple sets of numbers are shown entered in Table 16.

A two-way analysis of variance table is obtained with the command "TWO-WAY ANALYSIS DATA IN C1, FIRST VARIABLE IN C2, SECOND IN C3." The resulting table is shown at the foot of Table 16. The table does not include the F values. These have been entered in the table by hand. For example,

$$F_R = \frac{\text{MS}_{\text{rows}}}{\text{MS}_{\text{error}}} = \frac{14.523}{.927} = 15.67$$

From Table A8 with $\nu_1 = 2$ and $\nu_2 = 24$, $F_{2,24,.05} = 3.40$ and $F_{2,24,.01} = 5.61$. Since both 15.67 and 61.96 are significantly greater than 5.61, we conclude that there are significant differences among the mean "errors" for (1) the different age groups and (2) the different light groups. From Table A8 with $\nu_1 = 4$ and $\nu_2 = 24$, $F_{4,24,.05} = 2.78$ and $F_{4,24,.01} = 4.22$. Since $F_I = 1.37$ is below 2.78, we conclude that there is no significant interaction between the "light" groups and the age groups.

■

TABLE 16

Entering the data and obtaining the two-way analysis of variance

```
MTB > READ DATA INTO C1-C3
DATA> 5.2,1,1
DATA> 4.3,1,1
DATA> 5.8,1,1
DATA> 6.0,1,1
DATA> 5.2,2,1
DATA> 4.2,2,1
DATA> 6.0,2,1
```

TABLE 16

(*cont.*)

```
DATA> 5.4,2,1
DATA> 6.2,3,1
DATA> 7.1,3,1
DATA> 6.7,3,1
DATA> 6.5,3,1
DATA> 7.1,1,2
DATA> 6.4,1,2
DATA> 7.9,1,2
DATA> 5.6,1,2
DATA> 7.0,2,2
DATA> 7.1,2,2
DATA> 8.2,2,2
DATA> 6.0,2,2
DATA> 8.0,3,2
DATA> 6.7,3,2
DATA> 9.1,3,2
DATA> 9.2,3,2
DATA> 8.9,1,3
DATA> 10.4,1,3
DATA> 8.4,1,3
DATA> 7.9,1,3
DATA> 9.5,2,3
DATA> 9.0,2,3
DATA> 7.6,2,3
DATA> 10.7,2,3
DATA> 10.6,3,3
DATA> 12.1,3,3
DATA> 13.1,3,3
DATA> 12.4,3,3
DATA> TWOWAY ANALYSIS DATA IN C1 FIRST VARIABLE IN C2, SECOND IN C3
```

```
    36 ROWS READ

ANALYSIS OF VARIANCE ON C1
```

SOURCE	DF	SS	MS	F
C2	2	29.047	14.523	15.67
C3	2	114.872	57.436	61.96
INTERACTION	4	5.097	1.274	1.37
ERROR	27	25.032	0.927	
TOTAL	35	174.048		

PROBLEMS 13.5

38. The accompanying printout is a one-way ANOVA based on the nutrition data. In this case the dependent variable is the mother's weight gain (C10), and the independent variable is the mother's smoking history (1 = nonsmoker, 2 = light smoker, and 3 = heavy smoker). Complete the F test and interpret the result.

```
MTB > ONEWAY, DATA IN C10, LEVELS IN C3

ANALYSIS OF VARIANCE ON C10
SOURCE      DF        SS         MS         F
C3           2       514        257      1.53
ERROR       65     10938        168
T OTAL      67     11452
                                      INDIVIDUAL 95 PCT CI'S FOR MEAN
                                      BASED ON POOLED STDEV
LEVEL        N      MEAN      STDEV   ----+---------+---------+---------+--
    1       49     28.61      12.16          (----*----)
    2        9     28.78       7.51   (----------*-----------)
    3       10     36.40      19.41              (----------*---------)
                                      ----+---------+---------+---------+--
POOLED STDEV = 12.97                      22.5      30.0      37.5      45.0
```

39. A researcher in economics interviewed professors in the United States, Great Britain, and West Germany and recorded for each one the percentage of last year's income that was saved. The figures for American professors were entered into C1 and those for the British and German professors into C2 and C3, respectively. The accompanying printout contains the ANOVA table for these data.

(a) Complete the F test and interpret the result.

(b) Use the data in the printout to find a 90% confidence interval for the mean percentage of salary saved by all American professors.

```
MTB > SET DATA INTO C1
DATA> 6.2,9.4,8.1,7.5,3.2,1.9,4.4
DATA> SET DATA INTO C2
DATA> 8.3,13.3,6.4,10.5,4.7
DATA> SET DATA INTO C3
DATA> 11.7,10.0,8.3,11.5,12.4,8.3
DATA> AOVONEWAY ON DATA IN C1,C2,C3

ANALYSIS OF VARIANCE
SOURCE      DF        SS         MS         F
FACTOR       2     68.85      34.43      4.83
ERROR       15    106.93       7.13
TOTAL       17    175.79
```

```
                                               INDIVIDUAL 95 PCT CI'S FOR MEAN
                                               BASED ON POOLED STDEV
LEVEL       N      MEAN      STDEV    --------+---------+---------+--------
C1          7     5.814      2.746    (------*-------)
C2          5     8.640      3.385            (--------*-------)
C3          6    10.367      1.782                   (-------*------)
                                      --------+---------+---------+--------
POOLED STDEV = 2.670                         6.0       9.0      12.0
```

40. A sample of 12 full-term and 12 premature infants were given a neurological rating that can vary from 0 to 20. The 24 scores are broken down further by sex as follows:

	MALE						FEMALE					
FULL-TERM	17	10	13	14	13	11	12	19	14	12	15	17
PREMATURE	7	11	14	8	6	10	12	8	8	7	7	10

These data were entered into Minitab by using C1 for the scores, C2 for the male (1)/female (2) variable, and C3 for the full-term (1)/premature (0) variable. The accompanying printout contains (i) a one-way ANOVA with sex as the independent variable, (ii) a one-way ANOVA with full-term/premature as the independent variable, and (iii) a two-way ANOVA.

```
DATA> ONEWAY, DATA IN C1, LEVELS IN C2
    24 ROWS READ

ANALYSIS OF VARIANCE ON C1
SOURCE      DF        SS        MS         F
C2           1       2.0       2.0      0.16
ERROR       22     285.9      13.0
TOTAL       23     288.0
                                       INDIVIDUAL 95 PCT CI'S FOR MEAN
                                       BASED ON POOLED STDEV
LEVEL        N      MEAN     STDEV   ------+---------+---------+---------+
    1       12     11.17      3.21   (---------------*--------------)
    2       12     11.75      3.96       (--------------*--------------)
                                     ------+---------+---------+---------+
POOLED STDEV =  3.61                     9.8      11.2      12.6      14.0
MTB > ONEWAY, DATA IN C1, LEVELS IN C3
```

(a) Complete all the appropriate F tests.

(b) Using the original data, compute the four cell means and graphically check for interaction.

```
ANALYSIS OF VARIANCE ON C1
SOURCE      DF        SS        MS         F
C3           1    145.04    145.04     22.33
ERROR       22    142.92      6.50
TOTAL       23    287.96
                                     INDIVIDUAL 95 PCT CI'S FOR MEAN
                                     BASED ON POOLED STDEV
LEVEL        N      MEAN     STDEV   ---------+---------+---------+-------
    0       12     9.000     2.412   (-----*------)
    1       12    13.917     2.678                       (-----*-----)
                                     ---------+---------+---------+-------
POOLED STDEV = 2.549                          9.6      12.0      14.4
MTB > TWOWAY ANALYSIS DATA IN C1, FIRST VARIABLE IN C2, SECOND IN C3

ANALYSIS OF VARIANCE ON C1

SOURCE          DF        SS        MS
C2               1      2.04      2.04
C3               1    145.04    145.04
INTERACTION      1      9.38      9.38
ERROR           20    131.50      6.57
TOTAL           23    287.96
```

(c) Is your answer in part (b) consistent with the test for interaction in the two-way ANOVA?

(d) Find a 95% confidence interval for the difference in mean scores between full-term and premature infants.

(e) What assumptions were necessary in this problem?

SECTION 13.6
SUMMARY AND REVIEW

The **one-way analysis of variance** is a procedure used to test for significant differences among three or more population means.

We assume independently selected samples from r populations, the respective sample sizes being $n_1, n_2, \ldots, n_r$. The total number of values are $N = n_1 + n_2 + \ldots + n_r$. We call $\bar{x}_1, \bar{x}_2, \ldots, \bar{x}_r$ the sample means and $\bar{x}.$, the mean over all N values (or grand mean). The sample variances are $s_1^2, s_2^2, \ldots, s_r^2$.

The total variability among all N values can be broken down into (1) the variability within sample means and (2) the variability between sample means.

More formally, we can write

$$\sum^{N} (x_i - \bar{x}.)^2 = \sum^{r} n_i(\bar{x}_i - \bar{x}.)^2 + \sum^{r} (n_i - 1)s_i^2 \qquad \text{(Eq. 13.1)}$$

$$\text{SS}_{\text{total}} \qquad \text{SS}_{\text{between}} \qquad \text{SS}_{\text{within}}$$

Each of these terms is called a **sum of squares.** Associated with any sum of squares is a number of **degrees of freedom** (df). The df associated with these three sums of squares are $N - 1$ for SS_{total}, $r - 1$ for $\text{SS}_{\text{between}}$, and $N - r$ for $\text{SS}_{\text{within}}$.

A sum of squares divided by its corresponding df results in a variance-like quantity referred to as a **mean square,** denoted MS. The values for each SS and the corresponding MS and df are displayed in an analysis of variance table, or ANOVA table for short. The general format is as follows:

SOURCE OF VARIABILITY	SS	df	MS	F
Between sample means	$\text{SS}_{\text{between}}$	$r - 1$	$\text{MS}_{\text{between}}$	$F = \dfrac{\text{MS}_{\text{between}}}{\text{MS}_{\text{within}}}$
Within samples	$\text{SS}_{\text{within}}$	$N - r$	$\text{MS}_{\text{within}}$	
Total	SS_{total} $N-1$			

A computationally efficient method for computing these three sums of squares is provided by the following equations:

$$\text{SS}_{\text{between}} = \sum^{r} \frac{T_i^2}{n_i} - \frac{T^2}{N} \qquad \text{(Eq. 13.2)}$$

$$\text{SS}_{\text{within}} = \sum^{N} x_i^2 - \sum^{r} \frac{T_i^2}{n_i} \qquad \text{(Eq. 13.3)}$$

$$\text{SS}_{\text{total}} = \sum^{N} x_i^2 - \frac{T^2}{N} \qquad \text{(Eq. 13.4)}$$

where $T_1, T_2, \ldots, T_r$ are the sample totals and $T = \sum^{r} T_i$.
In practice, $\text{SS}_{\text{between}}$ is obtained as $\text{SS}_{\text{total}} - \text{SS}_{\text{within}}$.

The F test assumes that the r samples were drawn from r normal populations with unknown means $\mu_1, \mu_2, \ldots, \mu_r$ and common variance, σ^2. The null and alternative hypotheses are: $H_0: \mu_1 = \mu_2 = \cdots = \mu_r$ and H_1: these means are *not* all equal. The test statistic is $F = \text{MS}_{\text{between}}/\text{MS}_{\text{within}}$, which, if H_0 is true, has the $F_{r-1,N-r}$ distribution. The procedure for testing H_0 is given in Table 5.

ESTIMATING μ_i AND $\mu_i - \mu_j$

Provided that all assumptions underlying the F test are met, a $(100\gamma)\%$ confidence interval for μ_i is

$$\bar{x}_i - t_{N-r(\gamma)} \frac{s_*}{\sqrt{n_i}} \quad \text{to} \quad \bar{x}_i + t_{N-r(\gamma)} \frac{s_*}{\sqrt{n_i}} \qquad \text{(Def. 1)}$$

With these same assumptions, a $(100\gamma)\%$ confidence interval for $\mu_i - \mu_j$ is

$$\bar{x}_i - \bar{x}_j - t_{N-r(\gamma)} s_* \sqrt{\frac{1}{n_i} + \frac{1}{n_j}} \quad \text{to} \quad \bar{x}_i - \bar{x}_j + t_{N-r(\gamma)} s_* \sqrt{\frac{1}{n_i} + \frac{1}{n_j}}$$

(Def. 2)

In both definitions, s_* is $\sqrt{\text{MS}_{\text{within}}}$.

RELATIONSHIP BETWEEN F AND t

We pointed out in Section 13.3 that when there are only $r = 2$ groups or samples, the F test of $H_0: \mu_1 = \mu_2$ is equivalent to the two-sided t test of this same null hypothesis, outlined in Section 12.4. In fact, it can be shown that with $r - 1 = 2 - 1 = 1$ and $N - r = n_1 + n_2 - 2$, $F_{1, n_1+n_2-2} = t^2_{n_1+n_2-2}$.

Although the two tests are equivalent, the F test cannot be used with $r = 2$ groups when a *one-sided test* is appropriate. In this case the t test should be used.

TWO-WAY ANALYSIS OF VARIANCE

In a **two-way analysis of variance,** the data are classified on the basis of two, generally qualitative independent variables. The row variable has r levels and the column variable c levels. There will thus be rc experimental combinations (cells). The number of values in each cell is n. There will thus be rcn values in all.

We use the following notation: (1) M is the sum of the rcn squared values; (2) $C_1, C_2, \ldots, C_c$ are the column totals; (3) $R_1, R_2, \ldots, R_r$ are the row totals; (4) $T_1, T_2, \ldots, T_{rc}$ are the cell totals; and (5) G is the grand total.

The total variability in the rcn values around the grand mean (G/rcn) can be broken down into four components:

$$\text{SS}_{\text{total}} = \text{SS}_{\text{rows}} + \text{SS}_{\text{columns}} + \text{SS}_{\text{interaction}} + \text{SS}_{\text{within}} \qquad \text{(Eq. 13.5)}$$

These sums of squares can be computed using the formulas

$$\text{SS}_{\text{total}} = M - \frac{G^2}{rcn} \qquad \text{(Eq. 13.6)}$$

$$\text{SS}_{\text{rows}} = \frac{R_1^2 + R_2^2 + \cdots + R_r^2}{cn} - \frac{G^2}{rcn} \qquad \text{(Eq. 13.7)}$$

$$\text{SS}_{\text{columns}} = \frac{C_1^2 + C_2^2 + \cdots + C_c^2}{rn} - \frac{G^2}{rcn} \qquad \text{(Eq. 13.8)}$$

$$\text{SS}_{\text{within}} = M - \frac{T_1^2 + T_2^2 + \cdots + T_{rn}^2}{n} \qquad \text{(Eq. 13.9)}$$

and finally,

$$\text{SS}_{\text{interaction}} = \text{SS}_{\text{total}} - \text{SS}_{\text{rows}} - \text{SS}_{\text{columns}} - \text{SS}_{\text{within}} \qquad \text{(Eq. 13.10)}$$

The general format for a two-way analysis of variance table is as follows:

SOURCE OF VARIABILITY	SS	df	MS	F
Between row means	SS_{rows}	$r - 1$	$MS_{rows} = \dfrac{SS_{rows}}{r - 1}$	$F_R = \dfrac{MS_{rows}}{MS_{within}}$
Between column means	$SS_{columns}$	$c - 1$	$MS_{columns} = \dfrac{SS_{columns}}{c - 1}$	$F_C = \dfrac{MS_{columns}}{MS_{within}}$
Interaction	$SS_{interaction}$	$(r - 1)(c - 1)$	$MS_{interaction} = \dfrac{SS_{interaction}}{(r - 1)(c - 1)}$	$F_I = \dfrac{MS_{interaction}}{MS_{within}}$
Within cells	SS_{within}	$rc(n - 1)$	$MS_{within} = \dfrac{SS_{within}}{rc(n - 1)}$	
Total	SS_{total}	$rcn - 1$		

There are three pairs of hypotheses that we wish to test. They are:

1. $H_0: \mu_1 = \mu_2 = \cdots = \mu_r$; H_1: $\mu_1, \mu_2, \ldots, \mu_r$ are *not* all equal
2. $H_0: U_1 = U_2 = \cdots = U_c$; H_1: $U_1, U_2, \ldots, U_c$ are *not* all equal
3. H_0: there is no interaction between the two independent variables; H_1: there is some interaction

Here $\mu_1, \mu_2, \ldots, \mu_r$ are the true means for the r levels of the row variable, and $U_1, U_2, \ldots, U_c$ are the true means for the c levels of the column variable.

To test these hypotheses, we assume that the values in each cell are a random sample from a normal population and that all rc populations have the same standard deviation (Definition 4).

The procedures for performing an F test of these three hypotheses are set out in Table 11.

A major advantage of incorporating both independent variables into a single experiment is that we can investigate the interaction between the two variables. Formally: Two independent variables A and B are said to interact if the effect of A on the dependent variable is different for different levels of B (and vice versa) (Definition 3).

We have indicated how to test for significant interaction. Greater insight into the extent to which the two variables interact can be gained by compiling and displaying a table of the cell, row, and column means (see Tables 8 and 9 and Figure 6).

The two-way analysis of variance and the associated F tests are the appropriate tools for analyzing a randomized block design. In this case the blocking variable will be one of the independent variables. We introduced the randomized block design in Chapter 8 as a device for reducing bias. In Section 13.4 we indicated another advantage of the design is that it can eliminate an important source of variability in the dependent variable. This generally results in a far more sensitive F test of the effect of the independent variable of interest than would otherwise be possible.

REVIEW PROBLEMS

GENERAL

1. In which of the following situations is $SS_{Within} = 0$? Explain. In which is $SS_{between} = 0$? Explain.

	(a) I	(a) II	(a) III		(b) I	(b) II	(b) III		(c) I	(c) II	(c) III
	7.3	8.1	9.2		6.6	8.1	7.6		8.3	8.6	8.4
	8.2	9.4	8.4		9.4	7.7	9.0		8.3	8.6	8.4
	9.4	10.1	6.9		9.3	9.4	8.0		8.3	8.6	8.4
	8.3	6.8	9.1		8.3	8.4	9.0		8.3	8.6	8.4
$\bar{X}$	8.3	8.6	8.4	$\bar{X}$	8.4	8.4	8.4	$\bar{X}$	8.3	8.6	8.4

2. Construct the ANOVA table for the following data:

A	6	14	9	12
B	11	8	16	14
C	13	8	11	5

3. An experiment results in 12 values, each of which is classified on the basis of two variables. Construct the ANOVA table for these data:

		VARIABLE 1 A	VARIABLE 1 B	VARIABLE 1 C
VARIABLE 2	I	3, 4	7, 4	5, 4
	II	4, 4	6, 3	1, 2

4. Complete the following ANOVA table and test for significant differences among the three population means:

SOURCE OF VARIABILITY	SS	df	MS	F
Between group means	23			
Within groups				
Total	71	16		

5. Complete the following ANOVA table and test for (a) significant difference among variable A means; (b) significant differences among variable B means; (c) significant interaction between the two variables.

SOURCE OF VARIABILITY	SS	df	MS	F
Between levels of variable A	1104	4		
Between levels of variable B	817	4		
Interaction				
Within cells	3917	60		
Total	6455			

6. Early applications of analysis of variance were in agriculture. Here is an example typical of the early applications.

A large field was divided into 24 identical plots. Each of four fertilizers, I, II, III, and IV, were used on six randomly selected plots. The yields per plot (in bushels per acre of corn) were as follows:

							T_i	Σx_i^2
I	83	98	89	95	102	80	547	50,243
II	85	84	88	72	81	71	481	38,811
III	70	59	56	70	73	62	390	25,590
IV	52	90	69	81	76	64	432	31,998

(a) Construct an ANOVA table for these data.

(b) Perform the appropriate F test and explain the conclusion.

(c) Prior to performing the experiment there was special interest in comparing the mean yield for fertilizer I with both fertilizers III and IV. Find a 90% confidence interval for (i) $\mu_I - \mu_{III}$; (ii) $\mu_I - \mu_{IV}$.

(d) What assumptions were necessary in parts (a)–(c)?

7. A baseball team is interested in comparing the average speed of its four starting pitchers. The speed of the first 20 pitches made by each pitcher in successive games are recorded and the following summary data obtained:

PITCHER	ASKEY	RAY	HOWARD	HILL
Σx_i	1768	1854	1802	1876
Σx_i^2	156,409.64	171,980.76	163,046.20	176,337.39

(a) Construct the appropriate ANOVA

(b) Test whether the mean speeds for each pitcher are significantly different.

(c) What assumptions did you make about the data?

HEALTH SCIENCES

8. Twenty-four mice are inoculated with one of three strains of typhoid. The number of days until death for each mouse are given below:

STRAIN *A*			STRAIN *B*			STRAIN *C*			
5.9	6.7	3.7	7.9	9.4	10.7	5.7	6.9	8.5	
8.8	10.2	7.9	6.4	5.3	8.3	12.5	10.7	12.2	
			7.2	8.9		5.1	9.4	7.3	10.9

(a) Do these data suggest any significant differences in the mean survival time for the three strains?

(b) What assumptions did you make?

9. Zelazo et al. (1972) report on an experiment to determine the effect of special walking exercises on the age at which children begin to walk. Tweny-three infants were randomly divided into four groups. Those infants in groups *A* and *B* received various exercises, whereas those in groups *C* and *D* did not. The following data are the ages (in months) at which each of the 23 children first walked.

A	*B*	*C*	*D*
9.00	11.00	11.50	13.25
9.50	10.00	12.00	11.50
9.75	10.00	9.00	12.00
10.00	11.75	11.50	13.50
13.00	10.50	13.25	11.50
9.50	15.00	13.00	

Do these data suggest any significant differences in the mean age at walking for the four groups?

10. A clinic randomly assigns 24 patients suffering from blisters to receive one of three treatments, one of which is a placebo. The number of days for the blisters to completely heal are as follows:

PLACEBO	TREATMENT *A*	TREATMENT *B*
12	8	10
10	6	8
8	7	7
7	9	11
9	9	9
11	8	10
13	5	9
10	10	7

(a) Do these data suggest that the mean recovery times for the three treatments are significantly different?

(b) Find a 90% confidence interval for the difference between the mean number of days for healing for treatments *A* and *B*.

(c) What assumptions did you make in parts (a) and (b)?

SOCIOLOGY AND GOVERNMENT

11. A demographer is interested in the relationship between the birth of a family's first and second child and the eventual family size. She follows 29 families for 20 years, collecting the following data:

	NUMBER OF CHILDREN IN FAMILY					
	2	3	4	5	6	7
INTERVAL IN MONTHS BETWEEN FIRST AND SECOND CHILD	34	24	31	14	18	16
	22	18	19	24	14	
	18	19	24	20	10	
	49	23	21		24	
	39	24				

(a) Test whether the mean interval between first and second child differs significantly for the various family sizes.

(b) What assumptions were necessary in the analysis? (These data were given to the authors by Richard Harris.)

12. A random sample of 17 members of the New England Association of Psychiatric Social Workers is selected. The following figures are the caseloads for each worker arranged by state:

MASSACHUSETTS	CONNECTICUT	RHODE ISLAND	NEW HAMPSHIRE
25	30	18	41
29	21	16	37
31	22	19	
24	18		
28	27		
	24		
	28		

(a) Construct an ANOVA table for these data.

(b) Test for significant differences among the mean caseloads for the four states

(c) What assumptions were necessary here?

13. Refer to the data in Review Problem 12.

(a) Find an 80% confidence interval for the mean caseload in Massachusets.

(b) Find an 80% confidence interval for the difference between the mean caseloads for Massachusetts and Connecticut.

ECONOMICS AND MANAGEMENT

14. Allied Mills, a producer of breakfast cereals, is planning to advertise its least popular cereal in supermarkets. Three different advertisements under consideration will be tested in a sample of 12 supermarkets. Each advertisement is randomly assigned to appear in three supermarkets. The remaining three supermarkets will carry no advertising. After one week the increase in sales over the previous week is recorded for all 12 supermarkets:

NO ADVERTISING	ADVERTISEMENT *A*	ADVERTISEMENT *B*	ADVERTISEMENT *C*
14	14	28	19
−8	12	19	16
5	9	12	13

(a) Why did the company include in the experiment three supermarkets with no advertisements?

(b) Construct the ANOVA table for these data.

(c) Test for significant differences among the four advertising groups.

15. A movie company randomly divides 30 men and 30 women into two groups, *A* and *B*, each group consisting of 15 of each sex. Group *A* is shown a movie with a happy ending. Group *B* is shown the same movie except with a sad (but more realistic) ending. After the movie each person is asked to fill out a questionnaire, the answers to which are given a score. The more they enjoyed the movie, the higher the score. The mean scores are summarized as follows:

	MEN	WOMEN	MEAN
HAPPY ENDING	19.5	15.5	17.5
SAD ENDING	13.8	16.6	15.2
MEAN	16.65	16.05	

(a) Plot the means for each type of ending and comment on the nature of the interaction between sex and type of ending.

(b) Did you have to make any assumptions about the data in part (a)?

16. A company records the number of items sold by their salespersons on a sample of six Mondays, six Tuesdays, and so on. The results are as follows:

	MONDAY	TUESDAY	WEDNESDAY	THURSDAY	FRIDAY
	36	37	50	36	31
	37	36	32	39	40
	39	32	40	50	37
	30	43	34	49	40
	44	37	37	46	28
	24	47	47	44	28
Σx_i^2	7598	9116	9858	11,770	7098

(a) Construct the ANOVA table for these data.

(b) Test for a significant difference among the five daily means.

(c) What assumptions did you make in part (b)? Were these assumptions necessary in part (a)?

17. Refer to Review Problem 16. Of the six sales figures for each day, three referred to days in winter and three to days in spring. The sales on winter days were as follows:

MONDAY			TUESDAY			WEDNESDAY			THURSDAY			FRIDAY		
30	36	24	37	36	32	37	40	32	44	36	46	28	40	31

(a) Construct a two-way ANOVA for the 30 daily sales.
(b) Test for significant differences between days.
(c) How do your results in part (b) differ from those in Review Problem 16?
(d) Test for a significant difference between season means.
(e) Test for a significant interaction between days of the week and season of the year.

18. A bus company plans to begin service between two cities. Four routes, A, B, C, and D, are under consideration. To assess differences in the mean time for the four routes a bus makes the trip between the cities 32 times, taking each route eight times. The times (in hours) for each trip are as follows:

A	B	C	D
6.30	6.50	6.81	6.27
6.45	6.66	6.72	6.00
6.18	6.44	6.93	6.30
6.33	6.37	6.83	6.37
5.95	6.30	6.60	6.15
6.07	6.55	6.53	6.18
6.25	6.18	6.60	6.09
6.13	6.27	6.44	6.29

(a) Construct the ANOVA table for these data.
(b) Test whether the mean times for the four routes are significantly different.

EDUCATION AND PSYCHOLOGY

19. A psychologist is interested in the attitudes that young children have toward their school and how their attitudes change as they get older. A random sample of six children in each of the first, second, and third grades for each of the two elementary schools in a town are interviewed and given a score according to how much they are perceived to enjoy the school. The higher the score, the greater the enjoyment. The results are as follows:

		GRADE FIRST			GRADE SECOND			GRADE THIRD		
SCHOOL	WESSEX	9.4	12.5	11.3	10.6	11.2	9.8	10.0	6.7	8.8
		12.9	14.2	14.1	11.4	9.5	12.9	8.8	7.2	11.3
	ROSSMORE	12.8	14.7	11.9	13.4	11.9	13.9	8.2	5.9	5.0
		15.0	13.0	11.8	12.8	13.1	12.9	6.1	5.5	7.1

(a) Construct the two-way ANOVA for these data.
(b) Test for a significant difference in the mean score for the three grades.

(c) Test for a significant difference in mean scores for the two schools.

(d) Test for interaction between these two variables.

(e) What assumptions were necessary for parts (b)–(d)?

20. Refer to the data in Review Problem 19.

(a) Compute the mean scores for each cell.

(b) For each school plot the three grade means and connect the points.

(c) Explain any interaction you see between the two variables.

21. A college offers three semesters of introductory statistics, one meeting early in the morning (A), one at midday (M), and the other late in the afternoon (P). The following data are the number of absences for each section last semester. For convenience, these data are arranged in the form of a frequency distribution. For example, in Section P, four students missed exactly one class.

(A)

NUMBER OF ABSENCES	0	1	2	3	4	5	6	7	8	9
NUMBER OF STUDENTS	2	4	8	10	4	2	0	1	0	1

(M)

NUMBER OF ABSENCES	0	1	2	3	4	5	6	7	8	9
NUMBER OF STUDENTS	4	8	10	7	3	2	1	0	0	0

(P)

NUMBER OF ABSENCES	0	1	2	3	4	5	6	7	8	9
NUMBER OF STUDENTS	2	4	7	14	6	2	0	0	1	0

(a) Use the methods of Sections 1.3 and 1.4 to find the sum and the sum of squares of the number of absences for each section.

(b) Do these data suggest that there is any significant difference among the mean number of absences for the three sections?

22. The following data are the annual salaries (in thousands of dollars) of samples of 12 newly appointed assistant professors at three liberal arts colleges in a state. The three colleges differ in the extent to which the faculty are unionized. Compute the standard deviation of scores in each sample and hence explain why an F test for differences between mean starting salaries would be inappropriate here.

SUNSHINE COLLEGE (COMPLETELY UNIONIZED)			HAPPINESS COLLEGE (PARTIALLY UNIONIZED)			GLORY COLLEGE (NOT UNIONIZED)		
21.3	22.4	21.5	25.8	21.0	24.4	15.5	16.1	22.9
22.3	22.2	21.6	23.8	20.4	22.5	22.3	18.5	19.0
21.7	22.6	22.7	20.3	18.7	20.8	23.7	18.8	26.7
21.7	23.1	23.3	21.9	21.4	24.5	16.2	25.6	25.8
$\Sigma x_i^2 = 5918.72$			$\Sigma x_i^2 = 5923.09$			$\Sigma x_i^2 = 5435.87$		

SCIENCE AND TECHNOLOGY

23. The research and development division of an aircraft manufacturing company compares the hardness of three new alloys. Four specimens of each alloy are prepared and tested for hardness on the Moh scale, which runs from 1 to 10—the higher the score, the harder the material. The results are as follows:

ALLOY		
A	*B*	*C*
6.34	6.41	6.24
6.27	6.38	6.29
6.39	6.43	6.22
6.26	6.47	6.21

(a) Do these data suggest that the alloys differ significantly in their mean hardness?

(b) What assumptions were necessary in part (a)?

24. Refer to Review Problem 23.

(a) Find a 90% confidence interval for μ_C, the mean hardness of alloy C.

(b) With the obvious notation, find a 90% confidence interval for $\mu_A - \mu_C$.

25. The following data are the breaking strengths (in ounces) of 20 examples of yarn. Each value is classified by type of yarn (A, B, and C) and age (old/new).

	A		*B*		*C*	
NEW	16.2	15.7	19.1	17.5	17.4	16.7
OLD	15.0	15.9	18.1	17.8	17.1	16.3

(a) Construct the two-way ANOVA table for these data.

(b) Test for significant differences (i) between yarns; (ii) between age groups.

(c) Is there any significant interaction between the types of yarn and the age groups?

(d) Compute the mean breaking strength of the six cells and plot them. Do they indicate any interaction?

REFERENCES

ZELAZO, P. R., ZELAZO, N. A. and KOLB, S. "'Walking' in the newborn," *Science*, Vol. 176 (1972), pp. 314–315.

SUGGESTED READING

AFIFI, A. A. and AZEN, S. P. *Statistical Analysis: A Computer Oriented Approach*, 2nd ed. New York: Academic Press, 1979.

OTT, L. *An Introduction to Statistical Methods and Data Analysis*. 2nd Ed. North Scituate, Mass.: Duxbury, 1984.

KLEINBAUM, D. G. and KUPPER, L. L. *Applied Regression Analysis and Other Multivariate Methods*. North Scituate, Mass.: Duxbury, 1978.

14 LINEAR REGRESSION

PREDICTING SOME LIFE EXPECTANCIES

A group of 15 patients of different ages were observed from the date of onset of a disease. The age of the patient at this date (x) and the length of time (in years) that the patient survived (y) are as follows:

x	y
29	12.8
59	3.8
37	4.4
29	12.0
15	10.9
25	13.9
42	5.5
48	3.0
29	9.6
35	8.8
40	5.2
39	6.0
32	12.2
27	9.2
37	5.8

The researchers responsible for these data intend to compute the regression line relating survival time to age. The resulting line will be used to predict the survival time for other patients. But how accurate are such predictions? Review Problem 11 on page 571 takes up this question.

SECTION 14.1

INTRODUCTION TO LINEAR REGRESSION

In Chapter 2 we described how to find the correlation coefficient and the regression line for a set of bivariate (x, y) data. Usually, such a set of data is only a sample. In this chapter we explore how to use sample data to make inferences about the relationship between the two variables in the population. Before doing so, we review some of the techniques in Chapter 2 in the next example.

EXAMPLE 1 An accountant with a company owning a large chain of fast-food restaurants is interested in the relationship between monthly revenue (y) and advertising expenditures (x). As a pilot study the following values were obtained from a random sample of seven restaurants:

ADVERTISING EXPENDITURE, x (IN HUNDREDS OF DOLLARS)	REVENUE, y (TENS OF THOUSANDS OF DOLLARS)
2.7	5.0
2.0	2.5
2.4	3.3
2.8	3.8
2.2	4.5
2.0	2.1
2.3	2.8

(a) Plot these data on a scatterplot. (b) Find and plot the regression line predicting revenue from advertising. (c) Interpret b. (d) Compute the correlation coefficient, r. (e) Interpret r^2.

Solution (a) The scatterplot in Figure 1 suggests a moderate positive relationship between x and y.

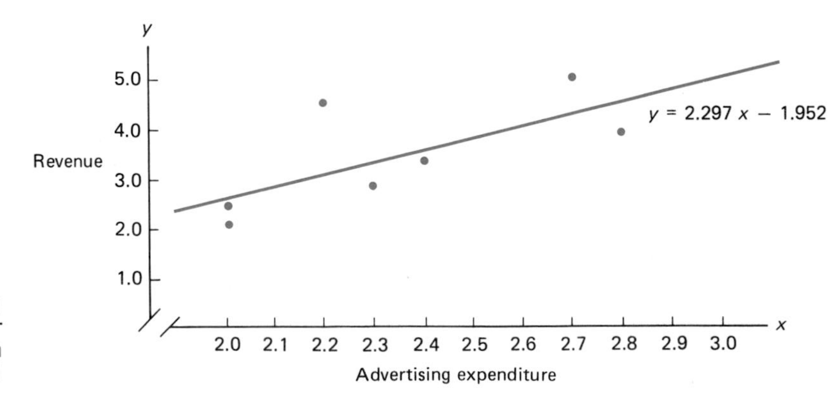

FIGURE 1

Scatterplot for the data in Example 1

(b) The regression line is of the form $y = a + bx$, where a and b are computed as

$$b = \frac{n \Sigma x_i y_i - (\Sigma x_i)(\Sigma y_i)}{n \Sigma x_i^2 - (\Sigma x_i)^2} \qquad \text{(Eq. 14.1)}$$

$$a = \frac{1}{n}\left(\sum y_i - b \sum x_i\right) \qquad \text{(Eq. 14.2)}$$

TABLE 1

Calculations necessary to compute a, b, and r in Example 1 ($n = 7$)

x	x^2	y	y^2	xy
2.7	7.29	5.0	25.00	13.50
2.0	4.00	2.5	6.25	5.00
2.4	5.76	3.3	10.89	7.92
2.8	7.84	3.8	14.44	10.64
2.2	4.84	4.5	20.25	9.90
2.0	4.00	2.1	4.41	4.20
2.3	5.29	2.8	7.84	6.44
16.4	39.02	24.0	89.08	57.60
$\bar{x} = 2.34$		$\bar{y} = 3.43$		

The components of a and b (and r) are computed in Table 1. Entering the appropriate column sums into Equations 14.1 and 14.2, we obtain

$$b = \frac{7(57.60) - (16.4)(24.0)}{7(39.02) - (16.4)^2} = \frac{9.6}{4.18} = 2.29665$$

$$a = \frac{1}{7}[24.0 - 2.29665(16.4)] = -1.95215$$

The regression line is then $y = -1.952 + 2.297x$ or $y = 2.297x - 1.952$. This line is drawn on the scatterplot in Figure 1.

(c) In this case $b = 2.297$ (or $22,970) is the increase in y (revenue) associated with an increase of 1 unit ($100) in x (advertising revenue).

Before continuing, it will be convenient to introduce a notation that will be helpful later.

$$T_{xx} = n \sum x_i^2 - \left(\sum x_i\right)^2 \qquad \text{(Eq. 14.3)}$$

$$T_{yy} = n \sum y_i^2 - \left(\sum y_i\right)^2 \qquad \text{(Eq. 14.4)}$$

$$T_{xy} = n \sum x_i y_i - \left(\sum x_i\right)\left(\sum y_i\right) \qquad \text{(Eq. 14.5)}$$

With this new notation, we have

$$b = \frac{T_{xy}}{T_{xx}} \qquad \text{(Eq. 14.6)}$$

$$\text{In this case } b = \frac{9.6}{4.18} = 2.297 \qquad \text{as before}$$

(d) with this notation the correlation coefficient becomes

$$r = \frac{n \Sigma x_i y_i - (\Sigma x_i)(\Sigma y_i)}{\sqrt{[n \Sigma x_i^2 - (\Sigma x_i)^2][n \Sigma y_i^2 - (\Sigma y_i)^2]}} = \frac{T_{xy}}{\sqrt{T_{xx}T_{yy}}}$$

We have already computed $T_{xy} = 9.6$ and $T_{xx} = 4.18$ when determining the value for b. From Table 1, $T_{yy} = 7(89.08) - (24.0)^2 = 47.56$, and thus

$$r = \frac{9.6}{\sqrt{(4.18)(47.56)}} = .6809$$

(e) $r^2 = .6809^2 = .4636$. For these data a little under one-half of the variability in revenue is associated with its linear relationship to advertising expenditure. ■

In order to draw inferences about a population based on a sample of bivariate data, we first summarize such data in what is called an **analysis of variance** (or ANOVA) **table.**[1] As in Chapter 2, we assume a sample of bivariate data of the form $(x_1, y_1), (x_2, y_2), \ldots, (x_i, y_i), \ldots, (x_n, y_n)$. We shall sometimes adopt the language of experimentation and call x the **independent variable** and y the **dependent variable.** As usual, $\bar{x}$ and $\bar{y}$ are the means of the x and y values.

We now introduce a symbol that we have not used before, $\hat{y}$ ("y hat"). We call $\hat{y}_i$ the **predicted** value for y if the observed value $x = x_i$ is inserted in the

[1]For most of this chapter we assume that the reader is *not* familiar with the material in Chapter 13. For those who have covered this material, we contrast it with linear regression at the end of Section 14.3.

equation for the regression line. In Example 1, for instance, the *predicted* revenue for a restaurant with an advertising expenditure of $x_i = 2.7$ would be $\hat{y}_i = a + b(2.7) = -1.952 + 2.297(2.7) = 4.25$ (or \$42,500). In the original data the *observed* value of y for the restaurant with $x_i = 2.7$ was $y_i = 5.0$ (or \$50,000). We indicate in Figure 2 the values $y_i = 5.0$ and $\hat{y}_i = 4.25$, corresponding to $x_i = 2.7$. We have also marked the value $\bar{y} = 3.43$ as well as $y_i - \hat{y}_i$, $\hat{y}_i - \bar{y}$, and $y_i - \bar{y}$.

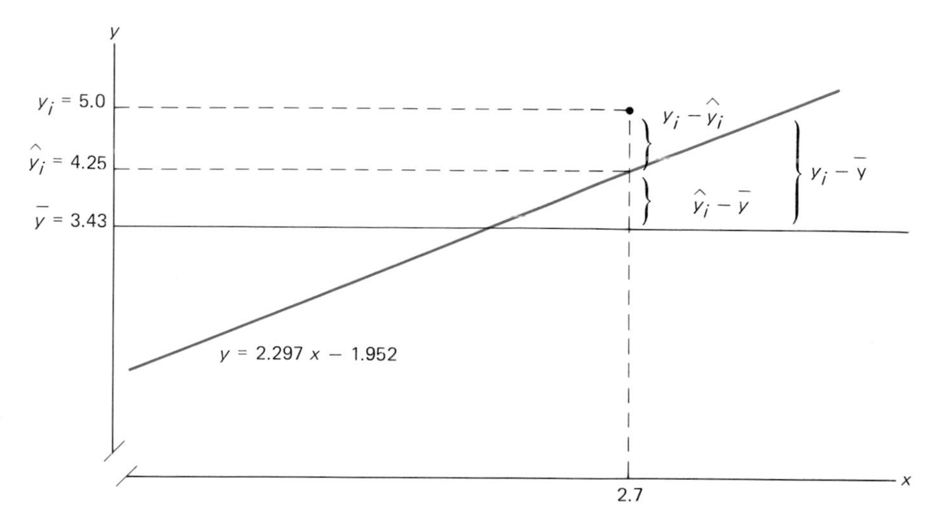

FIGURE 2

Illustrating the fact that $y_i - \bar{y} = (y_i - \hat{y}) + (\hat{y}_i - \bar{y})$

Notice from Figure 2 that the difference $y_i - \bar{y}$ can be written as the sum of two differences, $y_i - \hat{y}_i$ and $\hat{y}_i - \bar{y}$. That is,

$$y_i - \bar{y} = (y_i - \hat{y}_i) + (\hat{y}_i - \bar{y}) \qquad \text{(Eq. 14.7)}$$

It can be shown that if we square both sides of Equation 14.7 and sum over all n members of the sample, we will obtain the following equation involving *sums of squares* (SS):

$$\underbrace{\sum (y_i - \bar{y})^2}_{SS_{total}} = \underbrace{\sum (y_i - \hat{y}_i)^2}_{SS_{error}} + \underbrace{\sum (\hat{y}_i - \bar{y})^2}_{SS_{regression}} \qquad \text{(Eq. 14.8)}$$

On the left, the term $\Sigma\, (y_i - \bar{y})^2 = SS_{total}$ reflects the variability among the y's around their mean. This variability is broken into two components. The first, $\Sigma\, (y_i - \hat{y}_i)^2 = SS_{error}$, reflects the extent to which the values of the observed y's differ from those predicted by the regression line. It can also be viewed as the amount of variability in y that is not associated with its linear relationship to x. By contrast, $SS_{regression} = \Sigma (\hat{y}_i - \bar{y})^2$ represents the amount of variability in the y's that is associated with its linear relationship to x.

For ease of presentation we will, from now on, abbreviate SS_{total} to SST, SS_{error} to SSE, and $SS_{regression}$ to SSR.

As defined in Equation 14.8, the three sums of squares are messy to compute. More convenient computational forms are

$$\text{SST} = \frac{n \Sigma y_i^2 - (\Sigma y_i)^2}{n} = \frac{T_{yy}}{n} \qquad \text{(Eq. 14.9)}$$

$$\text{SSR} = \frac{[n \Sigma x_i y_i - (\Sigma x_i)(\Sigma y_i)]^2}{n[n \Sigma x_i^2 - (\Sigma x_i)^2]} = \frac{T_{xy}^2}{nT_{xx}} \qquad \text{(Eq. 14.10)}$$

The SSE can be found by subtracting SSR from SST. Notice that these forms eliminate the need to compute the values of $\hat{y}_i$.

For the advertising data in Example 1, $n = 7$, $T_{xx} = 4.18$, $T_{yy} = 47.56$, and $T_{xy} = 9.6$. Thus

$$\text{SST} = \frac{T_{yy}}{n} = \frac{47.56}{7} = 6.7943$$

$$\text{SSR} = \frac{T_{xy}^2}{nT_{xx}} = \frac{9.6^2}{7(4.18)} = 3.1497$$

and SSE $= 6.7943 - 3.1497 = 3.6446$.

Associated with a sum of squares is a number of *degrees of freedom*. The SST has $n - 1$, the SSR 1, and the SSE $n - 2$ degrees of freedom. Notice that the degrees of freedom for SST is the sum of the other two. Dividing a sum of squares by its corresponding number of degrees of freedom results in a variance-like quantity called a *mean square,* denoted MS. In the case of SST, dividing by $n - 1$ does, in fact, produce the variance (s_y^2) of the y's.

In Table 2 the three sums of squares, degrees of freedom, and mean squares for the advertising data are displayed in what is known as an analysis of variance (ANOVA) table.

TABLE 2

ANOVA table for the data in Example 1

SOURCE OF VARIABILITY	SS	df	MS
Regression on x	3.1497	1	3.1497
Error	3.6446	5	.7289
Total	6.7943	6	

In the following section we shall use the information in Table 2 to make inferences about the relationship between x and y in the population. For the present it is interesting to note the following.

As we indicated earlier, SSR is the amount of variability in the y's that is associated with their linear relationship to x. Thus the ratio SSR/SST is "the *proportion* of the total variability in y that is associated with their linear relationship to x." But this is precisely the way we defined r^2. In fact, from Table 2, SSR/SST = 3.1497/6.7943 = .4636, which is exactly the value for r^2 that we obtained in Example 1(e). In general, we write

$$r^2 = \frac{\text{SSR}}{\text{SST}} \quad \text{(Eq. 14.11)}$$

In Table 3 we provide the general format of an ANOVA table for regression.

TABLE 3

ANOVA table for simple linear regression

SOURCE OF VARIABILITY	SS	df	MS
Regression on x	$\text{SSR} = \dfrac{T_{xy}^2}{nT_{xx}}$	1	$\text{MSR} = \dfrac{\text{SSR}}{1}$
Error	SSE = SST − SSR	$n - 2$	$\text{MSE} = \dfrac{\text{SSE}}{n-2}$
TOTAL	$\text{SST} = \dfrac{T_{yy}}{n}$	$n - 1$	

EXAMPLE 2 Ten cars of the same model are equipped to record both speed (x), in mph, and gas consumption (y), in mpg. The cars are driven along a course at different speeds. The resulting data are as follows:

SPEED (x)	25	30	35	40	45	50	55	60	65	70
mpg (y)	34.5	32.4	35.0	30.6	31.7	32.2	30.4	28.5	25.7	26.6

(a) Compute the regression line relating y to x. (b) Construct the ANOVA table. (c) What proportion of the variability in gas consumption is associated with differences in speed? (The summary values are $\Sigma\, x_i = 475$, $\Sigma\, x_i^2 = 24{,}625$, $\Sigma\, y_i = 307.6$, $\Sigma\, y_i^2 = 9547.56$, and $\Sigma\, x_i y_i = 14{,}234.5$.)

Solution Using Equations 14.3, 14.4, and 14.5 with $n = 10$, we have

$$T_{xx} = 10(24{,}625) - (475)^2 = 20{,}625$$

$$T_{yy} = 10(9547.56) - (307.6)^2 = 857.84$$

$$T_{xy} = 10(14{,}234.5) - (475)(307.6) = -3765$$

(a) The slope and intercept of the regression line are

$$b = \frac{T_{xy}}{T_{xx}} = \frac{-3765}{20{,}625} = -.182545$$

$$a = \frac{1}{10}\left(\sum y_i - b \sum x_i\right)$$

$$= \frac{1}{10}[307.6 - (-.182545)(475)] = 39.4309$$

The regression line is thus

$$y = 39.4309 - .1825x$$

(b) From Equations 14.9 and 14.10 or Table 3,

$$\text{SSR} = \frac{T_{xy}^2}{nT_{xx}} = \frac{(-3765)^2}{10(20{,}625)} = 68.7284$$

$$\text{SST} = \frac{T_{yy}}{n} = \frac{857.84}{10} = 85.784$$

and SSE = 85.784 − 68.7284 = 17.0556. The resulting ANOVA table is shown in Table 4.

TABLE 4

ANOVA table for the data in Example 2

SOURCE OF VARIABILITY	SS	df	MS
Regression of mpg on speed	68.7584	1	68.7284
Error	17.0556	8	2.132
Total	85.784	9	

(c) In this case

$$r^2 = \frac{\text{SSR}}{\text{SST}} = \frac{68.7284}{85.784} = .801$$

Eighty percent of the variability in mpg is associated with its linear relationship to speed.

■

PROBLEMS 14.1

Save your computations, you will need them in later sections.

1. For the data in Problem 15 of Chapter 13 (page 484), treat x as the number of beers and y as the change in reaction time.
   (a) Compute T_{xx}, T_{yy}, and T_{xy}.

(b) Construct the ANOVA table.

(c) What proportion of variability in the change in reaction time is associated with its linear relationship to the number of beers drunk?

2. A plant nursery is interested in the relationship between the age and height of a variety of arborvitae. Four bushes aged 2, 3, 4, 5, 6, and 7 years old are selected and their heights recorded with the following results (in feet):

AGE	HEIGHT	AGE	HEIGHT	AGE	HEIGHT
2	5.6	4	6.2	6	7.2
2	4.8	4	6.7	6	7.5
2	5.3	4	6.4	6	7.8
2	5.7	4	6.7	6	7.4
3	6.2	5	7.1	7	8.9
3	5.9	5	7.3	7	9.2
3	6.4	5	6.9	7	8.5
3	6.1	5	6.9	7	8.7

(a) Compute T_{xx}, T_{yy}, and T_{xy}.

(b) Find the regression line predicting height from age.

(c) Interpret b in this case.

(d) Construct the ANOVA table.

(e) What proportion of the variability in height among the 24 bushes is due to their linear relationship to the ages of the bushes?

(f) Find the correlation coefficient, r, using your results from parts (a)–(d).

3. In a pilot study, a research assistant associated with the clinic of a health plan selects a sample of the records for 18 male members of the plan. The following data are the ages of the men and the number of visits to the clinic in the past year.

AGE, x	NUMBER OF VISITS, y	AGE, x	NUMBER OF VISITS, y	AGE, x	NUMBER OF VISITS, y
41	3	46	4	37	1
45	2	31	0	39	2
48	3	48	1	62	3
54	0	53	3	39	2
49	1	54	2	50	2
56	0	48	2	50	3

$$\sum x_i^2 = 41{,}128 \qquad \sum x_i y_i = 1630$$

(a) Find the regression line relating number of visits to the age of the plan member.

(b) Construct the ANOVA table.

(c) Find and interpret r^2 in this case.

4. The manager of the local outlet for the brand of beer sold at a ballpark is interested in the relationship between attendance and revenues from sales of beer. In a

random sample of 12 games played last season, the following data were obtained:

ATTENDANCE (THOUSANDS)	BEER SALES (THOUSANDS OF DOLLARS)
30.5	16.5
11.1	6.4
28.9	15.0
33.0	20.7
20.4	10.4
12.6	12.2
24.6	14.8
40.0	23.3
27.7	16.4
21.8	16.3
17.4	8.3
27.8	15.5

$$\sum x_i^2 = 8075.88 \qquad \sum y_i^2 = 2829.42 \qquad \sum x_i y_i = 4737.79$$

(a) Find the regression line predicting beer sales from attendance.
(b) Compute and interpret r^2.

5. A pediatrician interested in the relationship between the age of children and the number of hours of sleep asks the parents of 15 children (from 3 to 11 years old) to record the number of hours that their children slept on each of seven consecutive days. The following data are the age of each child and the average number of hours slept for each child:

AGE, x	HOURS OF SLEEP, y	AGE, x	HOURS OF SLEEP, y
5	11.1	11	8.8
7	9.8	5	11.1
3	11.4	7	10.2
10	8.3	11	9.5
8	11.0	5	10.7
11	7.7	5	11.1
3	11.4	8	9.2
8	9.2		

$$\sum x_i^2 = 871 \qquad \sum y_i^2 = 1530.27 \qquad \sum x_i y_i = 1032.6$$

(a) Compute (i) T_{xx}, T_{yy}, and T_{xy}, and (ii) construct the ANOVA table for these data.
(b) Compute and interpret r^2.
(c) Find the regression line relating y to x.
(d) Interpret b in this case.

6. The professor teaching the calculus sequence at a college is interested in the relationship between a student's score on the math competency test, which all students take, and the student's composite score in calculus. The following data on 24 students who had taken calculus last year were obtained:

TEST SCORE, x	CALCULUS SCORE, y	TEST SCORE, x	CALCULUS SCORE, y
29	98	15	59
21	67	16	59
27	72	18	74
29	86	25	91
21	83	18	70
15	53	23	72
19	67	21	88
25	86	28	78
21	69	29	100
19	74	25	90
20	76	15	63
33	85	27	79

$$\sum x_i^2 = 12{,}723 \qquad \sum y_i^2 = 144{,}435 \qquad \sum x_i y_i = 42{,}440$$

(a) Construct the ANOVA table for these data.

(b) What proportion of the variability in calculus scores is associated with their linear relationship to test scores?

(c) Obtain the regression line relating calculus scores to test scores.

7. For the data in Problem 22 in Chapter 13 (page 485), treat the educational level as x and the number of children as y.

(a) Compute T_{xx}, T_{yy}, and T_{xy}.

(b) Construct the ANOVA table.

(c) What proportion of the variability in the numbers of children per family is associated with educational level?

8. A psychologist gives a group of 20 students 30 minutes to memorize a list of 25 household items. The 20 students are then divided into five groups of four students each. One group was immediately asked to recall the number of items. The remaining groups were asked to recall the number of items after different lengths of time. The numbers of correct responses were as follows:

LENGTH OF TIME, x (MINUTES)	NUMBER OF CORRECT ITEMS, y			
0	19	23	17	21
30	16	20	15	18
60	16	14	19	20
90	20	12	15	15
120	14	16	19	17

(a) Compute T_{xx}, T_{yy}, and T_{xy}.

(b) Find the regression line relating y to x.

(c) Construct the ANOVA table.

(d) Compute and interpret r^2.

Solution to Problem 1

(a) Regarding x as the number of beers and y as the change in reaction time, the data can be written as follows:

x	y	x	y	x	y	x	y
0	−.21	2	.20	4	.45	6	.02
0	−.02	2	.36	4	.22	6	.43
0	−.17	2	.17	4	.12	6	.42
0	.03	2	.24	4	.34	6	.42
0	.03	2	.18	4	.35	6	.37
0	−.17	2	.08	4	.30	6	.27
		2	.34	4	.24	6	.55
		2	.06	4	.12		

Then we compute

$$\sum^{29} x_i = (6 \times 0) + (8 \times 2) + (8 \times 4) + (7 \times 6) = 90$$

$$\sum^{29} x_i^2 = (6 \times 0^2) + (8 \times 2^2) + (8 \times 4^2) + (7 \times 6^2) = 412$$

$$\sum^{29} y_i = -.21 - .02 + \cdots + .27 + .55 = 5.74$$

$$\sum^{29} y_i^2 = .1041 + .4141 + .6654 + 1.0504 = 2.234$$

(These values were given in Problem 15 on page 484. The x's there are the y's here.)

$$\sum x_i y_i = 0(-.21) + 0(-.02) + \cdots + 6(.55) = 26.7$$

Therefore,

$$T_{xx} = n \sum x_i^2 - \left(\sum x_i\right)^2 = 29(412) - 90^2 = 3848$$

$$T_{yy} = n \sum y_i^2 - \left(\sum y_i\right)^2 = 29(2.234) - 5.74^2 = 31.8384$$

$$T_{xy} = n \sum x_i y_i - \left(\sum x_i\right)\left(\sum y_i\right) = 29(26.7) - (90)(5.74) = 257.7$$

(b) From Table 3 we compute

$$\text{SST} = \frac{T_{yy}}{n} = \frac{31.8384}{29} = 1.0979$$

$$\text{SSR} = \frac{T_{xy}^2}{nT_{xx}} = \frac{257.7^2}{29(3848)} = .595108$$

and SSE $= 1.0979 - .595108 = .502792$. The ANOVA table is thus:

SOURCE OF VARIABILITY	SS	df	MS
Regression on x	.595108	1	.595108
Error	.502792	27	.0186
Total	1.0979	28	

(c) We require

$$r^2 = \frac{SSR}{SST} = \frac{.595108}{1.0979} = .542$$

(Notice that here we are treating the independent variable, number of beers, as quantitative. In Problem 15 of Section 13.2 we treated this variable as qualitative.)

SECTION 14.2

RELATIONSHIP BETWEEN x AND y IN THE POPULATION

The accountant in Example 1 would probably be interested in generalizing the relationship between advertising expenditure (x) and revenue (y) in the sample to the population of all the restaurants. In this section we outline the assumptions we need to make about the relationship between x and y in the population in order to make such a generalization.

We begin by noting that there is no reason to assume that all the points representing the pairs of (x, y) values in the population will lie along a straight line. To see what is a more realistic assumption, consider all the restaurants who spend \$200 ($x = 2.0$) on advertising. The amount of revenue (y) will vary for these restaurants, equaling 3.2 in some cases, 2.9 in others, 2.6 in others, and so on. We call the mean revenue for these restaurants $\mu_{y|2.0}$. In fact, for any specific value x (advertising expenditure) there will be a *distribution* of revenues (y values) which will have a mean value $\mu_{y|x}$. In Figure 3 we show what might be the distributions of y for $x = 1.8$, 2.0, and 2.3. In each case we indicate the corresponding $\mu_{y|x}$ on the vertical axis. We refer to $\mu_{y|x}$ as the mean of the y's for a specific x.

As indicated in Figure 3, we assume a linear relationship between the x's and the values for the corresponding means, $\mu_{y|x}$. Symbolically,

$$\mu_{y|x} = \alpha + \beta x \qquad \text{(Eq. 14.12)}$$

where α and β are unknown constants. This is the equation of a straight line

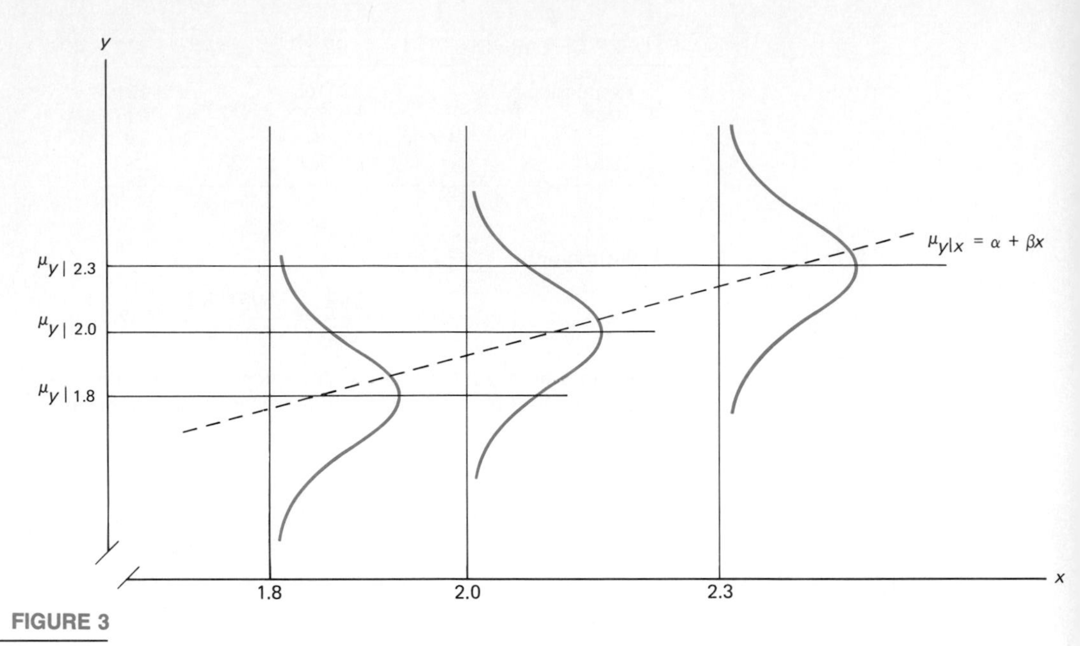

FIGURE 3

The relationship between x and $\mu_{y|x}$ is linear

which we call the **population line.** Here α is the intercept and β the slope of the population line. You may recall that the symbols α and β were also used in hypothesis testing. We use them here with a different meaning and trust that there will be no confusion.

In Figure 4 we have drawn the regression line $y = 2.297x - 1.952$ based on the sample in Example 1. We have also drawn a possible population line, dashed to indicate that its position is unknown.

In general, we regard the regression line $y = a + bx$ as an estimate for the

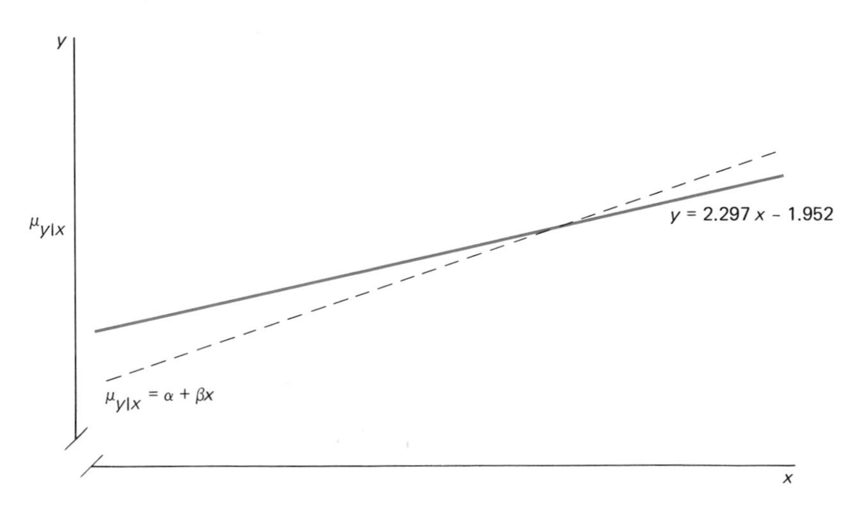

FIGURE 4

Population and regression lines

unknown population line. More specifically, we regard a as an estimate for α and b as an estimate for β.

In Figure 5 we show examples where (a) $\beta > 0$, (b) $\beta < 0$, and (c) $\beta = 0$. In the last case the population line is $\mu_{y|x} = \alpha + (0)x = \alpha$, so that the mean value of y for any x is *not related to* x.

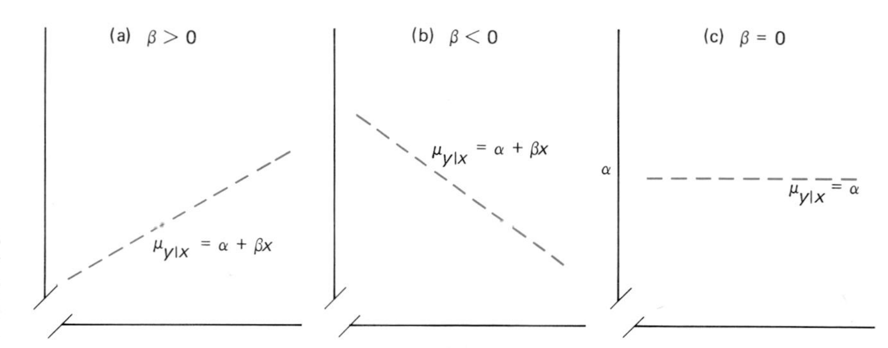

FIGURE 5

Examples where the slope of the population line is > 0, < 0, and = 0.

In addition to the linearity assumption we shall assume that for any *specific* x the distribution of y is normal with mean $\mu_{y|x}$ and standard deviation σ, and that *σ is the same for all x's*. These assumptions are illustrated in Figure 3. There, for each advertising expenditure, x, the distribution of revenue, y, is normal. Moreover, the standard deviation of revenue (y) is the same for each x. In Figure 6 we illustrate a case where the normality or equal-standard-deviation assumptions are violated. For $x = 1.8$ and $x = 2.0$ the distribution of y is normal but with very different standard deviations. When $x = 2.3$, the distribution of y is bimodal and far from normal.

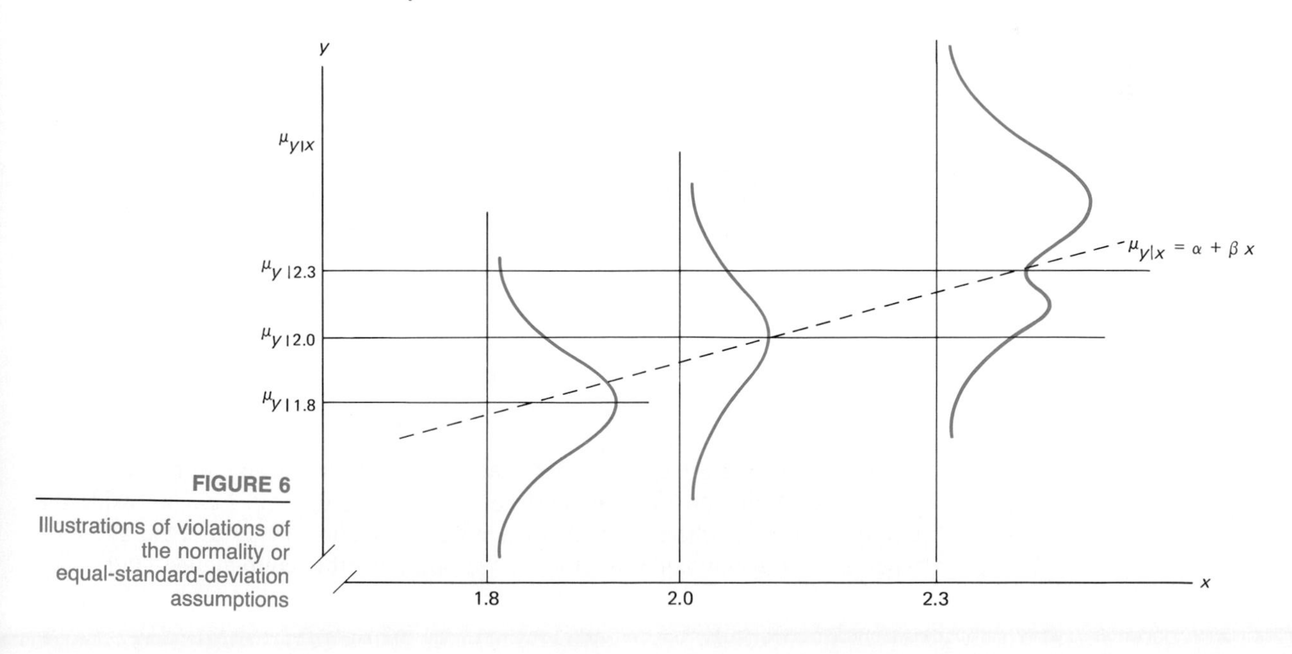

FIGURE 6

Illustrations of violations of the normality or equal-standard-deviation assumptions

The linearity, normality, and equal-standard-deviation assumptions are summarized in Definition 1.

DEFINITION 1

For any specific x we assume that the distribution of y in the population is $N[\mu_{y|x}, \sigma]$, where $\mu_{y|x} = \alpha + \beta x$.

Note: For completeness we add the assumption that the observations are selected independently.

In general, σ^2 is the variability in y among all members of the population having the *same* x value. Since all such members have the same value for x, σ^2 is the amount of the variability in y that is due to variables *other than* x. It may not come as a surprise, therefore, that an estimate for σ^2 is the MSE from the ANOVA table. More usefully, $\sqrt{\text{MSE}}$ is an estimate for σ. For instance, in Example 1 if the assumptions in Definition 1 are met, an estimate for σ, the standard deviation of revenue among all restaurants having the same advertising expenditure, is $\sqrt{\text{MSE}} = \sqrt{.7289} = .8538$ (or \$8538). (The value .7289 was taken from Table 2.)

EXAMPLE 3 Referring to Example 2, estimate the unknown α, β, and σ.

Solution We estimate α by $a = 39.4309$ and β by $b = -.1825$. We estimate σ, the standard deviation of miles per gallon for any specific speed, by $\sqrt{\text{MSE}} = \sqrt{2.132} = 1.46$ mpg.

■

THE POPULATION CORRELATION COEFFICIENT

To shift gears somewhat, we call ρ (the Greek lowercase letter rho) the unknown correlation coefficient based on all the (x, y) values in the *population* and we use the sample correlation coefficient, r, as an estimate for ρ.

It can be shown that if the assumptions in Definition 1 are met, then ρ and β, the slope of the population line, are related by the equation

$$\beta = \rho \frac{\sigma_y}{\sigma_x} \qquad \text{(Eq. 14.13)}$$

where σ_x and σ_y are the standard deviations for all the x's and all the y's in the population. Equation 14.13 is not true, however, if $\sigma_x = 0$. Leaving aside the unimportant case where $\sigma_y = 0$, we see from Equation 14.13 that the situation when $\rho = 0$ is equivalent to $\beta = 0$. We shall use this equivalence later.

FIXED AND RANDOM x'S

With bivariate data the y values are always regarded as randomly selected. The x values can be (1) fixed in advance by the experimenter, or (2) like the y's, randomly selected. In Example 2, for instance, the 10 speeds were fixed by the researcher. On the other hand, the resulting miles per gallon (the y's) were regarded as values selected at random from the distribution of mileages for each speed. By contrast, in Example 1 there was no attempt to pick restaurants having preselected advertising expenditures (x). The pair (x, y) is regarded as having been randomly chosen from all (x, y) pairs in the population.

Happily, provided that the assumptions in Definition 1 are met, the inferential methods developed in the next two sections are the same whether the x's are fixed or random. There is, nevertheless, an important difference between the two situations. If the x's are fixed, it makes sense to find the line predicting y from x *but not the reverse*. In Example 2, for instance, it makes sense to find the line relating mileage to speed, but it would hardly make sense to find the line relating speed to mileage. If in some other case, however, the pair (x, y) is randomly selected, the researcher may find either line.

PROBLEMS 14.2

9. (i)–(viii) For each of the eight problems at the end of Section 14.1, explain Definition 1 in the context of that problem.
10. For each of the eight problems at the end of Section 14.1, indicate whether the x values were "fixed" by the researcher or "randomly" chosen.
11. A population consists of the following 27 pairs of (x, y) values:

x	y	x	y	x	y
1	3	2	5	3	7
1	4	2	6	3	8
1	4	2	6	3	8
1	5	2	7	3	9
1	5	2	7	3	9
1	5	2	7	3	9
1	6	2	8	3	10
1	6	2	8	3	10
1	7	2	9	3	11

(a) Plot these values on a scatterplot.

(b) For each x value, do the y's seem approximately normal?

(c) For each value of x compute $\mu_{y|x}$, the mean value of the y's. Do these mean values lie on a straight line?

(d) For each value of x compute the standard deviation of the nine y's [use

$$s = \sqrt{\frac{n \Sigma y_i^2 - (\Sigma y_i)^2}{n(n-1}}\Bigg]$$

Do your results indicate that the equal-standard-deviation assumption in Definition 1 applies to this population?

12. A population consists of the following 45 pairs of (x, y) values:

x	y	x	y	x	y	x	y	x	y
0	−2	1	7	2	10	3	7	4	−2
0	−1	1	8	2	11	3	8	4	−1
0	−1	1	8	2	11	3	8	4	−1
0	0	1	9	2	12	3	9	4	0
0	0	1	9	2	12	3	9	4	0
0	0	1	9	2	12	3	9	4	0
0	1	1	10	2	13	3	10	4	1
0	1	1	10	2	13	3	10	4	1
0	2	1	11	2	14	3	11	4	2

(a) Plot these values on a scatterplot.

(b) For each x value do the y's seem approximately normal?

(c) For each value of x compute $\mu_{y|x}$, the mean value of the y's. Do these mean values lie along a straight line? If not, what relationship between x and $\mu_{y|x}$ might be more appropriate?

Solution to Problem 9(i)

(i) Refer to Problem 1 of Section 14.1 (page 532).

We assume that for any specific number of beers (x) the distribution of the change in reaction time (y) is normal with mean $\mu_{y|x} = \alpha + \beta x$ and with standard deviation σ, which is the same for each number of beers.

SECTION 14.3

INFERENCES ABOUT β, THE SLOPE OF THE POPULATION LINE

We indicated in Section 14.2 that b, the slope of the fitted regression line, can be regarded as an estimate for β, the unknown slope of the population line. But how precise an estimate? And, as important, can we be sure that β is *not* zero, in which case there would be no linear relationship between x and $\mu_{y|x}$?

It can be shown that over repeated samples the mean value of b is precisely β; that is, $E(b) = \beta$. The standard deviation of b, however, is quite different in form from standard deviations we have used up to this point. It is

$$\text{SD}(b) = \frac{\sigma}{\sqrt{\Sigma\,(x_i - \bar{x})^2}}$$

(*Note:* For clarity of exposition we will not distinguish between the random variable, the slope of the fitted line, and the observed value of that variable. We shall call both b.) Since σ, the standard deviation of y for any x, is generally unknown, we replace it by its estimate, $\sqrt{\text{MSE}}$. Also, for computational convenience we replace

$$\sum (x_i - \bar{x})^2 \quad \text{by} \quad \frac{n \sum x_i^2 - (\sum x_i)^2}{n} = \frac{T_{xx}}{n} \qquad \text{(from Equation 14.3)}$$

Thus

$$\text{SD}(b) = \sqrt{\frac{n\text{MSE}}{T_{xx}}}$$

Furthermore, if the assumptions in Definition 1 apply it can be shown that:

> DEFINITION 2
>
> $$t = \frac{b - \beta}{\text{SD}(b)} = \frac{b - \beta}{\sqrt{n\text{MSE}/T_{xx}}} \text{ has the } t_{n-2} \text{ distribution.}$$

CONFIDENCE INTERVALS FOR β Definition 2 provides the basis for a confidence interval for the unknown β. The details are given below.

> DEFINITION 3
>
> If the assumptions in Definition 1 apply, a $100(1 - \gamma)\%$ confidence interval for β is
>
> $$b - t_{n-2(\gamma)}\sqrt{\frac{n\text{MSE}}{T_{xx}}} \quad \text{to} \quad b + t_{n-2(\gamma)}\sqrt{\frac{n\text{MSE}}{T_{xx}}}$$

EXAMPLE 4 Use the data in Example 1 and Table 2 to find a 90% confidence interval for β, the slope of the population line relating advertising expenditure (x) to mean revenue ($\mu_{y|x}$).

Solution From Example 1, $b = 2.297$, $n = 7$, and $T_{xx} = 4.18$. From Table 2, MSE $= .7289$. Referring to Table A6, we have $t_{n-2(\gamma)} = t_{5(.9)} = 2.015$. Thus a 90% confidence interval for β is

$$2.297 \mp 2.015\sqrt{\frac{7(.7289)}{4.18}} = 2.297 \mp 2.226 = 0.71 \text{ to } 4.523$$

We can be 90% confident that β lies in this interval.

■

TESTING HYPOTHESES ABOUT β Definition 2 can be used to test whether or not $\beta = 0$. We emphasize that even if β were zero, sampling variability will ensure that the value for b will almost certainly *not* be zero.

We shall test $H_0: \beta = 0$ against $H_1: \beta > 0$, $H_1: \beta < 0$, or $H_1: \beta \neq 0$. If H_0 is true, $t = (b - 0)/\sqrt{n\text{MSE}/T_{xx}}$ has the t_{n-2} distribution. Evidence favoring H_0 will be a value of t close to zero. Evidence for $H_1: \beta > 0$ will be a value of t unusually *greater* than 0 and for $H_1: \beta < 0$ a value unusually *less* than 0. Evidence for $H_1: \beta \neq 0$ will be a value for t unusually different from 0.

The details of the t test of $H_0: \beta = 0$ are set out in Table 5.

TABLE 5

(a) Procedure for performing a one-sided test of H_0: $\beta = 0$

STEP	ALTERNATIVE HYPOTHESIS (a) $H_1: \beta > 0$	(b) $H_1: \beta < 0$
1. Compute:	$t = \dfrac{b - 0}{\sqrt{n\text{MSE}/T_{xx}}}$	
2. Find $t_{n-2,.05}$ and $t_{n-2,.01}$ from Table A6.		
3. *Conclusion:* Using $\alpha = .05$ or $.01$, reject H_0 at the α level of significance if:	$t \geq t_{n-2,\alpha}$	$t \leq -t_{n-2,\alpha}$
Assumption: As in Definition 1		

(b) Procedure for performing a two-sided test of H_0: $\beta = 0$

STEP	$H_1: \beta \neq 0$
1. Compute:	$t = \dfrac{b - 0}{\sqrt{n\text{MSE}/T_{xx}}}$
2. Find $t_{n-2,.025}$ and $t_{n-2,.005}$ from Table A6.	
3. *Conclusion:* Using $\alpha = .05$ or $.01$, reject H_0 at the α level of significance if:	$\lvert t \rvert > t_{n-2,\ \alpha/2}$
Assumption: As in Definition 1	

The procedures in Table 5 are similar to those in Tables 1 and 2 in Chapter 12 (pages 407 and 408) for testing $H_0: \mu = \mu_0$ in small samples. In Table 5, b replaces $\bar{x}$, 0 replaces μ_0, $\text{SD}(b) = \sqrt{n\text{MSE}/T_{xx}}$ replaces $\text{SD}(\bar{X}) = s/\sqrt{n}$, and the number of degrees of freedom for t is $n - 2$ here rather than the $n - 1$ in Tables 1 and 2 in Chapter 12.

EXAMPLE 5 Test $H_0: \beta = 0$ against $H_1: \beta > 0$ for the advertising data in Example 1.

Solution In this case, β, the slope of the population line, is the change in mean revenue associated with a one-unit (\$100) increase in advertising revenues among *all* the company's restaurants. From Example 4, $b = 2.297$, $n = 7$, $T_{xx} = 4.18$, and $\text{MSE} = .7289$. Following the procedure in Table 5 (*a*), we have

$$t = \frac{2.297 - 0}{\sqrt{7(.7289)/4.18}} = 2.079$$

Referring to Table A6 with $\nu = n - 2 = 5$, $t_{5,.05} = 2.015$ and $t_{5,.01} = 3.365$. Since $t = 2.079$ lies between 2.015 and 3.365, we reject H_0 at the .05 level of significance. The data suggest that β, the slope of the population line, is significantly greater than 0. This, in turn, suggests a positive linear relationship between the two variables in the population.

Prior to seeing the data, the accountant in Example 1 would reasonably expect that revenue would tend to rise with advertising expenditure and hence that if β were not zero, it would be greater than 0—hence the use of the alternative $H_1 : \beta > 0$ in Example 5. For illustrative purposes, however, suppose, conservatively, that the accountant preferred the two-sided alternative $H_1 : \beta \neq 0$. In this case $|t| = |2.079| = 2.079$. From Table A6, $t_{5,.025} = 2.571$ and $t_{5,.005} = 4.032$. Since 2.079 is less than 2.571, we cannot reject H_0 at the .05 level of significance. We conclude that β is not significantly different from 0 and hence that there is no significant linear relationship between $\mu_{y|x}$ and x. ■

Examples 4 and 5 may seem deceptively simple because most of the necessary calculations were completed in earlier examples. In the next example we test $H_0 : \beta = 0$ using the (x, y) totals, sums of squares, and sum of products.

EXAMPLE 6 In a recent study the age, x, and the systolic blood pressure (SBP), y, were recorded for each of 20 adult males. The following summary data were obtained:

$$\sum x_i = 833 \qquad \sum x_i^2 = 36{,}429 \qquad \sum y_i = 2710 \qquad \sum y_i^2 = 368{,}946$$

$$\sum x_i y_i = 113{,}156 \qquad n = 20$$

Test for a significant positive linear relationship between age and SBP among all adult men.

Solution We test $H_0 : \beta = 0$ against $H_1 : \beta > 0$, where β, the slope of the population line, is the change in mean blood pressure associated with a one-year increase in age. We first need to compute b, MSE, and therefore T_{xx}, T_{yy}, and T_{xy}.

From Equations 14.3, 14.4, and 14.5 with $n = 20$, we have

$$T_{xx} = 20(36{,}429) - 833^2 = 34{,}691$$

$$T_{yy} = 20(368{,}946) - 2710^2 = 34{,}820$$

$$T_{xy} = 20(113{,}156) - (833)(2710) = 5690$$

From Equation 14.6,

$$b = \frac{T_{xy}}{T_{xx}} = \frac{5690}{34{,}691} = .1640$$

From Table 3,

$$\text{SSE} = \text{SST} - \text{SSR} = \frac{T_{yy}}{n} - \frac{T_{xy}^2}{nT_{xx}} = \frac{34{,}820}{20} - \frac{5690^2}{20(34{,}691)}$$

$$= 1741 - 46.6635$$

$$= 1694.3365$$

and

$$\text{MSE} = \frac{\text{SSE}}{n-2} = \frac{1694.3365}{18}$$

$$= 94.1298$$

Now following the procedure in Table 5(a), we have

$$t = \frac{.1640 - 0}{\sqrt{20(94.1298)/34{,}691}} = \frac{.1640}{.2330} = .704$$

Referring to Table A6 with $n - 2 = 18$, we have $t_{18,.05} = 1.734$ and $t_{18,.01} = 2.552$. Since $t = .704$ is smaller than 1.734, we cannot reject H_0 at the .05 level of significance. The data suggest *no* significant positive linear relationship between age and mean (SBP) among all adult males.

■

In Section 14.2 we pointed out that when $\beta = 0$, ρ, the population correlation coefficient, is also 0. Thus testing $H_0: \beta = 0$ is equivalent to testing $H_0: \rho = 0$.

It is not hard to adapt Table 5 to test $H_0: \beta = \beta_0$, where β_0 is a value other than 0.

DEFINITION 4

To test $H_0: \beta = \beta_0$, use the procedure in Table 5 but with

$$t = \frac{b - \beta_0}{\sqrt{n\text{MSE}/T_{xx}}}$$

EXAMPLE 7 Use the information in Example 5 to test $H_0: \beta = 3.0$ against $H_1: \beta < 3.0$.

Solution In this case

$$t = \frac{2.297 - 3.0}{\sqrt{7(.7289)/4.18}} = -.6363$$

Since $-.6363$ is larger than -2.015 and -3.365, we cannot reject H_0 at the .05 level of significance. The data suggest that the slope of the population line is not significantly less than 3.0.

■

THE F TEST OF $H_0: \beta = 0$

There is an F test of $H_0: \beta = 0$ against $H_1: \beta \neq 0$. The test is based on the following result:

> DEFINITION 5
>
> If the assumptions in Definition 1 apply and $H_0: \beta = 0$ is true, the ratio $F = \text{MSR}/\text{MSE}$ has the $F_{1,n-2}$ distribution.

The test consists of rejecting H_0 if the value for F is unusually large. The details for $\alpha = .05$ and $\alpha = .01$ are given in Table 6.

TABLE 6

Procedure for Performing the F test of $H_0: \beta = 0$ and $H_1: \beta \neq 0$

1. Compute $F = \text{MSR}/\text{MSE}$.
2. Find $F_{1,n-2,.05}$ and $F_{1,n-2,.01}$ from Table A8.
3. *Conclusion:* Using $\alpha = .05$ or .01, reject H_0 at the α level of significance if $F \geq F_{1,n-2,\alpha}$.

Assumption: As in Definition 1

We illustrate this test using the advertising data.

EXAMPLE 8 Apply the F test to the data in Example 1 to check whether the slope of the population line relating advertising expenditure to mean revenue is significantly different from 0.

Solution In Table 7 we have reproduced from page 530 the ANOVA table for the data in Example 1 (Table 2). The last column on the right contains the ratio $F = \text{MSR}/\text{MSE}$.

TABLE 7

ANOVA table for the data in Example 1, including the F ratio

SOURCE OF VARIABIILTY	SS	df	MS	F
Regression on x	3.1497	1	3.1497	4.3212
Error	3.6446	5	.7289	
Total	6.7943	6		

The value for F is 4.3212. From Table A8 with $\nu_1 = 1$ and $\nu_2 = 5$, $F_{1,5,.05} = 6.61$ and $F_{1,5,.01} = 16.3$. Since $F = 4.3212$ is less than 6.61, we cannot reject H_0 at the .05 level of significance. This is the same conclusion that we came to earlier when we performed the two-sided t test on these data. ■

The advertising data illustrate an interesting relationship between the two-sided t test and the F test of $H_0: \beta = 0$. Recall from page 544 that the computed value for t was 2.079. But the square of this value, $2.079^2 = 4.32$, is the computed value for F above. Also, the squares of the tabulated values for t (i.e., $t^2_{5,.025} = 2.571^2 = 6.61$ and $t^2_{5,.005} = 4.032^2 = 16.3$) are precisely the tabulated values for F. The two test procedures are therefore precisely equivalent.

We emphasize, however, that the F test can only be used to test $H_0: \beta = 0$ against $H_1: \beta \neq 0$, whereas we can perform one- *or* two-sided t tests.

ONE-WAY ANALYSIS OF VARIANCE AND SIMPLE LINEAR REGRESSION

The reader familiar with the one-way analysis of variance in Chapter 13 will be interested in the relationship between that type of analysis and our work in this chapter. In both cases we were interested in the effect of an independent variable on a quantitative dependent variable (recovery times and pictures in Chapter 13, and monthly revenue and mileage here). One difference is that in Chapter 13 the independent variable in the analysis was generally qualitative (type of drug or particular technician), whereas in regression the independent variable is quantitative (advertising revenue and speed).

In the one-way analysis in Chapter 13, the null hypothesis was $H_0: \mu_1 = \mu_2 = \cdots = \mu_r$. In this chapter it was $H_0: \beta = 0$. In each case, if H_0 is true, there is no relationship between the two variables.

In both analyses the F test of H_0 is based on the ratio of two terms. The numerators ($MS_{between}$ and MSR) reflect the amount of variability in the dependent variable that is associated with the independent variable. The denominators (MS_{within} and MSE) reflect the amount of variability in the dependent variable due to variables *other* than the independent variable.

When drawing conclusions from a regression analysis the researcher should always keep in mind that a strong linear relationship between x and y should not, in general, be interpreted as x causing y. In fact, as we emphasized in Chapter 8, when dealing with people, a *causal* relationship between the independent and dependent variables can be established only with a randomized experiment. For example, suppose that a significant negative relationship were found between children's test scores and time spent watching TV. We could only conclude that watching more TV *results* in lower test scores if the children were randomly assigned to watch TV.

PROBLEMS 14.3

13. (a) Apply the t test of $H_0: \beta = 0$ against $H_1: \beta > 0$ using the results in Problem 1 on page 532.
    (b) Interpret β in this example.
    (c) What does the result in part (a) tell you about the relationship between the number of beers and the change (generally an increase) in reaction time?
14. (a) Define β in the context of Problem 2 on page 533.
    (b) Use the results in that problem to find a 95% confidence interval for β.
15. (a) Define β in the context of Problem 3 on page 533.
    (b) Use the results in the same problem to perform a t test of $H_0: \beta = 0$ against $H_1: \beta > 0$.

(c) Why is the F test of $H_0: \beta = 0$ not appropriate here?

16. (a) Define β in the context of Problem 4 on page 533.
    (b) Use the results in that problem to find an 80% confidence interval for β.
17. (a) Define β in the context of Problem 5 on page 534.
    (b) Do the results in that problem suggest a significant negative relationship between age and hours of sleep?
18. Find a 90% confidence interval for β using the results in Problem 6 on page 534.
19. (a) Define β in the context of Problem 7 on page 535.
    (b) Use the results to perform a t test of $H_0: \beta = 0$ against $H_1: \beta \neq 0$.
    (c) Show that the F test of $H_0: \beta = 0$ leads to the same conclusion as in part (b). Why will this always be so?
20. (a) Do the results in Problem 8 on page 535 suggest a significant negative relationship between x and y?
    (b) Perform a two-sided t test of $H_0: \beta = 0$ using the same results.
    (c) Show that the F test of $H_0: \beta = 0$ leads to the same conclusion as in part (b).
    (d) Why is a two-sided test not really appropriate for this problem?
21. (a) Test $H_0: \beta = 0$ against $H_1: \beta \neq 0$ using the following data:

x	22	29	28	24	22	20	33	28	21	27
y	14.7	12.1	11.6	10.2	11.8	9.2	8.3	8.0	11.3	11.0

$$\sum x_i^2 = 6612 \qquad \sum y_i^2 = 1206.56 \qquad \sum x_i y_i = 2719.7$$

(b) The data were selected by a computer from a population for which the slope of the population line β is 0. Did you make the correct decision in part (a)?

22. A manufacturer of fold-up umbrellas records the average weekly rainfall, x (in inches), for a 16-week period and the sale of their umbrellas, y, in the city during each of these weeks. The data are as follows

RAINFALL	SALES	RAINFALL	SALES	RAINFALL	SALES	RAINFALL	SALES
.19	588	.39	620	.29	680	.15	596
.05	492	.09	457	.37	604	.09	474
.15	487	.26	545	.21	505	.26	581
.38	625	.38	615	.06	479	.29	501

$$\sum x_i^2 = 1.0287 \qquad \sum y_i^2 = 4{,}964{,}237 \qquad \sum x_i y_i = 2089.08$$

(a) Define b and β in this case.
(b) Test $H_0: \beta = 0$ against $H_1: \beta > 0$.
(c) Find a 90% confidence interval for β.

23. The authors collected the following information on 20 Boston Red Sox games. Here x is the temperature (in degrees Fahrenheit) at game time and y the number of hits the Sox got in the game.

TEMPERATURE	HITS	TEMPERATURE	HITS	TEMPERATURE	HITS
93	13	87	9	89	14
76	1	78	12	90	11
99	6	77	8	75	9
87	10	81	12	83	12
99	10	81	2	77	7
71	10	79	9	84	9
85	12	70	13		

$$\sum x_i^2 = 139{,}227 \qquad \sum y_i^2 = 2009 \qquad \sum x_i y_i = 15{,}761$$

(a) Define b and β in this case.

(b) Test $H_0: \beta = 0$ against $H_1: \beta \neq 0$ using (i) the t test; (ii) the F test.

Solution to Problem 13

Refer to the solution to Problem 1 on page 536. From Equation 14.6,

$$b = \frac{T_{xy}}{T_{xx}} = \frac{257.7}{3848} = .067$$

(a) We wish to test $H_0: \beta = 0$ against $H_1: \beta > 0$. Following the procedure in Table 5(a), we have

$$t = \frac{.067}{\sqrt{29(.0816)/3848}} = 5.66$$

From Table A6, $t_{27,.05} = 1.703$ and $t_{27,.01} = 2.473$. Since $t = 5.66$ exceeds 2.473, we reject H_0 at the .01 level of significance. The data suggest that β, the slope of the population line relating change in reaction time to the number of beers consumed, is significantly greater than 0.

(b) In this case the unknown β is the increase in the *mean* change in reaction time corresponding to an increase in one beer consumed.

(c) The data suggest a positive linear relationship between the number of beers consumed and increased reaction time.

SECTION 14.4

PREDICTION

In this section we focus on methods for judging the *accuracy* of predictions based on the regression line. As in Section 14.3, we first assume that Definition 1 applies. In Figure 7 we show the population line together with the regression line $y = 2.297x - 1.952$ for the advertising data. As you can see when $x = 2.5$ ($2500), the predicted value for y from the regression line is $\hat{y} = 2.297(2.5) - 1.952 = 3.7905$ (or $37,905). If we knew the values of α and β, we could also compute $\mu_{y|2.5} = \alpha + \beta(2.5)$. Since we do not know α and β, we use the value $\hat{y} = 3.7905$ as an estimate for $\mu_{y|2.5}$, the unknown mean

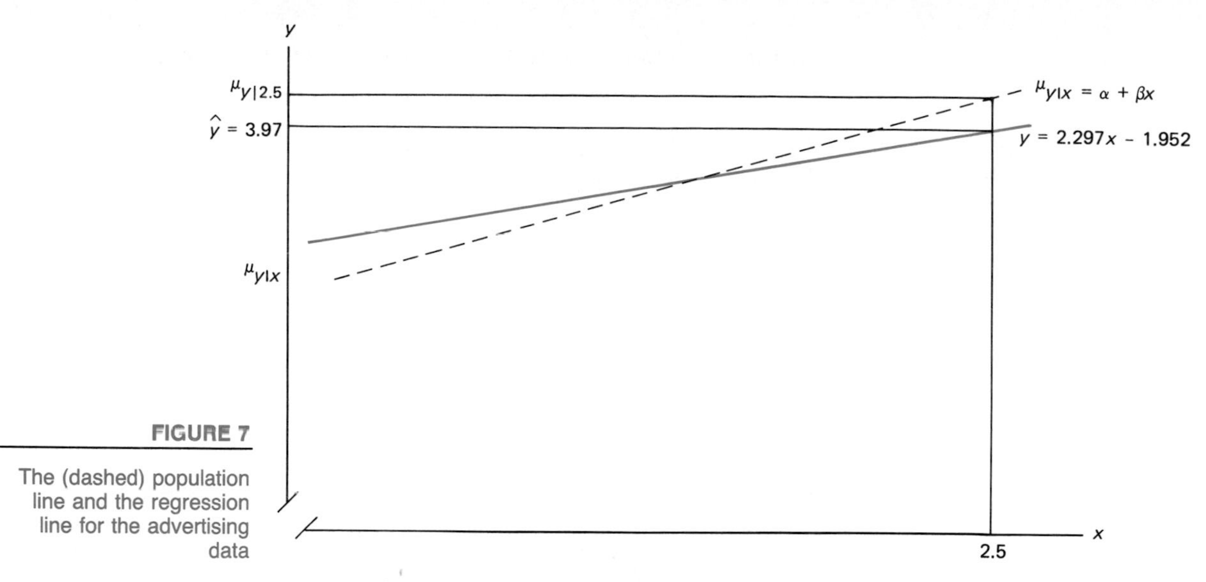

FIGURE 7

The (dashed) population line and the regression line for the advertising data

revenue for all restaurants spending 2.5 for advertising. But how accurate is this estimate?

In general, we regard the predicted value $\hat{y} = a + bx_0$ from the regression line as an estimate for $\mu_{y|x_0}$, the mean of y for all members of the population with $x = x_0$. As we indicated earlier, repeated sampling will result in different regression lines and hence different values for $\hat{y}_0$ for the same x_0. It can be shown that over repeated samples from a population the standard deviation of the distribution of $\hat{y}_0$ around $\mu_{y|x_0}$ is

$$\text{SD}(\hat{y}_0) = \sigma\sqrt{\frac{1}{n} + \frac{(x_0 - \bar{x})^2}{\Sigma(x_i - \bar{x})^2}}$$

where, as before, σ is the population standard deviation of the y's for each value of x. Since σ is usually unknown, we replace it by its estimate $\sqrt{\text{MSE}}$. Replacing $\Sigma(x_i - x)^2$ by T_{xx}/n, the $\text{SD}(\hat{y}_0)$ becomes

$$\text{SD}(\hat{y}_0) = \sqrt{\text{MSE}\left[\frac{1}{n} + \frac{n(x_0 - \bar{x})^2}{T_{xx}}\right]}$$

We can also specify the *shape* of the distribution of the standardized form of $\hat{y}_0$. Specifically:

> DEFINITION 6
>
> If the assumptions in Definition 1 apply, then
>
> $$\frac{\hat{y}_0 - \mu_{y|x_0}}{\sqrt{\text{MSE}\left[\frac{1}{n} + \frac{n(x_0 - \bar{x})^2}{T_{xx}}\right]}}$$
>
> has the t_{n-2} distribution.

This last result enables us to write a confidence interval for $\mu_{y|x_0}$:

> **DEFINITION 7**
>
> If the assumptions in Definition 1 apply, then a $(100\gamma)\%$ confidence interval for $\mu_{y|x_0}$, the mean value of y for all members of the population for which $x = x_0$, is
>
> $$\hat{y}_0 - t_{n-2(\gamma)}\sqrt{\text{MSE}\left[\frac{1}{n} + \frac{n(x_0 - \bar{x})^2}{T_{xx}}\right]} \quad \text{to} \quad \hat{y}_0 + t_{n-2(\gamma)}\sqrt{\text{MSE}\left[\frac{1}{n} + \frac{n(x_0 - \bar{x})^2}{T_{xx}}\right]}$$

EXAMPLE 9 Use the advertising data in Example 1 on page 526 to find a 90% confidence interval for $\mu_{y|2.5}$, the mean revenue for all restaurants spending 2.5 ($250) on advertising.

Solution The regression line in Example 1 is $y = 2.297x - 1.952$. When $x_0 = 2.5$, $\hat{y}_0 = 2.297(2.5) - 1.952 = 3.7905$.

From Example 1 and Table 2, $n = 7$, $T_{xx} = 4.18$, and MSE $= .7289$. From Table 1, $\bar{x} = 2.34$. Referring to Table A6 with $\nu = n - 2 = 5$, $t_{5(.9)} = 2.015$. With these values, for $x_0 = 2.5$, a 90% confidence interval for $\mu_{y|2.5}$ is

$$3.7905 \mp 2.015\sqrt{.7289\left[\frac{1}{7} + \frac{7(2.5 - 2.34)^2}{4.18}\right]} = 3.7905 \mp .74140$$

$$= 3.049 \text{ to } 4.532$$

We can be 90% confident that the mean revenue for all restaurants spending 2.50 ($250) on advertising lies between 3.049 and 4.532 (or $30,490 and $45,320). This confidence interval is shown in Figure 8.

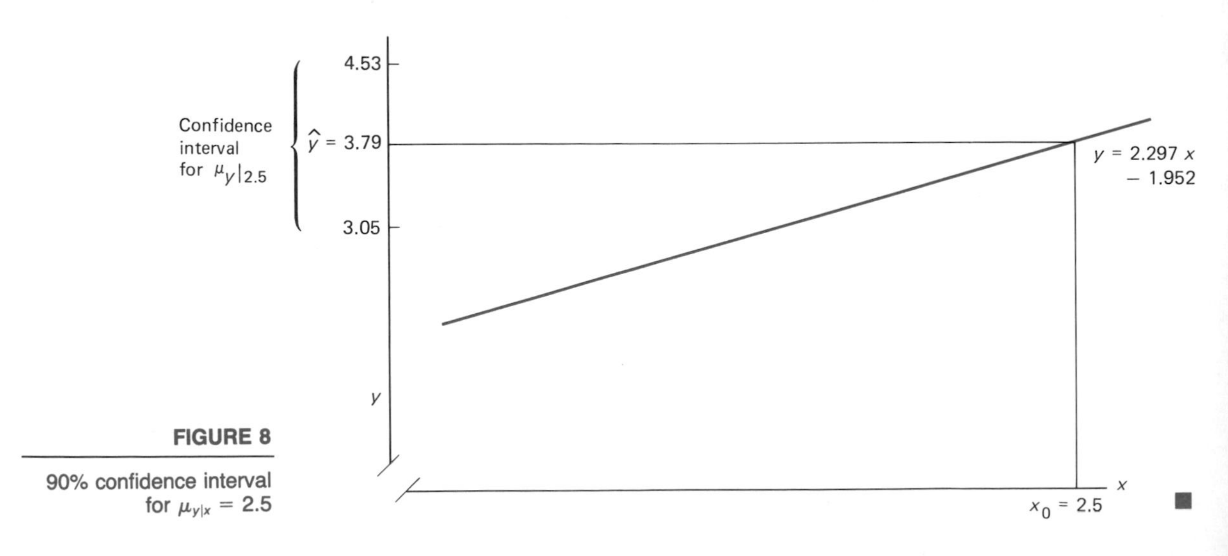

FIGURE 8

90% confidence interval for $\mu_{y|x} = 2.5$ ■

EXAMPLE 10 Use the results in Example 2 on page 531 to find an 80% confidence interval for the mean mileage per gallon for all cars of this model when they travel at (a) 50 mph; (b) 75 mph.

Solution From Example 2 we obtain the values $T_{xx} = 20{,}625$, $n = 10$, MSE $= 2.132$, and $\bar{x} = 475/10 = 47.5$. The regression line is $y = 39.4309 - .1825x$. Referring to Table A6 with $\nu = n - 2 = 8$, $t_{8(.8)} = 1.397$.

(a) With $x_0 = 50$, $\hat{y} = 39.4309 - .1825(50) = 30.3059$ and an 80% confidence interval for $\mu_{y|50}$ is

$$30.3059 \mp 1.397\sqrt{2.132\left[\frac{1}{10} + \frac{10(50 - 47.5)^2}{20{,}625}\right]} = 30.3059 \mp .65475$$

$$= 29.651 \text{ to } 30.961$$

(b) With $x_0 = 75$, $\hat{y}_0 = 39.4309 - .1825(75) = 25.7434$ and an 80% confidence interval for $\mu_{y|75}$ is

$$25.7434 \mp 1.397\sqrt{2.132\left[\frac{1}{10} + \frac{10(75 - 47.5)^2}{20{,}625}\right]} = 25.7434 \mp 1.3935$$

$$= 24.350 \text{ to } 27.1370$$

The two confidence intervals are shown in Figure 9.

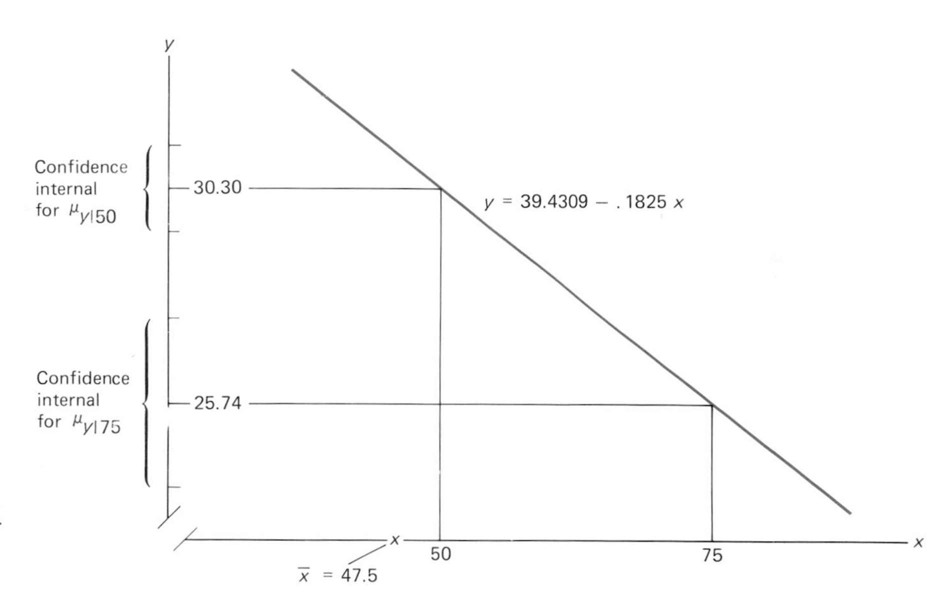

FIGURE 9

80% confidence interval for $\mu_{y|50}$ and $\mu_{y|75}$

Notice that the farther x_0 is from $\bar{x}$(47.5 above), the greater will be $(x_0 - \bar{x})^2$, the greater will be SD$(\hat{y}_0)$, and the wider the confidence interval for $\mu_{y|x_0}$. Thus the farther x_0 is from $\bar{x}$, the less accurately we will be able to predict $\mu_{y|x_0}$. In Example 10, for instance, where $\bar{x} = 47.5$, the confidence interval for $\mu_{y|75}$ is twice as wide as that for $\mu_{y|50}$. This difference is shown in Figure 9.

PREDICTION INTERVAL FOR AN INDIVIDUAL y

Instead of an estimate for the mean $\mu_{y|x_0}$, $\hat{y}_0$ can serve as an estimate for y_0, the y value for an *individual* with $x = x_0$. But again we ask how accurate this estimate is.

In this case we are concerned with the SD($\hat{y}_0$) for the distribution of $\hat{y}_0$ *around* y_0. Since the variability of $\hat{y}_0$ around y_0 reflects both the variability of $\hat{y}_0$ around $\mu_{y|x_0}$ and the variability of y around $\mu_{y|x_0}$, SD($\hat{y}_0$) will be larger than when $\hat{y}_0$ is used as an estimate of $\mu_{y|x_0}$. It is

$$\text{SD}(\hat{y}_0) = \sqrt{\text{MSE}\left[1 + \frac{1}{n} + \frac{n(x_0 - \bar{x})^2}{T_{xx}}\right]}$$

We can obtain a type of confidence interval for y_0 but, following custom, we refer to such an interval for an individual y value as a **prediction interval.**

DEFINITION 8

If the assumptions in Definition 1 apply, then a $(100\gamma)\%$ prediction interval for y_0, the unknown value of y for an *individual* with $x = x_0$ is

$$\hat{y}_0 - t_{n-2(\gamma)}\sqrt{\text{MSE}\left[1 + \frac{1}{n} + \frac{n(x_0 - \bar{x})^2}{T_{xx}}\right]} \quad \text{to}$$

$$\hat{y}_0 + t_{n-2(\gamma)}\sqrt{\text{MSE}\left[1 + \frac{1}{n} + \frac{n(x_0 - \bar{x})^2}{T_{xx}}\right]}$$

EXAMPLE 11

Using the results summarized in Example 9 on page 552, find a 90% prediction interval for the revenue for an individual restaurant spending 2.5 on advertising.

Solution

Using the information in Example 9, that is, $\hat{y}_0 = 3.7905$, MSE $= .7289$, $T_{xx} = 4.18$, $\bar{x} = 2.34$, and $t_{5(.9)} = 2.015$, a 90% confidence interval for y when $x_0 = 2.5$ is

$$3.7905 \mp 2.015\sqrt{.7289\left[1 + \frac{1}{7} + \frac{7(2.5 - 2.34)^2}{4.18}\right]} = 3.7905 \mp 1.8733$$

$$= 1.9172 \text{ to } 5.6638 \quad \text{or} \quad \$19{,}172 \text{ to } \$56{,}638$$

This prediction interval for an individual y when $x_0 = 2.5$ is approximately two and a half times as wide as that for $\mu_{y|2.5}$ in Example 9. This much greater width reflects the uncertainty associated with estimating the y value for an *individual* rather than the mean value of y for a specific x.

■

WHEN TO PREDICT

In reality it makes sense to use the regression line for prediction only when there is evidence of a significant relationship between x and y. This evidence may be a rejection of H_0: $\beta = 0$, although some practitioners prefer to use the regression line in this way if $r^2 \geq .5$ or some similar figure.

EXTRAPOLATION

One danger in prediction is *extrapolating* the regression line far beyond the range of values on which the line is based. The reason is that the linear relationship that seems appropriate for the observed set of x, y values may be entirely inappropriate beyond this range. In Example 1, for instance, the regression line $y = 2.297x - 1.952$ was based on a range of advertising expenditure from 2.0 (\$200) to 2.8 (\$280). There is no guarantee that this line represents the relationship between revenue and advertising expenditure if x were only 1.2 (\$120) or 3.5 (\$350).

This issue raises an interesting point. The researcher may wish to include a wide range of x values in the sample, but this may create a problem. Over a wide range of x values the relationship between x and $\mu_{y|x}$ may no longer be linear. In Figure 10 we show a situation where over all values of x the relationship between x and $\mu_{y|x}$ is curvilinear although in the range A–B the relationship can be well approximated by a straight line.

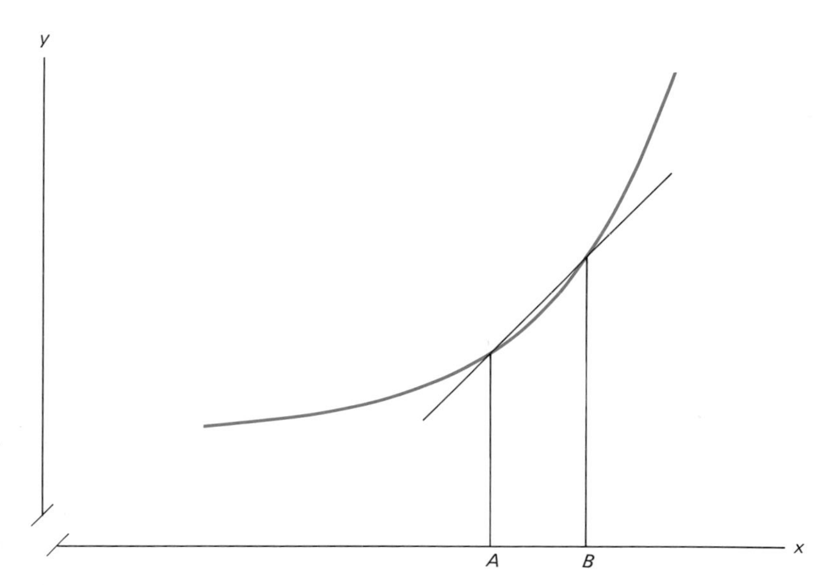

FIGURE 10

For a small range of x values the relationship between x and $\mu_{y|x}$ is approximately linear

GENERALIZING FROM THE SIMPLE LINEAR MODEL

There are two ways of generalizing beyond the simple linear regression line $y = a + bx$. More sophisticated models involving just x and y would involve the logarithm of x such as $y = a + b \log x$ or a quadratic in x of the form $y = a + bx + cx^2$. Researchers frequently have data on two or more independent variables $x_1, x_2, \ldots, x_m$ that may be used to predict the dependent variable y. In this case a **multiple regression line** of the form

$$y = a + b_1x_1 + b_2x_2 + \cdots + b_mx_m$$

may be used. We provide several references to this very important topic in the Suggested Reading at the end of the chapter.

PROBLEMS 14.4

24. Use your results in Problem 1 on page 532 and Problem 13 on page 548 to find (a) a 90% confidence interval for the mean change in reaction time for students drinking four beers between tests; (b) a 90% prediction interval for the change in reaction time for a specific student who drinks four beers between tests.

25. (a) Use your results in Problem 2 on page 533 and Problem 14 on page 548 to find an 80% confidence interval for the mean height of arborvitae that are (i) 5 years old; (ii) 9 years old. Why is your interval in (ii) wider than that in (i)?

    (b) Find an 80% prediction interval for the height of a specific 6-year-old arborvitae bush.

26. Use your results in Problem 4 on page 533 and Problem 16 on page 549 to find a 90% prediction interval for the beer sales at (a) a game in which 22,000 people are expected; (b) a game which is a sell-out with 42,050 people attending.

27. (a) Use your results in Problem 5 on page 534 and Problem 17 on page 549 to find a 95% confidence interval for (i) the mean number of hours of sleep for 7-year-old children; (ii) the mean number of hours of sleep for Eliot Mitchell, who is 7 years old.

    (b) In part (a), why are your answers in (i) and (ii) different?

28. Use your results in Problem 6 on page 535 and Problem 18 on page 549 to find an 80% confidence interval for the mean calculus score of students scoring (a) 22, and (b) 15, on the math competency test. Why would it be foolhardy to predict the calculus score for a student scoring 10 on the math competency test?

29. Why would it be inappropriate to use the results in Problem 23 on page 549 to find a prediction interval for the number of hits that the Red Sox will get in a game played in 84°F weather?

30. The admissions office of a large university rates applicants on a scale of 3 to 15. The *lower* the score, the better qualified the student. As a pilot study on the predictability of this system, y, the grade-point average (GPA), and x, the initial score of a random sample of 24 seniors, are recorded.

SCORE	GPA	SCORE	GPA	SCORE	GPA	SCORE	GPA
8	2.5	7	2.6	8	2.8	13	3.0
13	2.6	9	2.5	9	2.3	9	2.3
11	2.6	5	2.0	9	2.3	11	2.5
7	2.3	7	1.8	10	2.9	9	2.2
14	3.2	6	2.0	11	2.9	15	3.6
10	2.3	13	3.2	11	2.9	5	1.9

$$\sum x_i^2 = 2378 \qquad \sum y_i^2 = 160.68 \qquad \sum x_i y_i = 610.3$$

(a) Compute r.

(b) Find the regression line relating GPA to admissions score.

(c) Construct the ANOVA table.

(d) Find an 80% confidence interval for the mean GPA for students whose admission score is (i) 12; (ii) 4.

(e) In part (d), why is your interval in (i) narrower than that in (ii)?

Solution to Problem 24

(a) We have already computed $b = .067$. From Equation 14.2, $a = (1/29)[5.74 - (.067)90] = -.01$. Thus the regression line is $y = -.01 + .067x$ or $y = .067x - .01$. When $x_0 = 4$, $y_0 = .067(4) - .01 = .258$. Using MSE $= .0186$, $\bar{x} = \Sigma x_i/29 = 90/29 = 3.103$ and $t_{27(.9)} = 1.703$, a 90% confidence interval for $\mu_{y|4}$ is

$$.258 \mp 1.703\sqrt{.0186\left[\frac{1}{29} + \frac{29(4 - 3.103)^2}{3848}\right]} = .258 \mp .0468$$

$$= .211 \text{ to } .305 \text{ seconds}$$

We can be 90% confident that the mean increase in reaction time for people consuming four beers lies in this interval.

(b) A 90% prediction interval for the increase in reaction time for an individual consuming four beers is

$$.258 \mp 1.703\sqrt{.0186\left[1 + \frac{1}{29} + \frac{29(4 - 3.103)^2}{3848}\right]} = .258 \mp .237$$

$$= .021 \text{ to } .495 \text{ seconds}$$

We may be 90% confident that an individual increase in reaction time after four beers will lie between .021 and .495 second.

SECTION 14.5
ASSUMPTIONS IN REGRESSION

In Sections 14.2, 14.3, and 14.4 we made three assumptions about the relationship between x and y in the population. They are: (1) for any specific value of x, y, has a normal distribution, (2) there is a linear relationship of the form $\mu_{y|x} = \alpha + \beta x$ between x and $\mu_{y|x}$, and (3) for each value x, y has the same standard deviation, σ.

Some comments are appropriate concerning these assumptions:

1. In Chapter 12 we were able to check the assumption of normality using dot diagrams or histograms. With bivariate data it is more difficult to check for normality graphically. However, the t or F tests for β work well in the presence of moderate departures from normality.

2. The assumption of linearity *should* always be checked by sketching a scatterplot. If the relationship between x and y in the sample appears to be linear, we have some basis for assuming the same is true in the population. Figure 11 illustrates examples of scatterplots where the data support the assumption of a linear relationship between x and y (although with quite different error variances).

By contrast, Figure 12 illustrates two examples in which the scatterplots suggest that the relationship between x and y is more complex than in the linear case. More advanced techniques of analysis are required in these cases.

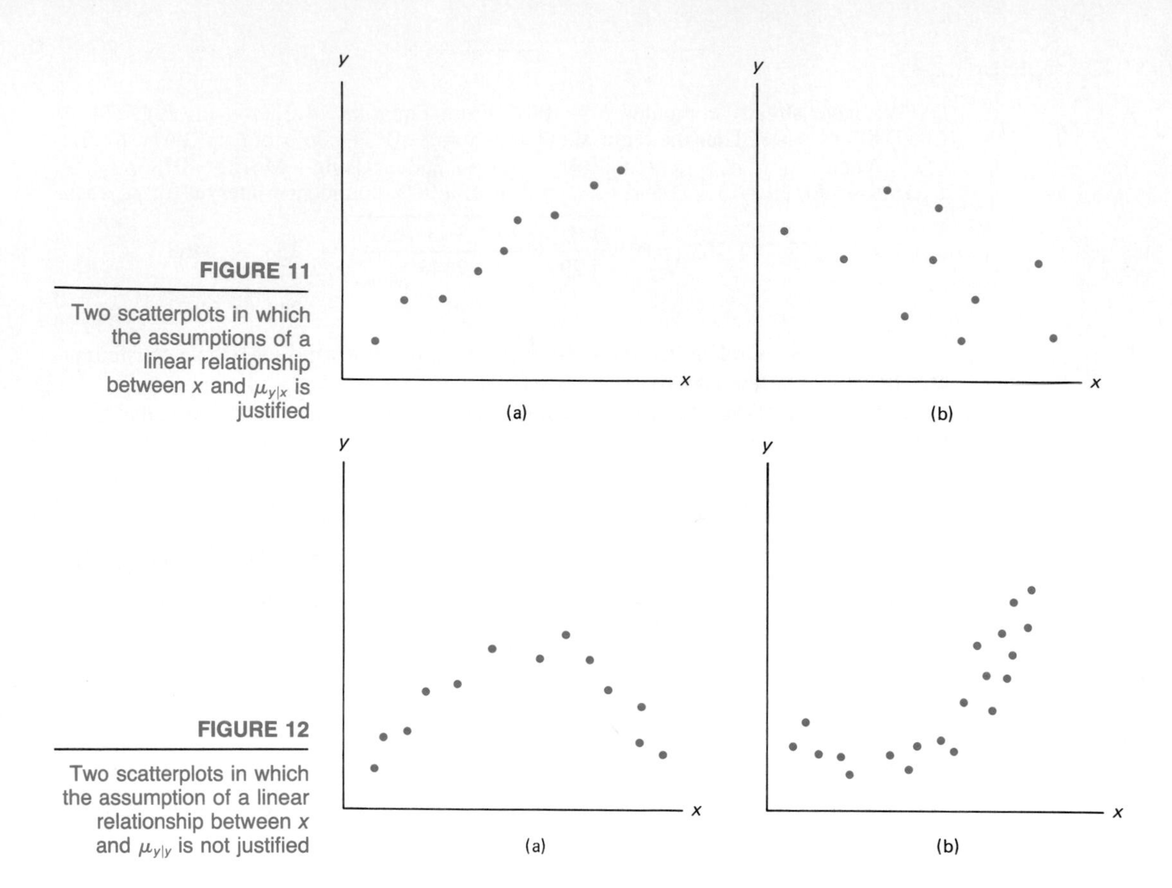

FIGURE 11

Two scatterplots in which the assumptions of a linear relationship between x and $\mu_{y|x}$ is justified

FIGURE 12

Two scatterplots in which the assumption of a linear relationship between x and $\mu_{y|y}$ is not justified

3. A scatterplot again provides a helpful way of checking the assumption of equal variances for the y's. In Example 6, for instance, we assumed that the variability of SBP was the same for 35-year-olds as for 55-year-olds. An example of the failure of this assumption is the situation where the variances of the y's increase as the value of x increases. This situation is shown in Figure 13. In this case the t or F test is not appropriate.

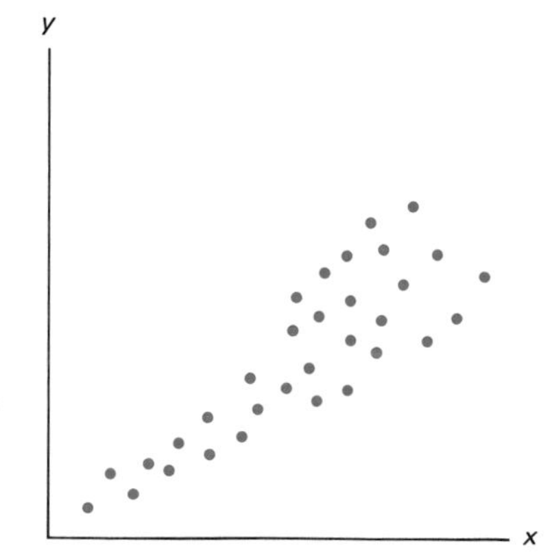

FIGURE 13

Scatterplot in which the variability of the y's increases as x increases

PROBLEMS 14.5

For each of the following scatterplots indicate which, if any, of the three assumptions (i) linearity, (ii) normality of y for each x, and (iii) equal standard deviations of y for each x appear to be violated.

31.

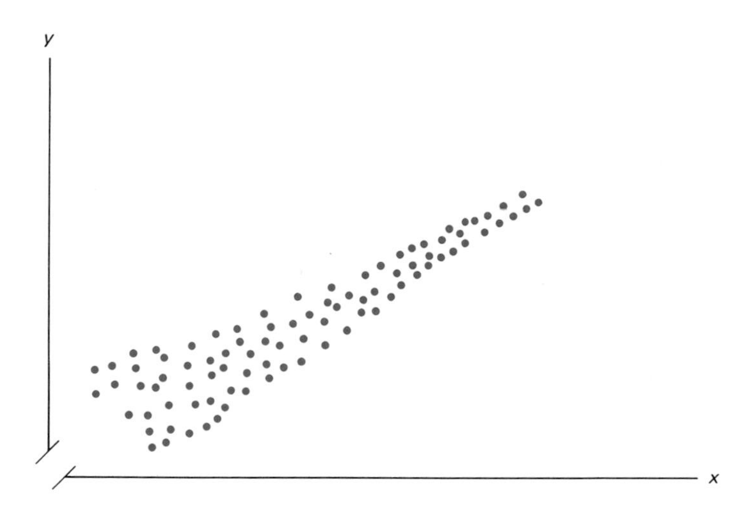

32.

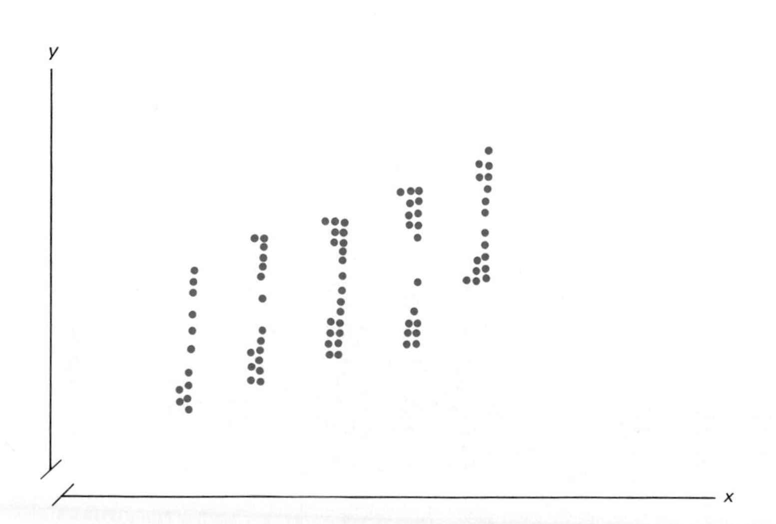

33.

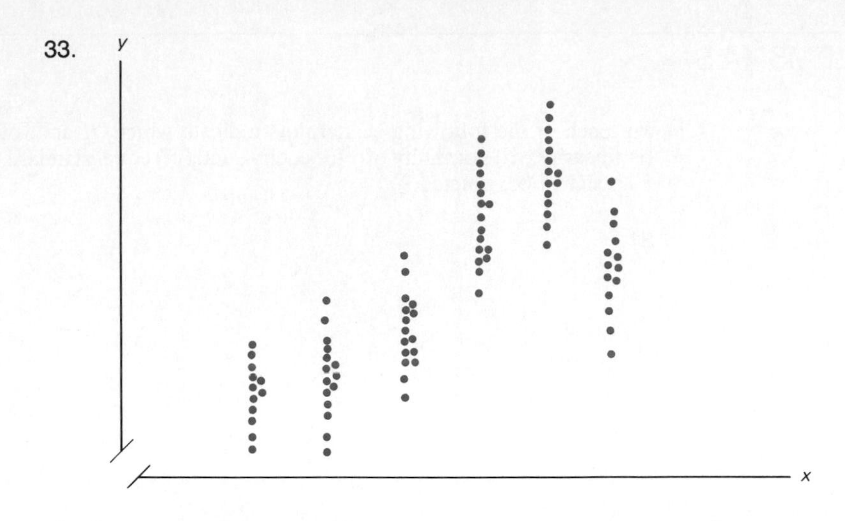

34.

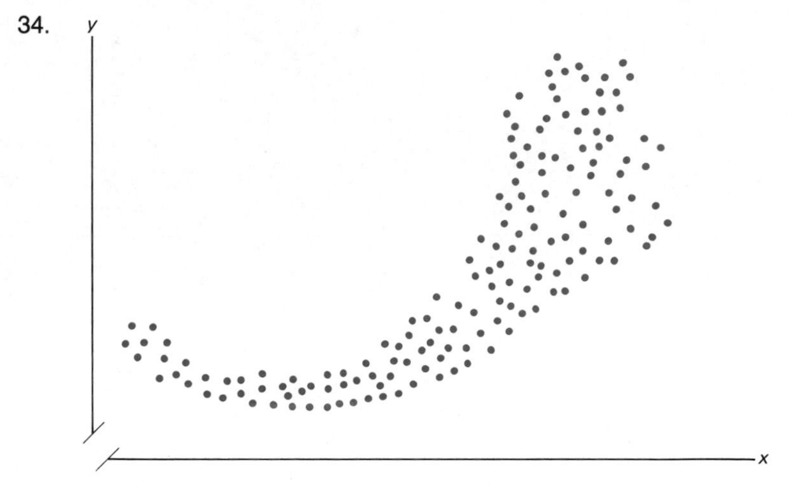

35.

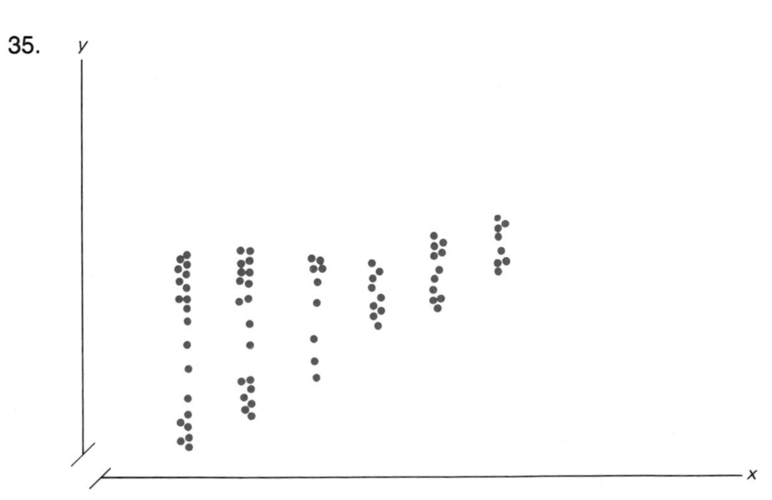

36.

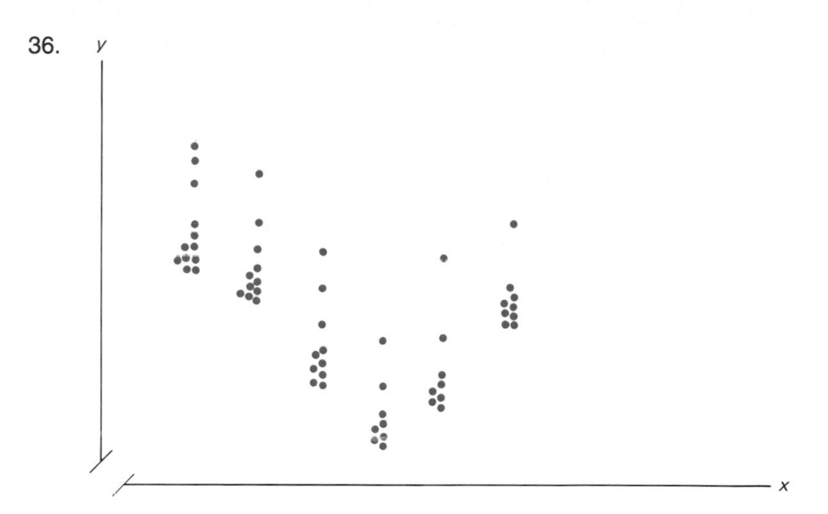

Solution to Problem 31

The assumption of equal variances certainly seems to be violated.

SECTION 14.6

USING COMPUTER OUTPUT: MINITAB AND LINEAR REGRESSION

In section 2.4 (page 99) we noted that the REGRESS command produces far more information than simply the regression line. The remaining output enables us to perform the analyses discussed in this chapter. An example will illustrate the procedure.

EXAMPLE 12 In Example 11 of Chapter 2, referring to the nutrition data, we asked for the regression line relating infant birthweight (in C7) to birth length in C8. The command "REGRESS C7 ON 1 PREDICTOR IN C8" produces the output shown in Table 8.

The first output is the regression line C7 = $-1233 + 89.6$ C8, or more conventionally, $y = -1233 + 89.6x$.

The values $a = -1233$ and $b = 89.6$ are sometimes referred to as **regression coefficients.** This explains their appearance under the heading "coefficients."

The ANOVA table (the general case is shown in Table 3) appears at the bottom of the printout. Notice that when referring to sources of variability Minitab uses the term "residual" rather than "error." In any event, MSE = $MS_{residual}$ = 136984. The F test of $H_0: \beta = 0$ against $H_1: \beta \neq 0$ is based on the ratio F = MSR/MSE = 4,576,595/136,984 = 33.41. (Minitab does not print

TABLE 8 Complete output obtained with the "REGRESS" command for the birthweight/birth length data

```
MTB > REGRESS C7 ON 1 PREDICTOR IN C8

THE REGRESSION EQUATION IS
C7 = - 1233 + 89.6 C8

                                  ST. DEV.    T-RATIO =
COLUMN       COEFFICIENT          OF COEF.    COEF/S.D.
                 -1233.2            774.2        -1.59
C8                 89.58            15.50         5.78

S = 370.1

R-SQUARED = 33.6 PERCENT
R-SQUARED = 32.6 PERCENT, ADJUSTED FOR D.F.

ANALYSIS OF VARIANCE

 DUE TO       DF          SS         MS=SS/DF
REGRESSION     1     4576595          4576595
RESIDUAL      66     9040912           136984
TOTAL         67    13617507
```

this ratio. We have provided it.) From Table A8 with $\nu_1 = 1$ and $\nu_2 = 60$ ($\nu_2 = 66$ is not in the table), $F_{1,60,.05} = 4.00$ and $F_{1,60,.01} = 7.08$. Since 33.41 is greater than 7.08, we reject H_0 at the .01 level of significance. The data suggest that β is signficantly greater than 0.

The statement "R-SQUARED = 33.6 PERCENT" right above the ANOVA table refers to the fact that, from Equation 14.11,

$$r^2 = \frac{\text{SSR}}{\text{SST}} = \frac{4{,}576{,}595}{13{,}617{,}507} = .336$$

Thus 100 $r^2 = 33.6$ or, in other words, only 33.6% of the variability in infant birthweight is associated with its linear relationship to birth length. The meaning of the "adjusted" r^2 is not important and we shall not refer to it again.

In the next column, 15.50 is the value for $\text{SD}(b) = \sqrt{n\text{MSE}/T_{xx}}$ from Section 14.3. The ratio $t = b/\text{SD}(b) = 89.6/15.50 = 5.78$ is given in the last column.[2] We can use this value for t to test $H_0: \beta = 0$ against the one-sided alternative $H_1: \beta > 0$. This is more realistic than $H_1: \beta \neq 0$ since we would expect the relationship, if any, between birthweight and birth length to be positive. Following the procedure in Table 5(a), we find that $t_{60,.05} = 1.671$ and $t_{60,.01} = 2.390$. Since $t = 5.78$ is far greater than 2.390, we reject H_0 at the .01 level of significance. The data suggest that β is significantly *greater* than 0.

[2]The corresponding values 774.2 and -1.59 for $a = -1233$ are SD(a) and $t = a/\text{SD}(a)$, respectively. We have not referred to these quantities in the text since they are rarely used.

Incidentally, notice that the square of the observed value for t, $5.78^2 = 33.41$, the observed value for F.

Finally, we note that the value $s = 370.1$ on the printout is $\sqrt{136{,}984}$, which is $\sqrt{MSE}$. In this context $s = 370.1$ is an estimate for σ, the standard deviation of birthweight for infants having the same length.

■

The printout obtained from the REGRESS command does not provide a confidence interval for β or prediction intervals, but the information in the printout can be used to obtain them. We indicate the procedure in the next example.

EXAMPLE 13 The printout in Table 9 was obtained for the age (x) and speed (y) data in Example 12 in Section 2.4. (a) Interpret β in this case. (b) Use the printout to (i) find and interpret a 90% confidence interval for β, and (ii) find and interpret a 90% confidence interval for $\mu_{y|35}$, the mean speed over all 35-year-old male marathoners.

TABLE 9

Complete printout using the regress command for the age/speed data

```
MTB > REGRESS C2 ON 1 PREDICTOR IN C1

THE REGRESSION EQUATION IS
C2 = 9.53 - 0.0680 C1

                                 ST. DEV.      T-RATIO =
COLUMN       COEFFICIENT         OF COEF.      COEF/S.D.
                  9.5274          0.3670          25.96
C1              -0.06804         0.01122          -6.06

S = 0.2796

R-SQUARED = 69.7 PERCENT
R-SQUARED = 67.8 PERCENT, ADJUSTED FOR D.F.

ANALYSIS OF VARIANCE

 DUE TO       DF           SS          MS=SS/DF
REGRESSION     1       2.8723            2.8723
RESIDUAL      16       1.2505            0.0782
TOTAL         17       4.1228
```

Solution (a) In this case, β is the unknown slope of the population line relating mean speed per person to age. As such it is the change in mean speed associated with an increase in age of one year.

(b) From Definition 3 a 90% confidence interval for β is

$$b - t_{n-2(\gamma)}\sqrt{\frac{n\text{MSE}}{T_{xx}}} \quad \text{to} \quad b + t_{n-2(\gamma)}\sqrt{\frac{n\text{MSE}}{T_{xx}}}$$

where $\sqrt{n\text{MSE}/T_{xx}} = \text{SD}(b)$.

From Table 9, $b = -.068$, $\text{SD}(b) = .01122$, and from Table A6 with $n - 2 = 16$ (the number of degrees of freedom for "residual" in the ANOVA table), $t_{16(.9)} = 1.746$. Thus a 90% confidence interval for β is

$$-.068 \mp 1.746(.01122) \quad \text{or} \quad -.088 \text{ to } 0.048$$

We can be 90% confident that the slope of the population line lies between $-.088$ and $-.048$.

(c) From Definition 7 a 90% prediction interval for $\mu_{y|x_0}$ is

$$\hat{y}_0 - t_{n-2(\gamma)}\sqrt{\text{MSE}\left[\frac{1}{n} + \frac{n(x_0 - \overline{x})^2}{T_{xx}}\right]} \quad \text{to}$$

$$\hat{y}_0 + t_{n-2(\gamma)}\sqrt{\text{MSE}\left[\frac{1}{n} + \frac{n(x_0 - \overline{x})^2}{T_{xx}}\right]}$$

where $\hat{y}_0 = a + bx_0$.

The only element of this interval, other than $t_{n-2(\gamma)}$, that cannot be found in the printout is $\overline{x}$. This can be found with the command "AVERAGE OF C1" (if the x's are in C1) or a simple calculation will show that the average of the 18 ages is $\overline{x} = 32.167$ years. The printout does not give T_{xx} alone but writing

$$\sqrt{\text{MSE}\left[\frac{1}{n} + \frac{n(x_0 - \overline{x})^2}{T_{xx}}\right]} = \sqrt{\frac{\text{MSE}}{n} + \frac{n\text{MSE}}{T_{xx}}(x_0 - \overline{x})^2}$$

You can see in part (b) that $n\text{MSE}/T_{xx} = \text{SD}(b)^2$, which is $.01122^2$ from the printout. Again from part (b), $t_{n-2(\gamma)} = t_{16(9)} = 1.746$. When $x_0 = 35$, $\hat{y}_0 = 9.53 - .068(35) = 7.15$ mph. Finally, $(x_0 - \overline{x})^2 = (35 - 32.167)^2 = 8.0259$ and $\text{MSE} = .0782$. A 90% prediction interval for $\mu_{y|35}$ is thus

$$7.15 \mp 1.746\sqrt{\frac{.0782}{18} + (.01122)^2(8.0259)} = 7.15 \mp .128$$

$$= 7.02 \text{ to } 7.28$$

We can be 90% confident that $\mu_{y|35}$ lies between 7.02 and 7.28 mph. ■

PROBLEMS 14.6

37. Refer to the output in Table 8.
    (a) Define β in this case.
    (b) Find an 80% confidence interval for β.

(c) Find an 80% confidence interval for (i) $\mu_{y|50}$; (ii) $\mu_{y|45}$. (Use the fact that $\bar{x} = 49.868$.) Why is your interval in part (a) considerably narrower than that in part (b)?

38. Refer to Example 13.

(a) Use the output in Table 9 to perform the F test of $H_0: \beta = 0$.

(b) Which of $H_1: \beta > 0$ and $H_1: \beta < 0$ would be the more appropriate alternative to $H_0: \beta = 0$?

(c) Perform the test you chose in part (b).

(d) What assumption did you make in performing this test?

39. The accompanying output was obtained, with the "REGRESS" command, for the additive data in Problem 23 of Section 2.4.

(a) Define β in this example.

(b) Test $H_0: \beta = 0$ against $H_1: \beta > 0$.

(c) Test $H_0: \beta = .5$ against $H_1: \beta > .5$.

(d) Find and interpret a 95% confidence interval for $\mu_{y|2.0}$ (compute $\bar{x}$ from the original data entries in Problem 23 in Section 2.4).

(e) What assumptions were necessary in parts (b), (c), and (d)?

```
MTB > REGRESS C2 ON 1 PREDICTOR IN C1

THE REGRESSION EQUATION IS
C2 = 45.8 + 0.729 C1

                                  ST. DEV.      T-RATIO =
COLUMN        COEFFICIENT         OF COEF.      COEF/S.D.
                  45.8214          0.2811        163.02
C1                 0.7286          0.1044          6.98

S = 0.2762

R-SQUARED = 90.7 PERCENT
R-SQUARED = 88.8 PERCENT, ADJUSTED FOR D.F.

ANALYSIS OF VARIANCE

 DUE TO       DF          SS         MS=SS/DF
REGRESSION     1      3.7157          3.7157
RESIDUAL       5      0.3814          0.0763
TOTAL          6      4.0971
```

40. The accompanying output was obtained, with the "REGRESS" command, for the father and sons' heights data in Problem 24 in Section 2.4.

(a) Define β in this example.

(b) Perform the F test of $H_0: \beta = 0$.

(c) Find a 90% confidence interval for β.

(d) What assumptions were necessary in parts (b) and (c)?

```
MTB > REGRESS C2 ON 1 PREDICTOR IN C1

THE REGRESSION EQUATION IS
C2 = 10.0 + 0.887 C1

                                      ST. DEV.     T-RATIO =
COLUMN         COEFFICIENT            OF COEF.     COEF/S.D.
                    10.031               7.682          1.31
C1                  0.8868              0.1103          8.04

S = 2.572

R-SQUARED = 69.8 PERCENT
R-SQUARED = 68.7 PERCENT, ADJUSTED FOR D.F.

ANALYSIS OF VARIANCE

 DUE TO        DF             SS          MS=SS/DF
REGRESSION      1         427.46            427.46
RESIDUAL       28         185.21              6.61
TOTAL          29         612.67
```

SECTION 14.7
SUMMARY AND REVIEW

In this chapter we indicated how to use the information in a sample of bivariate data to make inferences about the relationship between x and y in the population. We assume a set of bivariate data $(x_1, y_1), (x_2, y_2), \ldots, (x_n, y_n)$. We sometimes refer to x as the independent variable and y as the dependent variable.

We introduced the terms

$$T_{xx} = n \sum x_i^2 - \left(\sum x_i\right)^2 \qquad \text{(Eq. 14.3)}$$

$$T_{yy} = n \sum y_i^2 - \left(\sum y_i\right)^2 \qquad \text{(Eq. 14.4)}$$

$$T_{xy} = n \sum x_i y_i - \left(\sum x_i\right)\left(\sum y_i\right) \qquad \text{(Eq. 14.5)}$$

With this notation:

1. The slope and the intercept of the regression line $y = a + bx$ are, respectively,

$$b = \frac{T_{xy}}{T_{xx}} \qquad \text{(Eq. 14.6)}$$

$$a = \frac{1}{n}\left(\sum y_i - b \sum x_i\right) \qquad \text{(Eq. 14.2)}$$

2. The correlation coefficient is

$$r = \frac{T_{xy}}{\sqrt{T_{xx}T_{yy}}}$$

THE ANALYSIS OF VARIANCE

We distinguish between y_i, the *observed* value for y when $x = x_i$, and $\hat{y}_i = a + bx_i$, the *predicted* value of y when $x = x_i$ obtained from the regression line.

The analysis of variance table (ANOVA table) is based on the following equality:

$$\underset{\text{SST}}{\sum (y_i - \bar{y})^2} = \underset{\text{SSE}}{\sum (y_i - \hat{y}_i)^2} + \underset{\text{SSR}}{\sum (\hat{y}_i - \bar{y})^2} \qquad \text{(Eq. 14.8)}$$

where each sum of squares (SS) is given the indicated name. The computational formula for these sums of squares and their interpretation are as follows:

$$\text{SST} = \frac{T_{yy}}{n} \qquad \text{(Eq. 14.9)}$$

reflects the variability among the values for y.

$$\text{SSR} = \frac{T_{xy}^2}{nT_{xx}} \qquad \text{(Eq. 14.10)}$$

reflects the amount of variability in the y's that is associated with x, or more specifically, with the linear relationship between x and y.

$$\text{SSE} = \text{SST} - \text{SSR}$$

reflects the amount of variability in the y's that is associated with variables *other* than x.

Every sum of squares has a number of degrees of freedom (df) associated with it. The df for SST is $n - 1$, for SSR it is 1, and for SSE it is $(n - 1) - 1 = n - 2$. Dividing a sum of squares by the corresponding number of degrees of freedom results in a mean square (MS). The ANOVA table is the array showing the sums of squares, the degrees of freedom, and the mean squares. The general case for regression is shown in Table 3.

Since, from Chapter 2, r^2 is the *proportion* of the variability in y that is associated with its linear relationship to x, it can be found from the ANOVA table as

$$r^2 = \frac{\text{SSR}}{\text{SST}} \qquad \text{(Eq. 14.11)}$$

RELATIONSHIP BETWEEN x AND y IN THE POPULATION

We assume that in the population of x, y values: For any specific x the distribution of y is $N[\mu_{y|x}, \sigma]$ where $\mu_{y|x} = \alpha + \beta x$ (Definition 1) We estimate α by a, β by b, and σ^2 by MSE.

If the assumptions in Definition 1 apply, a $(100\gamma)\%$ confidence interval for β is

$$b - t_{n-2(\gamma)}\sqrt{\frac{n\text{MSE}}{T_{xx}}} \quad \text{to} \quad b + t_{n-2(\gamma)}\sqrt{\frac{n\text{MSE}}{T_{xx}}} \qquad \text{(Def. 3)}$$

Definition 2 can also be used for testing $H_0: \beta = 0$. The details are set out in Table 5. The contents of the ANOVA table also provide the elements for an F test of $H_0: \beta = 0$ against $H_1: \beta \neq 0$. The test is based on the fact that if the assumptions in Definition 1 apply and $H_0: \beta = 0$ is true, then the ratio

$$F = \frac{\text{MSR}}{\text{MSE}} \quad \text{has the } F_{1,n-2} \text{ distribution} \qquad \text{(Def. 5)}$$

The details of this F test are provided in Table 6.

PREDICTION

In general we regard the predicted value from the regression line $\hat{y}_0 = a + bx_0$ as an estimate for $\mu_{y|x_0}$, the mean of y for all members of the population with $x = x_0$.

If the assumptions in Definition 1 apply, then a $(100\gamma)\%$ confidence interval for $\mu_{y|x_0}$ is

$$\hat{y}_0 - t_{n-2(\gamma)}\sqrt{\text{MSE}\left[\frac{1}{n} + \frac{n(x_0 - \bar{x})^2}{T_{xx}}\right]} \quad \text{to}$$

$$\hat{y}_0 + t_{n-2(\gamma)}\sqrt{\text{MSE}\left[\frac{1}{n} + \frac{n(x_0 - \bar{x})^2}{T_{xx}}\right]} \qquad \text{(Def. 7)}$$

The predicted value $\hat{y}_0 = a + bx_0$ corresponding to $x = x_0$ can also be used as an estimate of the value of y for an *individual* with $x = x_0$. In this case if the assumptions in Definition 1 apply, then a $(100\gamma)\%$ prediction interval for y is

$$\hat{y}_0 - t_{n-2(\gamma)}\sqrt{\text{MSE}\left[1 + \frac{1}{n} + \frac{n(x_0 - \bar{x})^2}{T_{xx}}\right]} \quad \text{to}$$

$$\hat{y}_0 + t_{n-2(\gamma)}\sqrt{\text{MSE}\left[1 + \frac{1}{n} + \frac{n(x_0 - \bar{x})^2}{T_{xx}}\right]} \qquad \text{(Def. 8)}$$

When using the regression line for prediction it is important not to *extrapolate,* that is, predict far beyond the range of value in the sample. The reason is that the relationship between x and y may be very different beyond this range from what it is within the range.

REVIEW PROBLEMS

GENERAL

1. The following ANOVA table is based on a bivariate sample of 22 individuals.

SOURCE OF VARIABILITY	SS	df	MS	F
Regression on X	453.9			
Error				
Total	5671.9			

(a) Fill in the missing elements in the table.
(b) Compute r^2.
(c) Perform the F test of $H_0: \beta = 0$.

2. (a) Reproduce the accompanying diagram, and mark the position of y_4, $\hat{y}_4$, y_7, and $\hat{y}_7$.
(b) If the equation of the regression line is $y = 207 - 12x$, compute $\hat{y}_4$ if $x_4 = 6.5$ and $\hat{y}_7$ if $x_7 = 9.2$.

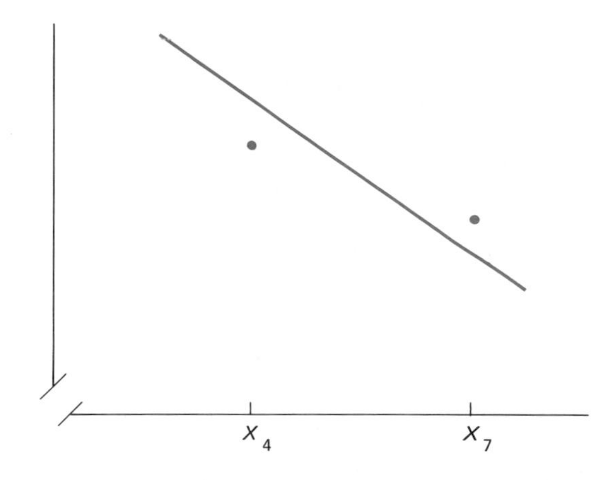

3. In a trade magazine it was reported that in California the correlation between the price of a secondhand car and its age is approximately $-.7$. According to the writer this meant that 70% of the variability in used-car prices can be associated with their linear relationship to the ages of the cars. Comment on this interpretation.
4. For the following bivariate data compute (a) Σx_i, Σx_i^2, Σy_i, Σy_i^2, and $\Sigma x_i y_i$; (b) The correlation coefficient; (c) the regression line relating y to x; (d) the ANOVA table.

x	−4	−2	0	2
y	16	13	10	4

5. Using a computer the authors generated a sample from a bivariate population in which $\beta = 0$. The following summary data were obtained:

$$n = 40 \quad \sum x_i = 1054 \quad \sum x_i^2 = 32{,}058 \quad \sum y_i = 1979$$
$$\sum y_i^2 = 100{,}463 \quad \sum x_i y_i = 51{,}536$$

(a) Find the regression line relating y to x.

(b) Construct the ANOVA table for these data.

(c) Test $H_0: \beta = 0$ against $H_1: \beta \neq 0$ using (i) the t test; (ii) the F test.

(d) Verify that the value for F is the square of the value for t.

(e) Did you make the correct decision in part (c)?

6. The data in Problem 15 of Chapter 2 (page 89) are the average temperature (y) and the latitude (x) for 13 U.S. cities. The summary data are as follows:

$$\sum x_i = 478 \qquad \sum x_i^2 = 18{,}018 \qquad \sum y_i = 745.4$$
$$\sum y_i^2 = 43{,}757.18 \qquad \sum x_i y_i = 26{,}746.5$$

(a) Use these data to construct the ANOVA table.

(b) Find a 95% confidence interval for β, the slope of the population line.

(c) Find a 95% prediction interval for the mean temperature at Kansas City (latitude 39°N).

HEALTH SCIENCES

7. The following data—from Review Problem 42 of Chapter 2 (page 120)—are the natural logarithms of the number of bacteria (y) per unit area on a piece of barley after various hours (x).

x	0	1	2	3	4	5	6	7
y	3.30	4.43	4.64	5.00	5.43	6.06	6.71	7.67

$$\sum x_i y_i = 174.68 \qquad \sum y_i = 43.24 \qquad \sum y_i^2 = 247.106$$

(a) Construct the ANOVA table for these data.

(b) Find the regression line relating y to x.

(c) Find an 80% confidence interval for β, the slope of the population line.

(d) Find an 80% confidence interval for the mean natural logarithm of the number of bacteria after (i) 4 hours; (ii) 8 hours.

(e) In part (d), why is your interval in (i) narrower than that in (ii)?

8. A researcher at the National Center for Health Statistics is interested in the relationship between the incidence of lung cancer and the number of years people have stopped smoking. The following data are hypothetical figures for men aged 55 to 65 years. Here y is the number of cases of lung cancer per 1000 such men who gave up smoking (x) years ago.

x	1	3	5	7	9	11	13	15
y	57	45	37	29	24	17	14	12

(a) Construct the ANOVA table for these data.

(b) Obtain the regression line relating y to x.

(c) Interpret b, the slope of the regression line.

(d) Find an 80% confidence interval for β.

(e) Interpret this interval.

(f) Predict the number of lung cancer deaths among men who had given up smoking 25 years ago.

(g) What is the danger in such extrapolation?

9. The data in Review Problem 16 of Chapter 2 (page 110) are the weight loss (y) and the initial weight (x) of 10 diabetic patients who have enrolled in a supervised weight-loss program. The summary data are as follows:

$$\sum x_1 = 1901 \qquad \sum x_i^2 = 372{,}249 \qquad \sum y_i = 240$$
$$\sum y_i^2 = 7394 \qquad \sum x_i y_i = 49{,}117$$

(a) Compute T_{xx}, T_{yy}, and T_{xy}.

(b) Construct the ANOVA table.

(c) Find the regression line relating y to x.

(d) Test H_0: $\beta = 0$ against H_1: $\beta \neq 0$ using (i) a t test; (ii) an F test.

(e) Verify that $F = t^2$.

10. The data in Review Problem 14 of Chapter 2, reproduced below, are the number of different behaviors (y) exhibited by a baby gibbon at different ages (x).

x	1	2	3	4	5	6
y	3	8	24	32	42	69

(a) Construct the ANOVA table for these data.

(b) Obtain the regression line relating y to x.

(c) Interpret β, the unknown slope of the population line.

(d) Find a 90% confidence interval for β.

(e) Find and interpret r^2 in this case.

11. A group of 15 patients of different ages were observed from the date of onset of a disease. The age of the patient at this date (x) and the length of time (in years) that the patient survived (y) are as follows:

x	y
29	12.8
59	3.8
37	4.4
29	12.0
15	10.9
25	13.9
42	5.5
48	3.0
29	9.6
35	8.8
40	5.2
39	6.0
32	12.2
27	9.2
37	5.8

(a) Construct the ANOVA table for these data.
(b) Find the regression line relating survival time to age at onset of the disease.
(c) Interpret b and the unknown β in this case.
(d) Set an 80% confidence interval for β.
(e) Predict with 80% confidence the mean survival time for people who have the disease at age (i) 20; (ii); 35 (iii) 50.
(f) What assumptions were necessary in parts (d) and (e)?

GOVERNMENT AND SOCIOLOGY

12. The data in Review Problem 28 of Chapter 2, reproduced below, are the median school years (x) and the unemployment rates (y) for 12 census tracts.

x	15.3	16.1	12.9	13.4	8.7	8.2	13.4	8.7	8.1	12.4	11.2	11.3
y	3.8	1.2	4.7	3.1	7.5	7.6	0.3	10.8	4.8	3.9	5.3	5.6

$$\sum x_i = 139.7 \qquad \sum x_i^2 = 1709.95 \qquad \sum y_i = 58.6$$
$$\sum y_i^2 = 376.02 \qquad \sum x_i y_i = 615.06$$

(a) Construct the ANOVA table for these data.
(b) Compute and interpret r^2 in this case.
(c) Interpret the unknown β.
(d) Obtain a 90% confidence interval for the mean unemployment rate for tracts in which the median years of school are (i) 10.5; (ii) 16.5.
(e) In part (d), explain why the interval in (i) is narrower than that in (ii).
(f) What assumptions were necessary in part (d)?

13. The data in Review Problem 11 of Chapter 13 (page 519) consisted of the interval (in months) between the first and second child (y) for 29 families classified by the eventual family size (x). In that problem we recommended the one-way analysis of variance. However, since the independent variable, x, is quantitative we can also apply the techniques of this chapter.
(a) Compute T_{xx}, T_{yy}, and T_{xy} for these data.
(b) Construct the ANOVA table.
(c) Interpret β in this case.
(d) Test $H_0: \beta = 0$ against $H_1: \beta < 0$.
(e) Why, prior to seeing the data, is it appropriate to use $H_1: \beta < 0$?

$$\sum x_i^2 = 540 \qquad \sum y_i^2 = 17{,}235 \qquad \sum x_i y_i = 2416$$

14. Refer to Review Problem 13. Do you think that the methods of this chapter or those of Chapter 13 are better suited to investigate the relationship between x and y? Explain.

ECONOMICS AND MANAGEMENT

15. The data in Review Problem 30 of Chapter 2 (page 116) are the number of mortgage applications at a bank (y) and the prevailing interest rate (x) for $n = 12$ months. The summary data are as follows:

$$\sum x_i = 232.75 \qquad \sum x_i^2 = 3407.5625 \qquad \sum y_i = 251$$
$$\sum y_i^2 = 4867 \qquad \sum x_i y_i = 3517.75$$

(a) Construct the ANOVA table for these data.

(b) Perform the F test of $H_0: \beta = 0$

(c) Find a 90% confidence interval for the mean number of applications in months when the interest rate is 16%.

(d) Find a 90% prediction interval for the number of applications in a specific month when the interest is 16%.

(e) Why do your results in parts (c) and (d) differ?

(f) What assumptions did you make in parts (b)–(d)?

16. The data in Review Problem 29 of Chapter 2 (page 115) are the number of sales representatives (x) and the sales (y), in thousands of dollars, for 20 of a company's sales areas. The summary data are as follows:

$$\sum x_i = 240 \qquad \sum x_i^2 = 3186 \qquad \sum y_i = 7795$$
$$\sum y_i^2 = 3{,}151{,}265 \qquad \sum x_i y_i = 96{,}654$$

(a) Construct the ANOVA table for these data.

(b) Find and interpret r^2 in this case.

(c) Interpret β in this case.

(d) Find a 95% confidence interval for β.

(e) Find the regression line relating sales to the number of representatives.

(f) Find a 95% confidence interval for the mean sales for sales areas in which there are 12 representatives.

(g) Find a 95% prediction interval for a particular sales area in which there are 12 representatives.

(h) Why is your interval in part (g) wider than that in part (f)?

17. The following summary data are for the monthly closing prices of the American Hospital Supply Corporation on the New York Stock Exchange (y) and the corresponding monthly Standard and Poor's index of 400 Industrials (x). The data are available for 15 months.

$$\sum x_i = 2398.93 \qquad \sum x_i^2 = 391{,}558.3487 \qquad \sum y_i = 607.25$$
$$\sum y_i^2 = 25{,}011.40625 \qquad \sum x_i y_i = 98{,}824.59375$$

(These data were provided by Pamela Jackson.)

(a) Construct the ANOVA table for these data.

(b) Test $H_0: \beta = 0$ against $H_1: \beta > 0$, where β is the slope of the population line.

(c) Why is the F test of $H_0: \beta = 0$ not appropriate here?

(d) Find and interpret a 95% confidence interval for β.

EDUCATION AND PSYCHOLOGY

18. As part of a senior project, a student investigated the relationship between the size of a group and the time it takes that group to reach a decision. Seven groups of various sizes (x) were formed and the number of hours necessary to reach a decision about a particular issue (y) recorded. The results were as follows:

SIZE OF GROUP, x	2	4	5	7	7	10	12
TIME TO REACH A DECISION, y	.5	.6	1.0	1.7	1.3	1.9	1.8

(These data were provided by John Millar.)

(a) Construct the ANOVA table for these data.

(b) Find the regression line relating y to x.

(c) Perform the t test of $H_0: \beta = 0$ against $H_1: \beta > 0$.

(d) Obtain a 90% confidence interval for the mean time to reach a decision by groups of size six.

(e) What assumptions did you have to make in parts (c) and (d)?

19. The data in Review Problem 34 of Chapter 2 (page 117) are the reading comprehension score (x) and a creativity index for each of 10 students (y). The summary data are:

$$\sum x_i^2 = 11{,}138 \qquad \sum y_i^2 = 1989 \qquad \sum x_i y_i = 4442$$

(a) Construct the ANOVA table for these data.

(b) Test $H_0: \beta = 0$ against $H_1: \beta > 0$.

(c) Compute and interpret r^2.

SCIENCE AND TECHNOLOGY

20. A research department of a company producing pesticides is concerned with the decline in potency of its product as the product ages. Fifteen containers of the product are aged for various numbers of months and the percentage loss in potency recorded. The results are as follows:

AGE, x	2	3	4	4	6	6	8
PERCENTAGE OF POTENCY LOST, y	2.4	2.7	3.1	3.6	4.5	3.6	4.5

AGE, x	8	10	10	12	12	15	17	18
PERCENTAGE OF POTENCY LOST, y	5.2	6.1	6.8	6.9	8.1	8.9	12.4	12.4

$$\sum y_i^2 = 699.52 \qquad \sum x_i y_i = 1042.4$$

(a) Construct the ANOVA table for these data.

(b) Find the regression line relating y to x.

(c) Find an 80% confidence interval for β, the slope of the population line.

(d) Use the regression line to predict the mean percentage loss of potency for (i) 7 months; (ii) 17 months.

(e) Find an 80% confidence interval for each of the means referred to in part (d).

SUGGESTED READING

Kleinbaum, D. G. and Kupper, L. L. *Applied Regression and Other Multivariate Methods*. North Scituate, Mass.: Duxbury, 1978.

Mendenhall, W. and McClave, J. T. A *Second Course in Business Statistics: Regression Analysis*. San Francisco: Dellen, 1981.

Ott, L. *An Introduction to Statistical Methods and Data Analysis 2nd ed.* North Scituate, Mass.: Duxbury, 1984.

Weisberg, S. *Applied Linear Regression*. New York: Wiley, 1980.

15 ANALYSIS OF COUNTS

HOW ACCURATE ARE POLLING PLACE SURVEYS?

The two leading vote getters in the primary election for mayor of the city are the candidates in a run-off election. Of considerable interest to candidate Collins are the results below, which are based on a random sample of voters as they left the polling place after voting in the primary. The data consist of a breakdown of the total number of voters in the sample by age group and whether or not they voted for Collins. For instance, 17 young voters voted for Collins.

	AGE GROUP		
	YOUNG	MIDDLE-AGED	OLDER
COLLINS	17	43	28
OTHER CANDIDATES	63	97	72
TOTAL	80	140	100

The proportion of young voters voting for Collins is 17/80 = .21. The corresponding proportions for the middle-aged group and the older group are 43/140 = .31 and 28/100 = .28, respectively. Are the differences among these three sample proportions due simply to chance (i.e., sampling variability), or do they indicate that the proportions of *all* young, middle-aged, and older voters favoring Collins are significantly different? (See Review Problem 11 on page 611.)

SECTION 15.1

TESTS OF FIT

In Section 10.3 we indicated how to use the number of successes in the sample to test hypotheses about π, the unknown proportion of successes in the population. In this chapter the methods of Section 10.3 are extended to problems involving three or more categories rather than just two (successes and failures). The next example illustrates this type of problem.

EXAMPLE 1 Data have been collected over a period of several months on the means of transportation used by most persons commuting into the city from a particular suburb. The results, in terms of the proportion of commuters using each means of transportation, are as follows:

MEANS OF TRANSPORTATION	Bus	Train	Car	Other	Total
PROPORTION OF ALL COMMUTERS	.25	.15	.50	10	1

For instance, 25% of the commuters used a bus. The category "other" primarily includes those who walked or cycled to work. In part as a result of this study, the city organized a campaign to encourage commuters to switch from their automobiles to public transportation.

Two months after the end of a three-month campaign, a random sample of 80 commuters from the suburb were selected and their current means of transportation recorded. The results are summarized as follows:

MEANS OF TRANSPORTATION	Bus	Train	Car	Other	Total
NUMBERS OF COMMUTERS	26	15	32	7	80

Do these data suggest that the campaign has been successful? We can answer this question by making use of the chi-square distribution. Before continuing you may want to review this distribution and the chi-square table, Table A7.

Table 1 contains all the information we need to answer the question. Column (3) lists the proportions of the population of commuters using various means of transportation *before* the campaign. Recall that data were collected on a sample of 80 commuters after the campaign. The number of commuters in this sample using each means of transportation are shown in column (5).

These sample counts are used to distinguish between two hypotheses, both of which relate to the postcampaign population of commuters. They are:

H_0: all the proportions in the postcampaign population (of commuters) are the same as those in the precampaign population as listed in column (3)

TABLE 1 Means of transportation used by suburban commuters

(1) CATEGORY	(2) MEANS OF TRANSPORTATION	(3) POPULATION PROPORTIONS	(4) EXPECTED COUNTS	(5) OBSERVED COUNTS
1	Bus	.25	20	26
2	Train	.15	12	15
3	Car	.50	40	32
4	Other	.10	8	7
	Total	1.00	80	80

(Thus, H_0 suggests that the campaign has had no effect on travel arrangements.)

H_1 : some of the proportions in the postcampaign population differ from the corresponding proportions in the precampaign population

We test H_0 by comparing the sample counts with those that would be "expected" if H_0 were true. Recall that for a sample of size n the expected number of successes is $n\pi$, where π is the probability of a success at each selection. For instance, if H_0 is true, the probability of selecting a commuter who takes the bus after the campaign is, from Table 1, still .25 and hence the expected number of such commuters in a sample of size 80 is $80(.25) = 20$. Similarly, the number that would be expected to take the train, travel by car, and fall in the "other" category are $80(.15) = 12$, $80(.5) = 40$, and $80(.10) = 8$, respectively. These expected counts are listed in column (4) of Table 1.

The observed counts in column (5) differ from the expected counts [in column (4)]. But even if H_0 were exactly true, sampling variability would almost guarantee some difference. How are we to decide whether (1) H_0 is true and the differences between the observed and expected counts are due to chance alone, or (2) the differences have occurred because the precampaign proportions in column (3) are no longer true for the postcampaign population (i.e., H_1 is true)?

A convenient method of summarizing the extent to which the observed and expected counts differ is to use the test statistic

$$Y = \sum_{\substack{\text{all four} \\ \text{categories}}} \frac{(\text{observed} - \text{expected})^2}{\text{expected}}$$

In words, Y is the sum over all four categories of the squared differences between the observed and expected counts each divided by the expected count. Notice that the closer the observed counts are to the expected counts, the closer Y will be to zero. On the other hand, the more they differ, the larger Y will be and the less support the data will offer for H_0.

For the counts in columns (4) and (5) in Table 1,

$$Y = \frac{(26 - 20)^2}{20} + \frac{(15 - 12)^2}{12} + \frac{(32 - 40)^2}{40} + \frac{(7 - 8)^2}{8}$$

$$= 1.8 + .75 + 1.6 + .125 = 4.275$$

The value 4.275 offers little help in deciding which hypothesis the data support. However, it can be shown that if H_0 is true, for repeated samples of size 80, Y will have approximately a χ^2_3 distribution. This suggests that we reject H_0 if Y is unusually large. In Figure 1 we show the χ^2_3 distribution on which we have indicated the value $Y = 4.275$ together with the values $\chi^2_{3,.05} = 7.81$ and $\chi^2_{3,.01} = 11.34$.

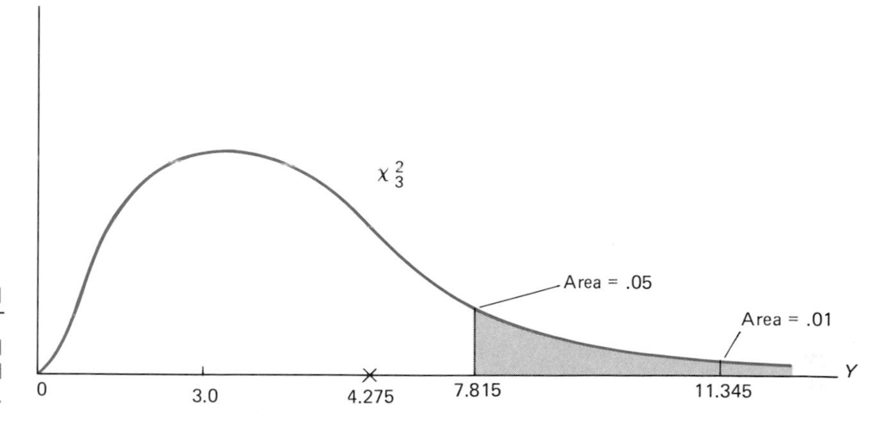

FIGURE 1

Value $Y = 5.05$ compared to $\chi^2_{3,.05} = 7.81$ and $\chi^2_{3,.01} = 11.34$.

Since $Y = 4.275$ is less than 7.81 we will not reject H_0 at the .05 level of significance and conclude that there has *not* been a significant shift in the proportion of commuters using the various means of transportation.

■

We can now generalize this procedure. Assume that a population is composed of M distinct categories. Each member of the population falls in one and only one category. The unknown proportion of the population in each category is $\pi_1, \pi_2, \ldots, \pi_M$, where $\pi_1 + \pi_2 + \cdots + \pi_M = 1$. A sample of size n is selected from this population. Either the sample is selected with replacement or n is small compared to the size of the population. As we pointed out in Chapter 5, the probability of selecting a member of the population in the ith category is simply the proportion of such members in the population (i.e., π_i). We call the number in the sample observed in each category $O_1, O_2, \ldots, O_M$, where $O_1 + O_2 + \cdots + O_M = n$.

The null hypothesis, H_0, is that $\pi_1 = k_1, \pi_2 = k_2, \ldots, \pi_M = k_M$, where $k_1, k_2, \ldots, k_M$ are hypothesized probabilities that sum to 1 (in Example 1, based on the precampaign figures $k_1 = .25$, $k_2 = .15$, $k_3 = .5$, and $k_4 = .10$). The alternative hypothesis H_1 is that some of the probabilities $k_1, k_2, \ldots, k_M$ are incorrect.

If H_0 is true, the number of observations that are expected to fall into each category are $E_1 = nk_1, E_2 = nk_2, \ldots, E_M = nk_M$. Table 2 shows a convenient way of displaying the k's, the O's, and the E's. Notice that the expected counts as well as the observed counts add to n. The reason is that $E_1 + E_2 + \cdots + E_M = nk_1 + nk_2 + \cdots + nk_M = n(k_1 + k_2 + \cdots + k_M) = n(1) = n$.

TABLE 2 Expected and observed counts

CATEGORY	HYPOTHESIZED POPULATION PROPORTIONS	EXPECTED COUNTS	OBSERVED COUNTS
1	k_1	$E_1 = nk_1$	O_1
2	k_2	$E_2 = nk_2$	O_2
3	k_3	$E_3 = nk_3$	O_3
⋮	⋮	⋮	⋮
M	k_M	$E_M = nk_M$	O_M
	1	n	n

The test of H_0 is based on the following result:

> DEFINITION 1
>
> The extent to which the observed and the expected counts differ is measured by
>
> $$Y = \sum_{i=1}^{M} \frac{(O_i - E_i)^2}{E_i}$$
>
> which, if H_0 is true will have approximately the χ^2_{M-1} distribution.

The details of how large Y needs to be before H_0 can be rejected are given in Table 3. Because we are interested in the extent to which the sample data are consistent with or "fit" a set of proportions or probabilities, this test is called a **chi-square test of fit.**

TABLE 3 Procedure for performing the chi-square test of fit

STEP	$H_0: \pi_1 = k_1, \pi_2 = k_2, \ldots, \pi_M = k_M$
1. Compute:	$Y = \sum_{i=1}^{M} \frac{(O_i - E_i)^2}{E_i}$ where $E_i = k_i n$
2. Find $\chi^2_{M-1,.05}$ and $\chi^2_{M-1,.01}$ from Table A7.	
3. *Conclusion:* Using $\alpha = .05$ or $.01$, reject H_0 at the level of significance if	$Y > \chi^2_{M-1,\alpha}$

EXAMPLE 2 A die is suspected of being biased. To test this, the die is rolled 300 times. The results are shown in column (4) of Table 4.

TABLE 4 Data for Example 2

(1) CATEGORY (NUMBER THROWN)	(2) PROBABILITIES UNDER H_0	(3) EXPECTED COUNTS	(4) OBSERVED COUNTS
1	1/6	50	68
2	1/6	50	55
3	1/6	50	40
4	1/6	50	45
5	1/6	50	53
6	1/6	50	39
Total	1	300	300

Thus the number 1 occurred 68 times; the number 2, 55 times; and so on. Do these counts suggest that the die is biased?

Solution In this case, there are $M = 6$ categories. The null hypothesis is that the die is fair, that is, that each of the numbers 1, 2, 3, 4, 5, and 6 are equally likely. We write H_0 and H_1 as:

$$H_0: \pi_1 = \pi_2 = \pi_3 = \pi_4 = \pi_5 = \pi_6 = 1/6$$

$$H_1: \text{some of the } \pi\text{'s are not } 1/6$$

These probabilities may be viewed as the assumed proportion of each number in an infinite population of throws. If H_0 is true, the expected numbers in each category are $E_1 = 300(1/6) = 50$, $E_2 = 300(1/6) = 50$, and so on. In this case:

$$Y = \frac{(68 - 50)^2}{50} + \frac{(55 - 50)^2}{50} + \frac{(40 - 50)^2}{50} + \frac{(45 - 50)^2}{50} + \frac{(53 - 50)^2}{50} + \frac{(39 - 50)^2}{50} = 12.08$$

From Table A7 with $M - 1 = 6 - 1 = 5$, $\chi^2_{5,.05} = 11.07$ and $\chi^2_{5,.01} = 15.09$.

Since $Y = 12.08$ lies between these two values, we reject H_0 at the .05 level of significance. The data suggest that the probabilities in column (2) are *not* appropriate and that the die is biased. From Table 4 you can see that the die seems to favor 1's at the expense of 3's and 6's.

■

Notice in this example that when we rejected H_0, we went further and found where the large discrepancies between the O_i and the E_i lay. We did this since an unusually large value for Y may lead us to reject H_0 but it does not indicate which of the M probabilities specified by H_0 seem to be incorrect. In general, we recommend this kind of examination whenever H_0 is rejected.

We indicated in Definition 1 that if H_0 is true, $Y = \sum^{M} (O_i - E_i)/E_i$ has only approximately a chi-square distribution. This approximation is, like the central limit theorem, applicable only for large samples. Conventional wisdom has it that the approximation can be used as long as the expected counts in each category all exceed five. In fact, there is some evidence that the chi-square test will be valid even if one (but only one) of the expected counts is as small as one.

One final note: In the context of chi-square tests the number of degrees of freedom for χ^2 can be defined as:

DEFINITION 2

df = the number of probabilities specified by H_0 that are free to vary

In the case where the number of probabilities are M, the sum of all the probabilities is known to be 1. Therefore, if any one probability is not known, *it* can be determined by subtracting the sum of the others from 1. The number of probabilities free to vary (the df) will therefore be $M - 1$. In the commuter problem, for instance, H_0 need only have specified three of the four probabilities, say $\pi_1 = .25$, $\pi_2 = .15$, and $\pi_3 = .5$. The remaining probability must be $1 -$ sum of the other three $= 1 - (.25 + .15 + .5) = .1$. In the next section we suggest another number of degrees of freedom based on Definition 2.

PROBLEMS 15.1

1. In a now famous experiment the founder of modern genetics, Gregor Mendel, crossed contrasting strains of garden peas. The details of the experiment are summarized below:

COMBINATION OF CHARACTERISTICS	PROBABILITY	E_i	O_i
Round and yellow	9/16		315
Round and green	3/16		101
Wrinkled and yellow	3/16		108
Wrinkled and green	1/16		32
			556

According to Mendel's theory the four combinations of second-generation peas should have occurred with the probabilities shown above. A total of $n = 556$ offspring were observed with the results shown in the column of O_i's above.

(a) What are H_0 and H_1 in this case?

(b) Compute the values for E_1, E_2, E_3, and E_4.

(c) Carry out the chi-square test of fit for these data.

(d) Do the data support Mendel's theory?

2. The results of a pre-election opinion poll of 1050 voters are summarized as follows:

PARTY	DEMOCRATS	REPUBLICAN	CONSERVATIVE	LIBERAL	OTHER
Number of Voters	425	406	105	104	10

Are these sample results consistent with the hypothesis that 40% of all voters intend to vote for the Democrats, 40% for Republicans, 8% for the Conservative Party, and 8% for the Liberal Party? If you find the data inconsistent with this hypothesis, discuss the discrepancies between the hypothesis and the data.

3. A company's products are classified as perfect, seconds, or defective. A random sample of 150 products were classified as follows:

CONDITION OF PRODUCT	NUMBER
Perfect	104
Second	38
Defective	8

Are these data consistent with the probabilities $P(\text{perfect}) = .8$; $P(\text{second}) = .15$?

4. A coin is suspected of being biased. To check this suspicion, the experiment of tossing the coin twice is repeated 100 times. In each case the number of heads is recorded with the following results:

NUMBER OF HEADS	0	1	2
NUMBER OF OCCURRENCES	16	44	40

(a) If the coin is fair, what is the probability of (i) no heads; (ii) one head; (iii) two heads?

(b) What are the null and alternative hypotheses in this case?

(c) Perform the chi-square test of H_0.

(d) If H_0 is rejected, explain whether the coin favors heads or tails.

5. Grade inflation occurs as an ever higher proportion of students is given high grades, particularly A's. A college registrar knows that of the students who have passed courses prior to this semester, 22% received A's, 35% B's, 27% C's, and 16% D's. In a random sample of 80 grades this semester, 19 were A's, 30 were B's, 20 were C's, and 11 were D's. Do these data indicate that there has been grade inflation?

Solution to Problem 1

(a) If π_1, π_2, π_3, and π_4 are the unknown probabilities for each combination of characteristics, then the two hypotheses are:

$$H_0: \pi_1 = 9/16, \quad \pi_2 = 3/16, \quad \pi_3 = 3/16, \quad \text{and} \quad \pi_4 = 1/16$$

$$H_1: \text{some of these probabilities are incorrect}$$

(b) $E_1 = 556(9/16) = 312.75$

$E_2 = E_3 = 556(3/16) = 104.25$

$E_4 = 556(1/16) = 34.75$

(c) $$Y = \sum^4 \frac{(O_i - E_i)^2}{E_i} = \frac{(315 - 312.75)^2}{312.75} + \frac{(101 - 104.25)^2}{104.25} + \frac{(108 - 104.25)^2}{104.25} + \frac{(32 - 34.75)^2}{34.75} = .47$$

From Table A7, $\chi^2_{3,.05} = 7.81$ and $\chi^2_{3,.01} = 11.34$. Since $.47 < 7.81$, we *cannot* reject H_0 at the .05 level of significance.

(d) The data *are* consistent with Mendel's theory.

SECTION 15.2
INDEPENDENCE IN CONTINGENCY TABLES

In Chapter 3 we discussed the idea of independence as it applied to two events A and B. They were said to be independent if the probability that one occurs is not changed by the occurrence of the other. A consequence of this definition was the rule that if two events A and B are independent, the probability of their joint occurrence is $P(A \text{ and } B) = P(A) \cdot P(B)$.

In this section we apply the concept of independence to the relationship among the counts in a *contingency table*. Such a table, introduced briefly in Chapter 2, can be illustrated by the following situation. Suppose that a population of voters in a small city were classified by political party and religion as shown in Table 5.

TABLE 5

Contingency table for voters' religions and political affiliations

		RELIGION			
		PROTESTANT (P)	CATHOLIC (C)	JEWISH (J)	TOTAL
POLITICAL PARTY	DEMOCRAT (D)	15,000	7,500	7,500	30,000
	REPUBLICAN (R)	10,000	5,000	5,000	20,000
	UNAFFILIATED (U)	5,000	2,500	2,500	10,000
	TOTAL	30,000	15,000	15,000	60,000

We call a table such as this a 3×3 contingency table. Each entry in the table is a count or a frequency. For example, there are 2500 politically unaffiliated Catholics, a total of 30,000 Democrats, and a total of 60,000 voters in the city. The three political groups are the levels of the qualitative variable, political party, and the three religious groups, the levels of the qualitative variable, religion.

Now suppose that we were to select a voter at random from this population. Following the notation used in Chapter 3, the probability of selecting a Catholic is $P(C) = 15{,}000/60{,}000 = .25$, the probability of a Democrat is $P(D) = 30{,}000/60{,}000 = .5$. Then $P(C) \cdot P(D) = .25 \times .5 = .125$. On the other hand, from the contingency table the probability of selecting a Catholic Democrat; that is, the probability of both C and D = $P(C \text{ and } D) = 7500/60{,}000 = .125$. Therefore, $P(C \text{ and } D) = P(C) \cdot P(D)$ and the events (or combination of levels) C and D *are* independent.

When working with contingency tables, we are interested primarily in determining whether two *variables* are independent but this requires that individual pairs of levels (from each of two variables) are independent. In this case, for instance, we would like to know whether religion and political party are independent. To check this, we require the following definition:

DEFINITION 3

Two qualitative variables are independent if *each* combination of levels of variable I and of variable II are independent.

In the present case these combinations for religion and political party are: D and P, D and C, D and J, R and P, . . . , U and J. Since in this case each combination of levels can be shown to be independent as we did in the case of C and D, the two variables political party and religion will also be independent. There is a simpler and more insightful way to check for independence of variables in contingency tables. This can be illustrated by referring again to Table 5.

Looking at the Protestant column you can see that one-half of the Protestants are Democrats, one-third are Republicans, and one-sixth are unaffiliated. Further, the proportions are exactly the same for the Catholics and the Jews. This indicates that the two variables are independent. You can also check that within each *party* one-half are Protestants, one-fourth are Catholics, and one-fourth are Jewish. In general:

DEFINITION 4

Two qualitative variables are independent if the proportion of observations in each level of variable I is identical for each level of variable II, and vice versa.

In Table 6 the voters in city B are also classified by political party and religion. They have the same row and column totals as those in city A, but for this city the two variables are *not* independent but *dependent*. For example, the proportion of Protestants who are Republican is $14{,}000/30{,}000 = .467$, whereas the proportion of Catholics who are Republican is $5000/15{,}000 = .333$, and for Jews it is only $1000/15{,}000 = .067$. This may be contrasted with the corresponding proportions in city A: $10{,}000/30{,}000$, $5000/15{,}000$, and $5000/15{,}000$, which all $= .333$, indicating that the variables are independent.

TABLE 6

Distribution of voters in city B by political party and religion

		RELIGION PROTESTANT (P)	 CATHOLIC (C)	 JEWISH (J)	TOTAL
POLITICAL PARTY	DEMOCRAT (D)	10,000	8,000	12,000	30,000
	REPUBLICAN (R)	14,000	5,000	1,000	20,000
	UNAFFILIATED (U)	6,000	2,000	2,000	10,000
	TOTAL	30,000	15,000	15,000	60,000

PROBLEMS 15.2

6. In a study of smoking habits, 800 people are classified by race and smoking habit as follows:

	RACE			
	WHITE	BLACK	HISPANIC	OTHER
NONSMOKER	131	69	39	8
LIGHT SMOKER	180	108	46	14
HEAVY SMOKER	147	27	18	13

Do these data indicate that race and smoking habits are independent?

7. Two groups of 400 persons are each classified by sex and whether or not they favor capital punishment. The results are summarized as follows:

	GROUP *A*: FAVOR CAPITAL PUNISHMENT		
	YES	NO	TOTAL
M	108	132	240
F	104	56	160
TOTAL	212	188	400

	GROUP *B*: FAVOR CAPITAL PUNISHMENT		
	YES	NO	TOTAL
M	132	108	240
F	88	72	160
TOTAL	220	180	400

In each case decide whether the two variables are independent.

8. The 180 students in an introductory statistics class are classified by class and area of concentration, as follows:

	CLASS				
	FRESHMAN	SOPHOMORE	JUNIOR	SENIOR	
SOCIAL SCIENCES	4	15	28	9	
NATURAL SCIENCES	15	38	16	3	
PROFESSIONAL STUDIES	6	15	23	8	
TOTAL	25	68	67	20	180

Are these two variables independent?

Solution to Problem 6

The complete table is as follows:

	RACE				
	WHITE	BLACK	HISPANIC	OTHER	TOTAL
NONSMOKER	131	69	39	8	247
LIGHT SMOKER	180	108	46	14	348
HEAVY SMOKER	147	27	18	13	205
TOTAL	458	204	103	35	800

The two variables are *not* independent. For example, the proportion of white persons who do not smoke is $131/458 = .286$. The corresponding proportion of black persons is $69/204 = .338$.

Similarly, of the nonsmokers a proportion $131/247 = .530$ are white. The corresponding proportion of heavy smokers who are white is $147/205 = .717$.

SECTION 15.3
TESTING FOR INDEPENDENCE

In Section 15.2 you saw how to determine whether qualitative variables in a population were independent. In reality it is most unlikely that we could construct contingency tables for *all* the voters in a city such as those in Tables 5 and 6. Suppose, however, that a sample of 200 voters were selected and their religion and political preferences obtained. The result might be a 3×3 contingency table such as that in Table 7.

TABLE 7

Contingency table for a sample of 200 voters

		RELIGION			
		PROTESTANT (P)	CATHOLIC (C)	JEWISH (J)	TOTAL
POLITICAL PARTY	DEMOCRAT (D)	49	26	23	98
	REPUBLICAN (R)	39	16	17	72
	UNAFFILIATED (U)	16	10	4	30
	TOTAL	104	52	44	200

The variables in this sample are *not* independent. For instance, among the 104 Protestants the proportion who are Democrats is $49/104 = .471$. The corresponding proportion among Jews is $23/44 = .523$. In most cases, as in this one, the variables in the sample will not be independent. However, we can ask whether this sample was drawn *from a population* such as that in Table 5, where the two variables were independent, or from a population such as that in Table 6, where the two variables were dependent. The remainder of this section is devoted to procedures for answering such questions.

Assume that we had a random sample arranged in the form of a contingency table on the basis of two variables. We want to know whether this sample could have come from a population in which the two variables were independent.

The two hypotheses would be:

H_0: the two variables in the population are independent

H_1: the two variables in the population are dependent

In Section 15.1 the null hypothesis specified the probabilities of a member of the sample falling into each of the M categories. In Table 7, for example, we have nine categories (or cells) but what should each of the nine probabilities be if H_0 is true?

To answer this question, consider the top middle cell in Table 7. If we knew the proportion of Catholics $P(C)$ and the proportion of Democrats $P(D)$ in the population as we, unrealistically, did earlier, the hypothesis of independence would indicate that the probability of a voter falling in this cell is the product $P(C) \cdot P(D)$. Since we do not know $P(C)$ or $P(D)$, this does not seem much help. But we can use the sample *estimates* for $P(C)$ and $P(D)$. The proportion of Catholics in the sample is 52/200, and the proportion of Democrats, 98/200. Thus an estimate of $P(C) \cdot P(D) = (52/200) \cdot (98/200)$. In the same way we can estimate the probability, if H_0 is true, of selecting a Jewish Republican, $P(J) \cdot P(R)$ by $(44/200) \cdot (72/200)$. These estimates and those for the other seven cells, assuming that H_0 is true, are shown in Table 8.

TABLE 8

Estimated probabilities of falling in each cell if H_0 is true

	PROTESTANT	CATHOLIC	JEWISH	TOTAL
DEMOCRAT	$\frac{104}{200} \cdot \frac{98}{200}$	$\frac{52}{200} \cdot \frac{98}{200}$	$\frac{44}{200} \cdot \frac{98}{200}$	$\frac{98}{200}$
REPUBLICAN	$\frac{104}{200} \cdot \frac{72}{200}$	$\frac{52}{200} \cdot \frac{72}{200}$	$\frac{44}{200} \cdot \frac{72}{200}$	$\frac{72}{200}$
UNAFFILIATED	$\frac{104}{200} \cdot \frac{30}{200}$	$\frac{52}{200} \cdot \frac{30}{200}$	$\frac{44}{200} \cdot \frac{30}{200}$	$\frac{30}{200}$
TOTAL	$\frac{104}{200}$	$\frac{52}{200}$	$\frac{44}{200}$	1

Notice that the *estimated* probabilities in any row or column add to the corresponding observed proportion in that row or column. For instance, the probabilities in the top row of Table 8 = $(104/200 + 52/200 + 44/200)$ $(98/100) = 1 \cdot 98/200$, which is the proportion of Democrats in the sample. We shall return to this point later.

As before, we find the *expected* number of observations in each cell by multiplying each probability by the total number of observations, $n = 200$ in this case. For the top middle cell, for instance, the expected number of observations if H_0 is true is $200(52/200 \cdot 98/200) = (52)(98)/200 = 25.48$. Similarly, the expected number of Jewish Republicans is $200(44/200 \cdot 72/200) =$

(44)(72)/200 = 15.84. This result suggests that we may obtain the expected values for each cell by multiplying the corresponding row and column totals of the contingency tables and dividing the product by n. We have listed the expected values for all nine cells in Table 9.

TABLE 9

Expected number of observations per cell

	PROTESTANT	CATHOLIC	JEWISH	TOTAL
DEMOCRAT	$\frac{(104)(98)}{200} = 50.96$	$\frac{(52)(98)}{200} = 25.48$	$\frac{(44)(98)}{200} = 21.56$	98
REPUBLICAN	$\frac{(104)(72)}{200} = 37.44$	$\frac{(52)(72)}{200} = 18.72$	$\frac{(44)(72)}{200} = 15.84$	72
UNAFFILIATED	$\frac{(104)(30)}{200} = 15.60$	$\frac{(52)(30)}{200} = 7.80$	$\frac{(44)(30)}{200} = 6.60$	30
TOTAL	104	52	44	200

Notice that the row and column totals of these expected values are exactly those for the observed row and column totals in Table 7. For instance, 50.96 + 25.48 + 21.56 = 98.

Again, as we did in Section 15.1, we compare the nine observed counts in Table 7 with the nine expected counts in Table 9 by means of the test statistic:

$$Y = \sum_{i=1}^{9} \frac{(O_i - E_i)^2}{E_i}$$

If H_0 is true, this has a χ^2 distribution but with how many degrees of freedom?

In Definition 2 we defined the number of degrees of freedom for χ^2 as the number of "free" probabilities given H_0. When referring to Table 7 earlier, we noted that with our method of estimating the probabilities for each cell the nine probabilities were constrained by the fact that they added to the observed row and column *proportions*.

If you now look at Table 10, you can see that once any *four* of the cell

TABLE 10

Once four of the cell probabilities have been estimated, the remaining five (shaded) are determined

	PROTESTANT	CATHOLIC	JEWISH	TOTAL
DEMOCRAT	$\frac{104}{200} \cdot \frac{98}{200}$	$\frac{52}{200} \cdot \frac{98}{200}$	$\frac{44}{200} \cdot \frac{98}{200}$	$\frac{98}{200}$
REPUBLICAN	$\frac{104}{200} \cdot \frac{72}{200}$	$\frac{52}{200} \cdot \frac{72}{200}$	$\frac{44}{200} \cdot \frac{72}{200}$	$\frac{72}{200}$
UNAFFILIATED	$\frac{104}{200} \cdot \frac{30}{200}$	$\frac{52}{200} \cdot \frac{30}{200}$	$\frac{44}{200} \cdot \frac{30}{200}$	$\frac{30}{200}$
TOTAL	$\frac{104}{200}$	$\frac{52}{200}$	$\frac{44}{200}$	1

probabilities have been estimated, the remaining five are determined. Thus only four of the nine cell probabilities can be independently estimated from H_0, and the number of degrees of freedom for Y is therefore four.

We can now test H_0 by following the procedure for performing a test of fit in Table 3. Using Tables 7 and 9, we have

$$Y = \frac{(49 - 50.96)^2}{50.96} + \frac{(26 - 25.48)^2}{25.48} + \frac{(23 - 21.56)^2}{21.56}$$
$$+ \frac{(39 - 37.44)^2}{37.44} + \frac{(16 - 18.72)^2}{18.72} + \frac{(17 - 15.84)^2}{15.84}$$
$$+ \frac{(16 - 15.6)^2}{15.6} + \frac{(10 - 7.8)^2}{7.8} + \frac{(4 - 6.6)^2}{6.6}$$
$$= 2.4144$$

From Table A7, $\chi^2_{4,.05} = 9.49$ and $\chi^2_{4,.01} = 13.28$. Since $Y = 2.4144$ is considerably less than 9.49, we do not reject H_0 at the .05 level of significance. The data suggest that the variables religion and political party are (at least) approximately independent.

The authors used a computer to select this sample from the population shown in Table 5, where the variables were independent. Thus in this case at least we know that our decision not to reject H_0: the two variables are independent, is correct.

We can now generalize the procedure for testing for independence with a contingency table. A sample of size n is selected from a population and classified into r categories for the row variable and c categories for the column variable. The sample will thus be composed of a total of rc observed counts (O's) as shown in the $r \times c$ contingency table in Table 11.

TABLE 11

The rc observed counts arranged in an $r \times c$ contingency table

		COLUMN VARIABLE					
		1	2	3	⋯	c	
ROW VARIABLE	1	O_{11}	O_{12}	O_{13}	⋯	O_{1c}	R_1
	2	O_{21}	O_{22}	O_{23}	⋯	O_{2c}	R_2
	3	O_{31}	O_{32}	O_{33}	⋯	O_{3c}	R_3
	⋮	⋮	⋮	⋮		⋮	⋮
	r	O_{r1}	O_{r2}	O_{r3}	⋯	O_{rc}	R_r
		C_1	C_2	C_3	⋯	C_c	n

Each observed count (O) has two subscripts. The first indicates the row it is in and the second the column. Thus O_{32} is the number of observations associated with the third level of the row variable and the second level of the column variable.

We call the total number of observations in each row $R_1, R_2, \ldots, R_r$ and in each column $C_1, C_2 \ldots, C_c$. The total number of observations is n, so that

$$R_1 + R_2 + \cdots + R_r = n$$

and

$$C_1 + C_2 + \cdots + C_c = n$$

The *rc* observed values will be used to test:

H_0: in the population, the two variables are independent

H_1: in the population, the two variables are dependent

We estimate the probabilities for each cell if H_0 is true as we did earlier by multiplying the corresponding row and column proportions as shown in Table 12. The *estimated probabilities* in each row then add to the *observed* proportion in that row and similarly for each column. For example, in row 1 the sum of the estimated probabilities is

TABLE 12

Estimated cell probabilities

		COLUMN VARIABLE					
		1	2	3	. . .	c	ROW PROPORTIONS
	1	$\frac{R_1}{n} \cdot \frac{C_1}{n}$	$\frac{R_1}{n} \cdot \frac{C_2}{n}$	$\frac{R_1}{n} \cdot \frac{C_3}{n}$	. . .	$\frac{R_1}{n} \cdot \frac{C_c}{n}$	$\frac{R_1}{n}$
ROW VARIABLE	2	$\frac{R_2}{n} \cdot \frac{C_1}{n}$	$\frac{R_2}{n} \cdot \frac{C_2}{n}$	$\frac{R_2}{n} \cdot \frac{C_3}{n}$	. . .	$\frac{R_2}{n} \cdot \frac{C_c}{n}$	$\frac{R_2}{n}$
		$\vdots \quad \vdots$	$\vdots \quad \vdots$	$\vdots \quad \vdots$		$\vdots \quad \vdots$	$\vdots$
	r	$\frac{R_r}{n} \cdot \frac{C_1}{n}$	$\frac{R_r}{n} \cdot \frac{C_2}{n}$	$\frac{R_r}{n} \cdot \frac{C_3}{n}$	. . .	$\frac{R_r}{n} \cdot \frac{C_c}{n}$	$\frac{R_r}{n}$
COLUMN PROPORTIONS		$\frac{C_1}{c}$	$\frac{C_2}{n}$	$\frac{C_3}{n}$	. . .	$\frac{C_c}{n}$	1

$$\frac{R_1}{n} \cdot \frac{C_1}{n} + \frac{R_1}{n} \cdot \frac{C_2}{n} + \cdots + \frac{R_1}{n} \cdot \frac{C_c}{n} = \frac{R_1}{n^2}(C_1 + C_2 + \cdots + C_c)$$

$$= \frac{R_1}{n^2} \cdot n = \frac{R_1}{n}$$

As we have shown, the expected number of observations for each cell is the product of the appropriate row and column totals divided by n. These expected values are shown in Table 13.

As a measure of how far the O's are from the E's we compute the test statistic

$$Y = \sum_{\text{all } rc \text{ cells}} \frac{(O - E)^2}{E}$$

Generalizing our earlier approach, once $(r - 1)(c - 1)$ of the cell probabilities have been obtained, the remainder can be obtained. The number of "free" probabilities is therefore $(r - 1)(c - 1)$. Thus if H_0 is true, Y has the $\chi^2_{(r-1)(c-1)}$ distribution.

TABLE 13

Estimated number of observations for each cell in the general case

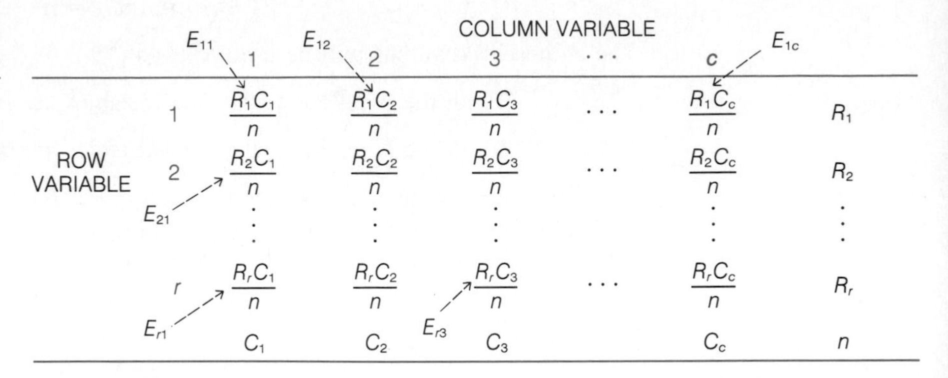

		COLUMN VARIABLE 1	2	3	$\cdots$	c	
ROW VARIABLE	1	$\frac{R_1C_1}{n}$ (E_{11})	$\frac{R_1C_2}{n}$ (E_{12})	$\frac{R_1C_3}{n}$	$\cdots$	$\frac{R_1C_c}{n}$ (E_{1c})	R_1
	2	$\frac{R_2C_1}{n}$ (E_{21})	$\frac{R_2C_2}{n}$	$\frac{R_2C_3}{n}$	$\cdots$	$\frac{R_2C_c}{n}$	R_2
	$\vdots$	$\vdots$	$\vdots$	$\vdots$		$\vdots$	$\vdots$
	r	$\frac{R_rC_1}{n}$ (E_{r1})	$\frac{R_rC_2}{n}$	$\frac{R_rC_3}{n}$ (E_{r3})	$\cdots$	$\frac{R_rC_c}{n}$	R_r
		C_1	C_2	C_3		C_c	n

The procedure for performing the chi-square test of independence can thus be summarized as follows:

> **DEFINITION 5**
>
> For a chi-square test of independence,
> (a) Compute the table of expected counts (as in Table 13). The expected count in each cell is RC/n where R and C are the row and column totals.
> (b) Compute $Y = \Sigma_{\text{all cells}} (O - E)^2/E$
> (c) Follow the procedure in Table 3 with $(r - 1)(c - 1)$ in place of $M - 1$.

EXAMPLE 3 A university classifies applicants whom they have accepted as excellent, very good, or good prospects. Some students accepted by the university in turn accept the university's offer, whereas others do not. A random sample of 700 students accepted by the university over the past 25 years were classified as follows:

		PROSPECTS EXCELLENT	VERY GOOD	GOOD	TOTAL
STUDENTS' ACTION	ACCEPTED	95	161	209	465
	REJECTED	85	81	69	235
	TOTAL	180	242	278	700

Do these data suggest that a student's decision is related to their evaluation or not?

Solution We test H_0: the variables, student's decision and their evaluation, are independent, against H_1: these two variables are not independent.

Following Definition 5 we first compute the table of expected counts. For example, for the middle top cell, $E_{12} = (465)(242)/700 = 160.76$. The complete table of expected counts is

	EXCELLENT	VERY GOOD	GOOD	TOTAL
ACCEPTED	119.57	160.76	184.67	465
REJECTED	60.43	81.24	93.33	235
TOTAL	180	242	278	700

Remember that we had to compute only $(r - 1)(c - 1) = (2 - 1)(3 - 1) = 2$ expected counts in this way. The remaining four can be obtained by subtracting from the (observed) row and column totals. In this case

$$Y = \frac{(95 - 119.57)^2}{119.57} + \frac{(161 - 160.76)^2}{160.76} + \frac{(209 - 184.67)^2}{184.67}$$

$$+ \frac{(85 - 60.43)^2}{60.43} + \frac{(81 - 81.24)^2}{81.24} + \frac{(69 - 93.33)^2}{93.33} = 24.59$$

From Table A7, $\chi^2_{(r-1)(c-1),\,.05} = \chi^2_{2,\,.05} = 5.99$, and $\chi^2_{2,\,.01} = 9.21$. Since $Y = 24.59$ is much greater than 9.21, we reject H_0 at the .01 level of significance.

The data strongly suggest that a student's decision is not independent of the evaluation. The nature of the dependence can be seen by examining the six contributions to Y in the calculations above. Of the students rated excellent, fewer than would be expected accepted the college. Of those rated good, far more than would be expected accepted the college.

■

At this point we note that the assumptions as to sample size that applied in Section 15.1 apply equally here. Specifically

> DEFINITION 6
>
> The chi-square test of independence is valid as long as no more than one cell has an *expected* count less than 5.

One final example will illustrate an interesting feature of 2×2 contingency tables.

EXAMPLE 4 A group of 150 executives were classified by sex and whether or not they were firstborn (see Table 14). Are these two variables independent?

TABLE 14

One hundred fifty executives classified by sex and birth order

	FIRSTBORN YES	NO	TOTAL
MALE	34	74	108
FEMALE	20	22	42
TOTAL	54	96	150

Solution The null and alternative hypotheses are:

H_0: among executives the two variables, sex and whether or not being firstborn, are independent variables

H_1: among executives the two variables are not independent

The table of expected values is

	FIRSTBORN YES	NO	TOTAL
MALE	38.88	69.12	108
FEMALE	15.12	26.88	42
TOTAL	54	96	150

We compute

$$Y = \sum^{4} \frac{(O_i - E_i)^2}{E_i}$$

$$= \frac{(34 - 38.88)^2}{38.88} + \frac{(74 - 69.12)^2}{69.12} + \frac{(20 - 15.12)^2}{15.12} + \frac{(22 - 26.88)^2}{26.88}$$

$$= .61251 + .34454 + 1.5750 + .88595$$

$$= 3.418$$

From Table A7, $\chi^2_{(r-1)(c-1),\,.05} = \chi^2_{1,\,.05} = 3.84$ and $\chi^2_{1,\,.01} = 6.63$. Since $Y = 3.418$ is less than 3.84 we do not reject H_0 at the .05 level of significance. The data are not inconsistent with the independence of these two variables. ■

In Examples 3 and 4 we assumed that a single sample of n was selected from a bivariate population and then classified on the basis of two variables. In the voter example at the beginning of this section, for instance, we assumed that the 200 voters were classified by political affiliation and religion after their selection. But suppose that *separate* samples of 98 Democrats, 72 Republicans, and 30 unaffiliated voters had been selected and then classified by religion. The resulting table would be indistinguishable from Table 7. Or, in Example 4, suppose that instead of dividing a single sample of 150 executives by sex and birth order, separate samples of men and women executives had been separately selected and then each sample divided by birth order. Again the resulting 2×2 table would be indistinguishable from Table 14.

In both these cases it would not be appropriate to test the hypothesis of independence in *the* population since samples have been drawn from a number of populations. Rather, in the voter example we would wish to test whether the proportion of Protestant, Catholic, and Jews are the same for each population (political party). In Example 4 we would want to test whether the proportions firstborn (and not firstborn) are the same for men and women.

Tests of such hypotheses are called **tests of homogeneity.** It is a remarkable fact that the method for testing the hypothesis of homogeneity of proportions in a number of populations leads to precisely the same χ^2 test as the hypothesis of independence. Only the conclusion would be slightly different. If a single sample is selected and H_0 not rejected, we say that the data are not inconsistent with independence. If the data consist of samples from a number of populations, not rejecting H_0 suggests that the proportions in the various populations are not significantly different.

If we view the data in Example 4 as consisting of separate samples of men and women executives, we can use the methods of Section 10.5 to test for a significant difference in the proportion of firstborns for women (π_1) and men (π_2) executives. The details follow.

We test $H_0: \pi_1 - \pi_2 = 0$ against $H_1: \pi_1 - \pi_2 \neq 0$. Following the procedure in Section 10.5, the corresponding sample difference in proportions is $\bar{p}_1 - \bar{p}_2 = 20/42 - 34/108 = .47619 - .31481 = .16138$ and the combined proportion $\bar{p}_c = (20 + 34)/(42 + 108) = 54/150 = .36$.

The value for the test statistic

$$z = \frac{\bar{p}_1 - \bar{p}_2}{\sqrt{\bar{p}_c(1 - \bar{p}_c)(1/n_1 + 1/n_2)}}$$

is

$$z = \frac{.16138}{\sqrt{(.36)(1 - .36)(1/42 + 1/108)}} = 1.8488$$

The P value is then

$$2P(Z > 1.8488) = 2(.0322) = .0644$$

which exceeds .05; therefore, as before, we cannot reject H_0 at the .05 level of significance. In this case, however, we conclude that the data suggest no significant difference in the proportion of firstborns among men and women executives.

Notice that the square of the computed value for Z, $1.8488^2 = 3.418$, is the computed value for Y in the χ_1^2 test of the same data. This is not a coincidence but an illustration of an interesting relationship between the χ_1^2 and the Z distribution. It can be shown that if the random variable Y has the χ_1^2 distribution, then $Y = Z^2$. A consequence of this result is that when frequencies occur in 2×2

contingency tables, the χ^2 test of independence is equivalent to the two-sided Z test of the difference between two proportions.

Readers familiar with the F test in the one-way analysis of variance and in simple linear regression will be interested in the following comparison. In linear regression the null hypothesis $H_0: \beta = 0$ is that two quantitative variables are (at least linearly) independent. In the one-way analysis of variance, H_0: $\mu_1 = \mu_2 = \cdots = \mu_r$ indicates that a qualitative and a quantitative variable are independent. Finally, in this section we tested for the independence of two qualitative variables.

PROBLEMS 15.3

In the following problems, write the appropriate null and alternative hypotheses whether or not you are asked to do so.

9. Each member of a sample of 166 persons taking the written part of the state driver's license test was classified by (i) whether or not they passed the test and (ii) socioeconomic level (the higher the score, the higher the level), as follows:

	SOCIOECONOMIC LEVEL				
	1	2	3	4	5
PASSED	2	13	35	40	40
FAILED	1	7	15	6	7

(a) State the appropriate null and the alternative hypotheses.
(b) Perform the corresponding chi-square test.
(c) State your conclusion.

10. A random sample of 80 women who have had a child within the past year are classified by whether or not they received nutritional counseling, and whether or not they are breastfeeding their child, with the following results:

		NUTRITIONAL COUNSELING	
		YES	NO
BREASTFEEDING	YES	24	21
	NO	10	25

(a) What are the appropriate null and alternative hypotheses for the chi-square test?
(b) Perform this test and state your conclusion.
(c) State the appropriate hypothesis in terms of two population proportions.
(d) Perform the Z test of H_0.
(e) Indicate how this test is equivalent to that in part (b).

11. Each school in a sample of 85 four-year colleges is classified by whether it is private or public, and the change in enrollment over the past five years, with the following results:

	ENROLLMENT INCREASED	DECREASED	UNCHANGED
PRIVATE	5	9	20
PUBLIC	14	5	32

(a) Do these data suggest any significant differences in the pattern of change in enrollment for the two types of school?

(b) If you find a significant difference, explain where the difference lies.

12. Regard the data in Problem 6 on page 586 and reproduced here as a random sample of U.S. adults.

		RACE WHITE	BLACK	HISPANIC	OTHER
SMOKING HABIT	NONSMOKER	131	69	39	8
	LIGHT SMOKER	180	108	46	14
	HEAVY SMOKER	147	27	18	13

Test whether smoking habits are independent of race.

13. In a survey of 220 adults in a city each respondent was classified by the newspaper read and their age group as follows:

		AGE YOUNG	MIDDLE-AGED	OLD
NEWSPAPER	RECORD	51	5	54
	TRAVELER	23	25	62

Do these figures suggest that the proportion of adults who read each newspaper is the same for each age group?

14. Regard the 180 students referred to in Problem 8 on page 586 and reproduced here as a sample of all students taking introductory statistics courses.

	CLASS FRESHMAN	SOPHOMORE	JUNIOR	SENIOR
SOCIAL SCIENCES	4	15	28	9
NATURAL SCIENCES	15	38	16	3
PROFESSIONAL STUDIES	6	15	23	8

Test whether class and concentration are independent.

15. A pilot survey of 67 faculty members at a large university included a question asking whether the member favored a unionized faculty. The responses, arranged by rank, are as follows:

		RANK			
		PROFESSOR	ASSOCIATE PROFESSOR	ASSISTANT PROFESSOR	INSTRUCTOR
FAVORING A UNION	YES	6	13	11	2
	NO	12	17	6	0

Do the proportions favoring a union differ significantly by rank? (You may have to combine the data for two ranks.) If they do differ explain how they differ.

16. A sample of 840 voters were asked to name the candidate they would like to see win the Democratic nomination for U.S. president. The results, arranged by sex, are as follows:

	CANDIDATES				
	LANIER	MINTON	NICHOLAS	OLIVIER	PRICE
MALE	97	117	38	68	75
FEMALE	116	126	44	70	89

Is there any relationship between the candidate favored and the sex of the voter?

Solution to Problem 9

(a) The two hypotheses are

H_0: whether or not a person fails the test is independent of socioeconomic status

H_1: these two variables are not independent

(b) The complete table is as follows, where the expected values are shown in parentheses [e.g., 39.16 = (50)(130)/166]:

	SOCIOECONOMIC LEVEL					
	1	2	3	4	5	TOTAL
PASSED	2(2.35)	13(15.66)	35(39.16)	40(36.02)	40(36.81)	130
FAILED	1(.65)	7(4.34)	15(10.84)	6(9.98)	7(10.19)	36
TOTAL	3	20	50	46	47	166

Notice that three of the expected values are less than 5. From Definition 6 the χ^2 test is therefore not appropriate. We can remedy this situation by *pooling* the observed and expected counts in the two lowest socioeconomic categories. The revised table is as follows:

	SOCIOECONOMIC LEVEL				
	1/2	3	4	5	TOTAL
PASSED	15(18.01)	35(39.16)	40(36.02)	40(36.81)	130
FAILED	8(4.99)	15(10.84)	6(9.98)	7(10.19)	36
TOTAL	23	50	46	47	166

We now compute

$$Y = \frac{(15 - 18.01)^2}{18.01} + \cdots + \frac{(7 - 10.19)^2}{10.19} = 7.659$$

In this case $(r - 1)(c - 1) = (2 - 1)(4 - 1) = 3$. From Table A7, $\chi^2_{3,.05} = 7.81$ and $\chi^2_{3,.01} = 11.34$.

(c) Since 7.659 is less than 7.81, we cannot reject H_0 at the .05 level of significance. The data suggest that these two variables are at least approximately independent.

SECTION 15.4

USING COMPUTER OUTPUT: MINITAB AND THE CHI-SQUARE TEST FOR INDEPENDENCE

In this section we indicate how Minitab can perform chi-square tests of independence as described in Section 15.3.[1] We begin with an example.

EXAMPLE 5 A sample of 240 students at a college are classified (1) by their area of concentration as follows: (a) natural sciences, (b) social sciences, (c) humanities, and (d) professional; and (2) according to whether they represent (a) their first, (b) second, or (c) third or higher generation in the family to go to college. The results are given in the following contingency table:

	CONCENTRATION			
	NS	SS	H	P
FIRST GENERATION	4	24	9	18
SECOND GENERATION	10	18	15	17
THIRD+ GENERATION	31	27	45	22

Do these data suggest that the hypothesis, H_0: generation and concentration are independent, is true? This hypothesis can be reformulated as H_0: each generation has approximately the same distribution of concentrators. The method for entering the 12 frequencies and requesting the chi-square test is shown in Table 15.

[1] Minitab can also perform tests of fit as described in Section 15.1, but the procedure is a bit complicated.

```
MTB > READ TABLE INTO C1-C4
DATA> 4,24,9,18
DATA> 10,18,15,17
DATA> 31,27,45,22
DATA> CHISQUARE FOR DATA IN C1-C4
     3 ROWS READ

EXPECTED FREQUENCIES ARE PRINTED BELOW OBSERVED FREQUENCIES
       I  C1    I  C2    I  C3    I  C4    ITOTALS
            NS        SS        H        P
-------I-------I-------I-------I-------I-------
     1 I     4 I    24 I     9 I    18 I     55
       I  10.3I   15.8I   15.8I   13.1I
1ST GEN
-------I-------I-------I-------I-------I-------
     2 I    10 I    18 I    15 I    17 I     60
       I  11.3I   17.3I   17.3I   14.3I
2ND GEN
-------I-------I-------I-------I-------I-------
     3 I    31 I    27 I    45 I    22 I    125
       I  23.4I   35.9I   35.9I   29.7I
3RD+GEN
-------I-------I-------I-------I-------I-------
TOTALS I    45 I    69 I    69 I    57 I    240

TOTAL CHI SQUARE =

          3.86 +  4.24 +  2.94 +  1.87 +
          0.14 +  0.03 +  0.29 +  0.53 +
          2.44 +  2.22 +  2.29 +  1.99 +

                  =  22.84

DEGREES OF FREEDOM = ( 3-1) X ( 4-1) =   6
```

TABLE 15

Requesting a chi-square test for the concentration data with printout

The command "READ TABLE INTO C1-C4" tells Minitab to expect four columns of frequencies corresponding to the four concentrations. The observed frequencies are then entered row by row. To obtain the value of $Y = \sum^{12} (O_i - E_i)^2/E_i$ in Section 15.3, we use the command "CHISQUARE FOR DATA IN C1-C4." The resulting output includes a copy of the contingency table, with each expected frequency printed beneath the corresponding observed frequency. In the top left-hand cell, for example, $10.3 = (55)(45)/240$ is the number of first-generation natural science concentrators to be expected if the two variables are independent. We have written in the titles, NS, SS, H, P, 1st GEN, 2nd GEN, and 3rd + GEN.

Minitab prints out the value for

$$Y = \sum^{12} \frac{(O_i - E_i)^2}{E_i} = 22.84$$

in this case. If H_0 is true, Y is χ^2_ν with $\nu = (3 - 1)(4 - 1) = 6$ degrees of freedom. From Table A7, $\chi^2_{6,.05} = 12.59$ and $\chi^2_{6,.01} = 16.81$. Since $Y = 22.84$ exceeds 16.81, we reject H_0 at the .01 level of significance. The data suggest that these two variables are *not* independent but are dependent. Minitab prints out each of the 12 components of $Y = \sum^{12} (O_i - E_i)^2/E_i$, so it is possible to search for particularly large discrepancies between the observed and expected frequencies. Notice that the two largest values for $(O_i - E_i)^2/E_i$, 3.86 and 4.24, are for first-generation natural and social scientists. Looking at the table that Minitab prints out, we see that there are fewer first-generation natural scientists and far more first-generation social scientists than would have been expected with independence.

■

Example 5 was similar to those in Section 15.3 in that the contingency table had already been constructed. But in reality the 12 frequencies in the table would have to be constructed from the original data. In the following example, we indicate how Minitab can do this as well as compute Y.

EXAMPLE 6 With regard to the nutrition data in Table 8 of Chapter 1 (page 48), it was of interest to examine whether the variables (1) breastfeeding or not, and (2) smoking habits, are independent, that is, to test H_0: these two variables are independent. In other words, does the proportion of women who breastfeed differ for the different smoking groups? In column (6) of Table 8 in Chapter 1, the mothers who breastfeed are identified by a 1 and the mothers who did not breastfeed by 0. Similarly, in column (3) nonsmokers are identified by a 1, light smokers by a 2, and heavy smokers by a 3. The command "CONTINGENCY TABLE OF C3 VS C6" instructs Minitab to construct a contingency table with smoking habits as the row variable and whether or not the mother was breastfeeding as the column variable. The resulting table is constructed and printed in Table 16. There are, for example, 21 nonsmoking mothers who were not breastfeeding. If the variables were independent, the corresponding expected frequency would be $(32)(49)/68 = 23.1$.

The "CONTINGENCY TABLE" command results not only in the table but also in the value $Y = 1.77$. From Table A7 with $\nu = (3 - 1)(2 - 1) = 2$ degrees of freedom, $\chi^2_{2,.05} = 5.99$ and $\chi^2_{2,.01} = 9.21$. Since $Y = 1.77$ is less than 5.991, we cannot reject H_0 at the .05 level of significance. The data suggest that the two variables are at least approximately independent.

```
MTB > CONTINGENCY TABLE OF C3 VS C6

EXPECTED FREQUENCIES ARE PRINTED BELOW OBSERVED FREQUENCIES

   ROW CLASSIFICATION - C3
   COLUMN CLASSIFICATION - C6
       I     0  I     1  I TOTALS
-------I--------I--------I-------
    1  I    21  I    28  I     49
       I   23.1I    25.9I
-------I--------I--------I-------
    2  I     6  I     3  I      9
       I    4.2I     4.8I
-------I--------I--------I-------
    3  I     5  I     5  I     10
       I    4.7I     5.3I
-------I--------I--------I-------
TOTALS I    32  I    36  I     68

TOTAL CHI SQUARE =

         0.18 +  0.16 +
         0.74 +  0.65 +
         0.02 +  0.02 +

                 =   1.77

DEGREES OF FREEDOM = ( 3-1 ) X ( 2-1 ) =   2

NOTE  3 CELLS WITH EXPECTED FREQUENCIES LESS THAN 5
```

TABLE 16

Instructing Minitab to construct a contingency table for the nutrition data and the printout

A word of caution is necessary here. At the foot of the printout in Table 16 is the statement "3 CELLS WITH EXPECTED FREQUENCIES LESS THAN 5." This is a warning that the chi-square test may not be appropriate in this case. In the circumstances we should accept the conclusion above only with some skepticism.

■

PROBLEMS 15.4

17. Refer to the nutrition data in Table 8 of Chapter 1 (page 48). The accompanying output consists of the contingency table with ethnic group as the row variable and whether or not the mother was breastfeeding as the column variable.
    (a) State H_0 and H_1 in this case.
    (b) Use the output to test H_0.

```
MTB > CONTINGENCY TABLE OF C1 VS C6

EXPECTED FREQUENCIES ARE PRINTED BELOW OBSERVED FREQUENCIES

   ROW CLASSIFICATION - C1
   COLUMN CLASSIFICATION - C6
        I     0  I     1  I TOTALS
-------I-------I-------I-------
     1  I    10  I    21  I      31
        I   14.6I   16.4I
-------I-------I-------I-------
     2  I    17  I     8  I      25
        I   11.8I   13.2I
-------I-------I-------I-------
     3  I     5  I     7  I      12
        I    5.6I    6.4I
-------I-------I-------I-------
TOTALS I    32  I    36  I      68

TOTAL CHI SQUARE =

         1.44 +   1.28 +
         2.33 +   2.07 +
         0.07 +   0.07 +

                  =     7.27

DEGREES OF FREEDOM = ( 3-1 ) X ( 2-1 ) =   2
```

18. Refer again to the nutrition data in Table 8 of Chapter 1. The accompanying output is the contingency table with ethnic group as the row variable and smoking habit as the column variable.
    (a) State H_0 and H_1 in this case.
    (b) Use the output of the table to test H_0.
    (c) Why should you view your results in part (b) with a good deal of skepticism?

```
MTB > CONTINGENCY TABLE OF C1 VS C3

EXPECTED FREQUENCIES ARE PRINTED BELOW OBSERVED FREQUENCIES

   ROW CLASSIFICATION - C1
   COLUMN CLASSIFICATION - C3
       I    1  I    2  I    3  I TOTALS
-------I-------I-------I-------I-------
    1  I   20  I    5  I    6  I     31
       I   22.3I    4.1I    4.6I
-------I-------I-------I-------I-------
    2  I   18  I    4  I    3  I     25
       I   18.0I    3.3I    3.7I
-------I-------I-------I-------I-------
    3  I   11  I    0  I    1  I     12
       I    8.6I    1.6I    1.8I
-------I-------I-------I-------I-------
TOTALS I    49 I     9 I    10 I     68

TOTAL CHI SQUARE =

         0.24 +  0.20 +  0.46 +
         0.00 +  0.14 +  0.12 +
         0.64 +  1.59 +  0.33 +

                   =    3.73

DEGREES OF FREEDOM = ( 3-1) X ( 3-1) =    4

NOTE  6 CELLS WITH EXPECTED FREQUENCIES LESS THAN 5
```

19. In a telephone survey conducted on behalf of a company producing coffee, respondents were classified as to whether or not they used the company's brand and by household income. The results were as follows:

		HOUSEHOLD INCOME (DOLLARS)			
		$< 15,000$	15,000–24,999	25,000–34,999	$> 35,000$
USED COMPANY'S BRAND	YES	15	22	10	3
	NO	34	42	42	22

In the accompanying output, the eight frequencies are entered and the "CHISQUARE" command given.

(a) What are the appropriate null and alternative hypotheses?

(b) Use the information in the output to test H_0.

```
MTB > READ TABLE INTO C1-C4
DATA> 15,22,10,3
DATA> 34,42,42,22
DATA> CHISQUARE FOR DATA IN C1-C4
     2 ROWS READ

EXPECTED FREQUENCIES ARE PRINTED BELOW OBSERVED FREQUENCIES
       I  C1    I  C2    I  C3    I  C4    ITOTALS
-------I-------I-------I-------I-------I-------
     1 I   15  I   22  I   10  I    3  I     50
       I   12.9I   16.8I   13.7I    6.6I
-------I-------I-------I-------I-------I-------
     2 I   34  I   42  I   42  I   22  I    140
       I   36.1I   47.2I   38.3I   18.4I
-------I-------I-------I-------I-------I-------
TOTALS I    49 I    64 I    52 I    25 I    190

TOTAL CHI SQUARE =

         0.34 +  1.58 +  0.99 +  1.95 +
         0.12 +  0.56 +  0.35 +  0.70 +

                 =    6.60

DEGREES OF FREEDOM = ( 2-1) X ( 4-1) =   3
```

20. After a change of ownership a magazine's format is altered to appeal to a more "up-beat" audience. Readers are invited to fill out and send in a short questionnaire about the changes that have been made. The questionnaire is contained in the latest edition of the magazine. The following table is a breakdown of the 420 replies by sex and whether or not they approved of the change.

		SEX	
		MALE	FEMALE
APPROVED	YES	53	90
OF CHANGE	NO	137	140

These four frequencies are entered and a chi-square test requested in the accompanying printout.

(a) State the null and alternative hypotheses.
(b) Use the output to test H_0.
(c) Perform a Z test of the difference between the two population proportions.
(d) In what way are the tests in parts (b) and (c) equivalent?
(e) Why should the publishers view these 420 respondents as a biased sample?

```
MTB > READ TABLE INTO C1,C2
DATA> 53,90
DATA> 137,140
DATA> CHISQUARE FOR  DATA IN C1,C2
     2 ROWS READ

EXPECTED FREQUENCIES ARE PRINTED BELOW OBSERVED FREQUENCIES
       I  C1    I  C2    ITOTALS
-------I-------I-------I-------
     1 I    53 I    90 I     143
       I   64.7I   78.3I
-------I-------I-------I-------
     2 I   137 I   140 I     277
       I  125.3I  151.7I
-------I-------I-------I-------
TOTALS I   190 I   230 I     420

TOTAL CHI SQUARE =

         2.11 +  1.75 +
         1.09 +  0.90 +

                 =    5.85

DEGREES OF FREEDOM = ( 2-1) X ( 2-1) =   1
```

SECTION 15.5

SUMMARY AND REVIEW

TESTS OF FIT A sample of size n is selected from a population each member of which is in one and only one of M categories. The (unknown) proportions of the population in each category are $\pi_1, \pi_2, \ldots, \pi_M$. The numbers in each category in the sample are $O_1, O_2, \ldots, O_M$, where $O_1 + O_2 + \cdots + O_M = n$.

In a test of fit we examine whether or not the O's are consistent with $H_0: \pi_1 = k_1, \pi_2 = k_2, \ldots, \pi_M = k_M$ when $k_1 \cdots k_M$ are hypothesized proportions. If H_0 is true, the number of observations that would be "expected" to fall in each category are $E_1 = nk_1, E_2 = nk_2, \ldots, E_M = nk_M$.

The extent to which the observed and the expected counts differ is measured by

$$Y = \sum^{M} \frac{(O_i - E_i)^2}{E_i}$$

which, if H_0 is true, will have approximately the χ^2_{M-1} distribution (Definition 1).

The procedure by which the value for Y is used to test H_0 is outlined in Table 3. This test is valid as long as no more than one expected count is less than five.

If H_0 is rejected, the $(O_i - E_i)$ differences should be examined to see where the large discrepancies lie.

In the context of chi-square tests the number of degrees of freedom for χ^2 can be defined as: df = the number of probabilities specified by H_0 that are "free" to vary (Definition 2).

TESTING FOR INDEPENDENCE

We frequently have to decide whether two (usually) qualitative variables *are* independent on the basis of only a sample from a population. The sample data can be arranged in a contingency table similar to that for a population.

Assume that a sample of size n is selected from a population and classified into an $r \times c$ contingency table involving r categories for the rows, variable I and c categories for the columns, variable II. The sample will thus consist of a total of rc observed counts $O_1, O_2, \ldots, O_{rc}$.

The number of observations in each row will be $R_1, R_2, \ldots, R_r$ and in each column $C_1, C_2, \ldots, C_c$. Since the total numbers of observation are n:

$$R_1 + R_2 + \cdots + R_r = n$$

$$C_1 + C_2 + \cdots + C_c = n$$

When we test for independence, the null and alternative hypotheses are:

H_0: in the population, the two variables are independent

H_1: in the population, the two variables are dependent

If H_0 is true, the expected count in each cell will be the product of the observed row and column totals divided by n, that is, $E = (R)(C)/n$. As before the test statistic is

$$Y = \sum_{\text{all cells}} \frac{(O - E)^2}{E}$$

which, if H_0 is true, has the $\chi^2_{(r-1)(c-1)}$ distribution. The procedure for testing H_0 is summarized in Definition 5.

If the contingency table results not from a sample from one population but from samples from each of r or c populations, we perform a **test of homogeneity.** This test is identical to the test of independence just mentioned but the interpretation will differ. As a consequence, when sample data are arranged in a 2×2 contingency table, the χ^2 test of independence is equivalent to the two-sided test of the difference between two sample proportions. This can be explained by noting that $Z^2 = Y$, where $Y = \chi^2_1 = \chi^2_{(2-1)(2-1)}$.

REVIEW PROBLEMS

GENERAL

1. In the following table (from Yule, 1900), 205 married couples were classified by the husband's and wife's height:

		WIFE		
		TALL	MEDIUM	SHORT
HUSBAND	TALL	18	28	14
	MEDIUM	20	51	28
	SHORT	12	25	9

Test whether the heights of husbands and wives are independent.

2. One of the authors was given a pair of dice, both of which contained the imprint of a famous Las Vegas gambling casino. To test whether they were fairly balanced, the author (and his family) rolled the pair of dice 1000 times, in each case recording the sum of the two numbers. The results are summarized as follows:

SUM	2	3	4	5	6	7	8	9	10	11	12
NUMBER OF TIMES SUM APPEARED	21	61	82	90	133	168	145	108	101	63	28

(a) If the two dice are fair, what are the probabilities of getting each of the values 2, 3, . . . , 12?

(b) Are these data consistent with the hypothesis that the dice are fair?

3. At a racetrack the horses are placed in one of eight starting stalls prior to the start of a race. The horses are assigned to the stalls at random. A racing enthusiast has collected the following data on which stalls the winners of the last 140 races have started from:

STALL	1	2	3	4	5	6	7	8
NUMBER OF WINNERS	13	13	16	18	17	19	23	21

(a) Use these data to test whether the winners are equally likely to come from any stall.

(b) If the null hypothesis is rejected, which numbered stalls seem to be the most likely to produce winners?

4. The following data are based on the records of 183 high-speed automobile accidents. Each accident was categorized by whether the driver was killed, and whether the driver was wearing a seat belt.

		WEARING A SEAT BELT	
		YES	NO
FATAL	YES	6	43
	NO	46	88

What do these data suggest about the effect of wearing seat belts on whether or not the driver survives an accident?

5. Assume at a set location on a highway, state police stop a random sample of motorists and measure the alcohol content of their blood. The same measurement is taken on those motorists stopped for speeding. During one holiday weekend 180 motorists were stopped at random and 110 were stopped for speeding. The following table contains a breakdown of each of these groups by whether there was no trace of alcohol, or an amount of alcohol below or above the legal limit.

	NO TRACE	BELOW LEGAL LIMIT	ABOVE LEGAL LIMIT	TOTAL
SPEEDERS	35	47	28	110
THOSE RANDOMLY STOPPED	91	67	22	180

(a) Do these data suggest that the distribution of alcohol content differs significantly for the two groups?

(b) If so, how do they differ?

(c) Find a 90% confidence interval for the proportion of all speeders who are stopped with an alcohol content above the legal limit.

6. A researcher with an automobile manufacturing company assumes that the distribution of city mileage per gallon (mpg)—on the road rather than in the laboratory—for a particular model is $N[24.2, .6]$. To examine whether actual road experience is consistent with this assumption, 120 cars of this model are driven over the same city course. The resulting mpg values are summarized in the following frequency distribution:

	INTERVAL	NUMBER OF CARS
1	Under 23.00	10
2	23.00–23.49	20
3	23.50–23.99	33
4	24.00–24.49	28
5	24.50–24.99	17
6	25.00–25.49	8
7	Above 25.50	4
		120

(a) Assuming that the distribution of mpg, X, *is* $N[24.2, .6]$, compute the probabilities of falling into each of the seven categories, that is, $k_1 = P(X < 23.00)$, $k_2 = P(23.00 < X < 23.50), \ldots, k_7 = P(X > 25.50)$.

(b) What are the corresponding expected number of cars in each interval?

(c) Compare these expected numbers with the observed numbers above by means of a chi-square test of fit.

(d) Do these results suggest that the researcher's assumption was appropriate?

HEALTH SCIENCES

7. Two antiulcer drugs, T and Z, are to be compared. A group of 200 persons suffering with duodenal ulcers are randomly divided into two groups of 100 each. The members of group I are given drug T and members of group II, drug Z. The numbers of cases in which the ulcers healed within a month were as follows:

		DRUG	
		T	Z
HEALED	YES	72	64
	NO	28	36

(a) Perform a χ^2 test to see whether the two drugs are equally likely to heal a person's ulcer.

(b) Suppose that, prior to the experiment, drug T is expected to have a higher probability of healing than drug Z. Perform the appropriate Z test of the difference between two proportions.

(c) Why would the χ^2 test in part (a) be inappropriate for the situation in part (b)?

8. Gordon et al. (1981) report on some of the outcomes of the Framingham heart study. The following table gives the number of deaths from heart attack or heart disease among men aged 45–64 broken down by daily caloric intake:

	DIED	
DAILY CALORIC INTAKE	YES	NO
Less than 2000	14	160
2000–2499	14	237
2500–2999	17	246
3000 or more	6	214

(a) Do these data suggest any significant difference in death rate for the various caloric groups?

(b) If yes, explain where the differences lie.

9. In a recent article (Helmrich, 1983) a group of 798 women with breast cancer were compared with a control group of 2019 women without breast cancer. In the following table we break down the numbers in each group by the age of the women at first birth.

		BREAST CANCER	CONTROL
	NEVER PREGNANT	180	475
AGE AT FIRST BIRTH	< 20	97	445
	20–24	229	664
	25–29	209	323
	30 +	83	112
	TOTAL	798	2019

(a) Do these data suggest that the distribution of age at first birth is the same for both groups?

(b) If not, explain how the two groups differ.

10. Pfeffer et al. (1983) classified 103 children receiving psychiatric help by race and whether they are inpatients or outpatients. The results were as follows:

	INPATIENT	OUTPATIENT
HISPANIC	25	8
BLACK	23	15
WHITE	14	18

Do these data suggest that the racial composition of inpatients is significantly different from that of outpatients?

SOCIOLOGY AND GOVERNMENT

11. After a primary election a random sample of voters is selected and classified according to age group and whether or not they voted for the incumbent candidate, Collins.

	AGE GROUP		
	YOUNG	MIDDLE-AGED	OLDER
COLLINS	17	43	28
OTHER CANDIDATES	63	97	72

Do these sample data suggest that the age distribution of Collins' voters differs from that for other candidates?

12. A random sample of 1315 policemen are classified by race (whether they are white or not) and by their assignment. The results are as follows:

		WHITE	OTHER
ASSIGNMENT	PATROL	508	123
	TRAFFIC	91	20
	DETECTIVE	271	40
	OTHER	225	37
		1095	220

Do these data suggest that the distribution of assignments differs significantly for the two racial groups?

13. In the following table we present the numbers of presidential appointees to U.S. District Court judgeships broken down by the president and by the type of undergraduate institution attended by the appointee. Those appointees not providing the necessary data are excluded (from *Sourcebook of Criminal Justice Statistics*, 1982).

		PRESIDENT			
		JOHNSON	NIXON	FORD	CARTER
UNDERGRADUATE INSTITUTION	PUBLIC SUPPORTED	47	74	25	116
	PRIVATE—NOT IVY LEAGUE	38	69	18	66
	IVY LEAGUE	20	35	9	20

Do these data indicate any significant difference in the distribution of type of institutions for the four presidents?

14. Those persons interviewed in an opinion poll are classified by whether they think the president should seek reelection and by their likelihood of voting.

	PRESIDENT SHOULD SEEK REELECTION YES	NO
UNLIKELY TO VOTE	30	34
LIKELY TO VOTE	150	169
CERTAIN TO VOTE	54	99

Perform the appropriate χ^2 test.

15. In an article by Black (1974) 87 U.S. cities were classified by their size (greater than or less than 30,000) and by turnout in municipal elections. The results were as follows:

		POPULATION SIZE	
		GREATER THAN 30,000	LESS THAN 30,000
TURNOUT	ABOVE 50%	3	26
	40–50%	6	23
	BELOW 40%	10	19
TOTAL		19	68

Do these data suggest that the distribution of turnout differs significantly for large and small cities?

16. There are theoretical reasons for believing that the probability distribution of X, the number of arrests of drug users admitted to drug-abuse treatment programs, is as follows:

NUMBER OF ARRESTS, x	0	1	2	3	4	5	6
$P(x)$	.52	.26	.12	.05	.03	.01	.01

As part of a study of the reasons for drug abuse a random sample of 2600 persons admitted to such programs within the past year is selected. The breakdown of their number of arrests is as follows:

NUMBER OF ARRESTS	0	1	2	3	4	5	6
NUMBER OF PERSONS	1406	702	276	110	75	14	17

(a) Are these data consistent with the probability distribution above?

(b) Referring to the sample data, compute the mean and standard deviation of the number of arrests and find a 95% confidence interval for the mean number of arrests per person.

(c) Find a 95% confidence interval for the proportion of all such persons who have no arrests.

ECONOMICS AND MANAGEMENT

17. A machine is designed to mix nuts so that 1/12 of the resulting product will consist of pecans, 1/4 will be cashews, 1/4 almonds, and 5/12 peanuts. In a recent random sampling of 144 nuts mixed by the machine 16 were pecans, 31 were cashews, 30 were almonds, and the remainder peanuts. Do these data suggest that the machine is mixing nuts in conformity with its design?

18. A survey organization selected a random sample of adults in a city. Each respondent is classified by annual income of the family and by daily newspapers read. (The few persons regularly receiving no or two or more daily newspapers were omitted.) The results are summarized as follows:

		NEWSPAPER			
		GLOBE	*HERALD*	*TIMES*	*U.S. NEWS*
FAMILY INCOME	LESS THAN $15,000	76	70	17	70
	$15,000–$24,999	128	78	33	71
	$25,000–$34,999	119	47	58	72
	GREATER THAN $35,000	103	17	64	40

(a) Do these data suggest that the distribution of newspaper readership differs by income level?

(b) Find a 95% confidence interval for the proportion of all such readers who read the *Globe*.

19. Each of a sample of 1450 currently unemployed adult males is classified on the basis of age and number of months unemployed. The results are summarized as follows:

		MONTHS UNEMPLOYED		
		LESS THAN 6 MONTHS	6–11 MONTHS	LONGER THAN 11 MONTHS
AGE	LESS THAN 25	158	90	64
	26–34	141	115	71
	35–44	149	102	58
	45–54	137	133	83
		100	35	14
		685	475	290

Do these data suggest that age and time spent unemployed are independent?

20. A random sample of 220 customers who had made purchases at a department store were asked whether they were satisfied with the service. The results were broken down by sex and by age group.

	SEX		AGE GROUP		
	MALE	FEMALE	LESS THAN 40	40–60	MORE THAN 60
SATISFIED	65	113	75	63	40
NOT SATISFIED	17	25	7	14	21

(a) Do these data suggest that whether a customer is satisfied or not is independent of (i) sex; (ii) age group?

(b) Find a 95% confidence interval for the proportion of all customers who have made purchases and are satisfied with the service.

EDUCATION AND PSYCHOLOGY

21. Dyson (1967) classified a sample of 568 seventh-grade students on the basis of two variables: (i) whether they were members of a homogeneous or heterogeneous classroom, and (ii) their self-concept. The results are as follows:

		SELF-CONCEPT	
		BELOW AVERAGE	ABOVE AVERAGE
CLASSROOM	HOMOGENEOUS	108	137
	HETEROGENEOUS	164	159

Perform the χ^2 test of independence for these data.

22. Erdmann (1983) reported on the results of a survey of 1100 high school seniors and 1100 high school counselors. A total of 401 seniors and 536 counselors responded. Test whether the proportion of the two groups responding is significantly different by (a) performing a Z test; (b) a χ^2 test.

23. A psychology student is interested in the behavior of prisoners as they pass through the various stages of their sentences. As part of his Ph.D. thesis he collects the following data on 200 attempted escapes (both successful and unsuccessful). Each attempt is classified by the quarter of the sentence when it was made; in the first, second, third, or fourth. The results are as follows:

QUARTER	1	2	3	4
NUMBER OF ATTEMPTS	59	38	42	61

(a) Do these results appear consistent with the hypothesis that attempted escapes are equally likely to occur in any quarter?

(b) If not, which quarters are favored by prisoners?

24. The following table shows the results of a survey of 150 colleges. Each college is categorized by level of highest degree offered and whether the faculty are unionized or not.

		TWO-YEAR COLLEGE	FOUR-YEAR COLLEGE	SOME MASTER'S PROGRAMS	DOCTORAL PROGRAMS
UNIONIZED	YES	10	15	9	5
	NO	21	52	23	15
	TOTAL	31	67	32	20

Do these data suggest that the proportion unionized differs among the four types of colleges?

25. A sample of 70 high school seniors were asked to fill out a lengthy questionnaire on the basis of which they were categorized as (i) either pessimistic or optimistic and (ii) favoring relatively stable low-risk careers or not. The results were as follows:

	PESSIMISTIC	OPTIMISTIC	TOTAL
LOW RISK	32	15	47
RISKY	13	10	23
TOTAL	45	25	70

(a) Perform a χ^2 test of whether pessimists and optimists are equally likely to favor low-risk careers.

(b) Perform a one-sided Z test of the hypothesis that pessimists are more likely than optimists to favor a low-risk career.

(c) Why is a χ^2 test not appropriate for testing the hypothesis in part (b)?

SCIENCE AND TECHNOLOGY

26. A statistician at an agricultural research station intends to compare the performance of three chemical pesticides in protecting plants from a specific disease. A total of 600 identical plants are divided into three batches, one of which is sprayed with chemical A, one with B, and one with C (the sizes of the batches differ because of the differing amounts of the chemicals that were available). The following table indicates the number of plants that were and were not infected by the disease.

	CHEMICAL		
	A	*B*	*C*
INFECTED	42	40	64
NOT INFECTED	108	160	186
TOTAL	150	200	250

(a) Which chemical *seems* to be the best at preventing infection?

(b) Perform the appropriate chi-square test.

27. The Air Force tests two early-warning radar systems. System A is tested for three months, after which system B is tested for four months. Among the data collected are the numbers of aircraft detected by the two systems. The results are summarized as follows:

		SYSTEM *A*	SYSTEM *B*
AIRCRAFT	DETECTED	699	924
	NOT DETECTED	81	100
	TOTAL	780	1024

(a) Perform the appropriate χ^2 test.

(b) Suppose that prior to the trials system *B* was expected to perform better than system *A*. Perform the appropriate one-sided *Z* test for the difference between two proportions.

(c) Why would the test in part (a) not be appropriate for the situation in part (b)?

28. In a now-classic experiment in radiation (Rutherford et al., 1951) scientists recorded the number of alpha particles hitting a screen for each of 2608 periods. The results are summarized as follows:

NUMBER OF ALPHA PARTICLES, x	0	1	2	3	4	5	6	7	8	9	10+
NUMBER OF PERIODS	57	203	383	525	532	408	273	139	45	27	16

For example, exactly three alpha particles hit in 525 of the 2608 periods. There are both physical and statistical reasons for believing that the probabilities associated with each number of particles are as follows:

NUMBER OF ALPHA PARTICLES, x	0	1	2	3	4	5
THEORETICAL PROBABILITIES	.021	.080	.156	.202	.195	.151

NUMBER OF ALPHA PARTICLES, x	6	7	8	9	10+
THEORETICAL PROBABILITIES	.097	.054	.026	.011	.007

(a) Compute the expected number of periods that each value for x will occur if these theoretical probabilities apply.

(b) Perform the appropriate χ^2 test of fit. (You should use M-2 rather than M-1 degrees of freedom in this case. Without going into detail, the reason for losing the degree of freedom is that the theoretical probabilities are based to some extent on the data.)

REFERENCES

BLACK, G. S. "Conflict in the community: a theory of the effects of community size," *The American Political Science Review,* Vol. LXVIII (September 1974), pp. 1245-1261.

DYSON, E. "A study of ability grouping and the self concept," *Journal of Educational Research,* 60 (1967), pp. 403-405.

ERDMANN, D. G. "An examination of factors influencing student choice in the college selection process," *Journal of College Admissions* (Summer, 1983), pp. 3-6.

GORDON, T., et al. "Diet and its relation to coronary heart disease and death in three populations," *Journal of Circulation,* Vol. 63, Number 3 (1981), pp. 500-515.

HELMRICH, S. P., et al. "Risk factors for breast cancer," *American Journal of Epidemiology,* Vol. 117 (January 1983), pp. 35-45.

PFEFFER, C. R., et al. "Predictors of assertiveness in latency age children," *American Journal of Psychiatry,* Vol. 140, Number 1, (Jan., 1983) pp. 31-35.

RUTHERFORD, Sir E., Chadwick, J., and Ellis, C. D. *Radiation from Radioactive Substances.* London: Cambridge University Press, 1951, p. 172.

YULE, G. U. "On the association of attributes in statistics: with illustration from the material of the childhood society," *Philosophical Transactions of the Royal Society,* Series A, 194 (1900), pp. 257-319.

SUGGESTED READING

ANDERSON, T. W., and Sclove, S. L. *An Introduction to the Statistical Analysis of Data.* Boston: Houghton Mifflin, 1978.

Two excellent guides to recent work in the area of contingency table analysis are:

FIENBERG, S. E. *The Analysis of Cross-Classified Categorical Data,* 2nd ed. Cambridge, Mass.: The M.I.T. Press, 1980.

UPTON, G. J. G. *The Analysis of Cross-Tabulated Data.* New York: Wiley, 1978.

16 NONPARAMETRIC METHODS

COMPARING COSTS

A federal auditor is interested in comparing the average amounts that two welfare offices approve for welfare recipients' medical expenses. Pilot samples of 15 cases are selected from each office and the last monthly medical payment recorded. The results are given below (rounded to the nearest dollar).

A: 135 0 0 72 1370 18 0 54 59 605 29 0 113 27 10

B: 79 64 27 299 0 17 0 30 122 25 0 8 1964 0 5

The auditor notices immediately that the two-sample *t* test is not appropriate in this case because both samples are so skewed. What alternative is there to the *t* test when the data are so evidently nonnormal? See Review Problem 14 on page 658.

SECTION 16.1

ONE-SAMPLE TESTS: THE SIGN TEST

NONPARAMETRIC TESTS

In Chapters 12 and 13 we noted that the one-sample t test assumed a sample from a population that was approximately normal. Further, in the two- and multiple-sample t and F tests the populations were assumed not only to be approximately normal but also to have equal variances.

There are other tests which may be used in place of the t and F tests but which assume little about the shape of the populations. These tests are usually referred to as being **nonparametric** or, less frequently but more appropriately, as **distribution free.**[1] There is, however, a price to be paid for relaxing the assumptions of normality and equality of variances. When a t or F test is appropriate, it will be more sensitive than the corresponding nonparametric test.

THE SIGN TEST

The simplest nonparametric alternative to the one-sample t test (Sections 12.3 and 12.5) is called the **sign test.** Unlike the t test, the sign test requires only that the population being sampled is continuous.

In the one-sample t test we tested a hypothesis about the population mean μ. By contrast the sign test is concerned with the unknown population median (M). Recall that m, the sample median as the middle observation of a *sample* of ordered data, divides the observations in half. Similarly, the population median, M, is that value which divides the *population* in half (see Figure 1). The median (M) and the mean (μ) will coincide if the population is symmetric. Otherwise, they will not (see also Figure 1). In either case, if X is a randomly selected observation:

$$P(X < M) = P(X > M) = .5 \qquad \text{(Eq. 16.1)}$$

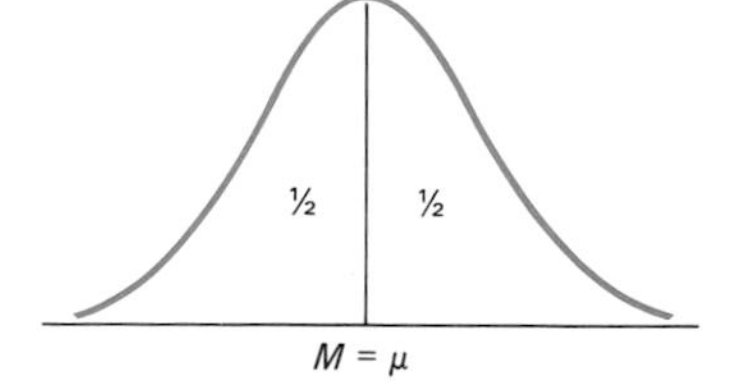

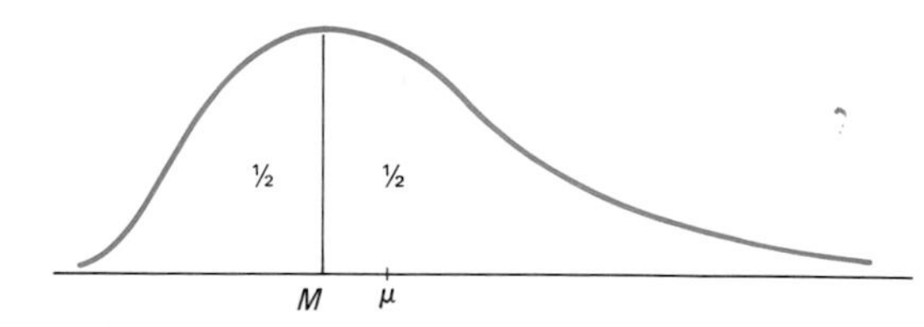

FIGURE 1

The median, M, of a continuous population

The following example illustrates the procedure used in the sign test.

EXAMPLE 1 The personnel director in a large plant believes that the median number of work days per employee lost due to illness during a specified year was 11. The days

[1]Although standard, the terminology "nonparametric" is misleading. Apparently, its use is based on the fact that nonparametric tests typically involve population medians rather than means. But such medians are as much parameters as are means.

lost that year for eight randomly selected employees were 1.5, 4.5, 2, 5, 12.5, 27, 3, and 7. Use these data to test $H_0: M = 11$ against $H_1: M < 11$.

Solution The (skewed) dot diagram for these data indicates that the one-sample t test is not appropriate here:

We begin our analysis by subtracting the null hypothesis median (11) from each of the observations, noting the sign of the difference in each case.

x_i	$x_i - 11$	SIGN OF DIFFERENCE
1.5	−9.5	−
4.5	−6.5	−
2.0	−9.0	−
5.0	−6.0	−
12.5	+1.5	+
27.0	+16.0	+
3.0	−8.0	−
7.0	−4.0	−
Number of positive differences, $y = 2$		

From Equation 16.1 and Figure 2(a), if $H_0: M = 11$ is true, there is an equal chance of an observation being *less* than 11 or *more* than 11, that is, $P(X < 11) = P(X > 11) = .5$. Equivalently, each difference $x_i - 11$ has the

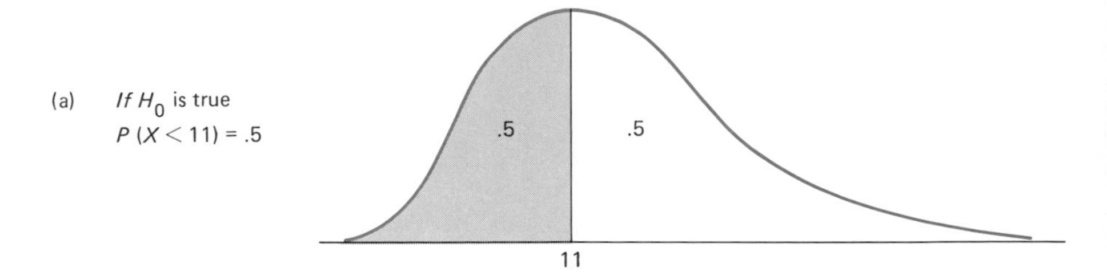

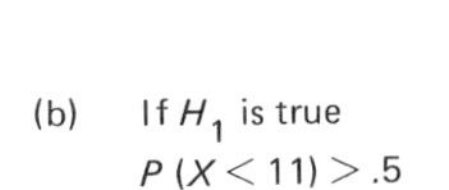

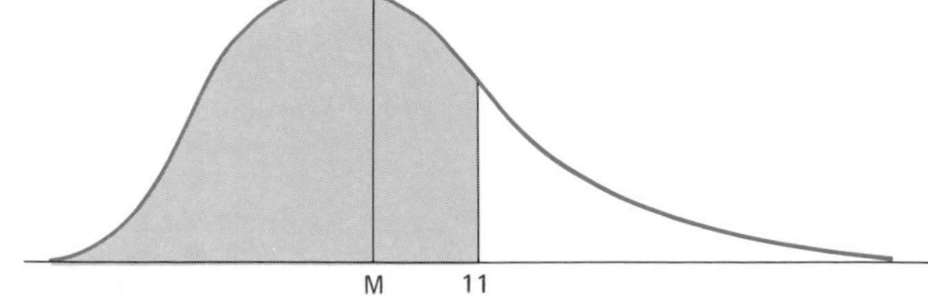

FIGURE 2

Situation in Example 1 if (i) H_0 is true and (ii) H_1 is true same probability .5 of being positive or negative. We may therefore regard the sample as a binomial experiment with the random variable Y, the number of positive differences $(x_i - 11)$, having the binomial distribution with $n = 8$ and $\pi = P(\text{positive difference}) = .5$. Recall that with $n = 8$ and $\pi = .5$, we would "expect" $n\pi = 8(.5) = 4$ positive differences.

The situation when $H_1: M < 11$ is true is shown in Figure 2(b). In this case, 11 is above the median so that *less* than half the population is above 11. As a result, the probability of an observation greater than 11 is less than .5. We would then expect less than half the $x_i - 11$ to be positive and consequently a value for y smaller than 4.

Our value $y = 2$ above is consistent with H_1, but is it sufficiently unusual for us to reject H_0? To answer this, we compute the P value and compare it with $\alpha = .05$ and .01. It is

$$P \text{ value} = P(Y \leq 2 \mid \pi = .5)$$

To find this probability we refer to the cumulative binomial probabilities in Table A2. With $n = 8$ and $\pi = .5$, $P(Y \leq 2) = .145$. Since the P value exceeds .05, we cannot reject H_0 at the .05 level of significance. The data suggest that the median time lost is not significantly less than 11 days. The P value is represented in Figure 3, the line histogram for the binomial distribution with $n = 8$ and $\pi = .5$. The P value is represented by the sum of the heights of the heavy lines above $y = 0$, 1, and 2.

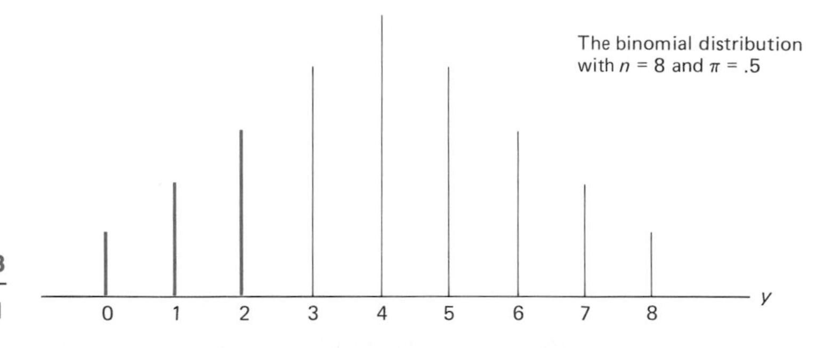

FIGURE 3

P value in Example 1

■

In Example 1 suppose that H_1 had been $H_1: M > 11$ instead of $H_1: M < 11$. Then, if H_0 were true, we would again expect four positive differences. But if H_1 were true, more than half the population would lie *above* 11, as shown in Figure 4. As a result, this time the probability of an observation greater than 11 would be greater than .5. We would thus expect more than half the $x_i - 11$ to be positive and thus a value for y greater than 4.

Suppose that the number of positive differences (y) were seven. This would be consistent with H_1. The P value would be $P(Y \geq 7 \mid \pi = .5) = 1 - P(Y \leq 6 \mid \pi = .5) = 1 - .965 = .035$, and we would therefore reject H_0 at the .05 level of significance. On the other hand, if Y were 2 as in Example 1, this would be contrary to our expectation of $y > 4$ and there would be no necessity to determine the P value; we would simply not reject H_0.

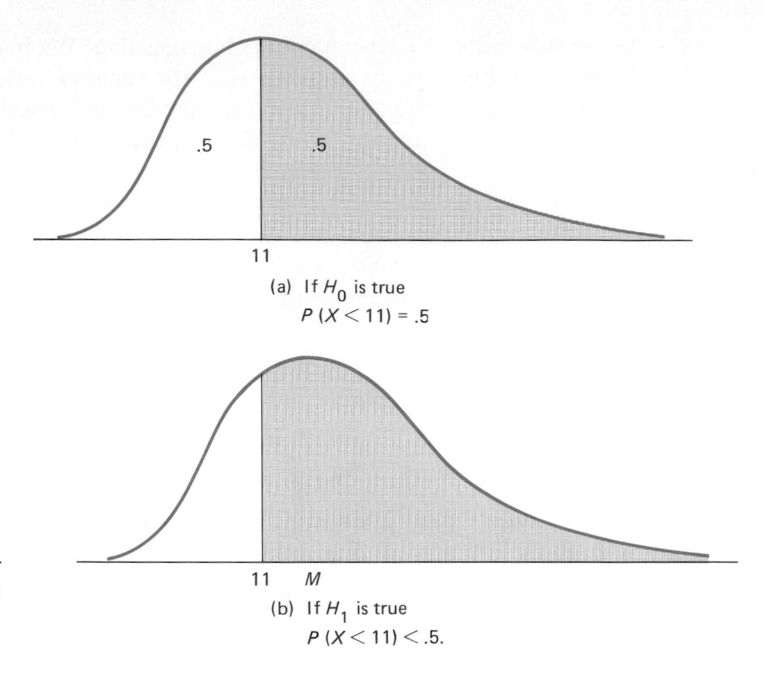

FIGURE 4

Situation if H_1 is $H_1: M > 11$ rather than $H_1: M < 11$

We can now summarize the general procedure for performing the sign test with $H_0: M = M_0$. Assume a random sample $x_1, x_2, \ldots, x_n$ from a population. We assume only that the population is continuous. If $H_0: M = M_0$ is true, we would expect half (or $n/2$) of the differences $x_i - M_0$ to be positive and half negative. We then ask whether the number of positive differences, y, is sufficiently far above or below $n/2$ to be inconsistent with H_0. The details are summarized in Table 1.

The rule in step 3(c, iii) needs some explanation. In Example 1 suppose that H_1 had been $H_1: M \neq 11$ and $y = 4$. In this case since $n = 8$, $y = n/2 = 4$

TABLE 1

Procedure for performing the sign test of $H_0: M = M_0$

1. Subtract M_0 from each observation.
2. Compute y, the number of the $x_i - M_0$ values that are > 0 (omit differences which $= 0$).
3. (a) $H_1: M > M_0$, P value $= P(Y \geq y \mid \pi = .5)$ $= 1 - P(Y \leq y - 1 \mid \pi = .5)$.
   (b) $H_1: M < M_0$, P value $= P(Y \leq y \mid \pi = .5)$.
   (c) $H_1: M \neq M_0$
      (i) If $y > n/2$, P value $= 2P(Y \geq y \mid \pi = .5)$ $= 2[1 - P(Y \leq y - 1 \mid \pi = .5)]$.
      (ii) If $y < n/2$, P value $= 2P(Y \leq y \mid \pi = .5)$.
      (iii) If $y = n/2$, do not compute a P value; do not reject H_0.
      (Find the probabilities in Table A2.)
4. Using $\alpha = .05$ or $.01$, reject H_0 at the α level of significance if the P value $\leq \alpha$.

Assumption: the population is continuous

and there would be no need to compute a P value. We would simply not reject H_0. No value for y is more supportive of H_0 than $y = 4$. You should verify that in this case both $2P(Y \leq 4 \mid \pi = .5)$ and $2P(Y \geq 4 \mid \pi = .5)$ exceed 1. A probability > 1 in such cases is merely an artifact of the statistical procedure and means that the P value is essentially 1.

The following example illustrates the use of the sign test with paired data.

EXAMPLE 2 A student compares the price of 15 items at two hardware stores, A and B. The difference (d) between price A and price B for each item has also been computed.

ITEM	*A*	*B*	*d*	ITEM	*A*	*B*	*d*	ITEM	*A*	*B*	*d*
1	2.72	2.68	.04	6	3.52	3.46	.06	11	.79	.72	.07
2	4.09	3.95	.14	7	3.22	3.31	−.09	12	.98	.90	.08
3	.48	.48	0	8	7.25	7.10	.15	13	1.77	1.72	.05
4	1.76	1.66	.10	9	1.04	1.04	0	14	1.99	2.05	−.06
5	2.19	2.23	−.04	10	4.60	4.48	.12	15	2.61	2.60	.01

Use the results above to test whether the median price difference for all items sold in both stores is (a) significantly greater than zero and (b) significantly different from zero.

Solution (a) We first test $H_0{:}M = 0$ against $H_1{:}M > 0$, where M is the (unknown) median difference in cost per item over all appropriate items. Using the 15 differences (d_i), we follow the steps in Table 1 with $M_0 = 0$. The signs of the $d_i - 0$ are $+ + 0 + - + - + 0 + + + + - +$. In practice we omit observations where the difference is 0, so the sample size here becomes $n = 13$ and y = number of positive differences = 10. From Table 1 the P value is $P(Y \geq 10 \mid \pi = .5) = 1 - P(Y \leq 9) = 1 - .954 = .046$. Since this value is less than .05, we reject H_0 at the .05 level of significance. The data suggest that the median difference in price is significantly greater than 0.

The P value, .046, is represented in Figure 5 as the sum of the heights of the lines over the values 10, 11, 12, and 13 of the histogram for the binomial distribution with $n = 13$ and $\pi = .5$.

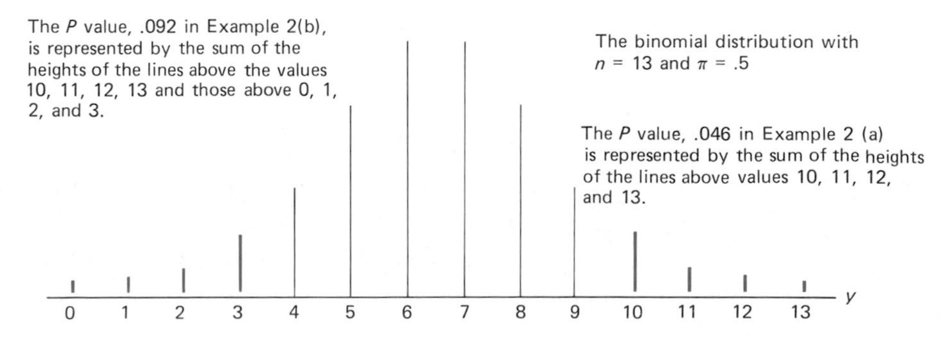

FIGURE 5

P values in Example 2

(b) If the student wished to test only for a significant difference in median prices, regardless of the direction, then $H_1: M \neq 0$ is appropriate. For the two-sided test we first check whether y is greater than or less than the expected number of positive differences $n/2 = 13/2 = 6.5$. Since $Y = 10 > 6.5$, we compute the P value as

$$P \text{ value} = 2P(Y \geq 10/\pi = .5) = 2(.046) = .092$$

Since this exceeds .05, we cannot reject H_0 at the .05 level of significance. The data suggest that the median difference in price is not significantly different from 0.

When $\pi = .5$ the binomial distribution is symmetric, in this case around $n\pi = 13(.5) = 6.5$. As a result, $P(Y \geq 10) = P(Y \leq 3)$ and the P value $= 2P(Y \geq 10)$ can also be written as $P(Y \geq 10) + P(Y \leq 3)$. These two probabilities are represented in Figure 5 by the sum of the heights of the lines above the values $y = 10, 11, 12, 13$ *and* those above the values $y = 0, 1, 2,$ and 3. ■

You may wonder why, although we use the binomial distribution, the sign test is called "distribution free." The point is that this test assumes nothing about the *shape* of the population being sampled; hence it is "distribution free." The test statistic Y is binomially distributed, not the population.

PROBLEMS 16.1

1. Twenty-one sections of oil cloth are each divided into two halves. One half, A, is treated with what it is hoped to be a strengthening agent. The other half, B, is left untreated. The breaking strength of each of the 42 halves is recorded with the following results (in hundreds of pounds per square inch).

A	B	A	B	A	B
5.9	5.2	6.0	5.0	2.9	3.5
5.1	5.2	6.0	5.2	4.7	4.1
4.8	4.4	3.8	3.3	6.8	6.8
4.6	4.5	5.2	6.3	8.2	7.5
3.7	4.6	6.6	7.4	6.6	5.7
7.2	7.0	4.8	5.4	3.9	3.8
8.0	8.1	7.4	5.9	7.1	8.4

Apply the sign test to these data to test whether the median difference in breaking strength $(A - B)$ is significantly greater than 0.

2. Test whether the following sample of nine incomes was drawn from a population with a median significantly less than \$23,800:

\$22,600	\$24,100	\$22,800
\$21,900	\$22,500	\$23,500
\$23,000	\$23,700	\$22,100

3. Use the data in Example 10 of Section 12.5 (page 421) to test whether the median decline in reaction times after drinking the beer is significantly greater than 0.
4. The following data are 20 times (in seconds) between telephone calls entering an exchange.

   12.9 10.7 14.5 3.7 2.4 7.2 10.2 15.2 9.2 18.0

   5.9 17.0 15.1 16.6 4.7 7.9 5.7 11.4 3.5 2.8

   Use the sign test to test $H_0: M = 12$ against $H_1: M < 12$.
5. Ten persons are each driven blindfolded over a predetermined course using two automobiles, A and B. In each case the person rates the quality of the ride on a scale of 0 to 20. The results are as follows:

PERSON	1	2	3	4	5	6	7	8	9	10
Automobile A	14	20	15	9	16	12	10	13	14	13
Automobile B	10	17	15	7	8	10	9	11	11	8

   Do the data suggest that the median difference in score $(A - B)$ over all potential participants is significantly (a) greater than 0; (b) different from 0?
6. The owner of a restaurant tells a potential buyer that the median number of customers per day is 550. The buyer feels that this figure is an overestimate and records the number of customers on each of 12 days with the following results:

   517 482 497 543 577 511

   555 522 580 529 479 536

   Do these data suggest that the owner was exaggerating?
7. A company is interested in comparing two new advertisements, one featuring a famous personality and one featuring a cartoon character. In a group of 40 consumers who were shown both ads, 26 preferred the ad featuring the cartoon character. Do these data suggest a significant difference in preference for the two ads?

Solution to Problem 1

We wish to test $H_0: M = 0$ against $H_1: M > 0$. One of the differences $(6.8 - 6.8)$ is zero, so we omit it. The 20 differences $(A - B) - 0$ and their signs are as follows:

$(A - B) - 0 = d_i - 0$	Sign
.7	+
−.1	−
.4	+
.1	+
−.9	−
.2	+
−.1	−
1.0	+
.8	+
.5	+
−1.1	−

$(A - B) - 0 = d_i - 0$	Sign
$-.8$	$-$
$-.6$	$-$
1.5	+
$-.6$	$-$
.6	+
0	0
.7	+
.9	+
.1	+
-1.3	$-$
	$y = 12$

Following the procedure in Table 1, the P value is $P(Y \geq 12 \mid \pi = .5) = 1 - P(Y \leq 11 \mid \pi = .5) = 1 - .748 = .252$. Since $.252 > .05$, we cannot reject H_0 at the .05 level of significance. The data suggest no significant differences in the median breaking strength for the treated and untreated sections.

SECTION 16.2
ONE-SAMPLE TESTS: THE WILCOXON SIGNED RANKS TEST

Another alternative to the one-sample t test is the **Wilcoxon signed ranks test,** which is almost as easy to apply as the sign test but makes greater use of the data. This test was developed in the 1940s by Frank Wilcoxon, an early advocate of nonparametric methods. Like the sign test, the Wilcoxon test also involves hypotheses about the population median, but unlike the sign test it assumes that the population is continuous and symmetric (although not necessarily normal).

For comparison purposes we shall use the data from Example 1 on page 619 to illustrate the Wilcoxon test though the dot diagram for these data suggests that the population may not be symmetric. The observations are the numbers of days lost because of illness for eight employees in a year. They are

$$1.5 \quad 4.5 \quad 2.0 \quad 5.0 \quad 12.5 \quad 27 \quad 3.0 \quad 7.0$$

We are to test $H_0: M = 11$ against $H_1: M < 11$.

We begin as we did with the sign test by subtracting 11 from each observation and recording the difference. This time, however, we *rank* the differences from smallest (1) to largest (8), temporarily ignoring their signs. In other words, we rank the absolute values of the differences. These steps are shown in Table 2. We now record the ranks in the last column again, but this time assigning to each rank the sign of $(x_i - 11)$ to obtain the *signed ranks*. The Wilcoxon test is based on w = sum of the positive rank values. In this case, from Table 2, $w = 1 + 8 = 9$.

Before continuing with this example we examine the behavior of w under more general conditions. In general, we consider a sample $x_1, x_2, \ldots, x_n$ of size

TABLE 2 Computing w for the data in Example 1

x_i	$x_i - 11$	RANK OF $\lvert x_i - 11 \rvert$	SIGNED RANKS	
1.5	−9.5	7		−7
4.5	−6.5	4		−4
2.0	−9.0	6		−6
5.0	−6.0	3		−3
12.5	+1.5	1	+1	
27.0	+16.0	8	+8	
3.0	−8.0	5		−5
7.0	−4.0	2		−2
			$w = 9$	

n drawn from a continuous and symmetric population with unknown median M. We wish to test $H_0: M = M_0$ against $H_1: M > M_0$, $H_1: M < M_0$, or $H_1: M \neq M_0$. We call W, the random variable "the sum of the positive rank values." It can be shown that if H_0 is true, the mean and standard deviation of W over repeated samples will be

$$\mu_W = \frac{n(n+1)}{4} \qquad \text{(Eq. 16.2)}$$

$$\sigma_W = \sqrt{\frac{n(n+1)(2n+1)}{24}} \qquad \text{(Eq. 16.3)}$$

For instance, for $n = 8$,

$$\mu_W = \frac{8(9)}{4} = 18 \qquad \text{and} \qquad \sigma_W = \sqrt{\frac{8(9)(17)}{24}} = 7.1414$$

For any n the actual distribution of W, the sum of the positive ranks of the $(x_i - M_0)$, can be derived in a straightforward manner, but the derivation is lengthy. For $n = 8$ we show the line histogram of the probability distribution of W in Figure 6. It is discrete and symmetric around its mean value 18. For later use we indicate that the sum of the probabilities represented by the lines above the values $w = 0, 1, 2, \ldots, 9$ is $P(W \leq 9) = .125$.

Table A9 contains various upper- and lower-tail probabilities for W for values of n from 4 to 12. Figure 7 contains a portion of Table A9 for $n = 8$. For example, the table shows that for $n = 8$, $P(W \leq 5) = P(W \geq 31) = .039$ and $P(W \leq 9) = P(W \geq 27) = .125$. [We represent the $P(W \leq 9) = .125$ in Figure 6.]

We now return to our analyis of the absentee data. We had computed $w = 9$. We now ask whether this value is consistent with $H_0: M = 11$ days lost.

If H_0 is true, then $M = 11$ and we would expect half the differences $(x_i - 11)$, and the corresponding ranks, to be positive and therefore a value w

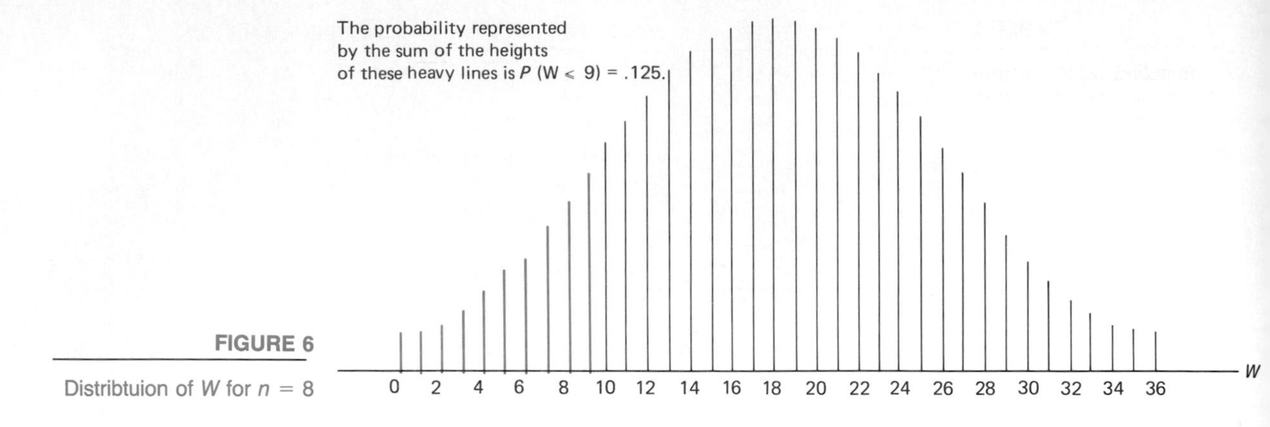

FIGURE 6

Distribtuion of W for $n = 8$

FIGURE 7

Portion of Table A9 for $n = 8$

	A	B	$P(W \leqslant A) = P(W \geqslant B)$
$n = 1$	.	.	.
.	.	.	.
.	.	.	.
$n = 8$	0	36	.004
	1	35	.008
	2	34	.012
	3	33	.020
	4	32	.027
	5	31	.039
	6	30	.055
	7	29	.074
	8	28	.098
	9	27	.125

close to its mean, $\mu_W = 8(9)/4 = 18$. If, on the other hand, $H_1: M < 11$ were true, we would expect *less* than half the $(x_i - 11)$'s and half the corresponding ranks to be positive, and therefore a value w less than 18.

We measure the extent to which the observed value $w = 9$ supports $H_0: M = 11$ by computing the P value, the probability of obtaining, by chance, a value for w as small or smaller than 9, that is, $P(W \leq 9)$. We have already noted that this probability is .125.

$$P \text{ value} = P(W \leq 9) = .125$$

Thus we cannot reject H_0 at the .05 level of significance. The data suggest that the median number of days lost to sickness is *not* significantly less than 11.

Notice from Table A9 that had the computed value for w exceeded 9, we could not have found the P value because the table contains only probabilities of the form $P(W \leq A)$ for values of A up to 9. But the mean of $W = n(n + 1)/4 = 18$ so the closer to 18 the more probable the value w. Had w been 11, for instance, we would have noted that the corresponding P value $- P(W \leq 11)$ *exceeds* .125.

Table 3 summarizes the procedure for performing the Wilcoxon signed ranks test of $H_0: M = M_0$ for the values of n from 4 to 12.

TABLE 3

Procedure for performing the Wilcoxon signed ranks test of $H_0: M = M_0$

1. Rank the n absolute differences $|x_i - M_0|$ from smallest (1) to largest (n). If any of the $|x_i - M_0|$'s = 0, omit those differences.
2. Assign the sign of $x_i - M_0$ to the rank.
3. Compute w, the sum of the positive signed rank values.
4. (a) $H_1: M > M_0$, P value $= P(W \geq w)$.
   (b) $H_1: M < M_0$, P value $= P(W \leq w)$.
   (c) $H_1: M \neq M_0$
      (i) If $w \geq n(n+1)/4$, P value $= 2P(W \geq w)$.
      (ii) If $w < n(n+1)/4$, P value $= 2P(W \leq w)$.
      (Obtain the probabilities from Table A9.)
5. Using $\alpha = .05$ or $.01$, reject H_0 at the α level of significance if the P value $\leq \alpha$.

Assumptions: the population is continuous and symmetric

We next apply the Wilcoxon test to paired data.

EXAMPLE 3 Twelve persons whose IQs had been measured in college between 15 and 20 years ago were recently tested with an equivalent IQ test. The results are as follows:

ORIGINAL SCORE	102	116	109	94	125	112	107	119	99	114	107	111
RECENT SCORE	115	122	122	97	116	112	113	134	112	118	106	122
IMPROVEMENT, d_i	13	6	13	3	−9	0	6	15	13	4	−1	11

Use the Wilcoxon test to decide if the median improvement in IQ is significantly greater than zero. In other words, do the results suggest that IQ improves with age?

Solution We test $H_0: M = 0$ against $H_1: M > 0$, where M is the unknown median improvement in IQ over all potential participants in the study. Since one of the differences $d_i - 0 = 0$, we omit this observation and analyze only the remaining 11. When ties occur, we give each difference the average of the ranks. We have indicated this by first using question marks to indicate ties and then listing the averages in parentheses. (See table at top of page 630.) For example, the two 6's are the fourth and fifth smallest differences, so they are each given the rank $(4 + 5)/2 = 4.5$. Similarly, the three 13's are given the rank $(8 + 9 + 10)/3 = 9$. Assigning the signs of the $(d_i - 0)$'s to the ranks and adding up the positive rank values, we obtain $w = 59$.

In this case the P value, the probability of getting a value for w as large or larger than 59 if H_0 is true, is $P(W \geq 59)$. From Table A9 with $n = 11$ (not 12), we see that $P(W \geq 59) = .009$. Accordingly, we reject H_0 at the .01 level of significance. The data suggest that the median improvement is significantly

$d_i - 0$	RANK $\lvert d_i - 0 \rvert$	SIGNED RANKS	
13	? (9)	+9	
6	? (4.5)	+4.5	
13	? (9)	+9	
3	2	+2	
−9	6		−6
6	? (4.5)	+4.5	
15	11	+11	
13	? (9)	+9	
4	3	+3	
−1	1		−1
11	7	+7	
		$w = 59$	

greater than 0 and therefore that IQ does increase with age, at least within the age range evaluated here.

■

Incidentally, had we been interested in the two-sided alternative $H_1: M \neq 0$, we would have followed the procedure in step 4(c) in Table 3. In that case we would first determine $\mu_W = n(n+1)/4 = (11)(12)/4 = 44$. Since $w = 59 > 44$, the P value $= 2P(W \geq 59) = .018$. We would again reject H_0, but this time at the .05 level of significance.

If the sample size is greater than $n = 12$ whether the data are paired or not, we assume that W is approximately normally distributed with mean $\mu_W = n(n+1)/4$ and standard deviation[2]

$$\sigma_W = \sqrt{\frac{n(n+1)(2n+1)}{24}}$$

As a consequence, the P value, which is a probability about W, can be standardized, becoming a probability about Z. We illustrate the technique below.

EXAMPLE 4 An educator has created a test for which it is hoped that the median score, M, will be 60. He is concerned, however, that M may be less than 60. The test is given to 27 persons. The following scores, listed and ordered for convenience, are obtained:

32 36 38 42 44 44 45 46 47 49 49 50 52 54

55 56 57 62 63 65 66 67 69 70 72 76 80

Do these data suggest that M is significantly below 60?

Solution We test $H_0: M = 60$ against $H_1: M < 60$, where M is the (potential) median score for the test. In Problem 9 at the end of this section you are asked to show that in this case $w = 110$. From Table 3 the P value is $P(W \leq 110)$. With

[2]If ties occur, σ_W is slightly smaller than this. The distinction is usually minor and we shall ignore it.

$n = 27$, $\mu_W = n(n + 1)/4 = (27)(28)/4 = 189$ and

$$\sigma_W = \sqrt{\frac{(27)(28)(55)}{24}} = 41.623$$

The test statistic is

$$z = \frac{w - \mu_W}{\sigma_W} = \frac{110 - 189}{41.623} = -1.90$$

and the P value is $P(Z < -1.90) = .0287$. Since this is less than .05, we reject H_0 at the .05 level of significance. The data suggest that the median score for this test is significantly less than 60.

■

The general procedure for performing the Wilcoxon test using the normal approximation is summarized in Table 4. Problem 8 at the end of this section offers another example of the application of the normal approximation of W.

TABLE 4

Procedure for testing $H_0: M = M_0$ using the normal approximation to W

1. Rank the n absolute differences $|x_i - M_0|$ from the smallest (1) to the largest (n). If any of the $(x_i - M_0)$'s $= 0$, omit those differences.
2. Assign the sign of $x_i - M_0$ to the ranks.
3. Compute w, the sum of the positive signed rank values.
4. Compute

$$\mu_W = \frac{n(n + 1)}{4} \quad \text{and} \quad \sigma_W = \sqrt{\frac{n(n + 1)(2n + 1)}{24}}$$

5. Compute the test statistic, $z = (w - \mu_W)/\sigma_W$.
6. (a) $H_1: M > M_0$ If $z > 0$, P value $= P(Z > z)$ (if $z < 0$, do not reject H_0).
   (b) $H_1: M < M_0$ If $z < 0$, P value $= P(Z < z)$ (if $z > 0$, do not reject H_0).
   (c) $H_1: M \neq M_0$ If $z > 0$, P value $= 2P(Z > z)$ if $z < 0$, P value $= 2P(Z < z)$.
7. Using $\alpha = .05$ or .01, reject H_0 at the α level of significance if the P value $\leq \alpha$.

Assumptions: the population is continuous and symmetric

One final note about the Wilcoxon test. Other versions of the test use (1) the sum of the negative ranks, (2) the sum of all the signed ranks, or (3) the smaller of the sums of positive and negative ranks.

COMPARING THE SIGN, WILCOXON, AND t TESTS

Studies have compared the performance of the sign and Wilcoxon tests to the one-sample t test. This is usually done on the basis of **relative efficiency.** Without going into detail, the relative efficiency of one test to another is the ratio of the sample sizes necessary so that the two tests will have the same probabilities of type I and type II errors. (See Section 10.6 for a brief discussion of these two types of errors.)

If the sampled population is normal, the mean (μ) and the median (M) coincide and for any given sample size, the t test is the most efficient. It can be shown, however, that in this case, while the sign test is only about 65% as efficient, the Wilcoxon test is almost (approximately 95%) as efficient as the t test. In the last case, for instance, if it requires a sample of $n = 20$ to perform the Wilcoxon test with specified probabilities of error, a t test based on a sample of $n = 19$ will have the same probabilities of error.

The other side of the coin is that if the sampled population is not normal but is symmetric, the Wilcoxon test generally performs better than the t test. The sign test is still inferior to both.

In the situation where the sample suggests a significantly nonsymmetric population, both the t test and the Wilcoxon test are inappropriate and the sign test should be used.

These comments explain the growing popularity of the Wilcoxon test among statistical practitioners. The authors recommend that if the data suggest an approximately symmetric population, the Wilcoxon test, not the t test, should be used.

PROBLEMS 16.2

8. Apply the Wilcoxon test to the data in Problem 1 on page 624. Test whether the median difference $(A - B)$ in breaking strength is significantly greater than 0.
9. Show that in Example 4 the sum of the signed ranks is $w = 110$.
10. Apply the Wilcoxon test to the data in Problem 2 on page 625.
11. Refer to the data in Example 10 of Chapter 12 (page 421). Use the Wilcoxon test to test whether the median decline in reaction time after drinking the beer is significantly greater than 0.
12. Refer to Problem 4 on page 625. Use the Wilcoxon test to test $H_0: M = 12$ against $H_1: M < 12$.
13. Refer to Problem 5 on page 625. Use the Wilcoxon test to decide whether the median differences in scores $(A - B)$ over all potential participants is significantly (a) greater than 0; (b) different from 0.
14. Why is the Wilcoxon test inappropriate in Problem 7 on page 625?

Solution to Problem 8

The differences $(d_i - 0)$, their ranks, and signed ranks are as follows:

$d_i - 0$	RANK $\|d_i - 0\|$	SIGNED RANK	
.7	11.5	+11.5	
−.1	2.5		−2.5
.4	6	+6	
.1	2.5	+2.5	
−.9	15.5		−15.5
.2	5	+5	
−.1	2.5		−2.5
1.0	17	+17	
.8	13.5	+13.5	
.5	7	+7	
−1.1	18		−18
−.8	13.5		−13.5

$d_i - 0$	RANK $\|d_i - 0\|$	SIGNED RANK	
−.6	9		−9
1.5	20	+20	
−.6	9		−9
.6	9	+9	
.7	11.5	+11.5	
.9	15.5	+15.5	
.1	2.5	+2.5	
−1.3	19		−19
		$w = 121$	

Following the procedure in Table 4, we compute

$$\mu_W = \frac{n(n+1)}{4} = \frac{20(21)}{4} = 105$$

$$\sigma_W = \sqrt{\frac{20(21)(41)}{24}} = 26.786$$

The value for z is $(121 - 105)/26.786 = .60$ and the P value $= P(Z > .60) = .2743$. Since the P value $> .05$, we cannot reject H_0 at the .05 level of significance. The data suggest that the median difference is not significantly greater than 0.

SECTION 16.3

COMPARING TWO INDEPENDENT SAMPLES: THE MANN–WHITNEY TEST

A number of nonparametric alternatives to the two-sample t test in Section 12.4 have been devised. The most popular is one published almost simultaneously by H. R. Mann and D. R. Whitney and by Frank Wilcoxon. To avoid confusion with the one-sample test, we refer to this as the **Mann–Whitney test.** This test is concerned with the *difference* between two population medians. An example will illustrate the test procedure.

EXAMPLE 5 Of the 17 babies born in a 24-hour period in a hospital, nine were born to nonsmoking mothers and eight to smoking mothers. The weights (in grams) of the infants are as follows:

NONSMOKING MOTHERS	SMOKING MOTHERS
3166	3204
2971	2808
3666	2771
3405	3133
2980	3476
3741	3222
2864	2732
3227	2934
3220	

Do these data suggest that the median weight of babies born to nonsmoking mothers is significantly larger than that for smoking mothers?

We assume that both sets of birthweights can be regarded as random samples from two populations of birthweights, one for children of nonsmoking and the other for children of smoking mothers. Call the unknown median birthweights in the two populations M_1 and M_2, respectively. We further assume that both populations have the same shape, although possibly differing in their medians (see Figure 8). The null and alternative hypotheses are $H_0: M_1 - M_2 = 0$ (or $M_1 = M_2$) and $H_1: M_1 - M_2 > 0$ (or $M_1 > M_2$). The situations if H_0 is true and if H_1 is true are shown in Figure 8.

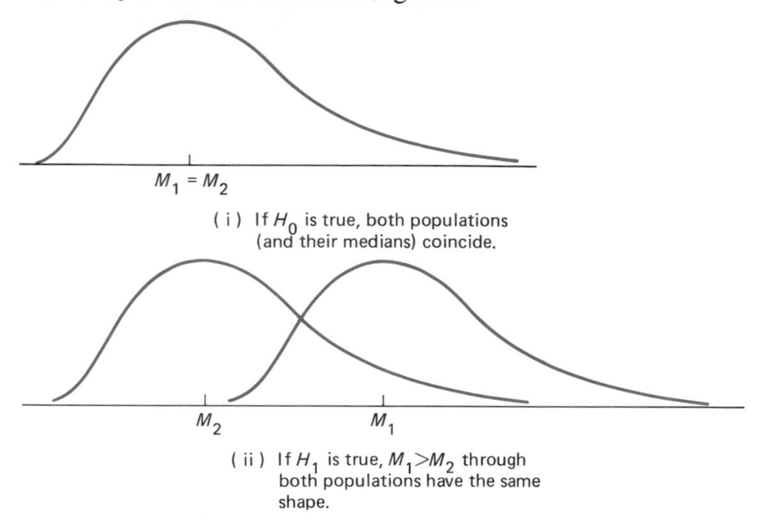

FIGURE 8

Distribution of birthweight for smoking and nonsmoking mothers assuming (i) H_0 is true; (ii) H_1 is true

We first rank all $8 + 9 = 17$ birthweights from smallest (1) to largest (17), being sure to record the sample with which each birthweight is associated. The following arrangement provides one way to do this:

NONSMOKERS	SMOKERS	RANK
	2732	1
	2771	2
	2808	3
2864		4
	2934	5
2971		6
2980		7
	3133	8
3166		9
	3204	10
3220		11
	3222	12
3227		13
3405		14
	3476	15
3666		16
3741		17
		153

We next compute the value for v, the sum of the ranks of the nonsmokers. In this case $v = 4 + 6 + 7 + 9 + 11 + 13 + 14 + 16 + 17 = 97$.

Stepping back for a moment, the sum of the 17 ranks will be $1 + 2 + 3 + \cdots + 17 = 153$. (A convenient way to represent the sum $1 + 2 + 3 + \cdots + n$ is $n(n + 1)/2$. For instance, $1 + 2 + 3 + \cdots + 17 = 17(18)/2 = 153$.)

If $H_0 : M_1 = M_2$ is true, we would expect a similar mix of high and low ranks in both samples. If both samples contained the same number of observations, the *expected* sum of ranks in each group, if H_0 were true, would be $153/2 = 76.5$. Since 9/17 of the birthweights belong to infants of nonsmoking mothers, the expected sum of ranks for *this* group, $E(V)$, is $(9/17)(153) = 81$. By contrast, if $H_1 : M_1 > M_2$ were true, we would expect the larger ranks to be concentrated in the nonsmoking group and hence v to take a value greater than 81.

The distribution of V, the random variable "the sum of the ranks in the first sample" has been tabulated for various combinations of sample sizes, but the tables tend to be tricky to use. We shall use the fact that even for sample sizes as small as 8 and 9, the distribution of V is approximately normal. In the present case, for example, if H_0 is true, V is approximately normal with a mean of 81 and a standard deviation of 10.392. (You will see the basis for this last value shortly.)

We compute the test statistic

$$z = \frac{v - \mu_V}{\sigma_V} = \frac{97 - 81}{10.392} = 1.54$$

The corresponding P value is $P(Z > 1.54) = .0618$.

We cannot reject H_0 at the .05 level of significance. Although the difference in birthweights appeared striking, the data suggest that the median birthweight of infants born to nonsmoking mothers is not (quite) significantly greater than that for infants born to smokers.

■

Generalizing the above procedures, we assume random samples of size n_1 from population I and n_2 from population II. We assume further that the two populations have the same shape although possibly different (but unknown) medians M_1 and M_2. Following our practice in Chapters 10 and 12, we use the subscript 1 for the median we believe to be the larger. We test $H_0 : M_1 - M_2 = 0$ (or $M_1 = M_2$) against $H_1 : M_1 - M_2 > 0$ (or $M_1 > M_2$).

The test procedure is identical to that followed in Example 5. We first rank all $n_1 + n_2$ observations from smallest (1) to largest $(n_1 + n_2)$, recording the sample from which each observation came. The test involves computing v, the sum of the ranks in sample 1. The total number of observations for the two samples is $n = n_1 + n_2$. The sum of the n ranks will then be $1 + 2 + 3 + \cdots + n = n(n + 1)/2$. As before, if H_0 is true, we would expect both samples to contain a similar mix of large and small ranks. Since a proportion n_1/n of the observations are in sample 1, the expected value for V is

$$E(V) = \mu_V = \frac{n_1}{n}\frac{n(n+1)}{2} = \frac{n_1(n+1)}{2} \qquad \text{(Eq. 16.4)}$$

We note without proof that if H_0 is true,

$$\mathrm{SD}(V) = \sigma_V = \sqrt{\frac{n_1 n_2(n+1)}{12}}$$

In the last example $n_1 = 9$, $n_2 = 8$; thus $n = 17$ and

$$\mu_V = \frac{9(17+1)}{2} = 81 \qquad \text{and} \qquad \sigma_V = \sqrt{\frac{9(8)(18)}{12}} = 10.392$$

Provided that n_1 and n_2 are both ≥ 8, V is approximately normal. That is, V is approximately

$$N\left[\frac{n_1(n+1)}{2}, \sqrt{\frac{n_1 n_2(n+1)}{12}}\right]$$

If, instead of H_0, $H_1: M_1 > M_2$ is true, we would expect the larger ranks to be concentrated in sample 1 and hence v to be greater than $\mu_V = n_1(n+1)/2$. The test is based on rejecting H_0 if the test statistic

$$z = \frac{v - n_1(n+1)/2}{\sqrt{\frac{n_1 n_2(n+1)}{2}}}$$

is sufficiently large. The above procedures are summarized in Table 5.

TABLE 5 Procedure for performing the one-sided Mann–Whitney test

STEP $\quad H_0: M_1 - M_2 = 0 \quad H_1: M_1 - M_2 > 0$
1. Designate M_1 as the population median thought to be larger.
2. Rank all $n_1 + n_2 = n$ observations from smallest (1) to largest (n). Ties are each given the average rank.
3. Compute v, the sum of the ranks in sample 1.
4. Compute $\mu_v = n_1(n+1)/2$ and $\sigma_V = \sqrt{n_1 n_2(n+1)/12}$.
5. Compute the test statistic, $z = (v - \mu_V)/\sigma_V$.
6. If $z \geq 0$, the P value $= P(Z > z)$ (If $Z < 0$, do not reject H_0.)
7. Using $\alpha = .05$ or $.01$, reject H_0 if the P value $\leq \alpha$.
Assumption: the populations have roughly the same shape.

EXAMPLE 6 Do "famous" persons live any longer than the rest of us? Sample 1 below indicates the ages at death of a sample of 15 famous people who died in 1980. Sample 2 includes the ages at death of a random sample of people whose death certificates were filed in New York City in 1980.

SAMPLE 1: FAMOUS PERSONS					SAMPLE 2: ORDINARY PEOPLE				
82	75	72	75	84	72	76	44	56	81
66	62	70	68	42	92	74	69	78	21
74	89	34	59	79	79	84	64	70	70

Do these data suggest that the median age at death of famous peoplc is greater than that for ordinary people?

Solution The question suggests that we refer to the famous persons as sample 1. We first list the 30 ages at death by appropriate sample, with the corresponding ranks in parentheses.

FAMOUS PERSONS			ORDINARY PERSONS		
34(2)	42(3)	59(6)	21(1)	44(4)	56(5)
62(7)	66(9)	68(10)	64(8)	69(11)	70(13)
70(13)	72(15.5)	74(17.5)	70(13)	72(15.5)	74(17.5)
75(19.5)	75(19.5)	79(23.5)	76(21)	78(22)	79(23.5)
82(26)	84(27.5)	89(29)	81(25)	84(27.5)	92(30)

The sum of the ranks for the famous persons is

$$v = 2 + 3 + 6 + \cdots + 26 + 27.5 + 29 = 228$$

We compute $\mu_V = 15(30 + 1)/2 = 232.5$, $\sigma_V = \sqrt{(15)(15)(30 + 1)/12} = 24.107$, and the test statistic $z = (228 - 232.5)/24.107 = -.19$. Since $z < 0$, we do not reject H_0. The data suggest that the median age at death for famous people is not significantly greater than that for ordinary people.

■

There is a two-sided version of the Mann–Whitney test, but it is quite complicated and we therefore omit it.

Studies have shown that even when the assumptions underlying the two-sample t test (Section 12.4) are valid, the Mann–Whitney test is almost as efficient. What is more important, when the t test is not valid, for instance, when the sampled populations are far from normal, the Mann–Whitney test generally performs better than the t test.

PROBLEMS 16.3

15. A sample of adoption records are selected at random from a state agency and a private agency. The time (in years) between applying for and receiving the child is noted for each record with the following results:

PRIVATE AGENCY	1.3	4.7	2.2	1.8	.7	1.9	2.4	2.0		
STATE AGENCY	2.4	2.1	2.7	1.7	4.4	4.9	2.8	3.1	4.0	2.8

Use the Mann–Whitney test to decide whether the median wait for adoption approval by the state agency is significantly greater than that for private agencies.

16. Twenty-four mice are involved in an experiment to measure the effect of a drug which is expected to retard the development of a type of cancer. The following data are the weights (in grams) of the tumors taken from 10 untreated mice and from 14 mice treated with the drug:

TREATED			UNTREATED	
1.24	1.47	1.55	1.56	1.29
1.52	1.44	.69	1.77	2.15
1.38	.82	1.32	2.07	1.84
1.17	1.20	1.46	1.40	1.69
1.04	1.28		1.64	1.99

Use the Mann–Whitney test to determine if the median weight of tumors taken from mice treated with the drug is significantly less than that for tumors taken from untreated mice.

17. A professional examination is taken by students who have studied certain material in graduate school and by students who have had experience with such material through full-time work. The scores for random samples of 10 students of each kind are as follows:

GRADUATE SCHOOL STUDENTS				WORKING STUDENTS			
374	378	338	340	374	370	344	432
380	399	360	394	383	372	378	358
349	402			327	392		

Do these data suggest that the median score for graduate school students is significantly greater than that for working students?

18. Draw a dot diagram for each sample in Problem 17 and explain why the two-sample t test would not be appropriate.

19. The OCRA gasoline company claims that following the elimination of credit cards, the retail cost of their gasoline has been lower than that of their competitors. In a survey of eight of their gas stations and 15 competitor stations in one state, the following prices per gallon of regular gasoline were posted (in dollars):

OCRA STATIONS		COMPETITOR STATIONS		
1.29	1.32	1.26	1.24	1.34
1.24	1.25	1.32	1.26	1.25
1.31	1.35	1.29	1.39	1.28
1.28		1.35	1.31	1.35
1.22		1.36	1.37	1.37

Use the Mann–Whitney test to check OCRA's claim.

20. For many years a university has leased 10 copying machines from company X and 8 from Company Y. Each machine gets roughly the same amount of use. The time (in weeks) between the last and the immediately preceding repair was recorded for each machine with the following results:

COMPANY X	2	38	20	17	31	21	24	10	15	22
COMPANY Y	62	19	24	36	39	34	48	14		

Do these data indicate that the median time between repairs for Y is significantly greater than that for X?

21. Apply the Mann–Whitney test to the age data in Problem 35 in Section 12.4 (page 420). Do these data suggest that the median age of men in the party exceeds that of women?

Solution to Problem 15

Following the procedure in Table 5, we designate the times for the state agency as sample 1. Then $n_1 = 10$, $n_2 = 8$, and $n = n_1 + n_2 = 18$. Each value is listed below together with its overall rank in parentheses.

PRIVATE AGENCY		STATE AGENCY	
1.3	(2)	2.4	(9.5)
4.7	(17)	2.1	(7)
2.2	(8)	2.7	(11)
1.8	(4)	1.7	(3)
.7	(1)	4.4	(16)
1.9	(5)	4.9	(18)
2.4	(9.5)	2.8	(12.5)
2.0	(6)	3.1	(14)
		4.0	(15)
		2.8	(12.5)
			$V = 118.5$

We compute $\mu_V = 10(18 + 1)/2 = 95$, $\sigma_V = \sqrt{10(8)(19)/12} = 11.2546$, and the test statistic $z = (118.5 - 95)/11.2546 = 2.09$. The P value is $P(Z > 2.09) = .0183$. Since this value lies between .01 and .05, we reject H_0 at the .05 level of significance. The data suggest that the median wait for the state agency is significantly greater than that for the private agency.

SECTION 16.4
MULTIPLE SAMPLES: THE KRUSKAL–WALLIS TEST

The Kruskal–Wallis test is a simple nonparametric alternative to the F test that was used in the one-way analysis of variance. This test is used to test for significant differences between r population medians—rather than means.

We illustrate the test procedure in the next example.

EXAMPLE 7 A business magazine selects samples of hotels from each of four major chains and records the occupancy rate for each hotel on a specified date, that is, the percentage of all available rooms in the hotel that were occupied the previous night. The results for each chain are as follows:

I	II	III	IV
74	81	63	74
76	86	77	66
59	94	68	58
79	84	69	57
82	80	62	67
75	72	67	63
		69	61
			64
			71

Do these data suggest any significant differences among the median occupancy rates for the four hotel chains?

In this example there are $r = 4$ samples. The sample sizes are $n_1 = n_2 = 6$, $n_3 = 7$, and $n_4 = 9$. The total number of observations is $n = n_1 + n_2 + n_3 + n_4 = 28$. We shall test $H_0: M_1 = M_2 = M_3 = M_4$ against $H_1: M_1, M_2, \ldots, M_4$ are not all equal. We begin by ranking all 28 observations from smallest (1) to largest (28). Ties are given the average rank. We then obtain the sum of the ranks in each of the four samples (R_1, R_2, R_3, and R_4). These tabulations are recorded below with the ranks in parentheses:

I	II	III	IV
74(17.5)	81(24)	63(6.5)	74(17.5)
76(20)	86(27)	77(21)	66(9)
59(3)	94(28)	68(12)	58(2)
79(22)	84(26)	69(13.5)	57(1)
82(25)	80(23)	62(5)	67(10.5)
75(19)	72(16)	67(10.5)	63(6.5)
		69(13.5)	61(4)
			64(8)
			71(15)
$R_1 = 106.5$	$R_2 = 144$	$R_3 = 82$	$R_4 = 73.5$

With four samples the Kruskal–Wallis test statistic is

$$Y = \frac{12}{n(n+1)}\left(\frac{R_1^2}{n_1} + \frac{R_2^2}{n_2} + \frac{R_3^2}{n_3} + \frac{R_4^2}{n_4}\right) - 3(n+1) \qquad \text{(Eq. 16.5)}$$

In this case

$$Y = \frac{12}{28(29)}\left(\frac{106.5^2}{6} + \frac{144^2}{6} + \frac{82^2}{7} + \frac{73.5^2}{9}\right) - 3(29)$$

$$= \frac{12}{28(29)}(6907.196) - 3(29) = 15.077$$

It can be shown that if H_0 is true, Y has approximately the χ_3^2 distribution. Recall from Section 12.6 that if Y is χ_3^2, then $\mu_Y = 3$. We will indicate in a moment that evidence in favor of H_1 will be a value for Y unusually far above 3. Table A7 of the χ_ν^2 distribution is not sufficiently detailed for us to compute P values associated with Y. Rather, we follow the procedure we adopted in Chapter 15 of comparing the value of Y to $\chi_{\nu,.05}^2$ and $\chi_{\nu,.01}^2$. From Table A7, $\chi_{3,.05}^2 = 7.81$ and $\chi_{3,.01}^2 = 11.34$. Since $Y = 15.077$ is greater than 11.34, we reject H_0 at the .01 level of significance. The data suggest that the median occupancy rates for the four chains differ significantly.

■

In the general case of the Kruskal–Wallis test we assume r random samples of size $n_1, n_2, \ldots, n_r$ selected independently from each of r populations all having roughly the same shape. The total number of observations is $n = n_1 + n_2 + \cdots + n_r$. We are to test $H_0: M_1 = M_2 = \cdots = M_r$ against $H_1: M_1, M_2, \ldots, M_r$ are not all equal, where $M_1, M_2, \ldots, M_r$ are the unknown population medians.

Before outlining the test procedure, we use the data in Example 7 to illustrate why we reject H_0 for sufficiently large values for Y.

As before, we begin by ordering all n observations and summing the rank values within each sample. The results are the rank sums $R_1, R_2, \ldots, R_r$. We may also compute the *grand mean rank* as the sum of *all* n rank values divided by n. Since the sum of all the ranks is $1 + 2 + 3 + \cdots + n = n(n+1)/2$, the grand mean rank is $n(n+1)/2n = (n+1)/2$. In Example 7 this was $(n+1)/2 = (28+1)/2 = 14.5$.

The *mean* rank value in each sample is the rank sum divided by the number of observations, or symbolically, $R_1/n_1, R_2/n_2, \ldots, R_r/n_r$. In Example 7 the four mean rank values were

$$\frac{R_1}{n_1} = \frac{106.5}{6} = 17.75 \qquad \frac{R_2}{n_2} = \frac{144}{6} = 24$$

$$\frac{R_3}{n_3} = \frac{82}{7} = 11.714 \qquad \frac{R_4}{n_4} = \frac{73.5}{9} = 8.167$$

The test of H_0 is based on the value of the test statistic

$$Y = \frac{12}{n(n+1)}\left[n_1\left(\frac{R_1}{n_1} - \frac{n+1}{2}\right)^2 + n_2\left(\frac{R_2}{n_2} - \frac{n+1}{2}\right)^2 + \cdots + n_r\left(\frac{R_r}{n_r} - \frac{n+1}{2}\right)^2\right] \quad \text{(Eq. 16.6)}$$

If H_0 is true, we would expect each sample to have a similar mix of high and low ranks. As a result, the mean rank value per sample $R_1/n_1, R_2/n_2, \ldots, R_r/n_r$ should all be close to the grand mean rank, $(n + 1)/2$, and Y will be small. In fact, if all the R_i/n_i values were equal to $(n + 1)/2$, each term within the large brackets in Equation 16.6 and hence Y itself would be 0.

By contrast, if H_1 is true, we would expect the high ranks to be concentrated in some samples and the low ranks in others. As a result, the sample mean ranks would differ substantially from the grand mean rank, $(n + 1)/2$, and Y would tend to be large. In Example 7 the sample mean ranks 17.75, 24, 11.714, and 8.167 differed sufficiently from the grand mean rank of 14.5 to result in a value of Y large enough for us to reject H_0.

Some rather messy algebra will show that the expression for Y in Equation 16.6 can be written equivalently and more conveniently as a generalized version of Equation 16.5:

$$Y = \frac{12}{n(n+1)}\left(\frac{R_1^2}{n_1} + \frac{R_2^2}{n_2} + \cdots + \frac{R_r^2}{n_r}\right) - 3(n+1) \quad \text{(Eq. 16.7)}$$

Provided that the sample sizes $n_1, n_2, \ldots, n_r$ are all ≥ 5, if H_0 is true, Y will have approximately the χ^2_{r-1} distribution. The exact procedure for performing the Kruskal–Wallis test is summarized in Table 6.

TABLE 6

Procedure for performing the Kruskal–Wallis test of $H_0: M_1 = M_2 = \cdots M_r$ against $H_1: M_1, M_2, \ldots, M_r$ are not all equal

1. Rank all $n = n_1 + n_2 + \cdots + n_r$ observations from smallest (1) to largest (n). Ties are given the average rank.
2. Compute $R_1, R_2, \ldots, R_r$, the sum of the rank values in each of the r samples.
3. Compute

$$Y = \frac{12}{n(n+1)}\left(\frac{R_1^2}{n_1} + \frac{R_2^2}{n_2} + \cdots + \frac{R_r^2}{n_r}\right) - 3(n+1)$$

4. Find $\chi^2_{r-1,.05}$ and $\chi^2_{r-1,.01}$ from Table A7.
5. *Conclusion:* Using $\alpha = .05$ or $.01$, reject H_0 at the α level of significance if $Y \geq \chi^2_{r-1,\alpha}$.

Assumption: the r populations have roughly the same shape.

EXAMPLE 8 Three treatments for a certain type of cancer are to be compared. A group of 24 patients having this type of cancer at the same stage are randomly divided into three treatment groups. The number of years of life from this point is recorded for each person. Do these data suggest any significant differences in the median survival time for the three treatments? (One of the patients receiving treatment C emigrated and was lost to the study.)

A	5.7	1.2	8.8	2.2	3.7	4.3	0.4	6.9
B	2.3	1.0	2.7	3.8	5.2	3.5	4.3	0.3
C	5.9	1.1	1.9	9.6	10.3	2.1	1.8	

Solution The following 23 observations are ranked from smallest (1) to largest (23):

A	B	C
5.7(18)	2.3(10)	5.7(19)
1.2(5)	1.0(3)	1.1(4)
8.8(21)	2.7(11)	1.9(7)
2.2(9)	3.8(14)	9.6(22)
3.7(13)	5.2(17)	10.3(23)
4.3(15.5)	3.5(12)	2.1(8)
.4(2)	4.3(15.5)	1.8(6)
6.9(20)	.3(1)	
$R_1 = 103.5$	$R_2 = 83.5$	$R_3 = 89$

Here $n_1 = n_2 = 8$, $n_3 = 7$, and $n = 23$. The value for Y is

$$Y = \frac{12}{23(24)}\left(\frac{103.5^2}{8} + \frac{83.5^2}{8} + \frac{89^2}{7}\right) - 3(24) = .655$$

From Table A7 with $\nu = r - 1 = 2$, $\chi^2_{2,.05} = 5.99$ and $\chi^2_{2,.01} = 9.21$. Since $Y = .655$ is much less than 5.99, we cannot reject H_0 at the .05 level of significance. We conclude that the data suggest no significant difference in the median length of survival for the three treatments. ■

Those of you familiar with the F test of the equality of r means in Chapter 13 will be interested in the fact that Y can be obtained by computing the ratio $F = \text{MS}_{\text{between}}/\text{MS}_{\text{within}}$ using the *ranks* rather than the actual observations.

PROBLEMS 16.4

22. The final scores in an economics course are classified by students' class as follows:

FRESHMAN	SOPHOMORE		JUNIOR		SENIOR	
64	81	69	73	80	76	96
78	82	85	81		84	
69	73		72		92	
54	74		76		88	
62	83		80		81	

(a) Use the Kruskal–Wallis test to investigate differences in the median scores for the four classes.

(b) Which class has the highest median score?

23. A consumer research laboratory tests eight 60-watt light bulbs for each of five brands and records their length of life. The results are as follows:

BRAND	*A*	1152	1052	1219	1098	1116	1191	1104	1203
	B	1300	1232	1268	1209	1197	1213	1278	1254
	C	1204	1108	1117	1134	1068	1092	1137	1146
	D	1312	1319	1288	1337	1412	1352	1302	1376
	E	1009	1050	1082	1121	1076	1094	1111	1027

(a) Using the Kruskal–Wallis test, determine whether these data suggest any significant differences among the median lengths of life of the five brands.

(b) Compute the median length of life for each of the five samples. Which brand has the longest median life? The shortest?

24. A sample of professional major league players are selected from the fields of hockey, basketball, baseball, and football. The age of each player sampled is as follows:

HOCKEY		BASKETBALL		BASEBALL		FOOTBALL	
27	29	32	30	31	41	31	26
22	30	25	29	19	34	26	26
37		29	22	22		22	23
19		32		24		29	29
24		27		28		30	30
19		26		35		27	

Do these data suggest any significant differences among the median ages of players for the four sports?

25. A manufacturer arranges to test three new ways of packaging one of its products in its 36 stores for one month. Each type of packaging is randomly assigned to 12 of the 36 stores. The mean number of sales per week for the trial month and the preceding month are compared and the "increase" for each store recorded as follows:

PACKAGING TYPE	A	0	14	−7	6	10	−2	5	12	6	19	−5	2
	B	13	27	41	37	29	17	20	10	38	26	20	29
	C	12	−9	14	6	−3	−5	8	−1	3	7	7	4

(a) Use the Kruskal–Wallis test to investigate differences among the median increases for the three types of packaging.

(b) Which type of packaging would you recommend that the company adopt?

26. An attempt was made to determine the extent to which different disciplines used statistical methods. Leading journals in psychology, sociology, government, and economics were studied. Eight copies of each journal were selected at random from among those for the past 10 years. The proportion of articles that used some aspect of statistics was recorded for each copy.

PSYCHOLOGY	SOCIOLOGY	GOVERNMENT	ECONOMICS
.84	.68	.80	.87
.75	.72	.74	.78
.89	.80	.63	.85
.75	.75	.74	.75
.69	.65	.71	.77
.84	.60	.68	.92
.81	.75	.85	.86
.72	.81	.77	.84

(a) Apply the Kruskal–Wallis test to these data.

(b) Compute the median for each discipline.

(c) Which discipline seems to make the most use of statistical methods?

27. Apply the Kruskal–Wallis test to the data in Problem 5 of Chapter 13 (page 475). Do the data suggest any significant differences among the median number of defective items per 200 items for the three machines? Compare your conclusions with those in the earlier problem.

Solution to Problem 22

(a) We are to test $H_0: M_1 = M_2 = M_3 = M_4$ against $H_1: M_1, M_2, M_3, M_4$ are not equal. The appropriate procedure is that in Table 6 with $r = 4$. The scores are given below with the overall rank in parentheses.

FRESHMAN	SOPHOMORE	JUNIOR	SENIOR
$n_1 = 5$	$n_2 = 7$	$n_3 = 6$	$n_4 = 6$
64 (3)	81 (16)	73 (7.5)	76 (10.5)
78 (12)	82 (18)	81 (16)	84 (20)
69 (4.5)	73 (7.5)	72 (6)	92 (23)
54 (1)	74 (9)	76 (10.5)	88 (22)
62 (2)	83 (19)	80 (13.5)	81 (16)
	69 (4.5)	80 (13.5)	96 (24)
	85 (21)		
$R_1 = 22.5$	$R_2 = 95$	$R_3 = 67$	$R_4 = 115.5$

Therefore,

$$Y = \frac{12}{24(25)}\left[\frac{22.5^2}{5} + \frac{95^2}{7} + \frac{67^2}{6} + \frac{115.5^2}{6}\right] - 3(25)$$

$$= \frac{1}{50}(4362.0774) - 75 = 12.24$$

From Table A7, $\chi^2_{3,.05} = 7.81$ and $\chi^2_{3,.01} = 11.34$. Since $Y = 12.24$ is greater than 11.34, we reject H_0 at the .01 level of significance. The data suggest that the median scores for the four classes are significantly different.

(b) The seniors have the highest median score (86).

SECTION 16.5

COMPARISON OF PARAMETRIC AND NONPARAMETRIC TESTS

Throughout this chapter we have noted that for many of the parametric tests described earlier there is an appropriate nonparametric counterpart which performs almost as well and involves less restrictive assumptions. As a consequence, in most cases we recommended that the nonparametric test be used in preference to the corresponding parametric one.

This being the case, the reader may wonder why we spent so much time discussing the t tests in Chapter 12 and the analysis of variance in Chapter 13. The fact is that t tests and the F test in analysis of variance are still widely taught and widely used in research. By contrast, their nonparametric counterparts are still viewed by some as insensitive, to be used, it at all, as a last resort—generally if the data are blatantly nonnormal. Although we generally recommend the Wilcoxon, the Mann–Whitney, and the Kruskal–Wallis tests, we emphasize that knowledge of the t and F tests is essential to the student who wishes to be statistically literate.

SECTION 16.6

THE RANK CORRELATION COEFFICIENT

This section is not concerned with a nonparametric test but rather with a nonparametric alternative to the correlation coefficient. It is called the **rank correlation coefficient.** Recall from Chapter 2 that r measures the strength of the relationship between two quantitative variables. On the other hand, the rank correlation coefficient, r_s, as the name suggests, is a measure of the extent to which the *ranks* of the two variables are in agreement. The s in r_s stands for Charles Spearman, the psychologist who developed this measure at the turn of the century. In the next example we show how to compute r_s.

EXAMPLE 9 The students in an intelligence testing course are interested in the relationship between IQ and scores on the college's mathematics placement test. The instructor obtains the following data on 12 anonymous students from the registrar's office:

IQ SCORE	112	115	121	113	117	108	106	128	124	133	116	109
SCORE ON MATH PLACEMENT TEST	24	29	25	27	25	20	28	50	37	45	38	32

The 12 IQs are listed below with their ranks in parentheses. The corresponding test scores are also listed, together with their ranks.

IQ SCORES	106 (1)	108 (2)	109 (3)	112 (4)	113 (5)	115 (6)	116 (7)	117 (8)	121 (9)	124 (10)	128 (11)	133 (12)
MATH TEST SCORES	(6) 28	(1) 20	(8) 32	(2) 24	(5) 27	(7) 29	(10) 38	(3.5) 25	(3.5) 25	(9) 37	(12) 50	(11) 45
d_i	−5	−1	−5	2	0	−1	−3	4.5	5.5	1	−1	1

Notice that those students who rank high in IQ tend to rank high on test scores. The value for r_s is based on the differences in ranks, which are labeled "d_i" above. With n pairs of observations,

$$r_s = 1 - \frac{6(d_1^2 + d_2^2 + \cdots + d_n^2)}{n(n^2 - 1)}$$

or

$$r_s = 1 - \frac{6 \Sigma d_i^2}{n(n^2 - 1)} \qquad \text{(Eq. 16.8)}$$

In this case, with $n = 12$,

$$\sum d_i^2 = (-5)^2 + (-1)^2 + \cdots + (-1)^2 + (1)^2 = 118.5$$

$$r_s = 1 - \frac{6(118.5)}{12(12^2 - 1)} = .586$$

You should check that if we had ranked the test scores first, the value for r_s would be the same.

Before interpreting this value, let us examine the expression for r_s in more detail.

■

Consider the two examples with $n = 6$ in Table 7. (For clarity only the ranks for the two variables have been provided.)

TABLE 7 Extreme values for r_s

EXAMPLE A				EXAMPLE B			
RANK OF VARIABLE I	RANK OF VARIABLE II	d_i	d_i^2	RANK OF VARIABLE I	RANK OF VARIABLE II	d_i	d_i^2
1	1	0	0	1	6	−5	25
2	2	0	0	2	5	−3	9
3	3	0	0	3	4	−1	1
4	4	0	0	4	3	1	1
5	5	0	0	5	2	3	9
6	6	0	0	6	1	5	25
		$\Sigma d_i^2 =$	0			$\Sigma d_i^2 =$	70
	$r_s = 1$				$r_s = -1$		

In Example A the ranks of the two variables are in complete agreement; $\Sigma d_i^2 = 0$ and from Equation 16.8, $r_s = 1 - 0 = 1$. In Example B the ranks of the two variables are in complete disagreement; $\Sigma d_i^2 = 70$ and $r_s = 1 - 6(70)/6(35) = 1 - 2 = -1$. This result is completely general; r_s, like r, lies between -1 and 1. (See Problem 35 at the end of this section.) The value $r_s = .586$ between IQ rank and math score rank indicates moderate agreement between the two sets of ranks.[3]

Values of r_s close to zero will occur when the ranks associated with one of the variables appear randomly mixed in relation to the ranks of the other. This would suggest that the two rankings are close to being independent.

The rank correlation coefficient r_s has several advantages over r:

1. The value for r_s is simpler to compute than the value for r.
2. On occasion researchers use measurement scales which are not well understood or trustworthy. For instance, sociologists sometimes construct and use an index of socioeconomic status. With such an index it is not clear whether, for example, people having scores of 80 and 90 differ to the same extent as those having scores of 10 and 20. In such a case the sociologist may well prefer a measure of correlation which requires only that the individuals be ranked on such an index.
3. In some cases the data on at least one of the variables may consist of ranks only. Example 10 is such a case.

EXAMPLE 10 The two candidates for governor are asked to rank seven issues in the order of their perceived importance. Their responses are listed beside the issues. Compute the rank correlation between the two sets of ranks.

[3]We use the same language to describe the magnitude of r_s as we did to describe the magnitude of r in Table 6 of Chapter 2 (page 85).

ISSUE	RANKING BY CANDIDATES *A*	*B*
Corruption in state government	1	2
Property taxes	4	3
Unemployment	2	1
Crime	3	4
Growth of state government	5	7
Social issues	6	5
Education	7	6

Solution The seven differences in ranks are -1, 1, 1, -1, -2, 1, and 1. Thus $\Sigma d_i^2 = 1 + 1 + 1 + 1 + 4 + 1 + 1 = 10$ and $r_s = 1 - 6(10)/7(48) = .821$. There is strong agreement between the two sets of rankings.

■

It is interesting to note that r_s is, in fact, simply the correlation coefficient, r, between the *ranks* of the observations. We will not prove this but will illustrate it with a simple example.

EXAMPLE 11 The price/sales data in Example 5 of Chapter 2 (page 85) are reproduced below. The ranks for both variables are included in parentheses.

PRICE, *x*		SALES, *y*	
\$41	(1)	435	(4)
\$45	(2)	375	(2)
\$50	(3)	410	(3)
\$54	(4)	354	(1)

From Equation 16.8, $\Sigma d_i^2 = (-3)^2 + 0^2 + 0^2 + 3^2 = 18$ and $r_s = 1 - 6(18)/4(15) = -.8$.

Now let us regard the two sets of ranks as the variables and compute the correlation coefficient r using Equation 2.2 (page 83). We need the following results:

$$\sum x_i = 1 + 2 + 3 + 4 = 10 \qquad \sum y_i = 4 + 2 + 3 + 1 = 10$$

$$\sum x_i^2 = 1^2 + 2^2 + 3^2 + 4^2 = 30 \qquad \sum y_i^2 = 4^2 + 2^2 + 3^2 + 1^2 = 30$$

$$\sum x_i y_i = (1)(4) + (2)(2) + (3)(3) + (4)(1) = 21$$

Then

$$r = \frac{n \Sigma x_i y_i - (\Sigma x_i)(\Sigma y_i)}{\sqrt{[n \Sigma x_i^2 - (\Sigma x_i)^2][n \Sigma y_i^2 - (\Sigma y_i)^2]}}$$

$$= \frac{4(21) - (10)(10)}{\sqrt{[4(30) - 10^2][4(30) - 10^2]}} = \frac{84 - 100}{\sqrt{(20)(20)}} = -.8 = r_s$$

Also note that the value $r_s = -.8$ indicates a strong tendency for the higher prices to be associated with lower sales.

■

PROBLEMS 16.6

28. A strike by major league baseball players divided the 1981 season into two halves. The standings in the American League, East, at the end of the first half and at the end of the second half were as follows:

STANDING	FIRST HALF	SECOND HALF
1	New York	Milwaukee
2	Baltimore	Boston
3	Milwaukee	Detroit
4	Detroit	Baltimore
5	Boston	Cleveland
6	Cleveland	New York
7	Toronto	Toronto

Compute r_s for the two sets of standings.

29. At the beginning of the 1981 season the three employees in a sporting goods store wrote down their projections of the final standings in the American League, East:

STANDING	*A*	*B*	*C*
1	New York	New York	Cleveland
2	Milwaukee	Boston	New York
3	Boston	Baltimore	Baltimore
4	Baltimore	Milwaukee	Cleveland
5	Detroit	Cleveland	Boston
6	Cleveland	Detroit	Detroit
7	Toronto	Toronto	Toronto

The person with the highest rank correlation with the final standings was to be taken to dinner by the other two. Which employee received a free meal? (Use the second-half standings in Problem 1 as the final standings.)

30. A consumer testing organization ranks five makes of imported televisions on the basis of the quality of the picture, listing beside each one its suggested retail price.

MAKE	RANK	PRICE
A	1	$547
B	2	572
C	3	527
D	4	550
E	5	539

Compute the rank correlation coefficient between quality and price.

31. A subcommittee consisting of two faculty members is to decide which of five students is to receive the prestigious alumnae award. Both instructors rank the five as follows (the *lower* the rank, the greater the regard for the student):

STUDENT	INSTRUCTOR *A*	INSTRUCTOR *B*
P. Alvarez	4	4
P. Brown	3	1
W. Chin	5	5
F. Drake	1	2
D. Epstein	2	3

(a) Compute the rank correlation coefficient for these two sets of ranks.

(b) Find the sum of the two ranks for each student.

(c) On thc basis of these sums, which student do you think should receive the award?

32. On one Sunday the Associated Press (AP) and United Press International (UPI) top five college football teams were as follows:

STANDING	AP	UPI
1	Georgia	Georgia
2	SMU	SMU
3	Arizona State	Penn State
4	Nebraska	Nebraska
5	Penn State	Arizona State

Compute the value of r_s for these ranks.

33. Compute the value of r_s for the height/weight data in Problem 2 of Chapter 2 (page 79).

34. Compute the value of r_s for the latitude/temperature data in Problem 15 of Chapter 2 (page 89).

35. Show that in the general case with n pairs of ranks (a) with complete agreement $r_s = 1$; (b) with complete disagreement $r_s = -1$. [To answer part (b) you should use the result that with complete disagreement $\Sigma\, d_i^2 = (1/3)n(n^2 - 1)$.]

Solution to Problem 28

The two corresponding sets of ranks are

TEAM	FIRST HALF	SECOND HALF	d_i	d_i^2
New York	1	6	−5	25
Baltimore	2	4	−2	4
Milwaukee	3	1	2	4
Detroit	4	3	1	1
Boston	5	2	3	9
Cleveland	6	5	1	1
Toronto	7	7	0	0
				44

Therefore,

$$r_s = 1 - \frac{6 \Sigma d_i^2}{n(n^2 - 1)}$$

$$= 1 - \frac{6(44)}{7(7^2 - 1)} = 1 - .786 = .214$$

There is only a small amount of agreement between the two sets of standings.

SECTION 16.7

SUMMARY AND REVIEW

There are nonparametric tests that can be used in place of the one- and two-sample t tests and the F test in the one-way analysis of variance. These nonparametric tests assume little about the sampled population(s) and are therefore sometimes called more appropriately "distribution free."

The Sign, Wilcoxon, Mann–Whitney, and Kruskal–Wallis all test hypotheses involving population medians rather than means. These last three tests do not *use* the sample medians but rather are based on sums of *ranks* of the values.

THE SIGN TEST

The simplest alternative to the one-sample t test is the **sign test,** which requires only that the population be continuous. The null and alternative hypotheses are $H_0: M = M_0$ and $H_1: M > M_0$ and $H_1: M < M_0$ or $H_1: M \neq M_0$. The test is based on y, the test statistic "the number of positive values for $x_i - M_0$." If H_0 is true, each such difference $(x_i - M_0)$ has a probability .5 of being positive or negative and the random variable, Y, has the binomial distribution with $\pi = .5$. P values associated with the value for y are found from Table A2. The details of the sign test are provided in Table 1.

THE WILCOXON TEST

The **Wilcoxon test** may also be used as an alternative to the one-sample t test. It makes more use of the data than does the sign test. The hypotheses are the same as those for the sign test. But this test requires that the population be continuous and symmetric, although not necessarily normal. For this test we rank the $|x_i - M_0|$'s and then obtain the signed ranks by combining the sign of $(x_i - M_0)$ with the rank. The test is based on the sum of the positive signed ranks (w).

For any n the actual distribution of the random variable W if H_0 is true is messy to obtain, but Table A9 contains upper- and lower-tail probabilities for w for values of n from 4 to 12. The procedure for using this table to compute the P value associated with a value for w is given in Table 3.

For $n > 12$ we compute the P value by assuming that W has approximately a normal distribution. The procedure in this case is given in Table 4.

THE MANN–WHITNEY TEST

The most popular nonparametric alternative to the two-sample t test is the **Mann–Whitney test.** The null hypothesis in this case is $H_0: M_1 - M_2 = 0$, where M_1 and M_2 are the two population medians. The Mann–Whitney test requires that the two populations have approximately the same shape (see Figure 8). For a one-sided test we use the subscript 1 for the median we believe to be the larger and then only $H_1: M_1 - M_2 > 0$. The test is based on independent samples of sizes n_1 and n_2, respectively. All $n = n_1 + n_2$ observations are ranked from smallest (1) to largest (n), recording the sample from which each observation came. Finally, v is the sum of the ranks for sample 1.

It can be shown that if H_0 is true and n_1 and n_2 are both ≥ 8, the random variable V is approximately $N[n_1(n + 1)/2, \sqrt{n_1 n_2(n + 1)/12}]$. The details of how this result is used to test H_0 are given in Table 5.

THE KRUSKAL–WALLIS TEST

The principal alternative to the F test for the one-way analysis of variance is the **Kruskal–Wallis test** for the difference in r population medians $M_1, M_2, \ldots, M_r$. We test $H_0: M_1 = M_2 \cdots = M_r$ against $H_1: M_1, M_2, \ldots, M_r$ are not all equal.

The respective sample sizes are $n_1, n_2, \ldots, n_r$ and their sum $n = n_1 + n_2 + \cdots + n_r$. We begin by ordering all n observations and summing the ranks for each sample. These sums of ranks are $R_1, R_2, \ldots, R_r$. The test is based on the value for

$$Y = \frac{12}{n(n + 1)}\left(\frac{R_1^2}{n_1} + \frac{R_2^2}{n_2} + \cdots + \frac{R_r^2}{n_r}\right) - 3(n + 1) \qquad \text{(Eq. 16.7)}$$

Provided that the $n_1, n_2, \ldots, n_r$ are all ≥ 5, if H_0 is true, Y will have approximately the χ^2_{r-1} distribution. Evidence in favor of H_1 will be an unusually large value for Y. The exact procedure for testing H_0 is given in Table 6.

THE RANK CORRELATION COEFFICIENT

The rank correlation coefficient, r_s, can be used as an alternative to the correlation coefficient, r, described in Chapter 2. The formula for r_s is

$$r_s = 1 - \frac{6 \Sigma d_i^2}{n(n^2 - 1)} \qquad \text{(Eq. 16.8)}$$

where the d_i's are the differences in ranks for the two variables. It can be shown that like r, r_s lies between 1 and -1.

REVIEW PROBLEMS (Many of these problems are selected from earlier chapters)

GENERAL

1. A landlord is considering the purchase of a device that the gas company claims will help the furnace work more efficiently and hence more cheaply than before. Before making a decision, the landlord asks six colleagues who have the device to provide her with their gas bills for the January before and the January after the device was installed. The results are as follows (in dollars):

COLLEAGUE	1	2	3	4	5	6
Before	182	110	215	129	117	176
After	136	112	184	100	102	145

Do these data suggest that the device significantly lowers the median January gas bill? Answer this question using (a) the sign test; (b) the Wilcoxon test.

2. The average number of points per game scored by the leading shooter in the National Basketball Association for the years 1947–1953 and 1976–1982 are

1947	23.2	1976	31.1
1948	21.0	1977	31.1
1949	28.3	1978	27.2
1950	27.4	1979	29.6
1951	28.4	1980	33.1
1952	25.4	1981	30.7
1953	22.3	1982	32.3

Use the Mann–Whitney test to determine whether there has been a significant increase in the median of the average points per game by the leading scorer.

3. Six contestants take part in a televised talent show. At the end of the show the contestants are ranked by the studio audience and by viewers at home who mail in their rankings. For one show the rankings were as follows:

CONTESTANT	*A*	*B*	*C*	*D*	*E*	*F*
Studio audience	1	2	3	4	5	6
Home audience	4	6	2	1	3	5

Compute the rank correlation coefficient for these two sets of ranks.

4. Assume at a set location on a highway, state police stop a random sample of motorists and measure the alcohol content of their blood. The same measurement is taken on those motorists who are stopped for speeding. The following data are the percent alcohol in the blood for all persons stopped on a Friday evening who had at least a trace of alcohol in their blood.

MOTORISTS STOPPED AT RANDOM					MOTORISTS STOPPED FOR SPEEDING				
.062	.025	.030	.051	.015	.062	.029	.012	.059	.150
.048	.057	.069	.069	.074	.034	.049	.042	.058	.060
.035	.039	.052	.042	.091	.058	.126	.079	.032	.068
.041	.018	.027	.046	.041	.038	.049	.052	.066	
.036									

Do these data suggest that the median percentage of alcohol among those motorists not speeding (the population from which, presumably, those chosen at random are selected) is significantly below that for all speeding motorists?

5. Apply the Kruskal–Wallis test to the following data:

A	6	14	9	12
B	11	8	16	14
C	13	8	11	5

6. A large field was divided into 24 identical plots. Each of four fertilizers, I, II, III, and IV, were used on six randomly selected plots. The yields per plot in bushels per acre of corn were as follows:

I	83	98	89	95	102	80
II	85	84	88	72	81	71
III	70	59	56	70	73	62
IV	52	90	69	81	76	64

Test for a significant difference in the median yields for the four fertilizers.

HEALTH SCIENCES

7. In an experiment involving 12 undergraduate women a dynamometer was used to record the isometric strength of the hamstring when the knee is flexed at (i) 5° and (ii) 45° to the horizontal. The force produced in each case (in pounds) is as follows:

	FORCE PRODUCED[a]	
SUBJECT	5°	45°
1	22.2	18.5
2	24.0	16.7
3	29.2	25.5
4	39.2	34.3
5	31.3	23.3
6	34.8	32.0
7	31.7	30.0
8	35.5	29.2
9	35.8	28.0
10	34.3	29.3
11	37.7	35.3
12	34.7	31.8

[a]These data were provided by Deanna DiTommaso.

Call M the median difference in force at 5° and at 45°. Use the Wilcoxon test to test whether M is significantly greater than 0.

8. Twenty-four mice are inoculated with one of three strains of typhoid. The number of days until death for each mouse are as follows:

STRAIN *A*			STRAIN *B*			STRAIN *C*			
5.9	6.7	3.7	7.9	9.4	10.7	5.7	6.9	8.5	
8.8	10.2	7.9	6.4	5.3	8.3	12.5	10.7	12.2	
			7.2	8.9		5.1	9.4	7.3	10.9

Test whether the median survival times with the three strains are significantly different.

9. Zelazo et al. (1972) report on an experiment to determine the effect of special walking exercises on the age at which children begin to walk. Twenty-three infants were randomly divided into four groups. Those infants in groups *A* and *B* received various exercises while those in groups *C* and *D* did not. The following data are the ages (in months) at which each of the 23 children first walked:

A	*B*	*C*	*D*
9.00	11.00	11.50	13.25
9.50	10.00	12.00	11.50
9.75	10.00	9.00	12.00
10.00	11.75	11.50	13.50
13.00	10.50	13.25	11.50
9.50	15.00	13.00	

Test whether the median ages of first walking for the groups are significantly different.

10. Weil et al. (1962) report on a study designed to measure some of the effects of smoking marijuana on intellectual functioning. Nine students were given two scores on a digit symbol substitution test, one 15 minutes after smoking a placebo cigarette (no marijuana) and one 15 minutes after smoking marijuana. The students were not told which cigarette was which. The following scores are the differences from the students' base-level score:

STUDENT	1	2	3	4	5	6	7	8	9
Placebo	−3	10	−3	3	4	−3	2	−1	−1
Marijuana	5	−17	−7	−3	−7	−9	−6	1	−3

(a) Test whether scores have declined to a greater extent with marijuana than with the placebo using (i) a one-sided t test for paired data; (ii) the sign test for paired data; (iii) the Wilcoxon test.

(b) Draw a dot diagram showing the nine (placebo-marijuana) differences.

(c) Based on the dot diagram, which of the three tests do you favor?

11. A clinic randomly assigns 24 patients suffering from blisters to receive one of three treatments, one of which is a placebo. The number of days for the blisters to completely heal were as follows:

PLACEBO	TREATMENT *A*	TREATMENT *B*
12	8	10
10	6	8
8	7	7
7	9	11
9	9	9
11	8	10
13	5	9
10	10	7

Do these data suggest that the median recovery times for the three treatments are significantly different?

SOCIOLOGY AND GOVERNMENT

12. A demographer is interested in the relationship between the birth of a family's first and second child and the eventual family size. She follows 29 families for 20 years, collecting the following data:

	NUMBER OF CHILDREN IN FAMILY					
	2	3	4	5	6	7
INTERVAL IN MONTHS BETWEEN FIRST AND SECOND CHILD	29	31	26	11	21	14
	34	24	31	14	18	16
	22	18	19	24	14	
	18	19	24	20	10	
	49	23	21		24	
	39	24				
	28					

(a) Test whether the median interval between the first and second child for the various family sizes differs significantly.

(b) For which group is the median interval the largest? (These data were given to the authors by Richard Harris.)

13. A sample of 17 members of the New England Association of Psychiatric Social Workers is selected. The following figures are the caseloads for each worker arranged by state:

MASSACHUSETTS	CONNECTICUT	RHODE ISLAND	NEW HAMPSHIRE
25	30	18	41
29	21	16	37
31	22	19	
24	18		
28	27		
	24		
	28		

Test whether the median caseloads for the four states differ significantly.

14. A federal auditor is interested in comparing the average amounts that two welfare offices approve for welfare recipients' medical expenses. Pilot samples of 15 cases are selected from each office and the last monthly medical payment recorded. The results are as follows (rounded to the nearest dollar):

A: 135 0 0 72 1370 18 0 54 59 605 29 0 113 27 10

B: 79 64 27 299 0 17 0 30 122 25 0 8 1964 0 5

Use the Mann–Whitney test to test whether the median payment in office B is significantly greater than that for office A.

15. In a recent survey men and women were asked to select which of six problem areas they felt were the primary cause of the increasing rate of crime. The aggregate rankings of these problem areas by men and by women were as follows:

PROBLEM AREA	MEN	WOMEN
Unemployment	1	1
Punishments not severe enough	2	3
Breakdown of family, society, morals, etc.	3	2
Courts too lenient	4	5
Drugs	5	4
TV violence	6	6

Compute the rank correlation between the men and the women rankings.

ECONOMICS AND MANAGEMENT

16. Two airlines, I and II, fly between New York and Los Angeles. For each of 12 consecutive months the percentage of flights between these two cities that arrived within 15 minutes of the scheduled arrival time were recorded for each airline.

MONTH	1	2	3	4	5	6	7	8	9	10	11	12
Airline I	75	74	66	49	54	64	73	81	88	72	75	72
Airline II	79	82	62	55	56	64	81	86	84	80	81	76

Use the (a) sign test and (b) Wilcoxon test to test whether the median difference in percentages is significantly different from 0.

17. In one town there are two real estate brokers, Allen and Baldwin. A group of 10 homeowners ask each broker for an approximate asking price for their houses. The results (in thousands of dollars) are as follows:

HOUSE	A	B
1	82.5	78.9
2	112.0	110.0
3	105.5	108.8
4	122.0	116.2
5	99.5	106.2
6	126.8	118.5
7	87.5	86.5
8	142.5	135.6
9	104.9	110.0
10	96.5	92.0

Use the (a) sign test and (b) Wilcoxon test to test whether the median difference in estimates for the two agencies is significantly different from 0.

18. Allied Mills, a producer of breakfast cereals, is planning to advertise its least popular cereal in supermarkets. Three different advertisements under consideration will be tested in a sample of 15 supermarkets. Each advertisement is randomly assigned to appear in three supermarkets. After one week the increase in sales over the previous week is recorded for all 15 supermarkets.

NO ADVERTISING	ADVERTISEMENT *A*	ADVERTISEMENT *B*	ADVERTISEMENT *C*
14	14	28	19
−8	12	19	16
5	9	12	13
6	10	17	20
−2	10	24	15

Test for significant differences among the median sales under the four conditions.

19. A company records the number of items sold by their salespersons in a sample of six Mondays, six Tuesdays, and so on. The results are as follows:

MONDAY	TUESDAY	WEDNESDAY	THURSDAY	FRIDAY
36	37	50	36	31
37	36	32	39	40
39	32	40	50	37
30	43	34	49	40
44	37	37	46	28
24	47	47	44	28

(a) Test whether the median sales for the five days are significantly different.
(b) Which day has the highest median sales?
(c) Which day has the lowest?

20. *Consumer Reports* (1983) ranked 22 makes of walkaround stereos. The rankings are given below with the price of each stereo (in dollars).

RANK	PRICE	RANK	PRICE	RANK	PRICE
1	120	9	150	17	50
2	180	10	140	18	60
3	130	11	75	19	70
4	120	12	90	20	200
5	170	13	170	21	220
6	70	14	100	22	70
7	170	15	110		
8	150	16	120		

(a) Compute the rank correlation coefficient for these data.
(b) Does the result surprise you?

(c) Approximately, what would you expect the value for r_s to be?

21. *Forbes Magazine* (1984) ranked 17 airlines on the basis of both five-year profitability and growth.

	PROFIT	GROWTH
Federal Express	1	1
Emery Air Freight	2	13
U.S. Air	3	6
Piedmont	4	4
Frontier	5	7
P.S.A.	6	8
Delta	7	9
Northwest	8	10
A.M.R.	9	12
Transworld	10	17
U.A.L.	11	15
Tiger International	12	11
Pan American	13	11
Eastern	14	14
Western	15	16
Texas Air	16	2
Republic	17	3

Compute the rank correlation coefficient between profitability and growth.

EDUCATION AND PSYCHOLOGY

22. The following data are the annual salaries (in thousands of dollars) of samples of 12 newly appointed assistant professors at three liberal arts colleges in a state. The three colleges differ in the extent to which the faculty are unionized.

SUNSHINE COLLEGE (COMPLETELY UNIONIZED)			HAPPINESS COLLEGE (PARTIALLY UNIONIZED)			GLORY COLLEGE (NOT UNIONIZED)		
21.3	22.4	21.5	25.8	21.0	24.4	15.5	16.1	22.9
22.3	22.2	21.6	23.8	20.4	22.5	22.3	18.5	19.0
21.7	22.6	22.7	20.3	18.7	20.8	23.7	18.8	26.7
21.7	23.1	23.3	21.9	21.4	24.5	16.2	25.6	25.8

Test whether the median salary for the three types of colleges differ significantly.

23. The two faculty members responsible for an art seminar rank the work of the six students in the class as follows:

	A	*B*	*C*	*D*	*E*	*F*
Faculty member (1)	3	1	4	2	6	5
Faculty member (2)	4	3	2	1	5	6

(a) Compute the rank correlation between the two sets of ranks.

(b) It is agreed that only the student with the lowest combined ranks will get an A for the course. Which student should get the A?

24. Webster (1983) compares the ranking of a large number of graduate schools by two publications. The rankings for 10 schools are as follows:

	GRADUATE SCHOOL	
	PUBLICATION *A*	PUBLICATION *B*
U. of California at Berkeley	1	2
Stanford U.	2	3
Harvard U.	3	1
Yale U.	4	4
M.I.T.	5	9
Princeton U.	6	8
U. of Chicago	7	6
U. of Wisconsin	8	7
U. of Michigan	9	5
U. of California at Los Angeles	10	10

Compute the rank correlation between these two sets of ranks.

25. A faculty advisor has six freshman advisees whom she ranks on the basis of their scores on the English and the mathematics competency tests.

STUDENT	*A*	*B*	*C*	*D*	*E*	*F*
English	1	2	3	4	5	6
Mathematics	4	2	5	1	6	3

Compute the rank correlation between these two sets of ranks.

26. Erdmann (1983) reported on a survey of high school seniors and high school counselors. Both groups were asked to rank the importance of eight factors in determining which colleges to apply to. The aggregate results for the two groups were as follows:

FACTORS	SENIORS	COUNSELORS
Academic program	1	2
Reputation	2	1
Location	3	3
Size	4	7
Parents' recommendation	5	4
Counselor's recommendation	6	5
Cost	7	6
Alumni contact	8	8

Compute the rank correlation coefficient between the ranks for the two groups. Would you say the two groups are in close agreement?

SCIENCE AND TECHNOLOGY

27. A chemist is concerned with the median percentage of impurities in a certain chemical product. In a random sample of 16 cases of the product, the percentage of impurities were:

6.4 5.5 6.0 6.2 6.7 6.2 5.3 5.8

6.1 6.5 6.6 5.5 5.9 6.1 6.9 5.9

Use the (a) sign test and (b) Wilcoxon test to test whether the median percentage of impurities differs significantly from 6.

28. The National Security Agency arranges an experiment to compare the precision of (i) a satellite-based and (ii) a land-based navigation system. Each system is used to estimate the location of 15 randomly selected points around the world whose locations are known precisely. For each point the error (in meters) in each system's estimate is recorded. The results are as follows:

	LAND SYSTEM	SATELLITE SYSTEM		LAND SYSTEM	SATELLITE SYSTEM
1	52	47	9	34	29
2	80	68	10	58	38
3	27	29	11	54	40
4	54	51	12	20	18
5	47	22	13	32	39
6	34	37	14	38	24
7	36	31	15	54	35
8	40	44			

Use the (a) sign test and (b) Wilcoxon test to test whether the median difference (land system error—satellite system error) is significantly greater than 0.

29. The research and development division of an aircraft manufacturing company compares the hardness of three new alloys. Four specimens of each alloy are prepared and tested for hardness on the Moh scale, which runs from 1 to 10 (the higher the score, the harder the material). The results are as follows:

ALLOY *A*	ALLOY *B*	ALLOY *C*
6.34	6.41	6.24
6.27	6.38	6.29
6.39	6.43	6.22
6.26	6.47	6.21

Do these data suggest that the median hardnesses for the three alloys are significantly different?

REFERENCES

ERDMANN, D. G. "An examination of factors influencing student choice in the college selection process," *Journal of College Admissions* (Summer 1983), pp. 3–6.

WEBSTER, D. S. "America's highest ranked graduate schools, 1925–1982," *Change* (May/June 1983), pp. 14–24.

WEIL, A. T., ZINBERG, N. E., and NELSON, J. M. "Effect of marijuana on naive subjects," *Science*, Vol. 162 (December 13, 1968), pp. 1234–1242.

ZELAZO, P. R., ZELAZO, N. A., and KOLB, S. "Walking in the newborn," *Science*, Vol. 176 (1972), pp. 314–315.

SUGGESTED READING

HOLLANDER, M., and WOLFE, D. A. *Nonparametric Statistical Methods*. New York: Wiley, 1973.

MOSTELLER, F., and ROURKE, R. E. K. *Sturdy Statistics*. Reading, Mass.: Addison-Wesley, 1973.

APPENDIX A STATISTICAL TABLES

TABLE A1

Individual binomial probabilities $P(Y = y) = \dfrac{n!}{y!(n-y)!}\pi^y(1-\pi)^{n-y}$

		π												
n	y	0.05	0.1	0.2	0.25	0.3	0.4	0.5	0.6	0.7	0.75	0.8	0.9	0.95
2														
	0	.903	.810	.640	.563	.490	.360	.250	.160	.090	.063	.040	.010	.002
	1	.095	.180	.320	.375	.420	.480	.500	.480	.420	.375	.320	.180	.095
	2	.002	.010	.040	.063	.090	.160	.250	.360	.490	.563	.640	.810	.903
3														
	0	.857	.729	.512	.422	.343	.216	.125	.064	.027	.016	.008	.001	.000
	1	.135	.243	.384	.422	.441	.432	.375	.288	.189	.141	.096	.027	.007
	2	.007	.027	.096	.141	.189	.288	.375	.432	.441	.422	.384	.243	.135
	3	.000	.001	.008	.016	.027	.064	.125	.216	.343	.422	.512	.729	.857
4														
	0	.815	.656	.410	.316	.240	.130	.063	.026	.008	.004	.002	.000	.000
	1	.171	.292	.410	.422	.412	.346	.250	.154	.076	.047	.026	.004	.000
	2	.014	.049	.154	.211	.265	.346	.375	.346	.265	.211	.154	.049	.014
	3	.000	.004	.026	.047	.076	.154	.250	.346	.412	.422	.410	.292	.171
	4	.000	.000	.002	.004	.008	.026	.063	.130	.240	.316	.410	.656	.815
5														
	0	.774	.590	.328	.237	.168	.078	.031	.010	.002	.001	.000	.000	.000
	1	.204	.328	.410	.396	.360	.259	.156	.077	.028	.015	.006	.000	.000
	2	.021	.073	.205	.264	.309	.346	.313	.230	.132	.088	.051	.008	.001
	3	.001	.008	.051	.088	.132	.230	.313	.346	.309	.264	.205	.073	.021
	4	.000	.000	.006	.015	.028	.077	.156	.259	.360	.396	.410	.328	.204
	5	.000	.000	.000	.001	.002	.010	.031	.078	.168	.237	.328	.590	.774
6														
	0	.735	.531	.262	.178	.118	.047	.016	.004	.001	.000	.000	.000	.000
	1	.232	.354	.393	.356	.303	.187	.094	.037	.010	.004	.002	.000	.000
	2	.031	.098	.246	.297	.324	.311	.234	.138	.060	.033	.015	.001	.000
	3	.002	.015	.082	.132	.185	.276	.313	.276	.185	.132	.082	.015	.002
	4	.000	.001	.015	.033	.060	.138	.234	.311	.324	.297	.246	.098	.031
	5	.000	.000	.002	.004	.010	.037	.094	.187	.303	.356	.393	.354	.232
	6	.000	.000	.000	.000	.001	.004	.016	.047	.118	.178	.262	.531	.735
7														
	0	.698	.478	.210	.133	.082	.028	.008	.002	.000	.000	.000	.000	.000
	1	.257	.372	.367	.311	.247	.131	.055	.017	.004	.001	.000	.000	.000
	2	.041	.124	.275	.311	.318	.261	.164	.077	.025	.012	.004	.000	.000
	3	.004	.023	.115	.173	.227	.290	.273	.194	.097	.058	.029	.003	.000
	4	.000	.003	.029	.058	.097	.194	.273	.290	.227	.173	.115	.023	.004

TABLE A1

(cont.)

n	*y*	0.05	0.1	0.2	0.25	0.3	0.4	0.5	0.6	0.7	0.75	0.8	0.9	0.95
								π						
	5	.000	.000	.004	.012	.025	.077	.164	.261	.318	.311	.275	.124	.041
	6	.000	.000	.000	.001	.004	.017	.055	.131	.247	.311	.367	.372	.257
	7	.000	.000	.000	.000	.000	.002	.008	.028	.082	.133	.210	.478	.698
8														
	0	.663	.430	.168	.100	.058	.017	.004	.001	.000	.000	.000	.000	.000
	1	.279	.383	.336	.267	.198	.090	.031	.008	.001	.000	.000	.000	.000
	2	.051	.149	.294	.311	.296	.209	.109	.041	.010	.004	.001	.000	.000
	3	.005	.033	.147	.208	.254	.279	.219	.124	.047	.023	.009	.000	.000
	4	.000	.005	.046	.087	.136	.232	.273	.232	.136	.087	.046	.005	.000
	5	.000	.000	.009	.023	.047	.124	.219	.279	.254	.208	.147	.033	.005
	6	.000	.000	.001	.004	.010	.041	.109	.209	.296	.311	.294	.149	.051
	7	.000	.000	.000	.000	.001	.008	.031	.090	.198	.267	.336	.383	.279
	8	.000	.000	.000	.000	.000	.001	.004	.017	.058	.100	.168	.430	.663
9														
	0	.630	.387	.134	.075	.040	.010	.002	.000	.000	.000	.000	.000	.000
	1	.299	.387	.302	.225	.156	.060	.018	.004	.000	.000	.000	.000	.000
	2	.063	.172	.302	.300	.267	.161	.070	.021	.004	.001	.000	.000	.000
	3	.008	.045	.176	.234	.267	.251	.164	.074	.021	.009	.003	.000	.000
	4	.001	.007	.066	.117	.172	.251	.246	.167	.074	.039	.017	.001	.000
	5	.000	.001	.017	.039	.074	.167	.246	.251	.172	.117	.066	.007	.001
	6	.000	.000	.003	.009	.021	.074	.164	.251	.267	.234	.176	.045	.008
	7	.000	.000	.000	.001	.004	.021	.070	.161	.267	.300	.302	.172	.063
	8	.000	.000	.000	.000	.000	.004	.018	.060	.156	.225	.302	.387	.299
	9	.000	.000	.000	.000	.000	.000	.002	.010	.040	.075	.134	.387	.630
10														
	0	.599	.349	.107	.056	.028	.006	.001	.000	.000	.000	.000	.000	.000
	1	.315	.387	.268	.188	.121	.040	.010	.002	.000	.000	.000	.000	.000
	2	.075	.194	.302	.282	.233	.121	.044	.011	.001	.000	.000	.000	.000
	3	.010	.057	.201	.250	.267	.215	.117	.042	.009	.003	.001	.000	.000
	4	.001	.011	.088	.146	.200	.251	.205	.111	.037	.016	.006	.000	.000
	5	.000	.001	.026	.058	.103	.201	.246	.201	.103	.058	.026	.001	.000
	6	.000	.000	.006	.016	.037	.111	.205	.251	.200	.146	.088	.011	.001
	7	.000	.000	.001	.003	.009	.042	.117	.215	.267	.250	.201	.057	.010
	8	.000	.000	.000	.000	.001	.011	.044	.121	.233	.282	.302	.194	.075
	9	.000	.000	.000	.000	.000	.002	.010	.040	.121	.188	.268	.387	.315
	10	.000	.000	.000	.000	.000	.000	.001	.006	.028	.056	.107	.349	.599
11														
	0	.569	.314	.086	.042	.020	.004	.000	.000	.000	.000	.000	.000	.000
	1	.329	.384	.236	.155	.093	.027	.005	.001	.000	.000	.000	.000	.000
	2	.087	.213	.295	.258	.200	.089	.027	.005	.001	.000	.000	.000	.000
	3	.014	.071	.221	.258	.257	.177	.081	.023	.004	.001	.000	.000	.000
	4	.001	.016	.111	.172	.220	.236	.161	.070	.017	.006	.002	.000	.000
	5	.000	.002	.039	.080	.132	.221	.226	.147	.057	.027	.010	.000	.000
	6	.000	.000	.010	.027	.057	.147	.226	.221	.132	.080	.039	.002	.000
	7	.000	.000	.002	.006	.017	.070	.161	.236	.220	.172	.111	.016	.001
	8	.000	.000	.000	.001	.004	.023	.081	.177	.257	.258	.221	.071	.014
	9	.000	.000	.000	.000	.001	.005	.027	.089	.200	.258	.295	.213	.087
	10	.000	.000	.000	.000	.000	.001	.005	.027	.093	.155	.236	.384	.329
	11	.000	.000	.000	.000	.000	.000	.000	.004	.020	.042	.086	.314	.569
12														
	0	.540	.282	.069	.032	.014	.002	.000	.000	.000	.000	.000	.000	.000
	1	.341	.377	.206	.127	.071	.017	.003	.000	.000	.000	.000	.000	.000
	2	.099	.230	.283	.232	.168	.064	.016	.002	.000	.000	.000	.000	.000
	3	.017	.085	.236	.258	.240	.142	.054	.012	.001	.000	.000	.000	.000
	4	.002	.021	.133	.194	.231	.213	.121	.042	.008	.002	.001	.000	.000

TABLE A1

(cont.)

		π												
n	*y*	0.05	0.1	0.2	0.25	0.3	0.4	0.5	0.6	0.7	0.75	0.8	0.9	0.95
	5	.000	.004	.053	.103	.158	.227	.193	.101	.029	.011	.003	.000	.000
	6	.000	.000	.016	.040	.079	.177	.226	.177	.079	.040	.016	.000	.000
	7	.000	.000	.003	.011	.029	.101	.193	.227	.158	.103	.053	.004	.000
	8	.000	.000	.001	.002	.008	.042	.121	.213	.231	.194	.133	.021	.002
	9	.000	.000	.000	.000	.001	.012	.054	.142	.240	.258	.236	.085	.017
	10	.000	.000	.000	.000	.000	.002	.016	.064	.168	.232	.283	.230	.099
	11	.000	.000	.000	.000	.000	.000	.003	.017	.071	.127	.206	.377	.341
	12	.000	.000	.000	.000	.000	.000	.000	.002	.014	.032	.069	.282	.540
13														
	0	.513	.254	.055	.024	.010	.001	.000	.000	.000	.000	.000	.000	.000
	1	.351	.367	.179	.103	.054	.011	.002	.000	.000	.000	.000	.000	.000
	2	.111	.245	.268	.206	.139	.045	.010	.001	.000	.000	.000	.000	.000
	3	.021	.100	.246	.252	.218	.111	.035	.006	.001	.000	.000	.000	.000
	4	.003	.028	.154	.210	.234	.184	.087	.024	.003	.001	.000	.000	.000
	5	.000	.006	.069	.126	.180	.221	.157	.066	.014	.005	.001	.000	.000
	6	.000	.001	.023	.056	.103	.197	.209	.131	.044	.019	.006	.000	.000
	7	.000	.000	.006	.019	.044	.131	.209	.197	.103	.056	.023	.001	.000
	8	.000	.000	.001	.005	.014	.066	.157	.221	.180	.126	.069	.006	.000
	9	.000	.000	.000	.001	.003	.024	.087	.184	.234	.210	.154	.028	.003
	10	.000	.000	.000	.000	.001	.006	.035	.111	.218	.252	.246	.100	.021
	11	.000	.000	.000	.000	.000	.001	.010	.045	.139	.206	.268	.245	.111
	12	.000	.000	.000	.000	.000	.000	.002	.011	.054	.103	.179	.367	.351
	13	.000	.000	.000	.000	.000	.000	.000	.001	.010	.024	.055	.254	.513
14														
	0	.488	.229	.044	.018	.007	.001	.000	.000	.000	.000	.000	.000	.000
	1	.359	.356	.154	.083	.041	.007	.001	.000	.000	.000	.000	.000	.000
	2	.123	.257	.250	.180	.113	.032	.006	.001	.000	.000	.000	.000	.000
	3	.026	.114	.250	.240	.194	.085	.022	.003	.000	.000	.000	.000	.000
	4	.004	.035	.172	.220	.229	.155	.061	.014	.001	.000	.000	.000	.000
	5	.000	.008	.086	.147	.196	.207	.122	.041	.007	.002	.000	.000	.000
	6	.000	.001	.032	.073	.126	.207	.183	.092	.023	.008	.002	.000	.000
	7	.000	.000	.009	.028	.062	.157	.209	.157	.062	.028	.009	.000	.000
	8	.000	.000	.002	.008	.023	.092	.183	.207	.126	.073	.032	.001	.000
	9	.000	.000	.000	.002	.007	.041	.122	.207	.196	.147	.086	.008	.000
	10	.000	.000	.000	.000	.001	.014	.061	.155	.229	.220	.172	.035	.004
	11	.000	.000	.000	.000	.000	.003	.022	.085	.194	.240	.250	.114	.026
	12	.000	.000	.000	.000	.000	.001	.006	.032	.113	.180	.250	.257	.123
	13	.000	.000	.000	.000	.000	.000	.001	.007	.041	.083	.154	.356	.359
	14	.000	.000	.000	.000	.000	.000	.000	.001	.007	.018	.044	.229	.488
15														
	0	.463	.206	.035	.013	.005	.000	.000	.000	.000	.000	.000	.000	.000
	1	.366	.343	.132	.067	.031	.005	.000	.000	.000	.000	.000	.000	.000
	2	.135	.267	.231	.156	.092	.022	.003	.000	.000	.000	.000	.000	.000
	3	.031	.129	.250	.225	.170	.063	.014	.002	.000	.000	.000	.000	.000
	4	.005	.043	.188	.225	.219	.127	.042	.007	.001	.000	.000	.000	.000
	5	.001	.010	.103	.165	.206	.186	.092	.024	.003	.001	.000	.000	.000
	6	.000	.002	.043	.092	.147	.207	.153	.061	.012	.003	.001	.000	.000
	7	.000	.000	.014	.039	.081	.177	.196	.118	.035	.013	.003	.000	.000
	8	.000	.000	.003	.013	.035	.118	.196	.177	.081	.039	.014	.000	.000
	9	.000	.000	.001	.003	.012	.061	.153	.207	.147	.092	.043	.002	000
	10	.000	.000	.000	.001	.003	.024	.092	.186	.206	.165	.103	.010	.001
	11	.000	.000	.000	.000	.001	.007	.042	.127	.219	.225	.188	.043	.005
	12	.000	.000	.000	.000	.000	.002	.014	.063	.170	.225	.250	.129	.031
	13	.000	.000	.000	.000	.000	.000	.003	.022	.092	.156	.231	.267	.135
	14	.000	.000	.000	.000	.000	.000	.000	.005	.031	.067	.132	.343	.366
	15	.000	.000	.000	.000	.000	.000	.000	.000	.005	.013	.035	.206	.463

TABLE A1

(cont.)

		π												
n	*y*	0.05	0.1	0.2	0.25	0.3	0.4	0.5	0.6	0.7	0.75	0.8	0.9	0.95
16														
	0	.440	.185	.028	.010	.002	.000	.000	.000	.000	.000	.000	.000	.000
	1	.371	.329	.113	.053	.023	.003	.000	.000	.000	.000	.000	.000	.000
	2	.146	.275	.211	.134	.073	.015	.002	.000	.000	.000	.000	.000	.000
	3	.036	.142	.246	.208	.146	.047	.009	.001	.000	.000	.000	.000	.000
	4	.006	.051	.200	.225	.204	.101	.028	.004	.000	.000	.000	.000	.000
	5	.001	.014	.120	.180	.210	.162	.067	.014	.001	.000	.000	.000	.000
	6	.000	.003	.055	.110	.165	.198	.122	.039	.006	.001	.000	.000	.000
	7	.000	.000	.020	.052	.101	.189	.175	.084	.019	.006	.001	.000	.000
	8	.000	.000	.006	.020	.049	.142	.196	.142	.049	.020	.006	.000	.000
	9	.000	.000	.001	.006	.019	.084	.175	.189	.101	.052	.020	.000	.000
	10	.000	.000	.000	.001	.006	.039	.122	.198	.165	.110	.055	.003	.000
	11	.000	.000	.000	.000	.001	.014	.067	.162	.210	.180	.120	.014	.001
	12	.000	.000	.000	.000	.000	.004	.028	.101	.204	.225	.200	.051	.006
	13	.000	.000	.000	.000	.000	.001	.009	.047	.146	.208	.246	.142	.036
	14	.000	.000	.000	.000	.000	.000	.002	.015	.073	.134	.211	.275	.146
	15	.000	.000	.000	.000	.000	.000	.000	.003	.023	.053	.113	.329	.371
	16	.000	.000	.000	.000	.000	.000	.000	.000	.003	.010	.028	.185	.440
17														
	0	.418	.167	.023	.008	.002	.000	.000	.000	.000	.000	.000	.000	.000
	1	.374	.315	.096	.043	.017	.002	.000	.000	.000	.000	.000	.000	.000
	2	.158	.280	.191	.114	.058	.010	.001	.000	.000	.000	.000	.000	.000
	3	.041	.156	.239	.189	.125	.034	.005	.000	.000	.000	.000	.000	.000
	4	.008	.060	.209	.221	.187	.080	.018	.002	.000	.000	.000	.000	.000
	5	.001	.017	.136	.191	.208	.138	.047	.008	.001	.000	.000	.000	.000
	6	.000	.004	.068	.128	.178	.184	.094	.024	.003	.001	.000	.000	.000
	7	.000	.001	.027	.067	.120	.193	.148	.057	.009	.002	.000	.000	.000
	8	.000	.000	.008	.028	.064	.161	.185	.107	.028	.009	.002	.000	.000
	9	.000	.000	.002	.009	.028	.107	.185	.161	.064	.028	.008	.000	.000
	10	.000	.000	.000	.002	.009	.057	.148	.193	.120	.067	.027	.001	.000
	11	.000	.000	.000	.001	.003	.024	.094	.184	.178	.128	.068	.004	.000
	12	.000	.000	.000	.000	.001	.008	.047	.138	.208	.191	.136	.017	.001
	13	.000	.000	.000	.000	.000	.002	.018	.080	.187	.221	.209	.060	.008
	14	.000	.000	.000	.000	.000	.000	.005	.034	.125	.189	.239	.156	.041
	15	.000	.000	.000	.000	.000	.000	.001	.010	.058	.114	.191	.280	.158
	16	.000	.000	.000	.000	.000	.000	.000	.002	.017	.043	.096	.315	.374
	17	.000	.000	.000	.000	.000	.000	.000	.000	.002	.008	.023	.167	.418
18														
	0	.397	.150	.018	.006	.002	.000	.000	.000	.000	.000	.000	.000	.000
	1	.376	.300	.081	.034	.013	.001	.000	.000	.000	.000	.000	.000	.000
	2	.168	.284	.172	.096	.046	.007	.001	.000	.000	.000	.000	.000	.000
	3	.047	.168	.230	.170	.105	.025	.003	.000	.000	.000	.000	.000	.000
	4	.009	.070	.215	.213	.168	.061	.012	.001	.000	.000	.000	.000	.000
	5	.001	.022	.151	.199	.202	.115	.033	.004	.000	.000	.000	.000	.000
	6	.000	.005	.082	.144	.187	.166	.071	.015	.001	.000	.000	.000	.000
	7	.000	.001	.035	.082	.138	.189	.121	.037	.005	.001	.000	.000	.000
	8	.000	.000	.012	.038	.081	.173	.167	.077	.015	.004	.001	.000	.000
	9	.000	.000	.003	.014	.039	.128	.185	.128	.039	.014	.003	.000	.000
	10	.000	.000	.001	.004	.015	.077	.167	.173	.081	.038	.012	.000	.000
	11	.000	.000	.000	.001	.005	.037	.121	.189	.138	.082	.035	.001	.000
	12	.000	.000	.000	.000	.001	.015	.071	.166	.187	.144	.082	.005	.000
	13	.000	.000	.000	.000	.000	.004	.033	.115	.202	.199	.151	.022	.001
	14	.000	.000	.000	.000	.000	.001	.012	.061	.168	.213	.215	.070	.009
	15	.000	.000	.000	.000	.000	.000	.003	.025	.105	.170	.230	.168	.047
	16	.000	.000	.000	.000	.000	.000	.001	.007	.046	.096	.172	.284	.168
	17	.000	.000	.000	.000	.000	.000	.000	.001	.013	.034	.081	.300	.376
	18	.000	.000	.000	.000	.000	.000	.000	.000	.002	.006	.018	.150	.397

TABLE A1 (*cont.*)

		π												
n	y	0.05	0.1	0.2	0.25	0.3	0.4	0.5	0.6	0.7	0.75	0.8	0.9	0.95
19														
	0	.377	.135	.014	.004	.001	.000	.000	.000	.000	.000	.000	.000	.000
	1	.377	.285	.068	.027	.009	.001	.000	.000	.000	.000	.000	.000	.000
	2	.179	.285	.154	.080	.036	.005	.000	.000	.000	.000	.000	.000	.000
	3	.053	.180	.218	.152	.087	.017	.002	.000	.000	.000	.000	.000	.000
	4	.011	.080	.218	.202	.149	.047	.007	.001	.000	.000	.000	.000	.000
	5	.002	.027	.164	.202	.192	.093	.022	.002	.000	.000	.000	.000	.000
	6	.000	.007	.095	.157	.192	.145	.052	.008	.001	.000	.000	.000	.000
	7	.000	.001	.044	.097	.153	.180	.096	.024	.002	.000	.000	.000	.000
	8	.000	.000	.017	.049	.098	.180	.144	.053	.008	.002	.000	.000	.000
	9	.000	.000	.005	.020	.051	.146	.176	.098	.022	.007	.001	.000	.000
	10	.000	.000	.001	.007	.022	.098	.176	.146	.051	.020	.005	.000	.000
	11	.000	.000	.000	.002	.008	.053	.144	.180	.098	.049	.017	.000	.000
	12	.000	.000	.000	.000	.002	.024	.096	.180	.153	.097	.044	.001	.000
	13	.000	.000	.000	.000	.001	.008	.052	.145	.192	.157	.095	.007	.000
	14	.000	.000	.000	.000	.000	.002	.022	.093	.192	.202	.164	.027	.002
	15	.000	.000	.000	.000	.000	.001	.007	.047	.149	.202	.218	.080	.011
	16	.000	.000	.000	.000	.000	.000	.002	.017	.087	.152	.218	.180	.053
	17	.000	.000	.000	.000	.000	.000	.000	.005	.036	.080	.154	.285	.179
	18	.000	.000	.000	.000	.000	.000	.000	.001	.009	.027	.068	.285	.377
	19	.000	.000	.000	.000	.000	.000	.000	.000	.001	.004	.014	.135	.377
20														
	0	.358	.122	.012	.003	.001	.000	.000	.000	.000	.000	.000	.000	.000
	1	.377	.270	.058	.021	.007	.000	.000	.000	.000	.000	.000	.000	.000
	2	.189	.285	.137	.067	.028	.003	.000	.000	.000	.000	.000	.000	.000
	3	.060	.190	.205	.134	.072	.012	.001	.000	.000	.000	.000	.000	.000
	4	.013	.090	.218	.190	.130	.035	.005	.000	.000	.000	.000	.000	.000
	5	.002	.032	.175	.202	.179	.075	.015	.001	.000	.000	.000	.000	.000
	6	.000	.009	.109	.169	.192	.124	.037	.005	.000	.000	.000	.000	.000
	7	.000	.002	.055	.112	.164	.166	.074	.015	.001	.000	.000	.000	.000
	8	.000	.000	.022	.061	.114	.180	.120	.035	.004	.001	.000	.000	.000
	9	.000	.000	.007	.027	.065	.160	.160	.071	.012	.003	.000	.000	.000
	10	.000	.000	.002	.010	.031	.117	.176	.117	.031	.010	.002	.000	.000
	11	.000	.000	.000	.003	.012	.071	.160	.160	.065	.027	.007	.000	.000
	12	.000	.000	.000	.001	.004	.035	.120	.180	.114	.061	.022	.000	.000
	13	.000	.000	.000	.000	.001	.015	.074	.166	.164	.112	.055	.002	.000
	14	.000	.000	.000	.000	.000	.005	.037	.124	.192	.169	.109	.009	.000
	15	.000	.000	.000	.000	.000	.001	.015	.075	.179	.202	.175	.032	.002
	16	.000	.000	.000	.000	.000	.000	.005	.035	.130	.190	.218	.090	.013
	17	.000	.000	.000	.000	.000	.000	.001	.012	.072	.134	.205	.190	.060
	18	.000	.000	.000	.000	.000	.000	.000	.003	.028	.067	.137	.285	.189
	19	.000	.000	.000	.000	.000	.000	.000	.000	.007	.021	.058	.270	.377
	20	.000	.000	.000	.000	.000	.000	.000	.000	.001	.003	.012	.122	.358
25														
	0	.277	.072	.004	.001	.000	.000	.000	.000	.000	.000	.000	.000	.000
	1	.365	.199	.024	.006	.001	.000	.000	.000	.000	.000	.000	.000	.000
	2	.231	.266	.071	.025	.007	.000	.000	.000	.000	.000	.000	.000	.000
	3	.093	.226	.136	.064	.024	.002	.000	.000	.000	.000	.000	.000	.000
	4	.027	.138	.187	.118	.057	.007	.000	.000	.000	.000	.000	.000	.000
	5	.006	.065	.196	.165	.103	.020	.002	.000	.000	.000	.000	.000	.000
	6	.001	.024	.163	.183	.147	.044	.005	.000	.000	.000	.000	.000	.000
	7	.000	.007	.111	.165	.171	.080	.014	.001	.000	.000	.000	.000	.000
	8	.000	.002	.062	.124	.165	.120	.032	.003	.000	.000	.000	.000	.000
	9	.000	.000	.029	.078	.134	.151	.061	.009	.000	.000	.000	.000	.000
	10	.000	.000	.012	.042	.092	.161	.097	.021	.001	.000	.000	.000	.000
	11	.000	.000	.004	.019	.054	.147	.133	.043	.004	.001	.000	.000	.000
	12	.000	.000	.001	.007	.027	.114	.155	.076	.011	.002	.000	.000	.000

TABLE A1 *(cont.)*

n	y	0.05	0.1	0.2	0.25	0.3	0.4	0.5	0.6	0.7	0.75	0.8	0.9	0.95
								π						
	13	.000	.000	.000	.002	.011	.076	.155	.114	.027	.007	.001	.000	.000
	14	.000	.000	.000	.001	.004	.043	.133	.147	.054	.019	.004	.000	.000
	15	.000	.000	.000	.000	.001	.021	.097	.161	.092	.042	.012	.000	.000
	16	.000	.000	.000	.000	.000	.009	.061	.151	.134	.078	.029	.000	.000
	17	.000	.000	.000	.000	.000	.003	.032	.120	.165	.124	.062	.002	.000
	18	.000	.000	.000	.000	.000	.001	.014	.080	.171	.165	.111	.007	.000
	19	.000	.000	.000	.000	.000	.000	.005	.044	.147	.183	.163	.024	.001
	20	.000	.000	.000	.000	.000	.000	.002	.020	.103	.165	.196	.065	.006
	21	.000	.000	.000	.000	.000	.000	.000	.007	.057	.118	.187	.138	.027
	22	.000	.000	.000	.000	.000	.000	.000	.002	.024	.064	.136	.226	.093
	23	.000	.000	.000	.000	.000	.000	.000	.000	.007	.025	.071	.266	.231
	24	.000	.000	.000	.000	.000	.000	.000	.000	.001	.006	.024	.199	.365
	25	.000	.000	.000	.000	.000	.000	.000	.000	.000	.001	.004	.072	.277
30														
	0	.215	.042	.001	.000	.000	.000	.000	.000	.000	.000	.000	.000	.000
	1	.339	.141	.009	.002	.000	.000	.000	.000	.000	.000	.000	.000	.000
	2	.259	.228	.034	.009	.002	.000	.000	.000	.000	.000	.000	.000	.000
	3	.127	.236	.079	.027	.007	.000	.000	.000	.000	.000	.000	.000	.000
	4	.045	.177	.133	.060	.021	.001	.000	.000	.000	.000	.000	.000	.000
	5	.012	.102	.172	.105	.046	.004	.000	.000	.000	.000	.000	.000	.000
	6	.003	.047	.179	.145	.083	.012	.001	.000	.000	.000	.000	.000	.000
	7	.000	.018	.154	.166	.122	.026	.002	.000	.000	.000	.000	.000	.000
	8	.000	.006	.111	.159	.150	.050	.005	.000	.000	.000	.000	.000	.000
	9	.000	.002	.068	.130	.157	.082	.013	.001	.000	.000	.000	.000	.000
	10	.000	.000	.035	.091	.142	.115	.028	.002	.000	.000	.000	.000	.000
	11	.000	.000	.016	.055	.110	.140	.051	.005	.000	.000	.000	.000	.000
	12	.000	.000	.006	.029	.075	.147	.081	.013	.000	.000	.000	.000	.000
	13	.000	.000	.002	.013	.044	.136	.112	.027	.001	.000	.000	.000	.000
	14	.000	.000	.001	.005	.023	.110	.135	.049	.004	.001	.000	.000	.000
	15	.000	.000	.000	.002	.011	.078	.144	.078	.011	.002	.000	.000	.000
	16	.000	.000	.000	.001	.004	.049	.135	.110	.023	.005	.001	.000	.000
	17	.000	.000	.000	.000	.001	.027	.112	.136	.044	.013	.002	.000	.000
	18	.000	.000	.000	.000	.000	.013	.081	.147	.075	.029	.006	.000	.000
	19	.000	.000	.000	.000	.000	.005	.051	.140	.110	.055	.016	.000	.000
	20	.000	.000	.000	.000	.000	.002	.028	.115	.142	.091	.035	.000	.000
	21	.000	.000	.000	.000	.000	.001	.013	.082	.157	.130	.068	.002	.000
	22	.000	.000	.000	.000	.000	.000	.005	.050	.150	.159	.111	.006	.000
	23	.000	.000	.000	.000	.000	.000	.002	.026	.122	.166	.154	.018	.000
	24	.000	.000	.000	.000	.000	.000	.001	.012	.083	.145	.179	.047	.003
	25	.000	.000	.000	.000	.000	.000	.000	.004	.046	.105	.172	.102	.012
	26	.000	.000	.000	.000	.000	.000	.000	.001	.021	.060	.133	.177	.045
	27	.000	.000	.000	.000	.000	.000	.000	.000	.007	.027	.079	.236	.127
	28	.000	.000	.000	.000	.000	.000	.000	.000	.002	.009	.034	.228	.259
	29	.000	.000	.000	.000	.000	.000	.000	.000	.000	.002	.009	.141	.339
	30	.000	.000	.000	.000	.000	.000	.000	.000	.000	.000	.001	.042	.215
40														
	0	.129	.015	.000	.000	.000	.000	.000	.000	.000	.000	.000	.000	.000
	1	.271	.066	.001	.000	.000	.000	.000	.000	.000	.000	.000	.000	.000
	2	.278	.142	.006	.001	.000	.000	.000	.000	.000	.000	.000	.000	.000
	3	.185	.200	.021	.004	.000	.000	.000	.000	.000	.000	.000	.000	.000
	4	.090	.206	.047	.011	.002	.000	.000	.000	.000	.000	.000	.000	.000
	5	.034	.165	.085	.027	.006	.000	.000	.000	.000	.000	.000	.000	.000
	6	.010	.107	.125	.053	.015	.000	.000	.000	.000	.000	.000	.000	.000
	7	.003	.058	.151	.086	.032	.001	.000	.000	.000	.000	.000	.000	.000
	8	.001	.026	.156	.118	.056	.004	.000	.000	.000	.000	.000	.000	.000
	9	.000	.010	.139	.140	.085	.010	.000	.000	.000	.000	.000	.000	.000

TABLE A1 (cont.)

n	y	π 0.05	0.1	0.2	0.25	0.3	0.4	0.5	0.6	0.7	0.75	0.8	0.9	0.95
	10	.000	.004	.107	.144	.113	.020	.001	.000	.000	.000	.000	.000	.000
	11	.000	.001	.073	.131	.132	.036	.002	.000	.000	.000	.000	.000	.000
	12	.000	.000	.044	.106	.137	.058	.005	.000	.000	.000	.000	.000	.000
	13	.000	.000	.024	.076	.126	.083	.011	.000	.000	.000	.000	.000	.000
	14	.000	.000	.011	.049	.104	.106	.021	.001	.000	.000	.000	.000	.000
	15	.000	.000	.005	.028	.077	.123	.037	.002	.000	.000	.000	.000	.000
	16	.000	.000	.002	.015	.052	.128	.057	.005	.000	.000	.000	.000	.000
	17	.000	.000	.001	.007	.031	.120	.081	.011	.000	.000	.000	.000	.000
	18	.000	.000	.000	.003	.017	.103	.103	.020	.001	.000	.000	.000	.000
	19	.000	.000	.000	.001	.009	.079	.119	.035	.002	.000	.000	.000	.000
	20	.000	.000	.000	.000	.004	.055	.125	.055	.004	.000	.000	.000	.000
	21	.000	.000	.000	.000	.002	.035	.119	.079	.009	.001	.000	.000	.000
	22	.000	.000	.000	.000	.001	.020	.103	.103	.017	.003	.000	.000	.000
	23	.000	.000	.000	.000	.000	.011	.081	.120	.031	.007	.001	.000	.000
	24	.000	.000	.000	.000	.000	.005	.057	.128	.052	.015	.002	.000	.000
	25	.000	.000	.000	.000	.000	.002	.037	.123	.077	.028	.005	.000	.000
	26	.000	.000	.000	.000	.000	.001	.021	.106	.104	.049	.011	.000	.000
	27	.000	.000	.000	.000	.000	.000	.011	.083	.126	.076	.024	.000	.000
	28	.000	.000	.000	.000	.000	.000	.005	.058	.137	.106	.044	.000	.000
	29	.000	.000	.000	.000	.000	.000	.002	.036	.132	.131	.073	.001	.000
	30	.000	.000	.000	.000	.000	.000	.001	.020	.113	.144	.107	.004	.000
	31	.000	.000	.000	.000	.000	.000	.000	.010	.085	.140	.139	.010	.000
	32	.000	.000	.000	.000	.000	.000	.000	.004	.056	.118	.156	.026	.001
	33	.000	.000	.000	.000	.000	.000	.000	.001	.032	.086	.151	.058	.003
	34	.000	.000	.000	.000	.000	.000	.000	.000	.015	.053	.125	.107	.010
	35	.000	.000	.000	.000	.000	.000	.000	.000	.006	.027	.085	.165	.034
	36	.000	.000	.000	.000	.000	.000	.000	.000	.002	.011	.047	.206	.090
	37	.000	.000	.000	.000	.000	.000	.000	.000	.000	.004	.021	.200	.185
	38	.000	.000	.000	.000	.000	.000	.000	.000	.000	.001	.006	.142	.278
	39	.000	.000	.000	.000	.000	.000	.000	.000	.000	.000	.001	.066	.271
	40	.000	.000	.000	.000	.000	.000	.000	.000	.000	.000	.000	.015	.129

TABLE A2

Cumulative binomial probabilities $P(Y \leq y)$

n	y	π 0.05	0.1	0.2	0.25	0.3	0.4	0.5	0.6	0.7	0.75	0.8	0.9	0.95
2														
	0	.903	.810	.640	.563	.490	.360	.250	.160	.090	.063	.040	.010	.002
	1	.998	.990	.960	.938	.910	.840	.750	.640	.510	.438	.360	.190	.097
	2	.999	.999	.999	.999	.999	.999	.999	.999	.999	.999	.999	.999	.999
3														
	0	.857	.729	.512	.422	.343	.216	.125	.064	.027	.016	.008	.001	.000
	1	.993	.972	.896	.844	.784	.648	.500	.352	.216	.156	.104	.028	.007
	2	.999	.999	.992	.984	.973	.936	.875	.784	.657	.578	.488	.271	.143
	3	.999	.999	.999	.999	.999	.999	.999	.999	.999	.999	.999	.999	.999
4														
	0	.815	.656	.410	.316	.240	.130	.063	.026	.008	.004	.002	.000	.000
	1	.986	.948	.819	.738	.652	.475	.313	.179	.084	.051	.027	.004	.000
	2	.999	.996	.973	.949	.916	.821	.688	.525	.348	.262	.181	.052	.014
	3	.999	.999	.998	.996	.992	.974	.938	.870	.760	.684	.590	.344	.185
	4	.999	.999	.999	.999	.999	.999	.999	.999	.999	.999	.999	.999	.999
5														
	0	.774	.590	.328	.237	.168	.078	.031	.010	.002	.001	.000	.000	.000
	1	.977	.919	.737	.633	.528	.337	.188	.087	.031	.016	.007	.000	.000

TABLE A2 (*cont.*)

n	y	π												
		0.05	0.1	0.2	0.25	0.3	0.4	0.5	0.6	0.7	0.75	0.8	0.9	0.95
	2	.999	.991	.942	.896	.837	.683	.500	.317	.163	.104	.058	.009	.001
	3	.999	.999	.993	.984	.969	.913	.813	.663	.472	.367	.263	.081	.023
	4	.999	.999	.999	.999	.998	.990	.969	.922	.832	.763	.672	.410	.226
	5	.999	.999	.999	.999	.999	.999	.999	.999	.999	.999	.999	.999	.999
6														
	0	.735	.531	.262	.178	.118	.047	.016	.004	.001	.000	.000	.000	.000
	1	.967	.886	.655	.534	.420	.233	.109	.041	.011	.005	.002	.000	.000
	2	.998	.984	.901	.831	.744	.544	.344	.179	.070	.038	.017	.001	.000
	3	.999	.999	.983	.962	.930	.821	.656	.456	.256	.169	.099	.016	.002
	4	.999	.999	.998	.995	.989	.959	.891	.767	.580	.466	.345	.114	.033
	5	.999	.999	.999	.999	.999	.996	.984	.953	.882	.822	.738	.469	.265
	6	.999	.999	.999	.999	.999	.999	.999	.999	.999	.999	.999	.999	.999
7														
	0	.698	.478	.210	.133	.082	.028	.008	.002	.000	.000	.000	.000	.000
	1	.956	.850	.577	.445	.329	.159	.063	.019	.004	.001	.000	.000	.000
	2	.996	.974	.852	.756	.647	.420	.227	.096	.029	.013	.005	.000	.000
	3	.999	.997	.967	.929	.874	.710	.500	.290	.126	.071	.033	.003	.000
	4	.999	.999	.995	.987	.971	.904	.773	.580	.353	.244	.148	.026	.004
	5	.999	.999	.999	.999	.996	.981	.938	.841	.671	.555	.423	.150	.044
	6	.999	.999	.999	.999	.999	.998	.992	.972	.918	.867	.790	.522	.302
	7	.999	.999	.999	.999	.999	.999	.999	.999	.999	.999	.999	.999	.999
8														
	0	.663	.430	.168	.100	.058	.017	.004	.001	.000	.000	.000	.000	.000
	1	.943	.813	.503	.367	.255	.106	.035	.009	.001	.000	.000	.000	.000
	2	.994	.962	.797	.679	.552	.315	.145	.050	.011	.004	.001	.000	.000
	3	.999	.995	.944	.886	.806	.594	.363	.174	.058	.027	.010	.000	.000
	4	.999	.999	.990	.973	.942	.826	.637	.406	.194	.114	.056	.005	.000
	5	.999	.999	.999	.996	.989	.950	.855	.685	.448	.321	.203	.038	.006
	6	.999	.999	.999	.999	.999	.991	.965	.894	.745	.633	.497	.187	.057
	7	.999	.999	.999	.999	.999	.999	.996	.983	.942	.900	.832	.570	.337
	8	.999	.999	.999	.999	.999	.999	.999	.999	.999	.999	.999	.999	.999
9														
	0	.630	.387	.134	.075	.040	.010	.002	.000	.000	.000	.000	.000	.000
	1	.929	.775	.436	.300	.196	.071	.020	.004	.000	.000	.000	.000	.000
	2	.992	.947	.738	.601	.463	.232	.090	.025	.004	.001	.000	.000	.000
	3	.999	.992	.914	.834	.730	.483	.254	.099	.025	.010	.003	.000	.000
	4	.999	.999	.980	.951	.901	.733	.500	.267	.099	.049	.020	.001	.000
	5	.999	.999	.997	.990	.975	.901	.746	.517	.270	.166	.086	.008	.001
	6	.999	.999	.999	.999	.996	.975	.910	.768	.537	.399	.262	.053	.008
	7	.999	.999	.999	.999	.999	.996	.980	.929	.804	.700	.564	.225	.071
	8	.999	.999	.999	.999	.999	.999	.998	.990	.960	.925	.866	.613	.370
	9	.999	.999	.999	.999	.999	.999	.999	.999	.999	.999	.999	.999	.999
10														
	0	.599	.349	.107	.056	.028	.006	.001	.000	.000	.000	.000	.000	.000
	1	.914	.736	.376	.244	.149	.046	.011	.002	.000	.000	.000	.000	.000
	2	.988	.930	.678	.526	.383	.167	.055	.012	.002	.000	.000	.000	.000
	3	.999	.987	.879	.776	.650	.382	.172	.055	.011	.004	.001	.000	.000
	4	.999	.998	.967	.922	.850	.633	.377	.166	.047	.020	.006	.000	.000
	5	.999	.999	.994	.980	.953	.834	.623	.367	.150	.078	.033	.002	.000
	6	.999	.999	.999	.996	.989	.945	.828	.618	.350	.224	.121	.013	.001
	7	.999	.999	.999	.999	.998	.988	.945	.833	.617	.474	.322	.070	.012
	8	.999	.999	.999	.999	.999	.998	.989	.954	.851	.756	.624	.264	.086
	9	.999	.999	.999	.999	.999	.999	.999	.994	.972	.944	.893	.651	.401
	10	.999	.999	.999	.999	.999	.999	.999	.999	.999	.999	.999	.999	.999

TABLE A2 (*cont.*)

		π												
n	y	0.05	0.1	0.2	0.25	0.3	0.4	0.5	0.6	0.7	0.75	0.8	0.9	0.95
11														
	0	.569	.314	.086	.042	.020	.004	.000	.000	.000	.000	.000	.000	.000
	1	.898	.697	.322	.197	.113	.030	.006	.001	.000	.000	.000	.000	.000
	2	.985	.910	.617	.455	.313	.119	.033	.006	.001	.000	.000	.000	.000
	3	.998	.981	.839	.713	.570	.296	.113	.029	.004	.001	.000	.000	.000
	4	.999	.997	.950	.885	.790	.533	.274	.099	.022	.008	.002	.000	.000
	5	.999	.999	.988	.966	.922	.753	.500	.247	.078	.034	.012	.000	.000
	6	.999	.999	.998	.992	.978	.901	.726	.467	.210	.115	.050	.003	.000
	7	.999	.999	.999	.999	.996	.971	.887	.704	.430	.287	.161	.019	.002
	8	.999	.999	.999	.999	.999	.994	.967	.881	.687	.545	.383	.090	.015
	9	.999	.999	.999	.999	.999	.999	.994	.970	.887	.803	.678	.303	.102
	10	.999	.999	.999	.999	.999	.999	.999	.996	.980	.958	.914	.686	.431
	11	.999	.999	.999	.999	.999	.999	.999	.999	.999	.999	.999	.999	.999
12														
	0	.540	.282	.069	.032	.014	.002	.000	.000	.000	.000	.000	.000	.000
	1	.882	.659	.275	.158	.085	.020	.003	.000	.000	.000	.000	.000	.000
	2	.980	.889	.558	.391	.253	.083	.019	.003	.000	.000	.000	.000	.000
	3	.998	.974	.795	.649	.493	.225	.073	.015	.002	.000	.000	.000	.000
	4	.999	.996	.927	.842	.724	.438	.194	.057	.009	.003	.001	.000	.000
	5	.999	.999	.981	.946	.882	.665	.387	.158	.039	.014	.004	.000	.000
	6	.999	.999	.996	.986	.961	.842	.613	.335	.118	.054	.019	.001	.000
	7	.999	.999	.999	.997	.991	.943	.806	.562	.276	.158	.073	.004	.000
	8	.999	.999	.999	.999	.998	.985	.927	.775	.507	.351	.205	.026	.002
	9	.999	.999	.999	.999	.999	.997	.981	.917	.747	.609	.442	.111	.020
	10	.999	.999	.999	.999	.999	.999	.997	.980	.915	.842	.725	.341	.118
	11	.999	.999	.999	.999	.999	.999	.999	.998	.986	.968	.931	.718	.460
	12	.999	.999	.999	.999	.999	.999	.999	.999	.999	.999	.999	.999	.999
13														
	0	.513	.254	.055	.024	.010	.001	.000	.000	.000	.000	.000	.000	.000
	1	.865	.621	.234	.127	.064	.013	.002	.000	.000	.000	.000	.000	.000
	2	.975	.866	.502	.333	.202	.058	.011	.001	.000	.000	.000	.000	.000
	3	.997	.966	.747	.584	.421	.169	.046	.008	.001	.000	.000	.000	.000
	4	.999	.994	.901	.794	.654	.353	.133	.032	.004	.001	.000	.000	.000
	5	.999	.999	.970	.920	.835	.574	.291	.098	.018	.006	.001	.000	.000
	6	.999	.999	.993	.976	.938	.771	.500	.229	.062	.024	.007	.000	.000
	7	.999	.999	.999	.994	.982	.902	.709	.426	.165	.080	.030	.001	.000
	8	.999	.999	.999	.999	.996	.968	.867	.647	.346	.206	.099	.006	.000
	9	.999	.999	.999	.999	.999	.992	.954	.831	.579	.416	.253	.034	.003
	10	.999	.999	.999	.999	.999	.999	.989	.942	.798	.667	.498	.134	.025
	11	.999	.999	.999	.999	.999	.999	.998	.987	.936	.873	.766	.379	.135
	12	.999	.999	.999	.999	.999	.999	.999	.999	.990	.976	.945	.746	.487
	13	.999	.999	.999	.999	.999	.999	.999	.999	.999	.999	.999	.999	.999
14														
	0	.488	.229	.044	.018	.007	.001	.000	.000	.000	.000	.000	.000	.000
	1	.847	.585	.198	.101	.047	.008	.001	.000	.000	.000	.000	.000	.000
	2	.970	.842	.448	.281	.161	.040	.006	.001	.000	.000	.000	.000	.000
	3	.996	.956	.698	.521	.355	.124	.029	.004	.000	.000	.000	.000	.000
	4	.999	.991	.870	.742	.584	.279	.090	.018	.002	.000	.000	.000	.000
	5	.999	.999	.956	.888	.781	.486	.212	.058	.008	.002	.000	.000	.000
	6	.999	.999	.988	.962	.907	.692	.395	.150	.031	.010	.002	.000	.000
	7	.999	.999	.998	.990	.969	.850	.605	.308	.093	.038	.012	.000	.000
	8	.999	.999	.999	.998	.992	.942	.788	.514	.219	.112	.044	.001	.000
	9	.999	.999	.999	.999	.998	.982	.910	.721	.416	.258	.130	.009	.000
	10	.999	.999	.999	.999	.999	.996	.971	.876	.645	.479	.302	.044	.004
	11	.999	.999	.999	.999	.999	.999	.994	.960	.839	.719	.552	.158	.030

TABLE A2

(cont.)

		π												
n	y	0.05	0.1	0.2	0.25	0.3	0.4	0.5	0.6	0.7	0.75	0.8	0.9	0.95
	12	.999	.999	.999	.999	.999	.999	.999	.992	.953	.899	.802	.415	.153
	13	.999	.999	.999	.999	.999	.999	.999	.999	.993	.982	.956	.771	.512
	14	.999	.999	.999	.999	.999	.999	.999	.999	.999	.999	.999	.999	.999
15														
	0	.463	.206	.035	.013	.005	.000	.000	.000	.000	.000	.000	.000	.000
	1	.829	.549	.167	.080	.035	.005	.000	.000	.000	.000	.000	.000	.000
	2	.964	.816	.398	.236	.127	.027	.004	.000	.000	.000	.000	.000	.000
	3	.995	.944	.648	.461	.297	.091	.018	.002	.000	.000	.000	.000	.000
	4	.999	.987	.836	.686	.515	.217	.059	.009	.001	.000	.000	.000	.000
	5	.999	.998	.939	.852	.722	.403	.151	.034	.004	.001	.000	.000	.000
	6	.999	.999	.982	.943	.869	.610	.304	.095	.015	.004	.001	.000	.000
	7	.999	.999	.996	.983	.950	.787	.500	.213	.050	.017	.004	.000	.000
	8	.999	.999	.999	.996	.985	.905	.696	.390	.131	.057	.018	.000	.000
	9	.999	.999	.999	.999	.996	.966	.849	.597	.278	.148	.061	.002	.000
	10	.999	.999	.999	.999	.999	.991	.941	.783	.485	.314	.164	.013	.001
	11	.999	.999	.999	.999	.999	.998	.982	.909	.703	.539	.352	.056	.005
	12	.999	.999	.999	.999	.999	.999	.996	.973	.873	.764	.602	.184	.036
	13	.999	.999	.999	.999	.999	.999	.999	.995	.965	.920	.833	.451	.171
	14	.999	.999	.999	.999	.999	.999	.999	.999	.995	.987	.965	.794	.537
	15	.999	.999	.999	.999	.999	.999	.999	.999	.999	.999	.999	.999	.999
16														
	0	.440	.185	.028	.010	.003	.000	.000	.000	.000	.000	.000	.000	.000
	1	.811	.515	.141	.063	.026	.003	.000	.000	.000	.000	.000	.000	.000
	2	.957	.789	.352	.197	.099	.018	.002	.000	.000	.000	.000	.000	.000
	3	.993	.932	.598	.405	.246	.065	.011	.001	.000	.000	.000	.000	.000
	4	.999	.983	.798	.630	.450	.167	.038	.005	.000	.000	.000	.000	.000
	5	.999	.997	.918	.810	.660	.329	.105	.019	.002	.000	.000	.000	.000
	6	.999	.999	.973	.920	.825	.527	.227	.058	.007	.002	.000	.000	.000
	7	.999	.999	.993	.973	.926	.716	.402	.142	.026	.007	.001	.000	.000
	8	.999	.999	.999	.993	.974	.858	.598	.284	.074	.027	.007	.000	.000
	9	.999	.999	.999	.998	.993	.942	.773	.473	.175	.080	.027	.001	.000
	10	.999	.999	.999	.999	.998	.981	.895	.671	.340	.190	.082	.003	.000
	11	.999	.999	.999	.999	.999	.995	.962	.833	.550	.370	.202	.017	.001
	12	.999	.999	.999	.999	.999	.999	.989	.935	.754	.595	.402	.068	.007
	13	.999	.999	.999	.999	.999	.999	.998	.982	.901	.803	.648	.211	.043
	14	.999	.999	.999	.999	.999	.999	.999	.997	.974	.937	.859	.485	.189
	15	.999	.999	.999	.999	.999	.999	.999	.999	.997	.990	.972	.815	.560
	16	.999	.999	.999	.999	.999	.999	.999	.999	.999	.999	.999	.999	.999
17														
	0	.418	.167	.023	.008	.002	.000	.000	.000	.000	.000	.000	.000	.000
	1	.792	.482	.118	.050	.019	.002	.000	.000	.000	.000	.000	.000	.000
	2	.950	.762	.310	.164	.077	.012	.001	.000	.000	.000	.000	.000	.000
	3	.991	.917	.549	.353	.202	.046	.006	.000	.000	.000	.000	.000	.000
	4	.999	.978	.758	.574	.389	.126	.025	.003	.000	.000	.000	.000	.000
	5	.999	.995	.894	.765	.597	.264	.072	.011	.001	.000	.000	.000	.000
	6	.999	.999	.962	.893	.775	.448	.166	.035	.003	.001	.000	.000	.000
	7	.999	.999	.989	.960	.895	.641	.315	.092	.013	.003	.000	.000	.000
	8	.999	.999	.997	.988	.960	.801	.500	.199	.040	.012	.003	.000	.000
	9	.999	.999	.999	.997	.987	.908	.685	.359	.105	.040	.011	.000	.000
	10	.999	.999	.999	.999	.997	.965	.834	.552	.225	.107	.038	.001	.000
	11	.999	.999	.999	.999	.999	.989	.928	.736	.403	.235	.106	.005	.000
	12	.999	.999	.999	.999	.999	.997	.975	.874	.611	.426	.242	.022	.001
	13	.999	.999	.999	.999	.999	.999	.994	.954	.798	.647	.451	.083	.009
	14	.999	.999	.999	.999	.999	.999	.999	.988	.923	.836	.690	.238	.050
	15	.999	.999	.999	.999	.999	.999	.999	.998	.981	.950	.882	.518	.208

TABLE A2 (*cont.*)

		π												
n	y	0.05	0.1	0.2	0.25	0.3	0.4	0.5	0.6	0.7	0.75	0.8	0.9	0.95
	16	.999	.999	.999	.999	.999	.999	.999	.999	.998	.992	.977	.833	.582
	17	.999	.999	.999	.999	.999	.999	.999	.999	.999	.999	.999	.999	.999
18														
	0	.397	.150	.018	.006	.002	.000	.000	.000	.000	.000	.000	.000	.000
	1	.774	.450	.099	.039	.014	.001	.000	.000	.000	.000	.000	.000	.000
	2	.942	.734	.271	.135	.060	.008	.001	.000	.000	.000	.000	.000	.000
	3	.989	.902	.501	.306	.165	.033	.004	.000	.000	.000	.000	.000	.000
	4	.998	.972	.716	.519	.333	.094	.015	.001	.000	.000	.000	.000	.000
	5	.999	.994	.867	.717	.534	.209	.048	.006	.000	.000	.000	.000	.000
	6	.999	.999	.949	.861	.722	.374	.119	.020	.001	.000	.000	.000	.000
	7	.999	.999	.984	.943	.859	.563	.240	.058	.006	.001	.000	.000	.000
	8	.999	.999	.996	.981	.940	.737	.407	.135	.021	.005	.001	.000	.000
	9	.999	.999	.999	.995	.979	.865	.593	.263	.060	.019	.004	.000	.000
	10	.999	.999	.999	.999	.994	.942	.760	.437	.141	.057	.016	.000	.000
	11	.999	.999	.999	.999	.999	.980	.881	.626	.278	.139	.051	.001	.000
	12	.999	.999	.999	.999	.999	.994	.952	.791	.466	.283	.133	.006	.000
	13	.999	.999	.999	.999	.999	.999	.985	.906	.667	.481	.284	.028	.002
	14	.999	.999	.999	.999	.999	.999	.996	.967	.835	.694	.499	.098	.011
	15	.999	.999	.999	.999	.999	.999	.999	.992	.940	.865	.729	.266	.058
	16	.999	.999	.999	.999	.999	.999	.999	.999	.986	.961	.901	.550	.226
	17	.999	.999	.999	.999	.999	.999	.999	.999	.998	.994	.982	.850	.603
	18	.999	.999	.999	.999	.999	.999	.999	.999	.999	.999	.999	.999	.999
19														
	0	.377	.135	.014	.004	.001	.000	.000	.000	.000	.000	.000	.000	.000
	1	.755	.420	.083	.031	.010	.001	.000	.000	.000	.000	.000	.000	.000
	2	.933	.705	.237	.111	.046	.005	.000	.000	.000	.000	.000	.000	.000
	3	.987	.885	.455	.263	.133	.023	.002	.000	.000	.000	.000	.000	.000
	4	.998	.965	.673	.465	.282	.070	.010	.001	.000	.000	.000	.000	.000
	5	.999	.991	.837	.668	.474	.163	.032	.003	.000	.000	.000	.000	.000
	6	.999	.998	.932	.825	.666	.308	.084	.012	.001	.000	.000	.000	.000
	7	.999	.999	.977	.923	.818	.488	.180	.035	.003	.000	.000	.000	.000
	8	.999	.999	.993	.971	.916	.667	.324	.088	.011	.002	.000	.000	.000
	9.	.999	.999	.998	.991	.967	.814	.500	.186	.033	.009	.002	.000	.000
	10	.999	.999	.999	.998	.989	.912	.676	.333	.084	.029	.007	.000	.000
	11	.999	.999	.999	.999	.997	.965	.820	.512	.182	.077	.023	.000	.000
	12	.999	.999	.999	.999	.999	.988	.916	.692	.334	.175	.068	.002	.000
	13	.999	.999	.999	.999	.999	.997	.968	.837	.526	.332	.163	.009	.000
	14	.999	.999	.999	.999	.999	.999	.990	.930	.718	.535	.327	.035	.002
	15	.999	.999	.999	.999	.999	.999	.998	.977	.867	.737	.545	.115	.013
	16	.999	.999	.999	.999	.999	.999	.999	.995	.954	.889	.763	.295	.067
	17	.999	.999	.999	.999	.999	.999	.999	.999	.990	.969	.917	.580	.245
	18	.999	.999	.999	.999	.999	.999	.999	.999	.999	.996	.986	.865	.623
	19	.999	.999	.999	.999	.999	.999	.999	.999	.999	.999	.999	.999	.999
20														
	0	.358	.122	.012	.003	.001	.000	.000	.000	.000	.000	.000	.000	.000
	1	.736	.392	.069	.024	.008	.001	.000	.000	.000	.000	.000	.000	.000
	2	.925	.677	.206	.091	.035	.004	.000	.000	.000	.000	.000	.000	.000
	3	.984	.867	.411	.225	.107	.016	.001	.000	.000	.000	.000	.000	.000
	4	.997	.957	.630	.415	.238	.051	.006	.000	.000	.000	.000	.000	.000
	5	.999	.989	.804	.617	.416	.126	.021	.002	.000	.000	.000	.000	.000
	6	.999	.998	.913	.786	.608	.250	.058	.006	.000	.000	.000	.000	.000
	7	.999	.999	.968	.898	.772	.416	.132	.021	.001	.000	.000	.000	.000
	8	.999	.999	.990	.959	.887	.596	.252	.057	.005	.001	.000	.000	.000
	9	.999	.999	.997	.986	.952	.755	.412	.128	.017	.004	.001	.000	.000
	10	.999	.999	.999	.996	.983	.872	.588	.245	.048	.014	.003	.000	.000

TABLE A2 (cont.)

n	y	0.05	0.1	0.2	0.25	0.3	0.4	0.5	0.6	0.7	0.75	0.8	0.9	0.95
	11	.999	.999	.999	.999	.995	.943	.748	.404	.113	.041	.010	.000	.000
	12	.999	.999	.999	.999	.999	.979	.868	.584	.228	.102	.032	.000	.000
	13	.999	.999	.999	.999	.999	.994	.942	.750	.392	.214	.087	.002	.000
	14	.999	.999	.999	.999	.999	.998	.979	.874	.584	.383	.196	.011	.000
	15	.999	.999	.999	.999	.999	.999	.994	.949	.762	.585	.370	.043	.003
	16	.999	.999	.999	.999	.999	.999	.999	.984	.893	.775	.589	.133	.016
	17	.999	.999	.999	.999	.999	.999	.999	.996	.965	.909	.794	.323	.075
	18	.999	.999	.999	.999	.999	.999	.999	.999	.992	.976	.931	.608	.264
	19	.999	.999	.999	.999	.999	.999	.999	.999	.999	.997	.988	.878	.642
	20	.999	.999	.999	.999	.999	.999	.999	.999	.999	.999	.999	.999	.999
25														
	0	.277	.072	.004	.001	.000	.000	.000	.000	.000	.000	.000	.000	.000
	1	.642	.271	.027	.007	.002	.000	.000	.000	.000	.000	.000	.000	.000
	2	.873	.537	.098	.032	.009	.000	.000	.000	.000	.000	.000	.000	.000
	3	.966	.764	.234	.096	.033	.002	.000	.000	.000	.000	.000	.000	.000
	4	.993	.902	.421	.214	.090	.009	.000	.000	.000	.000	.000	.000	.000
	5	.999	.967	.617	.378	.193	.029	.002	.000	.000	.000	.000	.000	.000
	6	.999	.991	.780	.561	.341	.074	.007	.000	.000	.000	.000	.000	.000
	7	.999	.998	.891	.727	.512	.154	.022	.001	.000	.000	.000	.000	.000
	8	.999	.999	.953	.851	.677	.274	.054	.004	.000	.000	.000	.000	.000
	9	.999	.999	.983	.929	.811	.425	.115	.013	.000	.000	.000	.000	.000
	10	.999	.999	.994	.970	.902	.586	.212	.034	.002	.000	.000	.000	.000
	11	.999	.999	.998	.989	.956	.732	.345	.078	.006	.001	.000	.000	.000
	12	.999	.999	.999	.997	.983	.846	.500	.154	.017	.003	.000	.000	.000
	13	.999	.999	.999	.999	.994	.922	.655	.268	.044	.011	.002	.000	.000
	14	.999	.999	.999	.999	.998	.966	.788	.414	.098	.030	.006	.000	.000
	15	.999	.999	.999	.999	.999	.987	.885	.575	.189	.071	.017	.000	.000
	16	.999	.999	.999	.999	.999	.996	.946	.726	.323	.149	.047	.000	.000
	17	.999	.999	.999	.999	.999	.999	.978	.846	.488	.273	.109	.002	.000
	18	.999	.999	.999	.999	.999	.999	.993	.926	.659	.439	.220	.009	.000
	19	.999	.999	.999	.999	.999	.999	.998	.971	.807	.622	.383	.033	.001
	20	.999	.999	.999	.999	.999	.999	.999	.991	.910	.786	.579	.098	.007
	21	.999	.999	.999	.999	.999	.999	.999	.998	.967	.904	.766	.236	.034
	22	.999	.999	.999	.999	.999	.999	.999	.999	.991	.968	.902	.463	.127
	23	.999	.999	.999	.999	.999	.999	.999	.999	.998	.993	.973	.729	.358
	24	.999	.999	.999	.999	.999	.999	.999	.999	.999	.999	.996	.928	.723
	25	.999	.999	.999	.999	.999	.999	.999	.999	.999	.999	.999	.999	.999
30														
	0	.215	.042	.001	.000	.000	.000	.000	.000	.000	.000	.000	.000	.000
	1	.554	.184	.011	.002	.000	.000	.000	.000	.000	.000	.000	.000	.000
	2	.812	.411	.044	.011	.002	.000	.000	.000	.000	.000	.000	.000	.000
	3	.939	.647	.123	.037	.009	.000	.000	.000	.000	.000	.000	.000	.000
	4	.984	.825	.255	.098	.030	.002	.000	.000	.000	.000	.000	.000	.000
	5	.997	.927	.428	.203	.077	.006	.000	.000	.000	.000	.000	.000	.000
	6	.999	.974	.607	.348	.160	.017	.001	.000	.000	.000	.000	.000	.000
	7	.999	.992	.761	.514	.281	.044	.003	.000	.000	.000	.000	.000	.000
	8	.999	.998	.871	.674	.432	.094	.008	.000	.000	.000	.000	.000	.000
	9	.999	.999	.939	.803	.589	.176	.021	.001	.000	.000	.000	.000	.000
	10	.999	.999	.974	.894	.730	.291	.049	.003	.000	.000	.000	.000	.000
	11	.999	.999	.991	.949	.841	.431	.100	.008	.000	.000	.000	.000	.000
	12	.999	.999	.997	.978	.916	.578	.181	.021	.001	.000	.000	.000	.000
	13	.999	.999	.999	.992	.960	.715	.292	.048	.002	.000	.000	.000	.000
	14	.999	.999	.999	.997	.983	.825	.428	.097	.006	.001	.000	.000	.000
	15	.999	.999	.999	.999	.994	.903	.572	.175	.017	.003	.000	.000	.000
	16	.999	.999	.999	.999	.998	.952	.708	.285	.040	.008	.001	.000	.000
	17	.999	.999	.999	.999	.999	.979	.819	.422	.084	.022	.003	.000	.000

Column headings 0.05–0.95 are values of π.

TABLE A2 *(cont.)*

		π												
n	y	0.05	0.1	0.2	0.25	0.3	0.4	0.5	0.6	0.7	0.75	0.8	0.9	0.95
	18	.999	.999	.999	.999	.999	.992	.900	.569	.159	.051	.009	.000	.000
	19	.999	.999	.999	.999	.999	.997	.951	.709	.270	.106	.026	.000	.000
	20	.999	.999	.999	.999	.999	.999	.979	.824	.411	.197	.061	.000	.000
	21	.999	.999	.999	.999	.999	.999	.992	.906	.568	.326	.129	.002	.000
	22	.999	.999	.999	.999	.999	.999	.997	.956	.719	.486	.239	.008	.000
	23	.999	.999	.999	.999	.999	.999	.999	.983	.840	.652	.393	.026	.001
	24	.999	.999	.999	.999	.999	.999	.999	.994	.923	.797	.572	.073	.003
	25	.999	.999	.999	.999	.999	.999	.999	.998	.970	.902	.745	.175	.016
	26	.999	.999	.999	.999	.999	.999	.999	.999	.991	.963	.877	.353	.061
	27	.999	.999	.999	.999	.999	.999	.999	.999	.998	.989	.956	.589	.188
	28	.999	.999	.999	.999	.999	.999	.999	.999	.999	.998	.989	.816	.446
	29	.999	.999	.999	.999	.999	.999	.999	.999	.999	.999	.999	.958	.785
	30	.999	.999	.999	.999	.999	.999	.999	.999	.999	.999	.999	.999	.999
40														
	0	.129	.015	.000	.000	.000	.000	.000	.000	.000	.000	.000	.000	.000
	1	.399	.080	.001	.000	.000	.000	.000	.000	.000	.000	.000	.000	.000
	2	.677	.223	.008	.001	.000	.000	.000	.000	.000	.000	.000	.000	.000
	3	.862	.423	.028	.005	.001	.000	.000	.000	.000	.000	.000	.000	.000
	4	.952	.629	.076	.016	.003	.000	.000	.000	.000	.000	.000	.000	.000
	5	.986	.794	.161	.043	.009	.000	.000	.000	.000	.000	.000	.000	.000
	6	.997	.900	.286	.096	.024	.001	.000	.000	.000	.000	.000	.000	.000
	7	.999	.958	.437	.182	.055	.002	.000	.000	.000	.000	.000	.000	.000
	8	.999	.985	.593	.300	.111	.006	.000	.000	.000	.000	.000	.000	.000
	9	.999	.995	.732	.440	.196	.016	.000	.000	.000	.000	.000	.000	.000
	10	.999	.999	.839	.584	.309	.035	.001	.000	.000	.000	.000	.000	.000
	11	.999	.999	.912	.715	.441	.071	.003	.000	.000	.000	.000	.000	.000
	12	.999	.999	.957	.821	.577	.129	.008	.000	.000	.000	.000	.000	.000
	13	.999	.999	.981	.897	.703	.211	.019	.000	.000	.000	.000	.000	.000
	14	.999	.999	.992	.946	.807	.317	.040	.001	.000	.000	.000	.000	.000
	15	.999	.999	.997	.974	.885	.440	.077	.003	.000	.000	.000	.000	.000
	16	.999	.999	.999	.988	.937	.568	.134	.008	.000	.000	.000	.000	.000
	17	.999	.999	.999	.995	.968	.689	.215	.019	.000	.000	.000	.000	.000
	18	.999	.999	.999	.998	.985	.791	.318	.039	.001	.000	.000	.000	.000
	19	.999	.999	.999	.999	.994	.870	.437	.074	.002	.000	.000	.000	.000
	20	.999	.999	.999	.999	.998	.926	.563	.130	.006	.001	.000	.000	.000
	21	.999	.999	.999	.999	.999	.961	.682	.209	.015	.002	.000	.000	.000
	22	.999	.999	.999	.999	.999	.981	.785	.311	.032	.005	.000	.000	.000
	23	.999	.999	.999	.999	.999	.992	.866	.432	.063	.012	.001	.000	.000
	24	.999	.999	.999	.999	.999	.997	.923	.560	.115	.026	.003	.000	.000
	25	.999	.999	.999	.999	.999	.999	.960	.683	.193	.054	.008	.000	.000
	26	.999	.999	.999	.999	.999	.999	.981	.789	.297	.103	.019	.000	.000
	27	.999	.999	.999	.999	.999	.999	.992	.871	.423	.179	.043	.000	.000
	28	.999	.999	.999	.999	.999	.999	.997	.929	.559	.285	.088	.000	.000
	29	.999	.999	.999	.999	.999	.999	.999	.965	.691	.416	.161	.001	.000
	30	.999	.999	.999	.999	.999	.999	.999	.984	.804	.560	.268	.005	.000
	31	.999	.999	.999	.999	.999	.999	.999	.994	.889	.700	.407	.015	.000
	32	.999	.999	.999	.999	.999	.999	.999	.998	.945	.818	.563	.042	.001
	33	.999	.999	.999	.999	.999	.999	.999	.999	.976	.904	.714	.100	.003
	34	.999	.999	.999	.999	.999	.999	.999	.999	.991	.957	.839	.206	.014
	35	.999	.999	.999	.999	.999	.999	.999	.999	.997	.984	.924	.371	.048
	36	.999	.999	.999	.999	.999	.999	.999	.999	.999	.995	.972	.577	.138
	37	.999	.999	.999	.999	.999	.999	.999	.999	.999	.999	.992	.777	.323
	38	.999	.999	.999	.999	.999	.999	.999	.999	.999	.999	.999	.920	.601
	39	.999	.999	.999	.999	.999	.999	.999	.999	.999	.999	.999	.985	.871
	40	.999	.999	.999	.999	.999	.999	.999	.999	.999	.999	.999	.999	.999

TABLE A3

Areas under the standardized normal distribution

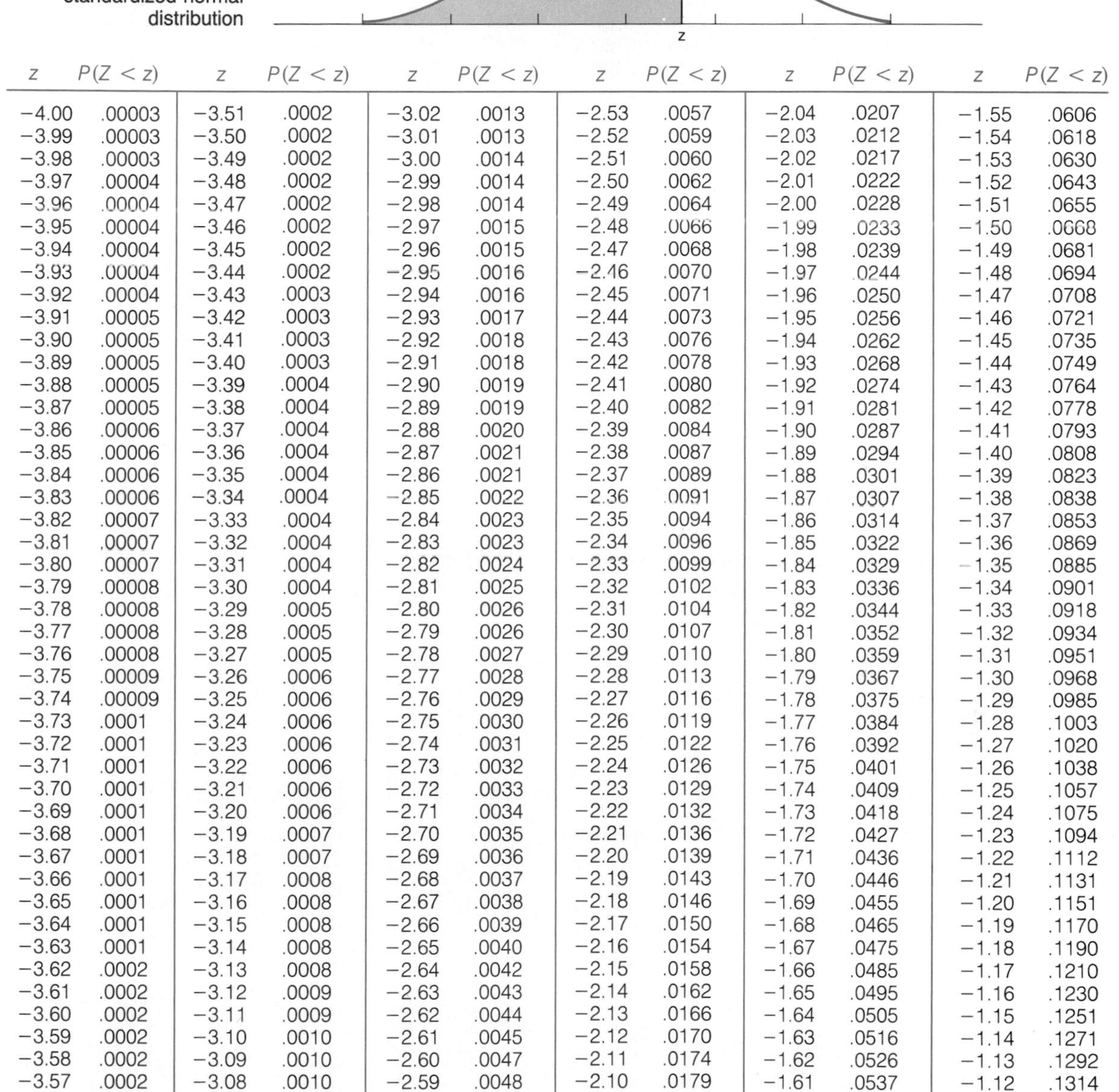

z	P(Z < z)	z	P(Z < z)	z	P(Z < z)	z	P(Z < z)	z	P(Z < z)	z	P(Z < z)
−4.00	.00003	−3.51	.0002	−3.02	.0013	−2.53	.0057	−2.04	.0207	−1.55	.0606
−3.99	.00003	−3.50	.0002	−3.01	.0013	−2.52	.0059	−2.03	.0212	−1.54	.0618
−3.98	.00003	−3.49	.0002	−3.00	.0014	−2.51	.0060	−2.02	.0217	−1.53	.0630
−3.97	.00004	−3.48	.0002	−2.99	.0014	−2.50	.0062	−2.01	.0222	−1.52	.0643
−3.96	.00004	−3.47	.0002	−2.98	.0014	−2.49	.0064	−2.00	.0228	−1.51	.0655
−3.95	.00004	−3.46	.0002	−2.97	.0015	−2.48	.0066	−1.99	.0233	−1.50	.0668
−3.94	.00004	−3.45	.0002	−2.96	.0015	−2.47	.0068	−1.98	.0239	−1.49	.0681
−3.93	.00004	−3.44	.0002	−2.95	.0016	−2.46	.0070	−1.97	.0244	−1.48	.0694
−3.92	.00004	−3.43	.0003	−2.94	.0016	−2.45	.0071	−1.96	.0250	−1.47	.0708
−3.91	.00005	−3.42	.0003	−2.93	.0017	−2.44	.0073	−1.95	.0256	−1.46	.0721
−3.90	.00005	−3.41	.0003	−2.92	.0018	−2.43	.0076	−1.94	.0262	−1.45	.0735
−3.89	.00005	−3.40	.0003	−2.91	.0018	−2.42	.0078	−1.93	.0268	−1.44	.0749
−3.88	.00005	−3.39	.0004	−2.90	.0019	−2.41	.0080	−1.92	.0274	−1.43	.0764
−3.87	.00005	−3.38	.0004	−2.89	.0019	−2.40	.0082	−1.91	.0281	−1.42	.0778
−3.86	.00006	−3.37	.0004	−2.88	.0020	−2.39	.0084	−1.90	.0287	−1.41	.0793
−3.85	.00006	−3.36	.0004	−2.87	.0021	−2.38	.0087	−1.89	.0294	−1.40	.0808
−3.84	.00006	−3.35	.0004	−2.86	.0021	−2.37	.0089	−1.88	.0301	−1.39	.0823
−3.83	.00006	−3.34	.0004	−2.85	.0022	−2.36	.0091	−1.87	.0307	−1.38	.0838
−3.82	.00007	−3.33	.0004	−2.84	.0023	−2.35	.0094	−1.86	.0314	−1.37	.0853
−3.81	.00007	−3.32	.0004	−2.83	.0023	−2.34	.0096	−1.85	.0322	−1.36	.0869
−3.80	.00007	−3.31	.0004	−2.82	.0024	−2.33	.0099	−1.84	.0329	−1.35	.0885
−3.79	.00008	−3.30	.0004	−2.81	.0025	−2.32	.0102	−1.83	.0336	−1.34	.0901
−3.78	.00008	−3.29	.0005	−2.80	.0026	−2.31	.0104	−1.82	.0344	−1.33	.0918
−3.77	.00008	−3.28	.0005	−2.79	.0026	−2.30	.0107	−1.81	.0352	−1.32	.0934
−3.76	.00008	−3.27	.0005	−2.78	.0027	−2.29	.0110	−1.80	.0359	−1.31	.0951
−3.75	.00009	−3.26	.0006	−2.77	.0028	−2.28	.0113	−1.79	.0367	−1.30	.0968
−3.74	.00009	−3.25	.0006	−2.76	.0029	−2.27	.0116	−1.78	.0375	−1.29	.0985
−3.73	.0001	−3.24	.0006	−2.75	.0030	−2.26	.0119	−1.77	.0384	−1.28	.1003
−3.72	.0001	−3.23	.0006	−2.74	.0031	−2.25	.0122	−1.76	.0392	−1.27	.1020
−3.71	.0001	−3.22	.0006	−2.73	.0032	−2.24	.0126	−1.75	.0401	−1.26	.1038
−3.70	.0001	−3.21	.0006	−2.72	.0033	−2.23	.0129	−1.74	.0409	−1.25	.1057
−3.69	.0001	−3.20	.0006	−2.71	.0034	−2.22	.0132	−1.73	.0418	−1.24	.1075
−3.68	.0001	−3.19	.0007	−2.70	.0035	−2.21	.0136	−1.72	.0427	−1.23	.1094
−3.67	.0001	−3.18	.0007	−2.69	.0036	−2.20	.0139	−1.71	.0436	−1.22	.1112
−3.66	.0001	−3.17	.0008	−2.68	.0037	−2.19	.0143	−1.70	.0446	−1.21	.1131
−3.65	.0001	−3.16	.0008	−2.67	.0038	−2.18	.0146	−1.69	.0455	−1.20	.1151
−3.64	.0001	−3.15	.0008	−2.66	.0039	−2.17	.0150	−1.68	.0465	−1.19	.1170
−3.63	.0001	−3.14	.0008	−2.65	.0040	−2.16	.0154	−1.67	.0475	−1.18	.1190
−3.62	.0002	−3.13	.0008	−2.64	.0042	−2.15	.0158	−1.66	.0485	−1.17	.1210
−3.61	.0002	−3.12	.0009	−2.63	.0043	−2.14	.0162	−1.65	.0495	−1.16	.1230
−3.60	.0002	−3.11	.0009	−2.62	.0044	−2.13	.0166	−1.64	.0505	−1.15	.1251
−3.59	.0002	−3.10	.0010	−2.61	.0045	−2.12	.0170	−1.63	.0516	−1.14	.1271
−3.58	.0002	−3.09	.0010	−2.60	.0047	−2.11	.0174	−1.62	.0526	−1.13	.1292
−3.57	.0002	−3.08	.0010	−2.59	.0048	−2.10	.0179	−1.61	.0537	−1.12	.1314
−3.56	.0002	−3.07	.0011	−2.58	.0049	−2.09	.0183	−1.60	.0548	−1.11	.1335
−3.55	.0002	−3.06	.0011	−2.57	.0051	−2.08	.0188	−1.59	.0559	−1.10	.1357
−3.54	.0002	−3.05	.0011	−2.56	.0052	−2.07	.0192	−1.58	.0571	−1.09	.1379
−3.53	.0002	−3.04	.0012	−2.55	.0054	−2.06	.0197	−1.57	.0582	−1.08	.1401
−3.52	.0002	−3.03	.0012	−2.54	.0055	−2.05	.0202	−1.56	.0594	−1.07	.1423

TABLE A3

(cont.)

z	$P(Z < z)$	z	$P(Z < z)$	z	$P(Z < z)$	z	$P(Z < z)$	z	$P(Z < z)$	z	$P(Z < z)$
−1.06	.1446	− .52	.3015	0.00	.5000	0.54	.7054	1.08	.8599	1.62	.9474
−1.05	.1469	− .51	.3050	0.01	.5040	0.55	.7088	1.09	.8621	1.63	.9485
−1.04	.1492	− .50	.3085	0.02	.5080	0.56	.7123	1.10	.8643	1.64	.9495
−1.03	.1515	− .49	.3121	0.03	.5120	0.57	.7157	1.11	.8665	1.65	.9505
−1.02	.1539	− .48	.3156	0.04	.5160	0.58	.7190	1.12	.8686	1.66	.9515
−1.01	.1563	− .47	.3192	0.05	.5199	0.59	.7224	1.13	.8708	1.67	.9525
−1.00	.1587	− .46	.3228	0.06	.5239	0.60	.7257	1.14	.8729	1.68	.9535
− .99	.1611	− .45	.3264	0.07	.5279	0.61	.7291	1.15	.8749	1.69	.9545
− .98	.1635	− .44	.3300	0.08	.5319	0.62	.7324	1.16	.8770	1.70	.9554
− .97	.1660	− .43	.3336	0.09	.5359	0.63	.7357	1.17	.8790	1.71	.9564
− .96	.1685	− .42	.3372	0.10	.5398	0.64	.7389	1.18	.8810	1.72	.9573
− .95	.1711	− .41	.3409	0.11	.5438	0.65	.7422	1.19	.8830	1.73	.9582
− .94	.1736	− .40	.3446	0.12	.5478	0.66	.7454	1.20	.8849	1.74	.9591
− .93	.1762	− .39	.3483	0.13	.5517	0.67	.7486	1.21	.8869	1.75	.9599
− .92	.1788	− .38	.3520	0.14	.5557	0.68	.7517	1.22	.8888	1.76	.9608
− .91	.1814	− .37	.3557	0.15	.5596	0.69	.7549	1.23	.8907	1.77	.9616
− .90	.1841	− .36	.3594	0.16	.5636	0.70	.7580	1.24	.8925	1.78	.9625
− .89	.1867	− .35	.3632	0.17	.5675	0.71	.7611	1.25	.8944	1.79	.9633
− .88	.1894	− .34	.3669	0.18	.5714	0.72	.7642	1.26	.8962	1.80	.9641
− .87	.1922	− .33	.3707	0.19	.5753	0.73	.7673	1.27	.8980	1.81	.9649
− .86	.1949	− .32	.3745	0.20	.5793	0.74	.7704	1.28	.8997	1.82	.9656
− .85	.1977	− .31	.3783	0.21	.5832	0.75	.7734	1.29	.9015	1.83	.9664
− .84	.2005	− .30	.3821	0.22	.5871	0.76	.7764	1.30	.9032	1.84	.9671
− .83	.2033	− .29	.3859	0.23	.5910	0.77	.7794	1.31	.9049	1.85	.9678
− .82	.2061	− .28	.3897	0.24	.5948	0.78	.7823	1.32	.9066	1.86	.9686
− .81	.2090	− .27	.3936	0.25	.5987	0.79	.7852	1.33	.9082	1.87	.9693
− .80	.2119	− .26	.3974	0.26	.6026	0.80	.7881	1.34	.9099	1.88	.9699
− .79	.2148	− .25	.4013	0.27	.6064	0.81	.7910	1.35	.9115	1.89	.9706
− .78	.2177	− .24	.4052	0.28	.6103	0.82	.7939	1.36	.9131	1.90	.9713
− .77	.2207	− .23	.4090	0.29	.6141	0.83	.7967	1.37	.9147	1.91	.9719
− .76	.2236	− .22	.4129	0.30	.6179	0.84	.7995	1.38	.9162	1.92	.9726
− .75	.2266	− .21	.4168	0.31	.6217	0.85	.8023	1.39	.9177	1.93	.9732
− .74	.2297	− .20	.4207	0.32	.6255	0.86	.8051	1.40	.9192	1.94	.9738
− .73	.2327	− .19	.4247	0.33	.6293	0.87	.8079	1.41	.9207	1.95	.9744
− .72	.2358	− .18	.4286	0.34	.6331	0.88	.8106	1.42	.9222	1.96	.9750
− .71	.2389	− .17	.4325	0.35	.6368	0.89	.8133	1.43	.9236	1.97	.9756
− .70	.2420	− .16	.4364	0.36	.6406	0.90	.8159	1.44	.9251	1.98	.9761
− .69	.2451	− .15	.4404	0.37	.6443	0.91	.8186	1.45	.9265	1.99	.9767
− .68	.2483	− .14	.4443	0.38	.6480	0.92	.8212	1.46	.9279	2.00	.9773
− .67	.2514	− .13	.4483	0.39	.6517	0.93	.8238	1.47	.9292	2.01	.9778
− .66	.2546	− .12	.4522	0.40	.6554	0.94	.8264	1.48	.9306	2.02	.9783
− .65	.2579	− .11	.4562	0.41	.6591	0.95	.8289	1.49	.9319	2.03	.9788
− .64	.2611	− .10	.4602	0.42	.6628	0.96	.8315	1.50	.9332	2.04	.9793
− .63	.2644	− .09	.4641	0.43	.6664	0.97	.8340	1.51	.9345	2.05	.9798
− .62	.2676	− .08	.4681	0.44	.6700	0.98	.8365	1.52	.9357	2.06	.9803
− .61	.2709	− .07	.4721	0.45	.6736	0.99	.8389	1.53	.9370	2.07	.9808
− .60	.2743	− .06	.4761	0.46	.6772	1.00	.8413	1.54	.9382	2.08	.9812
− .59	.2776	− .05	.4801	0.47	.6808	1.01	.8437	1.55	.9394	2.09	.9817
− .58	.2810	− .04	.4840	0.48	.6844	1.02	.8461	1.56	.9406	2.10	.9821
− .57	.2843	− .03	.4880	0.49	.6879	1.03	.8485	1.57	.9418	2.11	.9826
− .56	.2877	− .02	.4920	0.50	.6915	1.04	.8508	1.58	.9429	2.12	.9830
− .55	.2912	− .01	.4960	0.51	.6950	1.05	.8531	1.59	.9441	2.13	.9834
− .54	.2946	0.00	.5000	0.52	.6985	1.06	.8554	1.60	.9452	2.14	.9838
− .53	.2981			0.53	.7019	1.07	.8577	1.61	.9463	2.15	.9842

TABLE A3

(cont.)

z	$P(Z < z)$	z	$P(Z < z)$	z	$P(Z < z)$	z	$P(Z < z)$	z	$P(Z < z)$	z	$P(Z < z)$
2.16	.9846	2.47	.9932	2.78	.9973	3.09	.9990	3.40	.9997	3.71	.9999
2.17	.9850	2.48	.9934	2.79	.9974	3.10	.9990	3.41	.9997	3.72	.9999
2.18	.9854	2.49	.9936	2.80	.9974	3.11	.9991	3.42	.9997	3.73	.9999
2.19	.9857	2.50	.9938	2.81	.9975	3.12	.9991	3.43	.9997	3.74	.9999
2.20	.9861	2.51	.9940	2.82	.9976	3.13	.9991	3.44	.9997	3.75	.9999
2.21	.9864	2.52	.9941	2.83	.9977	3.14	.9992	3.45	.9997	3.76	.9999
2.22	.9868	2.53	.9943	2.84	.9977	3.15	.9992	3.46	.9997	3.77	.9999
2.23	.9871	2.54	.9945	2.85	.9978	3.16	.9992	3.47	.9997	3.78	.9999
2.24	.9875	2.55	.9946	2.86	.9979	3.17	.9992	3.48	.9998	3.79	.9999
2.25	.9878	2.56	.9948	2.87	.9980	3.18	.9993	3.49	.9998	3.80	.9999
2.26	.9881	2.57	.9949	2.88	.9980	3.19	.9993	3.50	.9998	3.81	.9999
2.27	.9884	2.58	.9951	2.89	.9981	3.20	.9993	3.51	.9998	3.82	.9999
2.28	.9887	2.59	.9952	2.90	.9981	3.21	.9993	3.52	.9998	3.83	.9999
2.29	.9890	2.60	.9953	2.91	.9982	3.22	.9994	3.53	.9998	3.84	.9999
2.30	.9893	2.61	.9955	2.92	.9983	3.23	.9994	3.54	.9998	3.85	.9999
2.31	.9896	2.62	.9956	2.93	.9983	3.24	.9994	3.55	.9998	3.86	.9999
2.32	.9898	2.63	.9957	2.94	.9984	3.25	.9994	3.56	.9998	3.87	1.0000
2.33	.9901	2.64	.9959	2.95	.9984	3.26	.9994	3.57	.9998	3.88	1.0000
2.34	.9904	2.65	.9960	2.96	.9985	3.27	.9995	3.58	.9998	3.89	1.0000
2.35	.9906	2.66	.9961	2.97	.9985	3.28	.9995	3.59	.9998	3.90	1.0000
2.36	.9909	2.67	.9962	2.98	.9986	3.29	.9995	3.60	.9998	3.91	1.0000
2.37	.9911	2.68	.9963	2.99	.9986	3.30	.9995	3.61	.9999	3.92	1.0000
2.38	.9913	2.69	.9964	3.00	.9987	3.31	.9995	3.62	.9999	3.93	1.0000
2.39	.9916	2.70	.9965	3.01	.9987	3.32	.9996	3.63	.9999	3.94	1.0000
2.40	.9918	2.71	.9966	3.02	.9987	3.33	.9996	3.64	.9999	3.95	1.0000
2.41	.9920	2.72	.9967	3.03	.9988	3.34	.9996	3.65	.9999	3.96	1.0000
2.42	.9922	2.73	.9968	3.04	.9988	3.35	.9996	3.66	.9999	3.97	1.0000
2.43	.9925	2.74	.9969	3.05	.9989	3.36	.9996	3.67	.9999	3.98	1.0000
2.44	.9927	2.75	.9970	3.06	.9989	3.37	.9996	3.68	.9999	3.99	1.0000
2.45	.9929	2.76	.9971	3.07	.9989	3.38	.9996	3.69	.9999	4.00	1.0000
2.46	.9931	2.77	.9972	3.08	.9990	3.39	.9997	3.70	.9999		

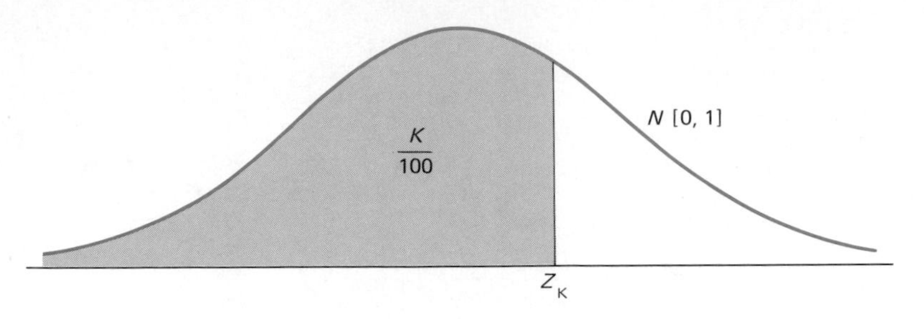

TABLE A4

Percentiles of the standardized normal distribution

K	$P(Z < Z_K) = \frac{K}{100}$	Z_K	K	$P(Z < Z_K) = \frac{K}{100}$	Z_K
.25	.0025	−2.807	50	.50	0
.5	.005	−2.576	55	.55	.1256
.75	.0075	−2.432	60	.60	.2533
1	.01	−2.326	65	.65	.3854
1.25	.0125	−2.241	70	.70	.5244
1.5	.015	−2.17	75	.75	.674
1.75	.0175	−2.108	80	.80	.842
2	.02	−2.054	85	.85	1.036
2.3	.023	−2.00	87.5	.875	1.15
2.5	.025	−1.960	90	.90	1.282
3	.03	−1.881	92.5	.925	1.44
4	.04	−1.751	95	.950	1.645
5	.05	−1.645	96	.960	1.751
7.5	.075	−1.440	97	.970	1.881
10	.10	−1.282	97.5	.975	1.96
12.5	.125	−1.150	97.7	.977	2.00
15	.15	−1.036	98	.98	2.054
20	.20	− .842	98.25	.9825	2.108
25	.25	− .674	98.5	.985	2.17
30	.30	− .5244	98.75	.9875	2.241
35	.35	− .3854	99	.99	2.326
40	.40	− .2533	99.25	.9925	2.432
45	.45	− .1256	99.5	.9950	2.576
50	.50	0	99.75	.9975	2.807

TABLE A5

Central Z values

γ	$Z_{[\gamma]}$
.995	2.807
.99	2.576
.98	2.326
.97	2.170
.96	2.054
.95	1.960
.94	1.881
.92	1.751
.90	1.645
.85	1.440
.80	1.282
.75	1.150
.70	1.036
.60	.842
.50	.674

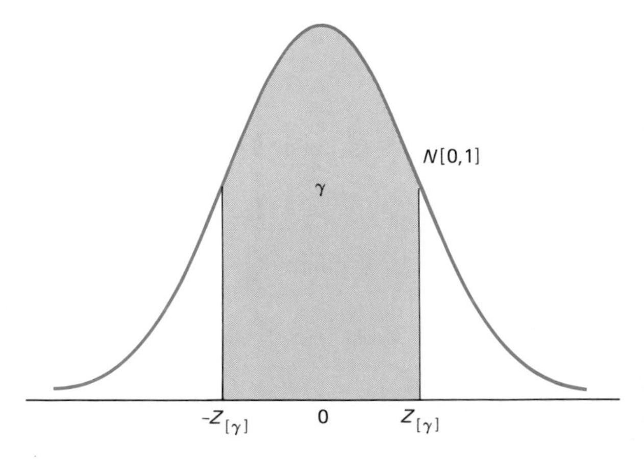

TABLE A6

The t distribution

	$t_{\nu,.25}$ $t_{\nu(.5)}$	$t_{\nu,.2}$ $t_{\nu(.6)}$	$t_{\nu,.15}$ $t_{\nu(.7)}$	$t_{\nu,.1}$ $t_{\nu(.8)}$	$t_{\nu,.05}$ $t_{\nu(.9)}$	$t_{\nu,.025}$ $t_{\nu(.95)}$	$t_{\nu,.01}$ $t_{\nu(.98)}$	$t_{\nu,.005}$ $t_{\nu(.99)}$
1	1.000	1.376	1.963	3.078	6.314	12.706	31.821	63.657
2	.817	1.061	1.386	1.886	2.920	4.303	6.965	9.925
3	.765	.978	1.250	1.638	2.353	3.183	4.541	5.841
4	.741	.941	1.190	1.533	2.132	2.776	3.747	4.604
5	.727	.920	1.156	1.476	2.015	2.571	3.365	4.032
6	.718	.906	1.134	1.440	1.943	2.447	3.143	3.707
7	.711	.896	1.119	1.415	1.895	2.365	2.998	3.500
8	.706	.889	1.108	1.397	1.860	2.306	2.896	3.355
9	.703	.883	1.100	1.383	1.833	2.262	2.821	3.250
10	.700	.879	1.093	1.372	1.813	2.228	2.764	3.169
11	.697	.876	1.088	1.363	1.796	2.201	2.718	3.106
12	.696	.873	1.083	1.356	1.782	2.179	2.681	3.055
13	.694	.870	1.079	1.350	1.771	2.160	2.650	3.012
14	.692	.868	1.076	1.345	1.761	2.145	2.624	2.977
15	.691	.866	1.074	1.341	1.753	2.132	2.602	2.947
16	.690	.865	1.071	1.337	1.746	2.120	2.583	2.921
17	.689	.863	1.069	1.333	1.740	2.110	2.567	2.898
18	.688	.862	1.067	1.330	1.734	2.101	2.552	2.878
19	.688	.861	1.066	1.328	1.729	2.093	2.539	2.861
20	.687	.860	1.064	1.325	1.725	2.086	2.528	2.845
21	.686	.859	1.063	1.323	1.721	2.080	2.518	2.831
22	.686	.858	1.061	1.321	1.717	2.074	2.508	2.819
23	.685	.858	1.060	1.319	1.714	2.069	2.500	2.807
24	.685	.857	1.059	1.318	1.711	2.064	2.492	2.797
25	.684	.856	1.058	1.316	1.708	2.060	2.485	2.787
26	.684	.856	1.058	1.315	1.706	2.056	2.479	2.779
27	.684	.855	1.057	1.314	1.703	2.052	2.473	2.771
28	.683	.855	1.056	1.313	1.701	2.048	2.467	2.763
29	.683	.854	1.055	1.311	1.699	2.045	2.462	2.756
30	.683	.854	1.055	1.310	1.697	2.042	2.457	2.750
31	.683	.854	1.054	1.310	1.696	2.040	2.453	2.744
32	.682	.853	1.054	1.309	1.694	2.037	2.449	2.739
33	.682	.853	1.053	1.308	1.692	2.035	2.445	2.733
34	.682	.852	1.053	1.307	1.691	2.032	2.441	2.728
35	.682	.852	1.052	1.306	1.690	2.030	2.438	2.724
36	.681	.852	1.052	1.306	1.688	2.028	2.434	2.720
37	.681	.852	1.051	1.305	1.687	2.026	2.431	2.716
38	.681	.851	1.051	1.304	1.686	2.024	2.428	2.712
39	.681	.851	1.050	1.304	1.685	2.023	2.426	2.708
40	.681	.851	1.050	1.303	1.684	2.021	2.423	2.705
50	.679	.849	1.047	1.299	1.676	2.009	2.403	2.678
60	.679	.848	1.046	1.296	1.671	2.000	2.390	2.660
70	.678	.847	1.044	1.294	1.667	1.995	2.381	2.648
80	.678	.846	1.043	1.292	1.664	1.990	2.374	2.639
90	.677	.846	1.043	1.291	1.662	1.987	2.368	2.632
100	.677	.845	1.042	1.290	1.660	1.984	2.364	2.626
∞	.674	.842	1.036	1.282	1.645	1.960	2.326	2.576

TABLE A7

The chi-square distribution

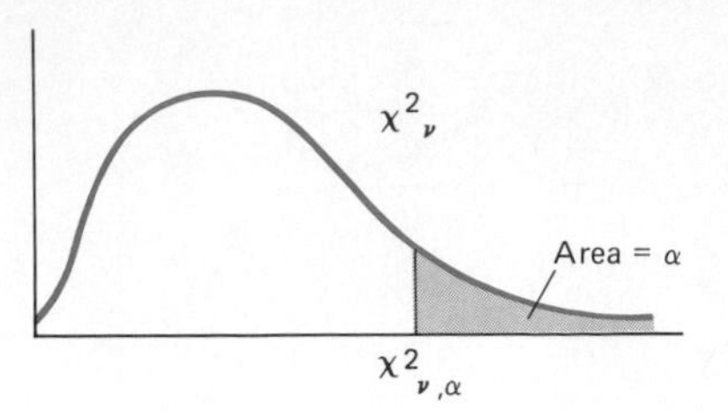

ν	$\chi^2_{\nu,.995}$	$\chi^2_{\nu,.99}$	$\chi^2_{\nu,.975}$	$\chi^2_{\nu,.95}$	$\chi^2_{\nu,.90}$	$\chi^2_{\nu,.10}$	$\chi^2_{\nu,.05}$	$\chi^2_{\nu,.025}$	$\chi^2_{\nu,.01}$	$\chi^2_{\nu,.005}$
1	.00	.00	.00	.00	.02	2.71	3.84	5.02	6.63	7.88
2	.01	.02	.05	.10	.21	4.61	5.99	7.38	9.21	10.60
3	.07	.11	.22	.35	.58	6.25	7.81	9.35	11.34	12.84
4	.21	.30	.48	.71	1.06	7.78	9.49	11.14	13.28	14.86
5	.41	.55	.83	1.15	1.61	9.24	11.07	12.83	15.09	16.75
6	.68	.87	1.24	1.64	2.20	10.64	12.59	14.45	16.81	18.55
7	.99	1.24	1.69	2.17	2.83	12.02	14.07	16.01	18.48	20.28
8	1.34	1.65	2.18	2.73	3.49	13.36	15.51	17.54	20.09	21.96
9	1.73	2.09	2.70	3.33	4.17	14.68	16.92	19.02	21.67	23.59
10	2.16	2.56	3.25	3.94	4.87	15.99	18.31	20.48	23.21	25.19
11	2.60	3.05	3.82	4.57	5.58	17.28	19.68	21.92	24.72	26.76
12	3.07	3.57	4.40	5.23	6.30	18.55	21.03	23.34	26.22	28.30
13	3.57	4.11	5.01	5.89	7.04	19.81	22.36	24.74	27.69	29.82
14	4.07	4.66	5.63	6.57	7.79	21.06	23.68	26.12	29.14	31.32
15	4.60	5.23	6.26	7.26	8.55	22.31	25.00	27.49	30.58	32.80
16	5.14	5.81	6.91	7.96	9.31	23.54	26.30	28.85	32.00	34.27
17	5.70	6.41	7.56	8.67	10.09	24.77	27.59	30.19	33.41	35.72
18	6.26	7.01	8.23	9.39	10.86	25.99	28.87	31.53	34.81	37.16
19	6.84	7.63	8.91	10.12	11.65	27.20	30.14	32.85	36.19	38.58
20	7.43	8.26	9.59	10.85	12.44	28.41	31.41	34.17	37.57	40.00
21	8.03	8.90	10.28	11.59	13.24	29.62	32.67	35.48	38.93	41.40
22	8.64	9.54	10.98	12.34	14.04	30.81	33.92	36.78	40.29	42.80
23	9.26	10.20	11.69	13.09	14.85	32.01	35.17	38.08	41.64	44.18
24	9.89	10.86	12.40	13.85	15.66	33.20	36.42	39.36	42.98	45.56
25	10.52	11.52	13.12	14.61	16.47	34.38	37.65	40.65	44.31	46.93
26	11.16	12.20	13.84	15.38	17.29	35.56	38.89	41.92	45.64	48.29
27	11.81	12.88	14.57	16.15	18.11	36.74	40.11	43.19	46.96	49.65
28	12.46	13.56	15.31	16.93	18.94	37.92	41.34	44.46	48.28	50.99
29	13.12	14.26	16.05	17.71	19.77	39.09	42.56	45.72	49.59	52.34
30	13.79	14.95	16.79	18.49	20.60	40.26	43.77	46.98	50.89	53.67
50	27.99	29.71	32.36	34.76	37.69	63.17	67.50	71.42	76.15	79.49
100	67.33	70.06	74.22	77.93	82.36	118.5	124.3	129.6	135.8	140.2
500	422.3	429.4	439.9	449.1	459.9	540.9	553.1	563.9	576.5	585.2
1000	888.6	898.8	914.3	927.6	943.1	1058	1075	1090	1107	1119

Source: Adapted from D. B. Owen, *Handbook of Statistical Tables.* Courtesy of the Atomic Energy Commission. Reading, Mass.: Addison-Wesley, 1962.

TABLE A8

The F distribution[a]

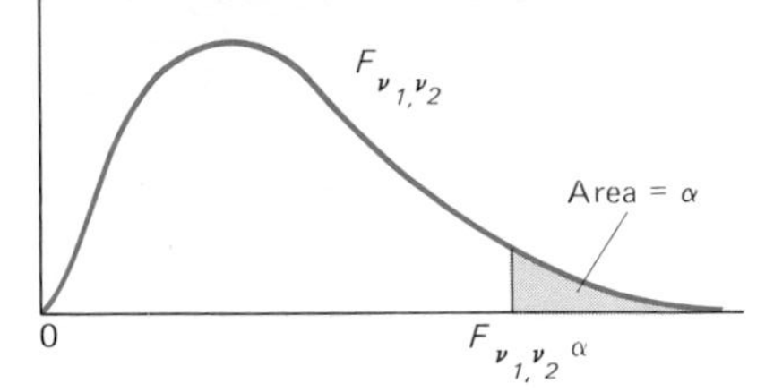

Entries in the table are $F_{\nu_1,\nu_2,\alpha}$

ν_2	α	ν_1 = 1	2	3	4	5	6	7	8	9	10	11	12	15	20	24	30
(1)	.25	5.83	7.50	8.20	8.58	8.82	8.98	9.10	9.19	9.26	9.32	9.36	9.41	9.49	9.58	9.63	9.67
	.1	39.9	49.5	53.6	55.8	57.2	58.2	58.9	59.4	59.9	60.2	60.5	60.7	61.2	61.7	62.0	62.3
	.05	161	200	216	225	230	234	237	239	241	242	243	244	246	248	249	250
	.025	648	800	864	900	922	937	948	957	963	969	973	977	985	993	997	100^1
	.01	405^1	500^1	540^1	562^1	576^1	586^1	593^1	598^1	602^1	606^1	608^1	611^1	616^1	621^1	623^1	626^1
	.005	162^2	200^2	216^2	225^2	231^2	234^2	237^2	239^2	241^2	242^2	243^2	244^2	246^2	248^2	249^2	250^2
(2)	.25	2.57	3.00	3.15	3.23	3.28	3.31	3.34	3.35	3.37	3.38	3.39	3.39	3.41	3.43	3.43	3.44
	.1	8.53	9.00	9.16	9.24	9.29	9.33	9.35	9.37	9.38	9.39	9.40	9.41	9.42	9.44	9.45	9.46
	.05	18.5	19.0	19.2	19.2	19.3	19.3	19.4	19.4	19.4	19.4	19.4	19.4	19.4	19.4	19.5	19.5
	.025	38.5	39.0	39.2	39.2	39.3	39.3	39.4	39.4	39.4	39.4	39.4	39.4	39.4	39.4	39.5	39.5
	.01	98.5	99.0	99.2	99.2	99.3	99.3	99.4	99.4	99.4	99.4	99.4	99.4	99.4	99.4	99.5	99.5
	.005	198	199	199	199	199	199	199	199	199	199	199	199	199	199	199	199
(3)	.25	2.02	2.28	2.36	2.39	2.41	2.42	2.43	2.44	2.44	2.44	2.45	2.45	2.46	2.46	2.46	2.47
	.1	5.54	5.46	5.39	5.34	5.31	5.28	5.27	5.25	5.24	5.23	5.22	5.22	5.20	5.18	5.18	5.17
	.05	10.1	9.55	9.28	9.12	9.01	8.94	8.89	8.85	8.81	8.79	8.76	8.74	8.70	8.66	8.63	8.62
	.025	17.4	16.0	15.4	15.1	14.9	14.7	14.6	14.5	14.5	14.4	14.4	14.3	14.3	14.2	14.1	14.1
	.01	34.1	30.8	29.5	28.7	28.2	27.9	27.7	27.5	27.3	27.2	27.1	27.1	26.9	26.7	26.6	26.5
	.005	55.6	49.8	47.5	46.2	45.4	44.8	44.4	44.1	43.9	43.7	43.5	43.4	43.1	42.8	42.6	42.5
(4)	.25	1.81	2.00	2.05	2.06	2.07	2.08	2.08	2.08	2.08	2.08	2.08	2.08	2.08	2.08	2.08	2.08
	.1	4.54	4.32	4.19	4.11	4.05	4.01	3.98	3.95	3.94	3.92	3.91	3.90	3.87	3.84	3.38	3.82
	.05	7.71	6.94	6.59	6.39	6.26	6.16	6.09	6.04	6.00	5.96	5.94	5.91	5.86	5.80	5.77	5.75
	.025	12.2	10.6	9.98	9.60	9.36	9.20	9.07	8.98	8.90	8.84	8.79	8.75	8.66	8.56	8.51	8.46
	.01	21.2	18.0	16.7	16.0	15.5	15.2	15.0	14.8	14.7	14.5	14.4	14.4	14.2	14.0	13.9	13.8
	.005	31.3	26.3	24.3	23.2	22.5	22.0	21.6	21.4	21.1	21.0	20.8	20.7	20.4	20.2	20.0	19.9
(5)	.25	1.69	1.85	1.88	1.89	1.89	1.89	1.89	1.89	1.89	1.89	1.89	1.89	1.89	1.88	1.88	1.88
	.1	4.06	3.78	3.62	3.52	3.45	3.40	3.37	3.34	3.32	3.30	3.28	3.27	3.24	3.21	3.19	3.17
	.05	6.61	5.79	5.41	5.19	5.05	4.95	4.88	4.82	4.77	4.74	4.71	4.68	4.62	4.56	4.53	4.50
	.025	10.0	8.43	7.76	7.39	7.15	6.98	6.85	6.76	6.68	6.62	6.57	6.52	6.43	6.33	6.28	6.23
	.01	16.3	13.3	12.1	11.4	11.0	10.7	10.5	10.3	10.2	10.1	9.96	9.89	9.72	9.55	9.47	9.38
	.005	22.8	18.3	16.5	15.6	14.9	14.5	14.2	14.0	13.8	13.6	13.5	13.4	13.1	12.9	12.8	12.7

[a] Read 562^1 as 5620, 231^2 as 23,100 and so on.

TABLE A8

(cont.)

ν_2	α	ν_1 = 1	2	3	4	5	6	7	8	9	10	11	12	15	20	24	30
(6)	.25	1.62	1.76	1.78	1.79	1.79	1.78	1.78	1.78	1.77	1.77	1.77	1.77	1.76	1.76	1.75	1.75
	.1	3.78	3.46	3.29	3.18	3.11	3.05	3.01	2.98	2.96	2.94	2.92	2.90	2.87	2.84	2.82	2.80
	.05	5.99	5.14	4.76	4.53	4.39	4.28	4.21	4.15	4.10	4.06	4.03	4.00	3.94	3.87	3.84	3.81
	.025	8.81	7.26	6.60	6.23	5.99	5.82	5.70	5.60	5.52	5.46	5.41	5.37	5.27	5.17	5.12	5.07
	.01	13.7	10.9	9.78	9.15	8.75	8.47	8.26	8.10	7.98	7.87	7.79	7.72	7.56	7.40	7.31	7.23
	.005	18.6	14.5	12.9	12.0	11.5	11.1	10.8	10.6	10.4	10.2	10.1	10.0	9.81	9.59	9.47	9.36
(7)	.25	1.57	1.70	1.72	1.72	1.71	1.71	1.70	1.70	1.69	1.69	1.69	1.68	1.68	1.67	1.67	1.66
	.1	3.59	3.26	3.07	2.96	2.88	2.83	2.78	2.75	2.72	2.70	2.68	2.67	2.63	2.59	2.58	2.56
	.05	5.59	4.74	4.35	4.12	3.97	3.87	3.79	3.73	3.68	3.64	3.60	3.57	3.51	3.44	3.41	3.38
	.025	8.07	6.54	5.89	5.52	5.29	5.12	4.99	4.90	4.82	4.76	4.71	4.67	4.57	4.47	4.42	4.36
	.01	12.2	9.55	8.45	7.85	7.46	7.19	6.99	6.84	6.72	6.62	6.54	6.47	6.31	6.16	6.07	5.99
	.005	16.2	12.4	10.9	10.0	9.52	9.16	8.89	8.68	8.51	8.38	8.27	8.18	7.97	7.75	7.65	7.53
(8)	.25	1.54	1.66	1.67	1.66	1.66	1.65	1.64	1.64	1.64	1.63	1.63	1.62	1.62	1.61	1.60	1.60
	.1	3.46	3.11	2.92	2.81	2.73	2.67	2.62	2.59	2.56	2.54	2.52	2.50	2.46	2.42	2.40	2.38
	.05	5.32	4.46	4.07	3.84	3.69	3.58	3.50	3.44	3.39	3.35	3.31	3.28	3.22	3.15	3.12	3.08
	.025	7.57	6.06	5.42	5.05	4.82	4.65	4.53	4.43	4.36	4.30	4.24	4.20	4.10	4.00	3.95	3.89
	.01	11.3	8.65	7.59	7.01	6.63	6.37	6.18	6.03	5.91	5.81	5.73	5.67	5.52	5.36	5.28	5.20
	.005	14.7	11.0	9.60	8.81	8.30	7.95	7.69	7.50	7.34	7.21	7.10	7.01	6.81	6.61	6.50	6.40
(9)	.25	1.51	1.62	1.63	1.63	1.62	1.61	1.60	1.60	1.59	1.59	1.58	1.58	1.57	1.56	1.56	1.55
	.1	3.36	3.01	2.81	2.69	2.61	2.55	2.51	2.47	2.44	2.42	2.40	2.38	2.34	2.30	2.28	2.25
	.05	5.12	4.26	3.86	3.63	3.48	3.37	3.29	3.23	3.18	3.14	3.10	3.07	3.01	2.94	2.90	2.86
	.025	7.21	5.71	5.08	4.72	4.48	4.32	4.20	4.10	4.03	3.96	3.91	3.87	3.77	3.67	3.61	3.56
	.01	10.6	8.02	6.99	6.42	6.06	5.80	5.61	5.47	5.35	5.26	5.18	5.11	4.96	4.81	4.73	4.65
	.005	13.6	10.1	8.72	7.96	7.47	7.13	6.88	6.69	6.54	6.42	6.31	6.23	6.03	5.83	5.73	5.62
(10)	.25	1.49	1.60	1.60	1.59	1.59	1.58	1.57	1.56	1.56	1.55	1.55	1.54	1.53	1.52	1.52	1.51
	.1	3.28	2.92	2.73	2.61	2.52	2.46	2.41	2.38	2.35	2.32	2.30	2.28	2.24	2.20	2.18	2.16
	.05	4.96	4.10	3.71	3.48	3.33	3.22	3.14	3.07	3.02	2.98	2.94	2.91	2.85	2.77	2.74	2.70
	.025	6.94	5.46	4.83	4.47	4.24	4.07	3.95	3.85	3.78	3.72	3.66	2.62	3.52	3.42	3.37	3.31
	.01	10.0	7.56	6.55	5.99	5.64	5.39	5.20	5.06	4.94	4.85	4.77	4.71	4.56	4.41	4.33	4.25
	.005	12.8	9.43	8.08	7.34	6.87	6.54	6.30	6.12	5.97	5.85	5.75	5.66	5.47	5.27	5.17	5.07
(11)	.25	1.47	1.58	1.58	1.57	1.56	1.55	1.54	1.53	1.53	1.52	1.52	1.51	1.50	1.49	1.49	1.48
	.1	3.23	2.86	2.66	2.54	2.45	2.39	2.34	2.30	2.27	2.25	2.23	2.21	2.17	2.12	2.10	2.08
	.05	4.84	3.98	3.59	3.36	3.20	3.09	3.01	2.95	2.90	2.85	2.82	2.79	2.72	2.65	2.61	2.57
	.025	6.72	5.26	4.63	4.28	4.04	3.88	3.76	3.66	3.59	3.53	3.47	3.43	3.33	3.23	3.17	3.12
	.01	9.65	7.21	6.22	5.67	5.32	5.07	4.89	4.74	4.63	4.54	4.46	4.40	4.25	4.10	4.02	3.94
	.005	12.2	8.91	7.60	6.88	6.42	6.10	5.86	5.68	5.54	5.42	5.32	5.24	5.05	4.86	4.76	4.65

(12)	.25	1.46	1.56	1.56	1.55	1.54	1.53	1.52	1.51	1.51	1.50	1.50	1.49	1.48	1.47	1.46	1.45
	.1	3.18	2.81	2.61	2.48	2.39	2.33	2.28	2.24	2.21	2.19	2.17	2.15	2.11	2.06	2.04	2.01
	.05	4.75	3.89	3.49	3.26	3.11	3.00	2.91	2.85	2.80	2.75	2.72	2.69	2.62	2.54	2.51	2.47
	.025	6.55	5.10	4.47	4.12	3.89	3.73	3.61	3.51	3.44	3.37	3.32	3.28	3.18	3.07	3.02	2.96
	.01	9.33	6.93	5.95	5.41	5.06	4.82	4.64	4.50	4.39	4.30	4.22	4.16	4.01	3.86	3.78	3.70
	.005	11.8	8.51	7.23	6.52	6.07	5.76	5.52	5.35	5.20	5.09	4.99	4.91	4.72	4.53	4.43	4.33
(13)	.25	1.45	1.55	1.55	1.53	1.52	1.51	1.50	1.49	1.49	1.48	1.48	1.47	1.46	1.45	1.44	1.43
	.1	3.14	2.76	2.56	2.43	2.35	2.28	2.23	2.20	2.16	2.14	2.12	2.10	2.05	2.01	1.96	1.96
	.05	4.67	3.81	3.41	3.18	3.03	2.92	2.83	2.77	2.71	2.67	2.64	2.60	2.53	2.46	2.42	2.38
	.025	6.41	4.97	4.35	4.00	3.77	3.60	3.48	3.39	3.31	3.25	3.20	3.15	3.05	2.95	2.89	2.84
	.01	9.07	6.70	5.74	5.21	4.86	4.62	4.44	4.30	4.19	4.10	3.98	3.96	3.82	3.67	3.59	3.51
	.005	11.4	8.19	6.93	6.23	5.79	5.48	5.25	5.08	4.94	4.82	4.73	4.64	4.46	4.27	4.17	4.07
(14)	.25	1.44	1.53	1.53	1.52	1.51	1.50	1.49	1.48	1.47	1.46	1.46	1.45	1.44	1.43	1.42	1.41
	.1	3.10	2.73	2.52	2.39	2.31	2.24	2.19	2.15	2.12	2.10	2.07	2.05	2.01	1.96	1.94	1.91
	.05	4.60	3.74	3.34	3.11	2.96	2.85	2.76	2.70	2.65	2.60	2.58	2.53	2.46	2.39	2.35	2.31
	.025	6.30	4.86	4.24	3.89	3.66	3.50	3.38	3.29	3.21	3.15	3.10	3.05	2.95	2.84	2.79	2.73
	.01	8.88	6.51	5.56	5.04	4.69	4.46	4.28	4.14	4.03	3.94	3.87	3.80	3.66	3.50	3.43	3.35
	.005	11.1	7.92	6.68	6.00	5.56	5.26	5.03	4.86	4.72	4.60	4.52	4.43	4.25	4.06	3.96	3.86
(15)	.25	1.43	1.52	1.52	1.51	1.49	1.48	1.47	1.46	1.46	1.45	1.44	1.44	1.43	1.41	1.41	1.40
	.1	3.07	2.70	2.49	2.36	2.27	2.21	2.16	2.12	2.09	2.06	2.04	2.02	1.97	1.92	1.90	1.87
	.05	4.54	3.68	3.29	3.06	2.90	2.79	2.71	2.64	2.59	2.54	2.51	2.48	2.40	2.33	2.39	2.25
	.025	6.20	4.76	4.15	3.80	3.58	3.41	3.29	3.20	3.12	3.06	3.01	2.96	2.86	2.76	2.70	2.64
	.01	8.68	6.36	5.42	4.89	4.56	4.32	4.14	4.00	3.89	3.80	3.73	3.67	3.52	3.37	3.29	3.21
	.005	10.8	7.70	6.48	5.80	5.37	5.07	4.85	4.67	4.54	4.42	4.33	4.25	4.07	3.88	3.79	3.69
(16)	.25	1.42	1.51	1.51	1.50	1.48	1.47	1.46	1.45	1.44	1.44	1.43	1.43	1.41	1.40	1.39	1.38
	.1	3.05	2.67	2.46	2.33	2.24	2.18	2.13	2.09	2.06	2.03	2.01	1.98	1.94	1.89	1.87	1.84
	.05	4.49	3.63	3.24	3.01	2.85	2.74	2.66	2.59	2.54	2.49	2.46	2.42	2.35	2.28	2.24	2.19
	.025	6.12	4.69	4.08	3.73	3.50	3.34	3.22	3.12	3.05	2.99	2.94	2.89	2.79	2.68	2.62	2.57
	.01	8.53	6.23	5.29	4.77	4.44	4.20	4.03	3.89	3.78	3.69	3.62	3.55	3.41	3.26	3.18	3.10
	.005	10.6	7.51	6.30	5.64	5.21	4.91	4.69	4.52	4.38	4.27	4.18	4.10	3.92	3.73	3.64	3.54
(17)	.25	1.42	1.51	1.50	1.49	1.47	1.46	1.45	1.44	1.43	1.43	1.42	1.41	1.40	1.39	1.38	1.37
	.1	3.03	2.64	2.44	2.31	2.22	2.15	2.10	2.06	2.03	2.00	1.98	1.96	1.91	1.86	1.84	1.81
	.05	4.45	3.59	3.20	2.96	2.81	2.70	2.61	2.55	2.49	2.45	2.42	2.38	2.31	2.23	2.19	2.15
	.025	6.04	4.62	4.01	3.66	3.44	3.28	3.16	3.06	2.98	2.92	2.87	2.82	2.72	2.62	2.56	2.50
	.01	8.40	6.11	5.19	4.67	4.34	4.10	3.93	3.79	3.68	3.59	3.52	3.46	3.31	3.16	3.08	3.00
	.005	10.4	7.35	6.16	5.50	5.07	4.78	4.56	4.39	4.25	4.14	4.06	3.97	3.79	3.61	3.51	3.41
(18)	.25	1.41	1.50	1.49	1.48	1.46	1.45	1.44	1.43	1.42	1.42	1.41	1.40	1.39	1.38	1.37	1.36
	.1	3.01	2.62	2.42	2.29	2.20	2.13	2.08	2.04	2.00	1.98	1.96	1.93	1.89	1.84	1.81	1.78
	.05	4.41	3.55	3.16	2.93	2.77	2.66	2.58	2.51	2.46	2.41	2.38	2.34	2.27	2.19	2.15	2.11
	.025	5.98	4.56	3.95	3.61	3.38	3.22	3.10	3.01	2.93	2.87	2.81	2.77	2.67	2.56	2.50	2.44
	.01	8.29	6.01	5.09	4.58	4.25	4.01	3.84	3.71	3.60	3.51	3.44	3.37	3.23	3.08	3.00	2.92
	.005	10.2	7.21	6.03	5.37	4.96	4.66	4.44	4.28	4.14	4.03	3.94	3.86	3.68	3.50	3.40	3.30
(19)	.25	1.41	1.49	1.49	1.47	1.46	1.44	1.43	1.42	1.41	1.41	1.40	1.40	1.38	1.37	1.36	1.35
	.1	2.99	2.61	2.40	2.27	2.18	2.11	2.06	2.02	1.98	1.96	1.93	1.91	1.86	1.81	1.79	1.76

TABLE A8

(cont.)

ν_2	α	ν_1 = 1	2	3	4	5	6	7	8	9	10	11	12	15	20	24	30
	.05	4.38	3.52	3.13	2.90	2.74	2.63	2.54	2.48	2.42	2.38	2.34	2.31	2.23	2.16	2.11	2.07
	.025	5.92	4.51	3.90	3.56	3.33	3.17	3.05	2.96	2.88	2.82	2.77	2.72	2.62	2.51	2.45	2.39
	.01	8.18	5.93	5.01	4.50	4.17	3.94	3.77	3.63	3.52	3.43	3.36	3.30	3.15	3.00	2.92	2.84
	.005	10.1	7.09	5.92	5.27	4.85	4.56	4.34	4.18	4.04	3.93	3.84	3.76	3.59	3.40	3.31	3.21
(20)	.25	1.40	1.49	1.48	1.47	1.45	1.44	1.43	1.42	1.41	1.40	1.39	1.39	1.37	1.36	1.35	1.34
	.1	2.97	2.59	2.38	2.25	2.16	2.09	2.04	2.00	1.96	1.94	1.91	1.89	1.84	1.79	1.77	1.74
	.05	4.35	3.49	3.10	2.87	2.71	2.60	2.51	2.45	2.39	2.35	2.31	2.28	2.20	2.12	2.08	2.04
	.025	5.87	4.46	3.86	3.51	3.29	3.13	3.01	2.91	2.84	2.77	2.72	2.68	2.57	2.46	2.41	2.35
	.01	8.10	5.85	4.94	4.43	4.10	3.87	3.70	3.56	3.46	3.37	3.29	3.23	3.09	2.94	2.86	2.78
	.005	9.94	6.99	5.82	5.17	4.76	4.47	4.26	4.09	3.96	3.85	3.76	3.68	3.50	3.32	3.22	3.12
(24)	.25	1.39	1.47	1.46	1.44	1.43	1.41	1.40	1.39	1.38	1.38	1.37	1.36	1.35	1.33	1.32	1.31
	.1	2.93	2.54	2.33	2.19	2.10	2.04	1.98	1.94	1.91	1.88	1.85	1.83	1.78	1.73	1.70	1.67
	.05	4.26	3.40	3.01	2.78	2.62	2.51	2.42	2.36	2.30	2.25	2.21	2.18	2.11	2.03	1.98	1.94
	.025	5.72	4.32	3.72	3.38	3.15	2.99	2.87	2.78	2.70	2.64	2.59	2.54	2.44	2.33	2.27	2.21
	.01	7.82	5.61	4.72	4.22	3.90	3.67	3.50	3.36	3.26	3.17	3.09	3.03	2.89	2.74	2.66	2.58
	.005	9.55	6.66	5.52	4.89	4.49	4.20	3.99	3.83	3.69	3.59	3.50	3.42	3.25	3.06	2.97	2.87
(30)	.25	1.38	1.45	1.44	1.42	1.41	1.39	1.38	1.37	1.36	1.35	1.35	1.34	1.32	1.30	1.29	1.28
	.1	2.88	2.49	2.28	2.14	2.05	1.98	1.93	1.88	1.85	1.82	1.79	1.77	1.72	1.67	1.64	1.61
	.05	4.17	3.32	2.92	2.69	2.53	2.42	2.33	2.27	2.21	2.16	2.13	2.09	2.01	1.93	1.89	1.84
	.025	5.57	4.18	3.59	3.25	3.03	2.87	2.75	2.65	2.57	2.51	2.46	2.41	2.31	2.20	2.14	2.07
	.01	7.56	5.39	4.51	4.02	3.70	3.47	3.30	3.17	3.07	2.98	2.91	2.84	2.70	2.55	2.47	2.39
	.005	9.18	6.35	5.24	4.62	4.23	3.95	3.74	3.58	3.45	3.34	3.25	3.18	3.01	2.82	2.73	2.63
(40)	.25	1.36	1.44	1.42	1.40	1.39	1.37	1.36	1.35	1.34	1.33	1.32	1.31	1.30	1.28	1.26	1.25
	.1	2.84	2.44	2.23	2.09	2.00	1.93	1.87	1.83	1.79	1.76	1.73	1.71	1.66	1.61	1.57	1.54
	.05	4.08	3.23	2.84	2.61	2.45	2.34	2.25	2.18	2.12	2.08	2.04	2.00	1.92	1.84	1.79	1.74
	.025	5.42	4.05	3.46	3.13	2.90	2.74	2.62	2.53	2.45	2.39	2.33	2.29	2.18	2.07	2.01	1.94
	.01	7.31	5.18	4.31	3.83	3.51	3.29	3.12	2.99	2.89	2.80	2.73	2.66	2.52	2.37	2.29	2.20
	.005	8.83	6.07	4.98	4.37	3.99	3.71	3.51	3.35	3.22	3.12	3.03	2.95	2.78	2.60	2.50	2.40
(60)	.25	1.35	1.42	1.41	1.38	1.37	1.35	1.33	1.32	1.31	1.30	1.29	1.29	1.27	1.25	1.24	1.22
	.1	2.79	2.39	2.18	2.04	1.95	1.87	1.82	1.77	1.74	1.71	1.68	1.66	1.60	1.54	1.51	1.48
	.05	4.00	3.15	2.76	2.53	2.37	2.25	2.17	2.10	2.04	1.99	1.95	1.92	1.84	1.75	1.70	1.65
	.025	5.29	3.93	3.34	3.01	2.79	2.63	2.51	2.41	2.33	2.27	2.22	2.17	2.06	1.94	1.88	1.82
	.01	7.08	4.98	4.13	3.65	3.34	3.12	2.95	2.82	2.72	2.63	2.56	2.50	2.35	2.20	2.12	2.03
	.005	8.49	5.80	4.73	4.14	3.76	3.49	3.29	3.13	3.01	2.90	2.82	2.74	2.57	2.39	2.29	2.19
(120)	.25	1.34	1.40	1.39	1.37	1.35	1.33	1.31	1.30	1.29	1.28	1.27	1.26	1.24	1.22	1.21	1.19
	.1	2.75	2.35	2.13	1.99	1.90	1.82	1.77	1.72	1.68	1.65	1.62	1.60	1.55	1.48	1.45	1.41
	.05	3.92	3.07	2.68	2.45	2.29	2.18	2.09	2.02	1.96	1.91	1.87	1.83	1.75	1.66	1.61	1.55
	.025	5.15	3.80	3.23	2.89	2.67	2.52	2.39	2.30	2.22	2.16	2.10	2.05	1.95	1.82	1.76	1.69
	.01	6.85	4.79	3.95	3.48	3.17	2.96	2.79	2.66	2.56	2.47	2.40	2.34	2.19	2.03	1.95	1.86
	.005	8.18	5.54	4.50	3.92	3.55	3.28	3.09	2.93	2.81	2.71	2.62	2.54	2.37	2.19	2.09	1.98

TABLE A9

Upper and lower values of the Wilcoxon signed ranks statistic

	A	B	$P(W < A) = P(W > B)$
$n = 4$	0	10	0.062
	1	9	0.125
$n = 5$	0	15	0.031
	1	14	0.062
	2	13	0.094
	3	12	0.156
$n = 6$	0	21	0.016
	1	20	0.031
	2	19	0.047
	3	18	0.078
	4	17	0.109
	5	16	0.156
$n = 7$	0	28	0.008
	1	27	0.016
	2	26	0.023
	3	25	0.039
	4	24	0.055
	5	23	0.078
	6	22	0.109
	7	21	0.148
$n = 8$	0	36	0.004
	1	35	0.008
	2	34	0.012
	3	33	0.020
	4	32	0.027
	5	31	0.039
	6	30	0.055
	7	29	0.074
	8	28	0.098
	9	27	0.125
$n = 9$	1	44	0.004
	2	43	0.006
	3	42	0.010
	4	41	0.014
	5	40	0.020
	6	39	0.027
	7	38	0.037
	8	37	0.049
	9	36	0.064
	10	35	0.082
	11	34	0.102
	12	33	0.125
$n = 10$	3	52	0.005
	4	51	0.007
	5	50	0.010
	6	49	0.014
	7	48	0.019
	8	47	0.024
	9	46	0.032
	10	45	0.042
	11	44	0.053
	12	43	0.065
	13	42	0.080
	14	41	0.097
	15	40	0.116
	16	39	0.138
$n = 11$	5	61	0.005
	6	60	0.007
	7	59	0.009
	8	58	0.012
	9	57	0.016
	10	56	0.021
	11	55	0.027
	12	54	0.034
	13	53	0.042
	14	52	0.051
	15	51	0.062
	16	50	0.074
	17	49	0.087
	18	48	0.103
	19	47	0.120
	20	46	0.139
$n = 12$	7	71	0.005
	8	70	0.006
	9	69	0.008
	10	68	0.010
	11	67	0.013
	12	66	0.017
	13	65	0.021
	14	64	0.026
	15	63	0.032
	16	62	0.039
	17	61	0.046
	18	60	0.055
	19	59	0.065
	20	58	0.076
	21	57	0.088
	22	56	0.102
	23	55	0.117
	24	54	0.133

TABLE A10

Table of random numbers

79202	05262	20275	92510	13420	93919	29308	41510	32444	40910
67528	71327	21233	78626	72170	70709	32042	33716	71447	30501
20006	68100	17493	07169	23554	23746	46797	31334	29685	82648
56544	09867	52453	19922	06764	05053	38750	93741	08612	17761
72133	50958	67464	62099	41524	71868	87521	41853	49573	88946
07206	41563	06723	78136	78707	14976	18118	40580	08629	77489
72732	99625	80605	39256	50222	75306	73591	92421	01328	21516
58164	39822	43322	26932	10137	74194	55359	84786	38249	24771
67040	82416	48892	46521	34050	83133	17968	98625	22505	10200
68022	76319	67673	07001	73609	51459	99184	04377	40914	53552
44207	07163	02596	45523	61532	65225	77314	46035	96285	32068
18216	93109	84314	73610	55776	08789	57301	48001	47548	76273
31526	93920	28157	54475	40650	37463	81075	11095	55696	79658
19802	17109	94245	79927	47329	88243	56729	09325	35403	83695
62809	29710	48265	06117	01548	37117	41624	05846	18689	52080
32870	28453	89693	71744	83703	87213	43342	56445	34288	39361
44193	89489	40402	29071	80350	81320	15279	04475	91987	13901
21324	81614	91174	89284	07094	70576	85344	25975	52665	29232
56639	00839	29588	98956	44283	96672	76663	85728	73185	71367
57413	65159	65250	60624	37859	94093	85117	87018	71494	71302
03285	61616	11296	48117	27691	70778	06216	22840	50570	45734
74597	65449	43357	42538	73834	83956	39205	34455	65969	97149
22793	47616	53262	61905	79450	52011	94870	35249	84240	14537
58122	71497	80003	50785	43979	84599	48168	59071	60045	21463
20604	25729	52308	58533	40642	10357	56121	62935	21134	98634
59054	45452	91682	06988	32471	28750	26484	96036	42981	18795
41895	63528	56920	62966	19363	60600	15845	44196	80048	96899
35444	50549	13225	21854	56122	66291	02487	56797	63182	16780
63553	10342	44458	23072	94262	14791	18329	25498	47756	98521
90002	77694	08221	43468	02605	73506	32916	71493	68821	77470
38049	75437	97365	20022	32651	16569	01447	09065	58520	68462
41883	46926	98172	55763	97239	30766	67282	69365	05667	36089
50119	11273	77428	74400	24977	98225	23765	08150	12500	87334
97924	02487	57219	15703	65713	64539	04555	18895	84297	29393
15514	82607	28791	91698	95723	87540	68143	72347	02233	84313
71347	92662	18684	69349	95089	22563	52281	70798	42678	32333
69731	14570	51552	99778	80221	61525	05975	37111	24919	85585
39754	20604	27803	44359	00067	20765	98893	99014	94439	36911
77426	67321	95734	25118	10308	33473	60967	21142	72688	54879
28278	61565	71308	66523	48015	54057	50013	93613	26085	94617
89958	15273	65471	77320	26869	07590	58947	56756	19942	77649
94498	45787	47339	54093	67702	14371	98794	79829	85231	56652
90064	79532	32310	04909	79451	54799	63684	66074	09663	34517
64183	90256	46529	89819	81235	60342	93451	96197	19896	23719
91280	15653	21120	05587	74555	21225	45354	39098	61860	17731
70235	74750	62918	72259	60667	50725	38933	97876	93306	22508
20377	60495	90821	08727	58695	14407	80989	60357	05988	68725
62434	49242	64310	46556	49671	68321	01114	26055	91145	97636
38189	29707	39476	16982	45555	59001	16716	71891	62747	33187
89691	65908	49013	39549	97002	50070	78398	19787	80370	04798

APPENDIX B SUMMATION NOTATION IN STATISTICS

For clarity, many of the formulas in probability and statistics which involve summing up numbers are written in summation notation. In this appendix we define and illustrate this notation. We begin with an example.

We want to represent the addition, or summing up, of the reciprocals of the whole numbers from 5 to 25. Without writing out all the terms, we can present this sum as $1/5 + 1/6 + \cdots + 1/24 + 1/25$. The three dots indicate the reciprocals of the numbers between (but not including) 6 to 24. Using summation notation, we can write this sum more conveniently as

$$\sum_{i=5}^{25} \frac{1}{i}$$

where Σ, the Greek uppercase sigma, is called the **summation sign;** $1/i$, the term that is to be added, is called the **summand;** and i is the number that is to be changed in each term. Five is the lower limit and 25 is the upper limit of the value of i. Thus, in words, we can describe $\Sigma\, 1/i$ as "the sum of all the $1/i$ for i from 5 to 25." Both the range of values and the form of the summand will vary from one situation to another.

In statistics we frequently represent numbers by subscripted letters. We might, for instance, represent the ages of six students as x_1, x_2, x_3, x_4, x_5, and x_6. We can write the sum of their ages in summation notation as $\sum_{i=1}^{6} x_i$, which in words is "the sum of all the x_i for i equals 1 to 6." In this case the summand is x_i. More generally, $\sum_{i=1}^{n} x_i$ represents the sum $x_1 + x_2 + \cdots + x_n$. The following four examples further illustrate the use of summation notation.

EXAMPLE A Given the values $x_1 = 70$, $x_2 = 48$, $x_3 = 74$, $x_4 = 62$, and $x_5 = 68$, evaluate (a) $\sum_{i=1}^{5} x_i$ (b) $\sum_{i=1}^{5} x_i^2$.

Solution (a) $\sum_{i=1}^{5} x_i = x_1 + x_2 + x_3 + x_4 + x_5 = 70 + 48 + 74 + 62 + 68 = 322.$

(b) In this case the summand is x_i^2 and $\sum_{i=1}^{5} x_i^2 = x_1^2 + x_2^2 + x_3^2 + x_4^2 + x_5^2 = 70^2 + 48^2 + 74^2 + 62^2 + 68^2 = 21{,}148.$

Incidentally, notice that the sum of the squared values, $\sum_{i=1}^{5} x_i^2 = 21{,}148$, is not the same as the square of the sum, which is $\left[\sum_{i=1}^{5} x_i\right]^2 = 322^2 = 103{,}684.$

■

EXAMPLE B The number of years that each of four instructors have taught at Upstate U. are $x_1 = 17$, $x_2 = 4$, $x_3 = 10$, and $x_4 = 9$. Evaluate $(1/3) \sum_{i=1}^{4} (x_i - 10)^2$.

Solution

$$\frac{1}{3}\sum_{i=1}^{4} (x_i - 10)^2 = \frac{1}{3}[(x_1 - 10)^2 + (x_2 - 10)^2 + (x_3 - 10)^2 + (x_4 - 10)^2]$$

$$= \frac{1}{3}[(17 - 10)^2 + (4 - 10)^2 + (10 - 10)^2 + (9 - 10)^2]$$

$$= \frac{86}{3} = 28.7$$

The value 28.7 is a measure of the variability among the numbers of years of service (see Section 1.4).

■

EXAMPLE C Using the values $f_1 = 4$, $f_2 = 10$, $f_3 = 8$, and $f_4 = 2$, evaluate $\sum_{x=1}^{4} xf_x$.

Solution In this case we arbitrarily use x instead of i. The summand is the product xf_x. Thus

$$\sum_{x=1}^{4} xf_x = (1)(f_1) + (2)(f_2) + (3)(f_3) + (4)(f_4)$$

$$= (1)(4) + (2)(10) + (3)(8) + (4)(2) = 56$$

■

In many cases where it is clear which set of values are being referred to, we omit the range of values from the summation sign. The following example illustrates this point.

EXAMPLE D The weights (in pounds) of five infants born in a hospital in one evening are 7.1, 6.3, 7.7, 10.5, and 5.9. Find $(1/5) \sum x_i$.

Solution There is the unwritten presumption here that $x_1 = 7.1$, $x_2 = 6.3$, and so on. When the range of values is omitted, assume that all the values are to be included. Thus

$$\frac{1}{5}\sum x = \frac{1}{5}(7.1 + 6.3 + 7.7 + 10.5 + 5.9) = \frac{37.5}{5} = 7.5 \text{ pounds}$$

The value 7.5—the sum of all the weights divided by the number of weights—is called the mean weight (see Section 1.3).

■

PROBLEMS

1. The number of credits that each of seven students needs for graduation are $x_1 = 24, x_2 = 12, x_3 = 16, x_4 = 16, x_5 = 9, x_6 = 10$, and $x_7 = 20$. Find $(1/7)\sum_{i=1}^{7} x_i$. (This is the mean number of credits needed for graduation.)
2. The total family incomes of the four families on a block are (in thousands of dollars) $x_1 = 24.8$, $x_2 = 32.4$, $x_3 = 28.6$, and $x_4 = 36.2$. Evaluate (a) $\sum_{i=1}^{4} x_i$; (b) $\sum_{i=1}^{4} (x_i - 20)$; (c) $\sum_{i=1}^{4} x_i^2$.
3. If $x_1 = 0$, $x_2 = 1$, $x_3 = 2$, and $x_4 = 3$, $p_1 = 1/8$, $p_2 = 3/8$, $p_3 = 3/8$, and $p_4 = 1/8$, evaluate (a) $\sum_{i=1}^{4} x_i p_i$; (b) $\sum_{i=1}^{4} x_i^2 p_i$.
4. The number of goals scored by the New York Rangers in a seven-game playoff series are 3, 4, 5, 4, 2, 4, and 6. Evaluate (a) $A = (1/7)\sum x_i$; (b) $\sum (x_i - A)^2$.

Solutions

1. $107/7 = 15.3$ credits
2. (a) 122.0; (b) 42.0; (c) 3793.2
3. (a) 1.5; (b) 3
4. (a) 4; (b) 10

ANSWERS TO THE ODD-NUMBERED PROBLEMS

If the first problem in a section has an odd number, the answer will be at the end of the problem set.

CHAPTER 1

3. (a) discrete (b) nominal (c) continuous (d) discrete (e) nominal (f) discrete (g) continuous (h) continuous (i) discrete
7. (a) and (b)

INTERVAL	FREQUENCY	RELATIVE FREQUENCY	CUMULATIVE FREQUENCY
850–999	4	.0500	4
1000–1149	9	.1125	13
1150–1299	20	.2500	33
1300–1449	13	.1625	46
1450–1599	15	.1875	61
1600–1749	10	.1250	71
1750–1899	5	.0625	76
1900–2049	3	.0375	79
2050–2199	1	.0125	80
	80	1.0000	

(c)

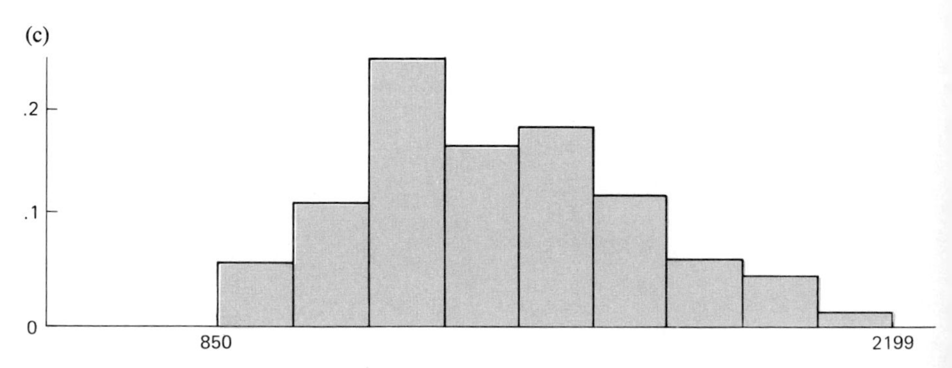

(d) (i) 71 (ii) 79 (e) (i) .25 (ii) .1875

(f)

8**	79
9**	05 36 95
10**	12 13 27 55 93
11**	17 22 36 47 57 73 76 77 82 92
12**	09 10 11 18 23 31 43 63 63 67 78 82 89 95
13**	29 33 34 43 43 44 52 53 58 65 73 83 92
14**	53 72 75 77 80 87 90 90
15**	00 61 71 76 86 86 97
16**	15 24 60 61 67 90 97
17**	13 28 35 61 81 83
18**	00 29
19**	04 17 60
20**	98

(g) The stem-and-leaf diagram is skewed slightly to the right.
(h) All the data are retained in order.

9. 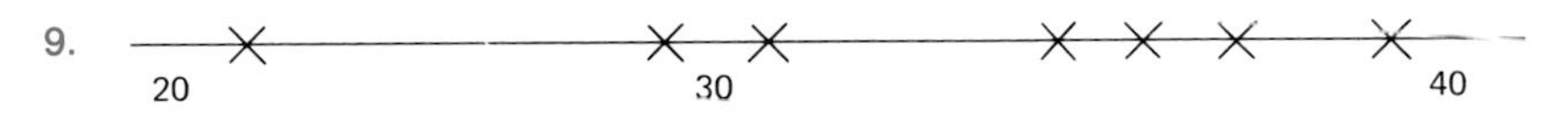

11. (a)

NUMBER OF GOALS	FREQUENCY	RELATIVE FREQUENCY	CUMULATIVE FREQUENCY
0	2	.0244	2
1	4	.0488	6
2	8	.0976	14
3	16	.1951	30
4	21	.2561	51
5	13	.1585	64
6	10	.1220	74
7	5	.0610	79
8	2	.0244	81
9	0	.0000	81
10	1	.0122	82
	82	1.0001	

(b)

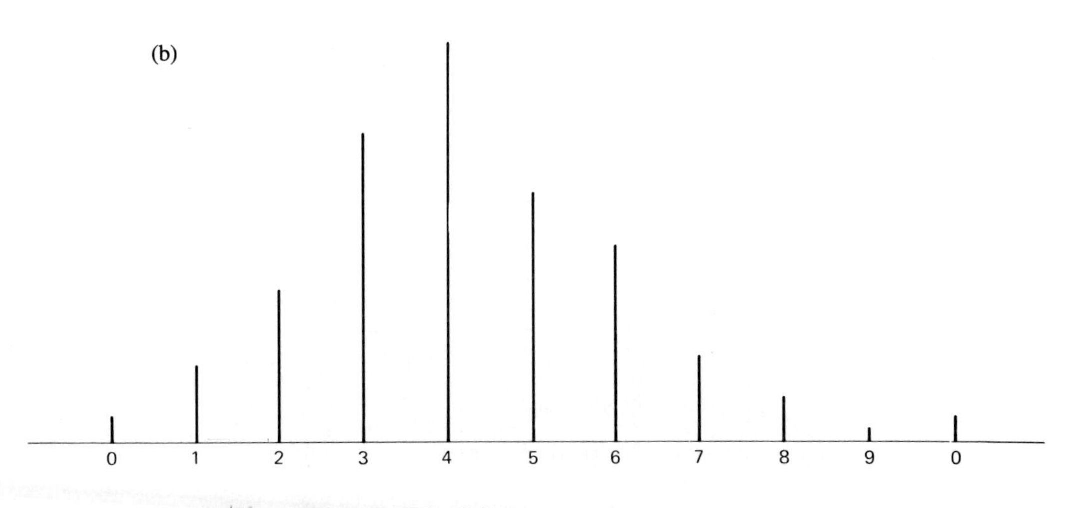

(c) $21/82 = .2561$ (d) (i) 51 (ii) 8

13. (a) and (b)

NUMBER OF DEPENDENTS	FREQUENCY	RELATIVE FREQUENCY	CUMULATIVE FREQUENCY
1	1	.0286	1
2	7	.2000	8
3	8	.2286	16
4	7	.2000	23
5	4	.1143	27
6	4	.1143	31
7	1	.0286	32
8	3	.0857	35
	35	1.0001	

(c) (i) 7 (ii) 12 (d) .1143

15. (a)

	FREQUENCY	RELATIVE FREQUENCY
M	192	.306
T	124	.197
W	152	.242
Th	96	.153
F	64	.102

(b)

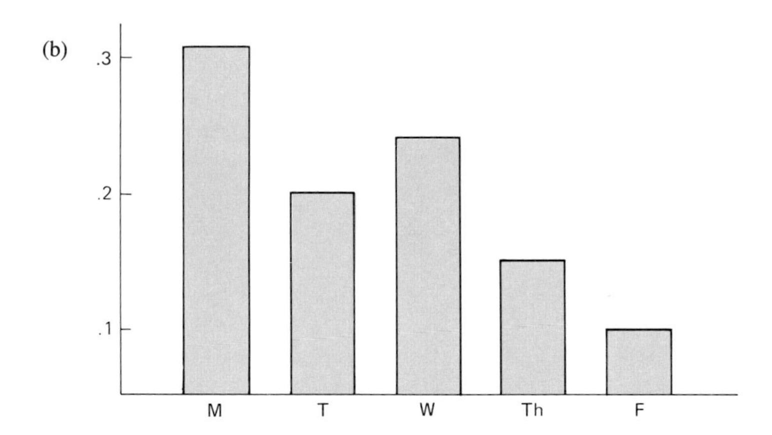

(c) ordinal 19. (a) (i) 7.59 lb (ii) 8.1 lb (b) no value is repeated

21. (a) 60.2 mph (b) 59 mph (c) 67 mph 23. (a) 4.1 (b) 4 (c) 4

25. (a) $16/25 = .64$ (b) the proportion of all hospitals in the city that are public

27. The distribution of age at death is skewed to the left (this is caused by the relatively few deaths of very young children).

29. (a) 13 (b) 4.4 31. (a) 24.7 (b) 48.1 (c) 6.9

33. (a) (i) 22 mph (ii) 5.5 mph (iii) 6.2 mph (b) (i) $28/40 = .7$ (ii) $40/40 = 1.0$

35. (a) 10 (b) 2.5 (c) 1.9

37. (a) 4.25 (b) The decimal point may have been misplaced.

41. (a)

94
84
80
76
60

(b)

43. (a) (i) The ranges are 1.47 for A and 2.97 for B. (ii) The interquartile ranges are .59 for A and 1.46 for B.
(b)

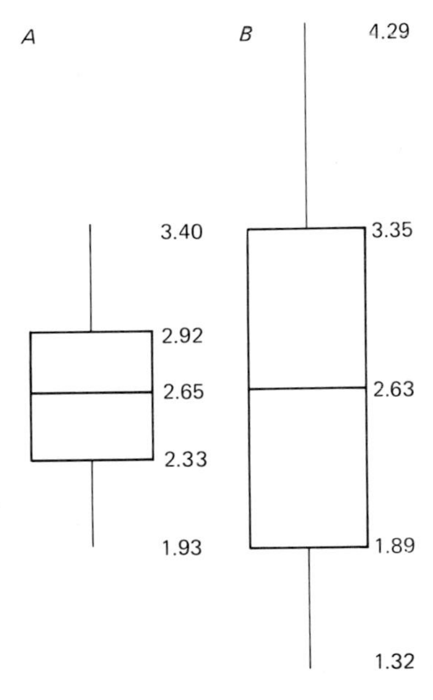

The median lengths are roughly the same but those in A are far less variable than those in B. Indeed, the interquartile range for B is roughly the range for A.

45. (a)

1955	1975
39	46
30.5	32.5
24.5	27.5
21	22.5
16	17

(b) Every one of the five numbers in 1975 is greater than the corresponding number for 1955. Women in 1975 are waiting longer to have children than did their counterparts in 1955.

47. The value .52941 is the proportion of the women who are breastfeeding.

49. (a)

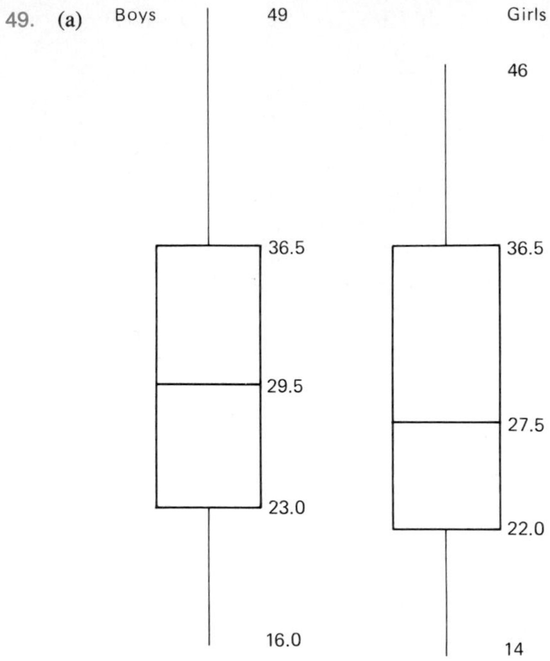

(b) The two distributions are not very different. The most ambitious girls, however, are not as ambitious as the most ambitious boys.

51. 15.8

REVIEW PROBLEMS

1. (a) 7.18 hours (b) .30 hour
3. (a)

4 *	2 3
4 □	6 7 8 9 9
5 *	0 0 1 1 1 1 2 2 4 4 4 4
5 □	5 5 5 5 6 6 6 7 7 7 7 8
6 *	0 1 1 1 2 4
6 □	5 8 9

(b) The distribution of ages is close to symmetric

(c) $\bar{x} = 52.8$ years, $s = 6.1$ years
(d) $Q_1 = 51$ years, $Q_2 = 55$ years, $Q_3 = 57.5$ years, IR = 6.5 years

5. (a) and (b)

NUMBER OF LETTERS	FREQUENCY	RELATIVE FREQUENCY	CUMULATIVE FREQUENCY
1	3	.0286	3
2	20	.1905	23
3	30	.2857	53
4	25	.2381	78
5	12	.1143	90
6	11	.1048	101
7	1	.0095	102
8	2	.0190	104
9	1	.0095	105
	105	1.000	

(c) (i) 78 (ii) 53 (d) $\bar{x} = 3.7$, $m = 3$, $m_o = 3$ (e) 1.6

7. (a)

	MEN	WOMEN
Instructor	.144	.165
Assistant Professor	.361	.505
Associate Professor	.288	.247
Professor	.207	.082
	1.000	.999

(b) A far greater proportion of men than women are in the upper ranks.
(c) ordinal

9. $\bar{p} = .6$ is the proportion of all those who started the race who finished it.

11. (a) and (b)

INTERVAL	FREQUENCY	RELATIVE FREQUENCY	CUMULATIVE FREQUENCY
70–84	3	.0375	3
85–99	4	.0500	7
100–114	19	.2375	26
115–129	21	.2625	47
130–144	23	.2875	70
145–159	5	.0625	75
160–174	2	.0250	77
175–189	3	.0375	80
	80	1.0000	

(c)

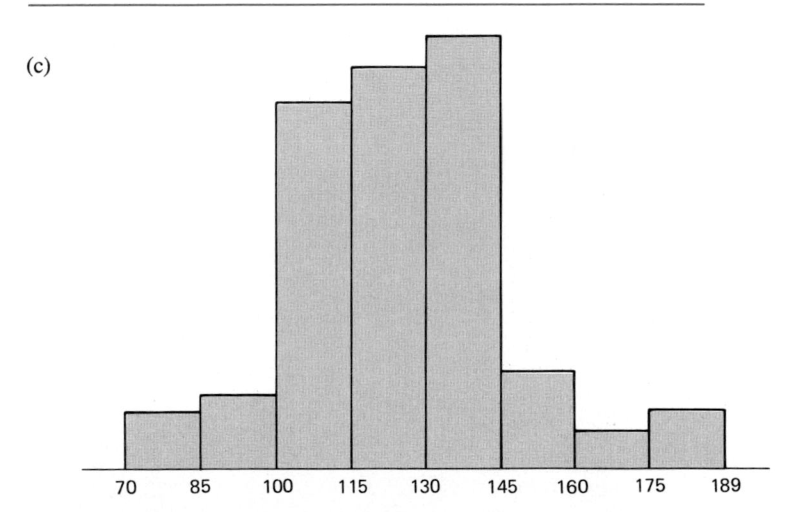

(d) Reasonably symmetric (e) around 133

13. (a)

12**	08 53 57 63 79
13**	03 03 09 28 32 39 42 42 51 51 54 80 82 87 96
14**	06 10 10 16 32 64 67 85
15**	05 07 08 08 11 15 16 18 30 32 49 49 53 67 78 88
16**	16 37 38
17**	03

(b) The data are bimodal in the broad sense of having two peaks and are slightly skewed to the right.

15. (a) $\bar{x} = .8038$ mg, $m = .7885$ mg (b) $R = .078$ mg, $R/4 = .0195$ mg (c) .0321 mg

17. (a)

SITE	RELATIVE FREQUENCY	SITE	RELATIVE FREQUENCY
1. Breast	.116	7. Kidney/bladder	.060
2. Colon/rectum	.172	8. Larynx	.012
3. Lung	.324	9. Prostate	.069
4. Oral	.030	10. Stomach	.046
5. Skin	.020	11. Leukemia	.050
6. Uterus	.036	12. Lymphomas	.066

(b)

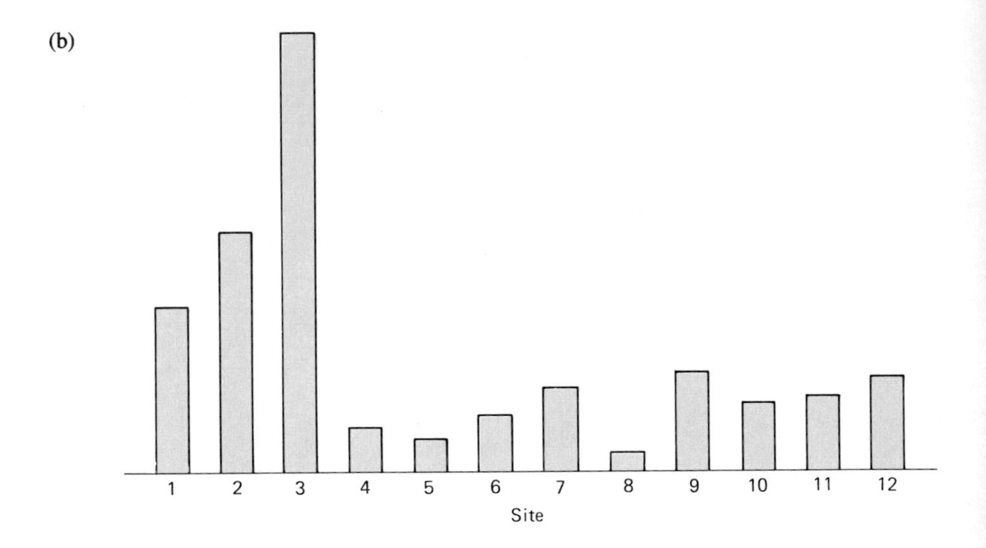

19. (a)

INTERVAL	FREQUENCY	RELATIVE FREQUENCY	CUMULATIVE FREQUENCY
180.00–199.99	2	.0333	2
200.00–219.99	4	.0667	6
220.00–239.99	16	.2667	22
240.00–259.99	13	.2167	35
260.00–279.99	11	.1833	46
280.00–299.99	9	.1500	55
300.00–319.99	3	.0500	58
320.00–339.99	2	.0333	60
	60	1.0000	

(b) .15 (c) 35

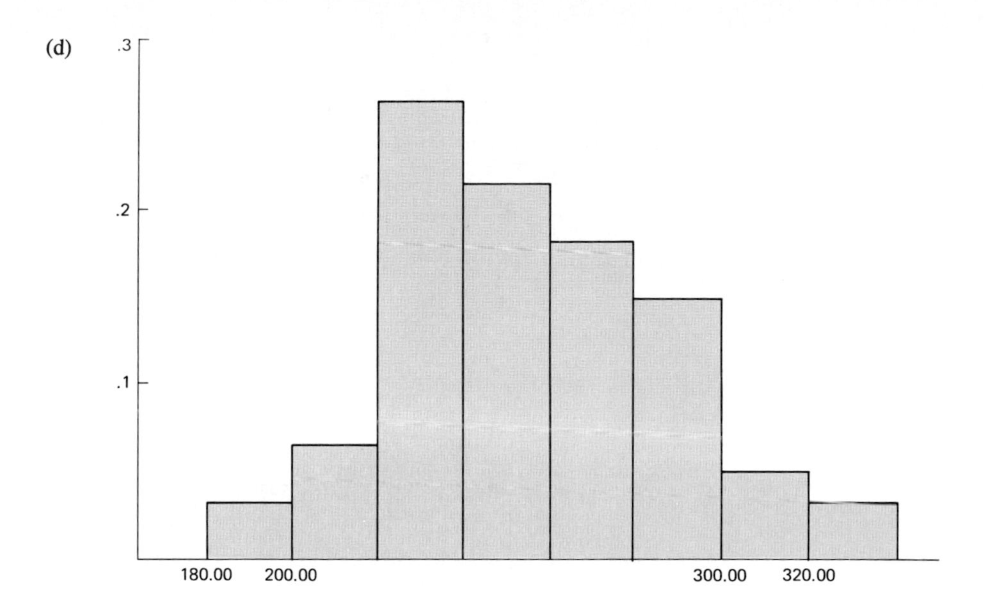

The distribution is skewed slightly to the right.

21. (a)

INTERVAL	FREQUENCY	RELATIVE FREQUENCY	CUMULATIVE FREQUENCY
0–9	1	.01	1
10–19	2	.02	3
20–29	3	.03	6
30–39	11	.11	17
40–49	14	.14	31
50–59	17	.17	48
60–69	18	.18	66
70–79	18	.18	84
80–89	10	.10	94
90–99	6	.06	100
	100	1.00	

(b) (i) .03 (ii) .16 (c) 31 (d)

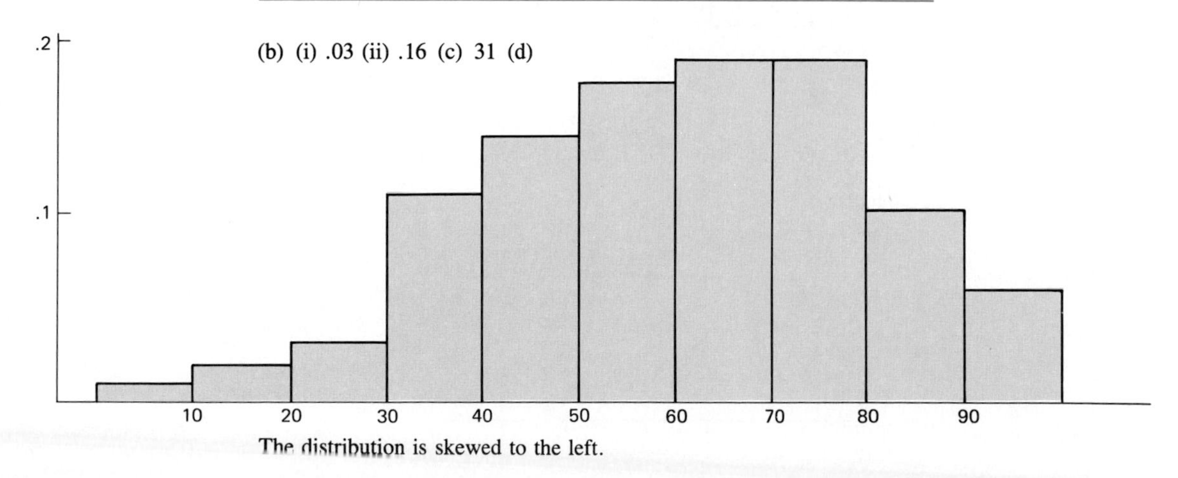

The distribution is skewed to the left.

23.

PARTY	$\bar{x}$	R	$R/4$	s
Dem.	46.3	70	17.50	15.8
Rep.	72.6	67	16.75	15.6

25. (a)

4.0 5.0 6.0

(b) $\bar{x} = 4.89$, $m = 4.85$ (c) .85

27. $\bar{p} = 211/450 = .469$ is the proportion of couples who received SSI payments.

29. (a) 427 (= 68% of 628) (b) 597 (= 95% of 628)

31. $\bar{x} = \$44.99$, $s = \$18.47$ 33. (a) 710 (= 68% of 1044) (b) 992 (= 95% of 1044)

35. (a)

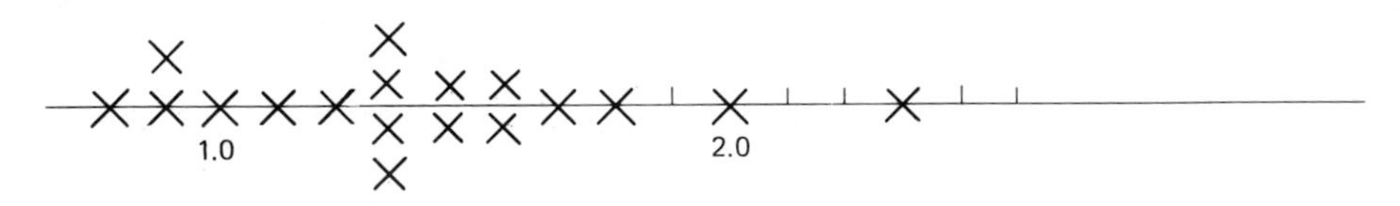

(b) $\bar{x} = 1.42$, $m = 1.40$ (c) $R = 1.5$, $s = .39$ (d) $Q_1 = 1.1$, $Q_3 = 1.6$

(e)

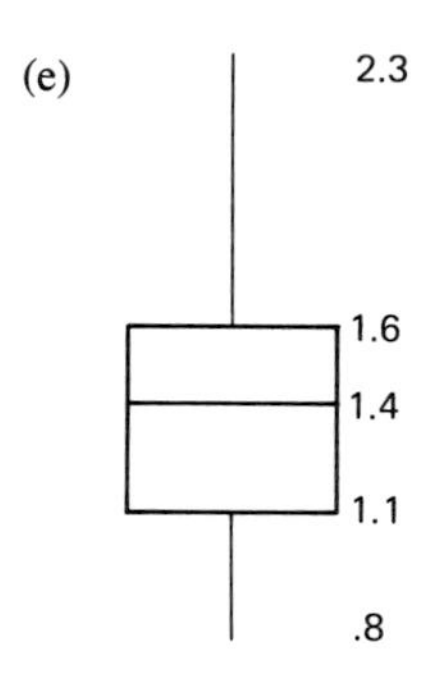

37. $\bar{p} = 26/70 = .371$ is the proportion of supermarkets using computerized checkout.

39. (a)

STATE	RELATIVE FREQUENCY
Arizona	.2122
California	.4162
Nevada	.0837
New Mexico	.1448
Utah	.1431

(b)

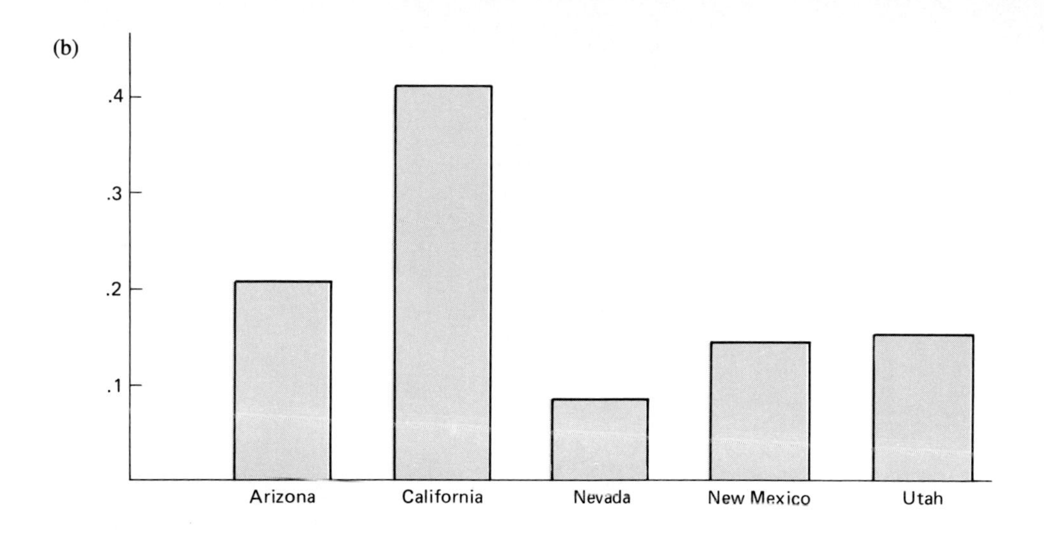

41. (a) 480.7 (b) $R = 382$, $R/4 = 95.5$ (c) 85.0

43. (a)

10 15 20

(b) 13.11 years (c) 12.7 years (This is due to the single, unusually large value, 18.8)
(d) $Q_1 = 12.1$, $Q_3 = 13.2$
(e)

Because of the unusually large value, 18.8, the top whisker is much longer than the bottom.
(f) 2.22

45. (a)

30 40 50

(b) 40.1 (c) 6.5

47. (a)

	RELATIVE FREQUENCY
A	.087
B	.452
C	.374
D	.061
F	.026

(b)

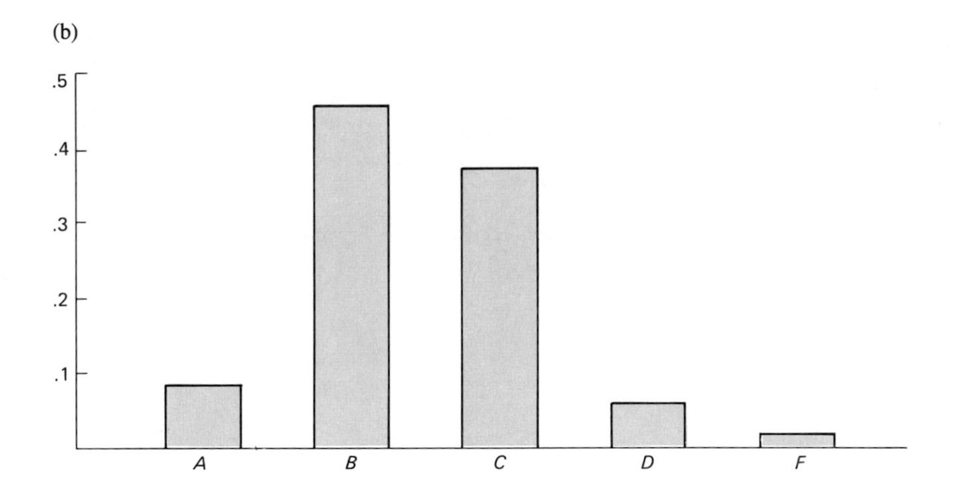

(c) ordinal

49. (a)

INTERVAL	FREQUENCY	RELATIVE FREQUENCY
5000–under 5500	1	.02
5500–under 6000	3	.06
6000–under 6500	8	.16
6500–under 7000	9	.18
7000–under 7500	15	.30
7500–under 8000	9	.18
8000–under 8500	1	.02
8500–under 9000	4	.08
	50	1.00

(b) (i) .16 (ii) .48

(c)

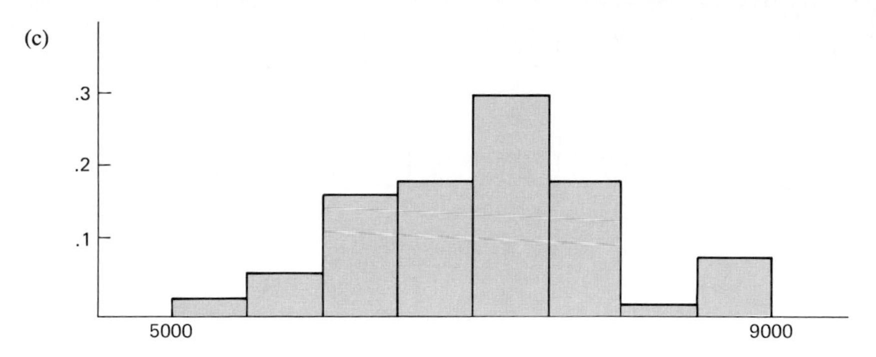

Histogram is skewed slightly to the left.

51. (a)

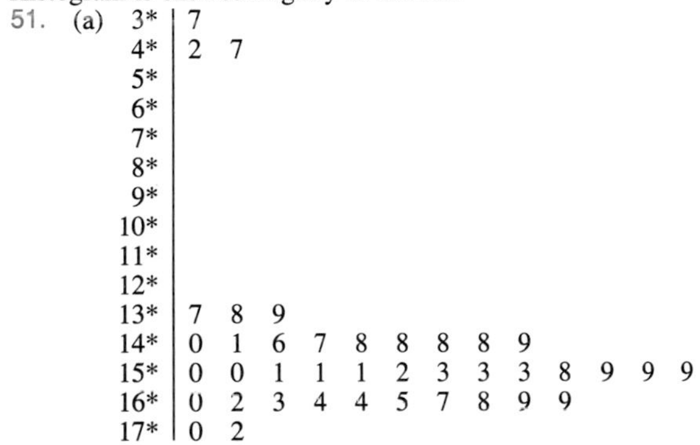

```
 3* | 7
 4* | 2 7
 5* |
 6* |
 7* |
 8* |
 9* |
10* |
11* |
12* |
13* | 7 8 9
14* | 0 1 6 7 8 8 8 8 9
15* | 0 0 1 1 1 2 3 3 3 8 9 9 9
16* | 0 2 3 4 4 5 7 8 9 9
17* | 0 2
```

(b) $\bar{x} = 146.66$, $m = 151.5$ (the three very small values pull down the mean) (c) 29.89

53. (a) [dot plot: × marks on axis labeled 1010, 1020, 1030]

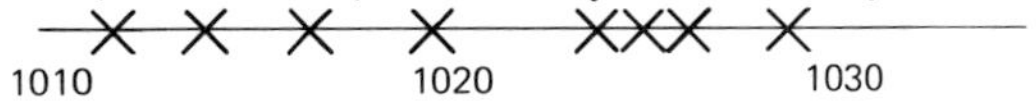

(b) 1021.98 (c) 1023.3 (d) 6.21

55. (a)

NUMBER OF PARTICLES	RELATIVE FREQUENCY
0	.005
1	.000
2	.055
3	.090
4	.125
5	.200
6	.150
7	.130
8	.100
9	.070
10	.050
11	.020
12	.000
13	.000
14	.005
	1.000

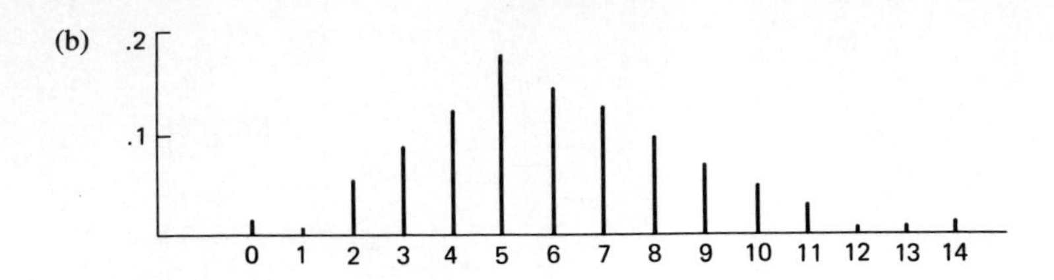

(c) 5.9 (d) $R = 14$, $R/4 = 3.5$ (e) 2.3; yes

CHAPTER 2

3.

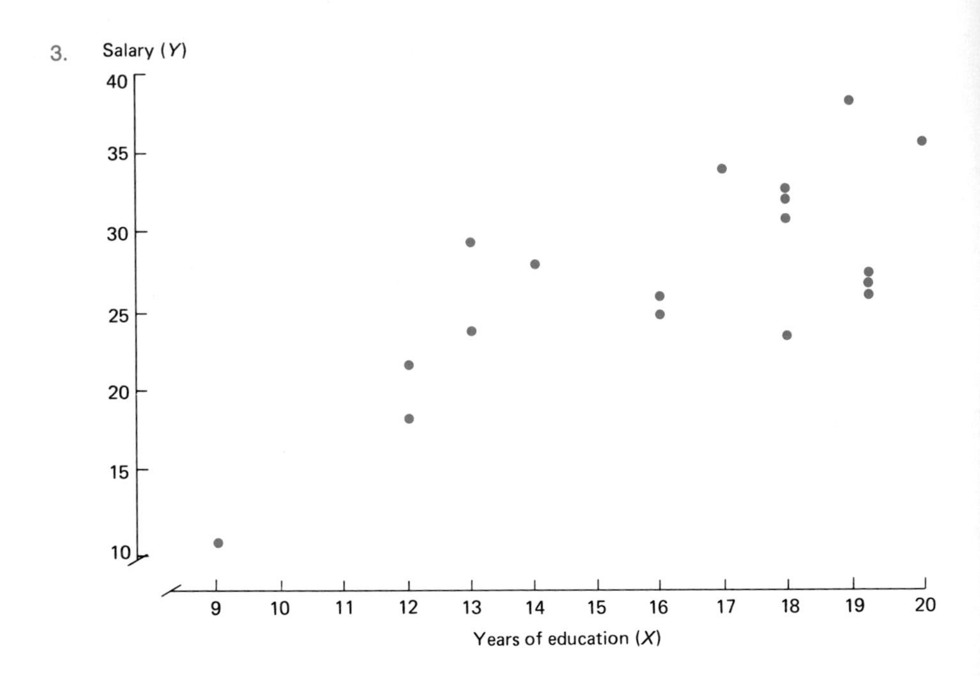

There seems to be a rough, positive relationship between salary and years of education.

5. (a)

M	T	W	Th	F
22	16	13	16	19
19	14.5	10.5	14	18
18	13	9	12	16
17	12	8.5	11.5	15
16	10	7	10	13

(b)

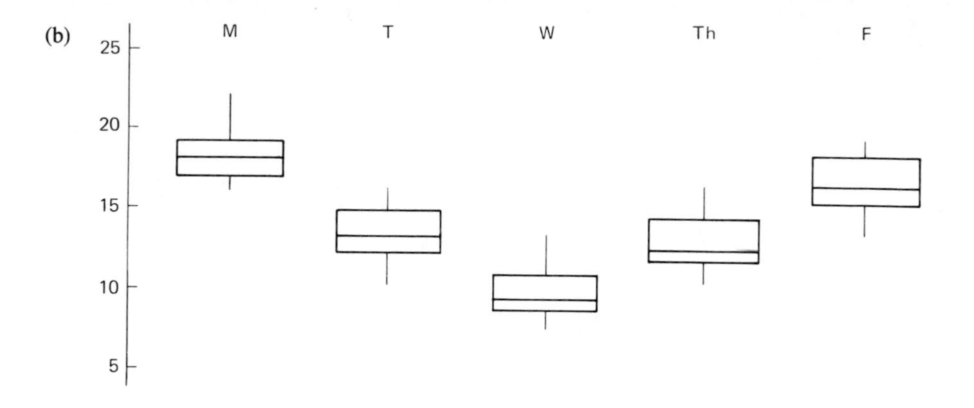

(c) The five distributions do not seem very different but their locations differ dramatically. Absences are highest on Monday, decline on Tuesday, are lowest on Wednesday, and then move up again although Friday is not as high as Monday.

7. (a) The number 241 is the number of women who approve of the president's economic program. The number 54 is the number of men who were neutral toward the program.

(b)

	APPROVE	NEUTRAL	DISAPPROVE	TOTAL
MEN	52%	9%	39%	100%
WOMEN	40%	14%	46%	100%

(c)

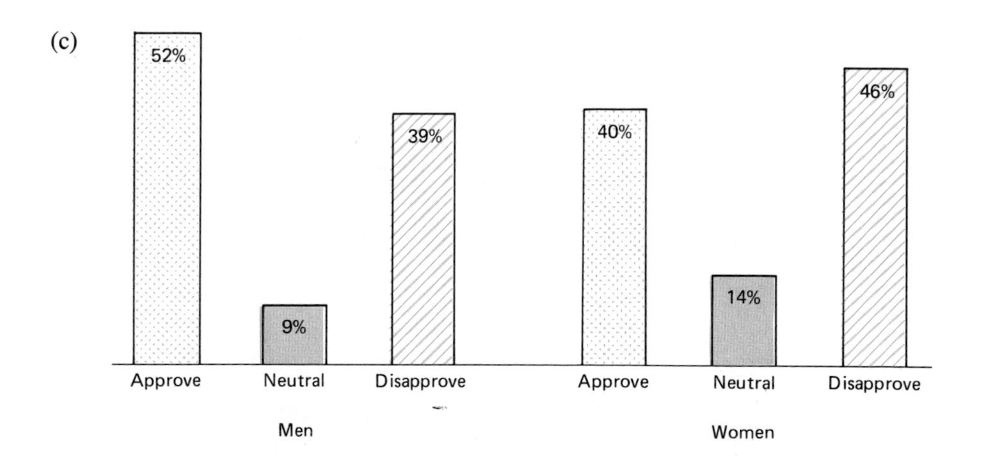

(d) A higher percentage of men (52) than women (40) approve of the program. Conversely, a higher proportion of women (46) than men (39) disapprove of the program.

9. (a)

	ABC	CBS	NBC	PBS	TOTAL
HIGH	33%	21%	14%	31%	99%[a]
MODERATE	31%	36%	19%	14%	100%
LOW	24%	30%	38%	8%	100%

[a]Due to rounding.

(b)

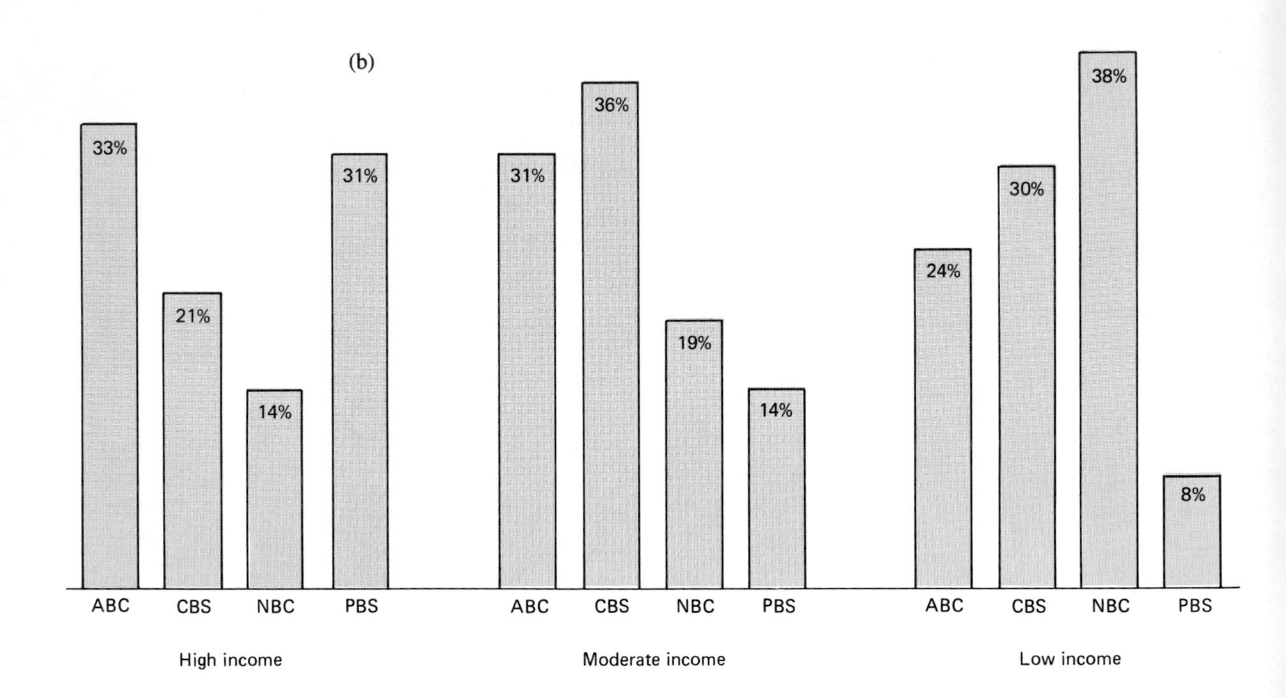

(c) Of the commercial networks, ABC is favored by the high-income group, CBS by the moderate-income group, and NBC by the low-income group. The higher-income group watches more of PBS (31%) than do the moderate-income group (14%) and the low-income group (8%).

11. .60 13. No; correlation does not imply causation.

 (a)

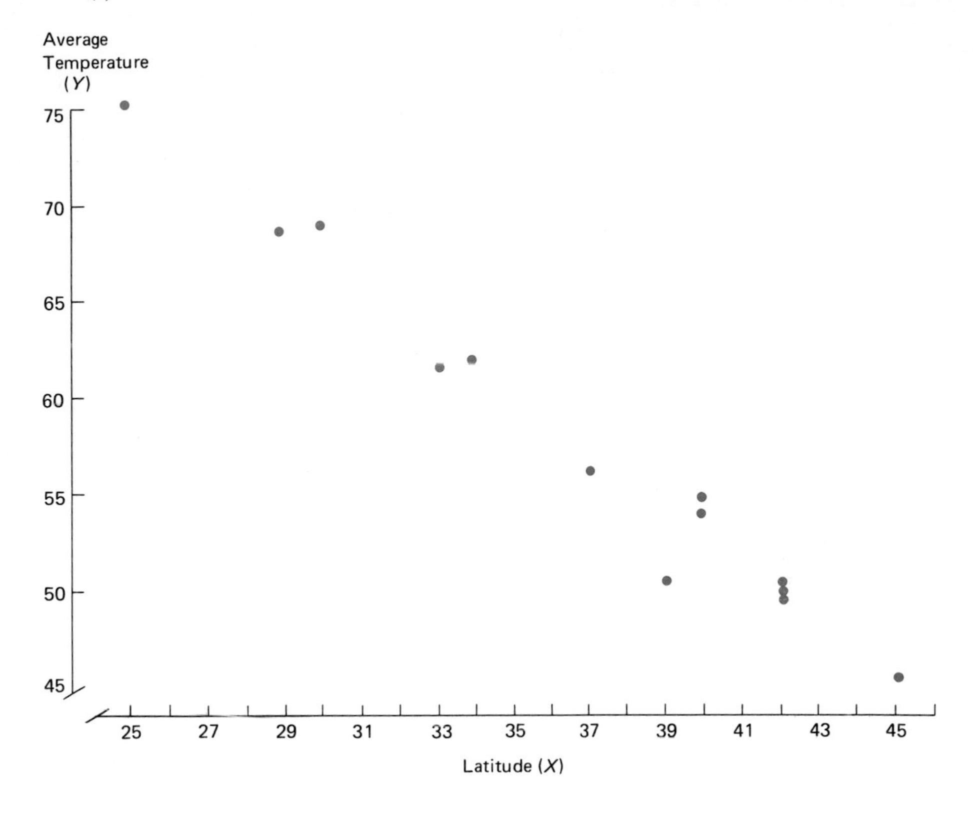

(b) −.99 17. $r = -.77$ indicates the strongest and −.18 the weakest.
19. (a) $y = 4.25x - 134.125$ (b) An increase of 1 inch in height is associated with an increase of 4.25 lb in weight. (c) .358. Only 36% of the variability in weight is associated with its linear relationship to height. (d) (i) 142.1 lb (ii) 163.4 lb
21. (a) $y = 13.915 - .582x$ (b) .764. Approximately 76% of the variability in unemployment rate is associated with its linear relationship to the inflation rate. (c) An increase of one point in the inflation rate is associated with a .58-point *decline* in the unemployment rate. (d) (i) 8.1% (ii) 11.0%
23. (a) An increase of 1% in the additive is associated with an increase of .729 in miles per gallon. (b) 47.805 mpg (c) .906
25. because there is no discernible linear relationship between the variables

REVIEW PROBLEMS

1. (a)

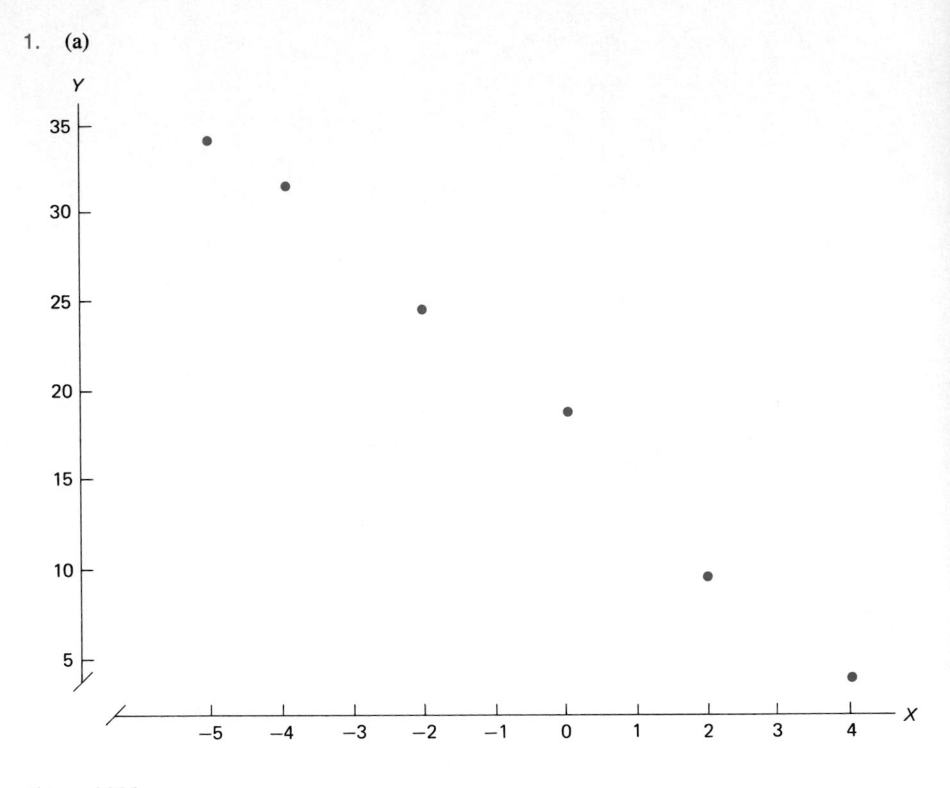

(b) −.9975

3. (a)

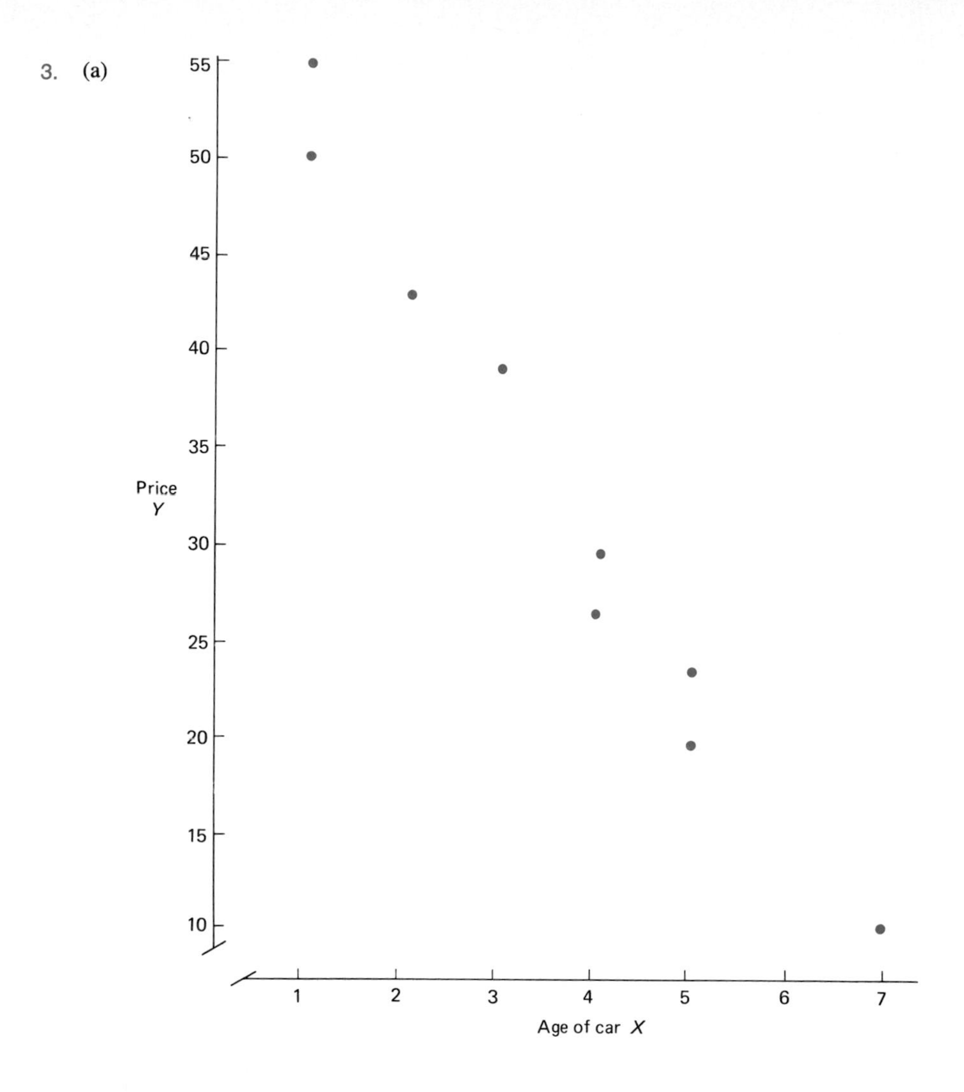

(b) $y = 58.88 - 7.25x$ (c) An increase of one year of age is associated with a 7.25 ($725) decrease in the price paid for the car. (d) $2175

5. (a) $y = 2x + 1$ (b) 1

9. $y = 60.275 + .228x$ (b) An increase of 1 point on the midterm is associated with an increase of .228 point on the final.

11. (a)

A	B	C
40	45	40
35	42	36
33	40	34
30	37	33
28	33	30

(b)

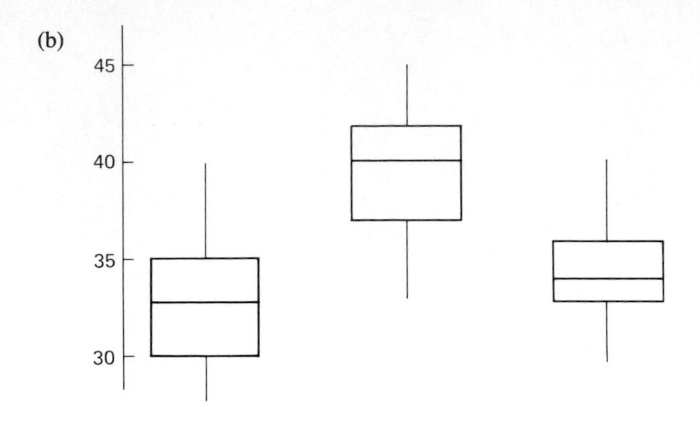

(c) The distribution of texture levels for the three batters are similar, but that for B is located at a higher level than that of A or C.

13. (a)

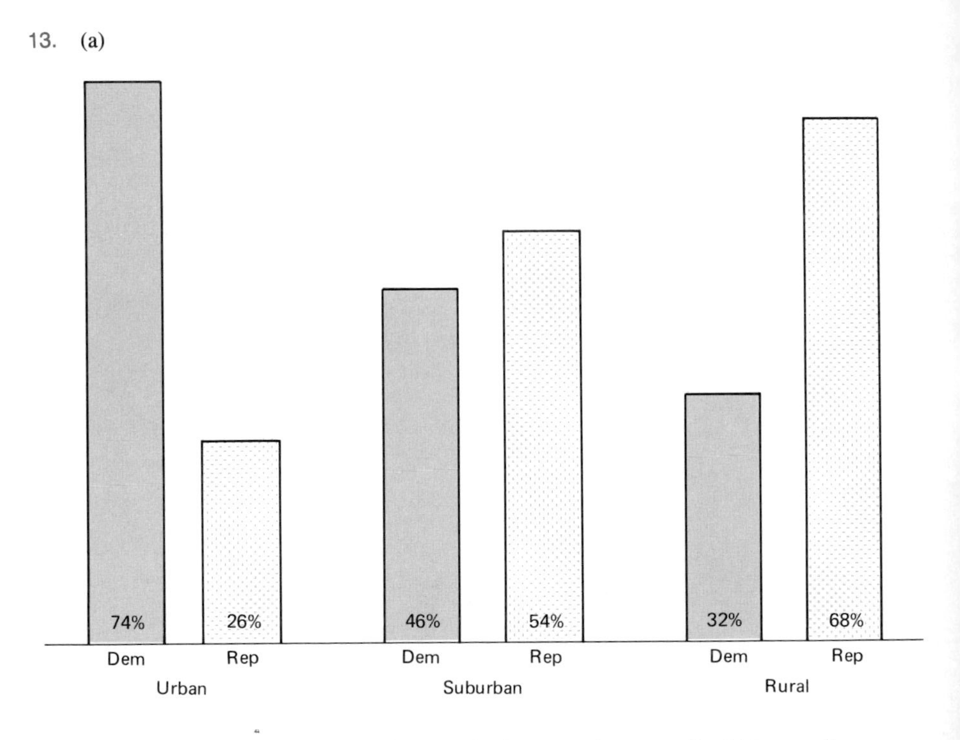

In urban counties Democrats predominate; suburban counties are split fairly equally among Democrats and Republicans, and in rural counties the Republicans predominate.

(b)

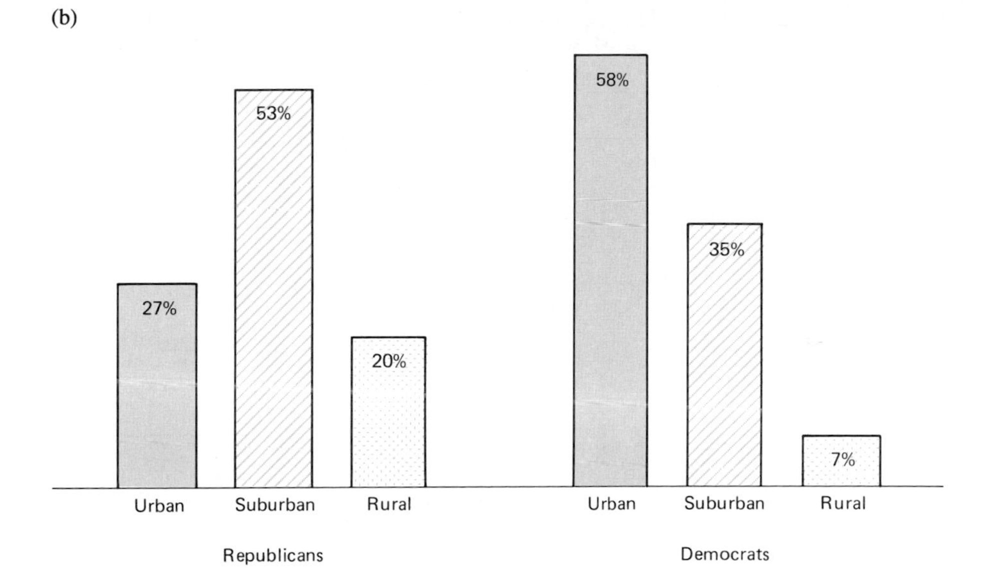

The majority of Republican counties are suburban. Among Democratic counties almost 60% are urban, with only a small percentage (7%) rural.

15. (a)

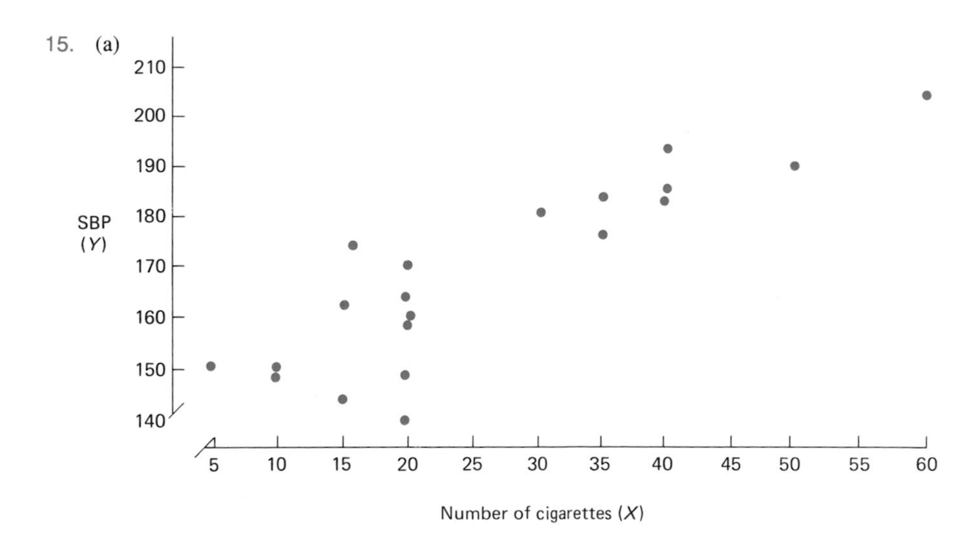

(b) .877 (c) $y = 139.212 + 1.107x$ (d) (i) 167 (ii) 150

17. (a) type of treatment and survival time
(b) and (c)

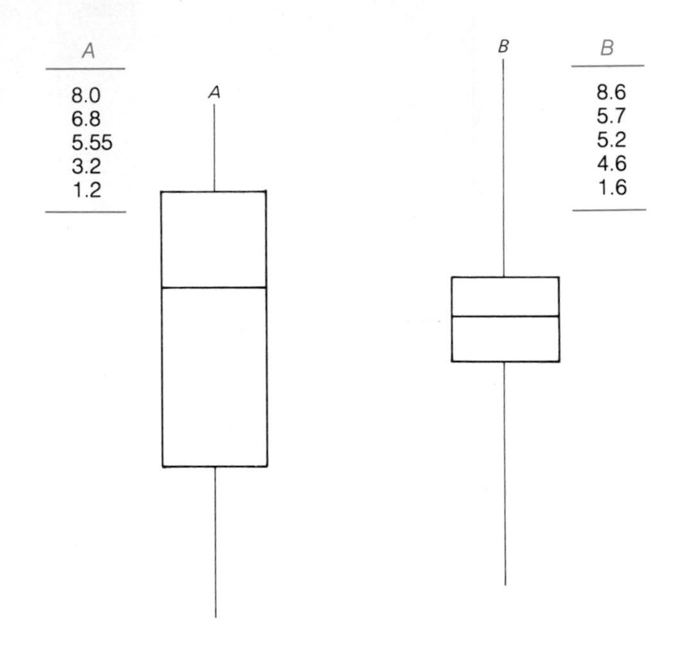

(d) The median survival time for *B* is slightly less than that for *A*, although the ranges of times are very similar. The two treatments differ considerably in that the interquartile range ($Q_3 - Q_1$) for *A* ($6.8 - 3.2 = 3.6$ months) is much greater than that for *B* ($5.7 - 4.6 = 1.1$ months).

19. (a) type of drug and time to relieve pain
(b)

C

Stem	Leaves
*	2
□	6 8 9
1*	0 0 0 0 1 1 3 3 4 4
1□	6

D

Stem	Leaves
*	4
□	5 5 7 8 9
1*	0 1 1 2 2 3 4
1□	6 7

P

Stem	Leaves
*	4 4
□	5 6 8 9 9
1*	0 1 1 3 3 4
1□	5 9

(c) It seems surprising that the distribution of the time to relieve pain for the placebo is not very different from the distributions for the two drugs.

21. (a)

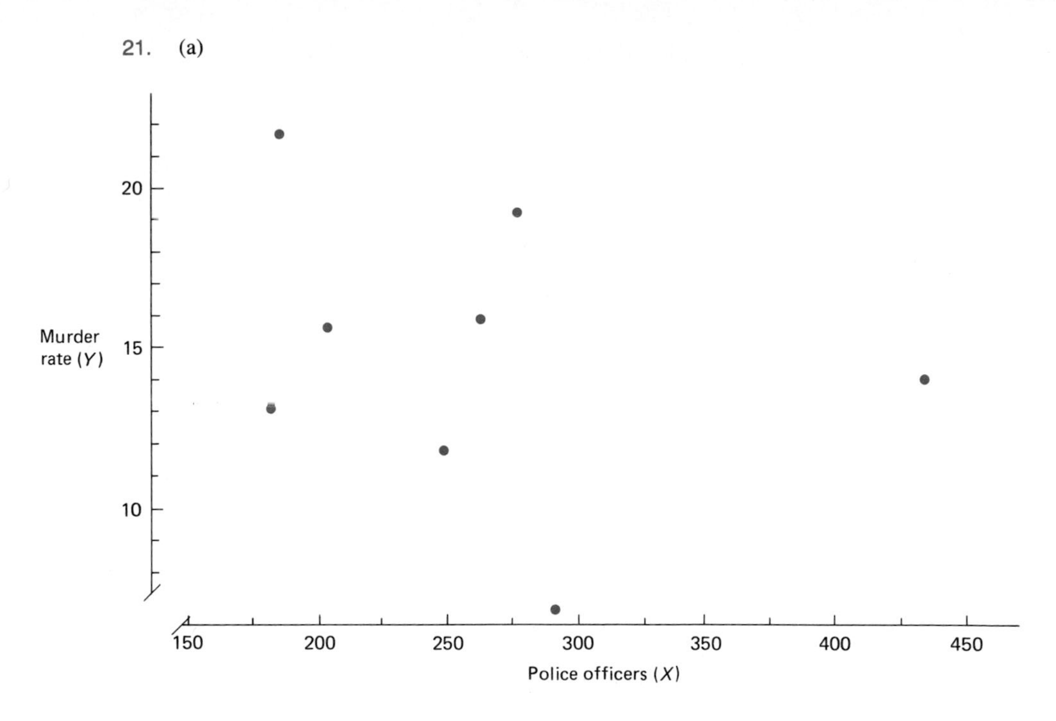

(b) −.279. This value is small and has a sign opposite to the one that might be expected.

23. (a)

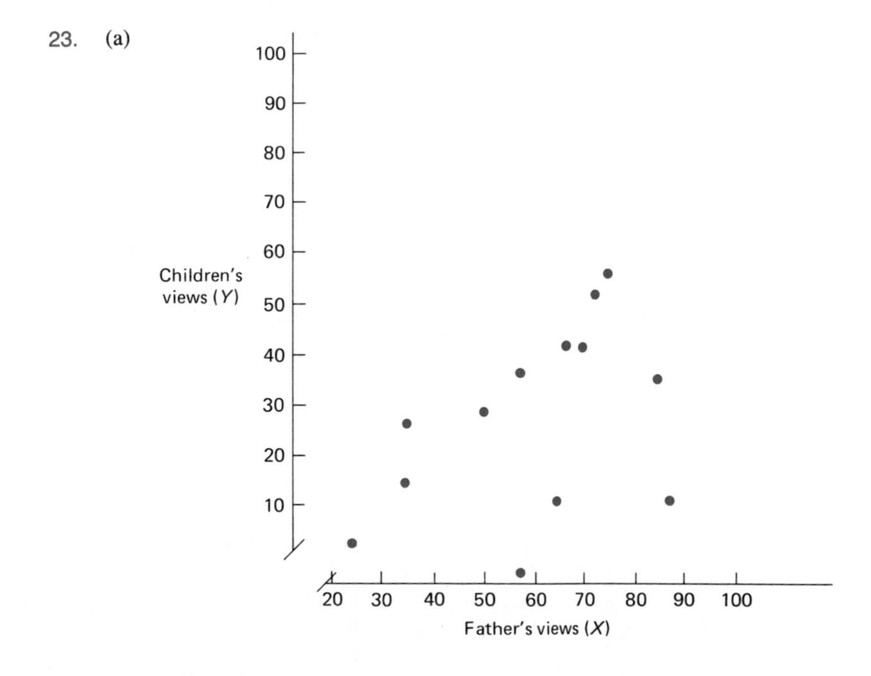

(b) .431 (c) $y = 16.077 + .421x$

25. (a) and (b)

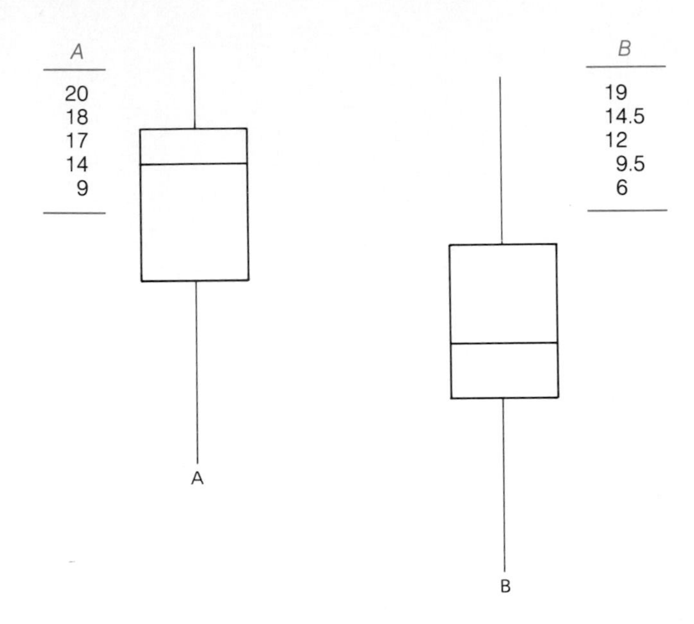

(c) The primary difference between the two sets of scores is that each of the five numbers for A is higher than the corresponding number for B. Recent immigrants appear more patriotic than others.

27. Yes; compare the percentage of men (44%) and the percentage of women (64%) who are firstborn or only children.
29. (a) $y = 267.6324 + 10.1765x$ (b) An increase of one sales representative in an area is associated with an increase of 10.1765 ($10,176.50) in sales. (c) 40.706 ($40,706)
31. (a) country and percentage of income that is saved
(b)

U.S.	GREAT BRITAIN	WEST GERMANY
9.4	13.4	16.3
7.5	10.8	11.7
5.7	8.85	9.85
4.4	4.7	8.3
1.9	1.3	3.9

(c)

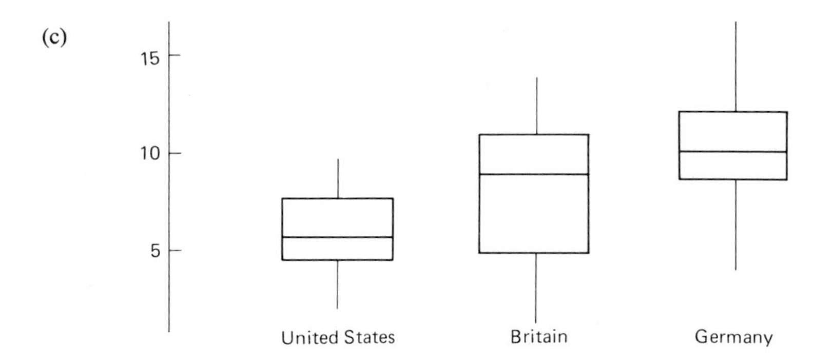

(d) The *range* of values in the U.S. group is smaller than that in Britain or West Germany. However, the data's major message is that German professors tend to save a larger percentage of their income than do their British colleagues, who in turn tend to save more than their U.S. counterparts.

33. (a)

	UNEMPLOYMENT	HIGH INTEREST RATES	INFLATION	TOTAL
LESS THAN $25,000	25%	23%	52%	100%
$25,000 OR MORE	11%	48%	41%	100%

(b)

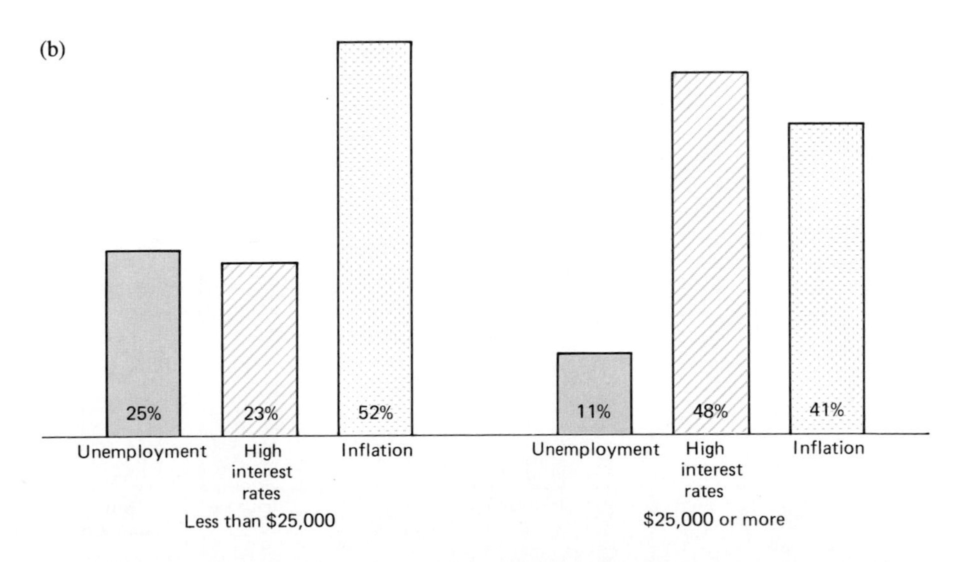

(c) The lower-income group is most concerned with the effects of inflation (52%), followed by unemployment (25%), and last, high interest rates (23%). The high-income group is *most* concerned with high interest rates (48%), followed by inflation (41%). 35. .9656

37. (a)

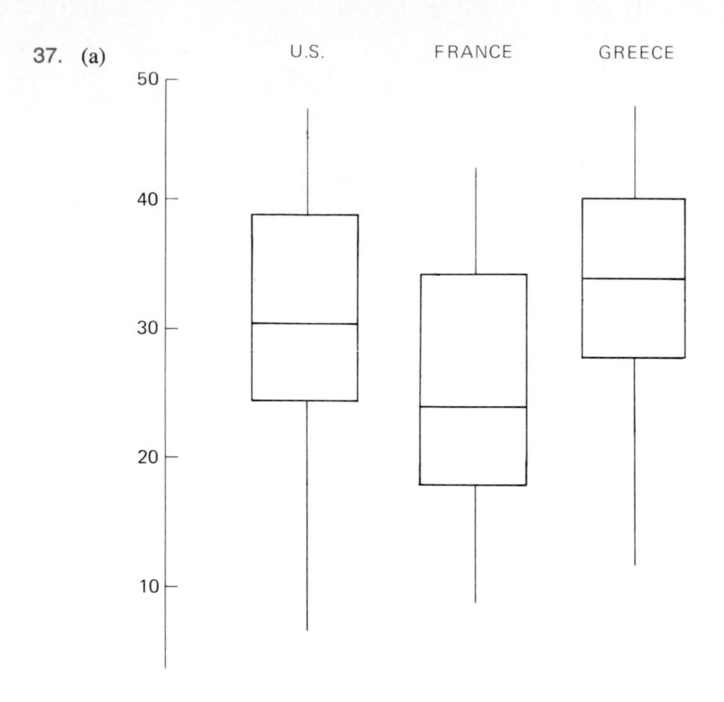

(b) The Greeks clearly tend to be the most self-confident group. With one exception, every one of the five Greek numbers are higher than the corresponding figures for the other two countries. The only exception is that the largest U.S. score exceeds the largest Greek score. The greatest range of self-confidence scores occurs in the U.S. group. The French scores, by contrast, are far less variable.

39. (a)

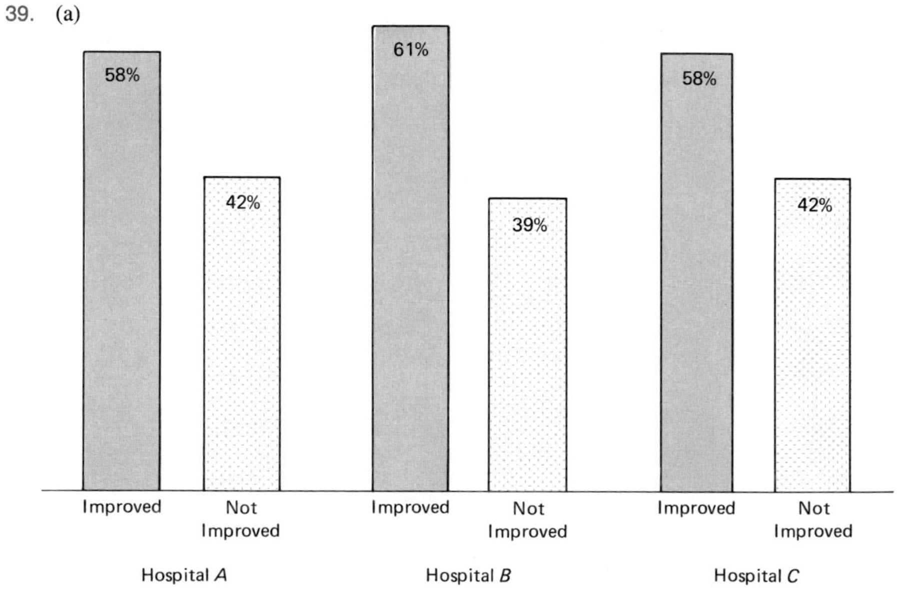

(b) The percentage who felt they had improved is marginally higher in hospital B than in A or C, but the therapies seem roughly equally effective.

41. (a)

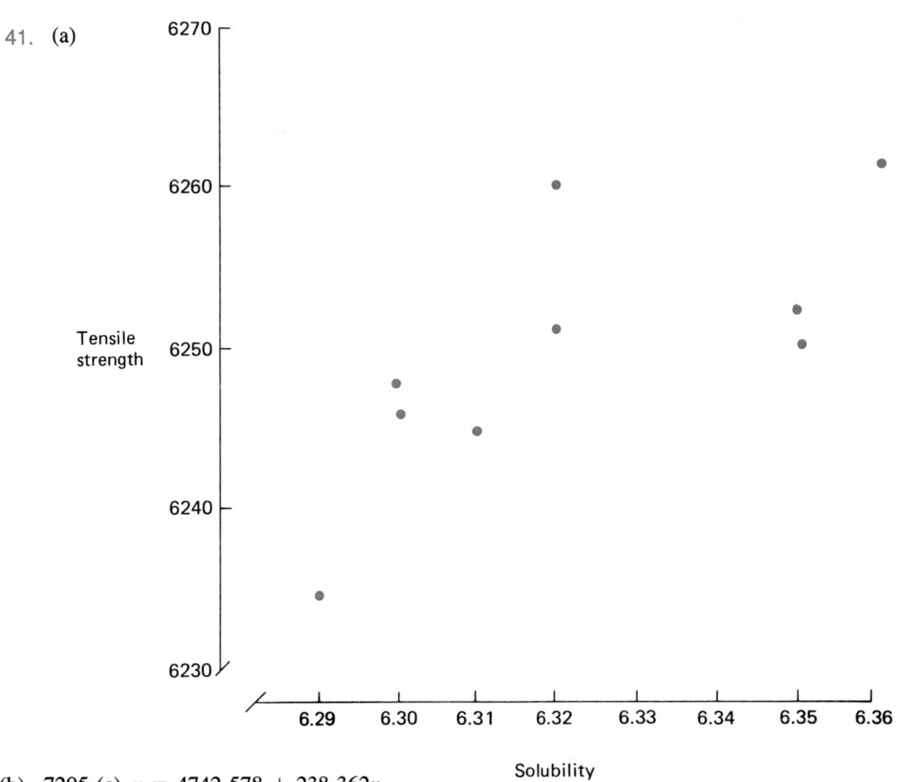

(b) .7205 (c) $y = 4742.578 + 238.362x$
(d) An increase of one unit of solubility is associated with an increase of 238.362 in tensile strength.

43. (a)

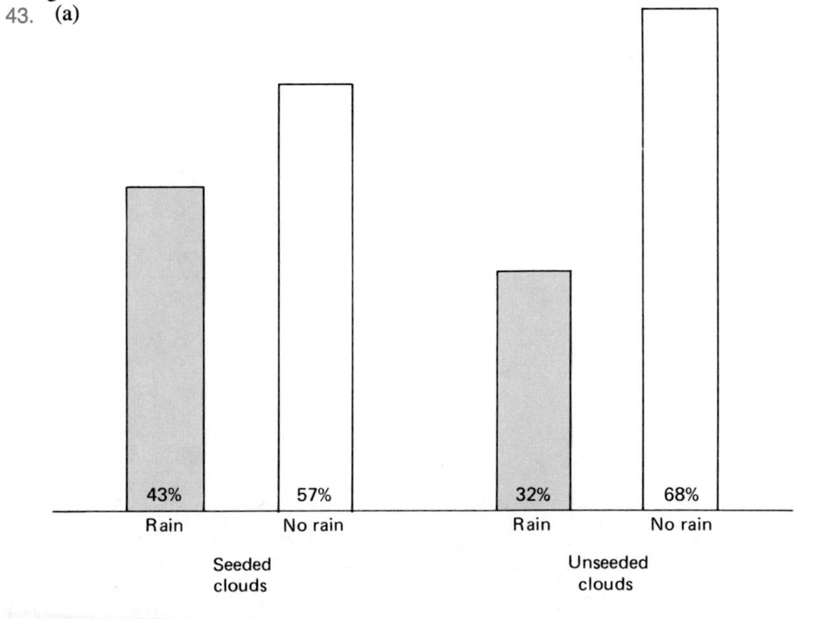

(b) Yes, it does.

CHAPTER 3

3. (a) .48 (b) .40 (c) .096 (d) .64 (e) .04
5. (a) SS = [AB, AC, AD, AE, AF, BC, BD, BE, BF, CD, CE, CF, DE, DF, EF]
(b) (i) 1/15 = .067 (ii) 2/15 = .133 (iii) 9/15 = .06
7. (a) SS = [ABC, ABD, ABE, ACD, ACE, ADE, BCD, BCE, BDE, CDE] (b) (i) .1 (ii) .3
9. The argument is false. Because there are three outcomes does not mean that they are equally likely. A sample space of equally likely outcomes would be [BB, BG, GB, GG].
11. Estimate $P(A)$ by 229/250 = .916.
15. (a) yes (b) (i) .25 (ii) .5 (iii) .45 (iv) .55 (v) .6 (c) (i) .8 (ii) .95 17. .1
19. (a) 1/52 (b) 16/52 (c) 6/52 (d) 32/52 (e) 16/52 (f) 28/52 (g) 51/52 (h) 48/52 (i) 39/52 (j) 3/52
21. (a) .094 (b) 0 (c) .168 (d) .858 (e) .367 (f) .906 (g) .775 25. (a) .484 (b) .721
27. (a) .167 (b) (i) .125 (ii) .125 (c) No; B and E are dependent since $P(B) = .400$ but $P(B|E) = .167$. (d) A and G *are* independent. $P(A) = .333 = P(A|G)$

29.

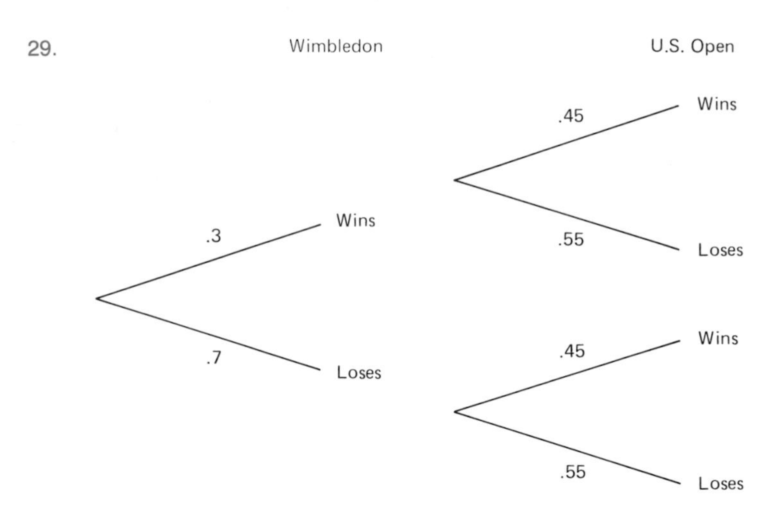

(a) .135 (b) .165 (c) .48 31. (a) .36 (b) .16 33. (a) .1255 (b) .2275 (c) .455
35. (a) .004 (b) .035 (c) .684
37.

	B	B'	TOTAL
E	25	75	100
E'	225	675	900
TOTAL	250	750	1000

REVIEW PROBLEMS

1. (a) .075 (b) 0 (c) .675 (d) .625 (e) .625 (f) .3
3. (a) .347 (b) .133 (c) .853 (d) .733 (e) .038 (f) .033 (g) .408
5. (a) .5 (b) .6 (c) .8 7. (a) .023 (b) .079
9. (a) .25 (b) .25 (c) .494 (d) .278 (e) .75

11. (a) $P(A|B) = .5$, $P(B|A) = .75$. No, A and B are dependent since $P(A) = .4$ but $P(A|B) = .5$. (b) $P(C|D) = .4$, $P(D|C) = .25$. Yes, C and D are independent since $P(C) = P(C|D) = .4$. 13. .333 15. (a) .5 (b) .5
17. (a) Yes (b) (i) .890 (ii) .110 (iii) .766 19. (a) .052 (b) .052 (c) .207
21. (a) .9544 (b) .0181 (c) .891 (d) .064 (e) .987 23. (a) .8 (b) .2
25. (a) .792 (b) .638 (c) .159 27. P(second generation pink) = 292/565 = .517
29. (a) .417 (b) .25 (c) .583 31. (a) .00007 (b) .986 (c) .032 (d) .968
33. (a) .02 (b) .36 (c) .018 35. .684 37. .093
39. (a) (i) .13 (ii) .87 (b) (i) .4 (ii) .2 41. (a) .25 (b) .75 43. .004
45. (a) .064 (b) .092 (c) .828 47. (a) (i) 30% (ii) 70% (b) 25%
49. (a) .316 (b) .684 51. (a) .99 (b) .81

CHAPTER 4

3.

g	0	1	2
$P(g)$	.09	.42	.49

5.

x	1	5	50
$P(x)$	29/36	6/36	1/36

7. (a) (b)

SS =

	(a)	(b) $\bar{x}$
1	(4, 3)	3.5
2	(4, 4)	4.0
3	(4, 4)	4.0
4	(4, 3)	3.5
5	(4, 2)	3.0
6	(3, 4)	3.5
7	(3, 4)	3.5
8	(3, 3)	3.0
9	(3, 2)	2.5
10	(4, 4)	4.0
11	(4, 3)	3.5
12	(4, 2)	3.0
13	(4, 3)	3.5
14	(4, 2)	3.0
15	(3, 2)	2.5

(c)

$\bar{x}$	2.5	3.0	3.5	4.0
$P(x)$	2/15	4/15	6/15	3/15

9. (a) This is *not* a probability distribution. The sum of the probabilities exceeds 1. (b) This is a probability distribution. (c) This is a probability distribution. (d) This is *not* a probability distribution. One of the $P(x)$ is negative. (e) This is *not* a probability distribution. The sum of the probabilities is less than 1. (f) This is a probability distribution.

11. (a) (b) (c) (d)

−1 0 1
y
$E(Y) = .5$

−3 −2 −1 0 1
z
$E(Z) = -1.25$

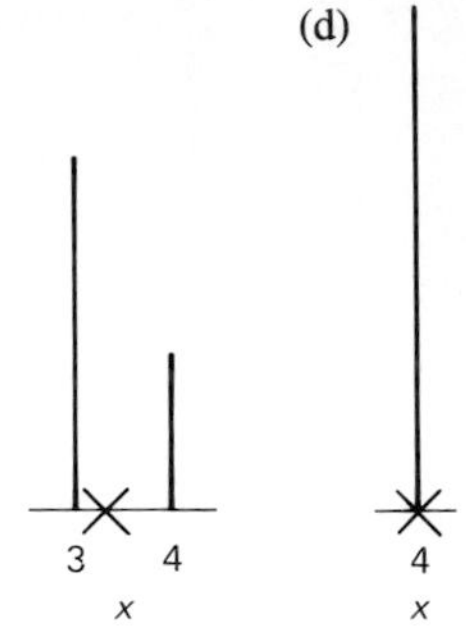

13. (a) .8 (b) 5.0

15. (a) \$1/12. On the average, over many repetitions of this game, the house will gain \$1/12, or \$1 per 12 games. (b) −\$1/12 (a loss) (c) \$11 17. \$.35

19. (a) 3.1 (b) 1.34 (c) .95 21. (a) (i) 3 cabs (ii) 1.14 cabs (b) (i) 5% (ii) 30%

23. (a) (b) (i) .6 (ii) .490

x	0	1
$P(x)$	.4	.6

REVIEW PROBLEMS

1. (a) This is *not* a probability distribution. The sum of the probabilities is less than 1. (b) This is *not* a probability distribution. One of the $P(x)$ is negative. (c) This is a probability distribution. (d) This is *not* a probability distribution. The sum of the probabilities exceeds 1.

3. (a) .25 (b) 2 (c) 5 (d) 2.4 (e) .35 (f) .75 5. (a) (i) 1.2 (ii) 1.1 (b) 46.5

7. (a) 3

(b) SS =	(c)	(d) M
1, 2	.1	1.5
1, 3	.1	2.0
1, 4	.1	2.5
1, 5	.1	3.0
2, 3	.1	2.5
2, 4	.1	3.0
2, 5	.1	3.5
3, 4	.1	3.5
3, 5	.1	4.0
4, 5	.1	4.5

(e)

m	1.5	2.0	2.5	3.0	3.5	4.0	4.5
$P(m)$	.1	.1	.2	.2	.2	.1	.1

(f) 3

9. (a) and (b)

	x		x		x		x		x		x
1,1	(1)	2,1	(2)	3,1	(3)	4,1	(4)	5,1	(5)	6,1	(6)
1,2	(2)	2,2	(2)	3,2	(3)	4,2	(4)	5,2	(5)	6,2	(6)
1,3	(3)	2,3	(3)	3,3	(3)	4,3	(4)	5,3	(5)	6,3	(6)
1,4	(4)	2,4	(4)	3,4	(4)	4,4	(4)	5,4	(5)	6,4	(6)
1,5	(5)	2,5	(5)	3,5	(5)	4,5	(5)	5,5	(5)	6,5	(6)
1,6	(6)	2,6	(6)	3,6	(6)	4,6	(6)	5,6	(6)	6,6	(6)

(c)

x	1	2	3	4	5	6
$P(x)$	1/36	3/36	5/36	7/36	9/36	11/36

(d) 4.472. Over an infinite number of replications of this experiment, the mean value of X is 4.472.
(e) 1.404. This is approximately the average amount by which the x values differ from 4.472.
(f) $1.11
11. (a) 4.9. Over all such patients the mean number of days spent in the hospital is 4.9 days.
(b) 1.179. This is approximately the average amount by which the d values differ from 4.9.
(c) .65
(d) $630
13. (a)

x	0	1	2	3
$P(x)$	.7	.03	.09	.18

(b) .75 aspirin per day
15. (a) $E(F)$ = 1.85 friends, $E(M)$ = 1.22 friends (b) .30 (c) .07
17. (a) $4.50 (b) No. In this case his average cost would be $5.00.
(c) Yes. In this case he should use the garage since his average cost ($4.00) would be less than that for the meter.
19. (a) .99. There will be an average of almost one complaint per day.
(b) 1.015. This is approximately the average amount by which the c values differ from .99.
21. $2,000,000
23.

	MEAN MPG	MEAN PROFIT
I	27.5	$2500
II	28.5	2300
III	29.0	2200
IV	29.25	2150

Plan 3 satisfies the law and produces the greatest profit.
25. (a) 3.2 (b) .1691

27. (a)

x	0	1	2	3
$P(x)$	.7073	.2	.09	.0027

(b) .3881

CHAPTER 5

5. No. The probability of a success *changes* with each trial (shot).
7. No. There are more than two outcomes per trial (selection).
9. No. This is not a number of Bernoulli trials.
11. No. A measurement rather than a 0 or a 1 is recorded for each family.
13. Yes. $n = 5$, $\pi = .1$
15.

	Equation 5.1	*Table A1*
(a)	.1115	.111
(b)	.384	.384
(c)	.3164	.316
(d)	.0439	.044
(e)	.0984	.098
(f)	.4096	.410
(g)	.1787	.179

17. (a) .197 (b) .267 (c) .789 (d) .862
19. (a) (i) .208 (ii) .100
(b)

	0	1	2	3	4	5	6	7	8
	.100	.267	.311	.208	.087	.023	.004	0	0

(c) 2
(d)

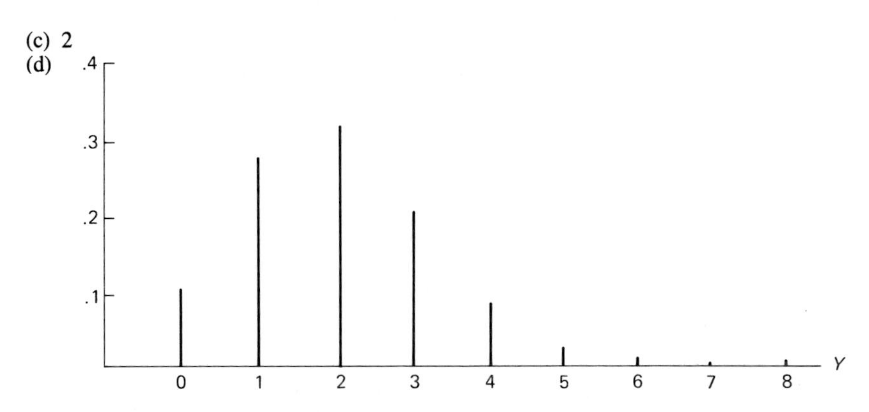

21. (a) .015 (b) .985 (c) .223 (d) .777
23. (a) $\frac{9!}{1!8!}(.0822)^1(.9178)^8$ (b) $1 - \frac{9!}{0!9!}(.0822)^0(.9178)^9$ where $.0822 = 30/365$.
25. $E(Y) = 16.667$, $SD(Y) = 3.909$
27. 26 29. 1985.046 31. $n = 5000$

REVIEW PROBLEMS

1. (a) This is *not* a binomial experiment. We are recording a measurement, not a 0 or a 1 for each teacher. (b) This is a binomial experiment
(c) This is *not* a binomial experiment. The number of trials (sample size) is not fixed.
(d) Same as part (c).
3. (a) .309 (b) .168 (c) .832 (d) .360 5. (a) .002 (b) .218 (c) .968
7. (a) 5 (b) 90 (c) 90
9. (a) 0, 1, 2, 3, 4
(b)

	0	1	2	3	4
	16/81	32/81	24/81	8/81	1/81

(c) $0\left(\frac{16}{81}\right) + 1\left(\frac{32}{81}\right) + \cdots + 4\left(\frac{1}{81}\right) = \frac{4}{3}$
11. $n = 100$, $\pi = .3$ 13. 15
15. (a) (i) .877 (ii) .125 (b) 8
17. (a) .202 (b) .415 (c) .041 (d) 15 (e) .617
19. (a) (i) .125 (ii) .134 (b) (i) .128 (ii) .130
21. (a) .444 (b) .556 (c) .138
23. .000325. A random process is very unlikely to result in three consecutive juries without blacks.
25. (a) .302 (b) .678 (c) .107 (d) .268
27. (a) .2 (b) (i) .410 (ii) .410 (iii) .590
29. (a) .391 (b) .495 31. (a) .011 (b) 6.25
33. (a) (i) .190 (ii) .133 (b) 2 35. .07

CHAPTER 6

3. (a) and (b)

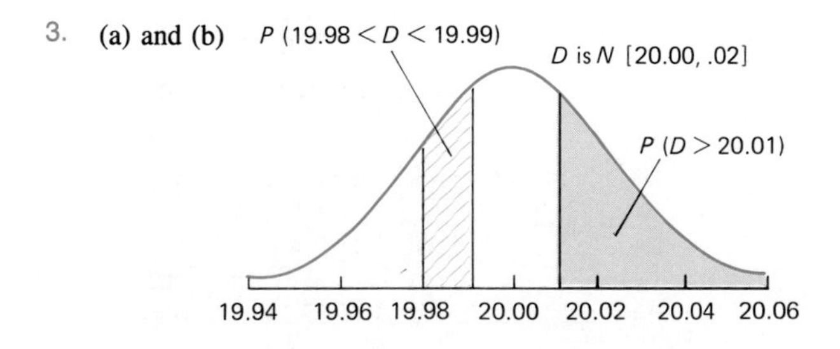

(c) (i) 68% (ii) 47.5% (iii) 2.5% (iv) 13.5% (v) 84%

5. (a) .5517

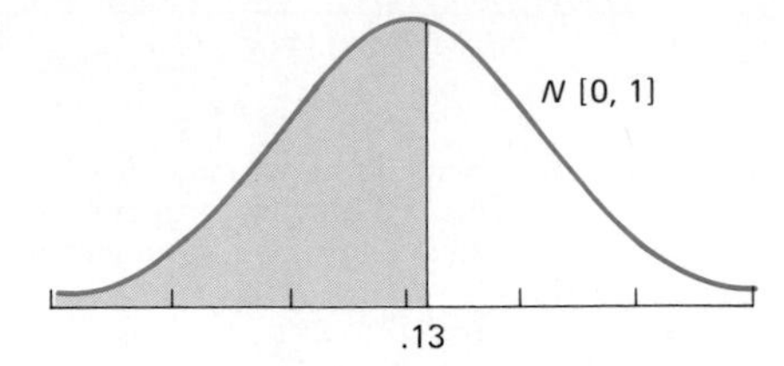

(b) .2946

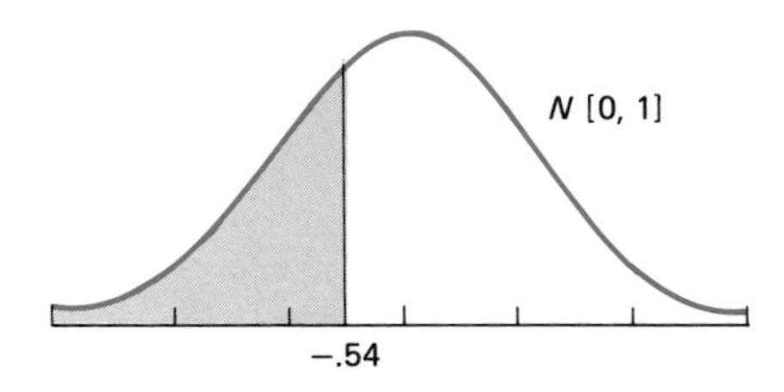

(c) 1

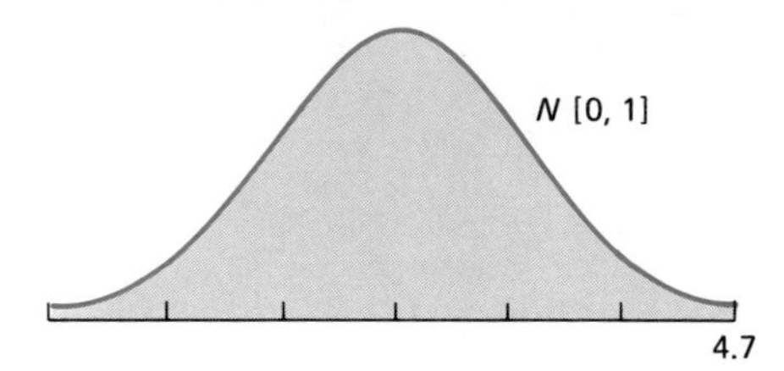

(d) .9591

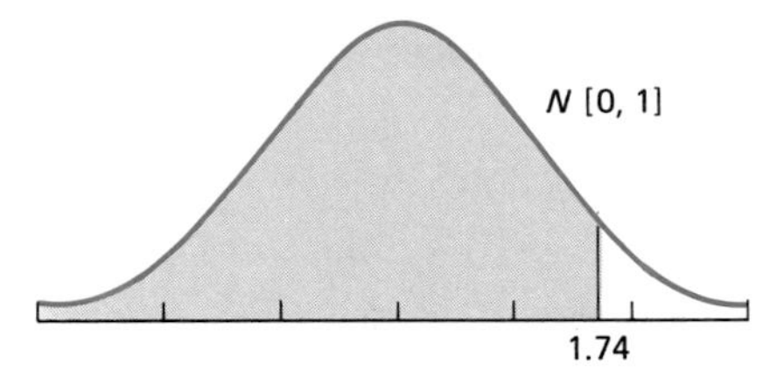

7. (a) .5477

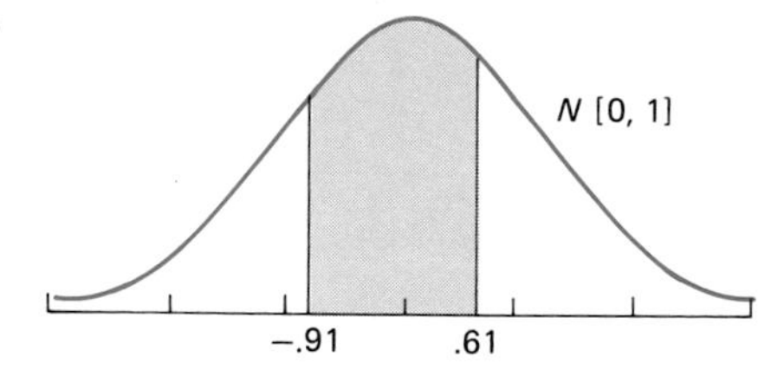

(b) .0895

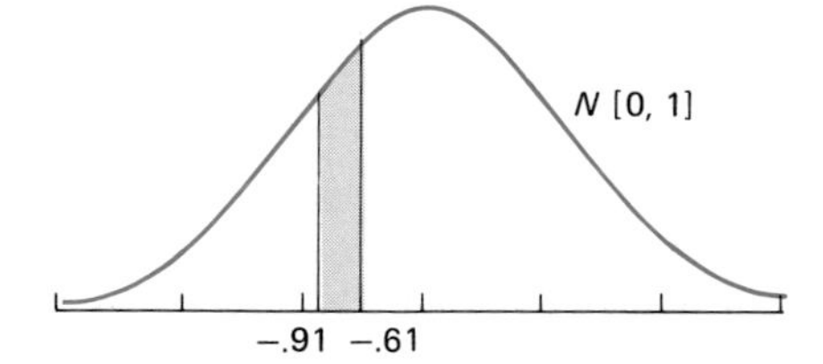

(c) .1865

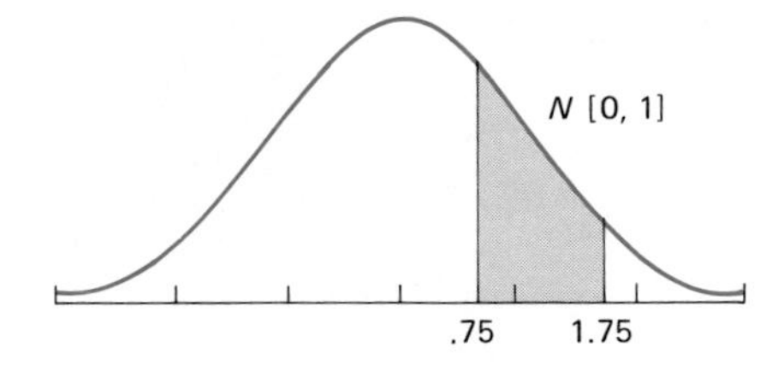

(d) .7864

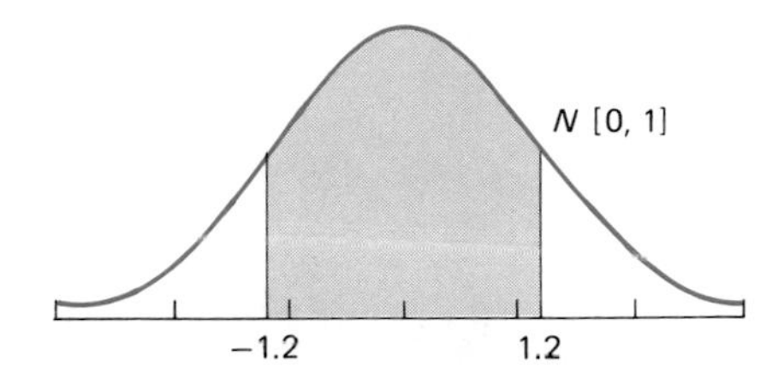

(e) .2642

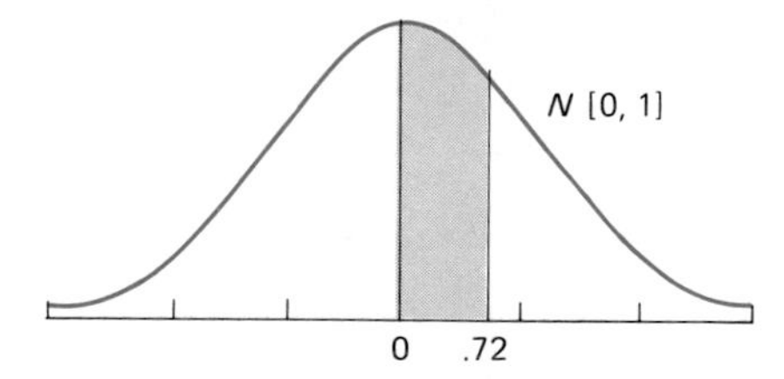

(f) .9495

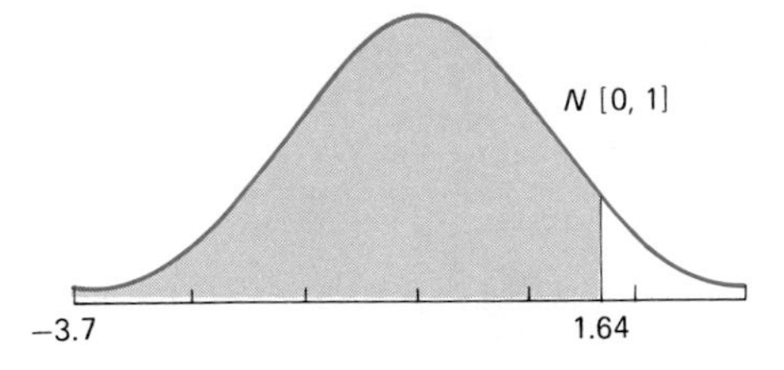

(g) .3594

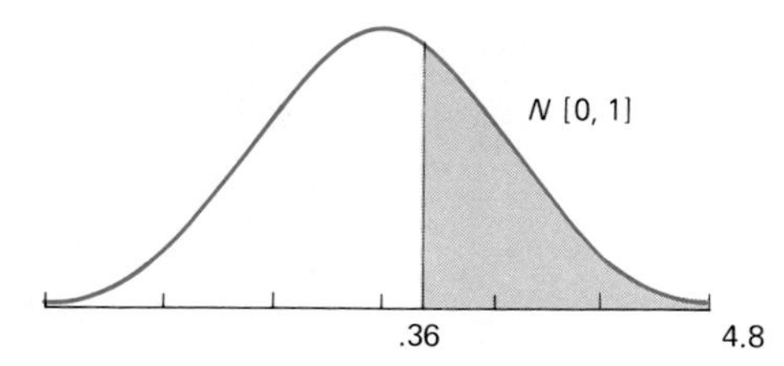

9. (a) −1.036

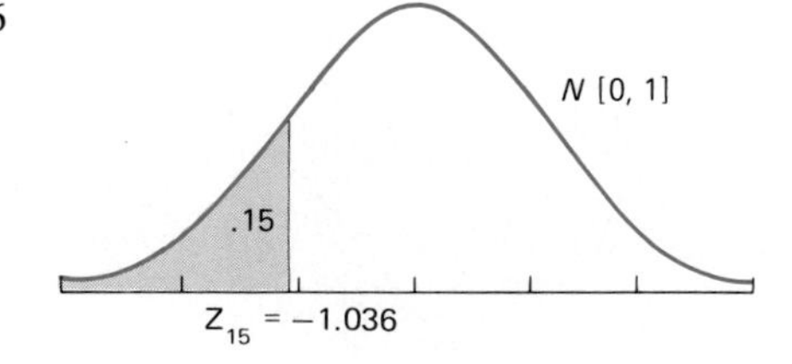

(b) 2.326

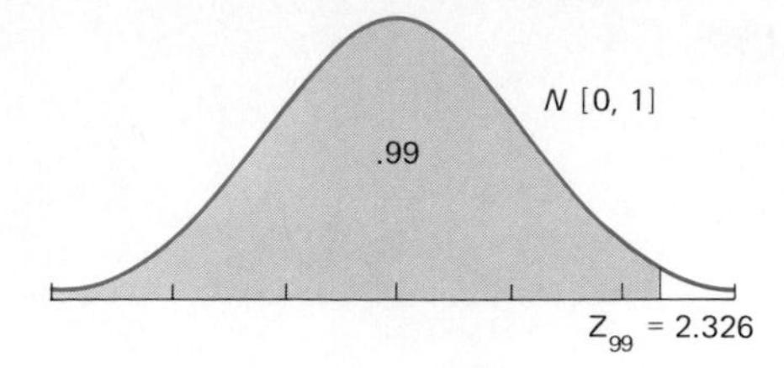

(c) −2.326

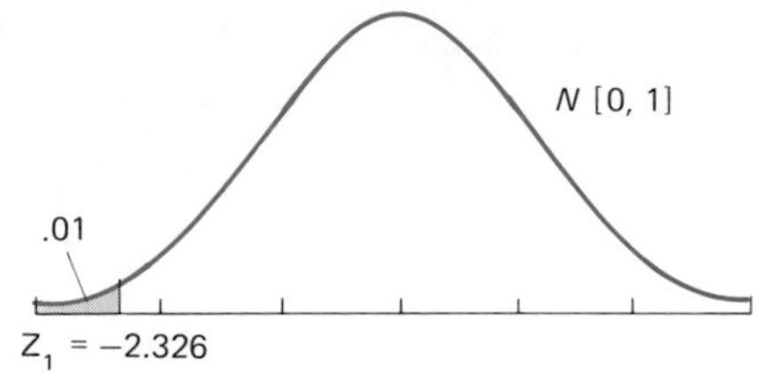

(d) .1256

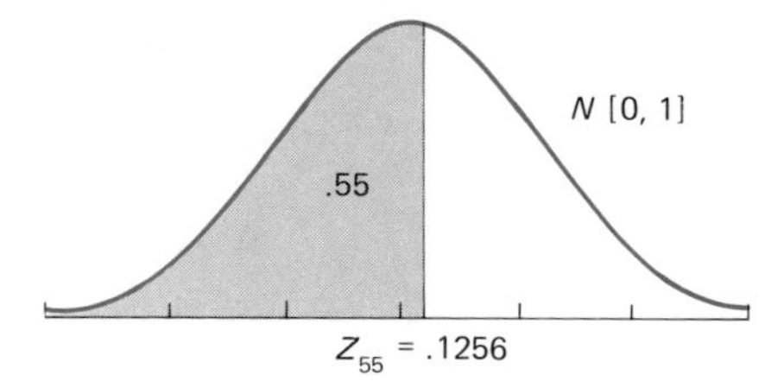

(e) −2.576

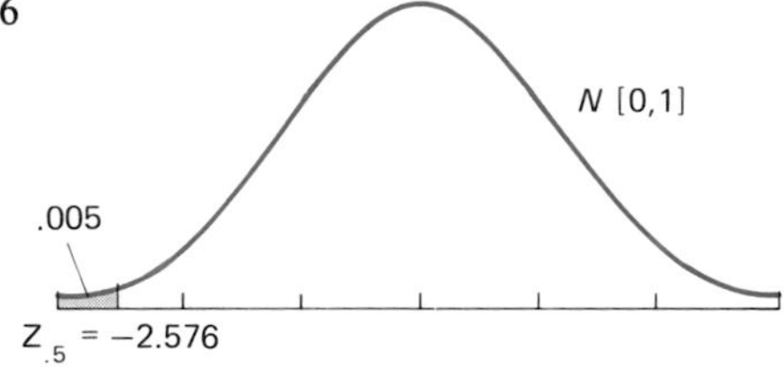

(f) 2.807

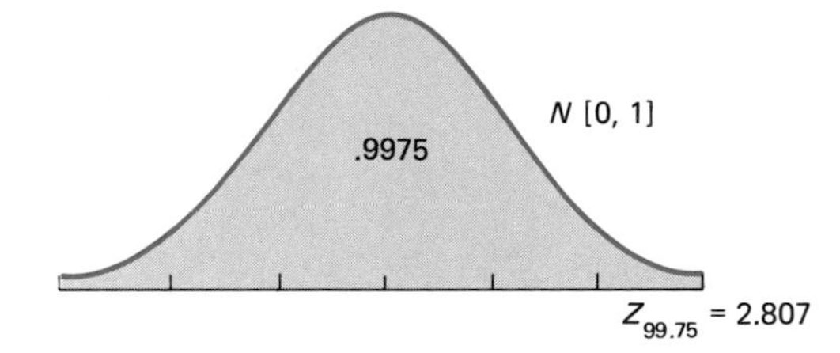

13. (a) .1335

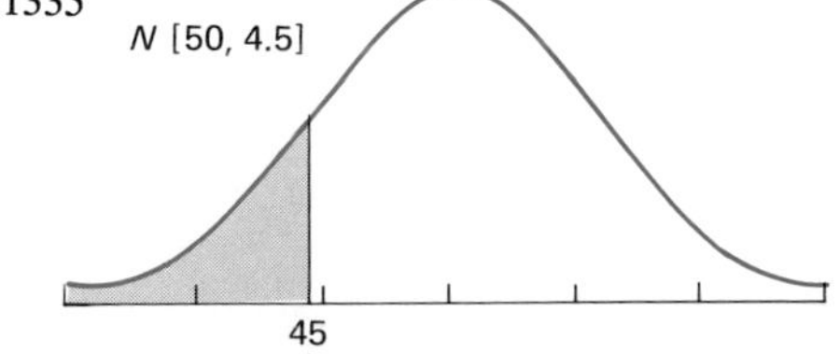

(b) .6266

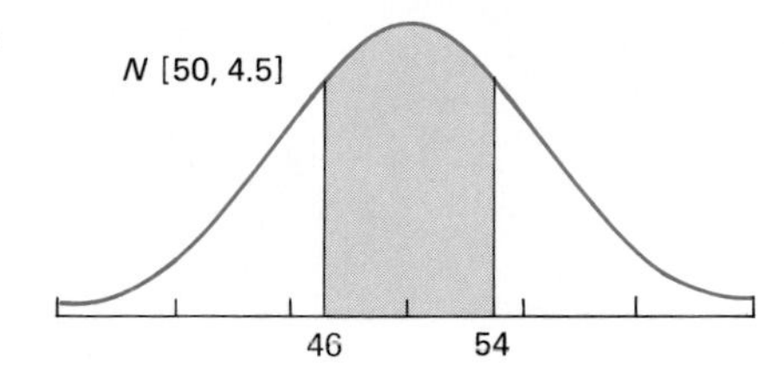

(c) .0132

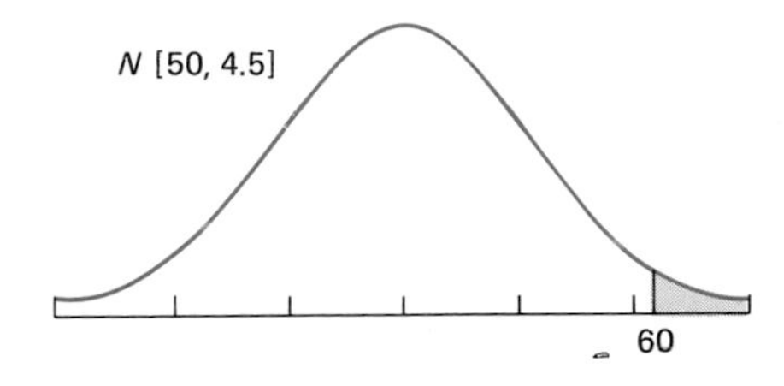

15. (a) $9536

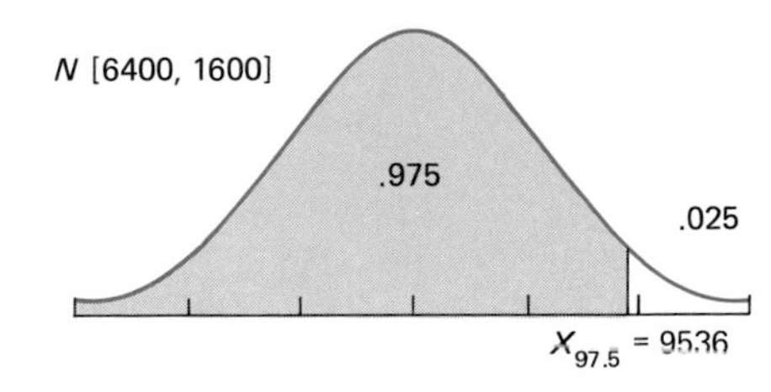

(b) $4096

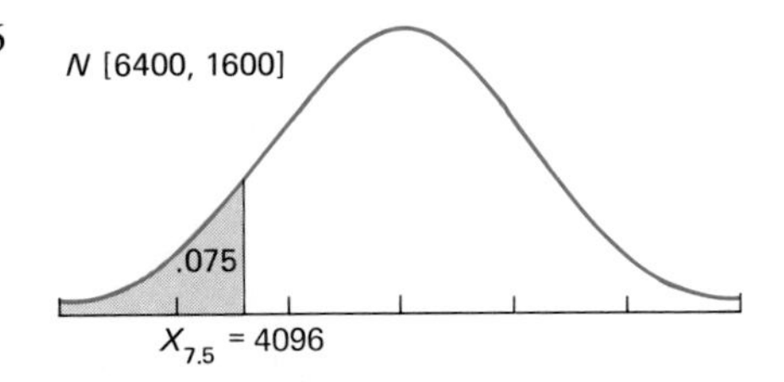

17. (a) .0084
(b) .42

19. (a) 552.56 (553)

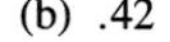

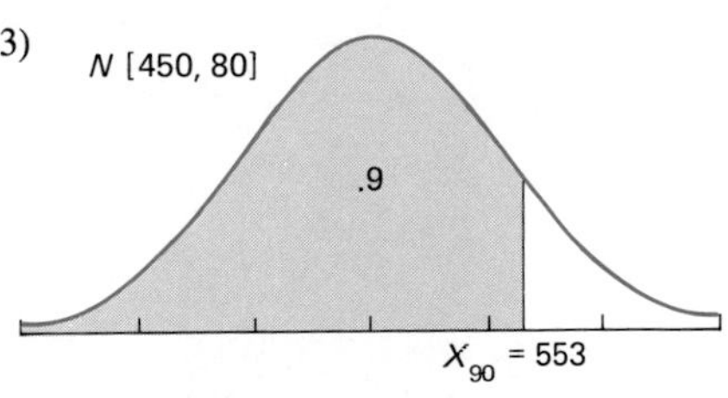

(b) (i) 408.05 (408)

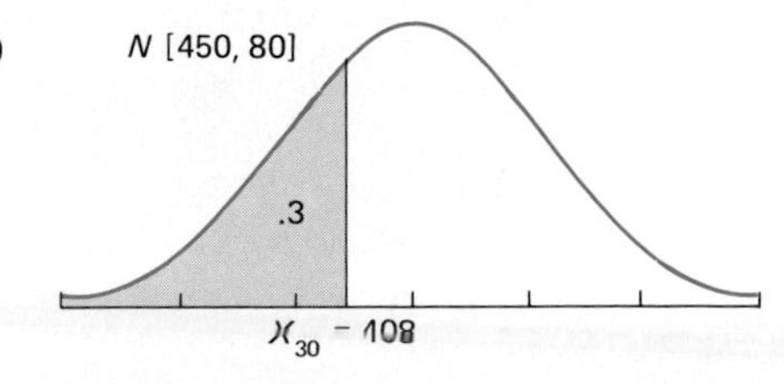

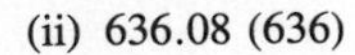
(ii) 636.08 (636)

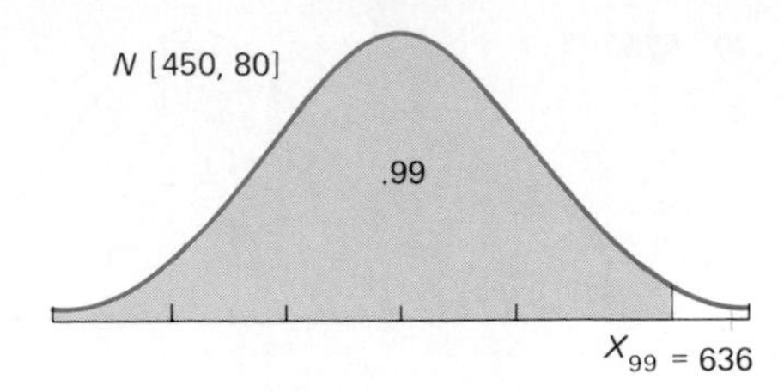

21. $17,990

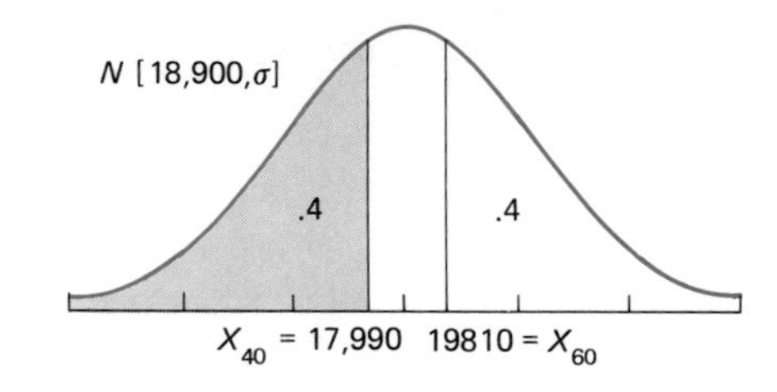

REVIEW PROBLEMS

1. (a) .8944

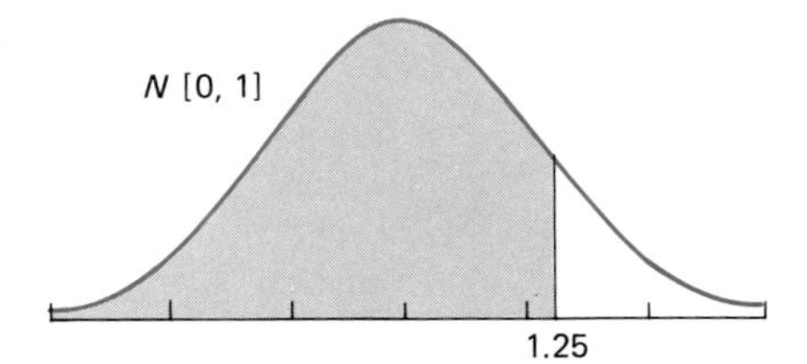

(b) .0778

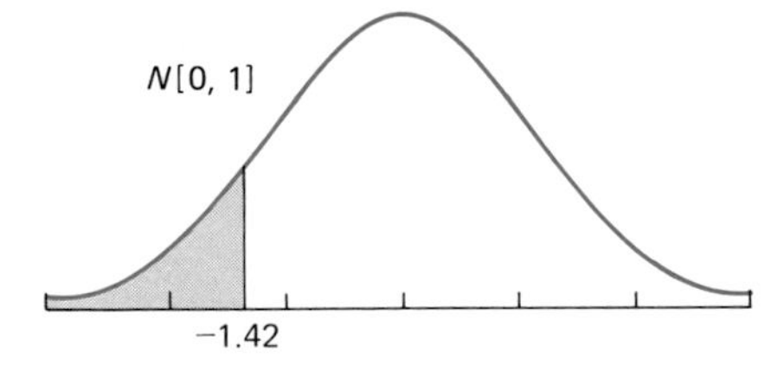

(c) 1

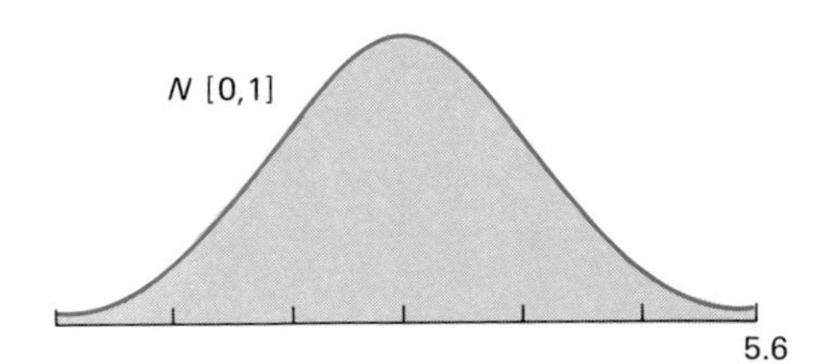

(d) .5803

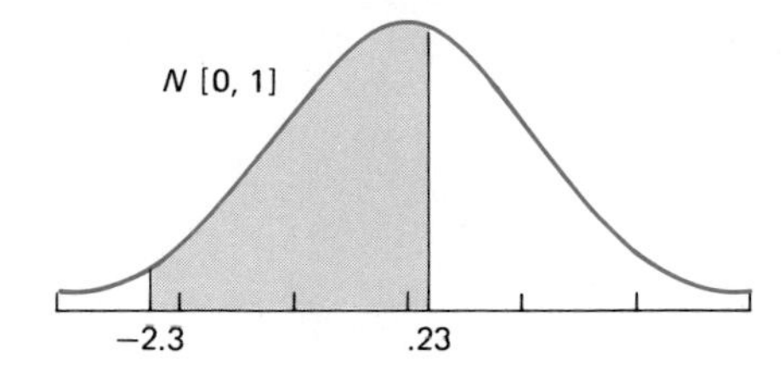

(e) .3599

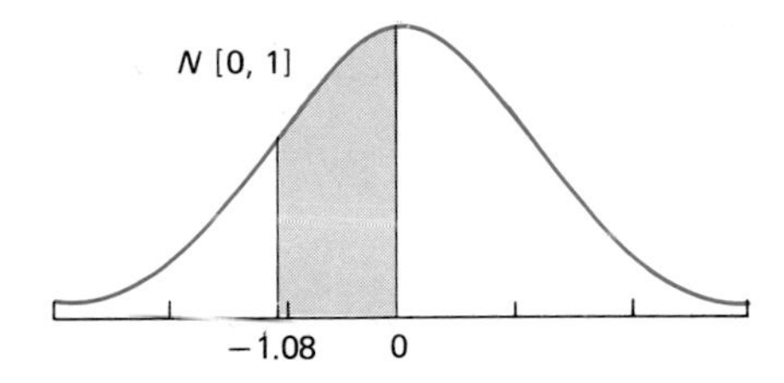

(f) .5239

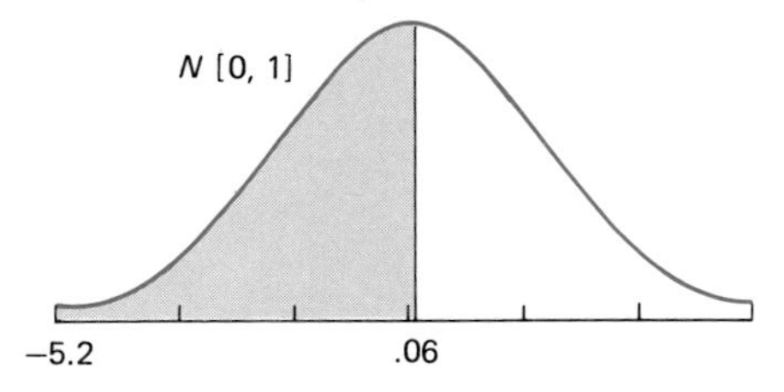

(g) .3409

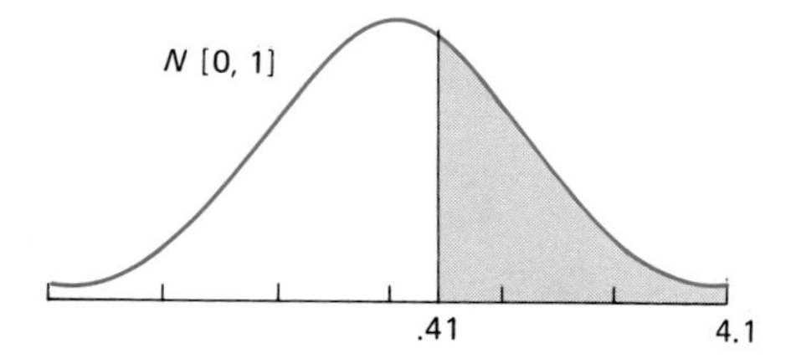

3. (a) (i) .3085

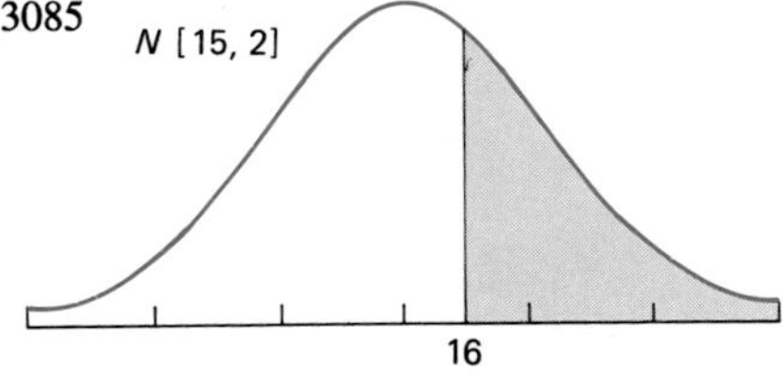

(ii) .4773

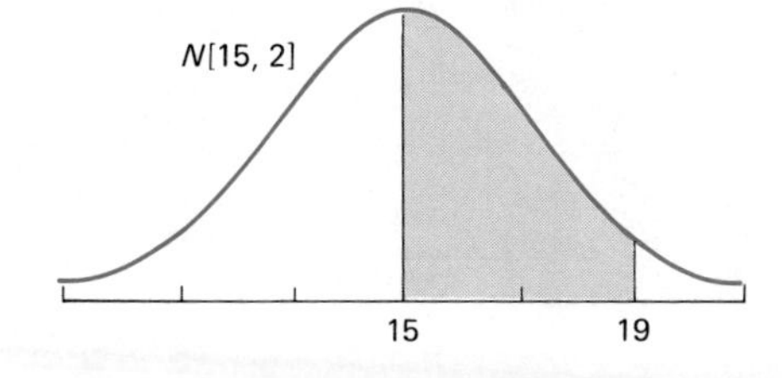

(iii) .8944

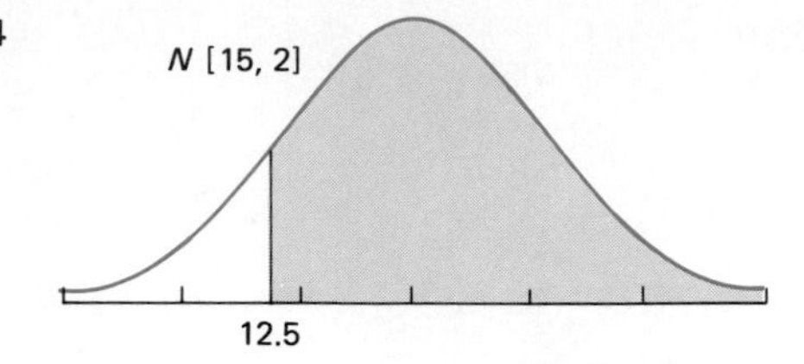

(b) (i) 17.564 hours

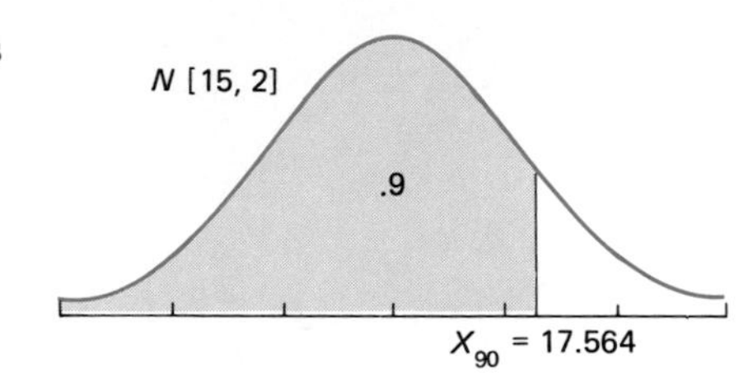

(ii) 15 hours

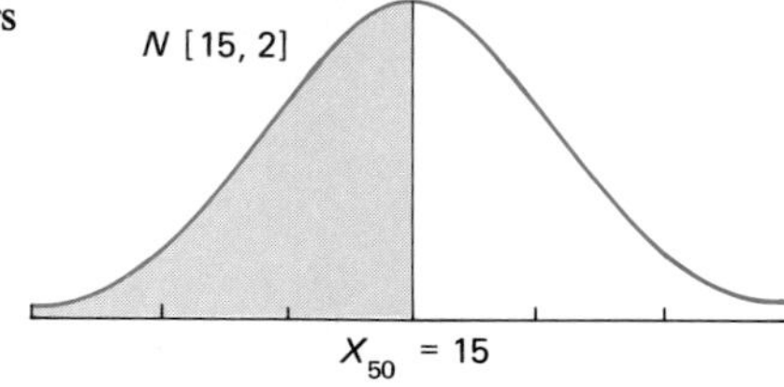

(iii) 17.072 hours

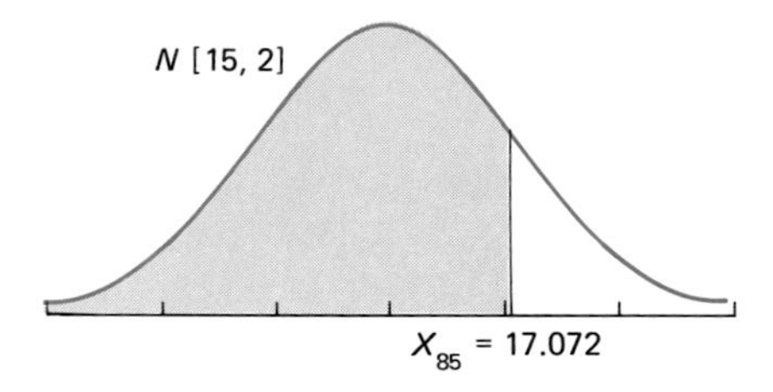

5. (a) (i) .1057

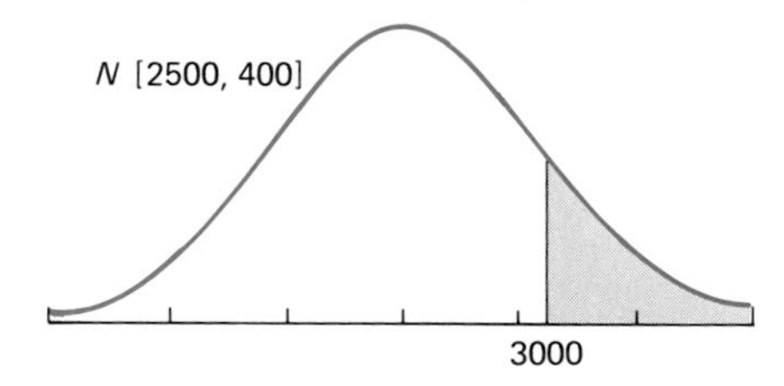

(ii) .00009

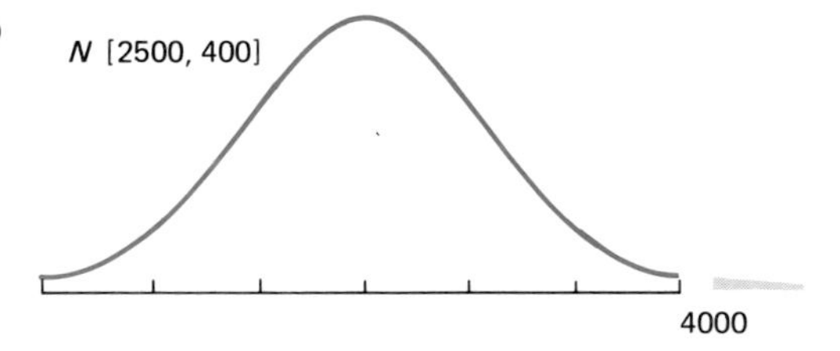

(iii) .3944

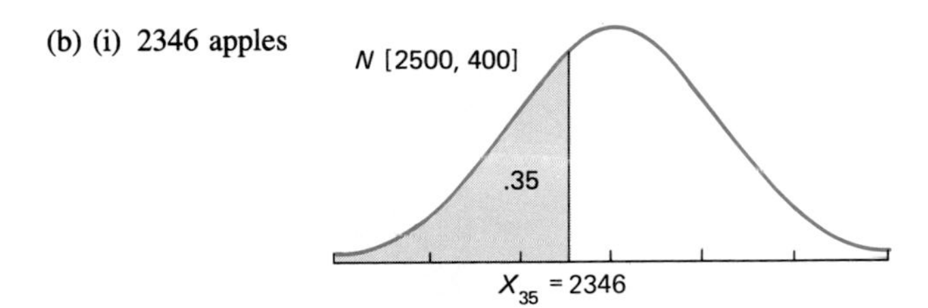

(b) (i) 2346 apples

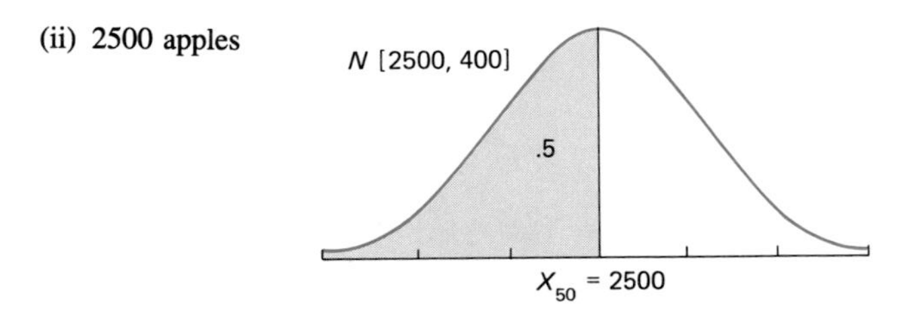

(ii) 2500 apples

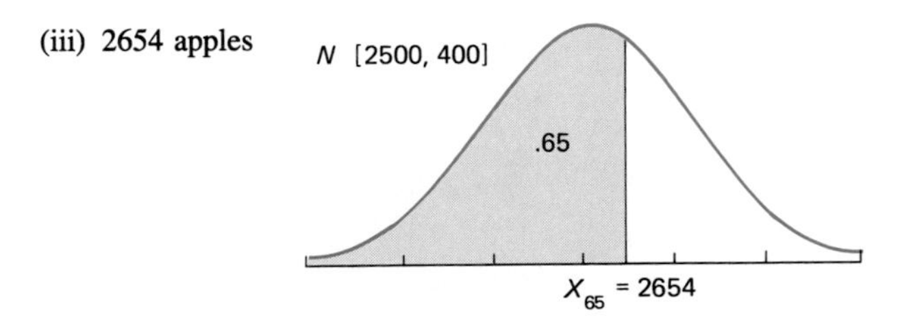

(iii) 2654 apples

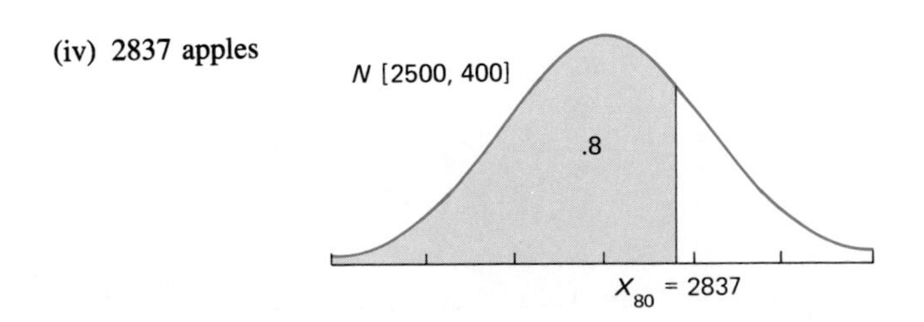

(iv) 2837 apples

N [2500, 400]

.8

X_{80} = 2837

7. (a) and (b)

	PROPORTION	EXPECTED
A	.0668	13.36
B	.2417	48.34
C	.383	76.60
D	.2417	48.34
F	.0668	13.36
	1.0000	200.00

9. $\mu = \$26{,}289.67$, $\sigma = \$7106.34$

11. (a) .9082

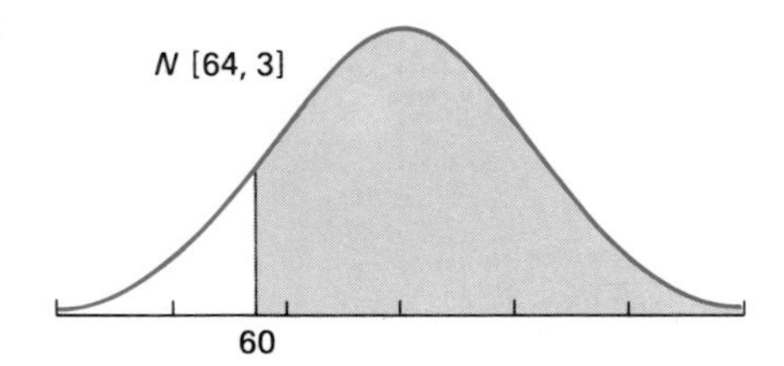

(b) .4082

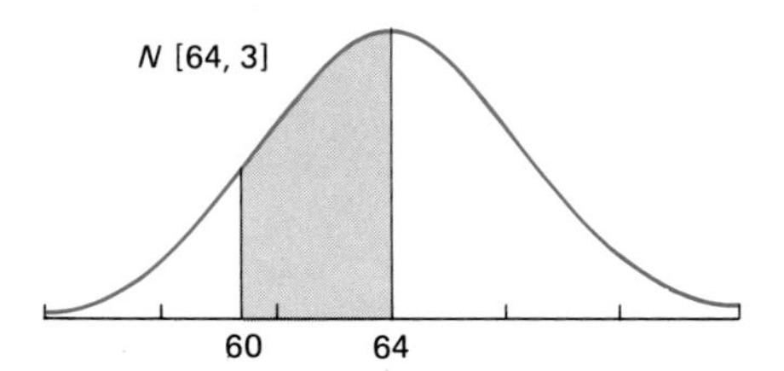

(c) .9962

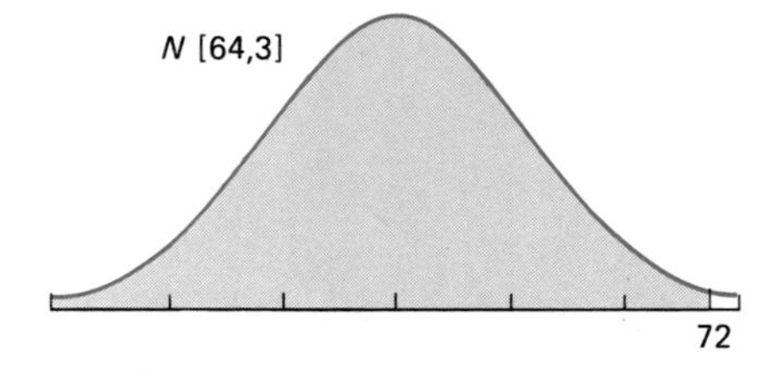

13. (a) $1 - t$

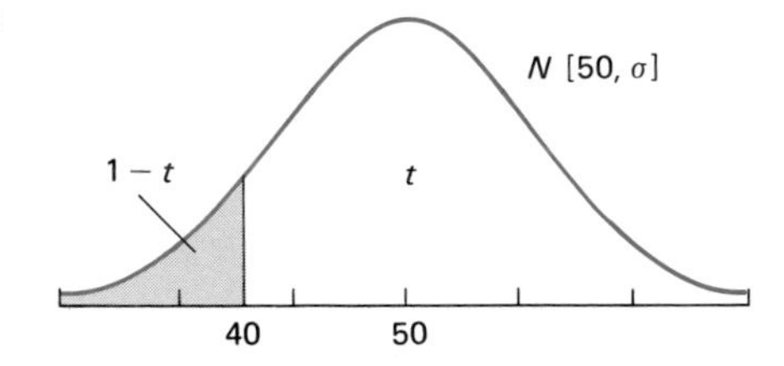

(b) t

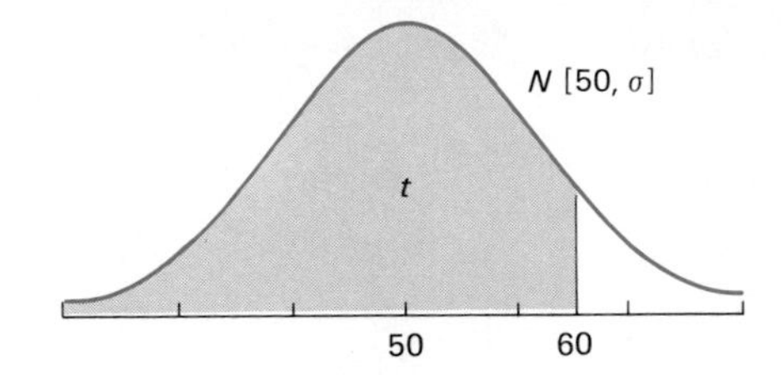

(c) $2t - 1$

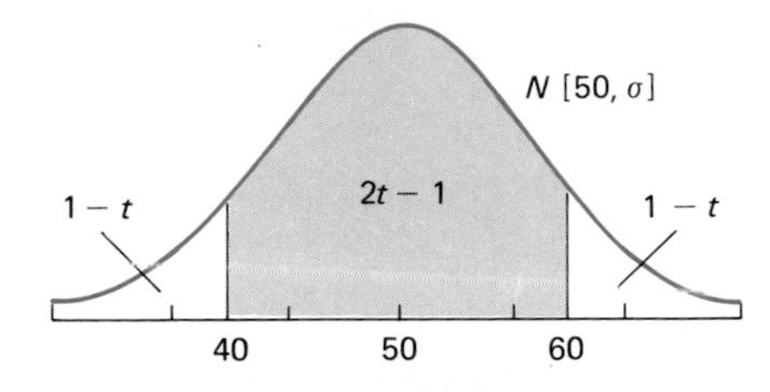

(d) $t - .5$

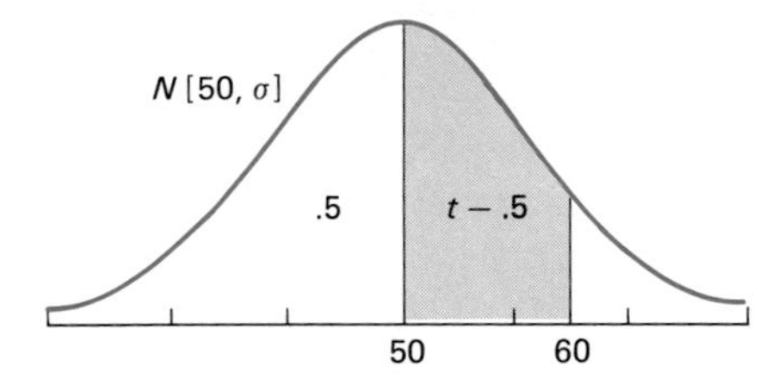

(e) $1.5 - t$

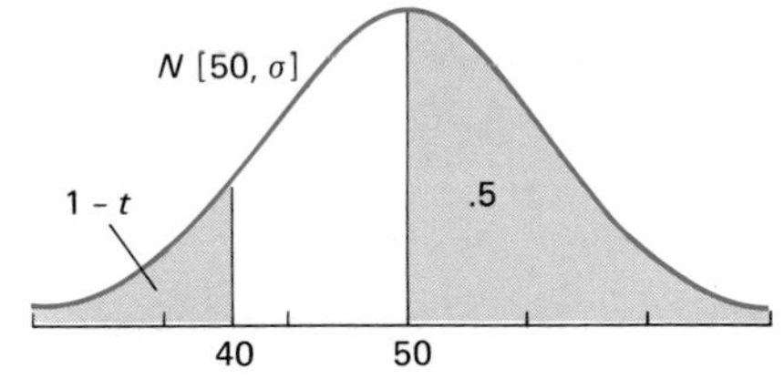

15. (a) (i) .2514

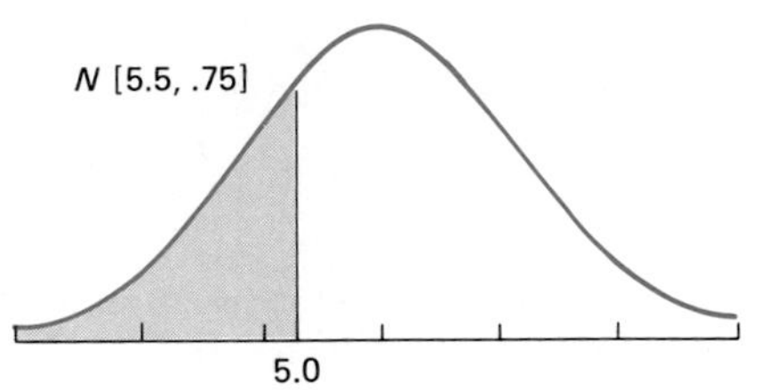

(ii) .4972

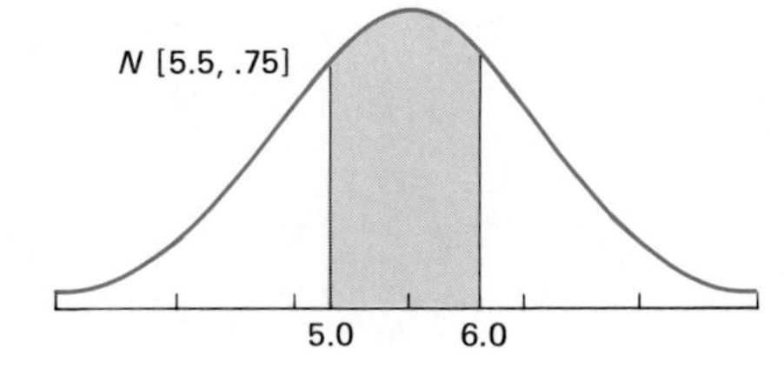

(b) 6.73 hours

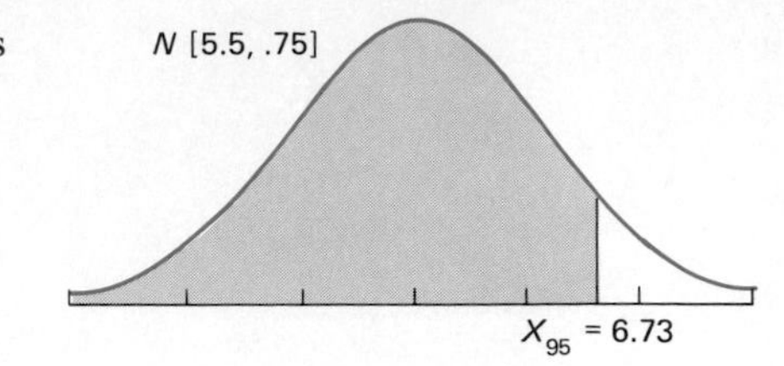

17. (a) .0554 (b) .0567
19. (a) (i) .2843

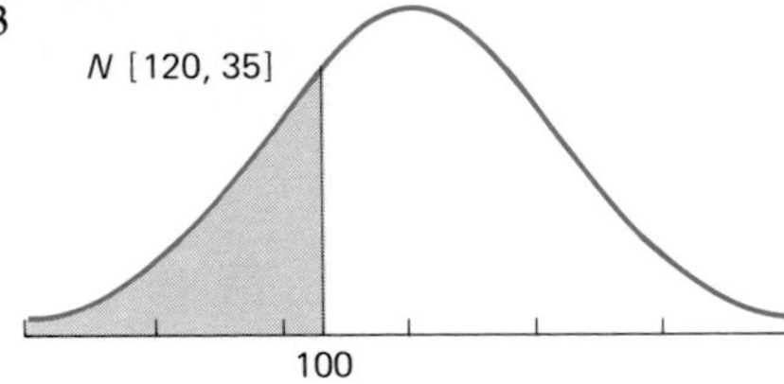

(ii) .2953

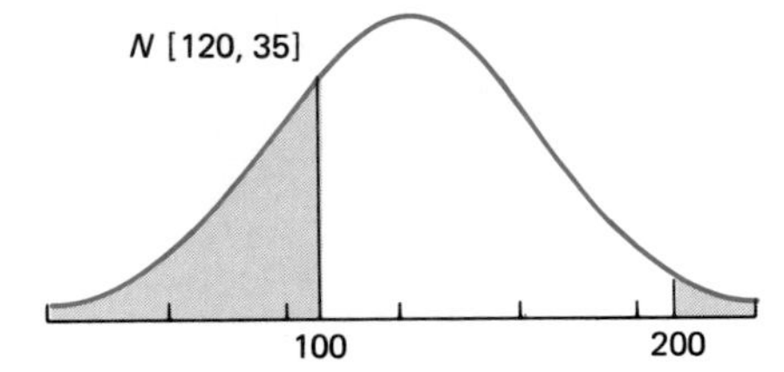

(b) $156.26

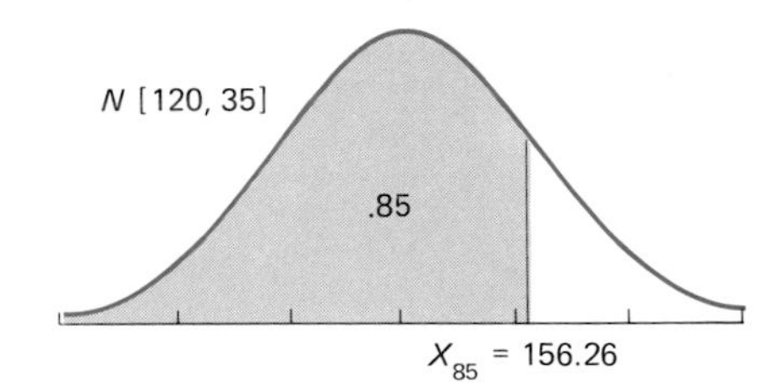

(c) Eighty-five percent of recipients received less than $156.26.

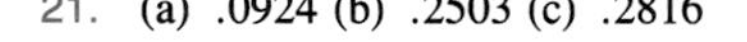
21. (a) .0924 (b) .2503 (c) .2816 23. (a) .0112 (b) .1891

25. (a) .4681 (b) .1026

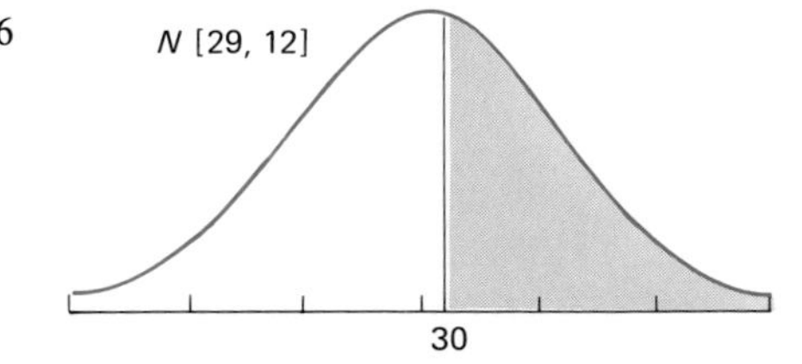

27. (a) (i) .2514

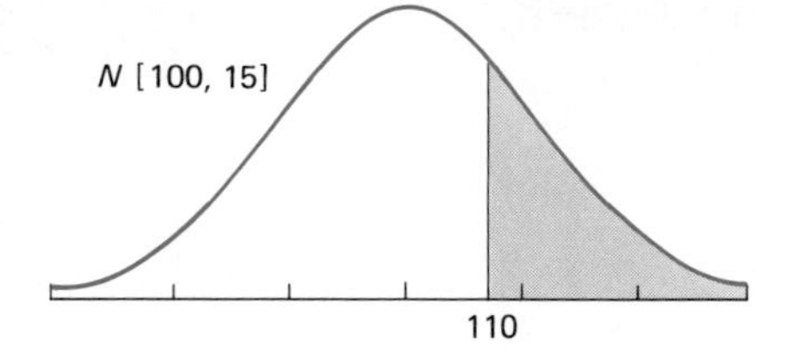

(ii) .2586

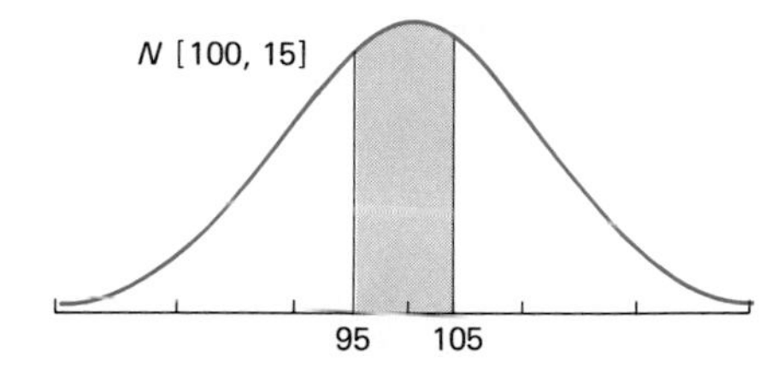

(iii) .1596

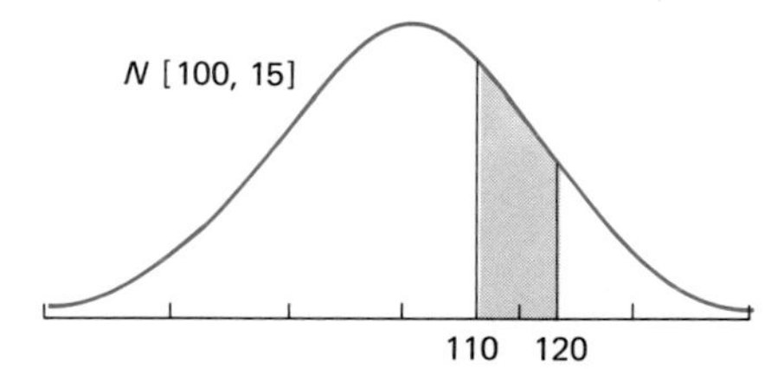

(b) The normal distribution does not assign equal probabilities to intervals of equal length. The area over an interval *close* to μ (i.e., the probability of falling in that interval) will be larger than the area over an interval of equal length far from μ.

(c) 138.64

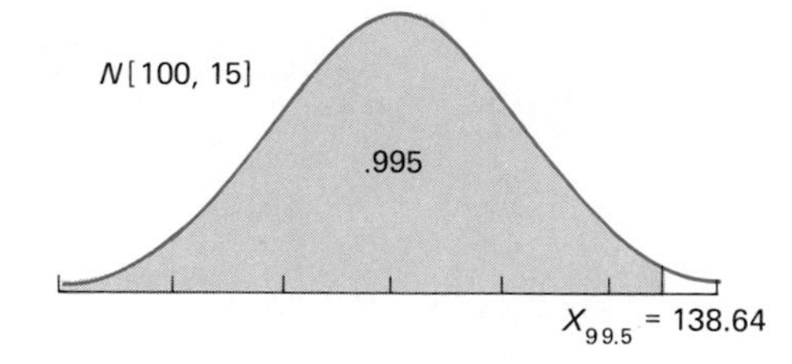

29. (a) .0359

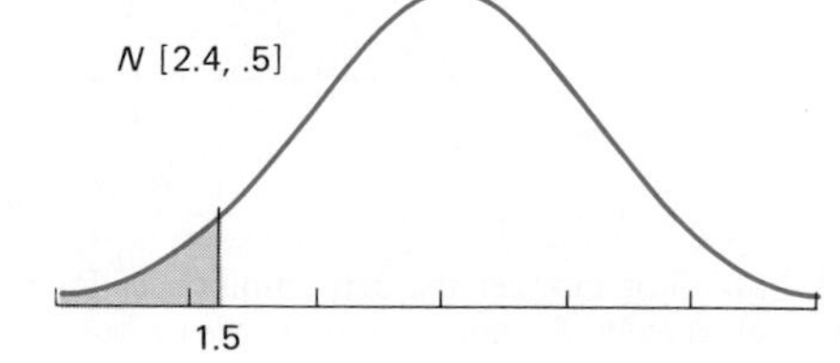

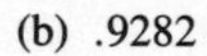

(b) .9282

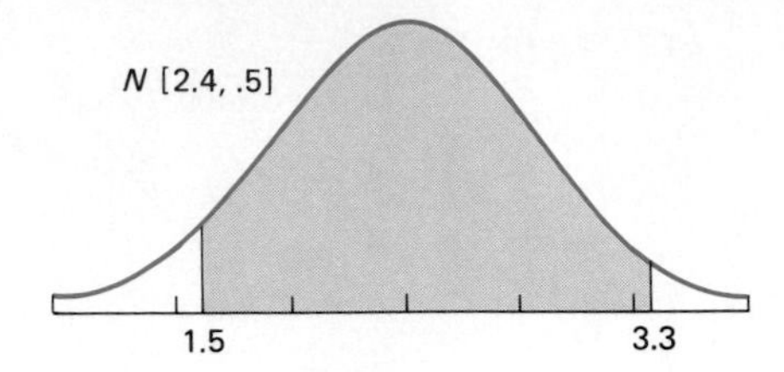

31. (a) (i) .9773

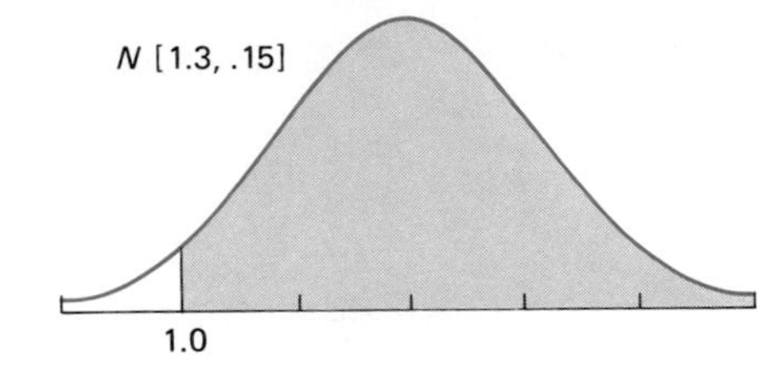

(ii) .0918

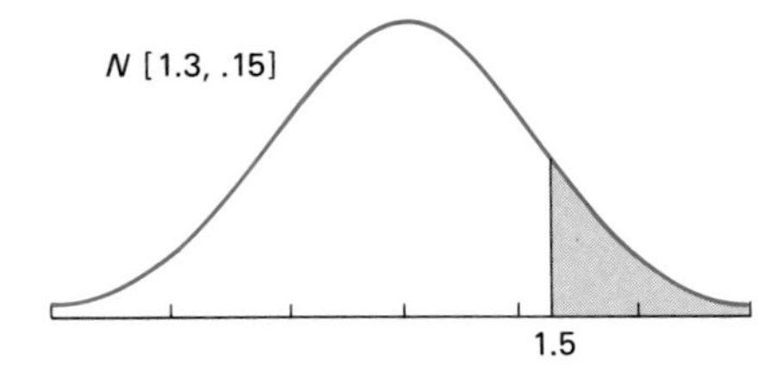

(b) (i) 1.426

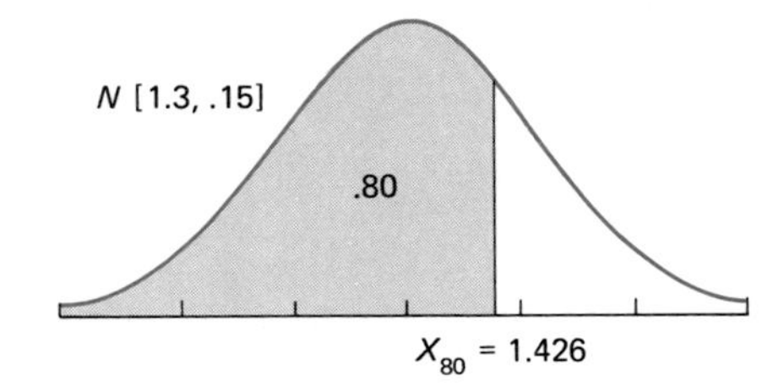

(ii) 1.262

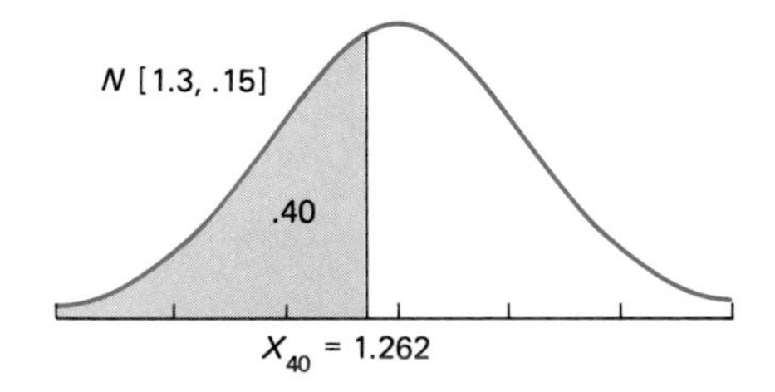

(c) In 80% of samples the concentration of fluoride will be below 1.426 parts per million. In 40% of samples the concentration will be below 1.262 parts per million.

33. (a) .0718

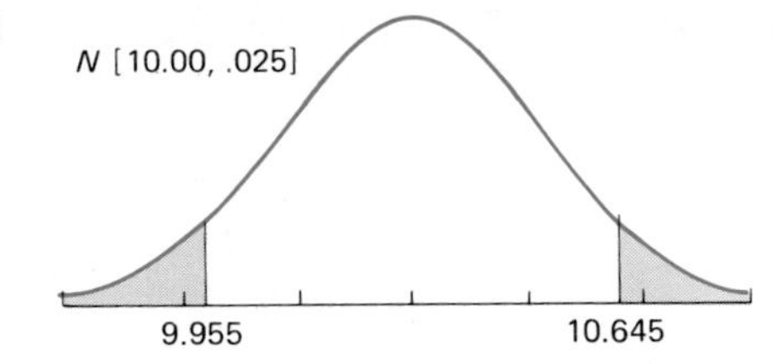

(b) .0028

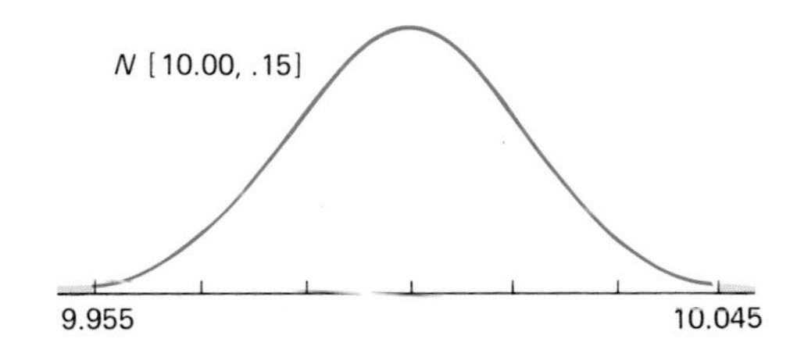

CHAPTER 7

3. (a) (i) .6 (ii) .02
(b) Over all possible samples of 600 adults, the average proportion who watch public TV is .6. Similarly, over all such samples, the sample proportions will *differ* from .6 by an average amount approximately equal to .02.
5. (a) 2.40 visits (b) .310 visit (c) .2 (d) .052 7. 2400

11. (a) .7324

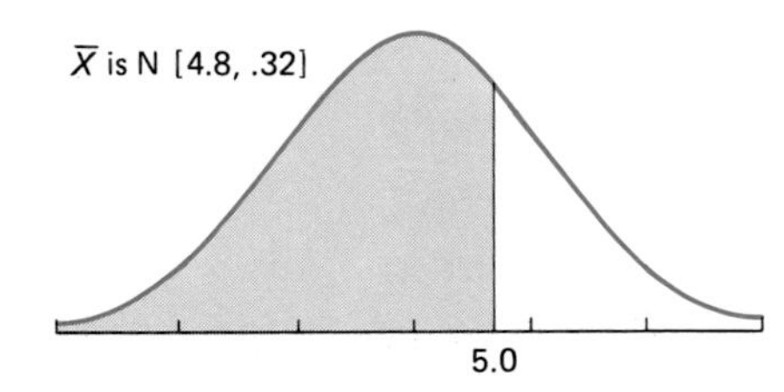

(b) .0009

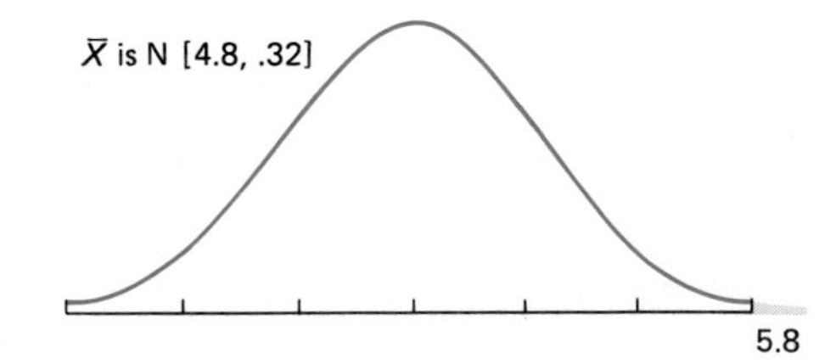

(c) .4648

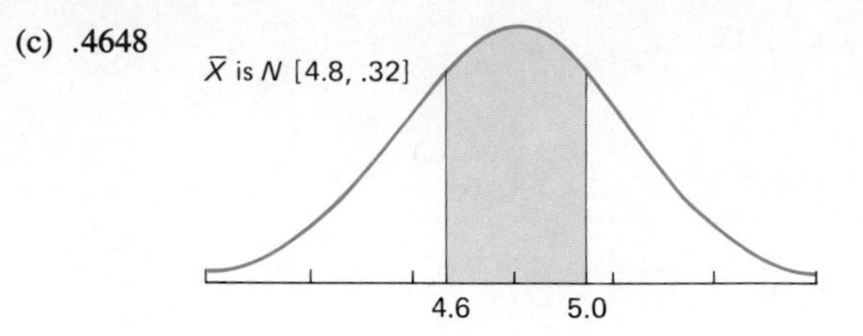

13. (a) .3085

(b) .1587

(c) .0475

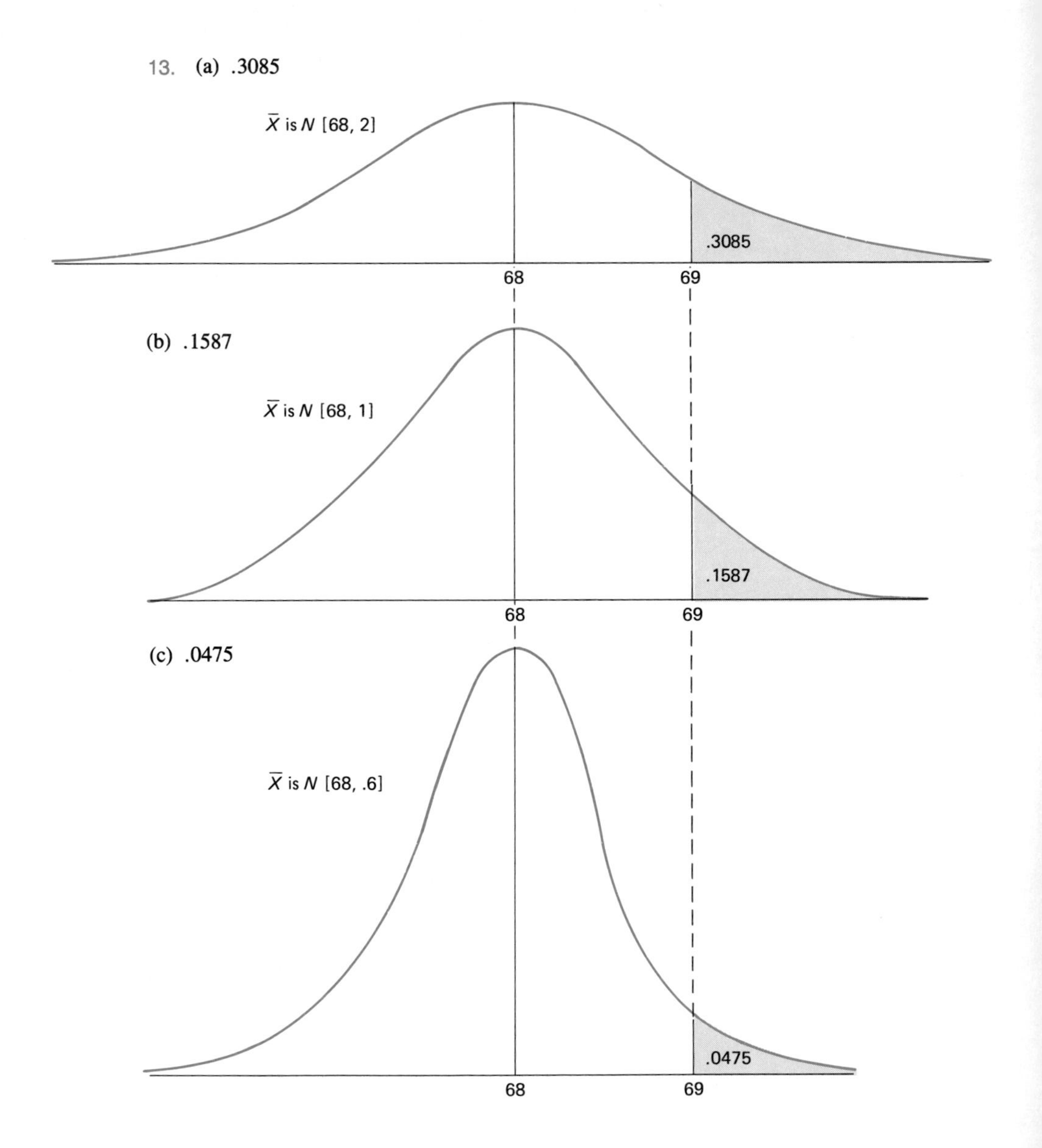

15. .0139

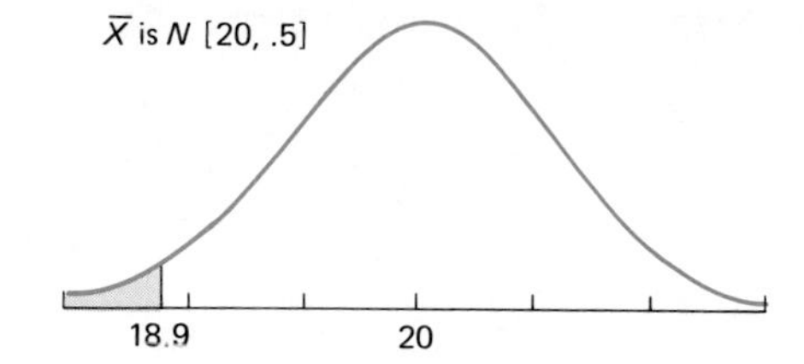

A sample mean below 18.9 would suggest that μ is not as large as 20. This is due to the fact that if μ were this large, a sample mean below 18.9 is unlikely (.0139) to occur.

17. (a) .0427

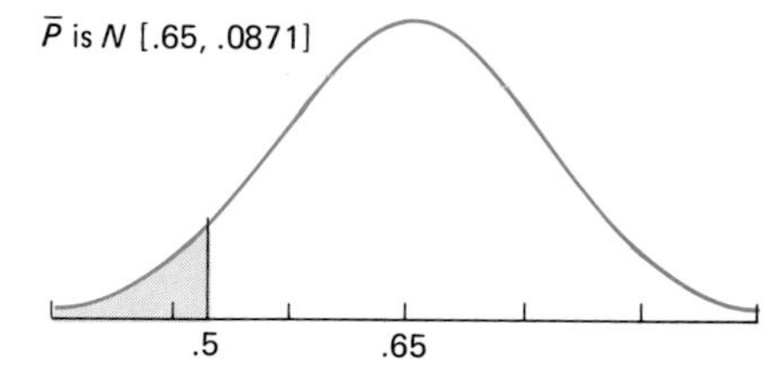

(b) .0837

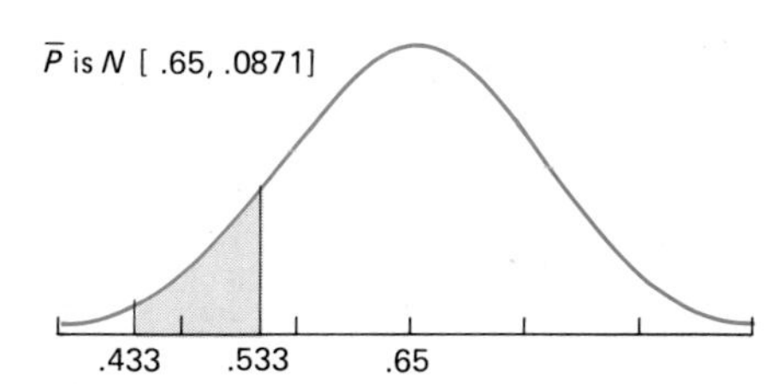

(c) .5

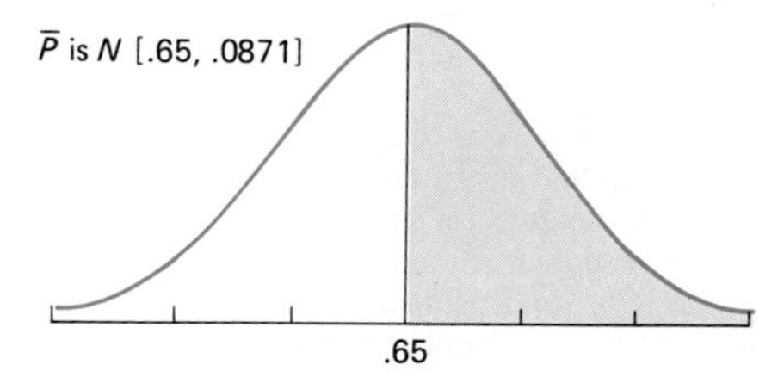

19. (a) .4448

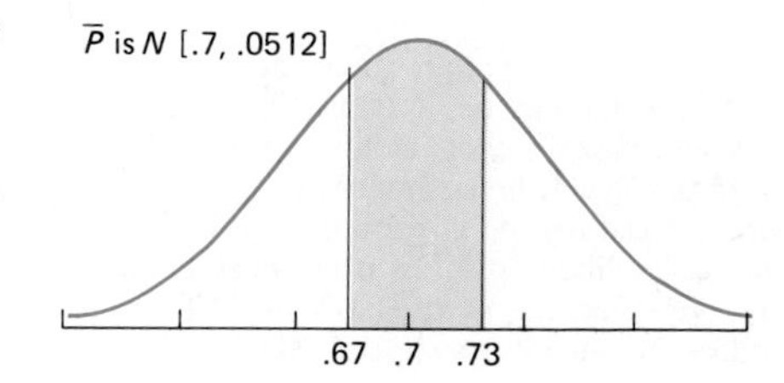

(b) .2776

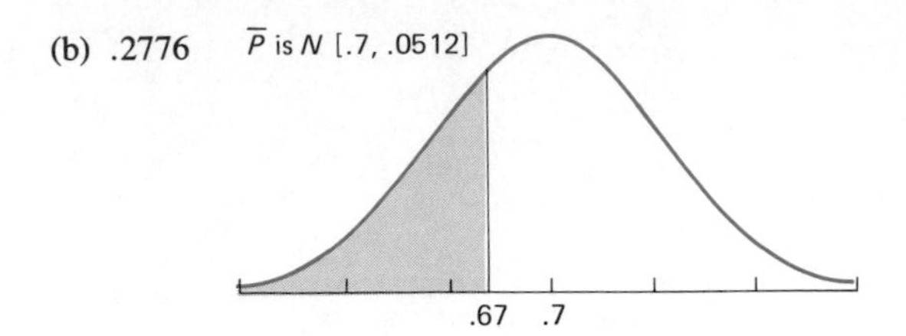

REVIEW PROBLEMS

1. (a) False. The CLT assumes only that the population has a mean and a standard deviation. (b) true
(c) False. The CLT is concerned with the *distribution* of $\overline{X}$ over all possible samples of a given size. (d) true
3. (a) .0110 (b) .9624
5. (a) (i) .5034 (ii) .6680 (iii) .7887
(b) As n increases, $SD(\overline{X}) = 8/\sqrt{n}$ gets smaller. Thus the values 62 and 64 are more standard deviations from 63 and $P(62 < \overline{X} < 64)$ increases.
7. (a) (i) .0436 (ii) .4766 (b) .200 9. (a) .0823 (b) .9997 11. 62
13. (a) (i) .0409 (ii) .1492
(b) (i) If $\overline{X} < 2250$ when, in fact, $\mu = 2500$, the Army will *not* accept the vehicles even though Brash has met the terms of the agreement ($\mu = 2500$). The probability of this error occurring is .0409.
(ii) Conversely, if $\overline{X} > 2250$ when, in fact, $\mu = 2100$, the Army will accept the vehicles even though Brash has in fact *not* met the terms of the agreement (μ is only 2100). The probability of this type of error is .1492.
15. (a) (i) .0392 (ii) .4608 (b) (i) .9231 (ii) .0753 17. .0683
19. (a) .0384 (b) .4616 21. (a) (i) .0146 (ii) .6157 (b) $n = 1421$
23. (a) .1190 (b) (i) .0384 (ii) .8485 25. (a) .6957 (b) .0102 27. .0985
29. (a) (i) .2090 (ii) .2017 (b) .3936 31. (a) (i) .2090 (ii) .1131 (b) .4139

CHAPTER 8

3. Assume that every freshman is given a number from 1 to 696. A table of random numbers can be used to select a sample of 25. A random starting point must be selected. Moving in some prearranged fashion select successive three-digit numbers. If a number is less than or equal to 696, the freshman with that number is included in the sample. Otherwise, skip to the next number. If a number has been selected previously, move to the next number. After 25 numbers (freshmen) have been selected, stop.
5. People choose whether or not to fill in and return such a ballot. There is no way of knowing or estimating how representative those returning ballots are of the entire voting population.
7. (a) Nursing home residents are *not* representative of *all* senior citizens. Further, the nursing home may be in an area quite atypical of the rest of the state. (b) By choosing the first 50 senior citizens encountered in the street, the sample will be biased against those who are institutionalized, house-bound, or in other ways unlikely to be in the street at that time. In both parts (a) and (b) the important point is that the needs and views of those excluded from the sample are likely to differ from those included.
11. (a) Blocking on the basis of SAT verbal scores would ensure that the three groups were (at least reasonably) comparable with respect to verbal ability. Blocking in this way would thus

eliminate the possibility of verbal ability acting as a confounding variable. (b) In the accompanying diagram we show four blocks—based on SAT verbal scores—each containing 15 students. We have assumed that the 60 students can be selected so that 15 will have scores below 400, 15 will have scores between 400 and 499, and so on. If this is not possible, block *T* would contain those 15 students with the lowest scores, block *V* those 15 students with the next lowest scores, and so on. The blocks are randomly divided into three groups of five.

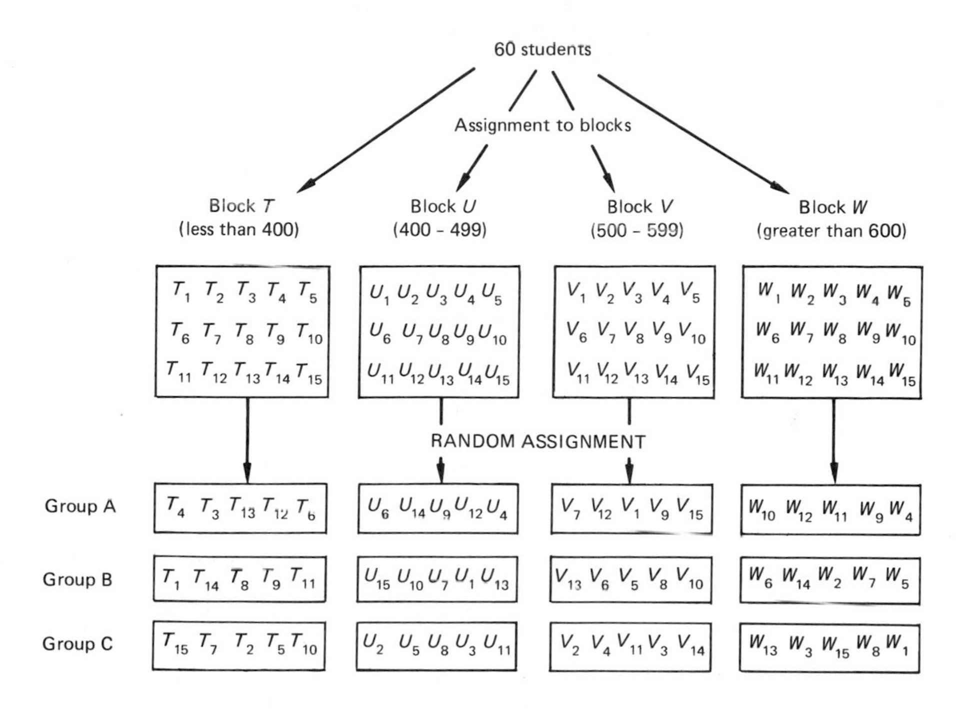

13. (a) The independent variable is "route used" and the dependent variable is "time to get to work." (b) The days (c) A randomized experiment (d) (i) The weather might be worse on *A* than *B* days. (ii) There might be greater traffic tie-ups on *A* than on *B* days. Even small differences between *A* and *B* days on those two variables may bias the result.

15. (a) The degree of success (b) women (c) a control group of women who are *not* successful (d) The researcher might find that, for example, 70% of the successful women were raised in families in which the father was at least 30 when the woman was born. But this in and of itself does not indicate that the age of the father is an important determinant of success. The 70% has to be *compared* to the corresponding frequency for either (i) the general population of all women or (ii) a control group of not-so-successful or unsuccessful women. (e) Include a control group of not-so-successful or unsuccessful women.

17. (a) The secretary should wait six months to allow drivers to get used to the new speed limit. The secretary should go back six months before the change to ensure that the 20 sites are examined at the same time of year. (b) The independent variable is the speed limit and the dependent variable is the accident rate. (c) the sites. (d) (i) The weather may be quite different at the two times. (ii) Any number of factors, such as road improvements or a gas shortage, may act as confounding variables.

19. The bias would be expected to work to the detriment of the self-paced section. The fact that the mean score in the self-paced section was 4.5 points higher than that in the lecture section *in spite of* the bias suggests that the self-paced method is greatly superior to the lecture method. In this case, knowing the direction of the bias strengthens the conclusion based on the data.

REVIEW PROBLEMS

3. Probabilities, whether about $\bar{X}$ or any other random variable, are always based on the assumption of random selection. If this assumption does not apply, it is impossible to estimate the likelihood of any particular event.
5. This claim is *not* correct. It is impossible to judge how representative the 5988 people are of even the newspaper's own readership. To take an extreme case, perhaps every single reader of the newspaper who opposed the scheme returned the coupon. These 4733 opponents are a large proportion of the sample but a tiny proportion of the readership.
7. Suggestion cards are likely to be filled out primarily by people upset with the service.
9. (a) The amount of alcohol consumed between tests (b) This is a randomized experiment (c) the student (d) The control group, D, provides a baseline indication of how reaction times change even without alcohol. (e) The 80 students should be made up of two blocks, one consisting of 40 males, the other of 40 females. Within each block 10 persons should then be randomly assigned to each of the groups A, B, C, and D.
11. Couples enrolled in childbirth classes are likely to be healthier, wealthier, and better educated than the general population of expectant couples. As such, their methods of preparing for childbirth are likely to be quite atypical of those for the rest of the population.
13. Not knowing which treatment they were receiving, the patients are less likely to react in the "expected" manner. Doctors, not knowing which patients have received which treatment, are unlikely to consciously or subconsciously bias progress reports in favor of, or against, particular treatments.
15. (a) the child (b) The independent variable was the kind of milk—raw, pasteurized, or none—that was provided. The dependent variables were weight gain and gain in height. (c) This interference in the randomization scheme introduced biases due to the judgments of the many school officials involved. In fact, it proved very difficult to measure the extent to which schools had "compensated" for too many well-fed or ill-nourished children.
17. (a) This is a quasi-experiment (b) The independent variable is whether or not a family is accepted for the new housing project. The dependent variables are degree of contentment and various health measures. (c) No. Those moving into the new public housing were "selected" by the housing authority. Perhaps they were selected for such reasons as financial security and/or health care awareness which are closely related to the dependent variable.
19. It is probable that a store's credit-card holders are generally the wealthier segment of the population of all customers. As such, their shopping habits are likely to be quite atypical of those of the population of all customers.
21. (a) This is a randomized experiment (b) The independent variable is type of packaging. The dependent variable is the number of items sold. (c) the store (d) by selecting the 120 participating stores at random
23. The fallacy is that students generally *choose* whether or not to take Latin. It may well be that Latin is taken primarily by the brighter students. In this case taking Latin does not "produce" the increase.
25. (a) Identical twins share exactly the same hereditary background. Therefore any differences in IQ must be due to the effect of the different environments. (b) We cannot randomly separate identical twins at birth. (c) We would select a random sample of 100 newborn identical twins. In each of the 100 cases one of the twins would be randomly selected and assigned to be raised in an affluent, nurturing family. The other twin would be raised in a poor, less nurturing family. At intervals the IQs of each twin would be recorded and compared in a matched pairs design. (d) Environment appears to have little effect on IQ.
27. Experiments in the natural sciences usually involve nonhuman experimental units. By contrast, experiments in the social sciences generally involve human beings. It is simple to assign mice or chemical samples at random to various treatments. However, it is frequently impossible to assign human beings to the treatments the experimenter wishes to study.

CHAPTER 9

5. (a) 56.76 to 61.59 mph
7. (a) 1.25 to 2.47 courses (b) We can be 90% confident that the mean number of math courses taken by all seniors lies in this interval. (c) 911 to 1805 courses

9. (a) (i) 683.4 to 713.6 apples per tree (ii) 2,597,110 to 2,711,490 apples (b) We can be 80% confident that the mean number of apples per tree lies in the interval 683.4 to 713.6. We can be 80% confident that the total number of apples in the orchard lies between 2,597,100 and 2,711,490. 11. 41.6 weeks 13. .257 to .383
15. Feeling optimistic about the future: .225 to .257; the United States is the best country in which to live: .941 to .958.
17. .314 to .345. We can be 80% confident that the proportion of all adults who watched the program lies between .314 and .345.
19. (a) .124 to .144 (b) 469,000 (c) 434,000 to 504,000 23. 7.5 to 13.9%
25. (a) −.036 to .256 (b) The fact that zero is in the interval suggests that it is plausible that $\pi_1 - \pi_2 = 0$ (i.e., that $\pi_1 = \pi_2$). 27. −.015 to .107 31. 139 33. 174
35. 72 37. 560 39. 89

REVIEW PROBLEMS

1. (a) (i) 78 − 1.67 to 78 + 1.67 (ii) 578 − 1.67 to 578 + 1.67 (b) The value for $\bar{x}$ has no effect on the accuracy of the confidence interval for μ.
3. Once the sample has been selected and the interval computed we cannot make a probability statement about whether or not μ is or is not in the interval.
5. (a) $10,350 to $13,650 (b) 304.47 to 343.53
7. (a) .302 to .448 (b) 5804 to 8596 9. .98 to 2.42 11. 640
13. (a) 174 (b) 44 15. 7.16 to 8.08 days 17. .057 to .183
19. −.164 to .064. We can be 90% confident that $\pi_1 - \pi_2$ lies in this interval.
21. 5.18 to 5.55 hours per week 23. .755 to .831
25. (a) 1068 (b) .599 − .029 to .599 + .029 or .570 to .628 (c) We can be 95% confident that π lies in this interval. (d) Yes; the accuracy is .029 instead of .03.
27. (a) .2824 (b) .2301
29. (a) 3.30 to 4.14 (b) 53,940 persons (c) 47,869 to 60,011 persons
31. Once the interval has been computed, it is fixed and therefore we cannot make a probability statement about whether or not μ is in the interval. 33. .095 to .245
35. 2.83 to 2.89 hours. We can be 80% confident that the mean time to complete this examination lies in this interval. 37. $232.50 39. (a) 1.96 to 4.44. (b) yes
41. (a) .818 to .941 (b) −.067 to .115 (c) No, because the interval stretches so far on either side of 0.

CHAPTER 10

3. (a) In this case μ is the mean concentration of pollutants in the river *after* the cleanup. (b) $H_0: \mu = 74$, $H_1: \mu < 74$ (c) P value $= P(Z < -9.48) \doteq 0$. Reject H_0 in favor of H_1.
5. (a) P value $= P(Z < -3.12) = .0009$. Reject $H_0: \mu = 1.7$ in favor of $H_1: \mu < 1.7$.

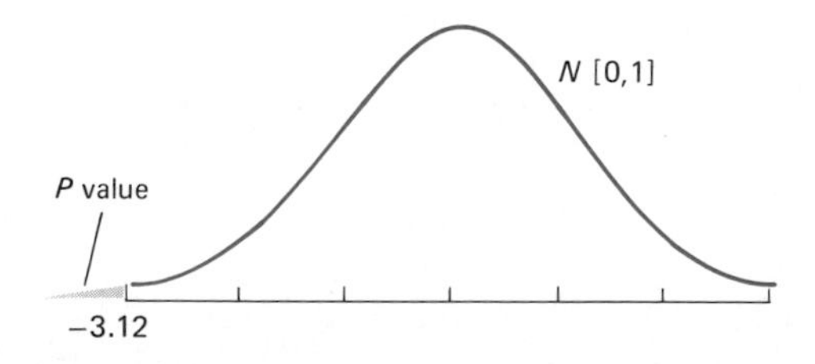

(b) .94 to 1.46 attempts
7. P value $= P(Z > 5.27) \doteq 0$. Reject $H_0: \mu = 55$ in favor of $H_1: \mu > 55$.
9. The P value is a probability about $\bar{X}$ and is based on the central limit theorem, which can be applied only if the sample size is greater than or equal to 30.

11. P value $= P(Z < -2.2) = .0132$. Reject $H_0: \mu = 20$ in favor of $H_1: \mu < 20$. Here μ is the true mean length of this brand of contact lens. The data suggest that the advertised claim is false.

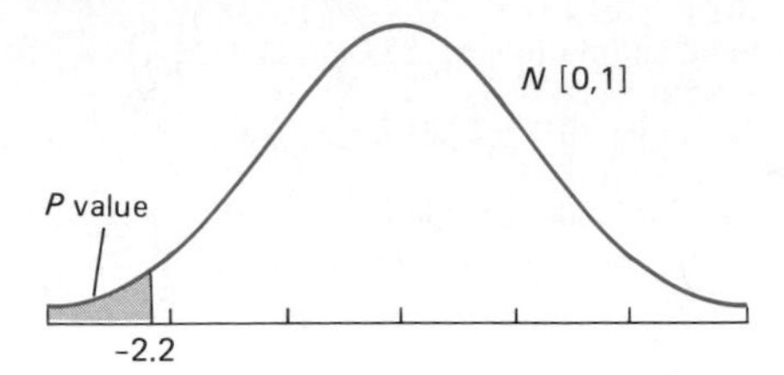

15. P value $= P(Z < -1.01) = .1563$. Do not reject H_0 at the .05 level of significance. The data suggest that the mean temperature for this date is not significantly less than 80°F.

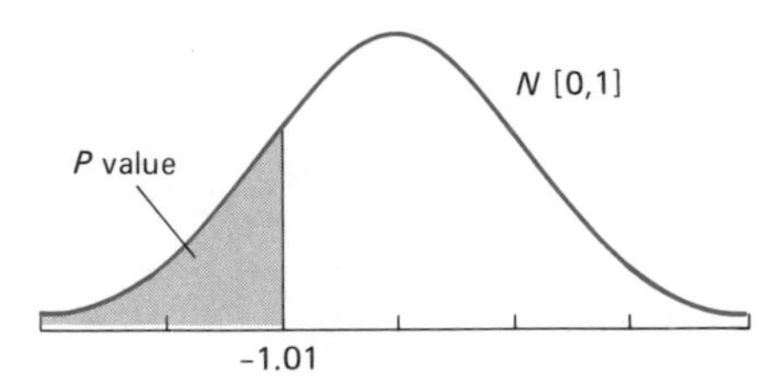

17. P value $= P(Z > .42) = .3372$. Do not reject H_0 at the .05 level of significance. The data suggest that the mean time to perform such operations is not significantly greater than 10 hours.

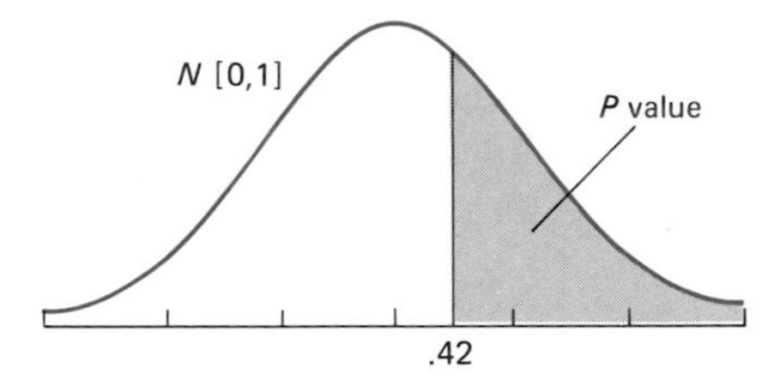

19. (a) $H_0: \mu = 19$, $H_1: \mu > 19$, where μ is the mean weight gain with the new feed. (b) (i) Do not reject H_0 ($\bar{x} = 18.29$ is *less* than 19). (ii) P value $= P(Z > 3.67) = .0001$. Reject H_0 at the .01 level of significance. The data strongly suggest that the mean weight gain with the new feed is significantly greater than 19 grams.

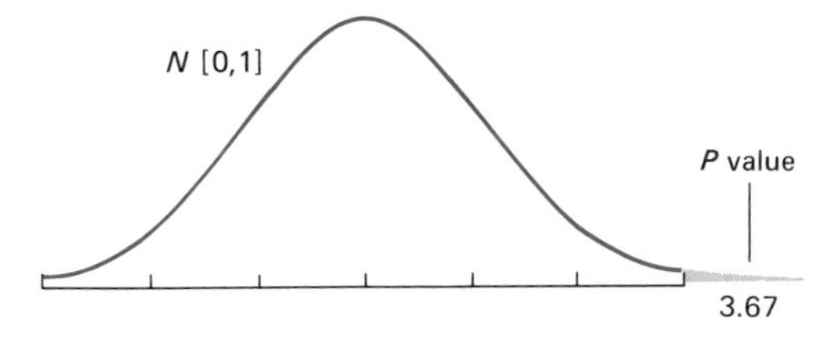

(iii) P value $= P(Z > 1.82) = .0344$. Reject H_0 at the .05 level of significance. The data suggest that the mean weight gain with the new feed is significantly greater than 19 grams.

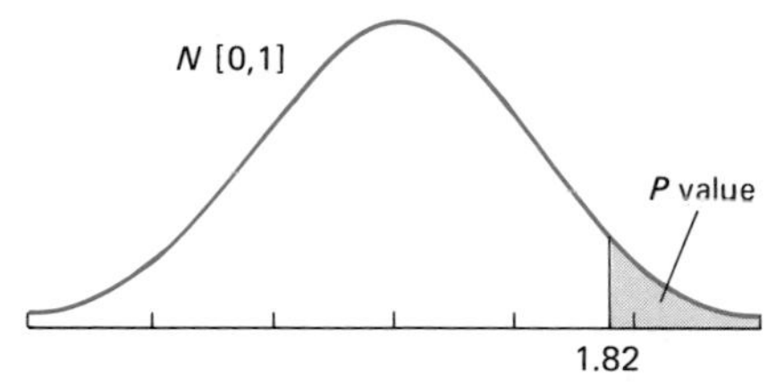

(iv) P value $= P(Z > .06) = .4761$. Do not reject H_0 at the .05 level of significance. The data suggest that the mean weight gain with the new feed is not significantly greater than 19 grams.

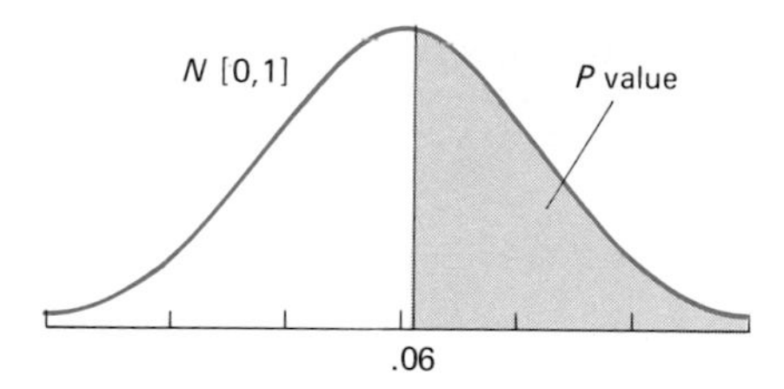

21. (a) P value $= P(Z > 6.03) \doteq 0$. Reject $H_0: \mu = 25$ at the .01 level of significance. The data strongly suggest that the mean percentage is significantly greater than 25%. (b) Do not reject H_0 since $\bar{x} = 22.1$ is *less* than 25.

23.

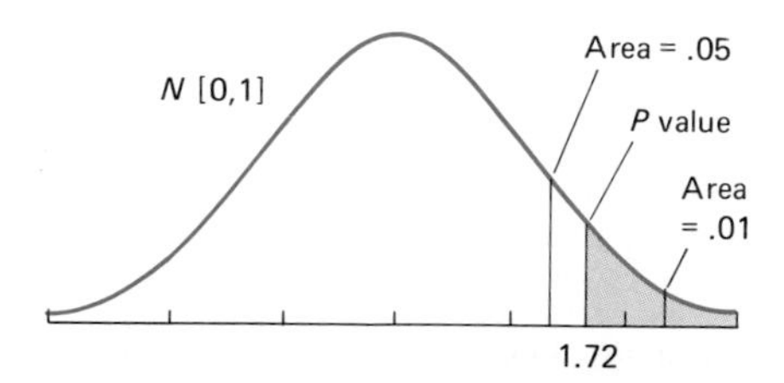

25. (a) In this case, π is the true proportion of those eligible who have enrolled.
(b) $H_0: \pi = .35$, $H_1: \pi < .35$ (c) P value $= P(Z < -3.71) = .0001$. Reject H_0 at the .01 level of significance. The data strongly suggest that the proportion of those eligible who have enrolled is significantly less than .35.

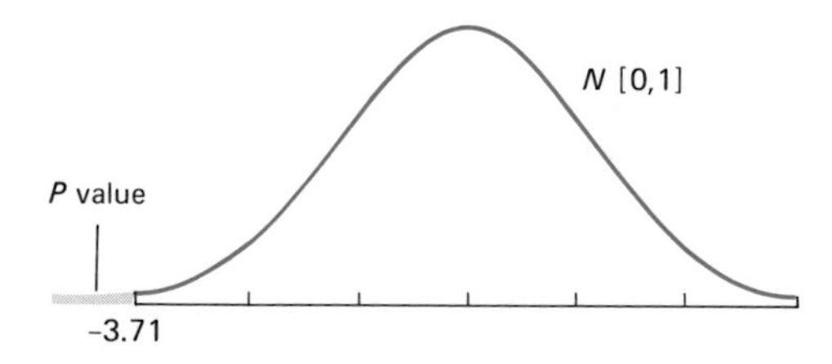

(d) .276 to .312

27. P value $= P(Z < -.71) = .2389$. Do not reject $H_0: \pi = .2$ at the .05 level of significance. The data suggest that the proportion of purchasers who are first-time buyers is not significantly less than .2.

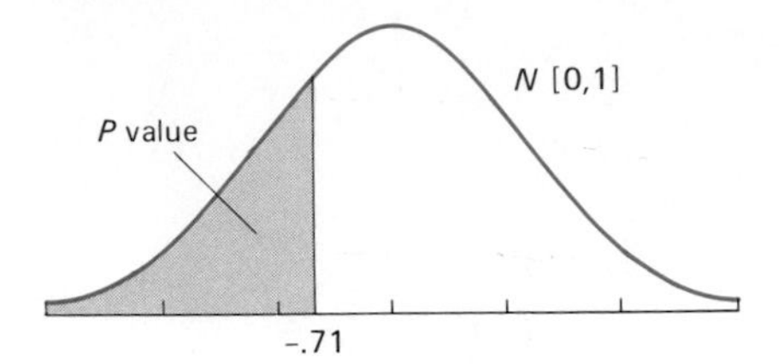

29. (a) \$36.66 to \$44.02 (b) .55 to .67

31. P value $= P(Z < -1.83) = .0336$. Reject $H_0: \pi = .4$ at the .05 level of significance. The data suggest that the proportion of undergraduates who are the first generation to go to college is significantly less than .4.

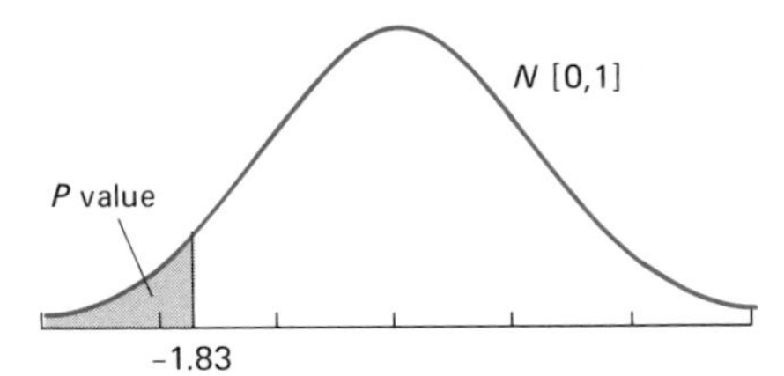

35. (a) In this case, μ_1 is the potential mean dexterity score over all women. Then μ_2 is similarly defined for men. (b) $H_0: \mu_1 - \mu_2 = 0$, $H_1: \mu_1 - \mu_2 > 0$ P value $=$ $P(Z > 7.54) \doteq 0$. Reject H_0 at the .01 level of significance. The data strongly suggest that the mean score for women is significantly greater than that for men.

37. (a) In this case μ_1 is the mean number of hours per week for all general practitioners. Then μ_2 is similarly defined for all specialists. (b) $H_0: \mu_1 - \mu_2 = 0$, $H_1: \mu_1 - \mu_2 > 0$ (c) P value $= P(Z > 2.72) = .0033$. Reject H_0 at the .01 level of significance. The data strongly suggest that the mean work week for GPs is significantly larger than that for specialists.

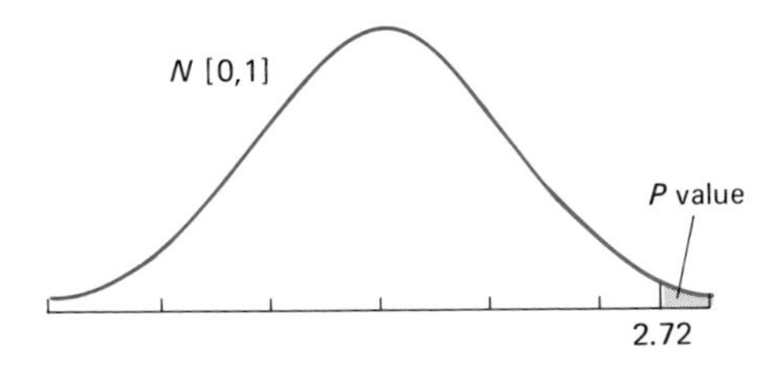

39. P value $= P(Z > 2.77) = .0028$. Reject $H_0: \mu_1 - \mu_2 = 0$ in favor of $H_1: \mu_1 - \mu_2 > 0$ at the .01 level of significance. The data strongly suggest that the mean number of occupants after the campaign is significantly greater than the mean number before the campaign.

41. (a) P value $= P(Z > 1.68) = .0465$. Reject $H_0: \pi_1 - \pi_2 = 0$ in favor of $H_1: \pi_1 - \pi_2 > 0$ at the .05 level of significance. The data suggest that the proportion using the bus has increased. (b) no

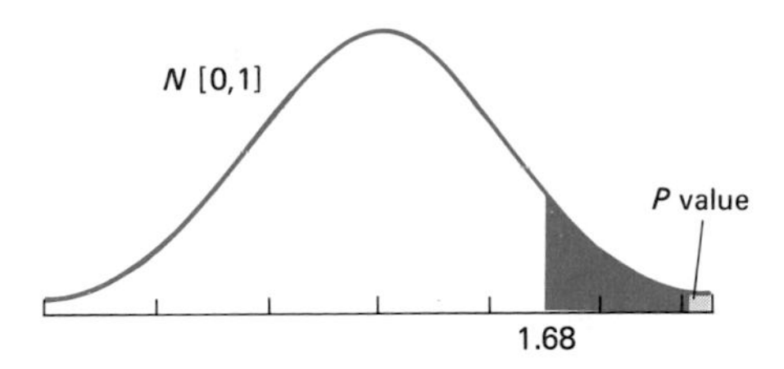

45.

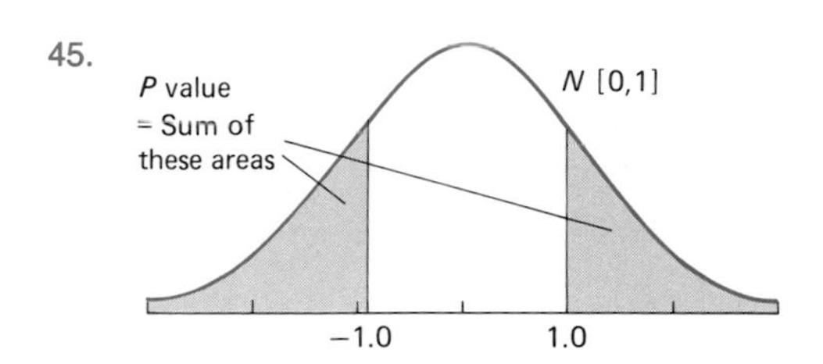

47. P value $= 2P(Z > 2.00) = .0454$. Reject $H_0: \mu = 64.5$ in favor of $H_1: \mu \neq 64.5$ at the .05 level of significance. The data suggest that the mean score for all women is significantly greater than 64.5.

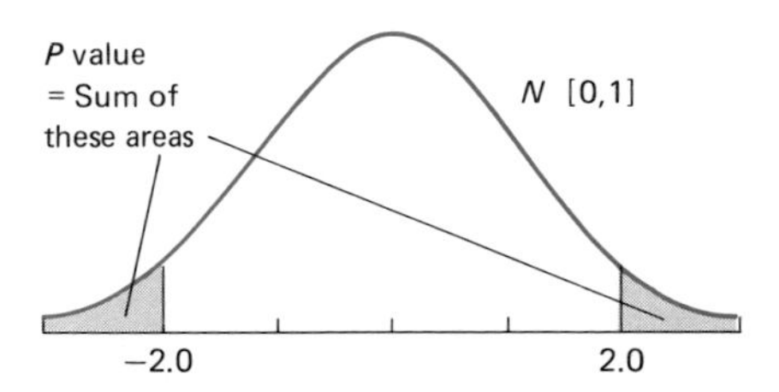

49. P value $= 2P(Z < -1.71) = .0872$. Do not reject H_0 at the .05 level of significance. The data suggest that the mean length of baseball games is not significantly different from 2.5 hours. Yes. Using $\alpha = .1$, we would reject H_0 at the .1 level of significance.

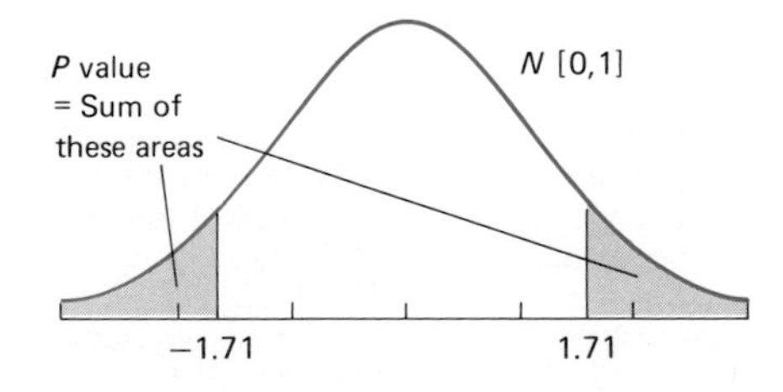

51. .258 to .392.

53. P value $= 2P(Z > .63) = .5286$. Do not reject H_0 at the .05 level of significance. The data suggest that the proportions not visiting the library in the populations are not significantly different.

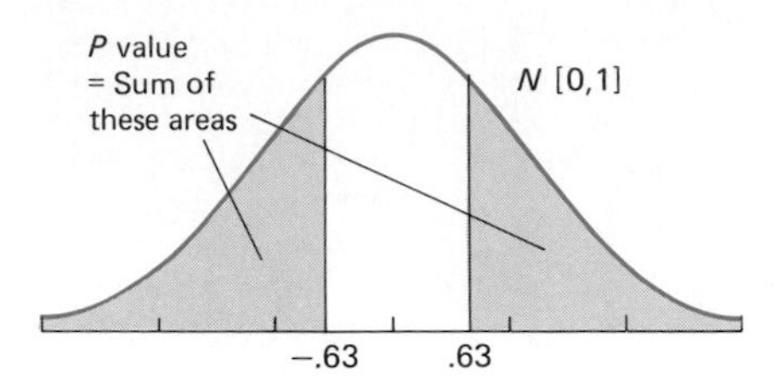

55. P value $= 2P(Z > 1.92) = .0548$. We cannot reject H_0 at the .05 level of significance. The data suggest that the mean times it takes for mail to travel in the two directions are not significantly different.

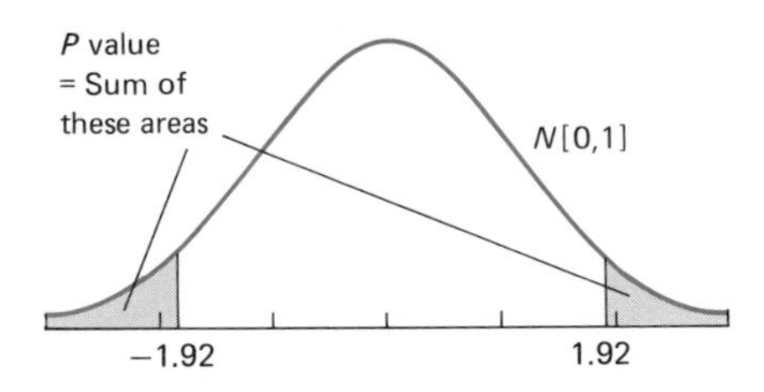

REVIEW PROBLEMS

3. (a) P value $= P(Z > .6) = .2742$. Do not reject H_0 at the .05 level of significance. (b) P value $= P(Z > 6) \doteq 0$. Reject H_0 at the .01 level of significance.
5. (a) P value $= P(Z < -1.13) = .1292$. Do not reject H_0 at the .05 level of significance. (b) P value $= P(Z < -2.55) = .0054$. Reject H_0 at the .01 level of significance. (c) Do not reject H_0 since $\bar{p} = 27/50 = .54$ *exceeds* .5.
7. P value $= P(Z > 1.68) = .0465$. Reject H_0 at the .05 level of significance. The data suggest that the mean age at death for "famous" people is significantly greater than 70.5.
9. P value $= 2P(Z > 4.54) \doteq 0$. Reject H_0 at the .01 level of significance. The data strongly suggest that the proportion of voters in Connecticut who think the president is doing a good job is higher than the corresponding proportion in Virginia.
11. (a) μ is the true mean amount of tar in this brand of cigarette. (b) $H_0: \mu = 9.8$, $H_1: \mu > 9.8$ (c) (i) P value $= P(Z > 1.22) = .1112$. Do not reject H_0 at the .05 level of significance. The data suggest that μ is not significantly greater than 9.8 milligrams. (ii) P value $= P(Z > 1.73) = .0418$. Reject H_0 at the .05 level of significance. The data suggest that μ is significantly greater than 9.8 milligrams. (iii) P value $= P(Z > 2.73) = .0032$. Reject H_0 at the .01 level of significance. The data strongly suggest that μ is significantly greater than 9.8 milligrams. (d) All other factors remaining equal the larger the value for n, the larger the value of the test statistic, z, and the smaller the P value.

13. P value $= 2P(Z > 1.32) = .1868$. Do not reject H_0 at the .05 level of significance. The data suggest that π_1 and π_2 are not significantly different.
15. P value $= P(Z > 3.27) = .0005$. Reject H_0 at the .01 level of significance. The data strongly suggest that the new treatment significantly increases the mean survival time.
17. (a) P value $= P(Z > .55) = .2912$. Do not reject H_0 at the .05 level of significance. The data suggest that the proportion of children with audio defects is not significantly greater than .08. (b) (i) Do not reject the H_0 since $\bar{p} = .075$ is less than .08. (ii) P value $= P(Z > 2.21) = .0136$. Reject H_0 at the .05 level of significance. The data suggest that the proportion of children with audio defects is significantly greater than .08.
19. P value $= P(Z > 1.72) = .0427$. Reject H_0 at the .05 level of significance. The data suggest that the mean interval between visits is higher for students than for staff members.
21. P value $= 2P(Z > 2.75) = .0060$. Reject H_0 at the .01 level of significance. The data strongly suggest that the mean absolute visual threshold for students in rural communities is greater than the corresponding mean for students in urban communities.
23. (a) $H_0: \mu = 80$, $H_1: \mu > 80$, where μ is the mean monthly expenditure on heating by welfare recipients in the region. (b) P value $= P(Z > 2.99) = .0014$. Reject H_0 at the .01 level of significance. The data strongly suggest that the mean expenditure for heating is significantly greater than \$80 per month.
25. P value $= P(Z < -1.69) = .0455$. Reject H_0 at the .05 level of significance. The data suggest that the proportion of all convictions involving first offenders this year is significantly less than .3.
27. P value $= P(Z > 3.27) = .0005$. Reject H_0 at the .01 level of significance. The data strongly suggest that, on average, girls have a significantly larger number of close friends than do boys.
29. P value $= P(Z > 2.18) = .0146$. Reject H_0 at the .05 level of significance. The data suggest that π_1 is significantly greater than π_2.
31. P value $= P(Z < -2.29) = .0110$. Reject H_0 at the .05 level of significance. The data suggest that the mean percentage of income saved by wealthy people is significantly less than 12%.
33. P value $= P(Z < -.53) = .2981$. Do not reject H_0 at the .05 level of significance. The data suggest that the proportion of customers who read this newspaper is not significantly less than .40.
35. P value $= P(Z > 1.99) = .0233$. Reject H_0 at the .05 level of significance. The data suggest that the mean cost of the "basket" is significantly higher in poor than in other sections of the city.
37. P value $= P(Z > .77) = .2206$. Do not reject H_0 at the .05 level of significance. The data suggest that the mean number of repairs for model A is not significantly less than that for model B.
39. P value $= 2P(Z > 4.68) \doteq 0$. Reject H_0 at the .01 level of significance. The data strongly suggest that the mean number of new employees has increased significantly.
41. P value $= P(Z < -1.27) = .1020$. Do not reject H_0 at the .05 level of significance. The data suggest that the proportion of private colleges offering the Master of Arts in Education degree is not significantly less than .45.
43. P value $= P(Z > 3.12) = .0009$. Reject H_0 at the .01 level of significance. The data strongly suggest that on average Greeks score higher on the test than do Americans.
45. P value $= P(Z > .44) = .3300$. Do not reject H_0 at the .05 level of significance. The data suggest that the proportion of classrooms with substitute teachers is not significantly greater than .15.
47. P value $= 2P(Z > 4.26) \doteq 0$. Reject H_0 at the .01 level of significance. The data strongly suggest that the mean time to complete checklist I is significantly greater than the mean time to complete checklist II.
49. P value $= 2P(Z > 1.89) = .0588$. Do not reject H_0 at the .05 level of significance. The data suggest that the mean time between malfunctions is not significantly different from 510 hours.
51. P value $= P(Z > 1.45) = .0735$. Do not reject H_0 at the .05 level of significance. The data suggest that the seeded clouds are not significantly more likely than the unseeded clouds to produce rain.
53. P value $= 2P(Z > .51) = .61$. Do not reject H_0 at the .05 level of significance. The data suggest that there is no significant difference in the probability that the systems will detect an aircraft.

CHAPTER 11

3. (a) In this case μ is the true mean salary for transit workers. (b) $H_0: \mu = 28{,}500$, $H_1: \mu < 28{,}500$ (c) Reject H_0 *if* $\bar{x} \leq 27{,}764.33$.

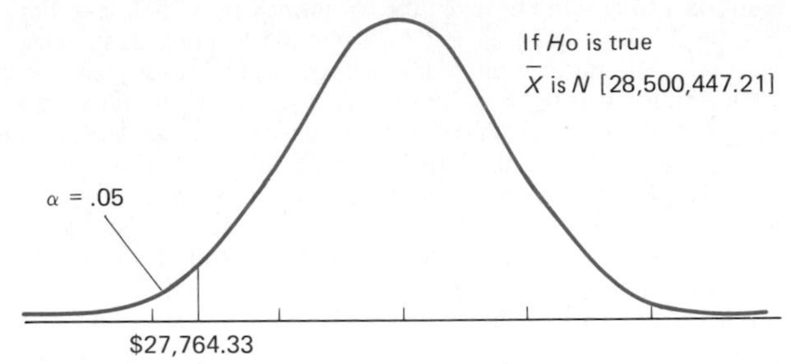

5.

	Rejection Region	Acceptance Region
(a) $n = 40$	$\bar{x} \geq 2.601$	$\bar{x} < 2.601$
(b) $n = 100$	$\bar{x} \geq 2.564$	$\bar{x} < 2.564$
(c) $n = 256$	$\bar{x} \geq 2.541$	$\bar{x} < 2.541$

The graphs are as follows.

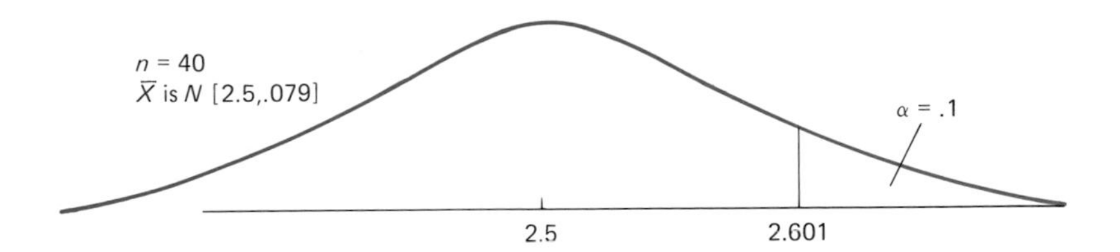

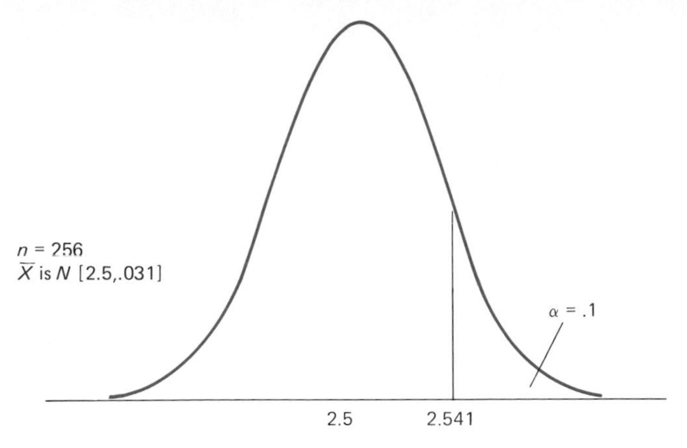

9. (a) true (b) true (c) No. The probability of rejecting H_0 if H_0 is true is $\alpha = .05$. (d) No. The probability of not rejecting H_0 when $\mu = 40$ is $\beta(40) = .05$. (e) No. The first probability is $1 - \alpha = .95$ and the second is $\beta(40) = .05$. (f) true

11. (a) In this case μ is the true mean IQ in the population. (b) $H_0: \mu = 115$, $H_1: \mu > 115$

(c)

μ_1	122	120	118	116
$\beta(\mu_1)$	.0027	.0643	.4013	.8437

(d) The farther the value for μ_1 is from 115 (i.e., in this case the larger μ_1 is), the smaller the value for $\beta(\mu_1)$. The situations for $\mu_1 = 120$ and $\mu_2 = 118$ are as shown (the acceptance region is $\bar{x} < 117.601$).

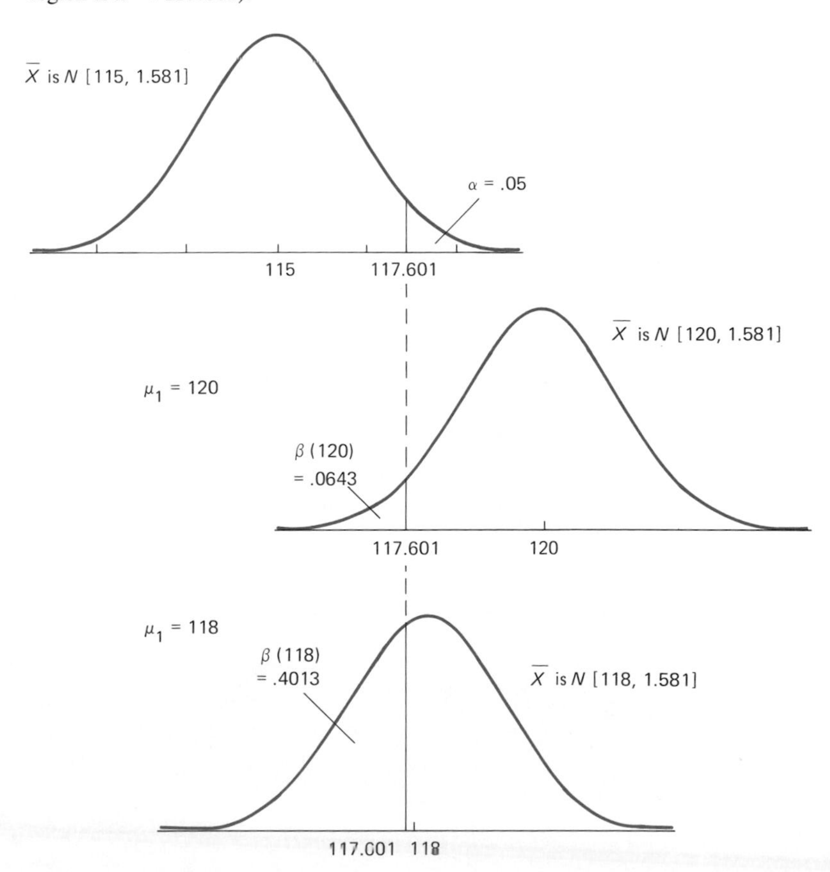

13. (a) In this case μ is the true mean number of years of education of the magazine subscribers. $H_0: \mu = 18.5$, $H_1: \mu < 18.5$.

(b)

	$\beta(18)$
(i) $n = 40$	.6064
(ii) $n = 100$	.2981
(iii) $n = 400$	.0035

(c) If $n = 400$, there is only a very small chance, .0035, of *not* rejecting H_0 if, in fact, $\mu = 18$ years. 17. 85

19. 171

21. 37

23. 230

REVIEW PROBLEMS

1. (a) Reject H_0 if $\bar{x} > 18.0232$. (b) $\beta(18.02) = .6064$

3. The reason is that once the sample is taken and $\beta(\mu_1)$ is found to be unacceptably large, it is usually too late to select a larger sample.

5. (a) Reject H_0 if $\bar{x} > K + .19\sigma$. (b) .7823

7. (a) proof (b) .226, .401, .923, .994 (c) The probability of at least one type 1 error increases as the number of tests increase.

9. (a) $H_0: \mu = 2.1$, $H_1: \mu < 2.1$, where μ is the mean number of workdays lost to colds over all potential users of the diet. (b) Reject H_0 if $\bar{x} < 1.937$ days. (c) $\beta(1.9) = .3557$

11. The sample size is so large that $\beta(205)$ is essentially 0. In fact, even $\beta(209)$ is close to 0. Rather than investigating whether μ is significantly less than 210, it makes sense to answer the question "what is μ" by finding a confidence interval for μ.

13. (a) In this case, μ is the mean length of sentence for defendants imprisoned for marijuana violations *since* the legislation. (b) $H_0: \mu = 38.6$, $H_1: \mu < 38.6$ (c) power $= 1 - \beta(35) = .9177$ (d) If μ is 35 months there is a .9177 probability of correctly rejecting H_0.

15. (a) $H_0: \mu = 655$, $H_1: \mu > 655$ (b) Reject H_0 if $\bar{x} > \$691.15$. (c) (i) $\beta(670) = .7734$ (ii) $\beta(685) = .5871$ (iii) $\beta(700) = .3783$ (iv) $\beta(715) = .1977$ (d) There is a large probability (.7734) of *not* rejecting $H_0: \mu = 655$ even if there has been an increase of \$15. On the other hand, there is a much smaller chance (.3783) of failing to reject H_0 if μ is as large as \$700.

17. (b) If there has been no change and $\mu = 12.5$, there is a probability $\alpha = .05$ of not rejecting H_0. If the mean interval has increased to 15 days, there is only a small probability, $\beta(15) = .1$, of failing to reject $H_0: \mu = 12.5$.

19. 88 cities 21. 109 students

23. (a) P value $= P(Z > 2.02) = .0217$. Reject H_0 at the .05 level of significance. The data suggest that the proportion of alumni favoring the new program is significantly greater than .6. (b) If $\pi = .7$, our decision in part (a) to reject $H_0: \pi = .6$ would be correct.

25. (a) $\beta(9.8) = .8079$ (b) The test is *insensitive* for the alternative, $\mu = 9.8$. There is a large probability, .8079, of failing to reject $H_0: \mu = 10$ if $\mu = 9.8$.

27. $1 - \beta(1185) = .9940$

CHAPTER 12

3. (a) .1

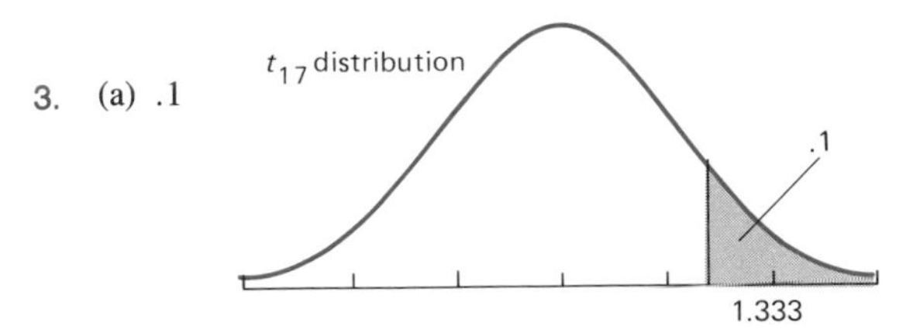

(b) .9

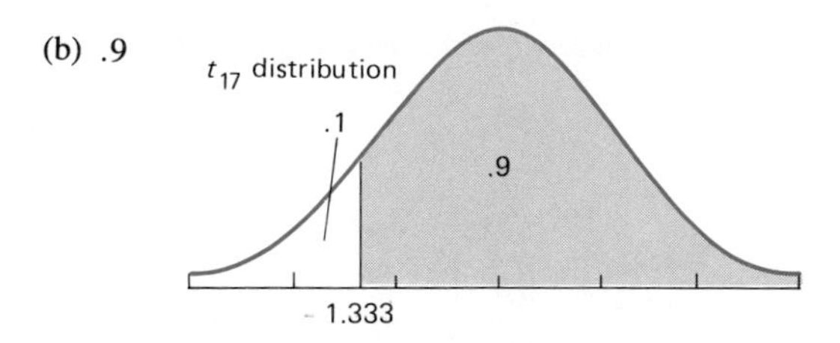

(c) .85

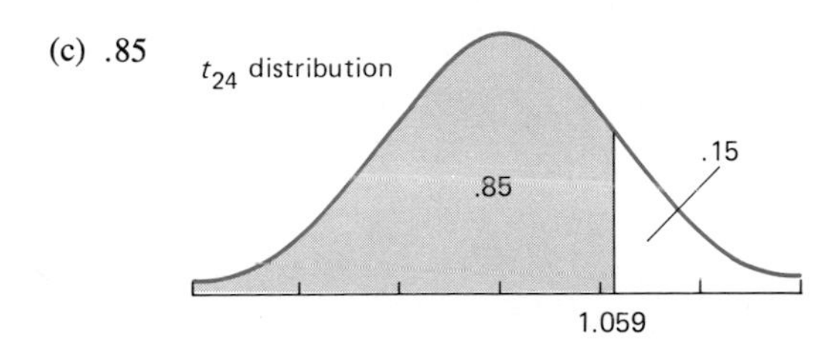

(d) .5

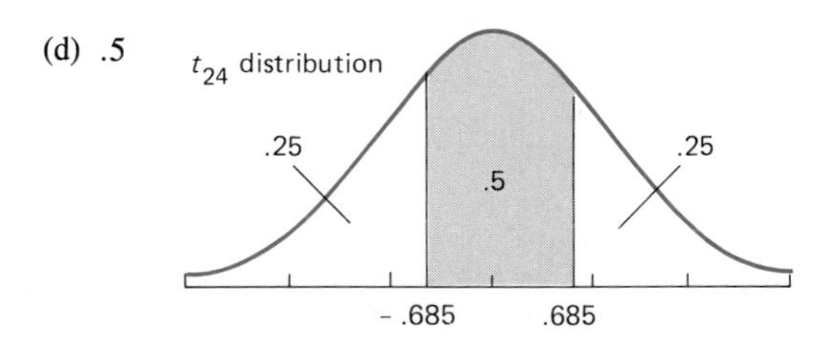

7. (a) \$76,515 to \$94,205 (b) We assume that the distribution of sale price is approximately normal. (c) We can be 90% confident that μ lies in this interval.
9. $\bar{x} = 3.143$, $s = 2.103$; 2.47 visits to 3.82 visits
11. (a) 1793.1 to 2231.7 (b) We assume that the distribution of the number of visitors is approximately normal.
13. The two very large values 23.6 and 31.8 suggest that the distribution of the number of years of marriage has a much larger right than left tail.
15. (a) In this case μ is the mean length of line *after* the introduction of the scheme.

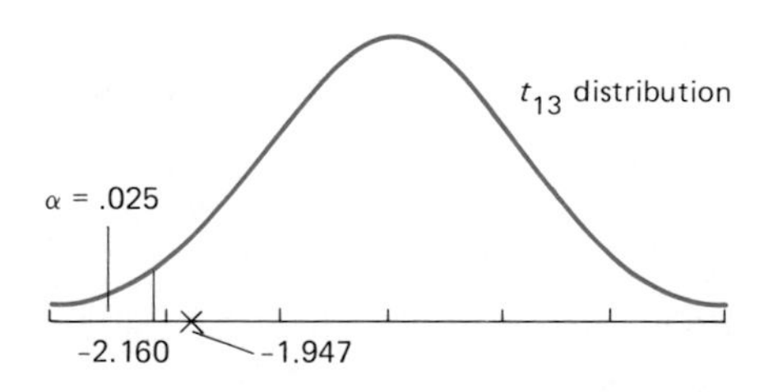

(b) $t = -1.947$. Do not reject H_0 at the .025 level of significance. The data suggest that the line length after the introduction of the scheme is not significantly less than 6.
(c) The distribution of long distance charges is approximately normal.
17. (a) $\bar{x} = 7.55$ parts per million. Since $\bar{x} = 7.55$ seems way below 9.4, without a test the data *appear* to suggest a significant decline. (b) $t = -3.39$. Reject H_0 at the .01 level of

significance. The data suggest that the mean level of carbon monoxide is significantly below 9.4 ppm.

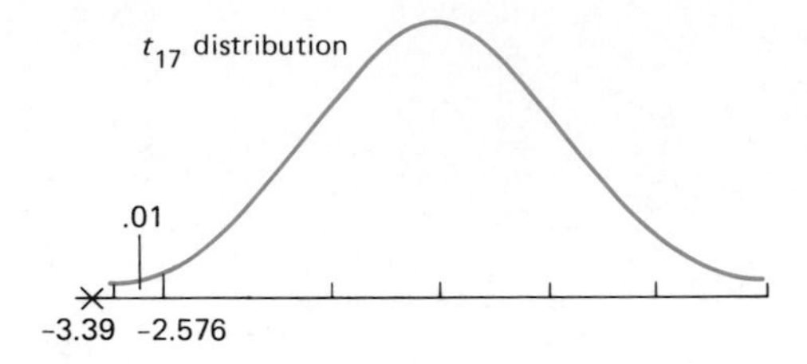

19. $t = 5.61$. Reject H_0 at the .005 level of significance. No, the conclusion would not change.

21. (a) $|t| = 3.156$. Reject H_0 at the .05 level of significance. The data suggest that the mean score on the test is significantly less than 32.

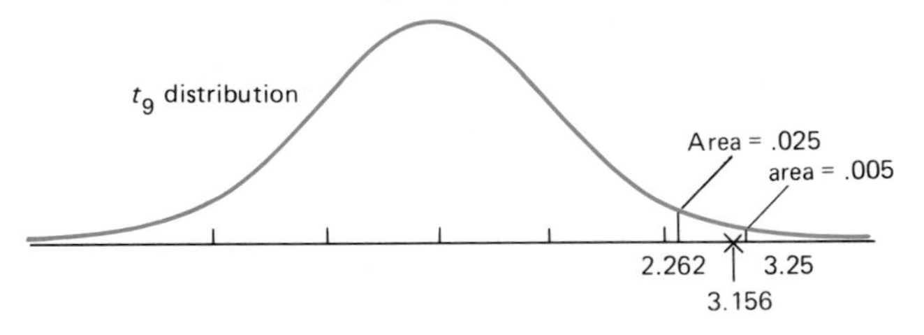

(b) We assume that the distribution of test scores is approximately normal.

23. $|t| = 1.923$. Do not reject H_0 at the .05 level of significance. The data suggest that the mean number of science courses taken by the present graduating class is not significantly different from 4.9.

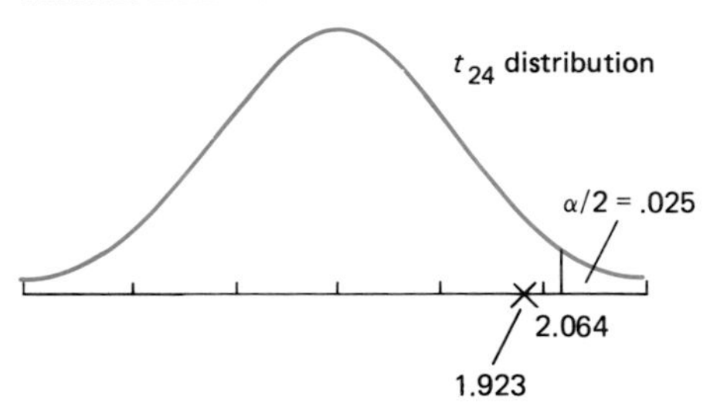

25. $t = -5.16$. Reject $H_0: \mu = 8$ at the .01 level of significance. The data strongly suggest that the mean number of hours of sleep for students at the college is significantly less than 8.0 hours.

29. .098 to .522. We can be 80% confident that $\mu_1 - \mu_2$ lies in this interval.

31. (a) $t = 1.926$. Do not reject H_0 at the .05 level of significance. The data suggest that there is no significant difference in the mean time to bloom for the two varieties of seeds.

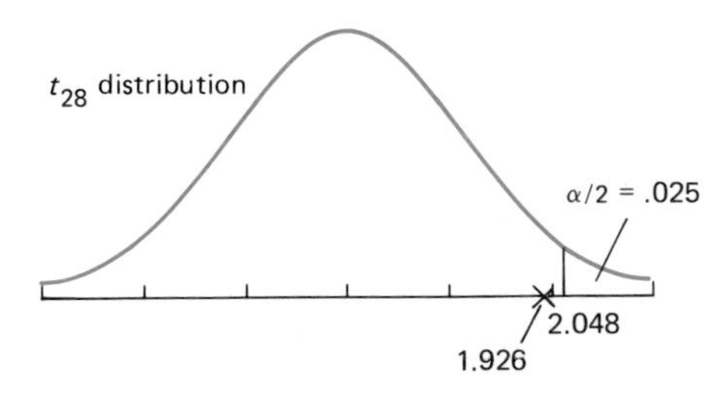

(b) We assume that, for both varieties, the distribution of the time to bloom is normal with a common variance.

33. $t = 1.453$. Do not reject H_0 at the .05 level of significance. The data suggest that there has been no significant reduction in the mean number of people waiting.

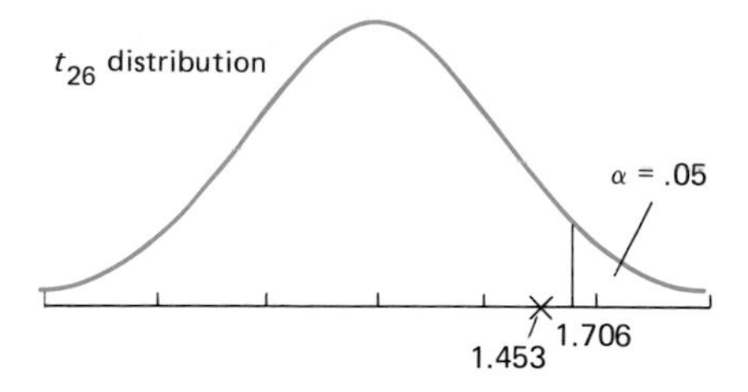

35. (a) $t = .694$. Do not reject H_0 at the .05 level of significance. The data suggest that the mean age of women in the party is not significantly less than that for men. (b) We assume that for both sexes the distribution of age is normal with the same variance.

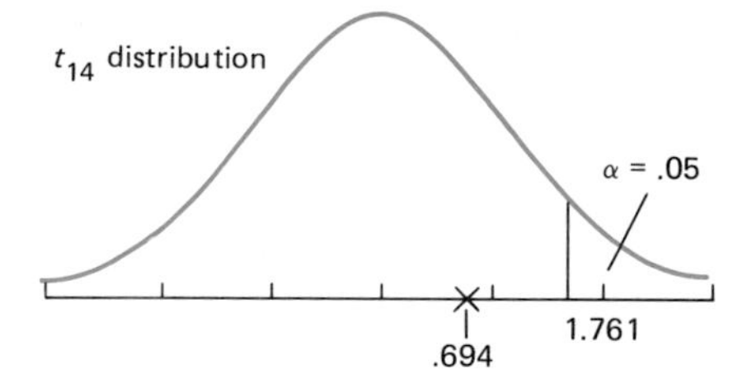

37. $t = 2.08$. Reject H_0 at the .05 level of significance. The data suggest that on average persons at the head of the line provide larger estimates of the length of the line than do persons at the end of the line.

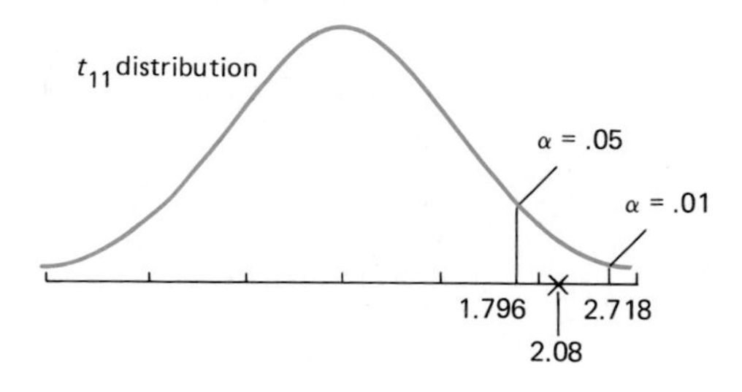

39. $t = 1.966$. Do *not* reject H_0 at the .025 level of significance.

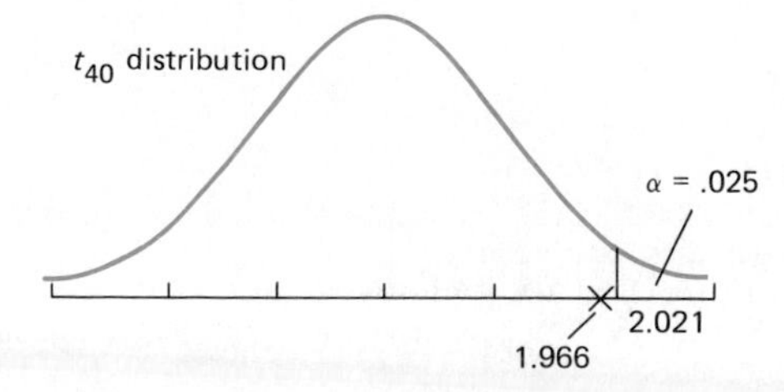

41. $t = 2.306$. (if you found PL67 − VS45). Do not reject H_0 at the .05 level of significance. The data suggest that the mean difference in computing times is not (quite) significantly different from 0.

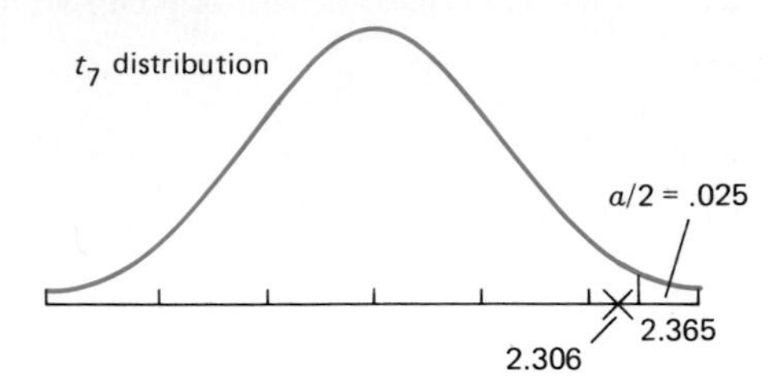

43. .009 to .231. **45.** (a) 10.22 to 13.14 lb. (b) We can be 90% confident that the mean decline in weight lies between 10.22 and 13.14 lb.
47. P value $= P(Z > 2.37) = .0089$. Reject H_0 at the .01 level of significance. The data suggest that the mean decline in scores is significantly greater than 0.

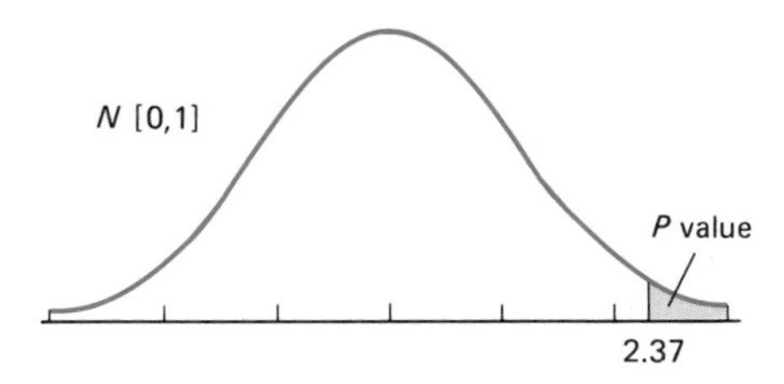

49. (a) (i) 15.99 (ii) 20.48 (iii) 3.25 (iv) 3.94 (b) 6 **51.** (a) .90 (b) .8
53. (a) .350 (b) .326 (c) .452 (d) .172 (e) .048 **55.** (a) .95 (b) .99
57. (a) 1.83 to 2.91 (b) We assume that the distribution of the length of time that the matches stay lighted is normal.
59. $y = 40.25$. Reject H_0 at the .01 level of significance. The data suggest that the standard deviation of times for mentally impaired persons is significantly greater than 2.3.

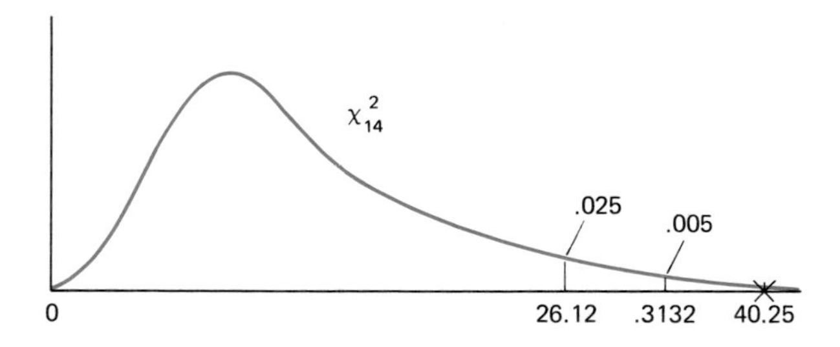

61. 8.92 to 24.17
63. (a) .0000082 to .0000255 (b) We can be 80% confident that the variance of plate thickness lies between these two values.
65. Do not reject H_0 provided that $.083 < s < .20$.
69. .123 to .631 We can be 90% confident that σ_1^2/σ_2^2 lies in this interval.

71. $F = .148$. Reject H_0 at the .01 level of significance. The data suggest that the variability in scores for form A is significantly greater than that for form B. (Use $F_{15,19}$ rather than $F_{19,19}$.)

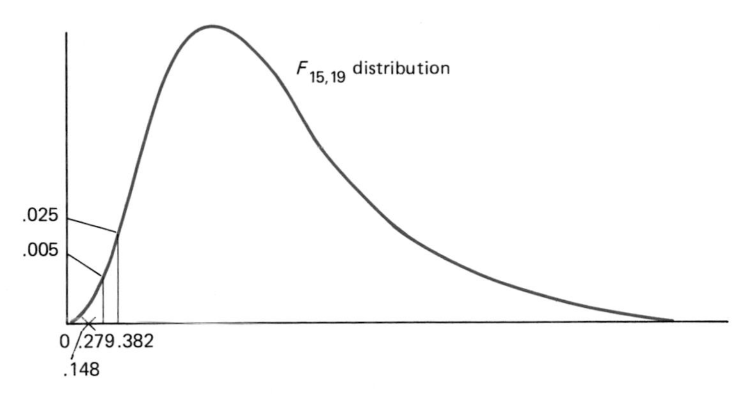

73. $F = 1.475$. Do not reject H_0 at the .05 level of significance. The data suggest that the variability of electricity use in City I is not significantly greater than that in City II.

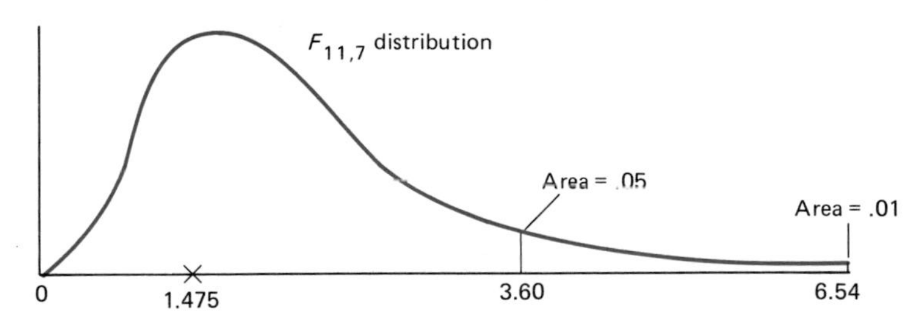

75. $t = .74$. Do not reject H_0 at the .05 level of significance. The data suggest that there is no significant difference in the mean amount of overtime for union members in the two plants.

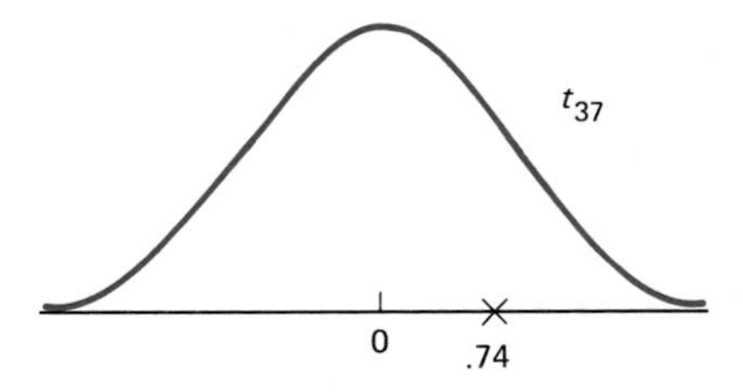

77. (i) No. The dot diagrams in (i) suggest "inverted" normal populations looking like this:

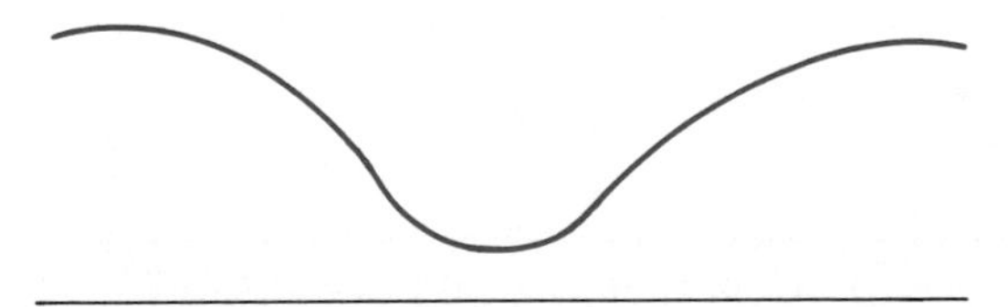

(ii) The dot diagrams in (ii), although not exactly what normal samples should look like, do not appear inconsistent with normal populations.

79.

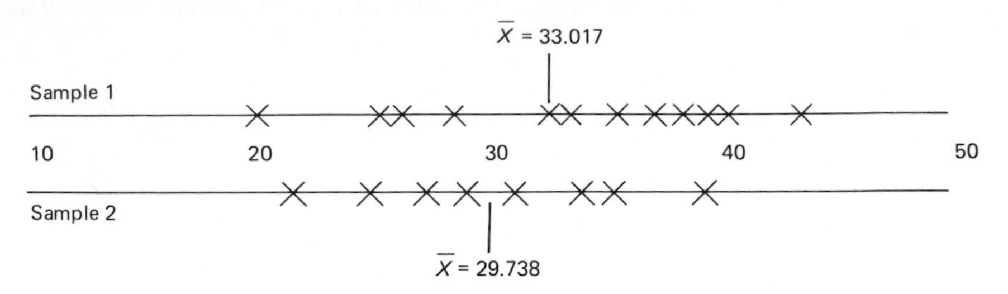

Yes, both samples appear consistent with having come from normal populations.
81. (b) Reject H_0 at the .01 level of significance (P value $< .01$). The data suggest that the potential mean number of larvae per test plot is significantly less than 18.
83. (b) In this case μ is the mean longest hitless streak for *all* major league baseball players. (c) We can be 80% confident that μ lies in the interval 18.4 to 21.0.
85. (a) $H_0: \mu = 86.5$, $H_1: \mu > 86.5$, where μ is the potential mean score with the new clubs. (b) The P value $< .01$, so reject H_0 at the .01 level of significance. The data strongly suggest that μ is significantly greater than 86.5.
87. (a) We performed a two-sided, two-sample t test, comparing the time spent with the nutritionist for those mothers who do and do not breastfeed. (b) The P value (.011) exceeds .01 but is less than .05. We therefore reject H_0 at the .05 level of significance. The data suggest that, on average, women who breastfeed spend significantly more time with the nutritionist than do those who do not breastfeed.

REVIEW PROBLEMS

1. (a) 37.56 to 50.64 seconds. (b) We assume that the distribution of the length of dives is approximately normal.
3. (a) $F = 4.329$. Reject H_0 at the .01 level of significance. The data strongly suggest that the variability of J's scores is significantly greater than the variability of L's. (Use the $F_{20,16}$ instead of the $F_{21,16}$ distribution.) (b) The fact that the median and the mean are fairly close suggests that the data are fairly close to symmetric.
9. $F = 2.478$. Do not reject H_0 at the .05 level of significance. The data suggest that there has been no significant change in the variability of the leading scorer's average.
11. 1.56. Do not reject H_0 at the .05 level of significance. The data suggest that the mean percent alcohol for those stopped at random is *not* significantly less than the corresponding mean for those stopped for speeding.
13. (a) $F = 1.92$. Do not reject H_0 at the .05 level of significance. The data suggest that σ_1^2 is not significantly greater than σ_2^2 (Use the $F_{12,16}$ rather than the $F_{14,16}$ distribution.) (b) We made the wrong decision. In fact, σ_1^2 is twice σ_2^2.
15. 1.88. Do not reject H_0 at the .05 level of significance. The data suggest no significant difference in mean assaultiveness for the two groups.

17.

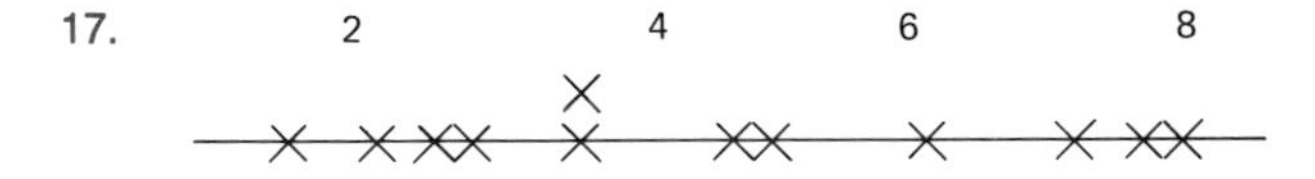

Although not exactly what we would expect from a normal population, the dot diagram does not suggest a significantly nonnormal population.

19. (a) $t = 2.79$. Reject H_0 at the .01 level of significance. The data suggest that the mean length of sentence for such violators this year is significantly greater than 37.2. (b) We assume that the distribution of the length of such sentences is approximately normal.
21. (a) $F = 2.674$. Reject H_0 at the .05 level of significance. The data suggest that the variance in service time for State is significantly greater than that for Mutual Insurance. (Use the $F_{20,20}$ instead of the $F_{23,23}$ distribution.) (b) $t = .97$. We cannot reject H_0 at the .05 level of significance. The data suggest that there is no significant difference between the mean time to service claims for the two companies. (c) In part (a) we assume that, for both companies, the distribution of service time is normal. In part (b) we assume not only that the distribution of service time for each company is normal but also that they have the same standard deviations. (d) P value $= 2P(Z > 2.20) = .0278$. Reject H_0 at the .05 level of significance. The data suggest that the mean time to service claims at Mutual Insurance is greater than at State Insurance. (e) No assumptions other than independent random samples were necessary in part (d).
23. (a) For *all* banks that contribute to nonprofit organizations, μ is the mean percentage of pretax net income that is given. (b) $H_0: \mu = 1.77$, $H_1: \mu < 1.77$ (c) $t = -2.73$. Reject H_0 at the .05 level of significance. The data suggest that μ is significantly less than 1.77%. (d) We assumed that the distribution of the percentage contribution over all banks that contribute is approximately normal.
25. $t = 1.038$. Do not reject H_0 at the .05 level of significance. The data suggest that the mean difference in estimates is not significantly different from zero.
27. $t = 1.36$. Do not reject H_0 at the .05 level of significance. The data suggest that the true mean student/teacher ratio in the district is not significantly greater than 23.5.
29. (a) 3.59 to 7.61 (b) $F = 1.940$. Do not reject H_0 at the .05 level of significance. The data suggest there is no significant difference between σ_1^2, and σ_2^2. 31. .350 to .649
33. (a) (i) 31.05 to 42.55 meters (ii) 2.8 to 11.6 meters (b) $t = 2.88$. Reject H_0 at the .01 level of significance. The data strongly suggest that the mean difference (land-based error − satellite-based error) is significantly greater than zero.

CHAPTER 13

3.

SOURCE OF VARIABILITY	SS	df	MS
Between sample means	1081	4	270.25
Within samples	3301	7	471.57
Total	4382	11	

5.

SOURCE OF VARIABILITY	SS	df	MS
Between machine means	65.408	2	32.704
Within machines	208.444	24	8.685
Total	273.852	26	

7.

SOURCE OF VARIABILITY	SS	df	MS
Between class means	2.685	3	.895
Within classes	27.889	76	.367
Total	30.574	79	

9.

SOURCE OF VARIABILITY	SS	df	MS
Between state means	45.732	2	22.866
Within states	51.589	21	2.457
Total	97.321	23	

11. (a)

SOURCE OF VARIABILITY	SS	df	MS	F
Between design means	44.882	4	11.220	5.82
Within designs	48.172	25	1.927	
Total	93.054	29		

(b) $H_0: \mu_1 = \mu_2 = \mu_3 = \mu_4 = \mu_5$ and $H_1: \mu_1, \mu_2, \mu_3, \mu_4, \mu_5$ are not equal. In this case the μ_i's are the potential mean times spent watching the design. (c) $F = 5.82$. Reject H_0 at the .01 level of significance. The data strongly suggest that the μ_i are significantly different.
13. (a)

SOURCE OF VARIABILITY	SS	df	MS	F
Between income means	3303.8	2	1,651.9	32.86
Within income levels	1357.4	27	50.274	
Total	4661.2	29		

(b) $F = 32.86$. Reject H_0 at the .01 level of significance. The data strongly suggest that the mean index of home safety for the three income levels is significantly different. (c) The mean index for high-income homes (67.1) is much higher than that for moderate-income homes (54.7), which, in turn, is considerably greater than that for low-income homes (41.4).
15. (a) In this case μ_1, μ_2, μ_3, and μ_4 are the mean differences in reaction times—over all potential participants—for the four different amounts of beer. (b) $F = 14.12$. Reject H_0 at the .01 level of significance. The data strongly suggest that the μ_i's are significantly different. (c) Given that it is a randomized experiment this is a reasonable interpretation of the data. (d) It would become more tentative. Allowing students to choose how much beer to drink introduces the possibility of many kinds of biases.
17. $F = 2.44$. Reject H_0 at the .05 level of significance. The data suggest that the mean GPA for the four classes is significantly different.
19. $F = 12.74$. Reject H_0 at the .01 level of significance. The data strongly suggest that the mean costs for the four hospitals are significantly different.
21. (a) $F = .502$. Do not reject H_0 at the .05 level of significance. The data suggest no significant differences in the four population means. (b) Yes, the four populations do all have the same mean (10).
25. (a) and (b) 1.43 to 89.77 [In both (a) and (b) we used $t_{40(.95)} = 2.021$ rather than $t_{48(.95)}$.]
27. (a) 5.29 to 7.27 lb. We can be 90% confident that the mean weight of infants born to heavy-smoking mothers lies in this interval. (b) $-.24$ to 2.18 lb. We can be 90% confident that the *difference* between the mean birthweight of infants born to heavy and light smoking mothers lies in this interval. [In both (a) and (b) we replaced $t_{56(.9)}$ with $t_{50(.9)} = 1.676$.]
31.

SOURCE OF VARIABILITY	SS	df	MS	F
Between transmission means	6.615	1	6.615	35.37
Between season means	11.235	3	3.745	20.03
Transmission/season interaction	.075	3	.025	.134
Within cells	3.000	16	.187	
Total	20.925	23		

(a) F(seasons) = 20.03. Reject H_0 at the .01 level of significance. The data strongly suggest that the mean mileages for the four seasons are significantly different. (b) F(transmission) = 35.37. Reject H_0 at the .01 level of significance. The data strongly suggest that the mean mileages for the two types of transmissions are significantly different. (c) F(interaction) = .134. Do not reject H_0 at the .05 level of significance. The data suggest that there is no significant interaction between season and type of transmission.

33. (a)

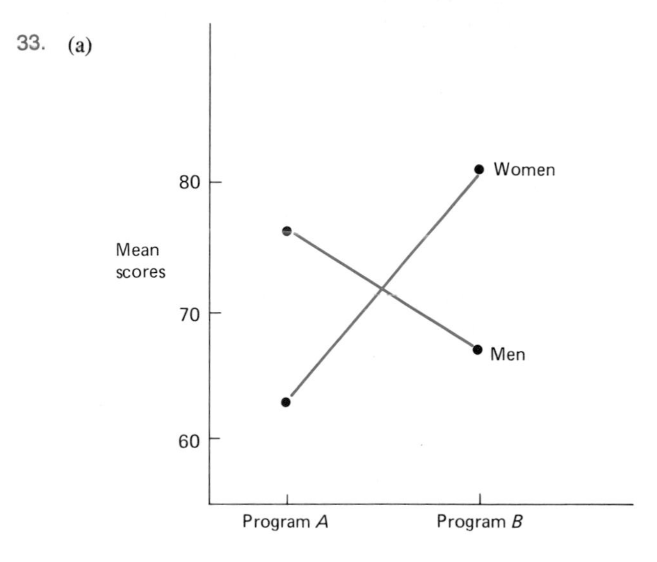

(b) There is substantial interaction. Men do much better in program A than in program B. Women do much better in B than in A.

35. (a)

		PANEL A	PANEL B	PANEL C
AIRCRAFT TYPE	I	7.8	9.2	8.0
	II	8.4	9.3	8.4
	BOTH	8.1	9.25	8.2

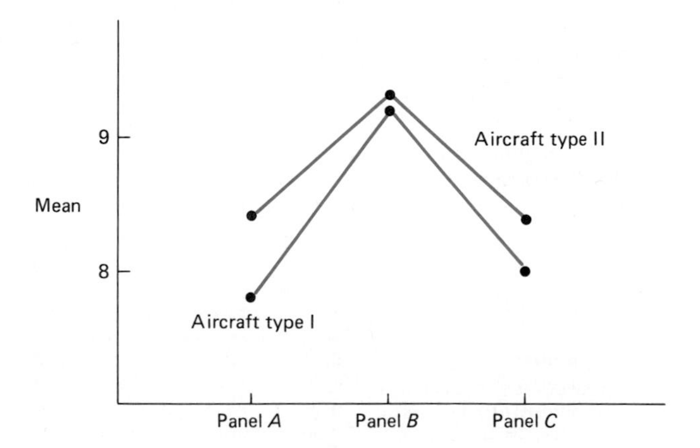

(b) panel A (8.1) (c) There is very little interaction (d) yes

37.

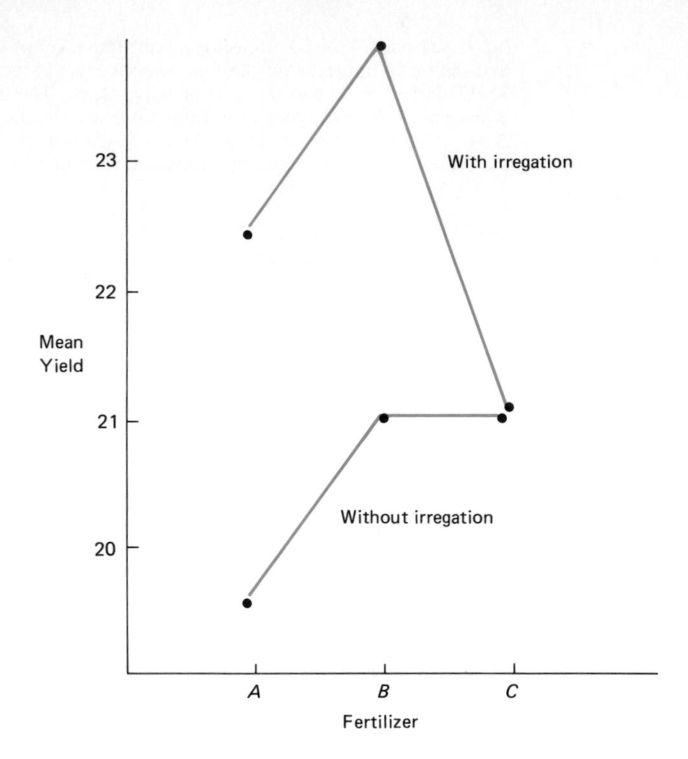

Most of the interaction is associated with fertilizer *C*. This fertilizer without irrigation does rather better than expected.

39. (a) $F = 4.83$. Reject H_0 at the .05 level of significance. The data suggest that the mean percentage of salary that is saved in the three counties differs significantly. (b) 4.04 to 7.58%

REVIEW PROBLEMS

1. $SS_{within} = 0$ in situation *C*. Within each group all the values are exactly the same. There is no variability within groups. $SS_{between} = 0$ in situation *B*. In this case all the group means are the same, so there is no variability between group means.

3.

SOURCE OF VARIABILITY	SS	df	MS	F
Variable 1	8.17	2	4.08	2.33
Variable 2	4.08	1	4.08	2.33
Interaction	6.17	2	3.08	1.76
Within cells	10.50	6	1.75	
Total	28.92	11		

5.

SOURCE OF VARIABILITY	SS	df	MS	F
Between levels of variable A	1104	4	276	4.23
Between levels of variable B	817	4	204.25	3.13
Interaction	617	16	38.56	.59
Within cells	3917	60	65.28	
Total	6455	84		

(a) Reject H_0: no difference between the means for variable A, at the .01 level of significance. (b) Reject H_0: no difference between the means for variable B, at the .05 level of significance. (c) Do not reject, at the .05 level of significance H_0: no interaction between variable A and variable B.

7. (a)

SOURCE OF VARIABILITY	SS	df	MS	F
Between pitchers	361	3	120.333	7.1
Within pitchers	1287.99	76	16.947	
Total	1648.99	79		

(b) Reject H_0: no difference in the mean speed for the four pitchers, at the .01 level of significance. (c) We assumed that for each pitcher the distribution of speed is approximately normal with the same standard deviation.

9. $F = 2.14$. Do *not* reject H_0: no difference in mean age of walking for the four groups.

11. (a) $F = 3.85$. Reject H_0: no difference in mean interval for the six family sizes, at the .05 level of significance. (We used values for the $F_{5,20}$ rather than the $F_{5,23}$ distribution.) (b) We assume that for each family size the distribution of this interval is approximately normal with the same standard deviation.

13. (a) 25.31 to 29.49 cases (b) .38 to 5.85 cases

15. (a)

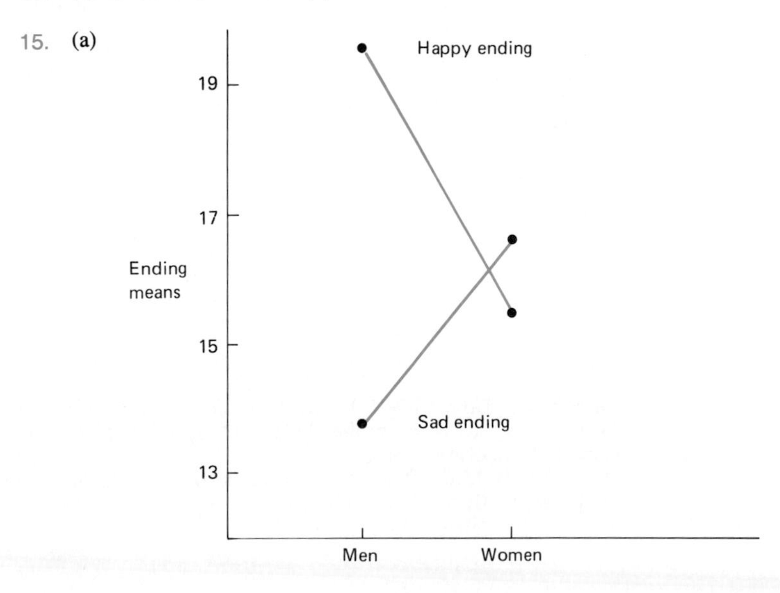

Since the lines are far from parallel, there is substantial interaction between the variables "type of ending" and "sex." On average, the happy ending scored slightly higher than the sad ending. Men, however, enjoyed the happy ending much more than the sad ending, whereas women, by a smaller amount, enjoyed the sad ending more than the happy ending. (b) No. The analysis in part (a) is purely descriptive.
17. (a)

SOURCE OF VARIABILITY	SS	df	MS	F
Between season means	294.5	1	294.5	9.44
Between day means	388.5	4	97.1	3.11
Season/day interaction	62.1	4	15.5	.497
Within cells	624.7	20	31.2	
Total	1369.8	29		

(b) Reject H_0: no difference between mean sales for the five days, at the .05 level of significance. (c) Yes. With the one-way analysis of variance we were unable to reject this hypothesis at the .05 level of significance. (d) Reject H_0: no difference in mean sales for the two seasons, at the .01 level of significance. (e) Do *not* reject, H_0: no interaction between day of the week and season, at the .05 level of significance.

SOURCE OF VARIABILITY	SS	df	MS	F
Between school means	.16	1	.16	.084
Between grade means	190.58	2	95.29	49.89
School/grade interaction	33.74	2	16.87	8.83
Within cells	57.22	30	1.91	
Total	281.70	35		

(b) Reject the hypothesis H_0: no difference in grade means, at the .01 level of significance. (c) Do not reject the hypothesis H_0: no difference in means for the two schools, at the .05 level of significance. (d) Reject the hypothesis H_0: no interaction between grade and school, at the .01 level of significance. (e) In parts (b)–(d) we assume that for each grade in each school the distribution of scores is approximately normal with the same standard deviation.
21. (a)

	T_i	Σx_i^2	n_i
A	92	370	32
M	77	245	35
P	102	368	36

(b) $F = 1.9$. Do not reject the hypothesis H_0: no difference in the mean number of absences for the three class times, at the .05 level of significance. (We used values for the $F_{2,60}$ rather than the $F_{2,100}$ distribution.)
23. (a) $F = 15.63$. Reject the hypothesis H_0: no difference in the mean hardness of the three alloys, at the .01 level of significance. (b) We assume that for each alloy the distribution of hardness is normal with equal variances.

25. (a)

SOURCE OF VARIABILITY	SS	df	MS	F
Between age means	.480	1	.480	1.19
Between yarn means	11.765	2	5.882	14.60
Yarn/age interaction	.015	2	.008	.02
Within cells	2.420	6	.403	
Total	14.680	11		

(b) (i) Reject the hypothesis H_0: no difference in mean strengths for the three types of yarn, at the .01 level of significance. (ii) Do not reject the hypothesis H_0: no difference in mean strength for the two age groups, at the .05 level of significance. (c) Do not reject, the hypothesis H_0: no interaction between age and type of yarn at the .05 level of significance.
(d)

		A	B	C
CELL	NEW	15.95	18.30	17.05
MEANS	OLD	15.45	17.95	16.70

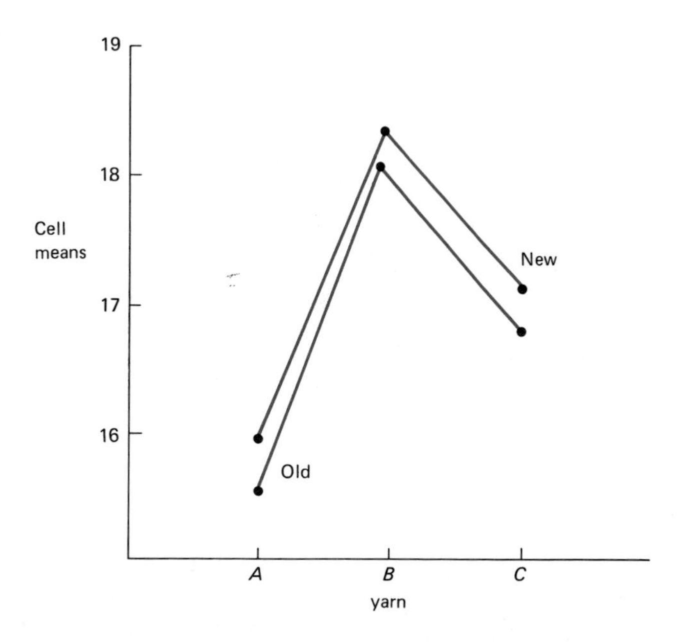

No, there seems to be very little interaction.

CHAPTER 14

3. (a) $y = .722 + .025x$.
(b)

SOURCE OF VARIABILITY	SS	df	MS
Regression of visits on age	.604	1	.604
Error	23.174	16	1.448
Total	23.778	17	

(c) $r^2 = .0254$. Only 2.5% of the variability in the number of visits is related to its linear relationship to age.

5. (a) (i) $T_{xx} = 1616$, $T_{yy} = 303.8$, $T_{xy} = -614.5$
(ii)

SOURCE OF VARIABILITY	SS	df	MS
Regression of sleep on age	15.578	1	15.578
Error	4.675	13	.36
Total	20.253	14	

(b) $r^2 = .769$. Almost 77% of the variability in hours of sleep is related to its linear relationship to age. (c) $a = .8496$ (d) $b = -.38$. An increase in age of one year is associated with a decrease of .38 hour of sleep.

7. (a) $T_{xx} = 1933.66$, $T_{yy} = 766$, $T_{xy} = -455.9$
(b)

SOURCE OF VARIABILITY	SS	df	MS
Regression of number of children on educational level	4.30	1	4.30
Error	26.34	23	1.145
Total	30.64	24	

(c) $r^2 = .1403$

11. (a)

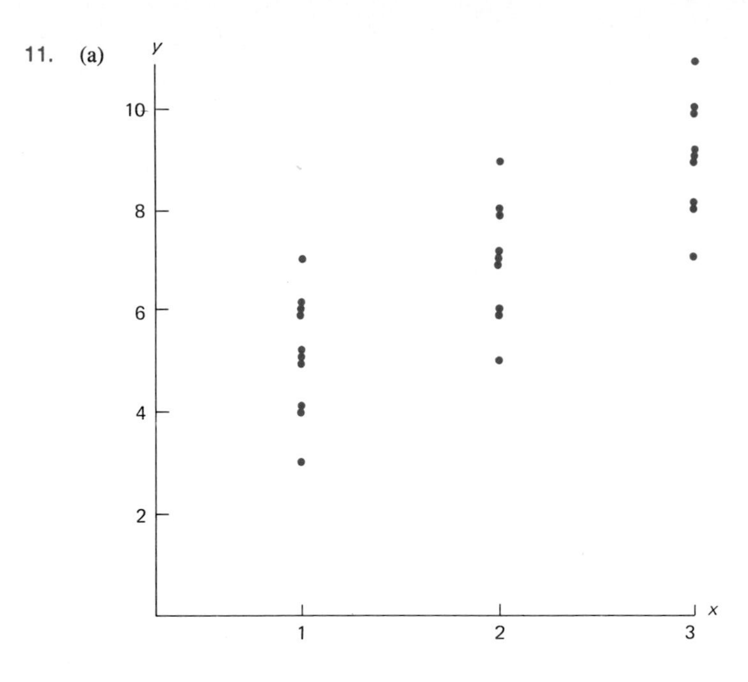

(b) Yes, they do. (c) $\mu_{y|1} = 5$, $\mu_{y|2} = 7$, $\mu_{y|3} = 9$ Yes, they lie on the line $y = 3 + 2x$. (d) Yes, since in each case $s = \sigma = 1.225$.

15. (a) In this case β is the unknown slope of the population line relating age to mean number of visits. (b) $t = .653$. Do not reject H_0 at the .05 level of significance. The data suggest that β is not significantly greater than 0, and hence that there is not a significant linear relationship between x and y. (c) Because the alternative hypothesis in the F test is $H_1: \beta \neq 0$.

17. (a) In this case β is the unknown slope of the population line relating age to mean number of hours of sleep. (b) $t = -6.57$. Reject H_0 at the .01 level of significance. The data strongly suggest that β is significantly less than 0, and hence that there is a significant linear relationship between x and y.

19. (a) In this case β is the unknown slope of the population line relating educational level to mean number of children. (b) $t = -1.94$. Do not reject H_0 at the .05 level of significance. The data suggest that β is not significantly different from 0 and hence that there is not a significant linear relationship between x and y. (c) $F = 3.755 = (-1.94)^2$. Do not reject H_0 at the .05 level of significance. The F test and the two-sided t test are equivalent.

21. (a) $t = -1.15$. Do not reject H_0 at the .05 level of significance. The data suggest that β is not significantly different from 0. (b) Yes, since in fact $\beta = 0$.

23. (a) In this case b (= .0504) is the slope of the regression line relating number of hits to temperature and β the unknown slope of the corresponding population line. (b) (i) $t = .516$. Do not reject H_0 at the .05 level of significance. The data suggest that β is not significantly different from 0, and hence that there is no significant linear relationship between temperature and mean number of hits. (ii) $F = .267$ ($= .516^2$). Do not reject H_0 at the .05 level of significance.

25. (a) (i) 7.11 to 7.30 ft (ii) 9.44 to 9.98 ft. Because 9 is farther from $\bar{x}$ (= 4.5 years) than is 5. (b) 7.35 to 8.31 ft

27. (a) (i) 9.75 to 10.42 hours (ii) 8.75 to 11.42 hours (b) In (a, i) we are estimating a mean value for y. In (a, ii) we are estimating a specific value of y.

29. Because in that problem no significant linear relationship was found between the number of hits and temperature.

33. The assumption of linearity may be violated.

35. Both the assumption of normality and equal variances appear to be violated.

37. (a) In this case β is the unknown slope of the population line relating infant birthweight to birthlength. (b) 69.49 to 109.67

39. (a) In this case β is the unknown slope of the population line relating the percentage of the additive to mean mileage. (b) From the printout, $t = 6.98$. Reject H_0 at the .01 level of significance. The data strongly suggest that β is significantly greater than 0, and hence that there is a significant positive linear relationship between x and y. (c) $t = 2.19$. Reject H_0 at the .05 level of significance. The data suggest that β is significantly greater than .5. (d) 45.6 to 48.1. We can be 95% confident that the mean mileage per gallon when 2% of the mixture is the additive lies in this interval. (e) We assumed that for any specific percentage of the additive (x) the distribution of mileage was normal with mean $\alpha + \beta x$ and standard deviation σ, which does not vary with x.

REVIEW PROBLEMS

1. (a)

SOURCE OF VARIABILITY	SS	df	MS	F
Regression on x	453.9	1	453.9	1.74
Error	5218	20	260.9	
Total	5671.9	21		

(b) $r^2 = .08$ (c) $F = 1.74$. Do not reject H_0 at the .05 level of significance. The data suggest that β is not significantly different from 0.

3. This interpretation is false. What the writer has given is almost a definition of r^2 not r.

5. (a) $y = 53.23 - .1425x$

(b)

SOURCE OF VARIABILITY	SS	df	MS	F
Regression on x	87.0204	1	87.0204	1.34
Error	2464.96	38	64.8675	
Total	2551.98	39		

(c) (i) $t = -1.158$. Do not reject H_0 at the .05 level of significance. The data suggest that β is not significantly different from 0, and hence that there is no significant linear relationship between x and y. (ii) $F = 1.34$. The same conclusion as in (i).
(d) $1.34 = (-1.158)^2$ (e) Yes, since in fact $\beta = 0$.

7. (a)

SOURCE OF VARIABILITY	SS	df	MS	F
Regression of log bacteria count on time	12.9704	1	12.9704	183.8
Error	.4234	6	.07057	
Total	13.3938	7		

(b) $y = 3.46 + .556x$ (c) .497 to .615 (d) (i) 5.54 to 5.82 (ii) 7.61 to 8.20
(e) Because 4 hours is closer to $\bar{x}$ (= 3.5 hours) than is 8 hours.

9. (a) $T_{xx} = 108{,}689$, $T_{yy} = 16{,}340$, $T_{xy} = 35{,}530$
(b)

SOURCE OF VARIABILITY	SS	df	MS	F
Regression of weight loss on initial weight	1161.462	1	1161.462	19.66
Error	472.538	8	59.067	
Total	1634	9		

(c) $y = .327x - 38.143$ (d) (i) $t = 4.434$. Reject H_0 at the .01 level of significance. The data strongly suggest that β is significantly greater than 0, and hence that there is a significant positive linear relationship between x and y. (ii) $F = 19.66$. The same conclusion as in (i). (e) $19.66 = 4.434^2$

11. (a)

SOURCE OF VARIABILITY	SS	df	MS	F
Regression of survival time on age	114.716	1	114.716	22.02
Error	67.713	13	5.2087	
Total	182.429	14		

(b) $y = 17.774 - .2744x$ (c) $b = -.2744$ is the slope of the regression line. An increase in age of one year is associated with a decrease of .2744 in years survived. β is the corresponding slope of the population line relating mean years survived to age at onset. (d) $-.353$ to $-.196$ (e) (i) 10.87 to 13.70 years (ii) 7.37 to 8.97 years (iii) 2.62 to 5.49 years (f) We assume that for any age (x) the distribution of survival times is normal with mean $\alpha + \beta x$ and standard deviation σ that does not vary with x.

13. (a) $T_{xx} = 2204$, $T_{yy} = 57{,}590$, $T_{xy} = -7076$
(b)

SOURCE OF VARIABILITY	SS	df	MS	F
Regression of interval on eventual family size	783.368	1	783.368	17.59
Error	1202.494	27	44.537	
Total	1985.862	28		

(c) In this case β is the unknown slope of the population line relating eventual family size to mean interval between the first and the second child. (d) $t = -4.19$. Reject H_0 at the .01 level of significance. The data strongly suggest that β is significantly less than 0 and hence that there is a significant negative linear relationship between x and y. (e) It seems reasonable to assume that, on average, the interval between the first and the second child will decline as family size increases.

15. (a)

SOURCE OF VARIABILITY	SS	df	MS	F
Regression of applicants on mortgage rate	784.828	1	784.828	75.98
Error	144.610	14	10.329	
Total	929.438	15		

(b) $F = 75.98$. Reject H_0 at the .01 level of significance. The data strongly suggest that β is significantly less than 0 ($b = -6.0$), and hence that there is a significant negative linear relationship between x and y. (c) 4.3 to 8.9 (d) .46 to 12.69 (e) In part (c) we were estimating a population mean value for y, while in part (d) we were estimating a specific value of y. (f) We assume that for any mortgage rate x the distribution of the number of applicants per month is normal with the mean $\alpha + \beta x$ and with standard deviation σ, which does not vary with x.

17. (a)

SOURCE OF VARIABILITY	SS	df	MS	F
Regression of closing price on Standard and Poor's	369.204	1	369.204	81.77
Error	58.698	13	4.515	
Total	427.902	14		

(b) $t = 9.04$. Reject H_0 at the .05 level of significance. The data strongly suggest that β is significantly greater than 0 and hence that there is a strong positive linear relationship between x and y. (c) because the alternative hypothesis for the F test is always $H_1\colon\ \beta \neq 0$ (d) .164 to .268

19. (a)

SOURCE OF VARIABILITY	SS	df	MS	F
Regression of creativity index on reading score	3.746	1	3.746	.277
Error	108.354	8	13.544	
Total	112.1	9		

(b) $t = .526$. Do not reject H_0 at the .05 level of significance. The data suggest that β is not significantly greater than 0, and hence that there is no significant linear relationship between x and y. (c) $r^2 = .0334$. Only a little over 3% of the variability in creativity index can be associated with a linear relationship to reading score.

CHAPTER 15

3. $Y = 12.84$. Reject H_0 at the .01 level of significance. The data appear inconsistent with the stated probabilities.

5. $Y = .626$. Do not reject H_0 at the .05 level of significance. The data appear consistent with the hypothesis of no grade inflation.

7. A: dependent. For example, $P(\text{yes}) = .53$ but $P(\text{yes} \mid M) = .45$. B: independent. For example, $P(\text{yes}) = .55 = P(\text{yes} \mid M)$.
11. $Y = 4.97$. Do not reject H_0 at the .05 level of significance. The data are consistent with the independence of these two variables.
13. $Y = 24.48$. Reject H_0 at the .01 level of significance. The data strongly suggest that the proportion of adults reading each newspaper differs significantly by age group.
15. Combine instructors and assistant professors. $Y = 4.99$. Reject H_0 at the .05 level of significance. The data suggest that a smaller proportion of the older faculty favor a union than do the younger.
17. (a) H_0: the likelihood of a woman breastfeeding is independent of ethnic group. H_1: these variables are dependent. (b) $Y = 7.27$. Reject H_0 at the .05 level of significance. The data suggest that the likelihood of a woman breastfeeding does vary significantly by ethnic group.
19. (a) H_0: the likelihood of a respondent using the company's brand is independent of income. H_1: these variables are dependent. (b) $Y = 6.60$. Reject H_0 at the .05 level of significance. The data suggest that the likelihood of a respondent using the company's brand does vary with income.

REVIEW PROBLEMS

1. $Y = 2.91$. Do not reject H_0 at the .05 level of significance. The data are consistent with the heights of husbands and wives being independent.
3. $Y = 5.03$. Do not reject H_0 at the .05 level of significance. The data are not inconsistent with the equally likely hypothesis.
5. (a) $Y = 12.98$. Reject H_0 at the .01 level of significance. (b) The data suggest that among speeders there is a higher proportion above the legal limit and a smaller proportion with no trace of alcohol than among those randomly stopped. (c) .186 to .323
7. (a) $Y = 1.47$. Do not reject H_0 at the .05 level of significance. The data suggest that the probabilities for the two drugs are not significantly different. (b) P value $= P(Z > 1.21) = .1131$. Do not reject H_0 at the .05 level of significance. The data suggest that the probability for T is not significantly greater than that for Z. (c) The alternative in part (a) is $H_1: \pi_T \neq \pi_Z$ rather than $H_1: \pi_T > \pi_Z$.
9. (a) $Y = 83.37$. Reject H_0 at the .01 level of significance. (b) As age at first birth increases the proportion with breast cancer increases.
11. $Y = 2.31$. Do not reject H_0 at the .05 level of significance. The data suggest that the age distribution of Collins' voters is not significantly different from that for other candidates.
13. $Y = 13.32$. Reject H_0 at the .05 level of significance. The data suggest that the distribution of type of institution does vary with the president.
15. $Y = 4.98$. Do not reject H_0 at the .05 level of significance. The data suggest that the distribution of turnout does not vary significantly with size of city.
17. $Y = 3.84$. Do not reject H_0 at the .05 level of significance. The data appear consistent with the stated proportions.
19. $Y = 39.64$. Reject H_0 at the .01 level of significance. The data strongly suggest that age and months unemployed are dependent.
21. $Y = 2.50$. Do not reject H_0 at the .05 level of significance. The data are not inconsistent with self-concept being independent of type of classroom.
23. (a) $Y = 8.2$. Reject H_0 at the .05 level of significance. (b) Prisoners seem to favor the first and last quarters rather than the middle two quarters.
25. (a) $Y = .90$. Do not reject H_0 at the .05 level of significance. The data suggest that these two variables are independent. (b) P value $= P(Z > .95) = .1711$. Do not reject H_0 at the .05 level of significance. The data suggest that pessimists are not significantly more likely than optimists to favor a low-risk career. (c) In the chi-square test in part (a) the alternative is $H_1: \pi_1 \neq \pi_2$ rather than $H_1: \pi_1 > \pi_2$ as in part (b).
27. (a) $Y = .19$. Do not reject H_0 at the .05 level of significance. The data are not inconsistent with both systems having the same probability of detection. (b) P value $= P(Z > .43) = .3336$. Do not reject H_0 at the .05 level of significance. The data suggest that the probability of detection for system B is not significantly greater than that for system A. (c) See answer to Review Problem 25(c).

CHAPTER 16

3. $n = 18$, $y = 16$. P value $= P(Y \geq 16 \mid \pi = .5) = .001$. Reject H_0 at the .01 level of significance. The data strongly suggest that the median decline in reaction time is significantly greater than 0.
5. $n = 9$, $y = 9$. (a) P value $= P(Y \geq 9 \mid \pi = .5) = .002$. Reject H_0 at the .01 level of significance. The data suggest that the median difference in scores $(A - B)$ is significantly greater than 0. (b) P value $= 2P(Y \geq 9 \mid \pi = .5) = .004$. Same conclusion as in part (a).
7. $n = 40$, $y = 26$. P value $= 2P(Y \geq 26 \mid \pi = .5) = 2(.04) = .08$. The data suggest no significant difference in preference for the two ads.
11. $n = 18$, $w = 163.5$. P value $= P(W \geq 163.5) = P(Z > 3.49) = .0002$. Reject H_0 at the .01 level of significance. The data strongly suggest that the median decline in reaction time is significantly greater than 0.
13. $n = 9$, $w = 45$. (a) P value $= P(W \geq 45) < .004$. Same conclusion as in Problem 5(a). (b) P value $= 2P(W \geq 45) < .008$. Same conclusion as in Problem 5(b).
17. $n_1 = n_2 = 10$, $v = 108$. P value $= P(Z > .09) = .4641$. Do not reject H_0 at the .05 level of significance. The data suggest that the median score for students in graduate schools is not significantly greater than that for working students.
19. $n_1 = 15$, $n_2 = 8$, $v = 204$. P value $= P(Z > 1.55) = .0606$. Do not reject H_0 at the .05 level of significance. The data suggest that the median price of competitors' gas is not significantly greater than that for OCRA.
21. $n_1 = n_2 = 8$, $v = 75.5$. P value $= P(Z > .79) = .2148$. Do not reject H_0 at the .05 level of signficance. The data suggest that the median age of men is not significantly greater than that for women.
23. (a) $n = 40$, $r = 5$, $R_1 = 130$, $R_2 = 224$, $R_3 = 118$, $R_4 = 291$, $R_5 = 57$, $Y = 31.51$. Reject H_0 at the .01 level of significance.
(b)

	A	B	C	D	E
Median	1110	1243	1125.5	1328	1079

Brand D has the largest median life, and E the lowest.
25. $n = 36$, $r = 3$, $R_1 = 161$, $R_2 = 356.5$, $R_3 = 148.5$, $Y = 20.43$. Reject H_0 at the .01 level of significance. The data suggest that there are significant differences among the three medians. (b) packaging B
27. $n = 27$, $r = 3$, $R_1 = 166$, $R_2 = 119$, $R_3 = 93$, $Y = 4.83$. Do not reject H_0 at the .05 level of significance. The data suggest that there are no significant differences among the median number of defectives for the three machines.
29.

	A	B	C
r_s	.429	.214	−.07

A wins.

31. (a) .7 (b) sum of ranks:

A	B	C	D	E
8	4	10	3	5

(c) D
33. −.938

REVIEW PROBLEMS

1. (a) $n = 6, y = 5$. P value $= P(Y \geq 5 \mid \pi = .5) = .109$. Do not reject H_0 at the .05 level of significance. The data suggest that the median decline in costs is not significantly greater than 0. (b) $n = 6, w = 20$. P value $= P(W \geq 20) = .031$. Reject H_0 at the .05 level of significance. The data suggest that the median decline in costs is significantly greater than 0.
3. $-.143$
5. $n = 12, r = 3, R_1 = 25.5, R_2 = 32.5, R_3 = 20, Y = 1.51$. Do not reject H_0 at the .05 level of significance. The data suggest that the three population medians are not significantly different.
7. $n = 12, w = 78$. P value $= P(W \geq 78) < .005$. Reject H_0 at the .01 level of significance. The data strongly suggest that the median difference is significantly greater than 0.
9. $n = 23, r = 4, R_1 = 39, R_2 = 71, R_3 = 82, R_4 = 84$, and $Y = 6.82$. Do not reject H_0 at the .05 level of significance. The data suggest that the median age when first walking is not significantly different for the four groups.
11. $n = 24, r = 3, R_1 = 130.5, R_2 = 68.5, R_3 = 101$, and $Y = 4.81$. Do not reject H_0 at the .05 level of significance. The data suggest that the median number of days for recovery for the three treatments is not significantly different.
13. $n = 17, r = 4, R_1 = 56, R_2 = 56.5, R_3 = 7.5, R_4 = 33$, and $Y = 10.57$. Reject H_0 at the .05 level of significance. The data suggest that the median caseloads for the four states are significantly different. 15. .886
17. (a) $n = 10, y = 7$. P value $= 2P(Y \geq 7 \mid \pi = .5) = 2(.172) = .344$. Do not reject H_0 at the .05 level of significance. The data suggest that the median difference in estimate is not significantly different from 0. (b) $n = 10, w = 38$. P value $= 2P(Y \geq 38 \mid \pi = .5) > 2(.138) = .276$. The same conclusion as in part (a).
19. (a) $n = 30, r = 5, R_1 = 70, R_2 = 93, R_3 = 104.5, R_4 = 133.5, R_5 = 64$, and $Y = 6.76$. Do not reject H_0 at the .05 level of significance. The data suggest that the median sales for the five days are not significantly different. (b) Thursday (45) (c) Friday (34)
21. .243 23. (a) .657 (b) D 25. .086
27. (a) $n = 15, y = 9$. P value $= 2P(Y \geq 9 \mid \pi = .5) = 2(.304) = .608$. Do not reject H_0 at the .05 level of significance. The data suggest that the median percentage of impurities is not significantly different from 6. (b) $n = 15, w = 71$. P value $= 2P(W \geq 71) = 2P(Z > .62) = .5352$. The same conclusion as in part (a).
29. $n = 12, r = 3, R_1 = 25, R_2 = 41, R_3 = 12$, and $Y = 8.115$. Reject H_0 at the .05 level of significance. The data suggest that the median hardness for the three alloys is significantly different.

INDEX

A

B

C

D

E

F

G

H

I

J

K

O

P

Q

R

S

T

U

V

W

Z

THE t DISTRIBUTION

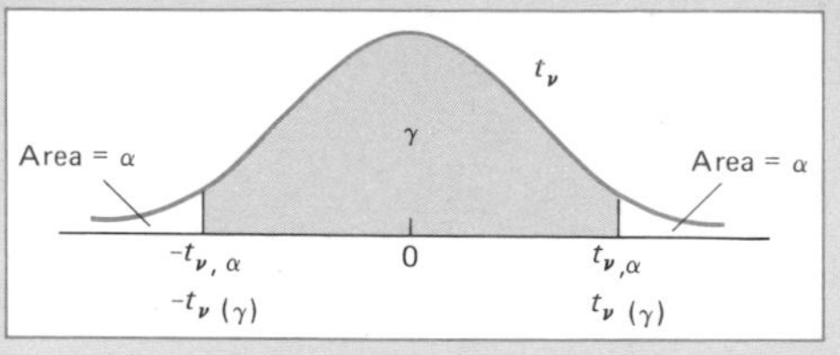

	$t_{\nu,.25}$ $t_{\nu[.5]}$	$t_{\nu,.2}$ $t_{\nu(.6)}$	$t_{\nu,.15}$ $t_{\nu[.7]}$	$t_{\nu,.1}$ $t_{\nu,(.8)}$	$t_{\nu,.05}$ $t_{\nu[.9]}$	$t_{\nu,.025}$ $t_{\nu[.95]}$	$t_{\nu,.01}$ $t_{\nu[.98]}$	$t_{\nu,.005}$ .99
1	1.000	1.376	1.963	3.078	6.314	12.706	31.821	63.657
2	.817	1.061	1.386	1.886	2.920	4.303	6.965	9.925
3	.765	.978	1.250	1.638	2.353	3.183	4.541	5.841
4	.741	.941	1.190	1.533	2.132	2.776	3.747	4.604
5	.727	.920	1.156	1.476	2.015	2.571	3.365	4.032
6	.718	.906	1.134	1.440	1.943	2.447	3.143	3.707
7	.711	.896	1.119	1.415	1.895	2.365	2.998	3.500
8	.706	.889	1.108	1.397	1.860	2.306	2.896	3.355
9	.703	.883	1.100	1.383	1.833	2.262	2.821	3.250
10	.700	.879	1.093	1.372	1.813	2.228	2.764	3.169
11	.697	.876	1.088	1.363	1.796	2.201	2.718	3.106
12	.696	.873	1.083	1.356	1.782	2.179	2.681	3.055
13	.694	.870	1.079	1.350	1.771	2.160	2.650	3.012
14	.692	.868	1.076	1.345	1.761	2.145	2.624	2.977
15	.691	.866	1.074	1.341	1.753	2.132	2.602	2.947
16	.690	.865	1.071	1.337	1.746	2.120	2.583	2.921
17	.689	.863	1.069	1.333	1.740	2.110	2.567	2.898
18	.688	.862	1.067	1.330	1.734	2.101	2.552	2.878
19	.688	.861	1.066	1.328	1.729	2.093	2.539	2.861
20	.687	.860	1.064	1.325	1.725	2.086	2.528	2.845
21	.686	.859	1.063	1.323	1.721	2.080	2.518	2.831
22	.686	.858	1.061	1.321	1.717	2.074	2.508	2.819
23	.685	.858	1.060	1.319	1.714	2.069	2.500	2.807
24	.685	.857	1.059	1.318	1.711	2.064	2.492	2.797
25	.684	.856	1.058	1.316	1.708	2.060	2.485	2.787
26	.684	.856	1.058	1.315	1.706	2.056	2.479	2.779
27	.684	.855	1.057	1.314	1.703	2.052	2.473	2.771
28	.683	.855	1.056	1.313	1.701	2.048	2.467	2.763
29	.683	.854	1.055	1.311	1.699	2.045	2.462	2.756
30	.683	.854	1.055	1.310	1.697	2.042	2.457	2.750
31	.683	.854	1.054	1.310	1.696	2.040	2.453	2.744
32	.682	.853	1.054	1.309	1.694	2.037	2.449	2.739
33	.682	.853	1.053	1.308	1.692	2.035	2.445	2.733
34	.682	.852	1.053	1.307	1.691	2.032	2.441	2.728
35	.682	.852	1.052	1.306	1.690	2.030	2.438	2.724
36	.681	.852	1.052	1.306	1.688	2.028	2.434	2.720
37	.681	.852	1.051	1.305	1.687	2.026	2.431	2.716
38	.680	.851	1.051	1.304	1.686	2.024	2.428	2.712
39	.681	.851	1.050	1.304	1.685	2.023	2.426	2.708
40	.681	.851	1.050	1.303	1.684	2.021	2.423	2.705
50	.679	.849	1.047	1.299	1.676	2.009	2.403	2.678
60	.679	.848	1.046	1.296	1.671	2.000	2.390	2.660
70	.678	.847	1.044	1.294	1.667	1.995	2.381	2.648
80	.678	.846	1.043	1.292	1.664	1.990	2.374	2.639
90	.677	.846	1.043	1.291	1.662	1.987	2.368	2.632
100	.677	.845	1.042	1.290	1.660	1.984	2.364	2.626
∞	.674	.842	1.036	1.282	1.645	1.960	2.326	2.576